SEISMIC DESIGN OF REINFORCED AND PRECAST CONCRETE BUILDINGS

SEISMIC DESIGN OF REINFORCED AND PRECAST CONCRETE BUILDINGS

ROBERT E. ENGLEKIRK
Consulting Structural Engineer
and
Adjunct Professor
University of California at San Diego

JOHN WILEY & SONS, INC.

This book is printed on acid-free paper. ♾

Published by John Wiley & Sons, Inc., Hoboken, New Jersey.
Published simultaneously in Canada.

For general information on our other products and services or for technical support, please contact our Customer Care Department within the United States at (800) 762-2974, outside the United States at (317) 572-3993 or fax (317) 572-4002.

Wiley also publishes its books in a variety of electronic formats. Some content that appears in print may not be available in electronic books. For more information about Wiley products, visit our web site at www.wiley.com.

Library of Congress Cataloging-in-Publication Data:

Englekirk, Robert E., 1936–
Seismic design of reinforced and precast concrete buildings / Robert Englekirk.
p. cm.
ISBN 0-471-08122-1
1. Earthquake resistant design. 2. Reinforced concrete construction. 3. Precast concrete construction. 4. Buildings, Reinforced concrete—Earthquake effects. I. Title.
TA658.44 .E56 2003
693.8'52—dc21 2002008561

Printed in the United States of America.

10 9 8 7 6 5 4 3 2 1

CONTENTS

PREFACE

"Man's mind, once stretched by a new idea, never regains its original dimensions."
—Oliver Wendell Holmes

Knowledge and imagination are essential components of the design process. Imagination without knowledge will quite often produce designs that are dangerous. Knowledge absent imagination can only produce designs of limited scope. The development and integration of these themes is the objective of this book.

My hope is to advance the reader's ability to design by reducing existing experimentally developed conclusions to design-relevant relationships and limit states. The reduction of experimental data to a usable form is essential to the design process because an engineer, faced with a design decision, cannot confidently develop a design approach from experiment data or basic principles as a part of each design, especially if the basic principle is not a part of his or her working vocabulary. Behavior models must also be available to and accepted by the designer. To this end Chapters 1 and 2 review experimental evidence and selected fundamental principles in search of appropriate design processes and limit states.

My objective in Chapters 1 and 2 is to stretch the reader's mind, not constrain a design to a particular approach or set of limit states, for I believe that all design procedures must be thoroughly understood and accepted by the user if they are to be appropriately applied. Algorithms, whether contained in a black box or reduced to napkin form, are an essential part of the designer's vocabulary. The algorithms developed herein are presented in sufficient detail so as to allow the user to adapt them to his or her predilection or interpretation of experimental evidence, and this is because it is the engineer, not the algorithm, black box, or experimental data, who is responsible for the building design and safety of the building's occupants.

Engineers are generally characterized as unimaginative or pedantic in their approach to problems. This generalization is not supported by historical evidence, which from a structural perspective includes the creation of ancient structures, medieval cathedrals, and the modern structures of today, because none of these structures were developed from scientifically supportable data. Even today, in our modern scientifically based society that probes the universe, the scientific data used to support building design are more speculative than scientific. The existence of codified relationships expressed in four significant figures only tends to suppress an engineer's imagination by creating a scientific illusion. Imagination cannot be taught, but it can be released and encouraged by removing suppressants, and this is the objective of Chapters 3 and 4.

Imagination can and must be effectively used to apply basic concepts to complex problems. It is also effectively used to extend experimental evidence. The exploitation of precast concrete as a seismic resisting system clearly demonstrates how imagination can be used to create better structural systems. Exploring all possible uses of imaginative thinking is impossible so I have tried to develop several imaginative approaches by example and where possible relate these specific examples to a generalization of the design objective.

Building codes, like dictionaries, are essential tools of the trade. One writes with the assistance of a dictionary—we do not use the dictionary to write. So it should be with design, and it is for this reason that I do not choose to describe or explain design using the building code as a basis. The thrust of this book is to produce structural systems that can be shown by rational analysis and experimental evidence to be capable of attaining performance objectives when subjected to design-level earthquakes.

ACKNOWLEDGMENTS

Absent the dedicated efforts of Joan Schulte, who not only typed but managed the processing of this manuscript; Dan Shubin, who converted my crude sketches to an art form; and my cardiology team, Drs. Kahn, Natterson, and Robertson, this book would not exist.

The concepts explored in this book are more simply applied than explained. The assistance of my associates is gratefully acknowledged and especially that of Nagi Abo-Shadi, Richard Chen, Robert Liu, and Michael Riddell.

My thanks also to Kimberly Tanouye for her design contribution to the cover.

A special expression of gratitude is also extended to those who have taken the time and made the effort to translate ideas and research into the written word on the subject of concrete and earthquake engineering—my good friends and colleagues, Bob Park, Tom Paulay, and Nigel Priestley.

My hope is that the material contained herein will encourage the development of a dialogue that will result in a more rational approach to the seismic design of concrete buildings. The comments of you, the reader, will be appreciated.

I dedicate this book to my family and, in particular, to my wife, Natalie, whose patience and understanding allowed for its development.

NOMENCLATURE

I have chosen to use both English and metric units so as not to alter the graphic description of experimental data. The following conversions are standard:

$$1 \text{ m} = 39.37 \text{ in.}$$

$$1 \text{ kN} = 0.2248 \text{ kips}$$

$$1 \text{ kN-m} = 0.737 \text{ ft-kips}$$

$$1 \text{ MPa} = 1000 \text{ kN/mm}^2$$

ADOPTED NOMENCLATURE

A Area, usually subscripted for definition purposes

A_j Effective cross-sectional area within a joint in a plane parallel to plane of reinforcement generating shear in the joint. The joint depth is the overall depth of the column. The effective width will depend to a certain extent on the size of the beams framing into the joint.

A_{ps} Area of prestressed reinforcement in tension zone

A_s Area of nonprestressed tension reinforcement

A'_s Area of compression reinforcement

A_{sh} Total cross-sectional area of transverse reinforcement (including crossties) within spacings

A_{st} Total area of longitudinal reinforcement

A_1 Loaded area

A_2 The area of the lower base of the largest frustum of a pyramid, cone, or tapered wedge contained wholly within the support and having for its upper base the loaded area and having side slopes of 1 vertical to 2 horizontal

C Compressive force—subscripted when qualification is required

C_d Force imposed on the compression diagonal

D Dead loads; depth of frame

DR Drift ratio (Δ_x/h_x) or (Δ_n/H)

E Load effects of seismic forces, or related internal moments and forces; modulus of elasticity usually subscripted to identify material

EI Flexural stiffness

F Loads attributable to strength of provided reinforcement, usually subscripted to identify condition

F_y Yield strength of structural steel

H Overall height of frame

I_{cr} Moment of inertia of cracked section transformed to concrete

I_e Effective moment of inertia

I_g Moment of inertia of gross concrete section about centroidal axis, neglecting reinforcement

L Live loads, or related internal moments and forces

M Moment in member, usually subscripted to identify loading condition, member, or stress state

M Mass subscripted when appropriate to identify (e) effective or (1) contributing mode

M_{bal} Nominal moment strength at balanced conditions of strain

M_{cr} Cracking moment

M_{el} Elastic moment

M_{pr} Probable flexural moment strength of members, with or without axial load, determined using the probable properties of the constitutive materials

N An integer usually applied to number of bays or number of connectors

P Axial load, usually subscripted to identify load type or strength state

P_b Nominal axial load strength at balanced conditions of strain

P_o Nominal axial load strength at zero eccentricity

P_{pre} Prestressing load applied to a high-strength bolt

Q Stability index for a story—elastic basis (see Section 4.3.1)

Q^* Stability index for a story—inelastic basis (see Section 4.3.1)

$\hat{R}$ Spectral reduction factor

S_a Spectral acceleration—in./sec

S_{ag} Spectral acceleration expressed as a percentage of the gravitational force g

S_d Spectral displacement

S_v Spectral velocity

SF	Square feet
U	Required strength to resist factored loads or related internal moments and forces
V	Shear force usually quantified to describe associated material or contributing load
V_c	Shear strength provided by concrete
V_{ch}	Nominal capacity of the concrete strut in a beam-column joint
V_N	Component of joint shear strength attributed to the axial load imposed on a column load
V_{sh}	Nominal strength of diagonal compression field
W	Wind load
W	Weight (mass) tributary to a bracing system
a	Depth of equivalent rectangular stress block, acceleration, shear span
b	Width of compression face of member
b_w	Web width
c	Distance from extreme compression fiber to neutral axis
c_c	Clear cover from the nearest surface in tension to the surface of the flexural tension reinforcement
d	Distance from extreme compression fiber to centroid of tension reinforcement
d	Displacement (peak) of the ground
$\dot{d}$	Velocity (peak) of the ground
$\ddot{d}$	Acceleration (peak) of the ground
d'	Distance from extreme compression fiber to centroid of compression reinforcement
d_b	Bar diameter
d_s	Distance from extreme compression fiber to centroid of tension conventional reinforcement
d_{ps}	Distance from extreme compression fiber to centroid of prestressed reinforcement
d_z	Depth of the plate
e	Eccentricity of axial load
f	Friction factor; measure of stress, usually subscripted to identify condition of interest
f'_c	Specified compressive strength of concrete
f'_{ci}	Compressive strength of concrete at time of initial prestress
$\sqrt{f'_{ci}}$	Square root of compressive strength of concrete at time of initial prestress
f_{cr}	Critical buckling stress
f_{ct}	Average splitting tensile strength of aggregate concrete

f_{cg} Stress in the grout

f_{pse} Effective stress in prestressed reinforcement (after allowance for all prestress losses)

f_{py} Specified yield strength of prestressing tendons

f_r Modulus of rupture of concrete

f_s Calculated stress in reinforcement

f_{sc} Stress in compression steel

f_y Specified yield strength of reinforcement

f_{yh} Specified yield strength in hoop reinforcing

h Overall thickness of member

h_c Cross-sectional dimension of column core measured center-to-center of confining reinforcement

h_n Height of the uppermost level of a frame

h_w Height of entire wall or of the segment of wall considered

h_x Maximum horizontal spacing of hoop or crosstie legs on all faces of the column; story height

k Effective length factor for compression members; system stiffness usually subscripted to identify objective

k_{el} Elastic stiffness

k_{sec} Secant stiffness

kd Depth of neutral axis—elastic behavior is assumed

ℓ Span length of beam center to center of supporting column

ℓ_c Clear span of beam from face to face of supporting column

ℓ_d Development length for a straight bar

ℓ_{dh} Development length for a bar with a standard hook

ℓ_w Length of entire wall or of segment of wall considered in direction of shear force

n An integer usually applied to number of floors

r Radius of gyration of cross section of a compression member

s Spacing of transverse reinforcement

t_g Thickness of grout

w Unit weight

w_z Width of steel plate

y_t Distance from centroidal axis of gross section, neglecting reinforcement, to extreme fiber in tension

α Factor in bar development length evaluation. 1.3 for top bars, 1.0 for bottom bars. See ACI,[2.6] Eq. 12.2.2

β Coating factor. See ACI,[2.6] Eq. 12.2.2

β_1	Factor that defines the relationship between the depth of the compressive stress block and the neutral axis depth, c[2.6]
γ_p	Postyield shearing angle
Γ	Participation factor
δ_u	Member or component displacement
Δ	An increment of force, stress, or strain
Δ_n	Relative lateral deflection between the uppermost level and base of a building
Δ_x	Relative lateral deflection between the top and bottom of a story
ε	Strain—usually subscripted to describe material or strain state
ζ	Structural damping coefficient expressed as a percentage of critical damping
$\hat{\zeta}$	Total damping coefficient expressed as a percentage of critical damping
θ	Rotation
λ	Lightweight aggregate concrete factor
λ_o	Component or member overstrength factor that describes overstrength expected in a member
μ	Ductility factor usually subscripted; bond stress; friction factor
μ_Δ	Displacement ductility factor
μ_ε	Strain ductility factor
μ_θ	Rotation ductility factor
μ_ϕ	Curvature ductility factor
ρ	Ratio of nonprestressed tension reinforcement, A_s/bd
ρ'	Ratio of nonprestressed compression reinforcement, A_s'/bd
ρ_b	Reinforcement ratio producing balanced strain conditions
ρ_g	Ratio of total reinforcement area to cross-sectional area of column
ρ_s	Ratio of volume of spiral reinforcement to total volume of core (out-to-out of spirals) of a spirally reinforced compression member
ρ_v	Ratio of area of distributed reinforcement perpendicular to the plane of A_{cv} to gross concrete area A_{cv}
ϕ	Curvature, rad/in.; capacity-based reduction factor; strength reduction factor
ϕ_e	Normalized elastic displacement (Δ_i/Δ_u)
ϕ_k	Stiffness reduction factor
ϕ_p	Probable overstrength of the steel
ω	Reinforcement index $\rho f_y/f_c'$
ω'	Reinforcement index $\rho' f_y/f_c'$
ω_p	Reinforcement index $\rho_p f_{ps}/f_c'$
Ω_o	System overstrength factor

SPECIAL SUBSCRIPTS

Special subscripts will follow a notational form to the extent possible. Multiple subscripts will be used where appropriate, and they will be developed as follows:

1. s, u, n, p, pr, y, i, max, and M will be used to describe member strength or deformation state:

   s, service or stress limit state (unfactored)

   u, ultimate or factored capacity (strength)

   n, nominal capacity

   p, postyield

   pr, probable

   i, idealized

   y, yield

   max, maximum permitted

   min, minimum permitted

   M, mechanism

2. c, b, s, f, and p will be used to describe a member category or characterize a system behavior condition:

   c, column

   b, beam

   s, shear component of deformation

   f, flexural component of deformation

   p, postyield component of deformation

3. e, i will be used to describe a location; i will also be used to identify an idealized condition such as yield:

   e, exterior beam or column

   i, interior beam or column

4. L, D, E will be used to describe a load condition:

   L, live load

   D, dead load

   E, earthquake load

5. A, B, C, L, R and 1, 2 will be used to locate an event with reference to a specific plan grid or point:

   L, left

   R, right

Example:

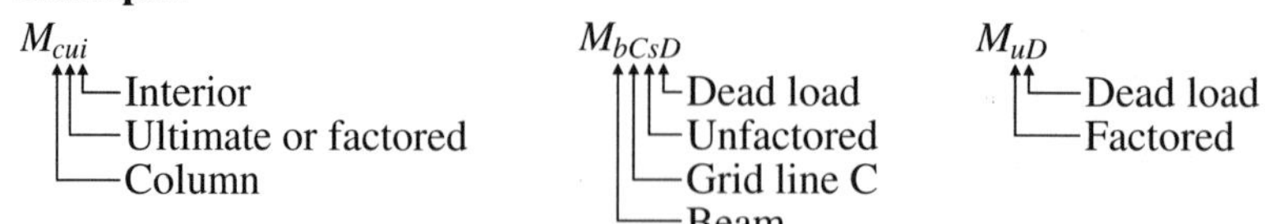

6. Capitalized subscripts will be used to describe the stress class and its location:
   B, bottom
   C, compression
   CB, compression bottom
   CT, compression top
   T, top, tension, transverse
   TB, tension bottom
   TT, tension top
7. Special subscripts will be used to identify the following:
   a, attainable or average
   d, design, as in design basis
   D, degrading or diaphragm
   ed, energy dissipater
   g, grout
   SDOF, single-degree-of-freedom system

INTRODUCTION

". . . the shoe that fits one person pinches another."
—Carl Gustav Jung

This book is primarily about design, which, as I use the term, is the creative process that seeks the proper blend of essential ingredients—specifically function, aesthetics, economy, and, in the context of this book, seismic behavior. There exists no single formula for creating a good design, for the design process involves making a set of decisions on issues for which no absolutely right answer exists. Thus the designer is continually seeking a comfortable rationally based design solution, and two identical solutions are not likely to be produced even successively by the same constructive designer.

Tools are essential to the completion of almost every task. I have tried to assemble, in as concise a form as possible, the tools necessary to the pursuit of a good design. From the extensive library of experimental efforts, I have selected representative works and demonstrated how both strength and deformation limit states might be predicted. Next, I review alternative design approaches and, in the process, simplify and adapt them to specific types of bracing systems. Finally I describe how designs might be comprehensively reviewed.

The focus of the book is concrete and the emphasis is on precast concrete. I have limited the scope to the satisfaction of seismic behavior objectives because the topic is complex and, though extensively studied and codified, not necessarily well understood by the structural design profession. The fact that seismic design can be reduced to an understandable level that can be creatively introduced into a building program makes it an ideal vehicle to study the design process.

Concrete as a composite material provides a medium that encourages freedom. The design of structures constructed using composite materials is not peculiar to the materials selected for any combination of dissimilar materials must satisfy the same basic fundamental laws and this is because equilibrium, compatibility, and adherence to the appropriate stress-strain relationship must always be attained. Accordingly, the choice of concrete as a vehicle should not be viewed as a constraint on the applicability of the material contained herein.

Precast concrete is but one creative extension of the use of concrete. It is an especially important extension because the prefabrication of structures can and will be required to meet the needs of society. The use of precast concrete has traditionally been viewed with skepticism in regions considered to have a potential for seismic activity. This is largely the result of a lack of understanding of the basic nature of seismic behavior and how the attributes of precast concrete can be exploited to improve behavior. The designer of a precast concrete structure, armed with the proper tools, can create a structure that will not only survive an earthquake, but do so with very little, if any, damage. To accomplish this lofty objective requires only that the designer take advantage of the jointed nature of the assemblage of precast elements.

To present the seismic design of precast concrete as a stand-alone topic would limit the usefulness of the treatment because a consistent base is critical to both explaining and understanding the behavior of precast concrete members and systems. Accordingly, the basic elements of both seismic behavior and the behavior and design of concrete must precede any treatment of precast concrete. The precast concrete seismic systems whose design is described in some detail herein are only intended to be examples of what can be accomplished with creative thinking. The objective then is to inspire creative applications of a versatile product.

The design process must be free and dynamic to be effective. Accordingly, a design must move aggressively to make the many decisions required in an orderly fashion with a minimum amount of distraction. The process usually starts by tackling the most difficult decision(s) first and, when necessary, looking quickly downstream in the decision-making process to confirm that potential problems do, in fact, have a solution.

I endeavor to place emphasis on the primary objectives of the design process and relieve, or at least loosen wherever possible, the ever-increasing number of prescriptive constraints being imposed on designs. This is especially important because the concept development (creative) part of a design must focus on the broader objectives and leave the details to the development of the concept. The importance of detail is not discounted by this apparent deferral, for the completed design package must be very clear on how the broader objectives are accomplished. It is this almost subliminal awareness of detail that will allow the focus essential to the creation of an excellent design.

Creative design clearly does not allow regimentation, and this makes it almost impossible to present design as a subject. My effort toward regimentation is limited to subdividing the presented material into four broad categories, but even this is not adhered to strictly. Chapter 1 discusses selected basic concepts. The objective is to provide the designer with the basic insight necessary to the effective development of a design. A comprehensive treatment of each topic would, in most cases, take volumes and tend to obscure the basic concepts and objectives. I have tried to identify references for the reader who is not satisfied with the brevity of treatment contained herein. Fortunately and unfortunately, the expanded treatment of many of the basic concepts presented herein has reached a level of development far beyond the technical capability of most of us. The fortunate aspect is that most of the theory is finding its way into computer applications that, if properly applied and understood, should help the designer make difficult decisions. The unfortunate aspect is that most of the material has been presented, at least to date, in a way that does not allow the rapid assimilation of concepts by the reader whose primary preoccupation is in an ancillary area. Perhaps a treatment along the lines of *Shakespeare for Dummies* would find a larger audience.

I considered including several additional topics in Chapter 1, but elected instead to scatter them throughout the book. They are covered in the discussions of the design processes where they may effectively be used. To compensate for the resulting scatter, I have tried to use the index as an effective locator.

The most important of these topics relates to understanding statics and indeterminate structures and how this understanding might quickly and reliably be reduced to design methodologies. It seems as though each passing year and each new software package causes us to become less facile in reducing complex structures to a level that allows us to make the appropriate design decisions. Design by iteration is becoming increasingly popular, but it will never be effective as a tool to create the desired balanced design. A learned mathematician once assured me that enough monkeys armed with typewriters would ultimately produce all of the works of Shakespeare. The problem from the structural design perspective is that we are given neither the time nor the money to follow this path. The question usually proposed to the designer is "Can I do this?" and the time frame allowed for coming up with an answer is measured in days. Such a time frame does not allow for extensive research or for time-consuming analytical procedures. The analytical reductions used in various example designs not only allow design insight and a quick means of evaluating the efficacy of a concept, but also a quick check of computer solutions.

Chapter 2 deals with the behavior and design of components of bracing programs. The approach to component design in Chapter 2 starts by reviewing selected experimental efforts and attempting to use the results of the experimental effort to support or propose design procedures for the component. One need only look at the tables of contents of the many technical journals to appreciate how much experimentation is being documented annually in universities around the world. Accordingly, it is impossible to reduce all available experimental data to a digestible form; thus I have been very selective.

Components can also be systems as, for example, in the case of shear walls. I have tried to draw the following distinction in the adopted approach. If a body of experimental work treats the subject, I have included the subject in Chapter 2. In the case of shear walls, Chapter 2 explores the experimental evidence and how these data might be effectively used to create a design approach. The element is then reintroduced as a part of a system in Chapter 3, expanding on the previously developed design procedures.

The seismic performance of a component is not exclusively concerned with its strength, for we know that seismically induced displacement demands will force members to deform well beyond their elastic limit states. The success with which a component responds to these postyield deformation demands can only be evaluated by understanding strain limit states and the damage that is likely to occur as a function of large ductility demands. Critical strains are those that define the inception of damage—shell spalling and strength degradation. The experimental efforts used to identify probable strain states is of necessity limited, and the reader is encouraged to continually review new or other pertinent experimental efforts in order to establish what he or she believes to be the appropriate limit state. Perhaps focused research will be undertaken to establish and confirm some of the more speculative limit states.

Having proposed a set of strength- and strain-based limit states, Chapter 2 next develops a design methodology for each component, and shows how limit states can be tested by example. Detailing considerations are discussed and details developed. Where appropriate, codification concepts are discussed and reduced to a level of analytical simplicity appropriate for design. Occasionally, the limited applicability of commonly held dogma is reviewed, as are procedures or behavior characteristics not commonly used by U.S. designers. The goals of this chapter are to reduce component design to as simple a process as possible and to provide insight into objectives often well disguised in the codification process.

Chapter 3 is the heart of the book. The focus of Chapter 3 is the design of bracing systems. The objective is to conceptually create a bracing program that is effective from both a cost and a behavior perspective. Building behavior must be controlled in the design process. The building must behave as you, the designer, intend it to, and only you can make this happen. I, as a grandfather, explain to my students that the behavior of a building is probably the only thing in your life that you have a chance of controlling.

One of the lessons I learned early in my career was that a design or a design concept must be less expensive and better than its alternative if it is to be accepted or adopted, and that the "better" part was a distant second consideration. Thus it is incumbent on the designer to create a cost-effective design in order for it to be realized and to almost subliminally include the "better" aspect into every design.

The appropriateness of a well-conceived design may defy codified dogma, and this will require courage on the part of the designer. An ethical issue is clearly raised, one that must be resolved by the responsible designer after careful study and consultation with peers.

Chapter 3 also presents a variety of design approaches. These include classical strength procedures as well as displacement-based approaches. I have advocated and used displacement-based procedures for more than thirty years, and I am convinced that they offer by far the best chance for producing a successful design. I support the development and acceptance of a performance-based approach wholeheartedly. Procedures currently proposed are not, in my opinion, easily applied to the conceptual design process. Further, they do not take advantage of a designer's understanding of the behavior characteristics inherent to structural systems. I, therefore, propose simpler procedures that include the basic philosophies of each displacement-based approach, specifically those that follow the equal-displacement tenet and those that treat structural damping as a system-dependent variable. When I first explored displacement-based procedures, design earthquakes were nowhere near as strong as they now are and, as a consequence, strain limit states that I established after analyzing experimental effort never were approached. Now with design earthquake intensities five or so times greater, these limit states are being approached, but who knows how strong the next decade's earthquakes might be? Therefore, I treat ductility much as a rich person treats money—there simply is no such thing as too much.

The precision suggested by most design procedures is illusory. Response modification factors now identify more than fifty categories with variations of 1.2% between vastly different types of structural systems. This suggests a solid technical basis that does not exist. When I work in three significant figures in examples, it is not because I believe it to be analytically appropriate to do so, but rather only to allow the reader a better chance to track the example. Typically, when I prepare a design, I work to two significant figures and try to constantly review or crosscheck my conclusions to make sure I get into the ballpark; once I'm in the ballpark, it is easy enough in the analysis phase to find the right seat.

Finding the ballpark is an essential part of the conceptual design process. As a designer you will soon learn that once a program is set it cannot be changed and the only real option is to mitigate mistakes in concept. On the other hand, if the first step is in the right direction and allows the latitude to properly consider potential contingencies, the design will flow smoothly. Early on I found that if I located my bracing systems in areas that would otherwise not be used, they could maintain their integrity. Whenever bracing systems can come into conflict with other building systems, rest assured a conflict will eventually occur. For years I have held the belief that mechanical/plumbing engineers, in spite of whatever lip service they may give during the conceptual design phase, do not start their designs in earnest until the concrete structure has progressed far enough to make it necessary to core or cut into it. So make sure your design, if it can be in harm's way, has some breathing room.

Chapter 4 introduces the reader to the elastic and inelastic time history procedures used to confirm or evaluate the efficacy of a design. The objective is not to support any particular design approach, but rather to better understand the messages time history analyses can convey. Presumably the designer will know the final answer before this type of analysis is undertaken, for it will be very painful if a major change has to be made at this stage. If the design procedures and checks presented in Chapter 3 are followed, it is unlikely that a significant change will be required. So why bother with a confirmation of the design? A retrospective review of any decision-making process extends the all-important experience base. It should, at the very least, add confidence to future designs or provide the courage necessary to take on more challenging designs. The architecture of structural expressionism is a thing of the past. The free form of today's architectural styles requires boldness on the part of the structural engineer, and this must be supported by knowledge, experience, and confidence.

Chapter 4 also explores the sensitivity of designs to parameters like strength and hysteretic damping. If parametric studies are used extensively in the retrospective review process, they should allow the designer to more effectively control the behavior of buildings in subsequent designs. Sufficient strength, for example, has traditionally been viewed as the key to a successful design. Some professional societies believe that buildings should be designed so as to respond almost exclusively in the elastic range. The parametric studies included herein, and those performed by others, suggest that the benefits associated with an increase in strength are small. This is certainly not intended to discount strength as an important design consideration, but rather to point out that associated negative impacts can be greater than the potential reductions in displacement. Remember, the strength of the yielding element will impose more demand on the brittle components along the lateral load path. Increases in system strength will also increase accelerations, and this too will tend to cause more damage to building contents.

The fullness of a hysteresis loop has always been considered a positive attribute. How full must it be to produce the desired control over building response? Parametric studies suggest that, like strength, there is only a vague link, provided that reasonable levels of both strength and energy dissipation are provided. It is possible that carefully designed shaking table tests will shed more light on these issues. Until such time, designers will have to use available tools and their intuition to produce the best possible building. I am convinced that, given today's knowledge base, successful designs can reliably be produced.

My hope is that the material contained in this book will make it possible for both the student and practitioner to effectively utilize the vast amount of material that has been developed over the past quarter century to develop designs with which they can be comfortable, designs that will serve society well.

1

BASIC CONCEPTS

"Nature to be commanded must be obeyed."
—Francis Bacon

To be an effective designer, an engineer must intuitively understand those elements that will significantly impact the behavior of a structure. The objective of this introductory chapter is to discuss, from a philosophical perspective, those elements whose intrinsic understanding and acceptance are essential to the effective seismic design of concrete and precast concrete structures.

The application of a scientific principle is traditionally accomplished by reducing it to a mathematical expression. Two topics are especially difficult to reduce to algorithmic form—ductility and shear. Absent an intuitive understanding of these topics, the imaginative component so essential to solving problems or promoting improved behavior cannot be exercised.

Ductility is an essential attribute of a structure that must respond to strong ground motions. Ductility serves as the shock absorber in a building, for it reduces the transmitted force to one that is sustainable. The resultant sustainable force has traditionally been used to design an hypothetically elastic representation of the building. This design simplification is deceptive because it deals with a fictitious structure and ductility becomes the incidental consequence of a design. Recently, an attempt has been made to appropriately direct the focus of the design process to the real structure and the region of anticipated behavior. The impact of ductility on system behavior and how it might be more rationally introduced into the design process will be discussed in Section 1.1.

Shear strength and the effective development of shear transfer mechanisms are important to the design of concrete structure and critical to the effective use of precast concrete as a seismic load path. The complex mechanisms of shear transfer that exist in cast-in-place concrete have been traditionally reduced to beam or pseudo beam behavior mechanisms, and this is not always appropriate, particularly in precast concrete structures. An understanding of shear transfer mechanisms and their limit states is essential to developing load paths in precast concrete systems. In Section 1.3 the mechanisms of shear transfer will be explored and limit states identified.

1.1 DUCTILITY—A SYSTEM BEHAVIOR ENHANCER

Elastic behavior serves as the basis for the development of most structural design procedures. In some cases the material is truly elastic in the region of design interest, and theory correctly predicts behavior. Where elastic behavior is not anticipated, we use idealizations (see Figure 2.1.3) to produce models that convert inelastic behavior to a form that fits analytical tools developed from the theory of elasticity. This idealization process is effective, provided that the idealization is not extreme and that the designer accounts for the idealization process in the development of the design, especially when strain states are a concern.

An elastic model must be used with caution and understanding in the design of structures that will be subjected to earthquake-induced ground motions. This is because the induced level of deformation will significantly exceed the idealized elastic limit of the system. The relationship between the anticipated level of displacement and the displacement at idealized yield (see Figure 1.1.1) is called ductility (Δ_u/Δ_{yi}), and this ductility demand can be on the order of 10.

Several design procedures have been explored in society's quest to produce structures that will survive earthquakes. These procedures can be categorized according to their basis as generally relating to

- Force
- Displacement
- Energy

Figure 1.1.1 idealizes the behavior of a ductile structure. At some point (F_{max}) the displacement increases with little or no increase in applied force. *Force*-based design procedures use elastic models to predict a hypothetical strength objective (μF_{max}),

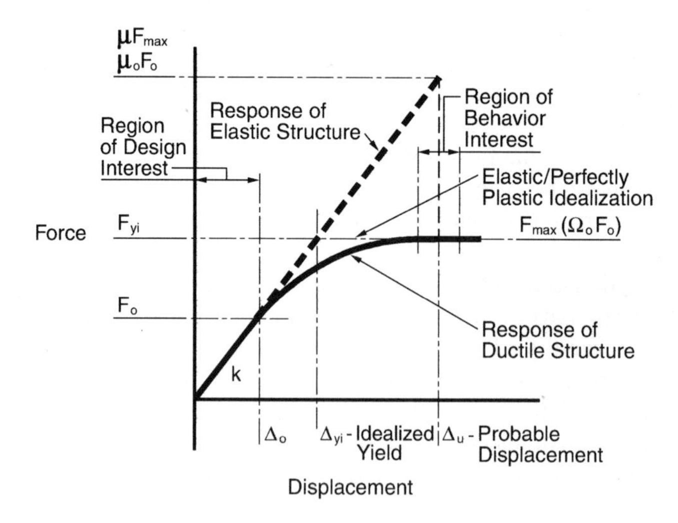

Figure 1.1.1 Force–displacement relationship of a ductile structure.

which is then reduced to an idealized strength objective (F_{max}) based on a perceived level of ductility. *Displacement*-based design procedures endeavor to predict the displacement a structure will experience during an earthquake. The impact of this displacement on components is then used to quantify elastic idealization objectives. *Energy* is dissipated by a ductile structure. Several design methodologies have been developed that consider dissipated energy, but the approach most recently proposed endeavors to convert dissipated energy to equivalent structural damping, which serves to reduce the level of system response.

These design alternatives will be philosophically discussed in this section, applied in Chapter 3, and to a limited extent, evaluated in Chapter 4.

Regardless of the adopted design approach, the designer must not only appreciate the importance of introducing ductility into a system but also be able to predict its impact and effectively develop a ductile structure. To this end, the following topics must be explored:

- How to predict the response of a ductile structure to ground motion
- How to create a linkage between the ductility available in a component with that available in the system
- How to effectively promote system ductility

1.1.1 Impact on Behavior

The designer of a building that is to be located in a seismically active area must understand how buildings respond to ground motion. One need not be an expert in

Photo 1.1 Ductile concrete, Northridge earthquake, 1994. (Courtesy of Englekirk Partners, Inc.)

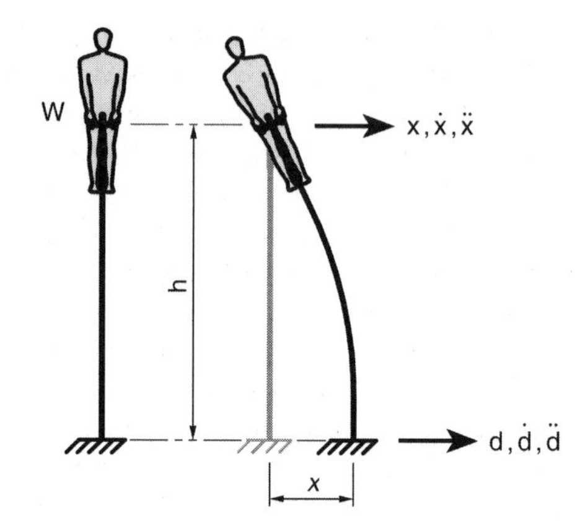

Figure 1.1.2 A single-degree-of-freedom system.

dynamics to intuitively understand the dynamic response of buildings and the impact ductility can have on this response. The material in this section develops the dynamic response of ductile structures using a single-degree-of-freedom model. The choice of this model is appropriate because almost all structures arrive at their displacement limit state when responding in a manner that can be analytically described in terms of the behavior of an equivalent single-degree-of-freedom system. Understanding the behavior of a single-degree-of-freedom system will not only provide the intuitive understanding we seek, but also provide a platform that can be extended to more complex systems.

Consider the single-degree-of-freedom system described in Figure 1.1.2. As the ground moves or displaces, you, at least for an infinitesimally small period of time, will not move. The characteristics of the ground to mass connection will play a major role in the response of the mass to ground motion. If the element that connects you to the ground is infinitely rigid, you will experience the same acceleration, velocity, and displacement as the ground. If, on the other hand, the linkage is very flexible, you will not move and the relative displacement (x) will be the displacement of the ground. Most buildings fall somewhere between these extremes.

If we treat the stiffness of the supporting element as the only variable, we can appreciate the impact stiffness characteristics and, of most importance, ductility will have on system behavior. At some point in the response, the distance between the mass and the base will be a maximum (x_{max}). The force imposed on the mass will be

$$F = kx_{max} \tag{1.1.1}$$

Following Newton, the mass, when subjected to this force, will experience an acceleration ($\ddot{x}$):

$$F = m\ddot{x} \tag{1.1.2}$$

It follows that the maximum acceleration is

$$\ddot{x}_{\max} = -\frac{k}{m} x_{\max} \tag{1.1.3a}$$

The introduction of the minus sign is easily understood by examining Figure 1.1.2, for the displacement and the sense of the acceleration are clearly opposite. This development is understood today by anyone who survives an elementary course in physics. The extension of Eq. 1.1.3a to structures that possess ductility is not universally appreciated.

Consider the elastic/perfectly plastic behavior described in Figure 1.1.1. Now force and displacement no longer maintain the proportional relationship defined by Eq. 1.1.1, for once the displacement reaches Δ_{yi}, the force applied to the mass (restoring force) will not exceed $F_{\max}$. This force limit also limits the restoring acceleration $(-\ddot{x})$ imposed on the mass. As a consequence, the maximum acceleration the mass can experience is

$$\ddot{x}_{\max} = \frac{F_{\max}}{m} \qquad \text{(see Eq. 1.1.2)}$$

The restoring force or acceleration imposed on the yielding structure (elastic/perfectly plastic) at a differential displacement of $2\Delta_{yi}$ (Figure 1.1.1) will, for example, be only half of that imposed on the comparable elastic structure, and this could be a positive attribute if for no other reason than the impact the reduced acceleration will have on the contents of the building.

Equation 1.1.3a describes the behavior of a single-degree-of-freedom system that is not acted upon by any external force. Were we to subject the system to any initial displacement and release it, we would create what is referred to as free undamped vibration. In other words, once released, the system would oscillate between plus and minus d_o. This (steady-state) motion can be analytically described using the differential equation developed from Eq. 1.1.3a.

$$m\ddot{x} + kx = 0 \tag{1.1.3b}$$

The solution of this differential equation is developed in most books on dynamics.[1.1–1.4] The solution so developed identifies the displacement of the mass, given an initial (impulsive) displacement of d_o as a function of time and the natural frequency of the system, (ω_n):

$$x(t) = d_o \cos \omega_n t \tag{1.1.4}$$

Equation 1.1.4 describes a sinusoidal motion with a natural frequency of ω_n, expressed in radians per second, where

$$\omega_n = \sqrt{\frac{k}{m}} \tag{1.1.5}$$

Velocity is the rate of change in displacement with respect to time and, as a consequence, is the (time) derivative of Eq.1.4 (dx/dt). Accordingly, a sine function is created. At maximum displacement ($\cos \omega_n t = 1.0$), it follows that the velocity of the mass will be zero ($\sin \omega_n t = 0$).

Acceleration, the time rate of change in velocity $(d\dot{x}/dt)$, is developed in the same manner. A negative cosine function describes acceleration. Accordingly, the acceleration of the mass will be a maximum value when the displacement is at a maximum, but it will be opposite in sense—essentially trying to restore the mass to its original relationship with the ground. These motions are described in Figure 1.1.3.

Observe also that the relationships between displacement and velocity (Eq. 1.1.5) and between velocity and acceleration are functions of the natural frequency (ω_n). This allows the peak responses described in Figure 1.1.3 to be related as follows:

$$\dot{x}(t)_{\max} = \omega_n x(t)_{\max} \tag{1.1.6a}$$

$$\ddot{x}(t)_{\max} = \omega_n \dot{x}(t)_{\max} \tag{1.1.6b}$$

$$\ddot{x}(t)_{\max} = \omega_n^2 x(t)_{\max} \tag{1.1.6c}$$

The usual philosophical concern with the dynamic response of a ductile structure is that the displacement experienced by the ductile structure will be significantly larger than that of a comparable elastic structure because the restoring force (deceleration rate) has been reduced.

To understand how this limited restoring force (Figure 1.1.1) impacts response, consider how a controlled level of acceleration and deceleration $(-\ddot{x})$ changes the motion of the mass. Obviously, a mass laterally supported by a system of limited strength or restoring force will initially experience a greater differential displacement (Figure 1.1.2) than it would if the deceleration rate were unlimited (elastic system).

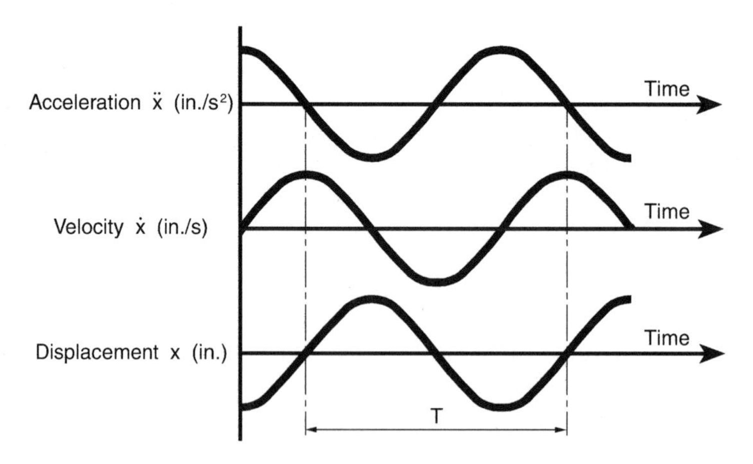

Figure 1.1.3 Steady-state vibration.

This happens to all ductile buildings, as we shall see in Chapter 4 when we compare the response of an elastic system to the response of a system of limited strength. In buildings, this initially higher level of relative displacement will occur during the first postyield excursion, but this initial postyield displacement level will, in all but very stiff buildings, be exceeded as the intensity of the earthquake builds. As the ground motion increases, a limited acceleration input will result in a lower velocity than that which would be caused by the input of an unlimited acceleration. A reduction in velocity will cause subsequent displacements to be smaller. It has generally been accepted[1.5] that the maximum displacement response of a ductile system will not be greater than the response of a comparable elastic system. From a design perspective the equal displacement hypothesis is very convenient, for it allows us to analyze an elastic model of the building and directly apply the conclusions to the design of a ductile structure. This favors the adoption of a displacement-based design approach for ductile structures.

Accordingly, we might reasonably conclude that ductility, properly introduced into a bracing system, will improve the behavior of a building—primarily by reducing experienced accelerations.

1.1.2 Impact of Strength Degradation on Response

The dynamic response of a structure will have a stability limit state (Section 4.3.1). Presuming that the level of provided strength is sufficient to prevent collapse at maximum levels of drift, we must concern ourselves with the impact a limited amount of strength degradation will have on the dynamic response of the structure. In essence, we must be satisfied that drift projections based on a nondegrading model will not be exceeded should the structure experience some degree of strength degradation.

System strength degradation can be caused by the loss of strength a component experiences as it is subjected to increasing deformation demands. Figures 2.1.47 and 2.3.8 describe components that experience a limited amount of strength degradation. System strength degradation or, more specifically, a reduction in restoring force, is also caused by $P\Delta$ effects. Figure 2.2.5 describes how the restoring force in a bracing system is reduced by $P\Delta$ forces as displacement levels are increased.

The impact of $P\Delta$ effects on building response is analogous to that caused by the introduction of ductility into the system. For taller, more flexible structures, peak displacements will, if anything, tend to be reduced by $P\Delta$ effects, and this will be demonstrated in Chapter 4 when example buildings are analyzed. The displacement response of shorter stiffer buildings, or of buildings whose response maxima are created by impulse-type displacements, will increase, and as a consequence system stability might become a concern.

$P\Delta$ effects are generally understood though too often inappropriately considered. Figure 1.1.4 describes a frame—beam *c* and column *b*—that provides lateral support for a vertical load-carrying system. Column *a* and beam *d* rely on the frame for stability and lateral support. That portion of the strength of the frame (F_{max}) available to restore the mass, given the displacement of the ground during an earthquake, will

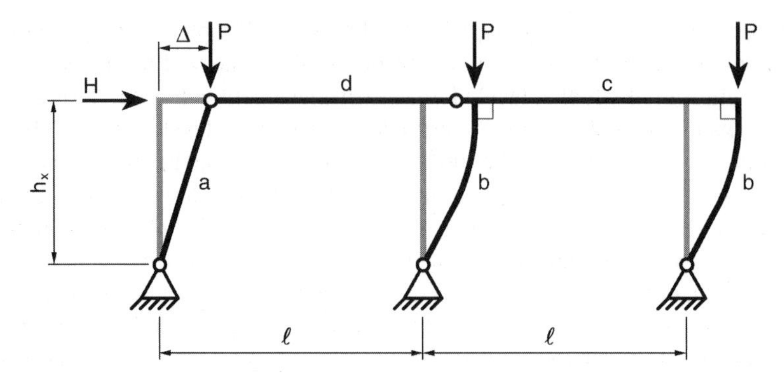

Figure 1.1.4 Framework subjected to $P\Delta$ sidesway.

be reduced by the force required to stabilize all of the tributary vertical load (mass). The force required to stabilize the vertical load in Figure 1.1.4 is

$$F_{P\Delta} = \frac{3P\Delta}{h_x} \quad \text{or} \quad F_{P\Delta} = \frac{\left(\sum W\right)\Delta}{h_x} \tag{1.1.7}$$

where W is the total mass (weight) whose lateral stability is provided by the frame.

This negative force, since it opposes the restoring force, will reduce the restoring force (deceleration) and, as a consequence, should impact the response of the structure in much the same way as does the introduction of a limited restoring force (ductility). At some point the restoring force or deceleration will be reduced to a point that no restoring force exists, and the structure will, as a consequence, collapse. Current system strength levels seem to significantly exceed those required to create a stable system. This subject will be addressed in more detail in Section 4.3.1. For now let us understand intuitively that neither the $P\Delta$ effect nor a limited amount of system strength degradation is likely to induce larger system displacements as buildings respond to earthquakes.

1.1.3 Quantifying the Response of Structures to Ground Motion

The objective of this section is to describe how one might determine the displacement and force a structure is likely to experience during an earthquake. Methodologies are presented in Sections 1.1.4 and 1.1.5 that allow this quantification of response to be converted into a design procedure appropriate for the development of a structure whose response is expected to be in the inelastic range. For conceptual design purposes the response of a system to earthquake ground motion is most easily developed from a response spectrum because this approach reduces a complex dynamic problem to one that may be treated as though it were static. A response spectrum quantifies the maximum response of an entire spectrum of single-degree-of-freedom systems to ground motion. The spectrum of a system is classified according to the system's level of damping. The response can be provided in the form of relative displacement, velocity, or absolute acceleration. Response spectra can be developed from a specific earthquake ground motion or generalized by combining specific spectra so as to create a design spectrum. Figure 4.1.1 describes the various earthquake-specific spectra and the generalized design spectra used in many examples developed in the book. It also shows how ground motion can be modified to fit the intensity of an adopted design spectrum. Design spectra are usually tailored to reflect an objective ground motion intensity.[1.1, 1.2, 1.6]

Elastic design response spectra are developed in a variety of ways. Newmark[1.5] developed a spectral form that is particularly well suited to the conceptual design process (Figure 1.1.5). Here, the elastic design spectrum is developed directly from the peak ground acceleration ($\ddot{d}_{go}$), velocity ($\dot{d}_{go}$), and displacement (d_{go}). Regardless of how a design spectrum is created, three behavior regions will evolve. The spectral acceleration will tend to be constant for stiff buildings ($T < 0.5$ second), the spectral velocity constant for most buildings (0.50 second $< T <$ 4 seconds), and the spectral displacement constant for buildings whose periods exceed about 4 seconds.

Damping, expressed as equivalent viscous damping, will affect the magnitude of the spectral response in each of these behavior regions. System damping is usually expressed as a percentage of critical viscous damping (ζ). Given this form of expressing damping and an estimate of peak ground motions, an elastic design spectrum can be created.

$$S_a = (4.38 - 1.04 \ln \zeta)\ddot{d}_{\max} \tag{1.1.8a}$$

$$S_v = (3.38 - 0.67 \ln \zeta)\dot{d}_{\max} \tag{1.1.8b}$$

$$S_d = (2.73 - 0.45 \ln \zeta)d_{\max} \tag{1.1.8c}$$

where

S_a is the design elastic spectral acceleration in the acceleration constant region (Figure 1.1.5, points b to c).

$\ddot{d}_{\max}$ is the (effective) peak ground acceleration (Figure 1.1.2).

S_v is the design elastic spectral velocity in the velocity constant region (Figure 1.1.5, points c to d).

$\dot{d}_{\max}$ is the peak ground velocity (Figure 1.1.2).

S_d is the design elastic spectral displacement in the displacement constant region (Figure 1.1.5, points d to e).

$d_{\max}$ is the peak ground displacement (Figure 1.1.2).

Understanding how spectral design values are impacted by damping is important in ductile structures because the experienced level of ductility is believed to be translatable to an equivalent level of structural damping, and this will have an impact on

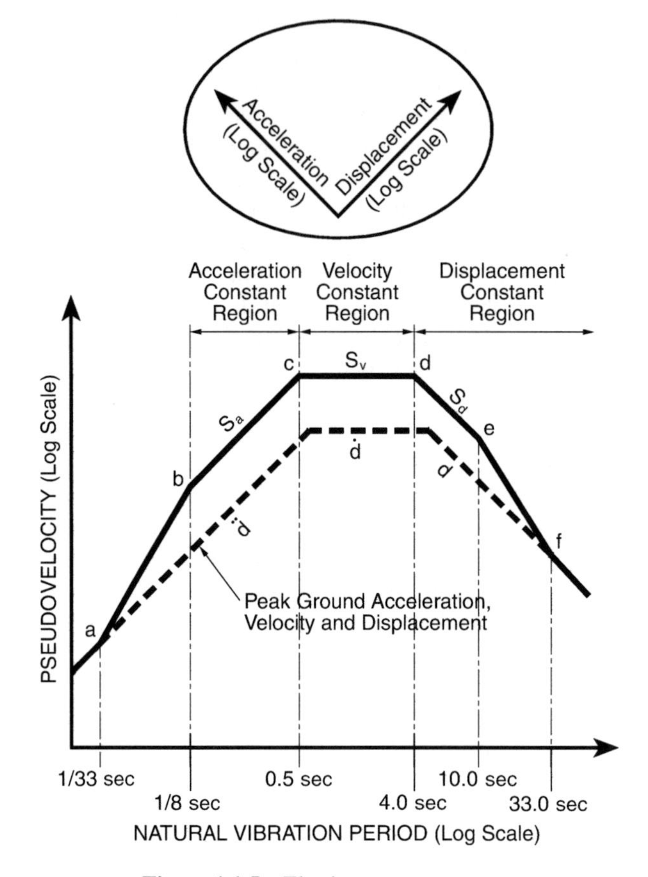

Figure 1.1.5 Elastic response spectra.

the response of a structure. Chopra and Goel[1.7] develop an expression for equivalent total viscous damping ($\hat{\zeta}_{\text{eq}}$) that includes the impact of the energy dissipated during a cycle of inelastic response:

$$\hat{\zeta}_{\text{eq}} = \zeta + \zeta_{\text{eq}} \tag{1.1.9a}$$

where the damping, which is equivalent to the hysteretic energy dissipated, is

$$\zeta_{\text{eq}} = \frac{2(\mu - 1)}{\mu\pi} \tag{1.1.9b}$$

and ζ is the nonstructural component of system damping.

This development (Eq. 1.1.9b) is based on the attainment of the "full" hysteretic response of a perfectly plastic member to a displacement of $\mu\Delta_y$. The energy dissipated during a full ductile excursion is described in Figure 1.1.6. An estimate of

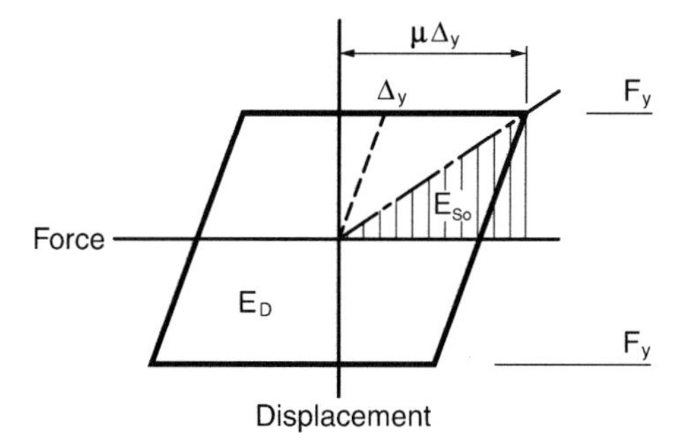

Figure 1.1.6 Energy dissipated by a "full" hysteretic response.

available structural damping produced by the hysteretic behavior (E_D) is developed by comparing the hysteretic energy dissipated to the maximum strain energy (E_{So}):

$$\zeta_{\text{eq}} = \frac{1}{4\pi}\frac{E_D}{E_{So}} \tag{1.1.9c}$$

The energy dissipated during a full hysteretic excursion as described in Figure 1.1.6 is

$$E_D = (\mu - 1)4\,\Delta_y F_y$$

while the maximum strain energy is

$$E_{So} = \frac{\mu\,\Delta_y F_y}{2}$$

Combining these relationships produces the estimate of equivalent structural damping proposed by Eq. 1.1.9b:

$$\zeta_{\text{eq}} = \frac{2(\mu - 1)}{\mu\pi} \qquad \text{(Eq. 1.1.9b)}$$

The damping equivalence proposed by Eq. 1.1.9b may be appropriate for a "full" hysteretic response as is typically approached in the response of a steel beam to cyclic motion,[1.6] but a full hysteretic response does not characterize the response of a concrete beam (Figure 2.1.2) and even less so the response of precast beams to cyclic postyield deformations (Figures 2.1.47 and 2.1.65).

Priestley[1.8] proposes a reduced level of equivalent structural damping (ζ_{eq}) for a concrete system responding in the inelastic range.

$$\zeta_{\text{eq}} = \frac{\sqrt{\mu} - 1}{\pi\sqrt{\mu}} \tag{1.1.9d}$$

For a system ductility of 4, the equivalent structural damping ζ_{eq}, according to Chopra[1.1] (Eq. 1.1.9b), would be

$$\zeta_{eq} = \frac{2(\mu - 1)}{\mu\pi} \qquad \text{(Eq. 1.1.9b)}$$

$$= \frac{2(4 - 1)}{4\pi}$$

$$= \frac{1.5}{\pi}$$

$$= 48\%$$

and according to Priestley[1.8], the level of equivalent structural damping would be

$$\zeta_{eq} = \frac{\sqrt{\mu} - 1}{\pi\sqrt{\mu}} \qquad \text{(Eq. 1.1.9d)}$$

$$= \frac{\sqrt{4} - 1}{\pi\sqrt{4}}$$

$$= \frac{0.5}{\pi}$$

$$= 16\%$$

Intuitively, it seems logical to somehow relate damping to the hysteretic energy dissipated. The relationship between the 16% developed from Eq. 1.1.9d and the 48% developed from Eq. 1.1.9b seems consistent and proportional to the area enveloped by the hysteresis loops of Figures 1.1.6 and 2.1.2. Whether or not these component-based relationships (Eq. 1.1.9b or 1.1.9d) can be extended to a system is reasonably questioned, especially in a frame braced structure where component ductility demands can vary significantly within the frame and be different from the ductility used to describe the response of the entire system. This caveat aside, elastic design spectra may be modified to account for varying levels of damping:

Example: Assume that the spectral velocity for a 5% damped elastic/perfectly plastic (Figure 1.1.1) structure is 60 in./sec. Assume further that we wish to modify this elastic spectral velocity to account for a system ductility of 2. First, determine the spectral velocity of the ground ($\dot{d}$) (see Figure 1.1.5):

$$\dot{d} = \frac{S_v}{3.38 - 0.67 \ln \zeta} \qquad \text{(see Eq. 1.1.8b)}$$

$$= \frac{60}{3.38 - 0.67 \ln 5}$$

$$= 26 \text{ in./sec}$$

Then determine the equivalent structural damping:

$$\zeta_{eq} = \frac{2(\mu - 1)}{\pi\mu} \quad \text{(Eq. 1.1.9b)}$$

$$= \frac{2(2-1)}{\pi(2)}$$

$$= 0.32 \quad \text{or} \quad 32\%$$

and the equivalent effective level of system damping:

$$\hat{\zeta}_{eq} = \zeta + \zeta_{eq} \quad \text{(Eq. 1.1.9a)}$$

$$= 5 + 32$$

$$= 37\%$$

The resultant design spectral velocity is

$$S_v = (3.38 - 0.67 \ln \hat{\zeta}_{eq})\dot{d} \quad \text{(Eq. 1.1.8b)}$$

$$= (3.38 - 0.67 \ln 37)(26)$$

$$= 25 \text{ in./sec}$$

Following Priestley,[1.8] the design spectral velocity would be developed as follows:

$$\zeta_{eq} = \frac{\sqrt{\mu} - 1}{\pi\sqrt{\mu}} \quad \text{(Eq. 1.1.9d)}$$

$$= \frac{\sqrt{2} - 1}{\pi\sqrt{2}}$$

$$= 0.09 \text{ or } 9\%$$

$$\hat{\zeta}_{eq} = \zeta + \zeta_{eq} \quad \text{(Eq. 1.1.9a)}$$

$$= 5 + 9$$

$$= 14\%$$

$$S_v = (3.38 - 0.67 \ln \hat{\zeta}_{eq})\dot{d}_{max} \quad \text{(Eq. 1.1.8b)}$$

$$= (3.38 - 0.67 \ln 14)(26)$$

$$= 42 \text{ in./sec}$$

Both procedures represent a significant reduction from the design elastic spectral velocity of 60 in./sec associated with a 5% damped system.

An inelastic design spectrum is most commonly created directly from the design elastic spectrum of Figure 1.1.5. Figure 1.1.7 describes a design inelastic spectrum following concepts proposed largely by Newmark.[1.5] It is consistent with the equal displacement concepts discussed in Section 1.1.1. Observe that the spectral velocity (S_v) and spectral displacement (S_d) are converted to force-based design values by dividing them by a ductility factor (μ).

This development is consistent with the force-displacement relationship described in Figure 1.1.1, where

$$\mu = \frac{\text{Elastic response } (\mu F_{\max})}{\text{Developable strength } (F_{\max})}$$

In the acceleration constant region, the reduction factor is attained by equating the elastic and inelastic strain energies. The resultant reduction factor is $\sqrt{2\mu - 1}$.

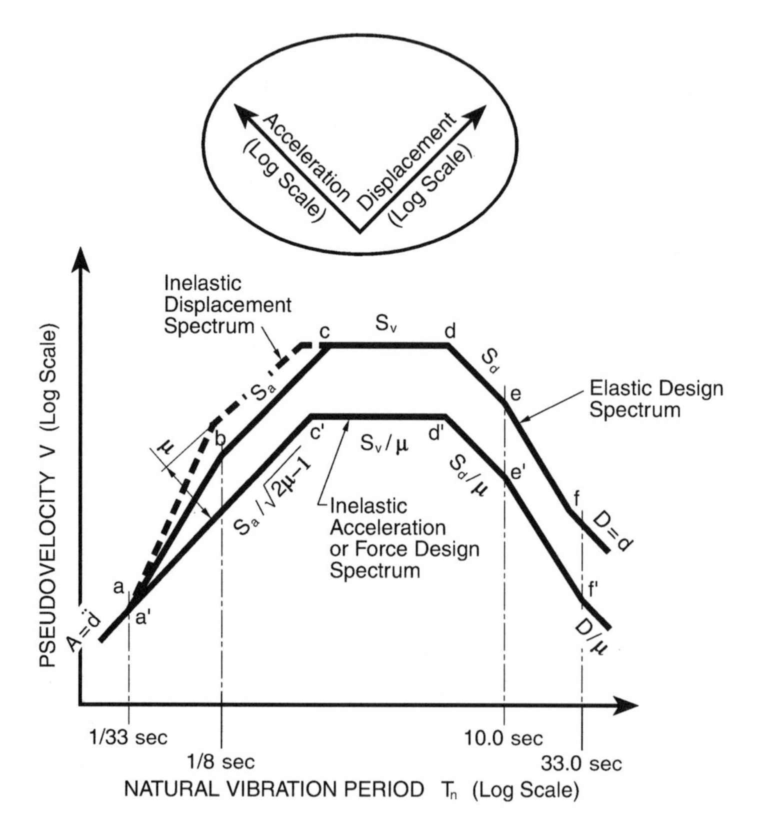

Figure 1.1.7 Construction of the design inelastic spectrum.

The inelastic displacement spectrum follows the elastic spectrum except in the acceleration constant region, where it is amplified by $\mu/\sqrt{2\mu - 1}$. This quantification of relative displacement maxima is usually referred to as "equal displacement" when incorporated into the description of a design process.

The design base shear response spectra of Figure 1.1.8 are commonly used in strength-based design procedures. The base shear spectrum provides the designer with a base shear coefficient that is a function of the fundamental period of the structure. It can be developed so as to include system ductility, and often includes subsurface soil characteristics. Base shear spectra also combine the effects of higher modes with that of the primary mode (T_1). The mode combining methodology adopted by current codes[1.9] is appropriate though conservative for estimating base shear, but inappropriate when the objective is to estimate displacement or the moment a shear wall is likely to experience. This inclusion of multimode effects differentiates the base shear spectra from the single-degree-of-freedom spectra described in Figures 1.1.5 and 1.1.7.

Regardless of the spectral approach used, the conceptual design process can be iterative, and the number of iterations will depend on the accuracy of the initial estimate of the period. The period (T) is the inverse of the natural frequency (Eq. 1.1.5) expressed in cycles per second:

$$T = 2\pi\sqrt{\frac{W}{kg}} \tag{1.1.10}$$

Observe that the period T is a function of the tributary mass or weight (W) and the stiffness of the bracing system (k). The designer is able to estimate the weight of the structure with reasonable accuracy during the conceptual design process.

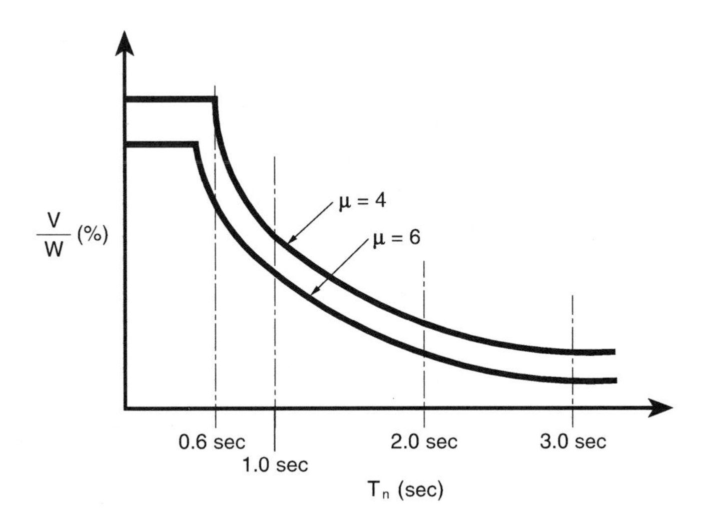

Figure 1.1.8 Design base shear response spectra.

Accordingly, Eq. 1.1.10 is reduced to a relationship between the period and the stiffness of the system. The period of the structure tends to be a function of the height of the structure and the characteristics of the bracing system. Fortunately, the period of most structures can be reasonably approximated through the use of simple relationships.

Current building codes[1.9] suggest that the period of a concrete frame braced building is

$$T = 0.03(h_n)^{3/4} \tag{1.1.11}$$

where h_n is the total height of the building in feet.

This relationship (Eq. 1.1.11) was developed for use in force-based design procedures. The objective was to create a conservative design force, so the predicted period (Eq. 1.1.11) is almost always less than the probable period. Were we interested in quantifying the ultimate displacement (Δ_u) the so-determined period (Eq. 1.1.11) is unconservative (see Figure 1.1.5). The use of a period that is double that suggested by Eq. 1.1.11 is recommended[1.6, Sec 4.7.4(a)] when displacement is the design parameter of interest. Hence, for displacement-based designs, Eq. 1.1.12 represents a better starting point than Eq. 1.1.11:

$$T \cong 0.06(h_n)^{3/4} \tag{1.1.12}$$

The period developed from Eq. 1.1.12 and the design response spectra (Figure 1.1.7) may be used to estimate the probable ultimate displacement (Δ_u, Figure 1.1.1), of a structure.

The concepts discussed very briefly in this section will be used to develop designs in Chapter 3, where they will be adjusted and converted to system-specific relationships.

1.1.4 Strength-Based Design

1.1.4.1 Identifying a Design Strength Objective The procedures that have traditionally been used to design buildings identify strength as the primary criterion or objective. The design focus is the idealized elastic behavior region described in Figure 1.1.1 ($F < F_{max}$). The quantification of the objective strength (F_{max}) assumes that the appropriate level of system ductility for the structure is known. The presumption is that sufficient system ductility will exist if the design follows a set of prescriptive rules in the development of both the bracing system and the components that create the bracing system to allow the system to reach the anticipated level of ultimate drift (Δ_u).

The strength-based design approach has worked well, especially given the limited scientific basis, but prescriptions often restrict the accomplishment of functional and aesthetic design objectives unnecessarily. A further, and probably more significant, drawback is the fact that the designer often does not understand the subtleties of a prescriptive requirement and, as we try to cover all conceivable conditions with

Photo 1.2 Exposed cast-in-place concrete, Lagoon Towers, Honolulu, HI, 1965. (Photo by Robert E. Englekirk.)

written dogma, conflicting propositions make compliance virtually impossible in all but the simplest buildings. As our knowledge base expands, prescriptive constraints can and should be replaced with more comprehensive design procedures.

The codified design version of the strength-based process and its objectives are described in Figure 1.1.1. The area of design interest is confined to the presumably subyield behavior region $F \leq F_o$. The deformation likely to be experienced by the structure is understood to be in the region of Δ_u—the ultimate or probable displacement. From a design perspective a design yield strength (F_o) and displacement (Δ_o) must be identified, and this process will not be easily standardized. The design yield displacement (Δ_o) depicted in Figure 1.1.1 is actually an idealization because, as we shall see in Chapters 2 and 3, the components are already operating in the inelastic behavior range, for the design strength (F_o) is associated with the factored nominal strength of critical components. The idealized yield strength level (F_{yi}) is developed from system response characteristics and, as a consequence, may require the development of a significant level of ductility in the critical components.

The ratio of the ultimate displacement (Δ_u) to the idealized yield displacement (Δ_{yi}) of the system (Δ_u/Δ_{yi}) is the most common definition of system displacement ductility (μ).

The definition of the displacement at yield can take at least two forms. From a design perspective, Δ_o has traditionally been used as a base line in spite of the fact that it too is an idealization; hence one definition of system ductility is

$$\mu_o = \frac{\Delta_u}{\Delta_o} \tag{1.1.13a}$$

The objective strength (F_o) will often significantly underestimate the real strength of a structure (F_{yi}) or the inception of inelastic behavior in the system as opposed to the yielding of a few critical components. Accordingly, the strength associated with system yield (F_{yi}) represents a more stable, identifiable definition of system strength, and this makes it better suited to the development of a design criterion. It is, therefore, more appropriate to define system ductility as

$$\mu = \frac{\Delta_u}{\Delta_{yi}} \tag{1.1.13b}$$

Standardizing the yield strength of a structure by identifying its ultimate strength eliminates one significant variable, but it does not resolve a much debated design relationship, specifically the relationship between the strength required of an equivalent elastic structure (μF_o) and that required of a ductile structure ($F_{\max}$). The subjectivity associated with establishing this relationship ($\mu_o F_o/F_{\max}$) will always lessen the confidence a designer has in force-based design procedures.

1.1.4.2 Creating a Ductile Structure Regardless of the design approach adopted, the relationship between the objective strength (F_o) and sustainable strength ($F_{\max}$), or the maximum lateral load the structure is likely to experience, must be established. This maximum lateral load is critical to the attainment of the design objective—creating a ductile structure—for it identifies the strength required of the nonductile or less ductile components along the lateral and vertical load paths. The design process used to create a structure that contains both ductile and brittle (or less ductile) elements is referred to as "capacity design" or "capacity-based design."

Consider the frame described in Figure 1.1.10. Assume that we can create a frame beam that possesses a large degree of ductility but that only a limited degree of ductility is possible in the column. If the probable flexural strength of the frame beam ($M_{b,\text{pr}}$), corresponding to the $F_{\max}$ in Figure 1.1.1, is known, then the probable ultimate demand on the column may be determined:

$$M_{c,\text{pr}} = M_{b,\text{pr}} \tag{1.1.14a}$$

$$P_{c,\text{pr}} = \pm\frac{2M_{b,\text{pr}}}{\ell} \tag{1.1.14b}$$

Photo 1.3 Precast clad concrete frame, Four Seasons, Newport Beach, CA. (Courtesy of Englekirk Partners, Inc.)

Thus we can attain the objective of creating a ductile structure using capacity-based procedures if we are able to create a column capable of sustaining the forces defined by Eq. 1.1.14a and 1.1.14b.

The extension of this process is not as easily accomplished on complex structures. Accordingly, codes[1.9–1.11] relate the objective design strength (F_o) to the sustainable strength ($F_{\max}$) through the inclusion of a system overstrength factor (Ω_o):

$$\Omega_o = \frac{F_{\max}}{F_o} \tag{1.1.15}$$

The level of "prescribed" overstrength for brittle member design purposes is on the order of 2.8. Accordingly, following the prescriptive approach to capacity-based design, the column of Figure 1.1.10 would be designed to the following criteria:

$$M_{co} = \Omega_o M_{bo} \tag{1.1.16a}$$

$$P_{co} = \pm\frac{2\Omega_o M_{bo}}{\ell} \tag{1.1.16b}$$

where M_{bo} is the objective flexural design strength related to the design shear (F_o) and Ω_o is the prescribed level of overstrength (Eq. 1.1.15). This design approach will usually produce a conservative design so long as the designer does not decide to

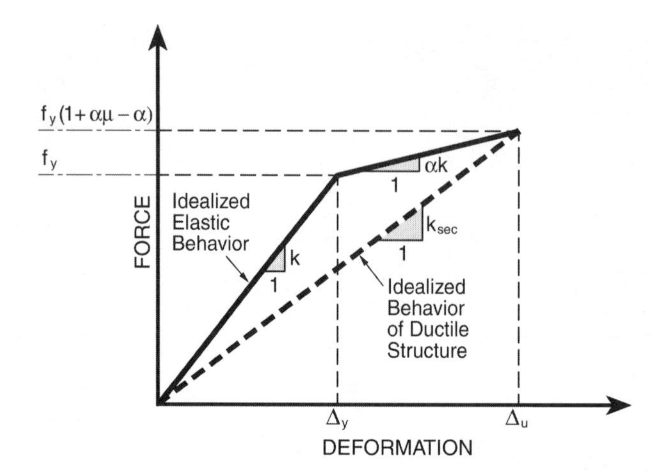

Figure 1.1.9 Idealized behavior of single-degree-of-freedom system with bilinear force–deformation relation.[1.8]

provide a frame beam whose real strength significantly exceeds the objective level of strength (M_{bo}).

The strength of more brittle elements along the lateral load path is extremely important. Accordingly, the overconservatism inherent to Eq. 1.1.16a and 1.1.16b is appropriate. In subsequent chapters, an alternative approach to defining strength and deformation objectives for more brittle elements will be developed, and this will allow the process described by Eq. 1.1.14a to be adapted to more complex structures.

1.1.5 Displacement-Based Design

More recently design methodologies based on displacement have been proposed as a part of performance-based design. This is logical since performance in a ductile structure (Figure 1.1.1) can only be evaluated based on estimates of deformation (Δ_u). Displacement-based design approaches start by identifying an objective system displacement (Δ_u) and ductility (μ) (see Eq. 1.1.13b and Figure 1.1.1). They then proceed to establish the system strength and stiffness necessary to the safe attainment of these objectives.

Displacement-based design procedures follow two conceptually logical paths. These methods are referred to as equal displacement and direct displacement design. They differ only in terms of how the ultimate displacement (Δ_u) is quantified and what idealization process is used to describe the system. The equal-displacement approach follows the Newmark–Hall[1.5] proposition described by Figure 1.1.7, as discussed in Section 1.1.3. The basic proposition is that the displacement response of a ductile structure can be developed from estimates of the response of an otherwise equivalent elastic structure. The direct displacement approach assumes that the ductility experienced by a structure can be converted to structural damping and that

Photo 1.4 Flat slab with shear head shear failure, Northridge earthquake, 1994. (Courtesy of Englekirk Partners, Inc.)

the response will be reduced by the amount of energy dissipated by the postyield deformation of system components (see Figure 1.1.6 and Section 1.1.3). The impact of ductility on system response is thereby directly considered. The period of an inelastically responding structure will be considerably different than that of the comparable elastic structure. A secant stiffness model is adopted to account for this period shift.[1.7,1.8] The force deformation described in Figure 1.1.9 can be used to envelope the behavior described in Figure 2.1.2. The fundamental period of the inelastic structure, presuming that its stiffness is reasonably represented by its secant stiffness, will increase, and it follows from Eq. 1.1.5 that the period of the structure responding inelastically (T_i) will be

$$T_i = T_e\sqrt{\mu} \tag{1.1.17}$$

where T_e is the fundamental period of the idealized elastic structure. Regardless of the conceptual design methodology, the final design should be tested to insure that drift objectives for the system have been attained and that the postyield deformation demands imposed on the components do not exceed their capacities.

Displacement-based design is most easily developed by example. The objective of the following examples is to demonstrate in the simplest manner possible how the required balance between strength, stiffness, and ductility can be attained. Accordingly, a single-degree-of-freedom model will be studied and several simplifying assumptions adopted.

The first simplifying assumption will be to assume that the structure will have a fundamental period that places it in the velocity-constant region (Figure 1.1.5). This establishes an equivalence between the peak deformations of the elastic and inelastic structures (see Figure 1.1.7).

Second, for conceptual design purposes, the behavior of the system will be presumed to be elastic/perfectly plastic, and the behavior characteristics of the frame components will be assumed identical.

1.1.5.1 Equal Displacement-Based Design Consider the single story frame of Figure 1.1.10 and assume that it will provide lateral support for a tributary building weight of 300 kips.

Step 1: Establish the Objective Drift Limit. A drift objective of 2.5% has traditionally been accepted for low-rise buildings.[1.9, Sec. 1630.10] The objective level of drift is

$$\Delta_u = 0.025(180)$$
$$= 4.5 \text{ in.}$$

Step 2: Determine the Objective Natural Frequency ω_n Based on a Criterion Spectral Velocity of 40 in./sec.

$$\omega_n = \frac{S_v}{\Delta_u} \qquad \text{(see Eq. 1.1.6a)}$$
$$= \frac{40}{4.5}$$
$$= 8.89 \text{ rad/sec}$$

Step 3: Determine the Objective Stiffness.

$$k = \omega_n^2 m \qquad \text{(Eq. 1.1.5)}$$
$$= (8.89)^2 \frac{300}{386.4}$$
$$= 61.3 \text{ kips/in.}$$

This stiffness becomes an objective minimum, for if it is not attained, the displacement of the structure will be greater than our objective deflection (Δ_u).

Step 4: Size the Frame Components. The relationship between force and displacement for the pin-based frame described in Figure 1.1.10 is

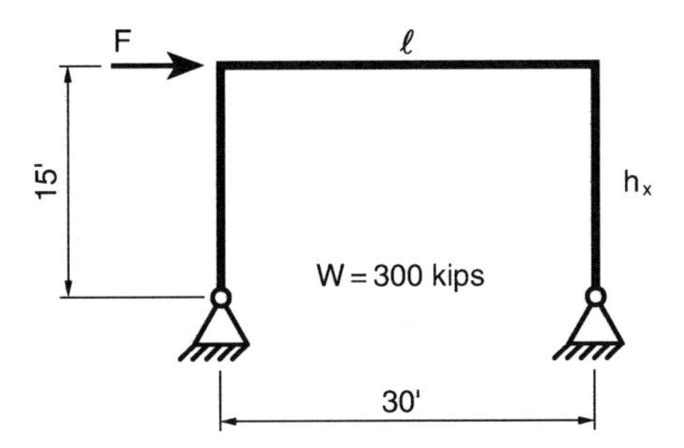

Figure 1.1.10 Single story frame.

$$\Delta = \frac{F(h_x)^2}{6E}\left(\frac{\ell}{2I_b} + \frac{h_x}{I_c}\right) \qquad \text{(Ref. 1.6, Sec. 4.2(a))} \quad (1.1.18)$$

$$\frac{6E}{k(h_x)^2} = \frac{\ell}{2I_b} + \frac{h_x}{I_c}$$

$$\frac{6(4000)}{61.3(180)^2} = \frac{360}{2I_b} + \frac{180}{I_c}$$

$$0.012 = 180\left(\frac{I_c + I_b}{I_b I_c}\right)$$

$$\frac{I_c + I_b}{I_b I_c} = 0.000067 \text{ in.}^4$$

A design assumption was that the effective moments of inertia (I_e) of the beam and column would be the same. The range required of the member moment of inertia is

$$\begin{aligned} I_e &= \frac{1}{0.0000335} \\ &= 29{,}850 \text{ in.}^4 \end{aligned}$$

If the effective moment of inertia of the members is 35% of the gross moment of inertia, then

$$\begin{aligned} I_g &= \frac{I_e}{0.35} \\ &= 85{,}300 \text{ in.}^4 \end{aligned}$$

This suggests a beam and a column that is 22 in. by 36 in. (85,500 in.^{4}).

Step 5: Determine the Strength Required of the Beam and Column. For this example it is assumed that the system ductility factor (μ) or the ratio of idealized yield displacement (Δ_{yi}) to ultimate displacement (Δ_u) is 4 (see Figure 1.1.1). Accordingly, this will produce an estimate of the ultimate strength required of the system ($F_{\max}$). The idealized elastic frame displacement is

$$\begin{aligned}\Delta_{yi} &= \frac{\Delta_u}{\mu} \qquad \text{(Eq. 1.1.13b)}\\ &= \frac{4.5}{4}\\ &= 1.125 \text{ in.}\end{aligned}$$

and the required ultimate strength, expressed as a mechanism shear force imposed on the frame, is

$$\begin{aligned}F_{\max} &= \Delta_{yi}k\\ &= 1.125(61.3)\\ &= 69 \text{ kips}\end{aligned}$$

Step 6: Consider the Impact of PΔ Forces. $P\Delta$ forces are created at a displacement of Δ_u of 4.5 in. in this case:

$$\begin{aligned}F_{P\Delta} &= \frac{\left(\sum W\right)\Delta_u}{h_x} \qquad \text{(Eq. 1.1.7)}\\ &= \frac{(300)(4.5)}{180}\\ &= 7.5 \text{ kips}\end{aligned}$$

The objective level of strength is

$$\Omega_o F_o = F_{\max} + F_{P\Delta} \qquad \text{(see Eq. 1.1.15)}$$

We have assumed that the system is elastic/perfectly plastic for design purposes. The required nominal strength of the beam must consider member and system overstrength. Adopt a system overstrength factor (see Eq. 1.1.15) for this example of 1.25.

$$M_{bu} = \frac{\Omega_o F_o h_x}{2\Omega_o} \qquad (1.1.19)$$

$$= \frac{(69 + 7.5)(15)}{(2)(1.25)}$$

$$= 459 \text{ ft-kips}$$

and this suggests that the beam be reinforced with about 3.5 in.^{2} of flexural reinforcement. This corresponds to a reinforcement ratio (ρ_s) of only 0.5% and, as we shall see in Chapter 2, is not very economical from a reinforcement perspective.

1.1.5.2 Direct Displacement-Based Design

Step 1: Proceed As in the Equal Displacement-Based Procedure to Establish a Drift Objective and Ultimate Drift ($\Delta_u = 4.5$ in.).

Step 2: Revise the Design Spectral Velocity to Reflect the Level of Provided Structural Damping.

$$\zeta_{\text{eq}} = \frac{\sqrt{\mu} - 1}{\pi\sqrt{\mu}} \qquad \text{(Eq. 1.1.9d)}$$

$$= \frac{\sqrt{4} - 1}{\pi\sqrt{4}}$$

$$= 0.16 \text{ or } 16\%$$

$$\dot{d}_{\max} = \frac{S_v}{3.38 - 0.67 \ \ln\ 5} \qquad \text{(see Eq. 1.1.8b)}$$

$$= \frac{40}{2.3}$$

$$= 17.4 \text{ in./sec}^2$$

$$\hat{\zeta}_{\text{eq}} = \zeta + \hat{\zeta}_{\text{eq}} \qquad \text{(Eq. 1.1.9a)}$$

$$= 5 + 16$$

$$= 21$$

$$S_v = (3.38 - 0.67 \ \ln\ 21)17.4 \qquad \text{(Eq. 1.1.8b)}$$

$$= 23.3 \text{ in./sec}$$

Step 3: Determine the Objective Natural Frequency of the Ductile Structure.

$$\omega_{n,\text{ductile}} = \frac{S_v}{\Delta_u} \qquad \text{(see Eq. 1.1.6a)}$$

$$= \frac{23.3}{4.5}$$

$$= 5.18 \text{ rad/sec}$$

Step 4: Determine the Objective Natural Frequency of the Elastic System.

$$\omega_{n,\text{elastic}} = \omega_{n,\text{ductile}} \sqrt{\mu} \qquad \text{(see Figure 1.1.9)}$$

$$= 5.18 \left(\sqrt{4}\right)$$

$$= 10.36 \text{ rad/sec}$$

Step 5: Determine the Stiffness Required of the Elastic Structure.

$$k = \omega_n^2 m$$

$$= (10.36)^2 \frac{300}{386.4}$$

$$= 83.3 \text{ kips/in.}$$

Step 6: Proceed to Develop the Stiffnesses Required of the Components. (Refer to Step 4 of the equal displacement procedure.)

$$0.0089 = 180 \left(\frac{I_c + I_b}{I_b I_c} \right) \qquad \text{(see Eq. 1.1.18)}$$

$$\frac{I_c + I_b}{I_b I_c} = 0.000049$$

$$I_c = \frac{2}{0.000049}$$

$$= 40{,}500 \text{ in.}^4 \qquad (I_e)$$

$$I_g = 115{,}700 \text{ in.}^4 \qquad (I_e = 0.35\, I_g)$$

Conclusion: Provide a 30-in. by 36-in. deep beam and column.

Step 7: Determine the Strength Required of the Beam and the Column.

$$\Delta_{yi} = \frac{\Delta_u}{\mu}$$

$$= \frac{4.5}{4}$$

$$= 1.13 \text{ in.}$$

$$F_{\max} = \Delta_{yi} k$$

$$= 1.13(83.3)$$

$$= 94 \text{ kips}$$

Add the $P\Delta$ shear:

$$\Omega_o F_o = F_{\max} + F_{P\Delta} \qquad \text{(see Eq. 1.1.15)}$$

$$M_{bu} = \frac{\Omega_o F_o h_x}{2\Omega_o} \qquad \text{(Eq. 1.1.19)}$$

$$= \frac{(94 + 7.5)(15)}{2(1.25)}$$

$$= 669 \text{ ft-kips}$$

Comment: Observe that, for this case, the design conclusions are significantly different. The relationship between the objective strengths will be quite sensitive to the presumed level of system ductility. For example, had we adopted a system ductility of 5, the resultant systems would have been essentially the same.

1.1.6 System Ductility

The objective of this section is to describe how system ductility factors can be rationally developed. The system ductility factors used in design must strive to control strain states and attain behavior limit states in system components. Chapter 2 will have as a focus the identification of component deformation limit states; so let us presume at this point that it is possible to identify the idealized yield rotation (θ_{yi}) and ultimate rotation limit state (θ_u) for a member. Let us first examine the impact of designer input into the design process by considering the design of the frame described in Figure 1.1.10. It was presumed in the example developed in Section 1.1.5 that the beam and the column would participate equally in the attainment of the system ductility objective. As a consequence, system and component ductility factors were the same. Creating this equivalence will not typically be possible, for it means that the development of the system must promote a sharing of the postyield demand between the beam and the column in order to validate the system ductility used to develop the design criterion for the beam. Accordingly, this sharing must become an objective of the design process.

The development of Eq. 1.1.18 was based on a separation of the components that contribute to the drift of the building. The idealized yield displacement (Δ_{yi}) of the system described in Figure 1.1.10 is then reasonably broken down into its two contributors—that of the beam (δ_{byi}) and that of the column (δ_{cyi}). These were presumed to participate equally in this example. Once idealized system yield (Δ_{yi}) has been reached, postyield drift is, or must be, accommodated in the members in the form of a postyield rotation. Design decisions can force this postyield rotation to occur in the beam, the column, or, idealistically, both in this case because they each should have the same level of available ductility. To differentiate the contributing elements, it is convenient to describe their deformations in terms of rotations. Accordingly,

$$\theta_{byi} = \frac{\delta_{byi}}{\ell/2} \qquad \text{(Hypothetically)} \tag{1.1.20a}$$

where δ_{byi} is the displacement required to restore the beam at midspan so as to be compatible with its symmetrical other half.[1.6, Sec. 4.2(a)]

$$\theta_{cyi} = \frac{\delta_{cyi}}{h_x} \tag{1.1.20b}$$

θ_{byi} and θ_{cyi} are the elastic components of story drift associated with the deformation of the beam and the column, respectively. It follows that

$$\frac{\Delta_{yi}}{h_x} = \theta_{byi} + \theta_{cyi} \tag{1.1.20c}$$

The postyield drift of the system can be described in a similar manner,

$$\frac{\Delta_p}{h_x} = \theta_{bp} + \theta_{cp} \tag{1.1.20d}$$

Now we are in a position to better understand the consequences of design decisions relating to the choice of member strengths. In the preceding development it was assumed that the deflection of the frame was contributed to equally by the beam and the column in both the elastic and postyield ranges. Had we adopted a system ductility of 5, the beam and the column would also have needed to possess a member ductility of 5 if our displacement objectives were to be attained.

In essence we created a structure whose system ductility (μ) is the same as that of its components. The objective idealized yield was developed from the ultimate deflection of the system.

$$\Delta_{yi} = \frac{\Delta_u}{\mu}$$

$$= \frac{4.5}{5}$$

$$= 0.9 \text{ in.}$$

Comment: A system ductility factor of 5 is used to simplify the mathematics.

This resulted in the identification of a postyield component of drift (Δ_p):

$$\Delta_p = \Delta_u - \Delta_{yi}$$

$$= 4.5 - 0.9$$

$$= 3.6 \text{ in.}$$

When combined these components confirm the adopted level of system ductility.

$$\mu = \frac{\Delta_p + \Delta_{yi}}{\Delta_{yi}}$$

$$= \frac{3.6 + 0.9}{0.9}$$

$$= 5$$

Given the assumed equal stiffness objective, we are in a position to identify the idealized displacement of the beam:

$$\delta_{byi} = \delta_{cyi} = \frac{\Delta_{yi}}{2}$$

$$= \frac{0.9}{2}$$

$$= 0.45 \text{ in.}$$

In this case, equal strength and stiffness, we get

$$\theta_{byi} = \frac{\delta_{byi}}{\ell/2}$$

$$= \frac{0.45}{180}$$

$$= 0.0025 \text{ radian}$$

Since the postyield behavior is presumed to be the same,

$$\theta_{bp} = \theta_{cp} = \frac{\Delta_p}{2h_x} \qquad \text{(since } h_x = \ell/2\text{)}$$

$$= \frac{3.6}{2(180)}$$

$$= 0.01 \text{ radian}$$

$$\mu_b = \frac{\theta_{bp} + \theta_{byi}}{\theta_{byi}}$$

$$= \frac{0.01 + 0.0025}{0.0025}$$

$$= 5$$

Had we elected to develop the design proposed in the preceding example following capacity-based objectives, the strength of the column would have been greater than that of the beam, for our objective would have been to force yielding into the beam—presumably because we believed that the beam would be capable of sustaining significantly larger postyield rotations than the column. The logical design procedure for the frame following capacity-based objectives would be as outlined below.

The strength required of the column, should the design be capacity-based, would be

$$M_{cu} = \Omega_o M_{bn} \qquad \text{(see Eq. 1.1.16a)}$$

$$= 1.25 M_{bn}$$

$$P_{cu} = \frac{2\Omega_o M_{bn}}{\ell_c} \qquad \text{(see Eq. 1.1.16b)}$$

$$= \frac{2(1.25)M_{bn}}{30 - 3}$$

$$= \frac{M_{bn}}{10.8}$$

where M_{bn} is the provided strength, not the strength suggested by Eq. 1.1.19.

The provided flexural strength of the column would now be greater than that of the beam and, as a consequence, all of the postyield rotation would be forced into the beam. The ductility demand imposed on the beam were the column designed to the strength dictated by capacity-based concepts, would be significantly larger than our objective component ductility of 5. To determine the ductility imposed on the component, describe the story drift in terms of the components of rotation:

$$\frac{\Delta_u}{h_x} = \theta_{cyi} + \theta_{cp} + \theta_{byi} + \theta_{bp} \tag{1.1.20e}$$

where θ_{cyi} and θ_{byi} are the elastic rotations of the column and beam, and θ_{cp} and θ_{bp} are the respective postyield rotations.

Our design to this point has balanced the elastic rotations of the beam and column such that at beam yield

$$\theta_{cyi} = \theta_{byi} = 0.0025 \text{ radian}$$

and, because the column is designed so as not to yield and yet have the same effective stiffness as the beam,

$$\begin{aligned}\theta_{bp} &= \frac{\Delta_u}{h_x} - \theta_{byi} - \theta_{cyi} \\ &= 0.025 - 0.0025 - 0.0025 \\ &= 0.02 \text{ radian}\end{aligned}$$

Now the ductility demand imposed on the beam is

$$\begin{aligned}\mu_b &= \frac{\theta_{bp} + \theta_{byi}}{\theta_{byi}} \\ &= \frac{0.02 + 0.0025}{0.0025} \\ &= 9\end{aligned}$$

and this approaches what may be viewed as the attainable level of member ductility (see Figure 2.1.2).

Conclusion: Were the system described in Figure 1.1.10 designed to a capacity-based criterion, the member ductility demand would be 9 if a system ductility of 5 were used in the design.

The system ductility that would produce a design that attains a component ductility of 5($\Delta = 4.5$ in.; $\mu_b = 5$) would be developed as follows:

$$\Delta_u = \Delta_{cyi} + \Delta_{byi} + (\mu_b - 1)\,\Delta_{byi} \tag{1.1.20f}$$

$$\Delta_u = \theta_{cyi} h_x + \theta_{byi}\left(\frac{\ell}{2}\right) + (\mu_b - 1)\,\theta_{byi}\left(\frac{\ell}{2}\right) \tag{1.1.20g}$$

$$\Delta_u = \Delta_{yi} + \Delta_p \tag{1.1.21a}$$

$$= (\Delta_{cyi} + \Delta_{byi}) + 4\Delta_{byi}$$

$$= 2\Delta_{byi} + 4\Delta_{byi}$$

$$\mu_s = \frac{\Delta_u}{\Delta_{yi}}$$

$$= \frac{2\Delta_{byi} + 4\Delta_{byi}}{2\Delta_{byi}}$$

$$= 3$$

Conclusion: Were a system ductility factor (μ) of 3 used to develop the design, the ductility demand imposed on the beam (μ_b) would be 5.

The strength of the member must be great enough so that it will remain essentially elastic until it reaches idealized yield (Figure 1.1.1). Hence, since $h_x = \ell/2$, it follows that

$$\theta_{byi} = \frac{\Delta_u}{6h_x} \qquad \text{(see Eq. 1.1.20f)}$$

$$= \frac{4.5}{6(180)}$$

$$= 0.0042 \text{ radian}$$

$$\phi_{byi} = \frac{6\theta_{byi}}{\ell_c}$$

$$= \frac{6(0.0042)}{324}$$

$$= 0.0000078 \text{ radian/in.}$$

$$M_{byi} = \phi_{byi} EI_e$$

$$= 0.0000078(3600)(30{,}000)$$

$$= 8400 \text{ in.-kips}$$

$$F_{\max} = \frac{2M_{byi}}{h_x}$$

$$= \frac{2(8400)}{180}$$

$$= 93.3 \text{ kips}$$

$$k = \frac{F_{\max}}{\Delta_{yi}}$$

$$= \frac{93.3}{0.9}$$

$$= 104 \text{ kips/in.}$$

$$F_o = \frac{F_{\max}}{\Omega_o}$$

$$= \frac{93.3}{1.25}$$

$$= 74.6 \text{ kips}$$

Accordingly, the design strength of the beam must be 35% greater than that proposed to attain our balanced yield design with a system ductility of 4.

The effective stiffness of the column can always be related to that of the beam. Equation 1.1.20f may be rewritten to reflect a generalized relationship as

$$\Delta_u = \frac{k_b}{k_c}\Delta_{byi} + \Delta_{byi} + (\mu_b - 1)\Delta_{byi}$$

$$\mu = \frac{(k_b/k_c)\Delta_{byi} + \mu_b\Delta_{byi}}{(k_b/k_c + 1)\Delta_{byi}}$$

$$= \frac{k_b/k_c + \mu_b}{k_b/k_c + 1}$$

or more elegantly, if the ratio of member stiffnesses is $A = k_b/k_c$, the design system ductility (μ) becomes

$$\mu = \frac{A + \mu_b}{A + 1} \tag{1.1.21b}$$

where μ_b is the available ductility in the beam and A is the component stiffness ratio (k_b/k_c). This relationship is described graphically in Figure 1.1.11.

An engineer confronted with Figure 1.1.11 would reasonably wonder why the system ductility factors (μ) described therein are so much lower than the reduction factor (R) used to develop current code[1.9] design forces.

Comment: The interested reader is referred to the Blue Book and Commentary[1.10] for a comprehensive treatment of the development of the code seismic design criterion. Briefly stated, Figure 1.1.12 describes the relationship between the elastic response spectrum and the inelastic-based design shear (V_S). The "typical deformation

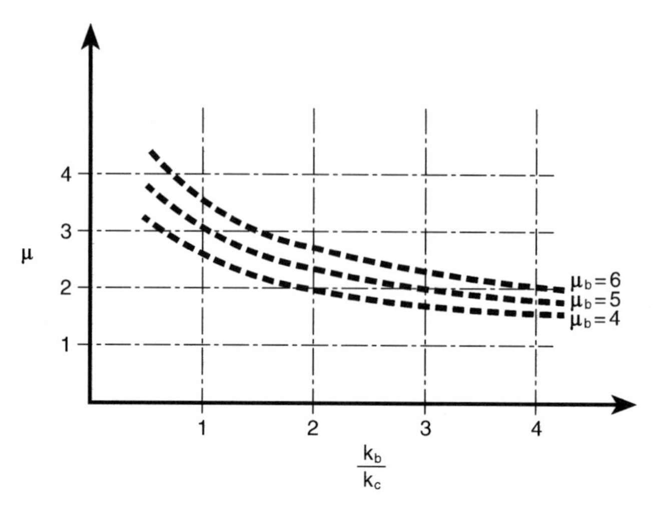

Figure 1.1.11 Relationship between member ductility (μ_b) and system ductility (μ). Design is presumed to be capacity based.

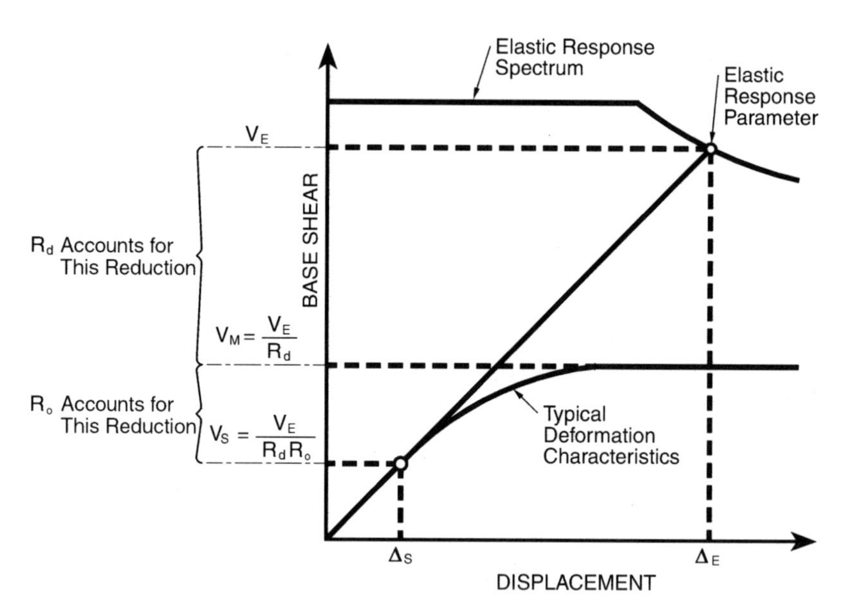

Figure 1.1.12 Relationship between base shear and displacement.[1.10]

characteristics" are identical to the response of the ductile structure described in Figure 1.1.1. The system ductility (μ), as defined by Eq. 1.1.13b, is the ratio of ultimate displacement (Δ_u) to the idealized design displacement (Δ_{yi}). The code development as described in Figure 1.1.12 is force based, and the relationship between the elastic response parameter (V_E) and the design force (V_S) is the same as the relationship that exists between similarly developed displacements on Figure 1.1.1 ($\mu_o F_o$ is V_E and F_o is V_S while Δ_u is Δ_E and Δ_o is Δ_S).

Current codes[1.9, 1.10] use R_d to define the "global ductility capacity of the lateral force resisting system" (see Figure 1.1.12). R_d for a concrete ductile frame is 3.4,[1.9 or 1.10, Table 104–6] and this is reasonably consistent with the values developed in Figure 1.1.11, especially if a beam ductility of 6 can be attained in a system (Fig. 1.1.10) of balanced stiffness ($k_b/k_c = 1.0$).

The design strength or basis (V_S of Figure 1.1.12) is not the maximum (mechanism) strength of the system (V_M or $F_{\max}$) that was used to quantify the global (R_d) or system (μ) ductility developed in Figures 1.1.12 and 1.1.1, respectively. To reach the design force V_s or F_o, the inherent overstrength of the system and its components must be considered. Current codes[1.9, 1.10] refer to the relationship between the maximum developable strength (V_M) and the design strength (V_s) as R_o. R_o speculatively identifies the relationship between the total overstrength of the system and the presumably factored nominal strength of the component. In Figure 1.1.12, R_o is described as

$$R_o = \frac{V_M}{V_S} \tag{1.1.22}$$

In Figure 1.1.1 this is the ratio $F_{\max}/F_o$, usually identified as Ω_o (Eq. 1.1.15). The code design force (V_s) is then related to the elastic response spectrum by the product $R_d R_o$.

For a concrete ductile frame, R_o is 2.5.[1.10, Table 104-6] This may be broken down into the constitutive components as follows, using the notation adopted by Paulay and Priestley[1.2, Sec. 1.3.3]:

$$\begin{aligned} R_o &= \frac{\phi_p}{\phi}\lambda_o\lambda_s \qquad (1.1.23) \\ &= \frac{1.25}{0.9}(1.25)(1.4) \\ &= 2.5 \end{aligned}$$

where

ϕ is the capacity reduction factor.

ϕ_p is the probable overstrength of the reinforcing steel.

λ_o describes the work hardening that is likely to occur.

λ_s takes us from member to system overstrength.

Observe that this notation is particular to Ref. 1.12 and not the notation used in this book.

It follows then that the code reduction factor is

$$R = R_d R_o \tag{1.1.24}$$

$$= 3.4(2.5)$$

$$= 8.5$$

What can be concluded from Figure 1.1.11?

- The code anticipates a member ductility of between 5 and 6 for the frame beam of Figure 1.1.10 were capacity-based concepts used to determine the strength of the columns and, the stiffness of the components are comparable.
- The relationship between the effective stiffness of the beam (k_b) and that of the column (k_c) will impact the ductility demand on the beam if the column strength is large enough to force all of the inelastic behavior into the beam.

If the two approaches, force and displacement, can be developed so as to have a similar basis, why not stay with the traditional force-based approach? The answer lies in the speculation required to generalize R_o and R_d, a speculation that can be refined in the design and controlled by the designer.

If we consider the one-story frame described in Figure 1.1.10, the inelastic deformation can occur in either the beam, the column, or the beam-column joint. In a multistoried building, the inelastic response will or can be directed toward a number of components largely at the dictate of the designer. A global relationship between member ductility and system ductility must be assumed, and this is best determined by the designer of a structure in whose control it resides. Further, a force-based approach bridges the gap between F_o and $F_{\max}$ (Figure 1.1.1) prescriptively, without any understanding of design methodologies or designer predilection. Only the designer will know the assumptions and level of conservatism used in the design of the structure.

A comparison between the strength objective of the code and the preceding development will add meaning to the selection of a seismic design process—force or displacement based. The equal-displacement based objective base shear (F_o), exclusive of the $P\Delta$ amplification, was 55.2 ($69/\Omega_o$) kips or $0.18W$ for the system described in Figure 1.1.10, assuming shared yielding between the beam and the column. Alternatively, the base shear for the same system, given an identical beam ductility objective (μ_b) of 5 and an elastic column, would have been 93.3 kips ($0.31W$). The code base shear coefficient could be developed as follows:

$$T = C_t(h_n)^{3/4} \qquad \text{(Ref. 1.9, Eq. 30-8)}$$

$$= 0.03(7.62)$$

$$= 0.23 \text{ second} \qquad \text{(Code period)}$$

The maximum period allowed by the Building Code[1.9] is

$$T_{\max} = 1.3T$$
$$= 1.3(0.23)$$
$$= 0.3 \text{ second}$$

The design base shear coefficient would be the lesser of

$$\frac{V}{W} = \frac{2.5C_a}{R} \qquad \text{(Ref. 1.9, Eq. 30-5)}$$
$$= \frac{2.5(0.44)}{8.5}$$
$$= 0.13 \qquad \text{(Governs)}$$

or

$$\frac{V}{W} = \frac{C_v}{RT} \qquad \text{(Ref. 1.9, Eq. 30-4)}$$
$$= \frac{0.64}{8.5(0.3)}$$
$$= 0.25$$

The design base shear (ultimate) is $0.13W$ (V_S of Figure 1.1.12).

$F_{\max}$ (V_M of Figure 1.1.12), the maximum developable strength, is

$$F_{\max} = \Omega_o F_o$$
$$= 2.5(0.13W)$$
$$= 0.33W$$
$$= 0.33(300)$$
$$= 100 \text{ kips}$$

The developable system strength ($F_{\max}$) required by the capacity-based design following the displacement-based approach was 93.3 kips. The designer following the code procedure is not required to produce a structure with a developable strength of 100 kips, for it is assumed by the code that this will be a normal consequence of the member design process (see Eq. 1.1.23). Hence, no direct consistent link exists between V_S and V_M (Figure 1.1.12). This is because the provided levels of system and component strength are likely to be significantly different than the stated code objectives (V_M), and this could adversely impact capacity-based design objectives.

Consider the dynamic consistency requirements imposed on a code design. Member sizes required to meet code stiffness objectives are

$$\Delta_S = \frac{0.025}{0.7(R)}(h_x)$$

$$= \frac{0.025}{0.7(8.5)}(180)$$

$$= 0.76 \text{ in.}$$

This corresponds to a required system stiffness of

$$k = \frac{V}{\Delta_S}$$

$$= \frac{0.13(300)}{0.76}$$

$$= 51 \text{ kips/in.}$$

The associated system stiffness would need be only 49% (51/104) of that required by the equal-displacement based design of Section 1.1.5. Since consistency is not required the designer may elect to use members that are larger ($k > 51$ kips/in.), and this would alter the ductility demand.

Conclusion: The designs produced are quite similar. The insight and consistency afforded the designer by a displacement-based approach is significant, and this will be considerably more important in the design of more complex structures.

1.1.7 Recommended Displacement-Based Design Procedure

Start by identifying a drift objective that will control or limit nonstructural damage and a member ductility objective that will preclude system failure or minimize structural damage. Member test data should be used to establish component ductility objectives. If the response falls in the velocity-sensitive region (Figure 1.1.5) where spectral velocity is constant, the spectral velocity will define system stiffness objectives. While this is the most common case, it is not a condition of the design process, as will be demonstrated.

The frame of Figure 1.1.10 will be designed to demonstrate the process. Select a design earthquake whose spectral velocity is 40 in./sec and assume that the acceptable member ductility is 5.

Step 1: Identify the Objective Ultimate Drift (Δ_u) and Select an Objective Drift Ratio of 2%.

$$\Delta_u = 0.02h_x$$

$$= 0.02(180)$$

$$= 3.6 \text{ in.}$$

Step 2: Find the Objective Frequency of the System.

$$\omega_n = \frac{S_v}{\Delta_u} \qquad \text{(see Eq. 1.1.6a)}$$

$$= \frac{40}{3.6}$$

$$= 11.1 \text{ rad/sec} \qquad (T = 0.57 \text{ second})$$

Step 3: Determine the Required Stiffness of the System.

$$\omega_n = \sqrt{\frac{k}{m}} \qquad \text{(Eq. 1.1.5)}$$

$$k = \omega_n^2 m$$

$$= (11.1)^2 \frac{300}{386.4}$$

$$= 95.7 \text{ kips/in.}$$

Step 4: Determine Objective Strength.

$$S_a = S_v\omega_n \qquad \text{(see Eq. 1.1.6a)}$$

$$= 40(11.1)$$

$$= 444 \text{ in./sec}^2$$

$$= 1.15g$$

Step 5: Select a System Ductility Factor.

$$\mu = 3$$

(See Figure 1.1.11, where $k_b/k_c = 1.0$.)

Step 6: Solve for $F_{\max}$. (See Figure 1.1.1 and V_M in Figure 1.1.12.)

$$F_{\max} = \frac{S_a(W)}{\mu}$$

$$= \frac{1.15}{3}(300)$$

$$= 115 \text{ kips}$$

Step 7: Solve for F_o. (See Figure 1.1.1 and V_S in Figure 1.1.12.) Assume $\Omega_o = 1.5$.

$$F_o = \frac{F_{\max}}{\Omega_o}$$

$$= \frac{115}{1.5}$$

$$= 76.7 \text{ kips}$$

Step 8: Determine Required Beam Moment Capacity.

$$M_{nb} = \frac{F_o}{2} h_x$$

$$= \frac{76.7}{2}(15)$$

$$= 575 \text{ ft-kips}$$

Step 9: Determine the Member Stiffness Objectives Required to Satisfy Our System Stiffness Objectives.

$$k_b = k_c \qquad \text{(Adopted objective)}$$

$$\theta_{byi} = \theta_{cyi} \qquad \left(h_x = \frac{\ell}{2}\right)$$

$$\frac{\Delta_u}{h_x} = \theta_{cyi} + \theta_{byi} + (\mu - 1)\theta_{byi} \qquad \text{(see Eq. 1.1.20g)}$$

For a member ductility factor of 5 ($\mu_b = 5$)

$$\frac{\Delta_u}{h_x} = 6\theta_{byi}$$

$$\theta_{byi} = \frac{\Delta_u}{6h_x}$$

$$= \frac{3.6}{6(180)}$$

$$= 0.00333 \text{ radian}$$

$$\phi_{byi} = \frac{6\theta_{byi}}{\ell_c}$$

$$\phi_{byi} = \frac{6(0.00333)}{324}$$

$$= 0.00006 \text{ rad/in.}$$

$$M = \phi E I_e$$

$$I_e = \frac{M}{\phi E}$$

$$= \frac{575(12)}{0.00006(4000)} \qquad (f_c' = 5 \text{ ksi})$$

$$= 28{,}750 \text{ in.}^4$$

$$I_b = \frac{I_e}{0.35} \qquad \text{(Assumed relationship)}$$

$$= 82{,}100 \text{ in.}^4$$

$$I_c = 0.5 I_b$$

(Column behavior will by design be in the idealized elastic range; hence $I_c \cong I_e/0.7$)

Step 10: Size the Beam and Column for Strength and Stiffness.

$$I_b \geq 82{,}100 \text{ in.}^4$$

Assume $b = 0.5h$. Then

$$I_b = \frac{0.5h(h)^3}{12}$$

$$h \geq \sqrt[4]{24 I_b}$$

$$= 37.5 \text{ in.}$$

Select a 20 in. by 37 in. beam ($I_b = 84{,}400 \text{ in.}^4$):

$$A_s = \frac{M}{\phi f_y(d - d')}$$

$$= \frac{575(12)}{0.9(60)(31)}$$

$$= 4.12 \text{ in.}^2 \qquad \rho_b = 0.6\%$$

Conclusion: Select a beam and column whose size is at least

Beam: 20 in. by 37 in.
Column: 26 in. by 26 in.

Design the strength of these members to attain the following objectives:

Beam: $M_{nb} > 575$ ft-kips
Column: $M_{nc} > \lambda_o M_{nb,\text{provided}}$

$$P_{nc} > \frac{2\lambda_o M_{nb,\text{provided}}}{\ell}$$

Should the structure place the response outside the velocity-sensitive region (Figure 1.1.5), a displacement-based design that adopts the criterion spectral velocity will be conservative for those structures that fall in the displacement-sensitive region and stronger than required for those structures in the acceleration-sensitive region.

Designs in the displacement-sensitive region (Figure 1.1.5) involve more considerations than those addressed by the criterion for this project. This aside, the recommended procedure is easily adjusted to accommodate a constant displacement objective.

Step 1: Identify the Objective Ultimate Drift (Δ_u). For a single-degree-of-freedom system, the ultimate drift is the spectral drift. Hence,

$$\Delta_u = S_d \tag{1.1.25}$$

Step 2: Find the Objective Minimum Natural Frequency of the System.

$$\omega_{n,\min} = \frac{S_v}{\Delta_u} \tag{1.1.26}$$

Continue the design based on the adopted or provided natural frequency.

Designs in the acceleration-sensitive region (Figure 1.1.5) can proceed directly from Step 4 of the last procedure, based on the fact that the system is a single-degree-of-freedom system and, as a consequence, S_a is known. This will produce appropriate designs so long as

$$\omega_n \leq \frac{S_a}{S_v} \tag{1.1.27}$$

The displacement of the ductile structure whose response falls in the acceleration-constant region will be

$$\Delta_u = \frac{S_a}{\omega_n^2} \frac{\mu}{\sqrt{2\mu - 1}} \tag{1.1.28}$$

1.1.8 Selecting Design Strength Objectives

Let us reexamine the behavior described in Figure 1.1.1 from the perspective of selecting the appropriate level of strength for the bracing system described in Figure 1.1.10. Following a displacement-based design approach, the desired strength described in Figure 1.1.13*a* would be based on attaining the objective drift (Δ_u). Since the drift objective must be attained in either sense (+ or −), it is logical that the strength be the same in either sense.

Now, consider the consequences associated with the existence of an initial vertical load on the beam. The beam and column will deform to some rotation given the existence of the nonseismic load. Call this imposed load M_D and its associated deformation θ_D. Given the symmetrical nature of the frame and most bracing systems, the existence of this moment (M_D) will not reduce the level of restoring force (F_R), which will remain the same. If the level of provided strength in each sense is the same we may quantify the restoring force from the following

$$\begin{aligned} h_x F_R &= M_{yi} - M_D + M_{yi} + M_D \\ &= 2M_{yi} \end{aligned}$$

The impact on deformation will be minimal, for the attainable deformation of the yielding components(s) will now be $\theta_u - \theta_D$ and $\theta_u + \theta_D$. The ductility and energy-dissipating characteristics of the system will not be materially altered. Consider the envelope of hysteretic behavior described in Figure 1.1.13*a*. Displacement cycles will begin at M_D, θ_D as opposed to the origin ($M = 0$, $\theta = 0$). From a behavior perspective, the attainable level of drift, Δ_u, will be a function of the strain imposed on the concrete, which in turn will be a function of the performance objectives for the system. In Chapter 2 these concrete strain limit states will be studied and found to be at least on the order of 0.007 in./in. (ε_{cu}). The strain state in the concrete at a rotation associated with θ_D will be by comparison quite small. Even if M_D is as much as $0.5M_{yi}$, an unlikely condition, ε_{cD} will be on the order of 0.0005 in./in. and this is only 7% of the probable strain demand.

It is for this reason that significant moment redistributions are permitted in the strength-based designs of structures, a feature that is particularly important in the design of precast concrete systems.

Strength-based design procedures usually adopt a first yield approach. Thus they combine seismic strength objectives (F_E) with sustained load strength objectives

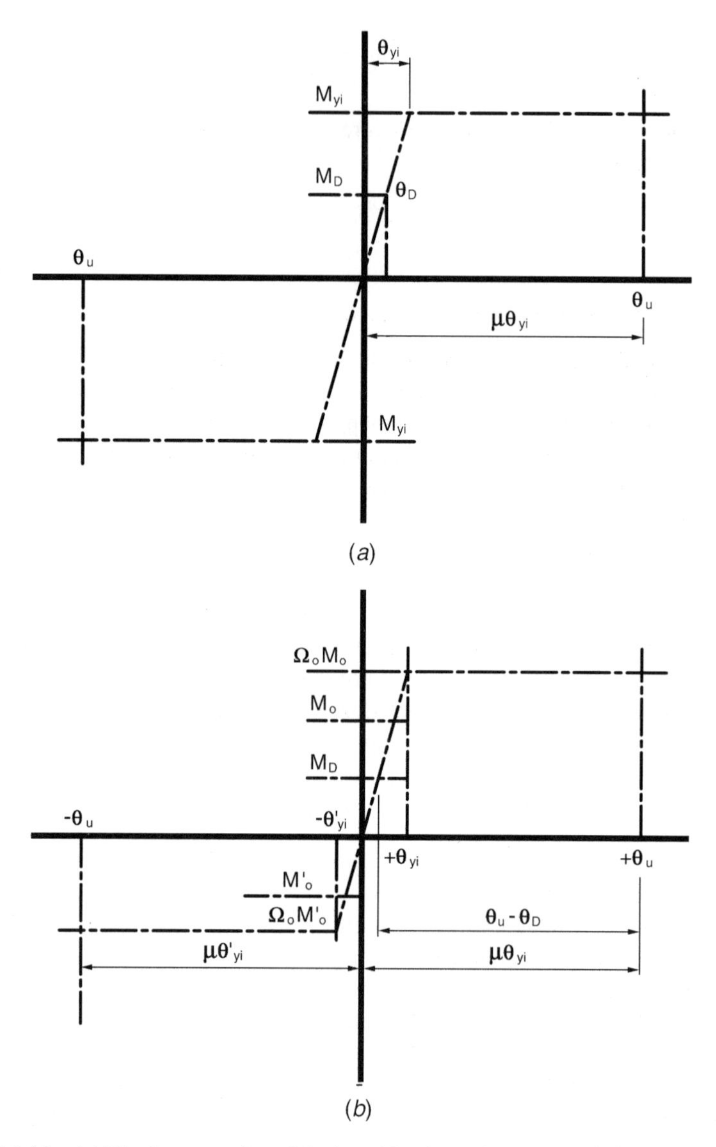

Figure 1.1.13 (*a*) Displacement-based design objectives. (*b*) Force-based design objectives.

(F_D) and factored live loads to arrive at an objective system strength. Neglecting the contribution of live loads, the design strength objectives would be (see Figure 1.1.13*b*)

$$F_o = F_E + F_D$$

$$F_o = F_E - F_D$$

The resultant envelope of hysteretic behavior is described in Figure 1.1.13*b*. The consequence of a strength imbalance is a reduction in the attainable ductility in the stronger beam ($+\Omega_o M_o$) as a direct consequence of the higher level of provided reinforcing and this loss will usually exceed the slight loss of deformability ($\theta_u - \theta_D$). Observe also that a higher strain demand will be imposed on the reinforcing provided to accommodate negative displacements at an ultimate drift of $-\theta_u$.

Further, from a constructibility perspective, congestion will be reduced by an equal distribution of reinforcement since the required total amount of reinforcing ($A_s + A_s'$) will be the same in either case.

Conclusion: The reinforcing provided in the top and bottom of a frame beam should be the same since this produces a system that will perform better and be more easily constructed—an attribute that should never be discounted. The interested reader is refered to Ref. 1.12, Sections 4.3.1–4.3.5.

1.1.9 Concluding Remarks

Ductility is an essential ingredient in structures that must withstand strong earthquakes. Its introduction into the strength based design process is difficult. Current force-based design procedures require the generalization of many features known to exist in systems that include ductile components arranged in a manner that allows their ductility to be exploited. The designer who uses strength-based procedures must understand the basis for the development of the system and member prescriptions contained therein. Design objectives are not always clear, especially to those who lack an intuitive understanding of system behavior, and this complicates the process.

The designer who uses a displacement-based procedure should not be bound by established or codified estimates of system ductility. Design displacement objectives can be related directly to performance goals. Component behavior characteristics are established in the laboratory, and this allows the designer significant insight and control over the behavior of a structure. The appropriate use of this insight should facilitate the development of a bracing program and a structure that will attain performance objectives.

The basic concepts discussed in this section should facilitate the design process by identifying simple relationships and design procedures that are easily remembered and applied. They are summarized as follows:

Equal Displacement: The response of a yielding or ductile structure to earthquake-induced ground motion will be essentially the same as that of a comparable elastic structure (Section 1.1.1).

PΔ Effects: A bracing system must be strong enough to equilibrate all of the mass that is tributary to it (Figure 1.1.4). The response of a ductile structure to earthquake ground motion will not be increased by the inclusion of $P\Delta$ effects (Section 1.1.2).

Response Spectra: Response spectra are the most effective conceptual design tool. They can be developed directly from estimates of peak ground motion (Figure

1.1.5). Spectral velocity is the most applicable quantification of earthquake intensity because it defines the response of most structures (Section 1.1.3). *Base shear spectra* attempt to include higher mode effects, but this is not appropriate when the objective is to quantify displacement or flexural strength in the design of a shear wall.

Damping: The response of structures to ground motion is reduced by increasing levels of damping. Energy dissipated by yielding (Figure 1.1.6) may be converted to structural damping and used in the design of ductile structures (Section 1.1.3).

Strength-Based Design: This is the traditional seismic design process. A singe-degree-of-freedom response spectrum is usually converted to an elastic base shear spectrum, which is then reduced to design loads through the introduction of a reduction factor that attempts to capture system ductility and overstrength (Figure 1.1.12).

Displacement-Based Design: Displacement-based design is a performance-based design procedure that endeavors to produce a system whose earthquake-induced displacement will be less than an objective level of drift. Thus the focus of the design is the ultimate drift (Δ_u) (Figure 1.1.1). Objective levels of drift can be developed directly from component strain states.

Capacity-Based Design: Capacity-based design is a design approach that strives to protect more brittle elements along both the lateral and vertical load paths so as to prevent the failure of essential components. The process must be applied with understanding because the extent to which members need to be protected is not identical. The magnitude of the load to which a component would be subjected is quantified by $F_{\max}$ (Figure 1.1.1). The process that is required to produce the desired result will be discussed in subsequent chapters, but the best example is the frame column. The axial demand imposed on a frame column can be developed from the system force $F_{\max}$ (Figure 1.1.1). The frame column need not be designed to the worst case moment that can result from the dynamic application of loads (multimode effects) so long as the created column section is ductile when subjected to the axial load associated with $F_{\max}$ (Section 2.2.3).

System Ductility: System ductility (μ) is an important design parameter that relates the ultimate drift (Δ_u) to an idealized estimate of system drift at yield (Δ_{yi}) (see Figure 1.1.1). Components along the lateral load path are expected to yield at system displacements that are significantly less than Δ_{yi} and, as a consequence, it is the strain state or *component ductility* of these elements that will determine the success of the design. The linkage between *component* and *system ductility* is building specific; it can be established analytically (Section 1.1.6) or by evaluating completed designs (Chapter 4).

Dynamic Characteristics of a Building: Analytical models of buildings must be created as a part of the design process. Unfortunately, dynamic characteristics are a function of the behavior characteristics of the ductile components located along the lateral load path, and these component characteristics will be neither repeatable nor stable (Figure 2.1.2). In fact, the dynamic characteristic of a building will change during an earthquake. Understanding this, the designer

Photo 1.5 Confined concrete, The Remington, Westwood, CA. (Courtesy of Morley Construction Company.)

must parameterically approach the design of a building utilizing worst case situations to help make a decision. If a strength-based design is being used, the estimate of building period should flow from component models that are conservatively stiffer than test data suggest (Figure 2.1.4). If a displacement-based approach is proposed, the design should be based on more flexible estimates of component behavior.

One truism bears repeating: The design of a building that must withstand earthquake-induced ground motion involves considerably more art than science. The engineer must use the scientific tools that are available to craft a building, but he or she should never lose sight of the many speculations required to create the illusion that earthquake engineering is an applied science.

1.2 CONFINEMENT—A COMPONENT BEHAVIOR ENHANCEMENT

The strength and deformability of concrete are significantly improved through the application of a confining pressure. Unfortunately, such confinement is too often used as a cure-all: "I am concerned about the behavior of this element—so I will confine it." To attain an appreciation for the impact and limitations of confinement, consider how the strength of a cylinder of sand is enhanced by a surrounding steel cylinder (tin can). Extend this analogy to a cemented sand or concrete cylinder. If the surrounding steel cylinder is assembled in a manner that produces a stress perpendicular to the direction of desired strength, as by radial prestressing, it will add to the lateral tensile strength of the contents that define the axial capacity of the cylinder. The increase in strength provided by the prestressing is caused by what is referred to as active confinement. Now imagine that the surrounding cylinder is integral with the concrete that it encloses but that at rest it does not apply a pressure to the concrete. In this case, as axial load is imposed on the concrete, the enclosing cylinder and the confinement it provides will have no effect on the strength of the concrete until the axial stress level in the concrete approaches its compressive strength ($\pm 0.85 f_c'$). At this stress level, the concrete in the cylinder expands laterally[1.13, Chapter 3, Fig 3.1]. Now the concrete cylinder will be constrained by the enclosing steel cylinder against dilatation, and the presence of the steel enclosing cylinder will, as the concrete expands laterally, apply a passive pressure to the contained concrete cylinder. This passively developed confining pressure will increase the strength of the concrete much as the tin can did for the sand. The confining pressure so applied is referred to as a passive pressure because it can only be activated by the expansion or dilatation of the concrete. Accordingly, absent an applied axial stress on the order of 75 to 80% of f_c', and the associated dilatation, passive confining reinforcement will not change the behavior of concrete.

1.2.1 Impact of Confining Pressure on Strength

Confinement and its impact on strength have been understood for many years. Confinement was first exploited in the design of spirally reinforced concrete columns. Spirally reinforced columns were found to be capable of sustaining imposed axial loads to strains well in excess of the accepted unconfined peak concrete strain of 0.0025 to 0.003 in./in. Tests performed on concrete cylinders confined by fluids (active confinement) in the 1920s[1.14] were used to establish a relationship between the confining pressure (f_ℓ) and the compressive strength of the confined concrete (f_{cc}'):

$$f_{cc}' = f_c' + 4.1 f_\ell \tag{1.2.1}$$

Equation 1.2.1 was used to develop an axial capacity for the core of a spirally reinforced column that exceeded the capacity of the same column were its shell and core fully effective. This development may be found in Ref. 1.13, Sec. 11-2. The resultant quantification of the spiral volumetric ratio (ρ_s) has long been a part of ACI codes[1.15]:

$$\rho_s = 0.45\left(\frac{A_g}{A_c} - 1\right)\frac{f'_c}{f_{yh}} \tag{1.2.2a}$$

where

A_g is the gross cross-sectional area of the column.
A_c is the area of the core measured to the outside diameter of the confining spiral.

Since Eq. 1.2.2a asymptotically approaches zero, a minimum volumetric ratio was established:

$$\rho_s = 0.12\frac{f'_c}{f_{yh}} \tag{1.2.2b}$$

where f_{yh} is the yield strength of the hoops.

This behavior enhancement was further acknowledged as being beneficial to spirally reinforced concrete columns through the acceptance of a higher strength reduction factor ($\phi = 0.75$ as opposed to 0.7).

The behavior enhancement provided by confining concrete subjected to postyield deformations (i.e., seismic events) was understood and acknowledged more than fifty years ago. The confinement of concrete is required in regions where postyield deformations and high concrete strains are anticipated.

Since most concrete sections were square or rectangular, the concepts developed for circular sections were extended to cover rectangular sections. The treatment of confining pressure for rectangular sections followed the approach developed for spirally reinforced columns. The column strength was maintained in the core, the objective of Eq. 1.2.2a, and a minimum level of confining pressure established. Rather than continue the volumetric ratio approach objective used in round shapes, objective confining pressures were stated in terms of required area of confinement reinforcing (A_{sh}):

$$A_{sh} = 0.3\left(sh_c\frac{f'_c}{f_{yh}}\right)\left[\frac{A_g}{A_c} - 1\right] \tag{1.2.3a}$$

The associated minimum value is

$$A_{sh} = 0.09\left(sh_c\frac{f'_c}{f_{yh}}\right) \tag{1.2.3b}$$

Volumetric ratios (ρ_s) and total transverse steel requirements (A_{sh}) tend to obscure a simple design objective, providing a passively developed internal transverse stress.

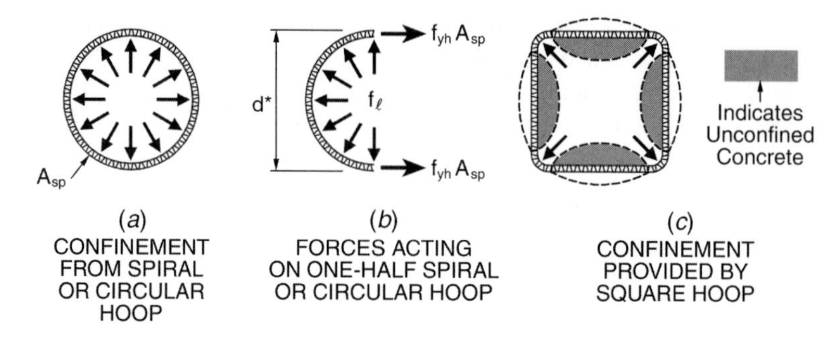

Figure 1.2.1 Confinement of concrete by circular and square hoops.[1.12]

Figure 1.2.1 describes how confining pressures (f_ℓ) or dilatational stresses load the reinforcing steel, which then develops the desired internal stresses. The confining pressure can be directly related to the steel that equilibrates it.

Consider the core of the spirally reinforced column shown in Figure 1.2.1. The confining pressure f_ℓ is a function of the strength that can be developed by the confining hoops at yield.

$$f_\ell = \frac{2 f_{yh} A_{sh}}{d_s s} \tag{1.2.4}$$

where

f_ℓ is the passively developed confining pressure at hoop or spiral yield.
f_{yh} is the yield strength of the hoop or spiral.
A_{sh} is the area of the spiral.
d_s is the center-to-center diameter of the spiral.
s is the spacing of the hoop or spiral along the length of the member.

The volumetric ratio (ρ_s) is defined[1.15] as the "ratio of volume of spiral reinforcement to total volume of core (measured out-to-out) of a spirally reinforced compression member." Hence

$$\rho_s = \frac{A_{sh} \pi d_s}{(\pi d^{*2}/4)s} \qquad \text{where } d^* = d_s + d_h \tag{1.2.5}$$

If we neglect d_h as being small in comparison with d_s, we may combine Eq. 1.2.2b, 1.2.4, and 1.2.5 and conclude that the ACI[1.15] objective confining pressure for a spirally confined round column is directly proportional to the strength of the concrete:

$$f_{\ell,\text{objective}} = 0.06 f'_c \tag{1.2.6}$$

This corresponds to 300 psi for 5000-psi concrete.

Equation 1.2.3b may also be restated in terms of an objective confining pressure:

$$f_\ell = 0.09 f_c' \tag{1.2.7}$$

The greater effectiveness of circular confining reinforcement is apparent from the stress flow described in Figure 1.2.1. Observe how the concrete in the core of the square column (Figure 1.2.1*c*) must "arch" to reach its equilibrating reaction.

The use of a volumetric limit (Eq. 1.2.2a and b) or its rectangular column equivalent (A_{sh} in Eq. 1.2.3a and b) as a design guide may be historically justified, but the design process would be simplified if they were replaced by the confining pressure objectives of Eq. 1.2.6 and 1.2.7. Equations 1.2.2a and 1.2.3a will typically establish the level of confinement for columns whose size is 24 in. or smaller. Since small frame columns are seldom used in regions of high seismicity Eq. 1.2.2a and 1.2.3a might be better dealt with by identifying the axial capacity objective, and this is especially true if high-strength concrete is used to reduce the size of the column. The quantification of confining reinforcement requirements using Eq. 1.2.2b or the determination of how to distribute the steel required by Eq. 1.2.3b is far more likely to produce design errors than are pressure-based relationships (Eq. 1.2.6 and 1.2.7).

The designer who understands confining pressure objectives will be able to handle situations that arise frequently, for example, boundary elements in walls, compression regions in beams and beam columns, or any area where concrete strains are expected to exceed 0.002 in./in.

Figure 1.2.2 describes how longitudinal bars and cross ties develop confining pressure in a rectangular shape. The typical situation is one in which the longitudinal bars are large and the crossties small. The longitudinal spacing of the hoops and crossties is usually much less than the space between longitudinal bars. The confining stresses tend to bridge the gap between the stiffer longitudinal bars that span from crosstie to crosstie. This will be true so long as the ratio of the longitudinal spacing of the ties to the reinforcing bar diameter (s/d_b) is less than that of the ties. Clearly, the effectiveness of a confined section will depend on the size and spacing of longitudinal bars and transverse reinforcing. It is, however, generally accepted that square tied confining reinforcement is 67% as effective as spirally confined cores. The ACI[1.15] confinement objectives clearly reflect this loss of efficiency (see Eq. 1.2.6 and Eq. 1.2.7).

Confinement provided in a shear wall or beam will be logically less effective than that provided in a column. This is because the level of confining pressure on the bottom face (see the cross section shown in Figure 2.1.1) will not typically be as effective because the deformation of the tie will be greater than that of a comparable tied column. In the design and detailing of wall sections, the tendency is to provide full hoops to produce more effective confinement. The interested reader is referred to Ref. 1.16. Paulay and Priestley[1.12, Eq. 3.62] suggest that the confining pressure should be a function of the axial load and this could result in a considerably higher level of confining pressure, almost twice that proposed by ACI[1.15] standards. Their objective is to significantly increase attainable levels of curvature ductility. Example 3.1 in Paulay and Priestley[1.12, Sec. 3.2.2e] analyzes a column confined to transverse stress levels more than twice that proposed by current ACI[1.15] standards ($0.09 f_c'$). They conclude

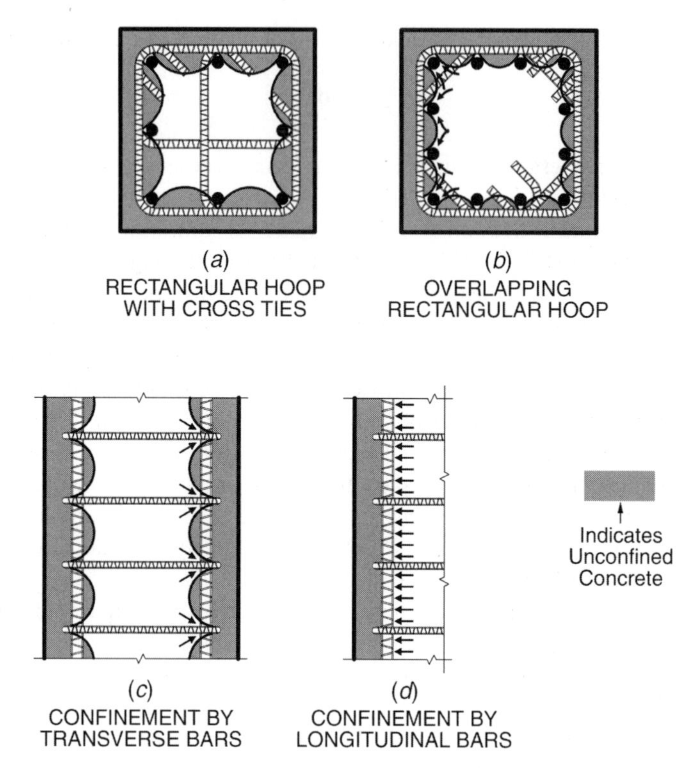

Figure 1.2.2 Confinement of column sections by transverse and longitudinal reinforcement.[1.12]

that the peak concrete strength will be 82% greater than the unconfined strength, and that the ultimate strain (ε_{cu}) in the confined concrete will be 0.055 in./in. This is consistent with experiments reported by Nawy,[1.17] but the placement of this amount of transverse reinforcing is impractical if not impossible. This topic will be a focal element in Section 1.2.2. Meticulosity is not justified based on available data so, ACI-recommended[1.15] confinement levels, used in conjunction with good engineering judgment, appear appropriate, at least for now.

The peak stress that can be attained by confined concrete is reasonably predicted by Eq. 1.2.1 or by the methods proposed by Pauley and Priestley.[1.12, Fig 3.6] At, for example, the code-prescribed level of confinement for a rectangular column (Eq. 1.2.7) whose concrete strength is 5000 psi, the attainable strength of the confined concrete (f'_{cc}) is 7000 psi following either procedure.

Summarizing Comments

- Confining pressure clearly enhances performance in concrete subjected to stresses at or near f'_c.
- A confining pressure of $0.09 f'_c$ is a reasonable design objective for a rectangular section.

- Constructibility is significantly impacted when higher levels of confinement are attempted, and the benefit is questionable because of the impact closely spaced large ties will have on concrete placement and consolidation.
- Load paths in the transverse reinforcing program must develop confining pressures. Member design procedures developed in Chapter 2 address this issue in more detail. Interior ties will clearly be subjected to larger loads (Figure 1.2.2) than exterior ties, and this is repeatedly confirmed by experiments (Chapter 2).
- Confining pressures of $0.09f_c'$ appear to be conservative (see Section 1.2.2) for concretes of higher strength. A tie spacing limit of 4 in. is too conservative, and may be relaxed to 6 in. in regions where the imposed loads are effectively controlled through the use of capacity-based design. (See Section 1.2.2 for comparative experimental data.)

1.2.2 High-Strength Concrete (HSC)

The use of high-strength concrete (HSC) does not appear to adversely affect the behavior of concrete members subjected to postyield deformations. This is contrary to our classical understanding, which is based on compression tests performed on cylinders. The traditional stress/strain relationship described in Figure 1.2.3 shows a significant decrease in ductility with an increase in concrete strength (f_c'). This brittleness is caused by a fracturing of the test specimen through the aggregate as opposed to, in the lesser strength cylinders, a propagation of the cracks through the paste. The micro- and macromechanics of concrete are discussed by Nawy[1.17, Ch.8] in some detail. We confine our discussion here to general characteristics of HSC and the identification of what seem to be desirable levels of confinement.

A considerable amount of research has been conducted in Southern California under the sponsorship of the Carpenters Contractors Cooperation Committee (C^4) to demonstrate that HSC can be developed and used in regions of high seismicity.[1.18] Professor James Anderson of the University of Southern California (USC) demonstrated that ductility could be attained in unreinforced HSC by controlling the relationship between the strength of the paste and the aggregate. The key to the creation of a more ductile concrete lies in attaining the objective level of strength and deformability at the time of the earthquake rather than at some arbitrary age such as 28 or 56 days. Figure 1.2.4 describes the behavior of a 12-ksi (f_c') mix. It will probably not reach the objective strength until it is one year old, but it is capable of attaining a strain of 0.005 in./in. at 91 days. Observe that this is higher than any of the concretes described in Figure 1.2.3. Testing of HSC components discussed in Chapter 2 clearly demonstrates that HSC, especially when designed to a controlled failure mode, is not likely to explode when subjected to large strains.

A major constructibility issue arises if, in fact, the level of confinement required by current codes (Eq. 1.2.7) is imposed on HSC. This is because the required level of confinement is directly proportional to the strength of the concrete. Consider, for example, the column described in Figure 2.2.3. The confinement reinforcing provided in the hinge region (#5 hoops at 4 in. on center) is that required to satisfy

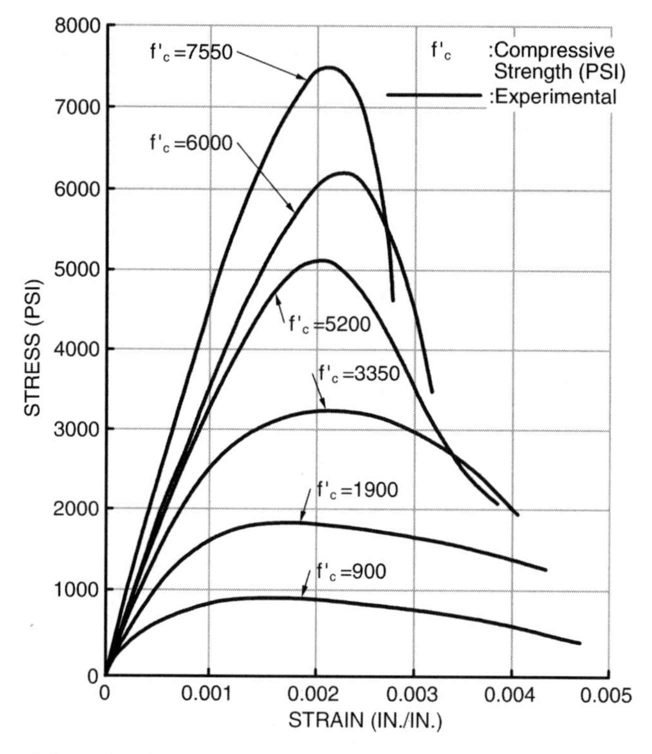

Figure 1.2.3 Classic stress/strain behavior of plain concrete.

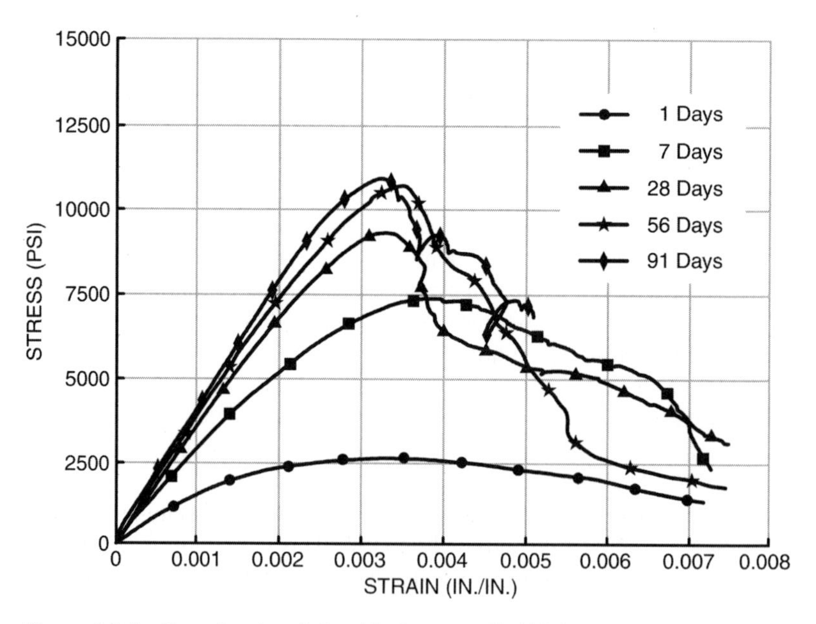

Figure 1.2.4 Stress/strain relationship for controlled high-strength concrete mix.

Eq. 1.2.7 for 8000 psi concrete. The described three-tie stack is 2 in. thick and the as-constructed reinforcing bar cage is one that approaches an impractical condition from a constructibility perspective (Figure 1.2.5). The amount of transverse reinforcement, and especially the development of this reinforcement (135° hooks), leads one to question the attainable quality of concrete placed in this member. Imagine the impact on the reinforcing bar cage were the concrete strength to be increased to 12 or 16 ksi.

There can be no denying the fact that increasing the confining pressure applied to HSC significantly improves its strength and ultimate strain. Nawy[1.17, Table 8.5] reports an increase in stress on the order of 80% when the confining pressure applied to 10- and 15-ksi concrete is on the order of twice that required by the ACI[1.15] (Eq. 1.2.7). The increase in attainable strain is on the order of 4 times. Clearly, this level of confining pressure is not practical from a constructibility standpoint. More recently Xiao[1.18 and 1.19, Sec. 1.8 & 6] tested a series of HSC columns subjected to various levels of axial load and controlled levels of postyield displacement. The level of confining pressure and the spacing and strength of the confining hoops were treated as variables. These tests are now summarized, and the conclusions below are drawn.

1.2.2.1 Ductility

Specimen HSC FF1-0.2 ($f_c' = 9$ ksi)

Axial load	$0.2f_c'A_g$
Confining pressure	$0.09f_c'$
Tie strength (f_{yh})	60 ksi
Vertical tie spacing	4 in.
Repeatable drift ratio	6%
Strength degradation	None

Figure 1.2.5 Preparation of steel cages.

SPECIMEN HSC FF5-0.2 ($f_c' = 9$ ksi)

Axial load	$0.2f_c'A_g$
Confining pressure	$0.06f_c'$
Tie strength (f_{yh})	60 ksi
Vertical tie spacing (s)	6 in.
Repeatable drift ratio	4%
Strength degradation	Slight (6.7%)

Conclusions

- The level of ductility available in an HSC column confined to $0.06f_c'$ is significant ($\mu = 4+$) and the attained drift ratio of 4% is more than might reasonably be expected to occur in a design that follows a capacity-based criterion.
- The attainable level of drift seems to be almost linear, and this appears to be consistent with the experimental data presented by Nawy.[1.19] A confinement pressure less than $0.06f_c'$ is not advisable.

Recommendations

- Confining pressure in HSC columns whose drift ratio is controlled through the application of capacity-based concepts need not exceed $0.06f_c'$. This does not comply with current code mandates (Eq. 1.2.7), and designer discretion is accordingly advised until such time as the code accepts lesser values.
- Where large postyield rotations might reasonably be expected in a column at the base of a building, for example, higher levels of confinement ($0.09f_c'$) are suggested.

1.2.2.2 High-Strength Ties One way to reduce congestion is to utilize higher strength ties at greater intervals. This alternative was considered by Xiao.[1.18, sec. 6]

SPECIMEN HSC FF3-0.2 ($f_c' = 9$ ksi)

Axial load	$0.22f_c'A_g$
Confining pressure	$0.09f_c'$
Tie strength (f_{yh})	75 ksi
Tie spacing (s)	5 in.
Repeatable drift ratio	6%
Strength degradation	None

Alternative levels of confining pressures were also tested along with greater tie spacing and hoop yield strength:

SPECIMEN HSC FF5-0.2 ($f_c' = 9$ ksi)

Axial load	$0.2f_c'A_g$
Confining pressure	$0.06f_c'$

Tie strength (f_{yh})	60 ksi
Tie spacing (s)	6 in.
Repeatable drift ratio	4%
Strength degradation	10%

SPECIMEN HSC FF6-0.2 ($f'_C = 9$ ksi)

Axial load	$0.2f'_cA_g$
Confining pressure	$0.075f'_c$
Tie strength (f_{yh})	75 ksi
Tie spacing (s)	6 in.
Repeatable drift ratio	4%
Strength degradation	None

Conclusions

- High-strength hoop tie reinforcing ($f_{yh} = 75$ ksi) may be used to provide confinement in HSC columns.
- Hoop tie spacings greater than the current maximum (4 in.) may be used without affecting attainable levels of drift provided tie spacing does not exceed 5 in. and confining pressures of $0.09f'_c$ are maintained (HSC FF1–HSC FF3).
- If greater hoop tie spacings are used ($s = 6$ in.) in regions where drift ratios are controlled through the use of capacity-based design concepts, the confining pressure should be increased from $0.06f'_c$ to $0.075f'_c$ (HSC FF5–HSC FF6).

Recommendations

- High-strength hoop reinforcement ($f_{yh} = 75$ ksi) may be used.
- Tie spacings that are greater than those currently used ($s = 4$ in.) are appropriate. This proposition is supported in the cited version of the ACI Code.[1.15] If currently recommended levels of confining pressure ($0.09f'_c$) are attained, tie spacing may be increased to 5 in. in regions where large postyield deformations are expected.
- If the confining pressure is at least $0.075f'_c$ the tie spacing may be increased to 6 in. in columns designed to a capacity-based criterion.

1.2.2.3 Higher Axial Loads As the axial load level is increased, the available level of ductility in a member subjected to both axial and flexural loads decreases. This loss of ductility (see Section 2.2.1.2) should be mitigated through the use of higher levels of confinement. It does, however, seem reasonable to use tie spacings larger than 4 in.

SPECIMEN HSC FF2-0.34 ($f'_c = 9$ ksi)

Axial load	$0.34f'_cA_g$
Confining pressure	$0.09f'_c$
Tie strength (f_{yh})	60 ksi

Tie spacing (s)	4 in.
Repeatable drift ratio	3% to 4%
Strength degradation	None

SPECIMEN HSC FF4-0.33 ($f'_c = 9$ ksi)

Axial load	$0.33 f'_c A_g$
Confining pressure	$0.09 f'_c$
Tie strength (f_{yh})	75 ksi
Tie spacing (s)	5 in.
Repeatable drift ratio	3% to 4%
Strength degradation	None

Conclusions

- Attainable levels of drift ratio are reduced as axial load levels are increased (HSC FF1–HSC FF2 and HSC FF3–HSC FF4).
- Tie spacings of up to 5 in. are reasonable.
- High-strength ($f_{yh} = 75$ ksi) hoop tie reinforcing may reasonably be used.

Recommendations

- Where axial load levels are high ($\pm 0.33 f'_c A_g$), hoop reinforcing should produce a confining pressure of $0.09 f'_c$, tie spacings may be as much as 5 in., and tie strengths of up to 75 ksi are appropriately used.

Concluding Remarks

- As the level of axial load approaches the balanced load (P_b, see Section 2.2), the available level of ductility will decrease and the level of confinement will become more important. At balanced conditions ($P_b \cong 0.33 f'_c A_g$), confining pressures should be at least $0.09 f'_c$.
- When capacity-based design is effectively used, lesser levels of confinement are justified, especially in HSC. Tie spacings of 6 in. on center are also reasonable, as is the use of higher strength ties.
- Welded wire fabric and continuous bend ties are becoming increasingly available. Their use promotes constructibility.

1.3 SHEAR

The effective transfer of shear is essential to the success of a concrete component—especially when it is subjected to cyclic loads and postyield deformations. Simple analytical models have traditionally been used to define strength limit states, and they have been reasonably effective. Unfortunately, the use of simple analytical models imposes constraints on their applicability, and these constraints can only be described by prescription. Prescriptions are difficult to write because, absent a thorough technical explanation, it is hard to clearly identify the condition of concern. The result is that prescriptions usually preclude the use of systems or load paths that are quite acceptable and often desirable. The price of simplification is not only the loss of design freedom, but in some cases the creation of undesirable behavior states.

In this section, an attempt is made to expand these simple models and identify limit states. This is necessary, especially in the design of precast concrete that, except in its emulative development form, does not fit the simple beam shear model that is the traditional basis for cast-in-place concrete design. The effective use of energy dissipation in precast concrete requires the development of strut and tie or truss models and the use of passively activated friction as a load path. The goal of this section is to expose the reader to the underlying theory of shear mechanisms, leaving example applications to later sections.

1.3.1 Shear Strength

Shear, once concrete has cracked, is transferred by two mechanisms—by the cracked concrete and by a truss mechanism. It is generally accepted that the concrete will transfer shear by an interlocking of aggregate along the cracked surface, friction across a flexurally activated compression region, and dowel action imposed on reinforcing crossing the created cracks. The extent to which each of these mechanisms is effective will depend on a number of conditions that the designer should be able to evaluate. This should allow the designer to subjectively allocate the appropriate portion of the currently accepted concrete shear (v_c) transfer mechanism $\left(\pm 2\sqrt{f'_c}\right)$ to the total shear transfer mechanism.

The reinforced shear (v_s) load path follows an internally developed truss or strut and tie model. Two basic assumptions are used to develop the codified mechanism strength for this load path: that truss panel points are square and that a uniform compression field is created. The truss described in Figure 1.3.1 is the basis for the codified load path and one may (analytically) visualize all of the shear reinforcing contained in a panel as being combined into the vertical tension member $\left(\Sigma A_v\right)$.

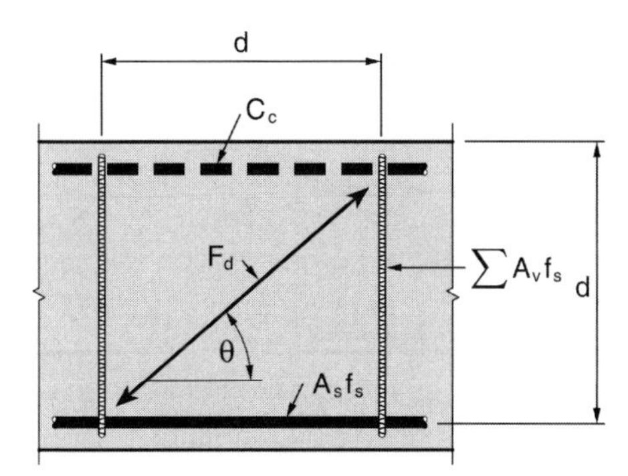

Figure 1.3.1 Truss analogy.

The shear strength provided by the reinforcing can be stated in terms of a uniform shear stress:

$$v_s = \frac{\sum A_v f_s}{bd} \tag{1.3.1}$$

As the amount of shear reinforcing (ΣA_v) is increased, the load on the compression diagonal (F_d) increases until the concrete along the compression diagonal reaches its limit state in compression. Were the truss of Figure 1.3.1 actually created, the capacity of the concrete diagonal would depend on the post-cracked condition of the concrete and the effective width of the compression diagonal. These variables are discussed extensively in Ref. 1.13, but for our purposes it is important to realize that the capacity of a concrete strut, in cyclically loaded elements, will diminish to a level of 30 to 35% of f_c' as cracking increases. The width of the compression strut is effectively increased by distributing the reinforcing, and it is for this reason that maximum spacing of shear reinforcing is reduced as the induced shear stress level increases.

The angle sustained (θ) by the compression diagonal in the development of shear as it applies to beam behavior is 45°, and this is typically conservative. The probable shear strength of a member whose strength in shear is controlled by the yielding of the shear reinforcing provided is determined by the "plastic truss" analogy described in Figure 1.3.2.

The plastic truss analogy assumes that all shear reinforcing will yield, and that the nodes will be capable of equilibrating imposed force levels. Given this yielding, the angle of the compression strut will flatten out to 25° to 30° provided the so-developed load path is maintained. Associated with this shear yielding is ductility in shear and this should be encouraged, especially when the attainment of flexural ductility is questionable.

The attainable strength of a reinforced section that can yield in shear will then be

$$V_s = \frac{A_v f_y d}{s \tan\theta} \tag{1.3.2}$$

where θ is the maximum developable angle but no less than 25°.

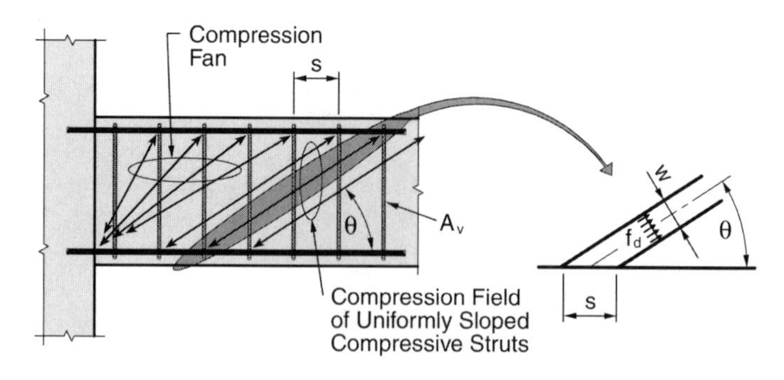

Figure 1.3.2 "Plastic truss" analogy.

Accordingly, so long as the compression diagonal does not fail, the codified version of the strength provided by the shear reinforcement may be conservative by as much as 2.14 (1/tan 25)

If the shear imposed on a concrete member is controlled by a capacity-based design, the strength provided in shear using beam shear theory will produce a member that meets our design objective provided that the shear imposed on the member does not cause the compression diagonal (F_d) to fail in compression. Hence, effective limit states are appropriately defined for shear reinforcement.

Since these effective limit states are based on the avoidance of a concrete compression failure along the compression diagonal, their exceedance will cause a brittle failure in shear. Accordingly, the designer should:

- Promote a flexural yield mechanism
- Not overreinforce a member in shear

Park and Paulay[1.20] describe the postyield behavior of a series of panels that are designed to promote flexural yielding, shear yielding, and one that is overreinforced in both flexure and shear. Flexural and shear yielding are accompanied by large ductilities, whereas the failure of the overreinforced panel is quite brittle. I have seen overreinforced shear walls that, during an earthquake, were overloaded. Individual, well-defined blocks of concrete were created by cracks that formed along the 45° diagonals in each direction. Unfortunately, extant codes tend to encourage overreinforcing in shear when a flexural mechanism cannot be attained. This is accomplished through the imposition of a low-capacity reduction factor ($\phi = 0.6$) in the development of objective shear reinforcing quantities.

The usual means for controlling the shear load on any member is to make sure that the flexural strength of the member limits the shear imposed on the member through the use of a capacity-based design. Two conditions need to be considered:

- What is the appropriate shear limit state?
- What is the appropriate shear limit state when it is impossible to control the induced shear level by a capacity-based design process—or otherwise stated when ductility in shear becomes a design objective or necessity?

To address these issues, we must evaluate the stress imposed on the compression diagonal and identify its probable strength limit state.

Figure 1.3.2 describes uniformly sloped load paths functioning at an angle θ. The load imposed on the compression diagonal is most easily visualized as a stress that is a function of the load carried by the steel tie at yield. The effective shear stress is a function of the tensile yield strength of the tension vertical

$$v_{sy} = \frac{A_v f_y}{bs} \tag{1.3.3a}$$

and the associated stress on the compression diagonal.

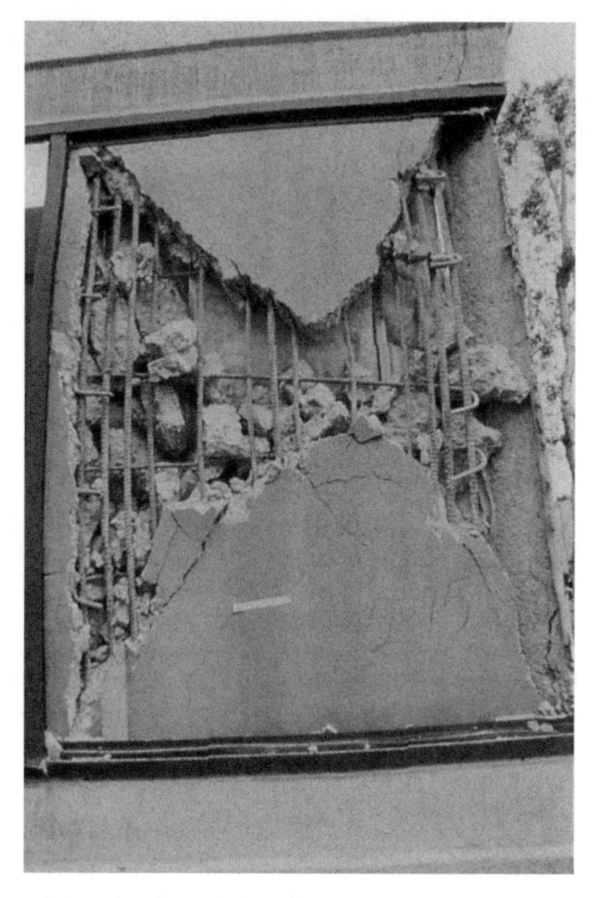

Photo 1.6 Overreinforced wall panel shear failure, Northridge earthquake, 1994. (Courtesy of Englekirk Partners, Inc.)

$$f_{cd} = \frac{v_{sy}}{\sin^2 \theta} \tag{1.3.3b}$$

Now if we choose to introduce a strength limit state that is a function of the compressive strength of the concrete,

$$f_{cd} \leq m f_c' \tag{1.3.4}$$

we can identify a shear strength limit state in terms of its effective strength factor:

$$v_{s,\max} = m \sin^2 \theta f_c' \tag{1.3.5a}$$

The probable worst case scenario (a severely cracked web) would, suggest the use of an effective strength factor (m), of 25%.[1.13] If the slope of the diagonal were 25°, the following shear stress limit state would be created:

$$v_{s,\max} = (0.25)(\sin^2 25) f_c'$$
$$= 0.045 f_c' \qquad (1.3.5b)$$

As the level of shear reinforcement is increased, the inclination of the angle θ would move toward 45° because the stirrups do not yield. This creates an overreinforced condition. The associated shear stress limit state for a severely cracked web could be as much as

$$v_{s,\max} = (0.45)(\sin^2 45) f_c'$$
$$= 0.225 f_c' \qquad (1.3.5c)$$

This migration of the compression diagonal is the direct consequence of the behavior of the tension vertical. In essence, the plastic elongation of the tension tie (stirrup) creates the load sharing characteristic of the "plastic truss" (Figure 1.3.2). Absent this yielding of the tension ties, a plastic truss will not be created.

Figure 1.3.3 describes a member loaded to its shear limit state. Observe that the angle of the cracking follows the plastic truss analogy described in Figure 1.3.2 and that the interior or constant inclination of the shear load path is about 35°. It appears that a shallower angle ($\theta \cong 25°$) will not form, for it certainly could have in the beam of Figure 1.3.3 given the close spacing of the stirrups.

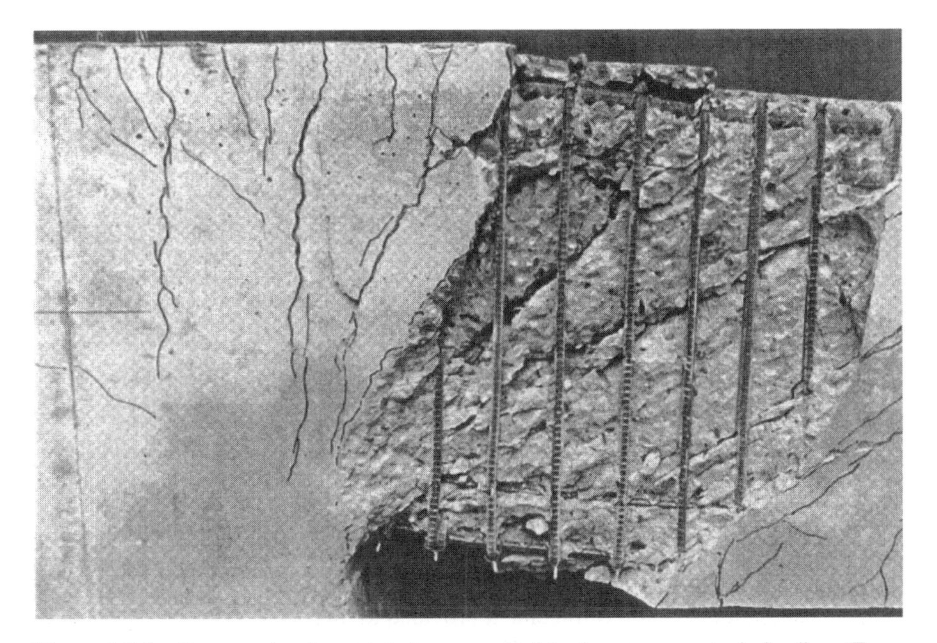

Figure 1.3.3 Compression fan at interior support of the beam—monotonic loading. (From *Reinforced Concrete: 3E, Mechanics and Design* by James G. MacGregor, © 1997. Reprinted by permission of Pearson Education, Inc., Upper Saddle River, NJ.)

It seems reasonable to conclude that the shear strength of a beam that will be cyclically loaded in the postyield domain and, as a consequence, cracked along both diagonals, will depend on the stress imposed on the stirrup. When the stress imposed on the shear reinforcement is certain to be within the elastic range, the limiting shear stress should be $0.225 f_c'$ (Eq. 1.3.5c), but ductility in shear will be severely limited in this case. Observe that an angle of 35° and an effective stress of 35% produce a shear stress limit state of $0.115 f_c'$, and this seems to be a more reasonable limit state provided it is used to limit the amount of shear reinforcement. Given this limit state, significant shear ductility will be realized.

Current codes identify the strength limit state as being a function of the square root of f_c' $(n\sqrt{f_c'})$. The preceding relationships can also be developed to this limit state characterization:

$$n\sqrt{f_c'} \leq m \sin^2 \theta f_c' \tag{1.3.6}$$

$$n = m \sin^2 \theta \sqrt{f_c'} \tag{1.3.7a}$$

And for the worst case ($m = 0.25$; $\theta = 25°$):

$$n = 0.25 \sin^2 25 \sqrt{f_c'} \qquad \text{(Eq. 1.3.7a)}$$

$$= 0.045 \sqrt{f_c'} \qquad \text{(Eq. 1.3.5b)}$$

For 5000 psi concrete this becomes

$$n = 0.045\sqrt{5000}$$

$$= 3.1$$

and for the more probable case ($m = 0.35$, $\theta = 35°$) and 5000-psi concrete,

$$n = 0.35 \sin^2 35° \sqrt{f_c'}$$

$$= 8.2$$

Hence, the shear reinforcing provided should be limited to

$$v_{s,\max} = 8\sqrt{f_c' f} \tag{1.3.7b}$$

Paulay and Priestley[1.12] recommend a shear stress limit of $0.16 f_c'$ and that the shear stress not exceed 870 psi. The ideal case would be a criterion that establishes and *limits* the amount of shear reinforcement provided to that associated with the attainment of the flexural overstrength ($\lambda_o M_n$) of a member and, further requires that the imposed level of shear not exceed $8\sqrt{f_c'}$.

In Chapter 2, when the behavior of frame beams is reviewed, we see that these limit states are probably too high. It is also suggested that when shear ductility is an objective, the level of shear reinforcement should be limited to something on the order of $0.1 f_c'$. Providing more reinforcement than this will cause the strength of the member to rapidly degrade as the compression field deteriorates.

Extending these concepts to shear walls requires the inclusion of member geometry, usually in the form of a shear span. The shear span (a) in the wall described in Figure 1.3.4 is h_w/ℓ_w and this is easily visualized. Hence

$$a = \frac{h_w}{\ell_w} \tag{1.3.8}$$

The identification of shear span as a variable essentially endeavors to include the contribution of arching action to shear strength. The action described in the interior portion of Figure 1.3.2 is referred to as beam shear. An arch is, however, developed in the region near the support. MacGregor[1.13]classifies shear behavior zones as being either beam behavior regions or regions of discontinuity. These regions of discontinuity are identified as being within a distance d of a change in load path from the uniformly sloped shear flow angle (θ) of Figure 1.3.2. The truss analogy (Figure 1.3.1 or 1.3.2) is appropriately used to design beam regions, but strut and tie models are often required to design regions of discontinuity.

The impact of shear span on shear strength is best described by Figure 1.3.5. This figure was developed from Park and Paulay,[1.20] who relied on the work of Leonhardt.[1.21] Two analytical projections of member strength are compared to experimentally attained strength (observed ultimate shear) in Figure 1.3.5. The flexural capacity of a member reinforced with a constant amount of flexural reinforcing and that associated with a constant "beam action" shear strength are related to the shear span ratio (a/d). The loading is applied by a pair of concentrated loads relocated so as to create various shear span ratios. Arch action begins to influence shear strength at a shear span ratio (a/d) of 2.5. This maximum arching shear span ratio of 2.5 corresponds to an

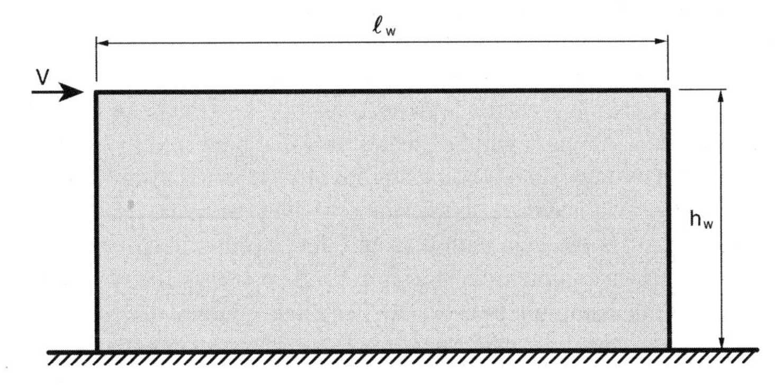

Figure 1.3.4 Shear wall.

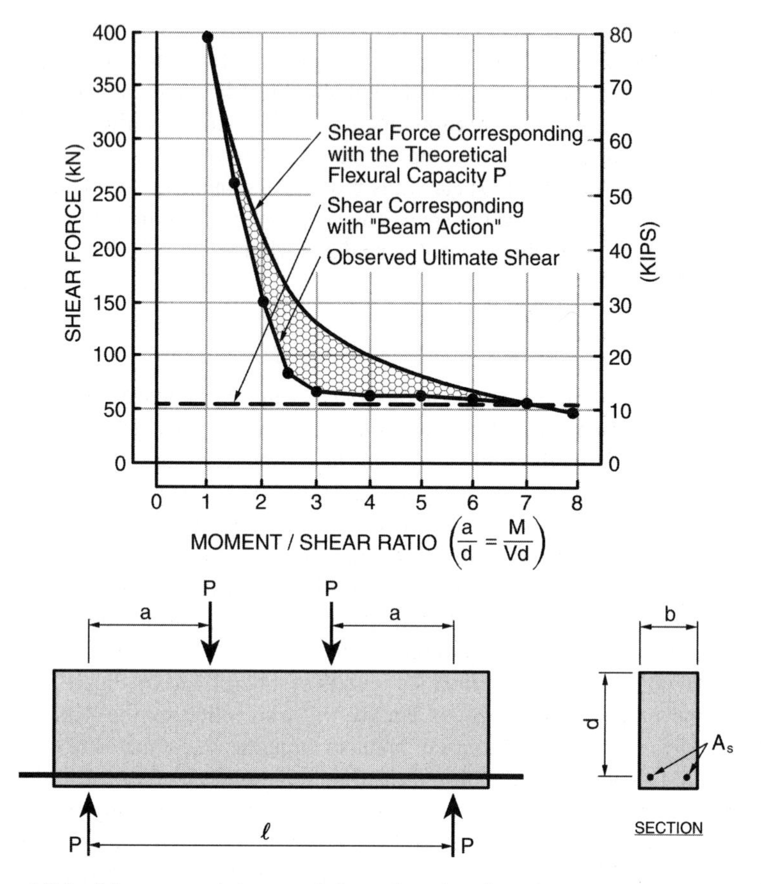

Figure 1.3.5 Moments and shears at failure plotted against shear span to depth ratio.[1.20]

angle (θ of Figure 1.3.2) of 22°. In essence, a partially effective arch is being formed between the load and the reaction at a shear span ratio of 2.5. Observe that arch action does not become fully effective until the shear span ratio reaches 1. An equilibrated load path (strut and tie model) must be used if one wishes to define shear strength for shear span ratios that are less than 1.5. In the transition area between 1.5 and 2.5, the attained level of shear strength will be greater than that predicted by beam theory.

Members that have very short shear spans are often referred to as deep beams, and this implies that the load is carried to the support by one or more compression struts. Figure 1.3.6*a* describes the load path that will develop in a deep beam. As can be seen in Figure 1.3.6*a*, a compression strut is created between nodes B and C and not exclusively at a depth that might be associated with the Whitney stress block (Node G of Figure 1.3.6*a*). Clearly beam theory will not describe the strength or deformation of a deep beam well.

The problem is complicated somewhat when distributed loads must be considered, as would be the case in a diaphragm. Now stress trajectories flow from the load points

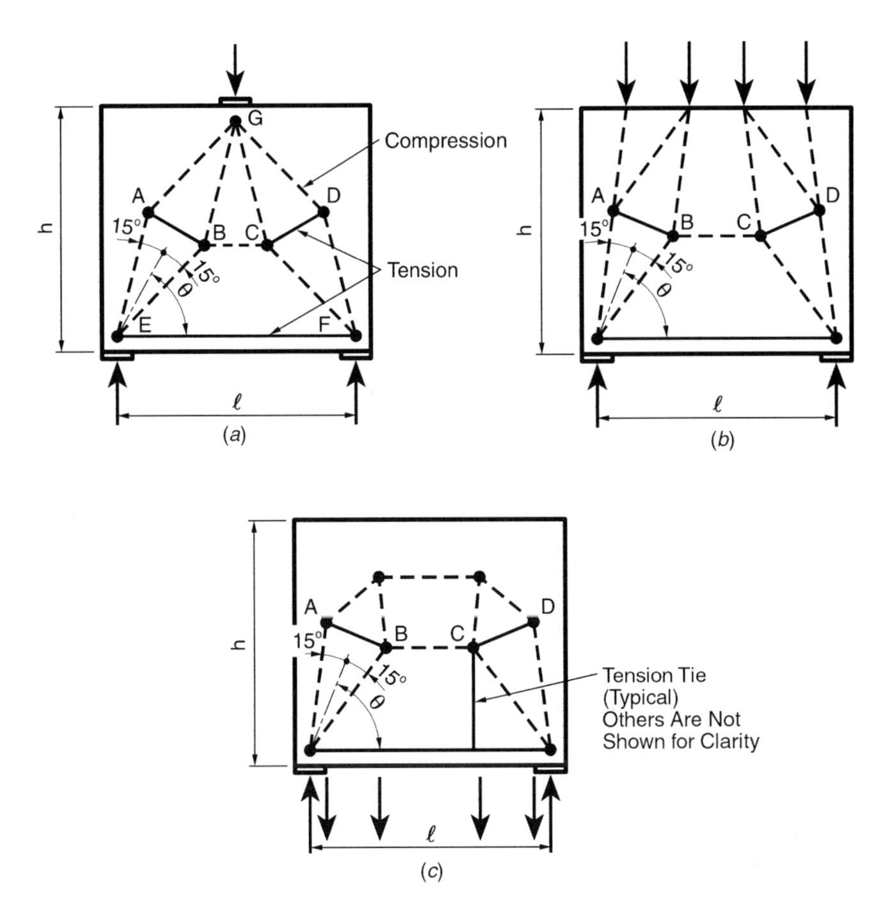

Figure 1.3.6 (*a*) Truss model. (*b*) Truss model subjected to distributed loads. (*c*) Strut and tie model subjected to tension loads.

to the support, but the basic model is little changed (Figures 1.3.6*b* and *c*). From a design perspective the existence of these interior compression struts, means that the developable moment in deep beams, or diaphragms whose shear spans are less than 1.25 (2.5 for continuous beams) will tend to be overestimated by beam theory. In seismic (capacity-based) designs, this is actually a positive attribute because the mechanism load will tend to be conservatively predicted by beam theory. It also means that a compressive force will be distributed across most of the member since a combination of the actions described in Figures 1.3.6*b* and *c* reasonably describe inertial loads as they are delivered to a diaphram.

Precast concrete diaphragms will not tend to develop stress flow patterns described in Figure 1.3.6 unless the induced secondary tensile stresses (i.e. tie *AB* in Figure 1.3.6*a*) are activated by reinforcing. The consequences associated with not providing tensile strength normal to the compression load path (Node *G* to *E* in Figure 1.3.6*a*)

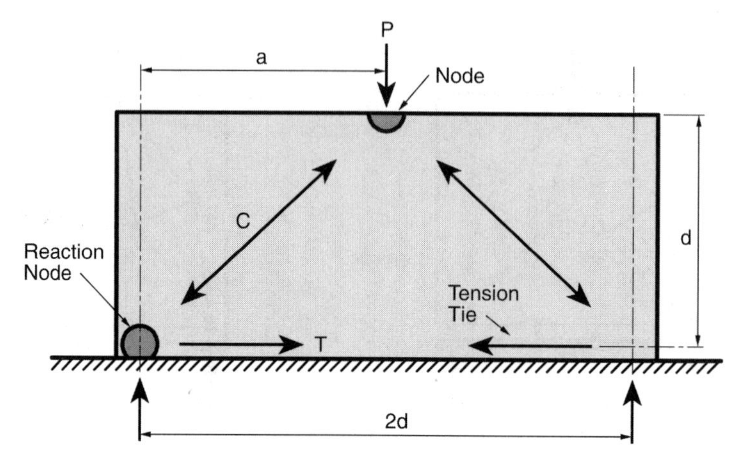

Figure 1.3.7 Arch action.

will be the creation of a more direct and narrower load path between the applied load and the reaction. Cracks will form parallel to the direction of the compression load (Figure 1.3.6*a*) whether or not reinforcing is provided because the principal tension stress is normal to the strut. The impact on design associated with the omission of reinforcing perpendicular to the load path is that broadening of the stress field will tend to be limited; in essence, the 15° spread described in Figure 1.3.6 will not tend to develop.

Figure 1.3.7 describes pure arch action. Essential to the realization of the load transfer is the development of effective nodes. In most cases the node will be the critical element, and this suggests that the load imposed on the node should be limited or controlled by ductile mechanisms. The effectiveness of the two nodes shown in Figure 1.3.7 will vary because they are not subjected to the same set of loads. The upper node is subjected to two compressive loads. Thus the node is "hydrostatically" stressed. This biaxial confinement imposed on the upper node will add strength, whereas the reaction node will be less effective because it is subjected to a tensile load and a compression load. MacGregor[1.13] suggests the node stress limit states described in Table 1.3.1.

The development of the node at the reaction (Figure 1.3.7) will depend on either the applied load or, given earthquake loads as a criterion, the strength (overstrength) of the yielding element.

Table 1.3.1 Effective Concrete Stresses in Nodes

Node Description	f_{ce}
Joints bounded by compression struts	$0.85 f_c'$
Joints anchoring one tension tie	$0.65 f_c'$
Joints anchoring more than one tension tie	$0.5 f_c'$

As in most connections, a precise evaluation of stress flow in the vicinity of a node is not possible, so the objective should be to develop a local equilibrated mechanism at the node. The conditions described for the tension (reaction) node of the element described in Figure 1.3.7 are those shown in Figure 1.3.8. This is an idealized condition, for all of the forces are shown to intersect at a centroidal point common to each force. This need not be a requirement provided secondary actions can be accommodated in the detailing.

For the case shown in Figure 1.3.8, the stress development might reasonably proceed as follows:

$$C = \frac{T}{\cos\theta} \tag{1.3.9a}$$

$$= 1.414T$$

$$w_{\min} = \frac{C}{bf_{ce}} \tag{1.3.9b}$$

where b is the width or thickness of the member and f_{ce}, the effective concrete strength, is $0.65f_c'$ (see Table 1.3.1).

The controlling element will tend to be the anchorage of the tension reinforcement, and the treatment of this must recognize the impact of the confining compressive forces R and C. The assignment of the probable width of the strut at the reaction is somewhat subjective but may reasonably be developed from ℓ_b, the width of bearing, or $2d'$. Examples of the application of this and other concepts will be developed where appropriate in later chapters.

As the compression force of Figure 1.3.7 distances itself from the node, it will tend to spread if, as previously discussed, it is reinforced to encourage this spreading more or less as described in Figure 1.3.6. The spread angle of 15° is widely accepted and conservatively often presumed to be 25% of the length of the diagonal. Accordingly,

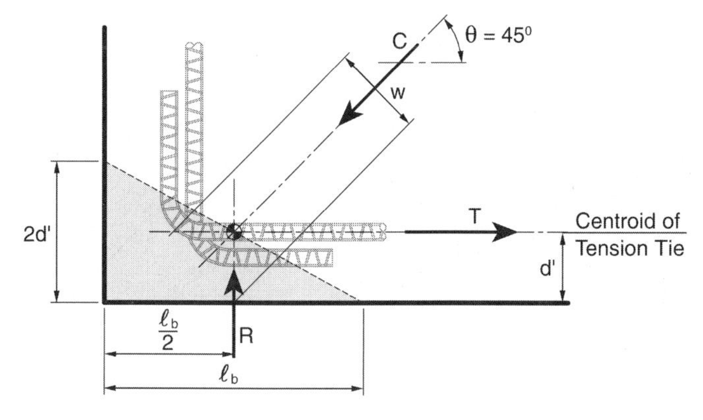

Figure 1.3.8 Reaction node of Figure 1.3.7.

the stress imposed on the diagonal for a tension controlled member like the one described in Figure 1.3.7 would be

$$F_{cd} = \frac{T}{\cos\theta} \tag{1.3.10a}$$

$$f_{cd} = \frac{F_{cd}}{bw}$$

$$= \frac{F_{cd}}{b(0.25)\ell_{cd}} \tag{1.3.10b}$$

The limiting stress on the compression diagonal will depend on the condition of the diagonal, cracked or uncracked, and the angle θ. For an uncracked diagonal, the effective stress is about $0.5f'_c$, while for severely cracked (as would be caused by postyield load reversals) webs or diaphragms, the sustainable stress would vary from $0.45f'_c$ when $\theta = 45°$ to $0.25f'_c$ when $\theta = 30°$.[1.13] It is important to remember that this load spreading will cause an orthogonal tension stress between points A and B of Figure 1.3.6a, and this simply means that a tension tie must be provided if the load spread is to be realized. Absent the tension tie, the width of the compression strut must remain at or close to its nodal width, and this will be an important consideration in the design of precast concrete diaphragms.

Squat shear walls ($h_x/\ell_w < 1.0$) are typically loaded in a fairly uniform manner, as shown in Figure 1.3.9. The consequence of this loading is the development of compression struts or a uniform stress field that activates the tension (vertical) reinforcing presumed to be uniformly distributed in Figure 1.3.9. As in all shear mechanisms, it is of paramount importance not to overstress the compression diagonal. Observe that the effective length of the compression diagonal can be the length of the wall if horizontal reinforcing is provided to develop the load path shown in Figure 1.3.9b. Accordingly, the previously discussed strength limit states are appropriately considered. Where uniformly distributed loads are not developed along the base as, for example,

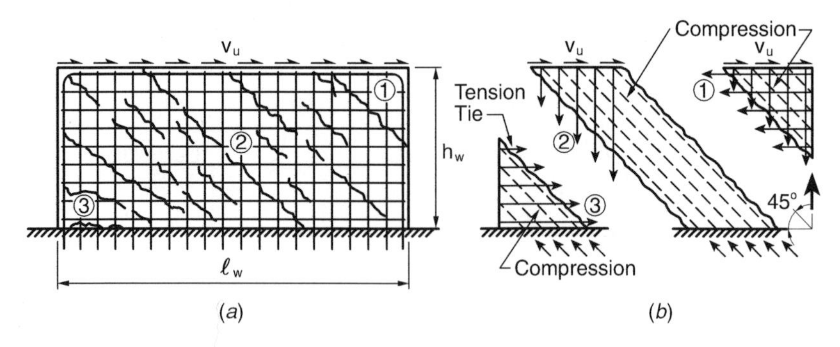

Figure 1.3.9 The shear resistance of low-rise shear walls.[1.20]

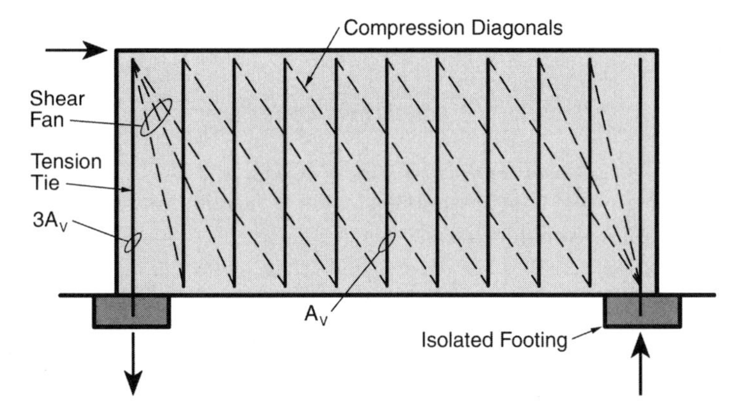

Figure 1.3.10 Precast wall panel with discrete support points.

in the development of a precast wall panel that is supported by individual footings, an activating tie and strut load path must be developed (Figure 1.3.10).

Shear ductility should be developed when it is impossible to develop a flexural mechanism, as is clearly the case in Figure 1.3.9. It is also advisable to develop shear ductility when the flexural mechanism is overly complex or unreliable.

Shear deformation must be estimated from simple models like the one shown in Figure 1.3.11. This simple model will not really develop because a shear fan will be produced with the yielding of each successive tension tie, as described in Figure 1.3.2. Nevertheless, an analytical model based on Figure 1.3.11 will reasonably and quickly predict the (vertical) sliding of planes that characterizes shear deformation. The shear deformation of a panel (δ_v) includes the elongation of the stirrup or tension tie (δ_s) and the vertical component of the shortening of the compression diagonal (δ_{vcd}).

$$\begin{aligned}\delta_v &= \delta_s + \delta_{vcd} \\ &= \delta_s + \sqrt{2}\delta_{cd}\end{aligned}$$

At stirrup yield the elongation of the tension tie will be

$$\delta_{sy} = \frac{f_y}{E_s} d$$

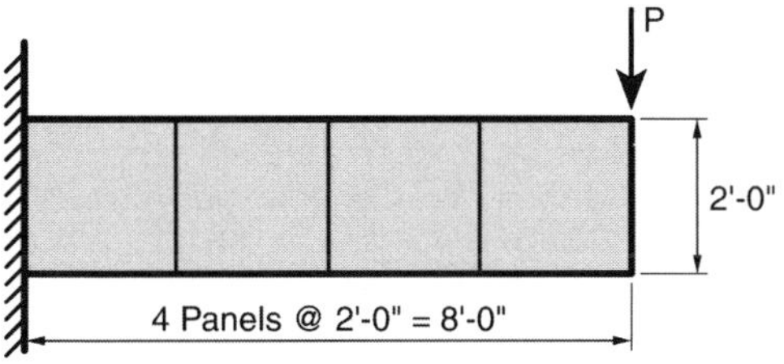

Figure 1.3.11 Beam shear deflection model.

and the stress in the concrete diagonal will be

$$f_{cd} = \frac{V_s}{b_w d \sin^2\theta \cot\theta}$$

For $\theta = 45°$, f_{cd} is $2v_s$, and for $\theta = 30°$, f_{cd} is $2.31v_s$.

The shortening of the compression diagonal is a function of the load carried by the steel (v_s) and, as a consequence, is

$$\delta_{cd} = \frac{f_{cd}}{E_c}\frac{d}{\sin\theta}$$

The vertical component of panel distortion at steel yield, given the adopted 45° angle (θ), becomes

$$\delta_{vy} = \delta_{sy} + \delta_{vcd} \tag{1.3.11a}$$

$$= \frac{f_y}{E_s}d + 4\frac{v_s}{E_c}$$

$$= \frac{f_y}{E_s}\left(d + \frac{4v_s d E_s}{f_y E_c}\right) \tag{1.3.11b}$$

Since v_s will typically be on the order of $4\sqrt{f_c'}$ and $E_c = 57{,}000\sqrt{f_c'}$, we can reasonably conclude that

$$\delta_{vy} \cong \varepsilon_{sy} d(1 + 0.14) \tag{1.3.12}$$

and this means that the contribution to shear deformation of the concrete strut at stirrup yield, given a shear stress on the order of $4\sqrt{f_c'}$, will only be 14%.

Since the preceding does not account for anchorage slip, cracking of the web, or an inclination of the diagonal strut greater than 45°, it is reasonable, from a design perspective, to adopt a panel yield distortion on the order of

$$\delta_{vy} = 1.25\varepsilon_{sy} d \tag{1.3.13a}$$

The postyield component of panel shear drift would be

$$\delta_{vp} = \mu\varepsilon_{sy} d \tag{1.3.13b}$$

and the total shear drift

$$\delta_{vu} = \delta_{vy} + \delta_{vp}$$

$$= (1.25 + \mu)\varepsilon_{sy} d \tag{1.3.14}$$

Consider how this might be applied to a beam. The beam described in Figure 1.3.11 has a span to depth ratio (shear span) of 4. The distortion of each panel at yield, assuming that

$$P = v_{sy}bd \tag{1.3.15}$$

will be

$$\begin{aligned} \delta_{vy} &= 1.25\varepsilon_{sy}d \qquad \text{(see Eq. 1.3.13a)} \\ &= 1.25(0.002)(24) \\ &= 0.06 \text{ in.} \end{aligned}$$

The total shear displacement of the beam at yield (Δ_{vy}) would be the sum of the panel shear displacements

$$\begin{aligned} \Delta_{vy} &= 4\delta_{vy} \\ &= 0.24 \text{ in.} \qquad \text{(see Figure 1.3.11)} \end{aligned}$$

and if we assume that the shear reinforcing is capable of a strain ductility of 10, the ultimate panel displacement would be

$$\begin{aligned} \delta_{vu} &= (1.25 + 10)\varepsilon_{sy}d \qquad \text{(see Eq. 1.3.14)} \\ &= 11.25(0.002)(24) \\ &= 0.54 \text{ in.} \end{aligned}$$

and the attainable level of postyield drift

$$\begin{aligned} \Delta_{vu} &= 4(\delta_{vu}) \\ &= 2.16 \text{ in.} \end{aligned}$$

This corresponds to a shear displacement ductility of 9, and this has reportedly been attained.[1.22]

Comment: Observe that the level of shear stress has little impact on displacement ductility because it only shortens the compression diagonal (Eq. 1.3.11b). This assumes that the stress in the compression diagonal will be controlled so as not to promote strength degradation. It also assumes that shear reinforcement will be distributed so as to create a uniform compressive stress field, and that the nodes will develop the strength of the tension verticals (stirrups).

The preceding assumes that a beam (shear) deflection mechanism will be realized. As the shear span becomes small, the beam (shear) deflection mechanism of Figure 1.3.11 will not be realized. Rather, deflection will be controlled by the tension tie that activates the arch mechanism (Figure 1.3.7). Several caveats bear repeating, for they apply to modeling the arch mechanism. Some shortening will occur along the compression diagonal, slippage will occur in the node, and the stress along the tension tie will be constant though variable along the length of the beam as the concrete carries part of the tension load between the cracks. Accordingly, an idealized behavior model is reasonably adopted. Figure 1.3.12 describes an idealized behavior that assumes that cracking will start at midspan and that postyield deformation will tend to accumulate in this region as the reinforcing bars debond. The behavior model assumes that two rigid blocks are created and that these blocks will rotate in proportion to the created hypothetical gap (δ_R).

If the tension tie is intentionally debonded over its entire length, it will be uniformly strained and presumed to have yielded in order to create the desired ductile behavior. The hypothetical gap created at yield can be expressed as

$$\delta_{Ry} = \varepsilon_{sy}\ell \tag{1.3.16a}$$

This will provide an idealized estimate of system deflection at yield:

$$\Delta_y = \frac{\delta_{Ry}}{2}\left(\frac{1}{h}\right)\left(\frac{\ell}{2}\right) \tag{1.3.16b}$$

$$= \frac{\varepsilon_{sy}\ell^2}{4h}$$

Observe that this relationship can also be developed from the constant curvature ($\phi_y = \varepsilon_{sy}/h$) that occurs along the beam.

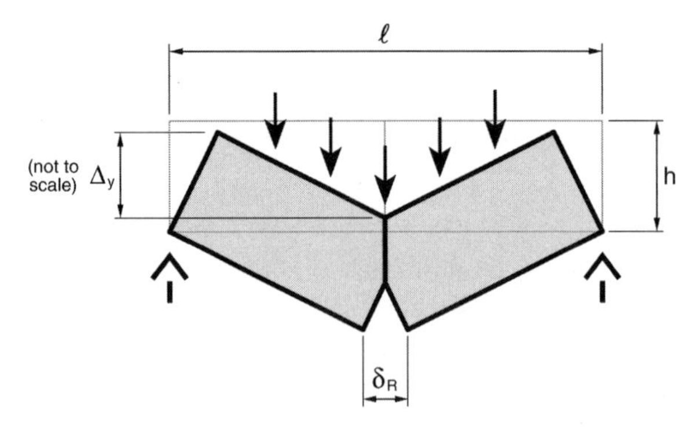

Figure 1.3.12 Shear deflection model—deep beam.

If the shear span ($a = \ell/2h$) is introduced, this becomes

$$\Delta_y = \frac{\varepsilon_{sy} a\ell}{2} \tag{1.3.17}$$

Comment: Debonding the reinforcing over its entire length will produce a very flexible structure, one that is not desirable, as we shall see when we develop design procedures for diaphragms in Section 3.3.

The deflection at yield (Δ_y) and the postyield deflection (Δ_p) will depend on the manner in which postyield straining manifests itself in the tension tie, and this can be effectively controlled by debonding the tension tie in the area in which the cracking will originate. If we neglect the tensile strain away from the debonded region, the so-created gap may be expressed as

$$\delta_{Ru} = (1 + \mu_s)\varepsilon_{sy}\ell_d \tag{1.3.18}$$

where ℓ_d is the developed debond length on each side of the centerline, and, accordingly

$$\Delta_y = \varepsilon_{sy} a\ell_d \qquad \text{(see Eq. 1.3.17)}$$

The total deflection is

$$\Delta_u = \Delta_y + \Delta_p$$

$$= (1 + \mu)a\varepsilon_{sy}\ell_d \qquad \text{(Eq. 1.3.18)}$$

Assume, for example, that a debond length of 0.1ℓ will be developed on each side of the centerline and that an ultimate strain of 5% is reasonably attainable. Then,

$$\Delta_u = (1 + 25)a\varepsilon_{sy}\ell_d \qquad \text{(see Eq. 1.3.18)}$$

$$= 0.0052a\ell$$

and this corresponds to a deflection ductility of 26.

Obviously the so-debonded region (ℓ_d) must be detailed in a manner that does not allow the bars that make up the tension tie to buckle on reverse cycle loading. This topic will be addressed in Section 3.3.

Comment: Experimental support for predicting the impact that arching action and debonding have on behavior is limited to that developed for the hybrid beam system (Section 2.1.4.4) and the coupling beam (Section 2.4.2.1). The load path developed in a coupling beam is basically that developed in the arch (see Figure 2.4.14b) and as such essentially the same as the load transfer mechanism of Figure 1.3.7. Accordingly, diaphragm behavior models must rely to a certain extent on experimental conclusions developed by testing coupling beams.

1.3.2 Shear Transfer across Concrete Discontinuities

Essential to the effective use of precast concrete is an understanding of how pure shear may be transferred across joined pieces of precast concrete. Most practical methods will rely on friction, and the axial force that creates the friction can be created either actively or passively. Friction as a load transfer mechanism has been slowly accepted over the last forty years, largely as a result of the work done by Mattock[1.22, 1.23] and the steel industry with the acceptance of high-strength slip critical bolts. Foerster, Rizkalla, and Heuvel[1.24] demonstrated the effectiveness of pure friction as a load path by performing a series of tests on the assembly described in Figure 1.3.13. The grout strength used was at least 6000 psi, and the levels of maintained preload were 290 and 580 psi. The initial slip or static friction load was 700 psi for both levels of maintained preload, while the sliding resistance was 86% of the applied load (Figure 1.3.14).

The introduction of castellations or shear keys increased the static friction sustained along the precast interface for the specimen preloaded to 580 psi by 50%. The degradation experienced by the castellated joint resulted in the same sliding friction coefficient as that of the uncastellated interface. Clearly, the texture of the interface

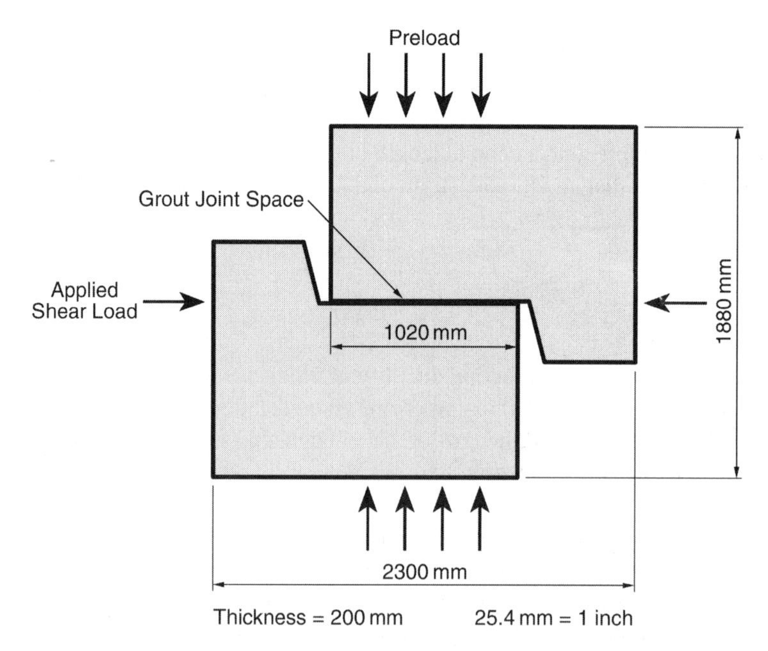

Figure 1.3.13 Specimen configuration—pure shear tests.[1.24]

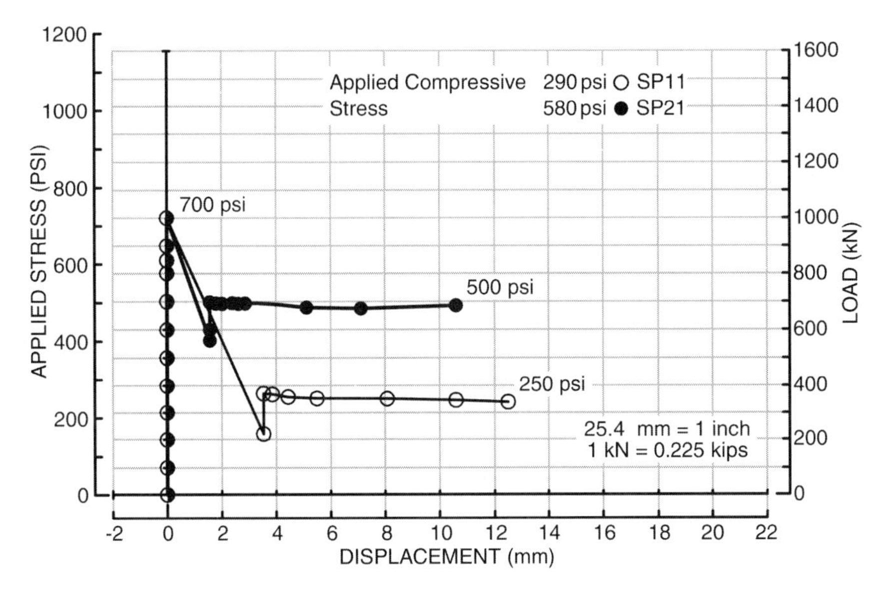

Figure 1.3.14 Load displacement relationship of a pure friction connection.[1.24]

surface will have a significant bearing on the precrack load, but not necessarily on the postcrack, load.

Engineers are typically concerned about the impact precracking will have on a shear transfer. Figure 1.3.15 describes a connection that might reasonably be provided between two stacked precast panels. In this test the interface was loaded in shear only, for no flexural action is possible (see Figure 1.3.13). From the standpoint of representing probable behavior, this reinforced specimen is more realistic than the unreinforced specimen described in Figure 1.3.13. Behavior of the reinforced specimen is described in Figure 1.3.16. In the static range resistance is directly proportional to the level of preload. Initial movement and cracking occur at 190% of the applied axial load (SP 12 and SP 22). The precracked interface (SP 12C) does not experience this "spike" and moves directly to a sustainable load of about 350 psi, the increase from 250 psi sustained by the Type I connection (SP 11 of Figure 1.3.14) being directly attributable to the contribution of the reinforcing as shear friction is achieved. The dowel load appears to be about 70 kips, and this corresponds to a stress of 88 ksi in the load side dowel if it is assumed the other dowel does not contribute to the shear transfer because of its proximity to the edge of the panel.

One might reasonably conclude that a grouted interface subjected to a maintained precompression load can transfer a shear load associated with a friction factor that is slightly less than unity even if the interface is precracked. An ultimate limit state obviously exists and has been suggested by Mattock.[1.23] An ultimate stress limit state is incorporated into shear friction theory.

Shear friction is a term used to describe the passive activation of friction. Mattock,[1.22] in the 1960s, performed a series of tests on specimens like the one described

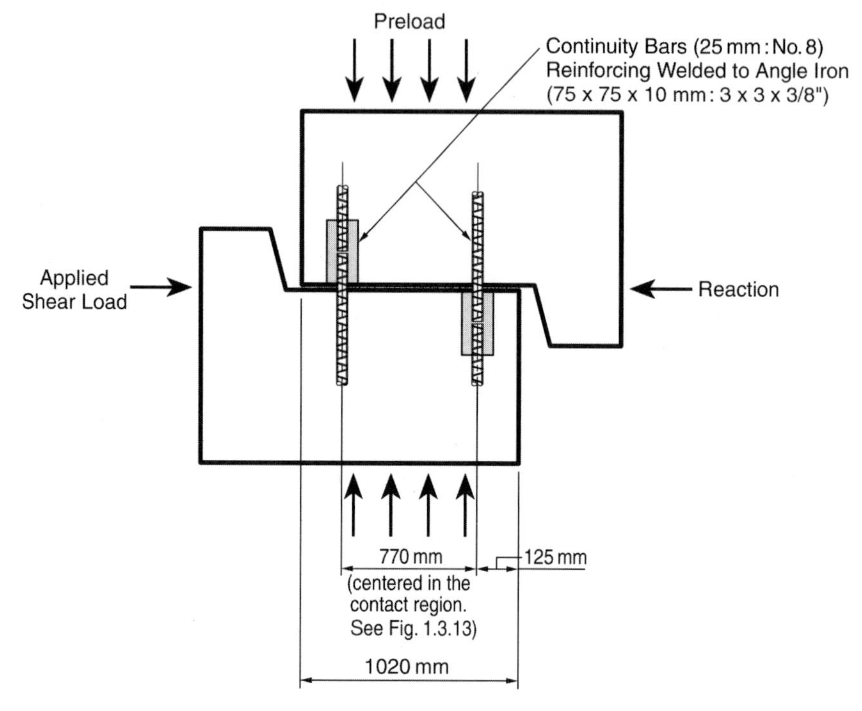

Figure 1.3.15 Reinforced friction connection configuration.[1.24]

in Figure 1.3.17. Uncracked and initially cracked specimens with various levels of transverse reinforcement were tested under conditions of pure shear. The results of these tests are described in Figure 1.3.18. The passively activated compression load imposed on the shear plane $(\Sigma A_s f_y)$ is created when the joined surfaces displace, causing the steel crossing this plane to yield.

The codified version of shear friction[1.15] develops "artificially high" values of the friction coefficient to model the behavior described in Figure 1.3.18 and places a limit of 800 psi on shear transfer regardless of the condition of the interface surfaces or the amount of reinforcing provided. Clearly this is a conservative interpretation of the experimental results reported in Figure 1.3.18. Shear friction experiments clearly support the use of an actively developed friction coefficient of at least 0.8. Passive and active shear resistance may be combined.[1.15, Paragraph 11.7.7] Accordingly, the nominal shear transfer across a precracked plane can be assumed to be

$$V_n = 0.8\frac{P}{A} + 1.0A_{vf}f_y \tag{1.3.19}$$

If the joined concrete surfaces are smooth, or if one is steel plated, the nominal shear transfer might reasonably be lessened to about 60% of that developed by Eq. 1.3.19.

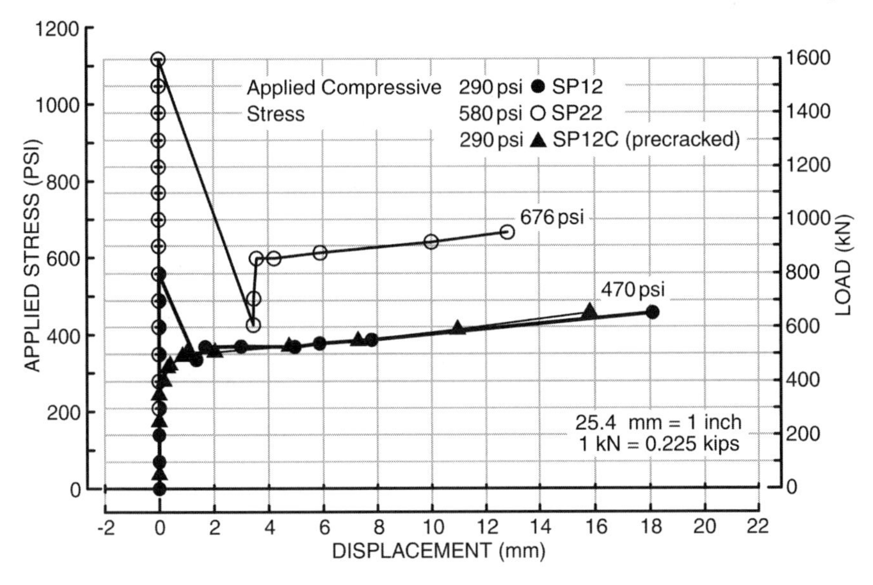

Figure 1.3.16 Load–displacement relationship of reinforced friction connection.[1.24]

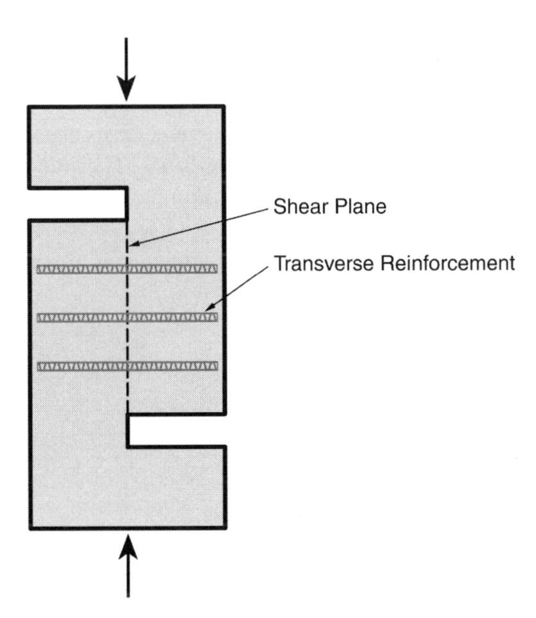

Figure 1.3.17 Shear friction load transfer specimen.[1.13]

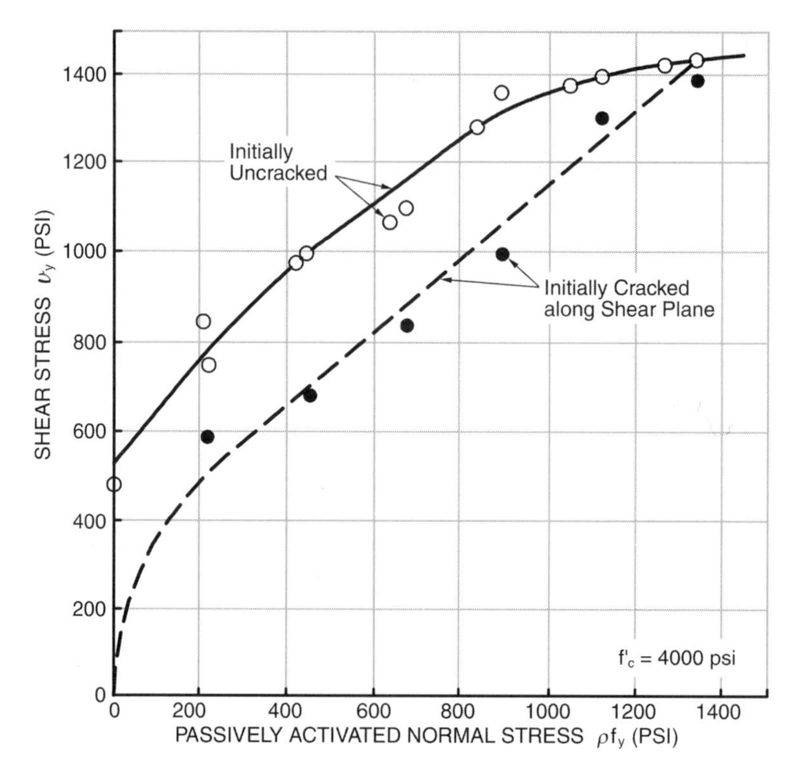

Figure 1.3.18 Shear friction behavior—specimen of Figure 1.3.17.[1.13]

1.3.3 Passively Activated Shear Transfer Mechanisms

Flexure creates a compression force. When a moment exists at a section, the activated tensile force is equilibrated by a compression force. The joined blocks of Figure 1.3.19 will transfer shear provided the activated compression force (C) creates a shear load path that is capable of transferring the shear (V).

The shear transfer will occur if the selected dimensions are appropriate:

$$Cf > V \tag{1.3.20a}$$

$$\frac{Va}{d - a/2} f > V$$

This is usually reduced for design purposes to

$$\frac{a}{d} > \frac{1}{f} \tag{1.3.20b}$$

Observe that slipping will not occur given the geometry constraint proposed by Eq. 1.3.20b because the compression is created by and is directly proportional to the applied shear.

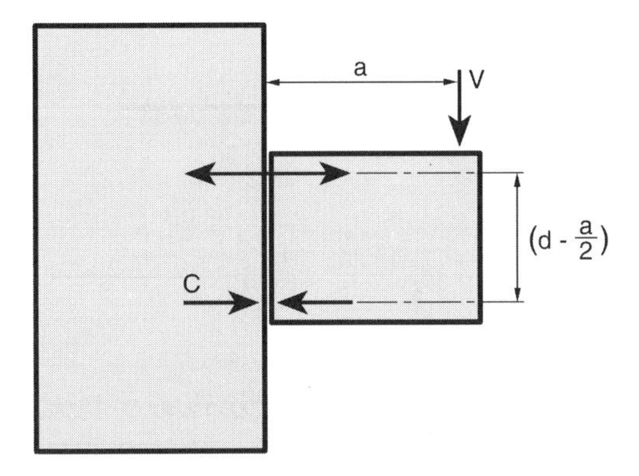

Figure 1.3.19 Flexurally activated pure friction.

In a continuous beam, the concept of shear transfer is somewhat more difficult to grasp. The shear span (a) may be viewed as the distance between the point of inflection ($M = 0$) and the point under consideration. Hence, for the usual seismic case where shear is essentially constant,

$$a = \frac{M}{V} \tag{1.3.21}$$

and the critical relationship we seek is developed from Eq. 1.3.20a

$$\frac{a}{d} > \frac{1}{f} \qquad \text{(see Eq. 1.3.20a)}$$

$$\frac{M}{Vd} > \frac{1}{f} \tag{1.3.22}$$

This passively developed shear transfer mechanism will be exploited in the development of precast concrete beam systems in subsequent chapters.

Diaphragms typically have span-to-depth ratios of 2 to 3. This means that their shear spans are on the order of 1 to 1.5, and this typically places them in the deep beam category. Accordingly, the shear transfer mechanism tends to be somewhat like that described in Figure 1.3.6*b* except for the location at which the load is applied, for this will be at the center of the mass or, in the cases of Figures 1.3.6*b* and *c*, distributed along the height of the member. An inertial load (w) must reach the support point or line (Figure 1.3.20). Arch action (Figure 1.3.7) and a shear fan (Figure 1.3.2) are activated. A load at point A (in Figure 1.3.20) will, given a secondary tensile load path, travel up to the shear fan and then to the support, while an inertial load at point B (Figure 1.3.20) will travel directly to the support. Loads located above the shear fan, point C (Figure 1.3.20), will either go directly to the reaction line or by compression down to the shear fan. The net result from a design perspective is the development of a uniformly loaded tied arch.

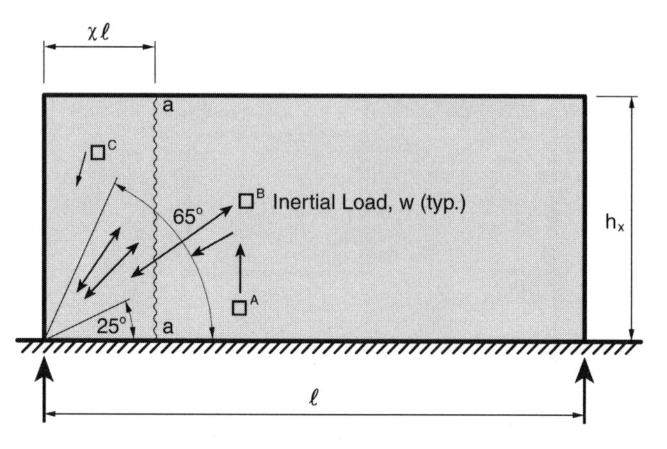

Figure 1.3.20 Diaphragm load paths.

Ductility should be developed in deep beams (diaphragms) that are subjected to earthquake-induced displacements. The only ductile element that can be developed is the tension tie (Figure 1.3.7). A capacity-based design approach is essential. All elements necessary to the development of the load paths described in Figure 1.3.20 must be activated and stresses, other than those in the tension tie, maintained within elastic limits. Chapter 3 will describe how one might arrive at the appropriate tensile tie demand and develop the details necessary to activate the (arch) load path. The focus here is to use capacity-based concepts to develop inertial loads and review critical sections so as to describe load transfer demands.

The first step is to quantify the unit inertial load (w). Regardless of the analytical model adopted, beam or rigid body (Figure 1.3.13), the relationship between the inertial load and the tension tie will be developed as

$$M = \frac{w\ell^2}{8}$$

$$T = \frac{w\ell^2}{8d_{\text{eff}}} \tag{1.3.23}$$

Observe that a fraction of h_x is more appropriately used to reflect the fact that the centroid of the compression load is not at the upper edge of the deep beam. Accordingly, use d_{eff} to describe the lever arm:

$$w = \frac{8Td_{\text{eff}}}{\ell^2} \tag{1.3.24}$$

If we adopt an effective depth of $0.75h_x$, we may reduce Eq. 1.3.24 to one that reflects the shear span:

$$w = \frac{6Th_x}{\ell^2} \tag{1.3.25}$$

With the introduction of the shear span ($a = 2h_x/\ell$), this becomes

$$w = \frac{3Ta}{\ell} \tag{1.3.26}$$

In these equations, w represents the inertial contribution of the mass or unit weight along each unit of length, and this should include the weight of exterior walls and any other tributary mass. From the standpoint of load path development, it is easiest to reduce the inertial force to a percentage of the line load (w).

The diaphragm will typically crack on a line parallel to the direction of the primary and secondary framing members where shrinkage strains combine with flexurally induced tension strains. Accordingly, load must travel across a cracked plane that, for our purposes, will be the line *a–a* shown in Figure 1.3.20.

Traditionally, in the design of diaphragms, it has been assumed that beam theory applies. In fact, this assumption is not truly appropriate, but it is usually conservative. In precast concrete it is usually assumed that all of the shear along a line like *a–a* of Figure 1.3.20 must be transferred by reinforcing activated shear friction, and this is very conservative because it neglects the passively activated compressive force developed by the tension tie (Figure 1.3.7). Observe that this compression force is constant and thus activated at every plane or section cut through the deep beam (i.e., *a–a* in Figure 1.3.20).

This understood, we may develop a capacity demand relationship for a plane passing through the deep beam of Figure 1.3.20.

The shear demand is

$$V = \frac{w\ell}{2} - wx\ell$$

$$= w\ell(0.5 - x) \tag{1.3.27}$$

The shear capacity may be expressed in terms of the contribution of passively (compression force) activated friction and reinforcing activated shear friction as developed in Eq. 1.3.19.

Consider the compression-activated component Cf where f is the adopted friction factor. Shear resistance will be provided by friction if

$$Cf > V \qquad \text{(see Eq. 1.3.20a)}$$

$$\frac{w\ell^2}{8d_{\text{eff}}} f > w\ell(0.5 - x) \qquad \text{(see Eq. 1.3.23)}$$

By replacing d_{eff} with $0.75h_x$, this can be transformed into

$$\frac{w\ell af}{3} > w\ell(0.5 - x)$$

$$\frac{af}{3} > (0.5 - x)$$

$$af > (1.5 - 3x) \tag{1.3.28}$$

Figure 1.3.14 suggests that the passive friction factor is a function of the level of applied compressive force. The slip producing shear stress (±700 psi) was developed for compression stresses of 290 and 580 psi. For this stress range, a friction factor of 1.25 might be assumed, and an even higher value for (lower) compression stress level as, for example, in the range of 250 psi.

If we adopt a friction factor of 1.25 on a diaphragm that has a shear span of 1, we might reasonably and conservatively assume that the compressive load would transfer all of the shear in the beam interior to $x\ell$.

$$af > (1.5 - 3x) \qquad \text{(Eq. 1.3.28)}$$

$$1(1.25) > 1.5 - 3x$$

$$x = \frac{0.25}{3}$$

$$= 0.083$$

Using these assumptions, the region between the support and 0.083ℓ would need to be supplemented by additional reinforcing provided in accordance with Eq. 1.3.19.

These concepts will be further developed and examples provided in Chapter 3. It is important to remember that these relationships should not be used for shear spans in excess of 1.5 because the shear transfer mechanism is changing (see Figure 1.3.5) and the load transfer mechanism is that developed by beam theory.

SELECTED REFERENCES

[1.1] A. K. Chopra, *Dynamics of Structures: Theory and Applications to Earthquake Engineering*, 2nd Edition. Prentice Hall, Upper Saddle River, New Jersey, 2001.

[1.2] A. Filiatrault, *Elements of Earthquake Engineering and Structural Dynamics*. Polytechnic International Press, Montreal, Canada, 1998.

[1.3] J. M. Biggs, *Introduction to Structural Dynamics*. McGraw-Hill Book Company, New York, 1964.

[1.4] M. F. Rubenstein and W. C. Hurty, *Dynamics of Structures*. Prentice Hall, Englewood Cliffs, New Jersey, 1964.

[1.5] N. M. Newmark, "Current Trends in the Seismic Analysis and Design of High-Rise Structures," in *Earthquake Engineering*, Chapter 16, R. L. Wiegel (Ed.), Prentice Hall, Englewood Cliffs, New Jersey, 1971, pp. 403–424.

[1.6] R. E. Englekirk, *Steel Structures: Controlling Behavior Through Design*. John Wiley & Sons, New York, 1994.

[1.7] A. K. Chopra and R. K. Goel, "Direct Displacement-Based Design: Use of Inelastic vs. Elastic Design Spectra," *Earthquake Spectra*, Vol. 17, No. 1, February 2001.

[1.8] M. J. N. Priestley and M. J. Kowalsky, "Direct Displacement-Based Seismic Design of Concrete Buildings," *Bulletin of the New Zealand Society for Earthquake Engineering*, Vol. 33, No. 4, December 2000.

[1.9] International Conference of Building Officials, *Uniform Building Code*, 1997 Edition, Whittier, California.

[1.10] Seismology Committee, Structural Engineers Association of California, *Recommended Lateral Force Requirements and Commentary*, Seventh Edition. Sacramento, California, 1999.

[1.11] International Code Council, *International Building Code*. Falls Church, Virginia, 200.

[1.12] T. Paulay and M. J. N. Priestley, *Seismic Design of Reinforced Concrete and Masonry Buildings*. John Wiley & Sons, New York, 1992.

[1.13] J. G. MacGregor, *Reinforced Concrete: Mechanics and Design*, 3rd Edition. Prentice Hall, Upper Saddle River, New Jersey, 1997.

[1.14] F. E. Richart, A. Brandtzaeg, and R. L. Brown, "A Study of the Failure of Concrete Under Combined Compressive Stresses," University of Illinois Engineering Experimental Station, Bulletin No. 185, 1928, 104 pp.

[1.15] American Concrete Institute, *Building Code Requirements for Structural Concrete (318-99) and Commentary (318R-99)*, ACI 318-99 and ACI 318R-99, Farmington Hills, Michigan, June 1999.

[1.16] J. B. Mander, M. J. N. Priestley, and R. Park, "Observed Stress-Strain Behavior of Confined Concrete," *Journal of Structural Engineering, ASCI*, Vol. 114, No. 8, August 1988, pp. 1827–1849.

[1.17] E. G. Nawy, *Fundamentals of High Performance Concrete*, 2nd Edition. John Wiley & Sons, New York, 2001.

[1.18] Carpenters/Contractors Cooperation Committee, Inc., *High Strength Concrete Research*. Los Angeles, December 2000.

[1.19] Y. Xiao and H. W. Yun, "Full-Scale Experimental Studies on High-Strength Concrete Columns," University of Southern California, Report No. USC-SERP 98/05, July 1998.

[1.20] R. Park and T. Paulay, *Reinforced Concrete Structures*. John Wiley & Sons, New York, 1975.

[1.21] F. Leonhardt, "Reducing the Shear Reinforcement in Reinforced Concrete Beams and Slabs," *Magazine of Concrete Research*, Vol. 17, No. 53, December 1965, pp. 187–198.

[1.22] J. A. Hofbeck, I. A. Ibrahim, and A. H. Mattock, "Shear Transfer in Reinforced Concrete," *ACI Journal, Proceedings,* Vol. 66, No. 2, February 1969, pp. 119–128.

[1.23] A. H. Mattock and N. M. Hawkins, "Shear Transfer in Reinforced Concrete—Recent Research," *Journal of the Prestressed Concrete Institute*, Vol. 17, No. 2, March–April 1972, pp. 55–75.

[1.24] H. R. Foerster, S. H. Rizkalla, and J. S. Heuvel, "Behavior and Design of Shear Connections for Loadbearing Wall Panels," *Journal of the Prestressed Concrete Institute*, Vol. 34, No. 1, January/February 1989, pp. 102–119.

2

COMPONENT BEHAVIOR AND DESIGN

"Wisdom is the daughter of experience."
—Leonardo da Vinci

"Experience in the design of concrete structures is required," the ad warned. In other words, you must have performed successful designs in the past if you wished to be considered for this position. When applied to vertical load carrying systems, a successful design is measured visually; cracking and deflection are minimal, ergo the design must have been appropriate. Seismic designs are seldom tested so this visual assurance link is not available. Where then does the designer go to develop the requisite level of experience in seismic design? The answer, at least insofar as component behavior is concerned, is the testing laboratory.

Reports of experimental efforts have increased exponentially these last decades and all profess to provide some design guidance. The sheer quantity of available data presents a significant problem to the design professional because reducing each experiment to a conclusion that can be confidently used in the design process leaves little time to design anything. The experiments reviewed in this chapter were selected because they focus on critical design issues and support limit states that define component capabilities. The summaries contained herein are, of necessity, reductions of reports that reduce extensive databases. So, the objective of each review cannot be viewed as technical support for an approach or suggested limit state in itself. The overviews presented do provide insight and allow the extrapolation of limit states and the development of design algorithms. Design is a very personal affair, and every designer has a methodology. Accordingly, the reader is encouraged to modify the included algorithms and proposed limit states to suit his or her predilection.

Building Code procedures are not specifically reviewed or commented on in this chapter; however, experimental evidence should at least add understanding to complex codifications. Further, an understanding of the experimental basis for a conclusion allows its extension to other issues for which the scientific basis is limited.

This chapter is devoted to demonstrating that a superior component design, from a performance perspective, can be produced in a cost-effective manner, both in terms of construction cost and design cost. Each section starts with an in-depth review of a selected test program designed to give the reader confidence in the ability of the component to attain the objective level of strength, and sustain levels of postyield deformation. A systematic approach to design is then developed and tested against demonstrated behavior. The design must comply with performance objectives; hence the conceptually designed component must be analyzed and detailed. An analysis procedure is developed and tested on the produced design. Critical construction details are developed, for the success of any design will depend on its execution.

Design and analysis procedures, if they are to be effective, must have focus and yet sufficient technical support to give the reader confidence in the adopted approach. This is also true of the explanation process. I do not attempt to create software ready algorithms, but rather to encourage independent original thought. I have never had confidence in a design procedure that I did not personally develop and thoroughly understand. I encourage the reader/designer to invest the same effort. The development of focused design procedures requires the use of efficient analytical reductions. I have elected to provide enough of a basis for the reader familiar with a supporting concept to understand the approach—supplementing this overview with a specific reference that contains a more thorough development of the methodology. Essentially, the supporting material contained herein should allow the "informed" reader to recall the technical basis without seeking out the reference. One clear drawback to this approach is that the user of the book will find it difficult to refer to topics of potential interest that are scattered throughout the text. I have compiled a subject index that should make it possible for the designer to quickly find a topic and the reference that develops the topic in more depth.

Cost-effectiveness is an essential part of the design process, in terms of both produced product and design time. Most buildings must make financial sense or they will not be built. If the cost of your design is excessive, another designer will get the job. There must be a balance between cost and seismic safety—a balance that respects functional and aesthetic objectives. The designer who cannot find this balance will not design many buildings.

2.1 BEAMS

Frames are used to provide lateral support for many buildings. In regions of moderate and high seismicity, frames must be capable of sustaining building drifts that are significantly greater than their elastic limit state. Frame beams provide the bulk of this postyield deformability; hence their appropriate design is essential to the success of the ductile frame.

The design of the frame beam is the most critical single step in the frame design process. This is because the strength of the frame beam will dictate the strength required of all of the other components of the bracing program, and this in turn will have an enormous impact on not only the cost, but also the constructibility of the building. The tendency is to provide a stronger beam than that suggested by the design criterion, believing that the additional strength will improve the behavior of the system. Stronger is, however, seldom better, especially in concrete beams where

Photo 2.1 Exposed cast-in-place concrete frame, Emerald Hotel, San Diego, CA, 1990. (Courtesy of Englekirk Partners, Inc.)

strength and ductility tend to be inversely proportional; the more reinforcement, the lower the level of provided ductility.

We start the frame design process with the beam, hence the overall strength of the system must be reduced to one that identifies the required strength of the beam. This reduction process will be developed in Chapter 3. An obvious question will be the appropriateness of an adopted strength or deformation criterion and this topic will be addressed in Chapter 4. This section will not only describe how the objective beam strength can be attained, but also how to avoid providing an undesirable level of component overstrength (λ_o).

Beam design procedures are developed in as simple a manner as possible. This is because a significant mistake, arithmetic or theoretical, will have an enormous impact on the behavior of the frame and on the design process, as the strength required of all frame components is a function of the provided strength of the beam. The adopted oversimplified approach to the beam design criterion development (Chapter 3) and the beam design process recognize that objective strength limits strive rather subjectively to develop acceptable behavior in a system whose real limit states are functions of induced strain states. (see Figure 1.1.1) Accordingly, the design strength limit state should not be venerated or viewed as necessarily producing the appropriate bracing program.

In this section then, the focus is on how to develop a beam that meets a strength objective and has a level of postyield deformability that will, upon examination, justify its selection.

2.1.1 Postyield Behavior—Flexure

2.1.1.1 Experimentally Based Conclusions—General Discussion The procedures we use today to predict the postyield behavior of beams were developed in the 1960s and early 1970s. An experimental program performed at the University of California at Berkeley by Popov, Bertero, and Krawinkler[2.1] has been chosen to explore both the analytical procedures proposed herein as well as strength and deformation limit states. This selection was made because the test specimens used were of reasonable size (15 in. by 29 in.), were reinforced at the upper limit of the optimal range ($A_s = 0.015bd$), and demonstrated the impact of shear reinforcement, shear stress, and confinement on flexural behavior.

The test program was developed so as to evaluate the behavior of a beam. Accordingly, column, beam-column joint, and anchorage deterioration were effectively eliminated. Figure 2.1.1 describes the test setup as well as beam reinforcing and geometry.

First, we predict the strength of the beam described in Figure 2.1.1 using design and analysis procedures that are commonly used to design beams.

From a design perspective, the strength of the beam of Figure 2.1.1 could most easily be developed as follows:

$$\begin{aligned} M_n &= A_s f_y (d - d') \qquad (2.1.1) \\ &= 6(60)(25.25 - 3.75) \\ &= 7740 \text{ in.-kips} \end{aligned}$$

The supportable nominal load (P_n) is 99.2 (see Figure 2.1.1).

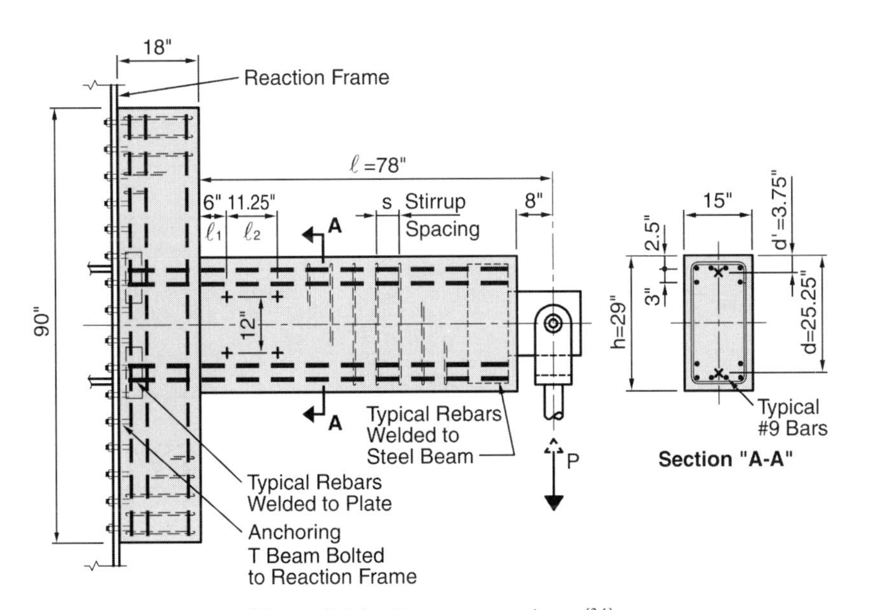

Figure 2.1.1 Beam test specimen.[2.1]

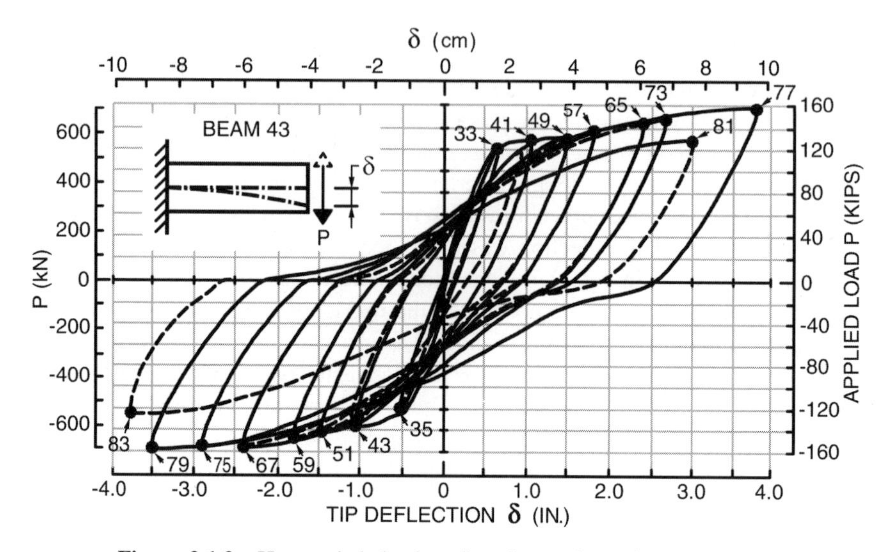

Figure 2.1.2 Hysteretic behavior of test beam shown in Figure 2.1.1.

This is admittedly a conservative estimate (see Figure 2.1.2), and one might alternatively follow the methods attributed to Whitney, but that approach would be complicated by the existence of a significant amount of compression reinforcement ($A_s' = A_s$).[2.2, 2.3] An iterative approach to locating the centroid of the compression for the neutral axis will often be helpful:

$$a = \frac{T - C_s}{0.85 f_c' b}$$

$$f_c' = 5 \text{ ksi}$$

$$f_y = 60 \text{ ksi}$$

The maximum possible depth of the compressive stress block ($C_s = 0$) would be

$$a = \frac{360}{0.85(5)(15)}$$

$$= 5.65 \text{ in.}$$

The use of a 5.65-in. compressive stress block is obviously too conservative, so a better "design" estimate of the nominal moment capacity would be based on a depth of the compressive stress block on the order of 4 in.

$$M_n = 6(60)(25.25 - 2) \qquad \text{(see Eq. 2.1.1)}$$

$$= 8370 \text{ in.-kips} \qquad (P_n = 107 \text{ kips})$$

The measured yield force (P) corresponding to Point 33 on Figure 2.1.2 is 120 kips. The maximum load that the beam could support was 160 kips (Point 77).

Comment: In subsequent analyses that endeavor to predict concrete strain states, the depth of the neutral axis will become an important factor. The iterative process reasonably starts by selecting a trial neutral axis and confirming it by satisfying equilibrium:

$$\begin{aligned} c &= 4.5 \text{ in.} \qquad \text{(Trial value)} \\ a &= \beta_1 c \\ &= 0.8(4.5) \\ &= 3.6 \text{ in.} \\ C_c &= 0.85 f_c' ab \\ &= 0.85(5)(3.6)(15) \\ &= 230 \text{ kips} \\ \varepsilon_s &= \frac{c - d'}{c}(0.003) \end{aligned}$$

where d' is now the depth of the top four #9 bars (see Figure 2.1.1). Thus

$$\begin{aligned} \varepsilon_s &= \frac{4.5 - 2.5}{4.5}(0.003) \\ &= 0.00133 \text{ in./in.} \\ f_s &= \varepsilon_s E_s \\ &= 0.00133(29{,}000) \\ &= 38.6 \text{ ksi} \\ C_s &= f_s A_s' \\ &= 38.6(4) \\ &= 154 \text{ kips} \\ C_c + C_s &= 230 + 154 \\ &= 384 \text{ kips} \cong T \end{aligned}$$

Conclusion: The neutral axis depth at the nominal flexural strength of the beam will be about 4.5 in.

Several idealizations of the load displacement behavior are described in Figure 2.1.3. An idealization that might be proposed by the research team would quantify

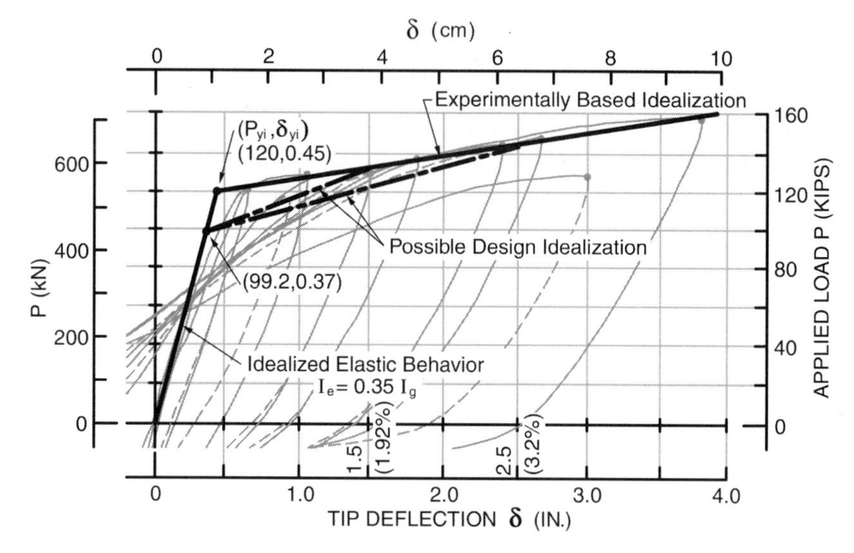

Figure 2.1.3 Behavior idealization—beam of Figure 2.1.1.[2.1]

the yield load (P_{yi}) as being 120 kips. The designer, on the other hand, would develop a strength idealization of 99.2 kips or 107 kips from design equations, as for example Eq. 2.1.1.

An important design relationship is the ratio of the idealized yield strength (P_{yi}) to the maximum attainable strength ($P_{\max}$). This relationship is identified by the term "overstrength factor." The overstrength factor for a member (λ_o) and a system (Ω_o) are important because they are used to develop the basic design criterion in a capacity-based design process (Chapter 1).

For the beam of Figure 2.1.1, several quantifications of member overstrength are possible. If the design basis is Eq. 2.1.1,

$$P_n = \frac{M_n}{\ell}$$

$$= \frac{7740}{78}$$

$$= 99.2 \text{ kips}$$

and λ_o is

$$\lambda_o = \frac{P_{\max}}{P_n} \tag{2.1.2a}$$

$$= \frac{160}{99.2}$$

$$= 1.61$$

If, on the other hand, the overstrength is based on an experimental idealization of the yield strength (P_{yi}), the associated overstrength factor would be

$$\lambda_o = \frac{P_{\max}}{P_{yi}} \tag{2.1.2b}$$

$$= \frac{160}{120}$$

$$= 1.33$$

Both of these quantifications are considerably more than the 25% advocated by current codes.

Another consideration in the development of an overstrength factor involves the selection of $P_{\max}$. The selected peak load (160 kips) corresponds to a beam rotation of almost 5% (Figure 2.1.2), and this level of deformation is not usually anticipated, especially when attributed solely to beam rotation as is the case here. Furthermore the work hardening developed by the many postyield excursions imposed on this specimen are not to be expected during an earthquake. A more reasonable rotation demand on the beam (exclusively) might be on the order of 2%, and this corresponds to Point 49 on Figure 2.1.2. The maximum load, as defined by Point 49, is only 130 kips and this creates an overstrength factor that is more consistent with the overstrength factors commonly used in design (1.25).

Conclusion: Overstrength factors are a function of the design process (identification of yield strength) and behavior objective generalizations, therefore, they must be made with care.

Behavior idealization also requires the identification of the deflection associated with yield. Point 33 on Figure 2.1.2 would be the logical experimental choice, but its analytical prediction is not easy, for it will depend on many factors, not the least of which is the level of applied load. Consider the alternatives.

Elastic Beam Model In this approach, moment, curvature, and deflection are all related by familiar equations because the system is presumed to be elastic. In the case of the pseudoelastic test beam (Figure 2.1.1), curvature and moment would vary linearly from beam tip to point of support:

$$\phi_y = \frac{M}{EI}$$

$$\delta_y = \frac{\phi_y \ell^2}{3}$$

$$= \frac{M\ell^2}{3EI}$$

$$= \frac{P\ell^3}{3EI}$$

The elastic prediction of yield displacement for the beam of Figure 2.1.1 is, for P_{33} (Figure 2.1.2),

$$\delta_y = \frac{120(78)^3}{3(4000)(30{,}500)}$$
$$= 0.156 \text{ in.}$$

This is clearly inconsistent with experimental evidence (Figure 2.1.2). So a flexural idealization is adopted in the form of an idealized or effective moment of inertia (I_e). For (seismic) frame beams the effective moment of inertia is usually assumed to be $0.35I_g$. This would analytically suggest an idealized yield deflection δ_{yi} of 0.45 in. (see Figure 2.1.4) and this is about midway between the tangent and secant modulus developed from Figure 2.1.2.

First Yield of Steel Model This model develops the curvature at yield from the strain in the steel at yield and the depth of the neutral axis. At first yield of the steel, the concrete will still be reasonably elastic and the depth of the neutral axis will be at a distance kd from the compression face (Figure 2.1.5). A reasonably accurate idealization of curvature at first yield of the steel is

$$\phi_y = \frac{\varepsilon_{sy}}{0.60h}$$
$$= \frac{0.0033}{h}$$

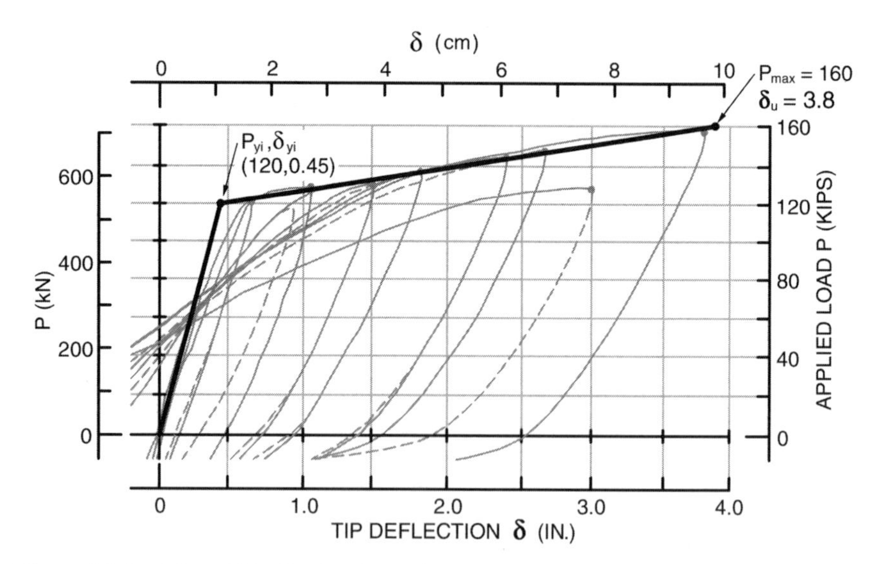

Figure 2.1.4 Adopted idealization describing behavior of the beam shown in Figure 2.1.1.

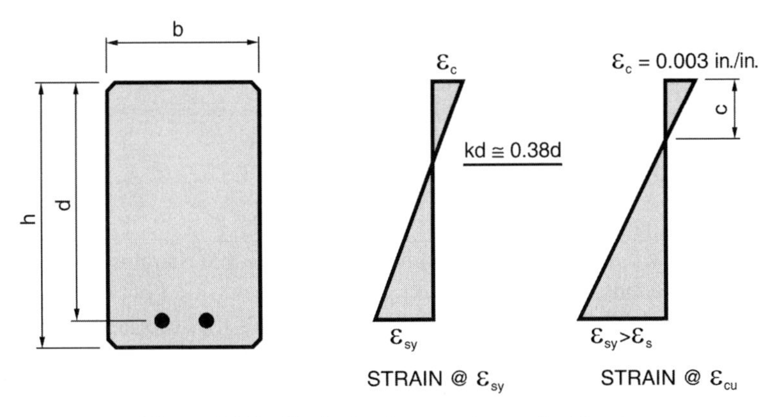

Figure 2.1.5 Strain development in a typical beam.

$$= \frac{0.0033}{29}$$

$$- 0.000114 \text{ rad/in.}$$

The prediction of flexural strength using the curvature model is quite close to the nominal strength predicted by Eq. 2.1.1:

$$M_y = A_s f_y \left(d - \frac{kd}{3} \right)$$

$$= 6.0(60)(25.25 - 3.2)$$

$$= 7938 \text{ in.-kips} \qquad (P_y = 102 \text{ kips})$$

But the projection of displacement using the linear curvature model is not consistent with observed behavior:

$$\delta_y = \frac{\phi \ell^2}{3}$$

$$= \frac{0.000114(78)^2}{3}$$

$$= 0.23 \text{ in.}$$

This identification of yield curvature is often used in the design of shear walls, but it is not used in conjunction with the development of a linear curvature model for a beam.

Inelastic Curvature Model The strength limit state in concrete is associated with the strain limit state in concrete (Figure 2.1.5). The curvature associated with the strain limit state of 0.003 in./in. (nominal strength of beam) is

$$\begin{aligned}\phi_{yn} &= \frac{0.003}{c} \\ &\cong \frac{0.003}{4.5} \\ &= 0.00067 \text{ rad/in.}\end{aligned}$$

Curvature idealizations will soon be compared with measured curvatures and analysis procedures proposed. For now, focus on the physical transition between steel first yield and the strain states at nominal flexural strength of the section. The steel will plastically deform over what may be referred to as a plastic hinge region and create a curvature diagram much like that shown loading a conjugate beam in Figure 2.1.11. If we adopt a plastic hinge region of $0.5h$ and an average postyield curvature in the plastic hinge region, we can estimate the deflection at idealized yield:

$$\begin{aligned}2\phi_{p,\text{avg}} &= \phi_{yn} - \phi_y \\ &= 0.00067 - 0.000114 \\ &= 0.00056 \text{ rad/in.} \\ \phi_{p,\text{avg}} &= 0.00028 \text{ rad/in.} \\ \delta_p &= \phi_{p,\text{avg}} \ell_p \left(\ell - \frac{\ell_p}{2}\right) \\ &= 0.00028(14.5)(78 - 7.25) \\ &= 0.29 \text{ in.} \\ \delta_{yn} &= \delta_y + \delta_p \\ &= 0.23 + 0.29 \\ &= 0.52 \text{ in.}\end{aligned}$$

This is reasonably consistent with experimental conclusions.

Comment: Since moment and curvature are no longer directly related, estimates of member deformation must be developed directly from predicted curvatures along the beam. My preference is to use the conjugate beam method,[2.4] which is a simplified version of the area-moment method.

Summary: Clearly the elastic beam model when used in conjunction with an effective stiffness produces reasonable conclusions quickly. As a consequence, it is the most popular approach.

My preference is the behavior idealization described in Figure 2.1.4. This is most easily attained by calculating the nominal strength of the beam (P_n) and directly amplifying it by an overstrength factor (λ_o), thereby producing an idealized or probable strength of the section (P_{yi}). The overstrength factor would endeavor to capture the material overstrength ($\cong$ 1.15) as well as the design conservatism ($\lambda_o = 1.2$). This is somewhat different than the classical approach (Eq. 2.1.2a), but easier to use in design because an elastic/perfectly plastic model can be reasonably produced:

$$\lambda_o = \frac{P_{yi}}{P_n} \tag{2.1.2c}$$

Comment: Member and system overstrength factors must be tailored to the design situation. They are not universally quantifiable, as we see when we study the behavior of other components.

Another important design quantification is member ductility. From a displacement perspective this too will depend on the selection of the displacement at yield. On Figure 2.1.3 a distinction was made between the displacement associated with P_n, the nominal design strength, and that associated with idealized yield P_{yi} ($\lambda_o P_n$—Eq. 2.1.2c). If the practical deflection limit state (δ_u) is identified with a tip displacement of 2.5 in., the level of idealized beam displacement ductility would be

$$\mu_{\delta i} = \frac{\delta_u}{\delta_{yi}} \qquad \text{(see Eq. 1.1.8)} \tag{2.1.3}$$

$$= \frac{2.5}{0.45}$$

$$= 5.6$$

while the displacement ductility associated with the nominal strength (P_n) and effective stiffness ($I_e = 0.35 I_g$) would be

$$\mu_{\delta d} = \frac{2.5}{0.37} \qquad \text{(see Figure 2.1.3)}$$

$$= 6.8$$

Observe that these analytical quantifications of member ductility are larger than those that would be developed from the experimentally established idealization of member yield ($\delta_{y33} = 0.65$ in.).

The idealized displacement ductility associated with Point 77 (Figure 2.1.2) would be

$$\mu_{\delta i} = \frac{\delta_{\max}}{\delta_{yi}}$$

$$= \frac{3.8}{0.45} \qquad \text{(see Figure 2.1.4)}$$

$$= 8.4$$

The condition of the beam at this displacement (3.8 in.) is shown in Figure 2.1.9.

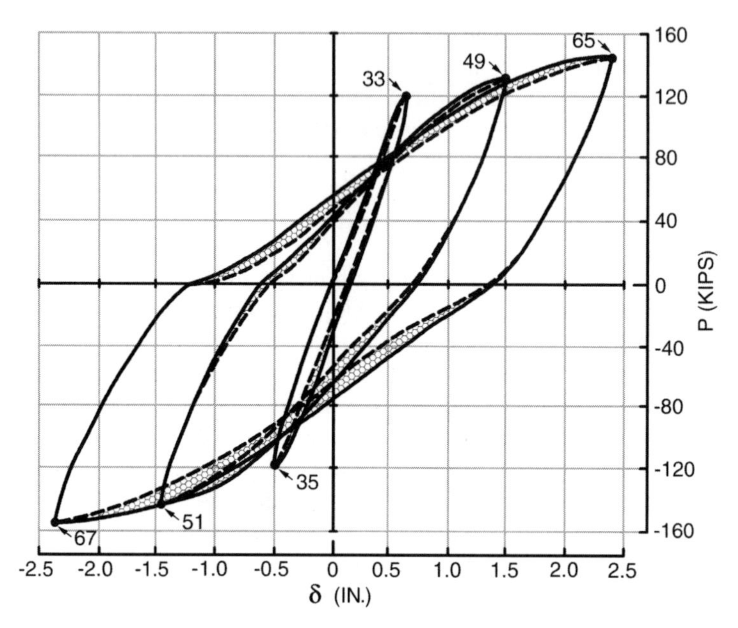

Figure 2.1.6 Stiffness deterioration from first to second cycle of the same deflection amplitude (see Figure 2.1.2).

Strength and stiffness degradation will, of course, impact the response of any building to earthquake ground motion. Figure 2.1.6 describes the stiffness at cycle 65 as well as the deterioration that was observed on the next loading cycle to the same level of deflection. Observe that the stiffness is now 56 kips/in. (140/2.5), but cyclic stiffness deterioration (to the same level of drift) can be neglected, especially given the assumptions used to develop an idealization of behavior. This is considerably less than the idealized stiffness of 267 kips/in. (120/0.45).

More important to the designer is how the computer treats stiffness degradation in an inelastic time history analysis. This will be discussed in Section 2.1.1.5, and its impact on behavior is assessed in Chapter 4.

Postyield deformation and especially deformation limit states must be understood by the designer if for no other reason than to develop confidence in the bracing programs he or she develops. The test specimen selected (Figure 2.1.1) offers excellent insight. Figure 2.1.7 identifies the tip displacement associated with shear deformation (δ_{shear}) and flexural deflection $\left(\delta^1_{\text{flex}} \text{ and } \delta^2_{\text{flex}}\right)$. Measured curvatures are described in Figure 2.1.8. The average curvatures for regions ℓ_1 and ℓ_2 (see Figure 2.1.1) represent an attempt to quantify fixed end rotation and postyield rotation, respectively. Together they create δ^1_{flex} in Figure 2.1.7. The deflection of the beam beyond this "hinge" region ($\ell_1 + \ell_2$) is referred to as δ^2_{flex}. Observe that deflection component δ^2_{flex} is constant between load Point 33 (idealized yield) and load Point 65 (see Figure 2.1.7). This confirms the fact that all postyield deformation is generated by a postyield rotation in the plastic hinge region.

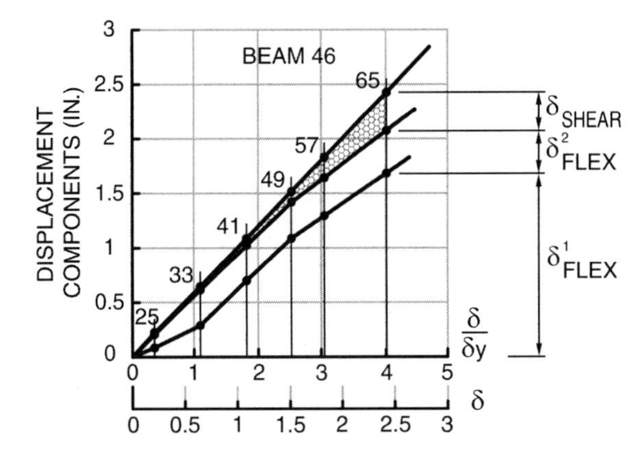

Figure 2.1.7 Components of tip deflection beam of Figure 2.1.1.

Consider the behavior at Point 41 of Figure 2.1.2 and 2.1.8; the objective is to predict the probable compressive strain states in the concrete. The flexure associated deflection is about 1 in. and the shear component is negligible. The research team[2.1] developed the flexural components of deflection (δ_{flex}) from the following relationship:

$$\delta_{\text{flex}}^1 = (6\phi_1)75 + (11.25\phi_2)66.375 \tag{2.1.4}$$

The products $\ell_1\phi_1$ and $\ell_2\phi_2$ develop rotations

$$\theta_1 = \ell_1\phi_1$$

and

$$\theta_2 = \ell_2\phi_2$$

It was these rotations that were measured (see, e.g., Figure 2.1.13) in region $\ell_1(6\phi_1)$ and region $\ell_2(11.25\phi_2)$ and then converted to estimates of curvature (Figure 2.1.8). δ_{flex}^2 was the deduced flexural deformation in the remaining portion of the beam. Assuming that the section was cracked δ_{flex}^2, using our idealized stiffness ($I_e = 0.35I_g$), turns out to be

$$\begin{aligned}\delta_{\text{flex}}^2 &= \frac{P(\ell - \ell_1 - \ell_2)^3}{3EI_e} \\ &= \frac{120(60.75)^3}{3(4000)(10{,}700)} \\ &= 0.21 \text{ in.}\end{aligned}$$

and this is almost half of the value suggested by Figure 2.1.7.

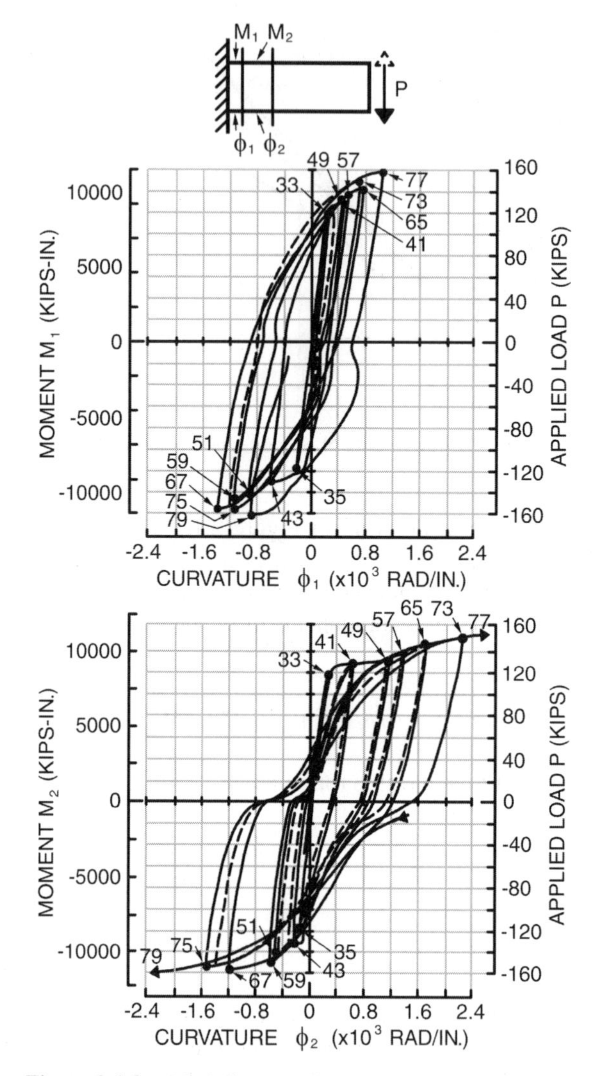

Figure 2.1.8 *M–ϕ* diagrams for the beam of Figure 2.1.1.

Curvatures suggested in Figure 2.1.8 for Point 41 are

$$\phi_1 = 0.00050 \text{ rad/in.}$$

$$\phi_2 = 0.00065 \text{ rad/in.}$$

The rotation experienced at a point can be used to predict deflection by multiplying the rotation (θ_1) by the length between it and the end of the beam.

Accordingly,

$$\delta^1_{\text{flex}} = 6(0.0005)(75) + 11.25(0.00065)(66.375) \qquad \text{(Eq. 2.1.4)}$$
$$= 0.22 + 0.49$$
$$= 0.71 \text{ in.}$$

and this is the value suggested by Figure 2.1.7.

As we have seen, an "estimate" of the depth to the neutral axis is required to convert these curvatures to strains. This involves a certain amount of speculation, which is not necessarily improved by a detailed analysis. The probable depth of the neutral axis at displacement Point 41 (Figure 2.1.5) is 4.5 in.

The associated levels of compressive strain are

$$\text{Region } \ell_1 \qquad \varepsilon_{c1} = 0.0005(4.5) = 0.0023 \text{ in./in.}$$
$$\text{Region } \ell_2 \qquad \varepsilon_{c2} = 0.00065(4.5) = 0.0029 \text{ in./in.}$$

The differentiation of strain states between regions ℓ_1 and ℓ_2 is not realistic. Rather, it seems more reasonable to conclude that the compressive strain level realized in the plastic hinge region $(\ell_1 + \ell_2)$ is consistent with the strains used to estimate the nominal moment capacity of the section ($\varepsilon_c = 0.003$ in./in.).

The practical deformation limit state is reasonably associated with Point 73 in Figure 2.1.2. The neutral axis will drop as the concrete near the face of the beam becomes less effective. If a neutral axis depth of 5 in. is assumed, the probable strain state is

$$\phi_2 = 0.00232 \text{ rad/in.} \qquad \text{(see Figure 2.1.8)}$$
$$\varepsilon_{c2} = 0.00232(5)$$
$$= 0.0116 \text{ in./in.}$$

In summary, at a drift angle (δ/ℓ) of 0.014 radian (Point 41), concrete strains in the plastic hinge region are on the order of 0.003 in./in. When the drift angle reaches 0.035 radian, the concrete strain in the plastic hinge region is on the order of 0.012 in./in., and the shell should spall, as it did on this beam (Figure 2.1.9). Observe that the extent of shell spalling and length of the plastic hinge (ℓ_p) is at least 17 in. Accordingly, our assumed plastic hinge length $(\ell_p = 0.5h)$ appears to be slightly conservative.

2.1.1.2 Predicting Postyield Deformation Limit States The development of a rational design methodology for predicting compressive strain states in flexurally loaded members requires that we address and quantify key components of the deformation algorithm. Specifically, the plastic hinge length (ℓ_p) and the curvature at yield are required if we wish to estimate postyield curvature (ϕ_p) in the plastic hinge region. The conjugate beam of Figure 2.1.10 describes one relationship between these elements.

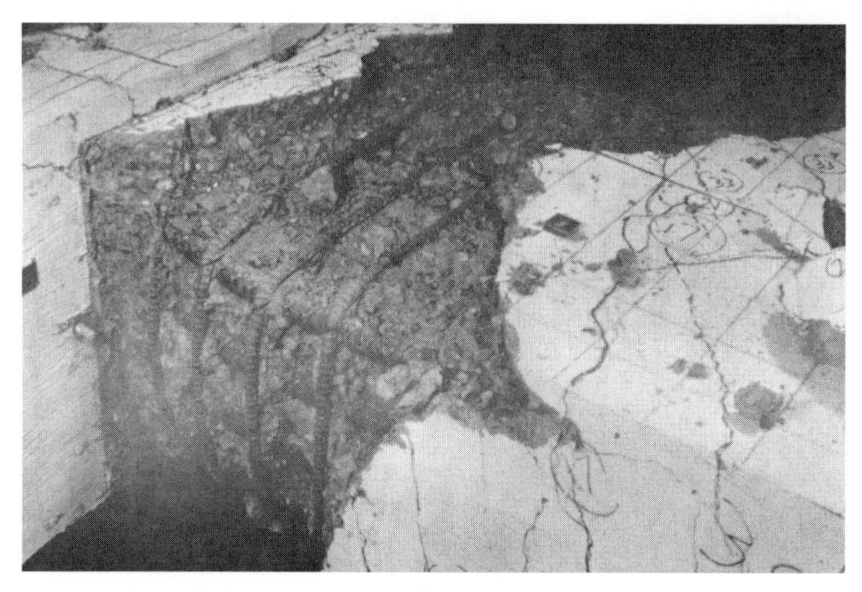

Figure 2.1.9 Beam 43 after failure. (Courtesy of Earthquake Engineering Research Center, University of California, Berkeley, EERC 72-5, 1972.[2.11])

(See Ref. 2.4, Sec. 2.2(b)) for a development of the conjugate beam method for calculating deflection.)

Comment: The conjugate beam described in Figure 2.1.10 is easily used in design but not unique or necessarily appropriate for all situations. Inelastic time history analyses, for example, endeavor to predict postyield rotations. These postyield rotations may be converted to curvatures and then concrete strains, as was done when the frame beam of Figure 2.1.1 was analyzed. To this postyield concrete strain must be added the elastic(!) strain in the hinge region. This process corresponds to predicting deformations using the conjugate beam described in Figure 2.1.11.

Theoretical values for ℓ_p and δ_y may be developed, but they must include complex behavior characteristics that are usually influenced by other factors not easily generalized. Consider, for example, the extension of the flexural reinforcing into the column in Figure 2.1.1. The stress in these bars inside the column will exceed yield, and the resultant elongation must manifest itself in what is usually referred to as fixed end rotation (ϕ_1). This is commonly referred to as tensile strain penetration, and it will have a greater impact when reinforcing bar anchorage is not as effective as it was in this experiment. Reinforcing bar size and bond deterioration will impact fixed end rotation. These factors will be discussed when we study the beam-column joint. The use of simple analytical models can be reasonably effective if calibrated to test data developed from members representative of those used in the design of proposed structures.

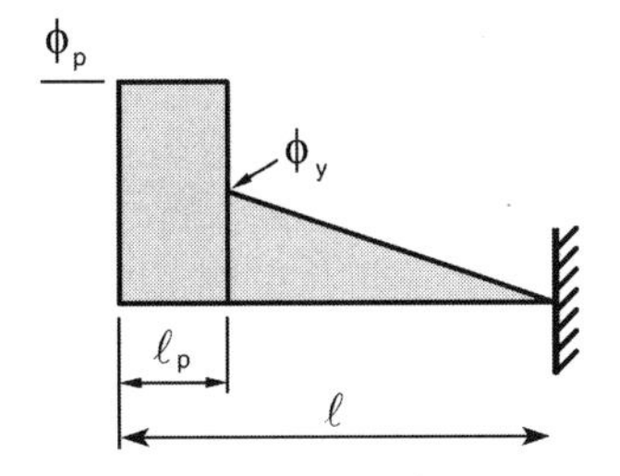

Figure 2.1.10 Conjugate beam (curvature diagram)—design model.

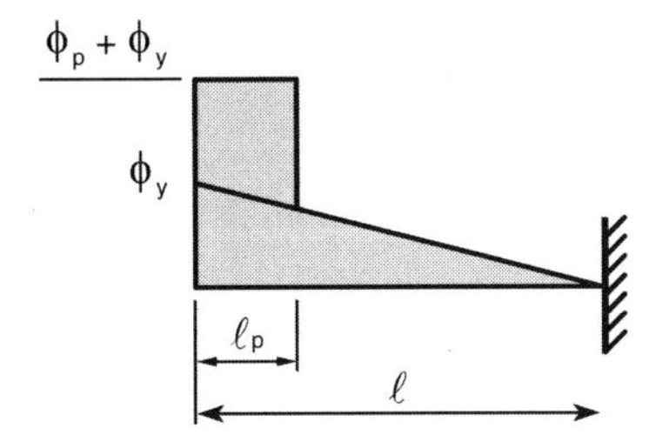

Figure 2.1.11 Conjugate beam (curvature diagram)—analysis model.

Paulay and Priestley[2.5] suggest that a good estimate of the plastic hinge length (ℓ_p) may be obtained from the following:

$$\ell_p = 0.08\ell + 0.15 d_b f_y \tag{2.1.5}$$

where ℓ is, in their development, a cantilever length or the distance between the point of maximum moment and the point of inflection. Alternatively, they suggest that a plastic hinge length (ℓ_p) of $0.5h$ produces adequate accuracy.

In the test beam (Figure 2.1.1) Eq. 2.1.5 suggests that

$$\begin{aligned} \ell_p &= 0.08(78) + 0.15(1.125)60 \\ &= 16.4 \text{ in.} \end{aligned}$$

However, the use of a hinge length of $0.5h$ produces concrete strain estimates that are consistent with these tests:

$$\begin{aligned} \ell_p &= 0.50(29) \\ &= 14.5 \text{ in.} \end{aligned}$$

The deflection of the beam at yield (δ_y), when used to quantify postyield strain states, is best associated with the First Yield of Steel Model because subsequent displacements tend to manifest themselves in postyield rotation in the plastic hinge region. This suggests the adoption of the curvature model of Figure 2.1.11.

One procedure for quantifying δ_y is to determine the moment at either the face of the plastic hinge (Figure 2.1.10) or the face of the support (Figure 2.1.11). Then convert this moment to a curvature, assuming elastic behavior and a cracked section.

An estimate of the concrete strain and curvature at yield may also be developed directly from the steel strain at first yield (Figure 2.1.5).

The depth of the neutral axis will lie somewhere between $kd(\pm 0.38d)$ and c as developed from the Whitney Stress Block. For convenience assume that kd is reasonably on the order of $0.33d$. Then

$$\varepsilon_c = \frac{\varepsilon_{sy}}{2} \tag{2.1.6}$$

or about 0.001 in./in., and

$$\phi_y = \frac{\varepsilon_{sy}}{0.67d} \tag{2.1.7a}$$

Consider, for example, behavior of the test beam at a drift of 3.5%. This corresponds to Point 73 of Figure 2.1.2:

$$\phi_y = \frac{0.002}{0.67(25.25)} \qquad \text{(Eq. 2.1.7a)}$$

$$= 0.00012 \text{ rad/in.}$$

$$\delta_y = \frac{\phi_y \ell^2}{3} \qquad \text{(See conjugate beam Figure 2.1.11)} \tag{2.1.7b}$$

$$= \frac{0.00012(78)^2}{3}$$

$$= 0.24 \text{ in.}$$

Observe that this estimate of δ_y is reasonably consistent with δ^2_{flex} (Figure 2.1.7) yet conservative.

The postyield deflection (δ_p) is developed as follows:

$$\delta_u = 0.035\ell \qquad \text{(Point 73, Figure 2.1.2)}$$

$$= 0.035(78)$$

$$= 2.73 \text{ in.}$$

$$\delta_p = \delta_u - \delta_y \tag{2.1.7c}$$

$$= 2.49 \text{ in.}$$

From this the rotation, curvature and concrete strain in the plastic hinge region can be developed:

$$\theta_p = \frac{\delta_p}{\ell - \ell_p/2} \tag{2.1.8}$$

$$= \frac{2.49}{78 - 7.25} \qquad \left(\ell_p = \frac{h}{2}\right)$$

$$= 0.035 \text{ radian}$$

$$\phi_p = \frac{\theta_p}{\ell_p} \tag{2.1.9}$$

$$= \frac{0.035}{14.5}$$

$$= 0.0024 \text{ rad/in.}$$

$$\varepsilon_{cp} = \phi_p c \tag{2.1.10a}$$

$$= 0.0024(5)$$

$$= 0.012 \text{ in./in.}$$

$$\varepsilon_{cu} = \varepsilon_{cy} + \varepsilon_{cp} \tag{2.1.10b}$$

$$= 0.001 + 0.012$$

$$= 0.013 \text{ in./in.}$$

Observe that this slightly overestimates the strain experienced in region ℓ_2 of the test.

$$\phi_2 = 0.00232 \text{ rad/in.}$$

$$\varepsilon_{c2} = 000232(5)$$

$$= 0.0116 \text{ in./in.}$$

Procedure for Estimating the Probable Postyield Concrete Strain State

Step 1. Estimate the Plastic Hinge Length.

$$\ell_p = 0.5h$$

Step 2. Estimate the Curvature Associated with First Yield in the Steel.

$$\phi_y = \frac{\varepsilon_{sy}}{0.67d} \qquad \text{(Eq. 2.1.7a)}$$

Step 3. Calculate the Deflection at Yield.

$$\delta_y = \frac{\phi_y \ell^2}{3} \qquad \text{(Eq. 2.1.7b)}$$

Step 4. Develop a Conjugate Beam (Figure 2.1.11) Whose Length Is Half of the Cleat Span of the Beam.

Step 5. Calculate the Postyield Component of Deflection.

$$\delta_p = \delta_u - \delta_y \qquad \text{(Eq. 2.1.7c)}$$

Step 6. Calculate the Rotation in the Hinge Region.

$$\theta_p = \frac{\delta_p}{\ell - \ell_p/2} \qquad \text{(Eq. 2.1.8)}$$

Step 7. Calculate the Average Curvature in the Plastic Hinge Region.

$$\phi_p = \frac{\theta_p}{\ell_p} \qquad \text{(Eq. 2.1.9)}$$

Step 8. Estimate the Postyield Strain in the Concrete.

$$\varepsilon_{cp} = \phi_p c \qquad \text{(Eq. 2.1.10a)}$$

Step 9. Estimate the Probable Strain in the Concrete.

$$\varepsilon_{cu} = \varepsilon_{cy} + \varepsilon_{cp} \qquad \text{(Eq. 2.1.10b)}$$

2.1.1.3 Impact of Shear and Confinement on Behavior To understand the impact of shear on the postyield behavior of a beam, we can compare the behavior of two specimens tested by the researchers[2.1]: Beam 43 and Beam 46. These beams are identical (Figure 2.1.1) with the exception of the amount of shear reinforcement provided. Beam 43 was reinforced with #4 stirrups at 3 in. on center while Beam 46 used #4 stirrups at 6 in. on center. Accordingly,

$$\begin{aligned} v_{s43} &= \frac{A_v f_y}{sb} \\ &= \frac{0.4(60)}{3(15)} \\ &= 0.53 \text{ ksi} \\ v_{s46} &= 0.26 \text{ ksi} \end{aligned}$$

The induced level of shear stress at a tip displacement of 1 in. ($\theta = 0.013$ radian) was generated by a tip load of 122 kips in Beam 46 (Figure 2.1.12) and 130 kips in Beam 43 (Figure 2.1.2). This variation in developed strength may be attributable to the difference in concrete strength ($f_c' = 4$ ksi for Beam 46 and 5 ksi for Beam 43). In any event, it is reasonable to conclude that the induced level of shear stress was on the order of

$$v = \frac{V}{bd}$$

$$= \frac{126}{15(25.25)}$$

$$= 0.333 \text{ ksi} \qquad \pm 5\sqrt{f_c'}$$

and this exceeds v_{s46} but is considerably less than v_{s43}.

Beam 46 could only attain a tip deflection of 2 in. (Figure 2.1.12), and this immediately preceded failure. The failure is clearly one of shear in the hinge region (Figure 2.1.13). This characteristic behavior pattern is explained by the deterioration of the concrete and the resultant inability of the concrete to participate in transferring shear ($V_c = 0$) in the hinge region. Recall that the concrete shear transfer mechanism (v_c) relies on aggregate interlock, dowel action, and friction within the compressive stress block, all of which are diminished when the load is cyclically reversed and the concrete cracks become wide, as they clearly did on Beam 46 (Figure 2.1.13).

One might observe that the plastic truss analogy described in Section 1.3.1 (Figure 1.3.2) is not activated because the compression fan cannot attain an angle of 25°, its identified static limit. Figure 2.1.14 describes the shear load path required to transfer

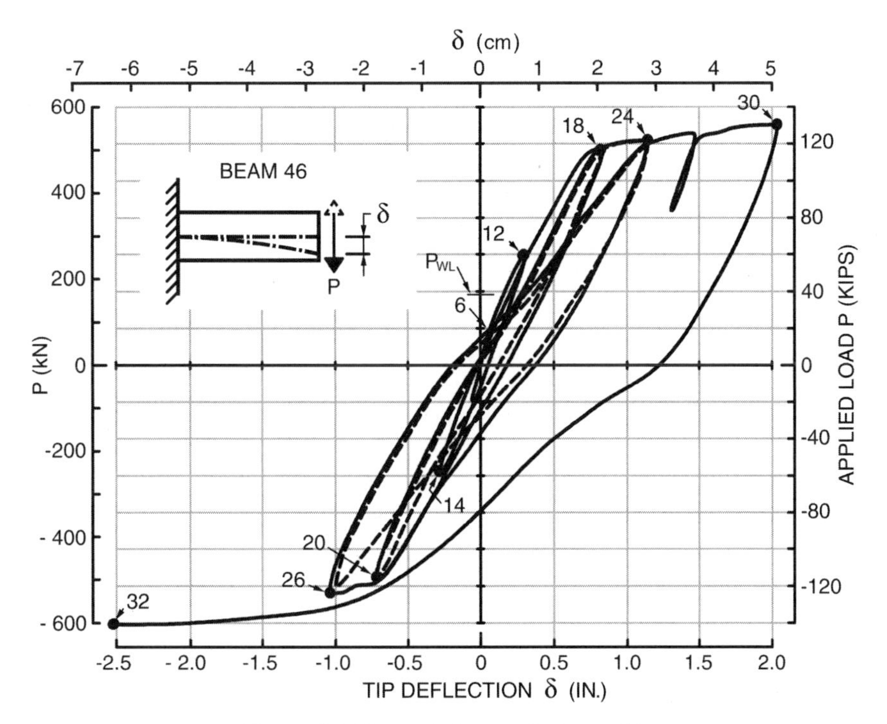

Figure 2.1.12 P–δ diagram for beam 46.

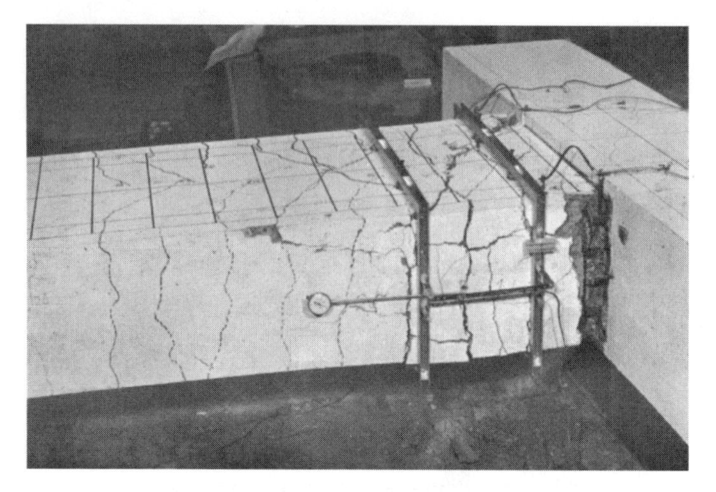

Figure 2.1.13 Beam 46 after failure. (Courtesy of Earthquake Engineering Research Center, University of California, Berkeley, EERC 72-5, 1972.[2.1])

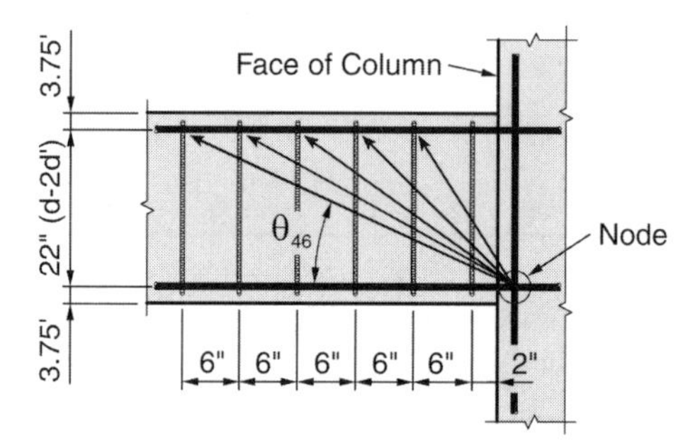

Figure 2.1.14 Shear fan—plastic truss analogy.

a shear force of about 125 kips (Figure 2.1.12). Five stirrup groups must participate in transferring the shear once the concrete component (V_c) is lost.

The compressive strut must reach an inclination of

$$\theta_{46} = \tan^{-1} \frac{22}{34}$$
$$= 33^\circ$$

and this it clearly was not able to do, probably because a large crack formed on a 45° angle (Figure 2.1.13) and this allowed only three or at most four stirrups to be activated.

In contrast, the angle of the shear fan in Beam 43 must engage six stirrup groups if it is to support a load of 150 kips:

$$\theta_{43} = \tan^{-1}\left(\frac{22}{19}\right)$$
$$= 49^\circ$$

and this is consistent with the idealized truss mechanism used to develop v_s (see Section 1.3.1). Further, it does not require a compressive strut to cross a principal tension stress induced crack line (45°).

Design Rule: Stirrups within the plastic hinge region must be designed to carry the probable shear assuming no participation from the concrete mechanism ($V_c = 0$). The plastic truss mechanism should not require a compressive strut angle greater than 45°.

Comment: The building code[2.6] requires that this level of shear reinforcement extend for a distance of twice the member depth ($2d$) from the face of the supporting member.

Confinement undoubtedly increases the deformation limit state of a frame beam. Observe that in spite of the shell having spalled completely (Figure 2.1.9), the strength sustained at load cycle 77 (Figure 2.1.2) exceeds that of load Point 73. The confining pressure provided by the stirrups in Beam 43 exceeds our objective level of 450 psi. See Section 1.2.1.

$$f_\ell = \frac{2A_{sh}f_{yh}}{d_c s} \qquad \text{(see Eq. 1.2.4)}$$
$$= \frac{2(0.2)(60)}{(15-3)(3)}$$
$$= 0.67 \text{ ksi}$$

Laterally this objective level of confinement is also exceeded.

Consider how the flexural strength developed during the test was attained at load Point 77 (Figure 2.1.2). The tip load is 160 kips. This means that

$$M_{77} = 78(160)$$
$$= 12{,}480 \text{ in.-kips}$$

The force in the tension reinforcement is

$$T_{77} \cong \frac{M_{77}}{d - d'}$$
$$\cong \frac{12{,}480}{25.25 - 3.75}$$
$$= 580 \text{ kips}$$

and this corresponds to a stress of

$$f_{s77} = \frac{T_{77}}{\sum A_s}$$

$$= \frac{580}{6}$$

$$= 97 \text{ ksi}$$

which is very near the reported ultimate strength of the bar.[2.1]

The compressive force is developed in the reinforcing bars and the confined concrete within the shell (Figure 2.1.9). It is unlikely that the compressive force developed in the bars (A_s') will exceed yield.

$$C_{s77} \cong 60(6)$$

$$= 360 \text{ kips}$$

$$C_{c77} = T - C_{s77}$$

$$= 580 - 360$$

$$= 220 \text{ kips}$$

If we assume that the confined compressive strength of the concrete f_{cc}' is on the order of 7.5 ksi (Eq. 1.2.1), the depth of the compressive stress block must be

$$a = \frac{C_{c77}}{b_e f_{cc}'}$$

$$= \frac{220}{12(7.5)}$$

$$= 2.4 \text{ in.}$$

This is consistent with the lever arm (d–d') assumed in development of the flexural strength of the member.

Accordingly, we have confirmed analytically that both a flexural and shear load path capable of supporting a tip load of 160 kips exist. Most importantly, this load is supported at a drift angle (θ) of 4.8%.

2.1.1.4 Importance of Detailing Ties or stirrups perform another important function—they prevent the compression bars from buckling. A close examination of Figure 2.1.9 reveals that the middle two flexural bars have caused the ties to deform outwardly. This may have been caused by confining induced stresses imposed on the ties or bar buckling. Clearly, the observed behavior demonstrates the importance of detailing.

As a precursor to the PRESSS Program,[2.7] which was instrumental in developing the precast frames discussed in Section 2.1.4, a one-third-scale model of a cast-in-place seismic frame beam designed to the then current ACI criterion was tested at the National Institute of Standards and Technology (NIST).[2.8] The behavior of this test specimen is described in Figure 2.1.15. The attained level of story drift was 3.5%. At this level of drift there was a significant loss of strength accompanied by major damage (Figure 2.1.16). The amount of damage is attributable to two major factors—shear transfer and bar buckling.

The beam size was $6\frac{2}{3}$ in. by 10 in. and the flexural reinforcing consisted of #3 bars. They were laterally supported by 0.2-in. diameter hoops spaced at $3\frac{1}{3}$ in. on center. By way of comparison this corresponds to an ℓ/d_b ratio of 16.5, while the ℓ/d_b ratio of the bars in the test specimen described in Figure 2.1.1 was only 2.67. The detrimental behavior is explained by considering the critical buckling stress of the #3 reinforcing bar. The stress required to buckle the previously overstrained tension bars (A_s) is

$$f_{\text{cr}} = \frac{\pi^2 EI}{\ell^2} \tag{2.1.11a}$$

$$= \frac{\pi^2 (29{,}000)(0.00097)}{(3.33)^2}$$

$$= 25 \text{ ksi}$$

The stability of the bar is further lessened by a reduced modulus of elasticity (Bauschinger effect). Accordingly, lateral support of these cyclically loaded bars must be provided at closer spacing than bars stressed in compression only. Buckling of compression bars is evident in Figure 2.1.16, for it has forced the shell concrete to bulge outwardly and spall.

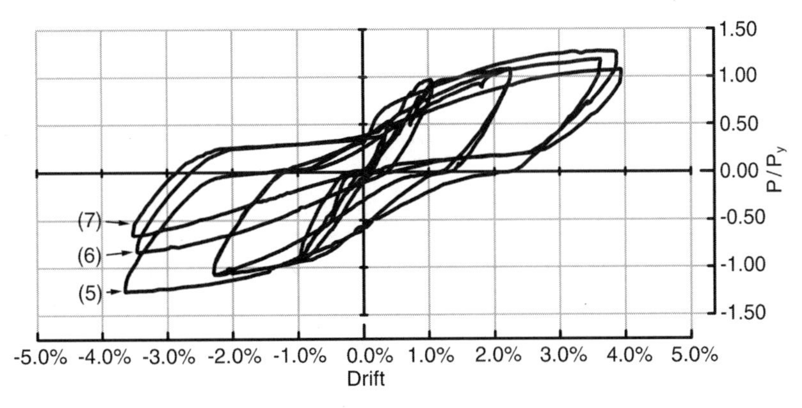

Figure 2.1.15 NIST load-displacement history.

Figure 2.1.16 NIST cast-in-place subassembly at 3.5% drift, cycle 3.

Shear slip will also occur in significantly larger amounts when small flexural reinforcing bars are used. This shear deterioration is evinced by the complete deterioration of the concrete at the beam column interface. Thus in spite of the fact that the beam described complied with code[2.9] strength objectives and was not overreinforced ($\rho = 0.01$), and that capacity-based design considerations were satisfied, the beam failed prematurely and after only four cycles of postyield drift, albeit at high levels of story drift. This clearly demonstrates the importance of detailing.

Prescriptive detailing rules are contained in buildings codes,[2.6] but they should be supplemented when and where significant deformations are anticipated. Pertinent rules for frame beam tie spacing contained in current codes[2.9] are

- Hoop reinforcement is required in the hinge region. The identified length of the hinge region in a frame beam is twice the depth of the member ($2d$).
- Spacing between hoops longitudinally is limited to the lesser of $d/4$, $8d_b$, $24d_h$, or 12 in.
- Longitudinal bars must be transversely supported as described in Figure 2.1.17.

A review of Figure 2.1.1 shows that hoop reinforcing does not comply with the transverse reinforcing required by Figure 2.1.17 because alternative bars are not supported and, as previously pointed out, it appears as though buckling is imminent (Figure 2.1.9). Further, the longitudinal spacing of the hoops is 3 in. in the test beam and this is much less than the 6.3 in. minimum ($d/4$) required by current codes.[2.9]

A review of the stirrup arrangement used to produce the NIST beam (Figure 2.1.16) indicates compliance with all but the $8d_b$ rule ($8(0.375) = 3$ in. < 3.33 in.). Paulay and Priestley[2.5] recommend closer tie spacings ($6d_b$) than those required by ACI 318–99.[2.6] Wallace,[2.10] in the testing of boundary elements of walls, reports

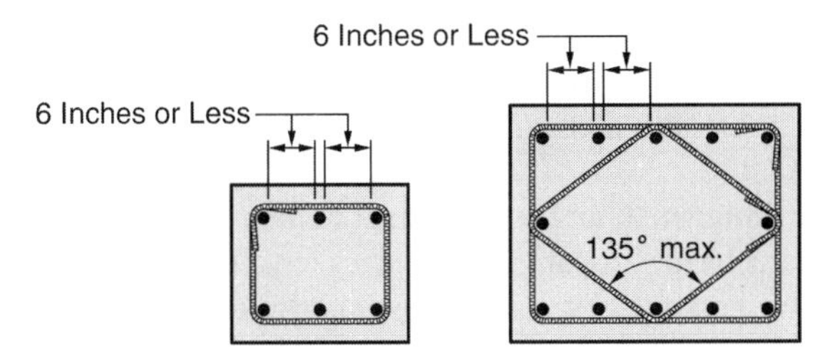

Figure 2.1.17 Sketch that describes maximum spacing between laterally supported column bars.[2.6]

the buckling of compression reinforcement in the plastic hinge region. The failure is attributed to a misplaced tie producing an s/d_b radio of 7. In spite of the fact that the buckled bars were small (#3 and #4) and therefore not representative of actual frame beam construction, prudence, especially in the hinge region, seems warranted.

Further, Paulay and Priestley[2.5] impose a strength requirement on the ties based upon the force imposed on the longitudinal reinforcing:

$$A_t = \frac{A_s f_y}{16 f_{yh}} \left(\frac{s}{4}\right) \tag{2.1.11b}$$

This relationship is intended to provide a lateral restraining force equivalent to $\frac{1}{16}$ of the force in the restrained longitudinal bar and presumes that support will be provided at 4-in. on center ($s = 4$ in.) longitudinally. A_s, in this case, is the sum of the areas of the longitudinal bars being supported by one tie A_t.

If the NIST beam is tested using this criterion (Figure 2.1.16),

$$A_{t-\text{req'd}} = \frac{2(0.11)(60)3.33}{16(60)4} \qquad \text{(Eq. 2.1.11b)}$$

$$= 0.0114 \text{ in.}^2 \ll \left(0.0314 \text{ in.}^2\right)$$

Accordingly, the provided tie size (0.2 in.) complies with the strength objective defined by Eq. 2.1.11b, but its longituding spacing significantly exceeds the objective ℓ/d_b ratio:

$$6d_b = 2.25 \text{ in.} < 3.33 \text{ in.}$$

The experienced buckling is undoubtedly attributable to the aspect ratio (s/d_b) of the bar. It should be noted that the span-to-bar diameter ratio (ℓ/d_b) does not uniformly define elastic stability (Eq. 2.1.11a), which is more critical for smaller bars.

Spacing between beam longitudinal bars in the transverse direction is also critical. The ACI[2.6] limits this to 14-in. centers, while Paulay and Priestley[2.5] suggest that the transverse spacing between bars be limited to 8 in. It appears that the 14 in. recommended by the ACI is too liberal for bars that are subjected to large postyield strains and then recompressed, while that recommended by Paulay and Priestley is probably too conservative for use as a general requirement. Observe that the detailing of Beam 43 exceeded the 8-in. limit and yet was capable of sustaining a drift angle of 4.8%.

Comment: The designer should be cognizant of his or her objectives when detailing transverse reinforcement. Probable strain states must be considered. When it is clear that large (> 2%) postyield reversible rotations are to be expected, the buckling of overstrained longitudinal bars must be inhibited. If concrete compressive strains are expected to cause shell spalling ($\varepsilon_c > 0.01$ in./in.), the core of the beam should be confined so as to minimize strength degradation. The need to provide special detailing over a region exceeding $0.75d$ is questionable. To state this succinctly, be conservative where conservatism is called for. The interested reader is referred to Priestley et al.[2.11]

2.1.1.5 Modeling Considerations Frame beam behavior in the postyield range is modeled through the use of a variety of simple elements. The most commonly used element model is the bilinear element incorporated into the computer program commonly referred to by its acronym DRAIN-2DX,[2.12] which is shown in Figure 2.1.18. When this element is used in computer programs, the designer inputs the "idealized" elastic stiffness (k_i), strength limit state (M_{yi}), and the strength hardening rate in the postyield range usually expressed as a function of the idealized elastic stiffness (γk_i). The use of idealized stiffness and strength hardening parameters is appropriate, and in the force–deformation relationship described in Figure 2.1.18, an envelope of the cyclic behavior is produced. This model also reasonably models monotonic behavior.[2.3] Some computer programs will provide bilinear options in the elastic range (k_i) and, whereas this is consistent with the initial behavior described in Figure 2.1.3, this bilinear elastic bilinear behavior is not repeated on subsequent cycles because the section is now cracked. Accordingly, refinement of the model in the elastic range is not necessary.

The bilinear model described in Figure 2.1.18 is more appropriate for steel beams than for concrete beams. Observe that it does not reproduce the behavior described in Figure 2.1.2. This is because after a yielded concrete section is unloaded, the

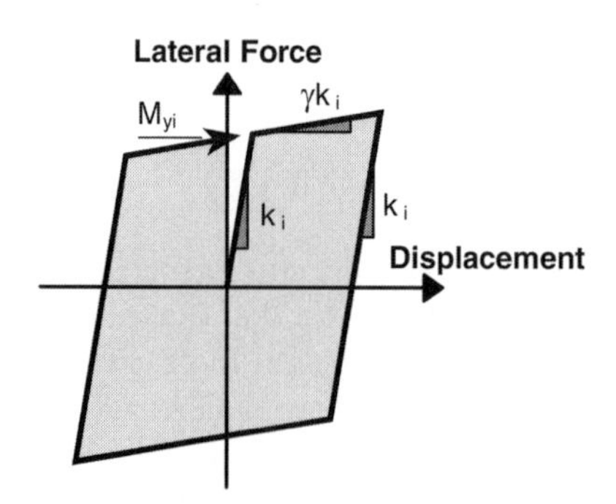

Figure 2.1.18 Force–deformation relationship for beam element, DRAIN-2DX.[2.12]

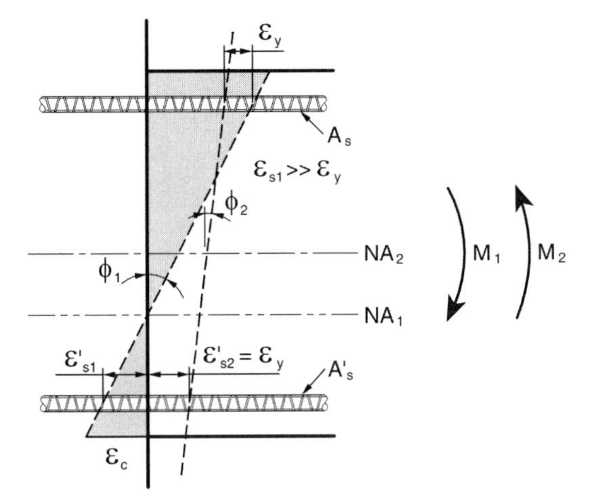

Figure 2.1.19 Curvature change from peak load to tension yielding of A_s' bars.

force applied to the member will initially operate on a section that consists of the reinforcing bars acting alone. The stiffness of this section will be considerably less than the idealized stiffness, and it is this change in stiffness coupled with shear slippage along the now opened cracks that causes the characteristic pinching of the hysteresis loop in a concrete member (Figure 2.1.2). Consider the behavior described in Figure 2.1.19. At peak load (M_1) the tension steel A_s has reached a strain state that considerably exceeds ε_y. When the load that has created M_1 is relieved, behavior is quasi-elastic, but upon reverse loading (M_2), the cracking that occurred during the previous loading (M_1) will not close until the top bars (A_s), now in compression, yield or debond. The reinforcing bars (A_s and A_s') now provide the stiffness, and this stiffness will be considerably less than the k_i assumed in the model of Figure 2.1.18. This reduced stiffness is contributed to by the absence of a stiff shear transfer mechanism. Observe that all of the shear imposed on the section must be transferred by dowel action before the cracks on the now compression side are closed. This pinching or reduced stiffness becomes more pronounced at higher levels of postyield deformation because the concrete on which these dowels bear has been crushed.

A critical review of the behavior of the test specimen (Figure 2.1.2) also suggests that the use of a constant value for the "elastic" stiffness (k_i) of a beam is probably not appropriate, for the stiffness of the section is a function of the level of deformation. A beam element that more closely models the behavior described in Figure 2.1.2 is incorporated into a computer program known by the acronym IDARC.[2.13] This element model contains a bilinear initial stiffness model, strength hardening, hysteretic pinching, and stiffness degradation. The moment curvature diagram described in Figure 2.1.20 can be internally generated by the computer program (IDARC2D 4.0). The beam cross section is divided into a number of horizontal strips that also include the reinforcing. The model (beam section) is then subjected to increasing levels of curvature. Strain distribution is obtained from compatibility, equilibrium, and the stress-strain models for the materials.

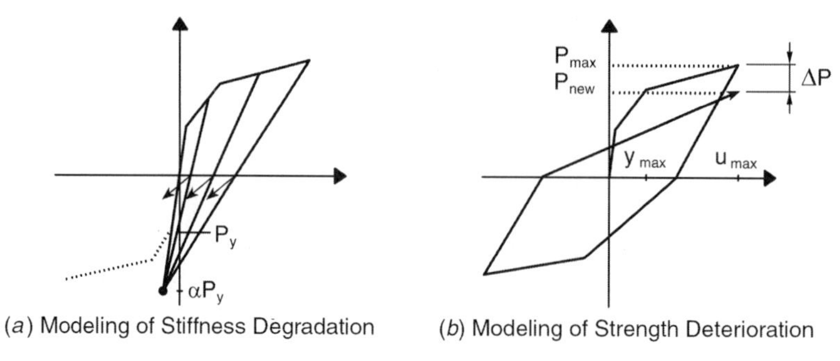

Figure 2.1.20 Control parameters for stiffness degrading hysteretic model.[2.13]

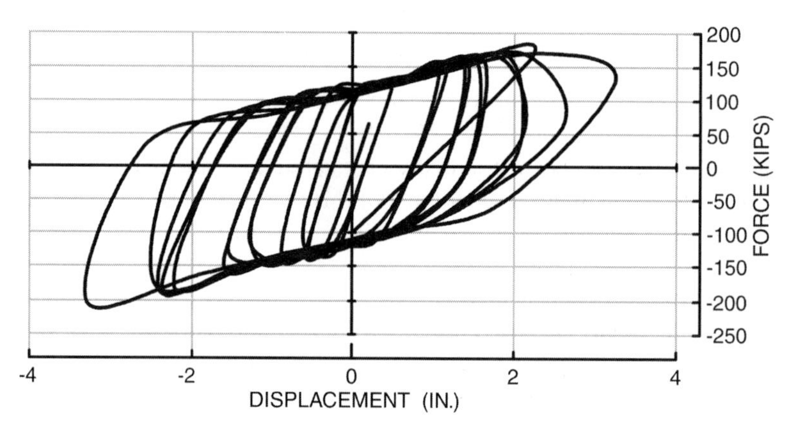

Figure 2.1.21 Computer-generated force versus displacement curve—DRAIN-2DX.

Both the DRAIN and IDARC programs were used to predict the behavior observed in the testing of the beam described in Figure 2.1.1. Figure 2.1.21 shows the hysteretic behavior of the beam element predicted by the DRAIN program when the beam tip is subjected to the displacements used to drive the test program. A similar analysis using the IDARC program is shown in Figure 2.1.22. The IDARC model more closely describes the behavior of the model (Figure 2.1.2), especially at higher levels of deformation. The error associated with the use of the DRAIN model in the analysis of a frame appears to be minimal, especially if large postyield deformations are not anticipated. The implications associated with the use of these two element models is contained in Reference 2.14, and will be discussed further when the behavior of systems is explored in Chapter 4.

2.1.2 Designing the Frame Beam

The design of the beam in a ductile frame is critical, not only to the survival of the building should it be subjected to the design earthquake, but also to the design process

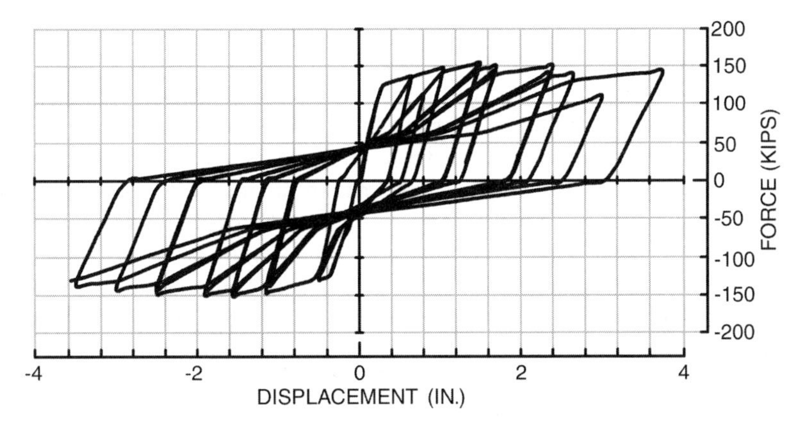

Figure 2.1.22 Computer-generated force versus displacement curve—IDARC.

itself. It is the strength of the beam that will impact, if not control, the design of every other structural component of the building. A proper design of the beam will cause the design process to flow smoothly and efficiently. An improper frame beam design will make the bracing system design process chaotic, disorganized, painful, and very time consuming.

The dynamic design process briefly described in the introduction to this book requires each design decision to be made in its proper order. It also requires that each design step consider the impact it will have on subsequent design decisions so as to avoid revisiting design decisions previously made. The careful consideration of subsequent impacts is especially critical in the design of a frame beam. The reinforcing program, for example, must be one that can be accomplished in congested regions. Stresses imposed on beam-column joints cannot exceed limit states. Too often a designer will propose a frame configuration or frame beam that cannot be detailed or that requires, after the analysis has been completed, a significant change in member size. The consequence is design chaos or a compromising of limit states. It is imperative that the designer of a frame beam look ahead without losing focus. Methodologies are developed in this section that allow the designer to efficiently consider the impact of a proposed design on subsequent steps and to produce a beam design that will, in the final analysis, optimize system behavior.

Good designers will resist the temptation to overreinforce concrete. The higher the level of developed tensile strength, the greater the compressive strain imposed on the concrete and the lower the level of available ductility. An increase in system strength does not assure improved system behavior. It can, in fact, produce a strength degrading system. This topic will be treated in Chapter 4. You will see as you study this and subsequent sections that I try to attain the strength objective through what might, by the "purist," be viewed as devious means. This is not solely an economic objective, but rather one that understands the consequence and compromises that typically are produced by introducing arbitrarily excessive amount of flexural reinforcement. With each design and analysis I become more convinced that the attainment of current basic strength objectives[2.97] produces building that will behave well if subjected to design level earthquakes.

Three elements need to be determined in the conceptual design of a beam—its height (h), width (b), and the amount of flexural reinforcing required. Design decisions must precede the design of the beam, but they all must be made with a clear understanding of the feasibility of creating a beam that will validate preceding decisions. Establishing the number of frame bays to be used and determining the strength objectives of each frame bay must precede designing the beam. These topics will be discussed in Chapter 3. We start our beam design process by presuming that we understand our strength objectives and any aesthetic constraints that must be considered. Aesthetic constraints usually include the frame geometry and story height as well as beam height, width, and column size limitations.

The factors that must be considered in the conceptual sizing of the frame beam are

- Beam-column joint capacity (Section 2.1.2.1)
- Reinforcing details (Section 2.1.2.2)
- Beam shear demand (Section 2.1.2.3)
- Column shear strength (Section 2.1.2.4)
- Available ductility (Section 2.1.2.5)

All of these considerations are examined in turn in the following subsections. Not all of them will affect every design, but if they are overlooked they may cause a significant redesign effort or compromise the behavior of the bracing program.

2.1.2.1 Beam-Column Joint Considerations The strength and deformation limit state of the beam-column joint will be discussed in Section 2.3, but the impact of the beam-column joint on the variables of interest must be quickly assessed by the designer of the frame beam. This may be accomplished by appropriately reducing the basic relationship between the amount of flexural reinforcing ($A_s + A_s'$) and joint area (A_j). The basic relationship between joint size and reinforcing is developed from the subassembly model described in Figure 2.1.23 by equating capacity and demand:

$$V_j = f_{j-\text{allow}} A_j \qquad \text{(Capacity)} \tag{2.1.12}$$

$$V_j = (A_s + A_s') f_y \lambda_o - V_c \qquad \text{(Demand)}$$

It may be conveniently reduced for an interior joint (Figure 2.1.23) in a plane frame to

$$A_j = \left[A_s \left(1 - \frac{\ell_1 (d - d')}{\ell_{c1} h_x} \right) + A_s' \left(1 - \frac{\ell_2 (d - d')}{\ell_{c2} h_x} \right) \right] \frac{\lambda_o f_y}{15\phi\sqrt{f_c'}} \tag{2.1.13}$$

Fortunately the general case (Eq. 2.1.13) may be reduced for the typical design. Beam spans are usually equal, and the clear span may be reduced in the conceptual design phase to a percentage of the center-to-center distance (about 90%):

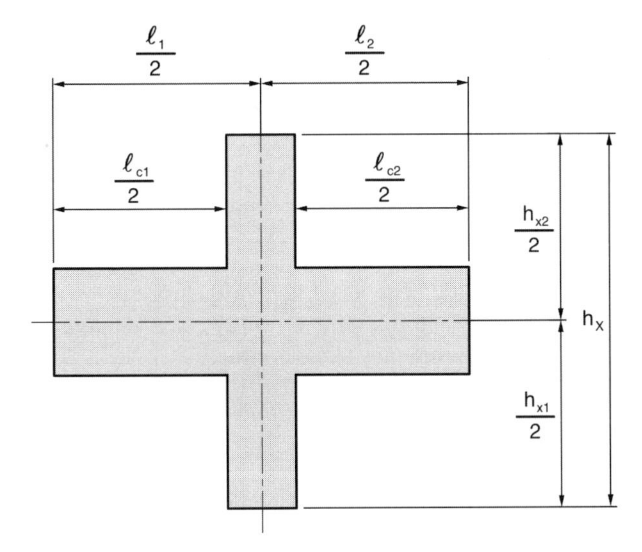

Figure 2.1.23 Beam-column subassembly.

$$A_j = \left(A_s + A_s'\right)\left(1 - \frac{1.1}{h_x}(d - d')\right)\left[\frac{\lambda_o f_y}{15\phi\sqrt{f_c'}}\right] \tag{2.1.14}$$

where h_x is the story height.

Most of the elements of the last term are office or regionally codified standards. Accordingly, the last term of Eq. 2.1.14 may be reduced for Grade 60 steel and 5000-psi concrete to

$$\frac{\lambda_o f_y}{15\phi\sqrt{f_c'}} = \frac{1.25(60)}{15(0.85)(0.071)} = 83$$

When higher strength concretes are to be used, the constant may be increased by the ratio $\sqrt{f_c'/5000}$. The maximum value, a topic that will be addressed in Section 2.3, of this constant should be 108.

$$A_j = \left(A_s + A_s'\right)\left(1 - \frac{1.1(d - d')}{h_x}\right)(83) \tag{2.1.15}$$

Aesthetic and functional constraints will usually allow reasonable estimates of h_x and d. For example, if the story height is 12 ft and the ceiling height is 8 ft, 6 in., the second term becomes

$$1 - \frac{1.1(d - d')}{h_x} = 1 - \frac{1.1(36 - 3)}{12(12)} = 0.75$$

Observe that a change in the depth of the beam will have little impact on the reduction proposed.

Now the relationship between the amount of reinforcing proposed and the area of the joint becomes one that is easily assessed:

$$\boxed{A_j = 62(A_s + A_s')} \quad \text{Interior joint} \tag{2.1.16}$$

Equation 2.1.16 is easily used. The usual known quantity is the (constrained) size of the column, say 36 in. by 36 in.; thus A_j is 1296 in.^{2}. Applying Eq. 2.1.16, the maximum amount of reinforcing may be determined:

$$\begin{aligned} A_s + A_s' &\leq \frac{1296}{62} \\ &\leq 20.9 \text{ in.}^2 \end{aligned}$$

Alternatively, if the design proposes eight #11 bars in the top and bottom of a frame beam,

$$\begin{aligned} A_s + A_s' &= 2(8)(1.56) \\ &= 25 \text{ in.}^2 \\ A_j &\geq 25(62) \qquad \text{(see Eq. 2.1.16)} \\ &\geq 1550 \text{ in.}^2 \end{aligned}$$

The column must be at least 36 in. by 43 in., but a 42-in. deep column would probably work.

Questions will arise as to effective area of the beam-column joint (A_j) in all but the simplest cases. The mechanism of load transfer within the beam-column joint is complex (see Section 2.3), and the use of shear stress is only a gauge that may be used reliably when calibrated by comparable tests. Caution in the conceptual design phase is advised when the frame beam width is considerably smaller than that of the column, when the frame beam is placed eccentrically, or when it is wider than the column.

Exterior beam-column joints are typically not a problem because frame beam reinforcement is usually about the same or slightly more than that used at an interior beam-column joint. In the case of a single bay frame, exterior joint size may govern, in which case, given the reductions discussed:

$$A_j \cong 96A_s \qquad \text{Exterior joint} \tag{2.1.17}$$

2.1.2.2 Reinforcing Details The experienced designer of seismic frames will carefully consider how frame beam reinforcing is efficiently integrated with column reinforcing and develop the reinforcing program from these considerations. The size of the beam will logically follow. The best place to start is by considering the relationship between the corner beam bar and the exterior column bar. It is desirable, from a joint behavior perspective, to have the beam bars pass within the confined area of the column, as shown in Figure 2.1.24. Column bars must pass between beam bars; accordingly, a layering will naturally develop. If a minimum beam width is desired, spacing dimensions need to be determined.

The clear distance between bars is a function of the bar diameter (d_b). For beams the minimum clear distance between parallel bars is one bar diameter; for columns it is $1\frac{1}{2}$ bar diameters. Since at this stage we know little about the reinforcing that may be required for the column, it is advisable to allow some degree of flexibility. The largest beam bar from a practical perspective is a #11 bar. Two bar bundles may be required, and the effective diameter of two bundled #11 bars is 2 in.: [$\sqrt{4(2)(1.56)/\pi}$]. A similar requirement may be developed in reinforcing the columns. Hence, a $4\frac{1}{2}$-in. center-to-center spacing between beam bars is comfortable and will allow for the minimum spacing between bundled #11 column bars. If the four beam bars shown in Figure 2.1.24 are a likely condition, the minimum beam width will be developed as follows:

Four #11 bars	4 (1.375)	=	5.5 in.
Three spaces @ 3 in.	3 (3)	=	9.0 in.
#5 hoop ties	2 (0.625)	=	1.25 in.
Fire cover	2 (1.5)	=	3.0 in.
	b	>	18.75 in.

Some detailing points are worth considering. In multifloor construction the beam size should remain constant over the height of the building in order to maintain a constant form size. Beam sizes are commonly set in 2-in. increments. In the preceding case a 20-in. wide beam would be comfortable and allow for the placement of eight

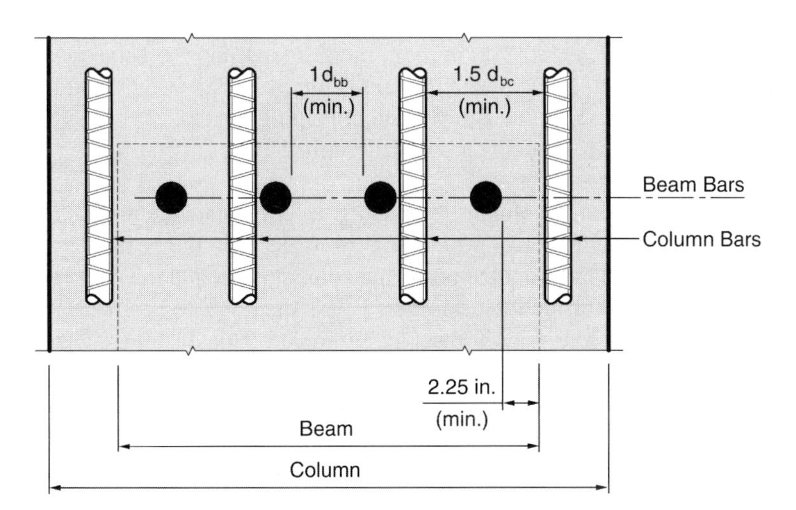

Figure 2.1.24 Beam and column reinforcing program.

Photo 2.2 Upturned frame beam, Emerald Hotel, San Diego, CA, 1990. (Courtesy of Englekirk Partners, Inc.)

#11 bars either in bundles or in two layers. A smaller beam could be used, but future flexibility would be limited. For example, beam bars and column bars of a smaller diameter could be used, but this is more expensive simply because more bars must be placed to attain the required area of steel. Bundled bars (two bars) are most economical from a placing perspective because a second layer if not bundled, must be tied off to a chair bar and supporting ties, while bundled bars are tied together and supported off of the same chair bar or tie. Detailing is common sense—imagine that you had to place the reinforcing you propose and don't get too carried away by esoteric arguments. Remember that a 60-ft long #11 bar weighs 300 lb. It does not belong in a watch so don't create a reinforcing program that can't be constructed. The beam and column reinforcing program described in Figure 2.1.24 will be difficult enough to build.

The minimum width of columns required to accommodate the reinforcing placement described in Figure 2.1.24 would be 3 in. wider than the beam. Aesthetically it may be desirable to flush out the beam and column edge, and this is acceptable so long as the concrete outside the beam bars is less than 4 in. When the thickness of the concrete shell exceeds 4 in., it must be reinforced. This flushing out is not done for economic reasons and, as can be seen in Figure 2.1.9, will only produce more concrete to spall.

We can conclude then that convenient beam sizes are 20 in. (4 bars in one layer), 16 in. (3 bars), and 24 in. (5 bars in one layer). Since we will use the same beam form throughout the height of the frame, a minimum practical size should be selected for the lower floors. The most logical choice is the 20-in. beam.

2.1.2.3 Beam Shear Demand In Section 2.1.1.3 we discussed the impact of shear stress on behavior in the plastic hinge region. The contribution of concrete to shear transfer (v_c) is not allowed or reasonably considered in the plastic hinge region. All of the induced shear in the plastic hinge region must be carried by the shear reinforcement (v_s). ACI 318[2.6] places a limit of $8\sqrt{f_c'}$ on the shear stress that may be carried by shear reinforcement, but this will produce a significant amount of shear deformation. A lower level, one more consistent with the behavior described in Figure 2.1.2, should be the design objective and this "optimum" shear stress is on the order of $5\sqrt{f_c'}$. It is important that this reduced objective shear capacity be applied to probable shears, and the probable shear should include realistic estimates of dead and live load.

The design procedure requires a capacity-based approach, specifically, that the shear strength of the beam exceed the maximum shear that can be delivered to the beam. This condition will occur when plastic hinges form at each end of the beam and they reach their probable strength ($\lambda_o P_{yi}$ of Figure 2.1.3).

The shear force (V_u) induced on the beam described in Figure 2.1.25, as a result of seismic actions, is a function of this idealized or probable moment capacity (M_{pr}) of the beam:

$$M_{\text{pr1}} = A_s' \lambda_o f_y \left(d - d'\right) \tag{2.1.18a}$$

and

$$M_{\text{pr2}} = A_s \lambda_o f_y \left(d - d'\right) \tag{2.1.18b}$$

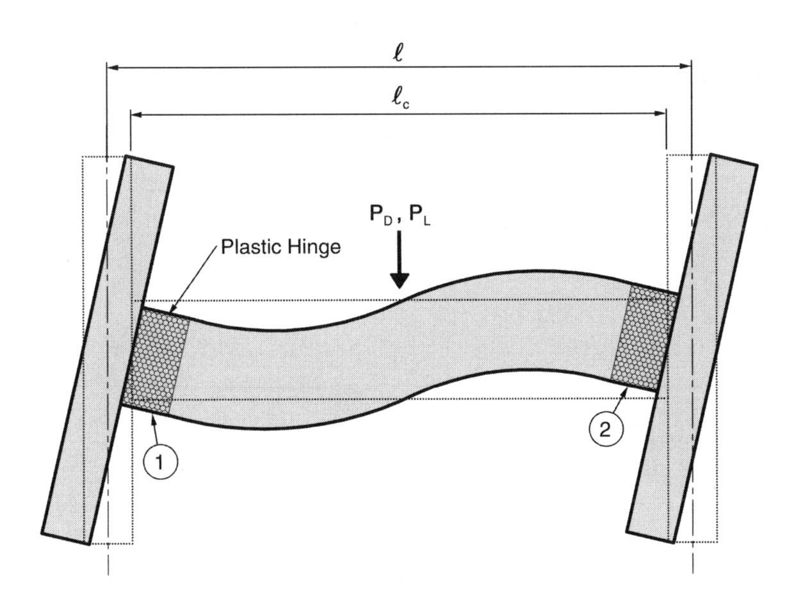

Figure 2.1.25 Frame beam at deformation limit state.

The seismic component of the shear force is

$$V_{u,\text{seismic}} = \frac{M_{\text{pr1}} + M_{\text{pr2}}}{\ell_c} \tag{2.1.19}$$

To this, from a code satisfying perspective, must be added the factored shears produced by the dead and live loads. At the strength limit state of the beam, the imposed shear will be dominated by seismic actions. Accordingly, given our rather arbitrarily selected limit state for ϕv_n of $5\sqrt{f_c'}$, real dead and live loads may often be neglected in the conceptual design process. Our design objective ($A_s = A_s'$) will cause M_{pr1} and M_{pr2} to be equal. We can then estimate the shear demand on the beam:

$$V_u = \frac{2A_s\lambda_o f_y(d - d')}{\ell_c} + \frac{P_D + P_L}{2} \tag{2.1.20}$$

From this we can size the beam as

$$bd = \frac{V_u}{\phi v_n} \tag{2.1.21}$$

Let us presume that eight #11 bars (A_s and A_s') are indicated by our design to this point, that d is 36 in., and that the clear span (ℓ_c) is 26 ft.

$$V_{u,\text{seismic}} = \frac{2(8)(1.56)(1.25)(60)(36 - 3)}{26(12)} = 198 \text{ kips} \qquad \text{(see Eq. 2.1.19)}$$

If the factored dead and live loads are on the order of 100 kips (quite high), the shear demand imposed on the frame beam would be 298 kips.

If we set our objective shear strength (ϕv_n) at $5\sqrt{f_c'}$ and presume that the probable dead and live loads will be less than 60% of the factor values, our suggested beam area is

$$bd = \frac{198 + 0.6(100)}{5(0.07)} = 737 \text{ in.}^2 \qquad \text{(see Eq. 2.1.21)}$$

and

$$b = \frac{737}{36} = 20 \text{ in.}$$

Observe that this is well within code limit states.

$$\begin{aligned} v_u &= \frac{298}{20(36)} \\ &= 0.41 \text{ ksi} \end{aligned}$$

$$8\phi\sqrt{f_c'} = 8(0.85)\sqrt{5000} \qquad \text{Code limit}$$
$$= 0.481 \text{ ksi} > 0.41 \text{ ksi}$$

The suggested sizing procedure is

- Estimate the maximum probable reinforcing program.
- Combine it with the likely dead and live loads to create a shear demand.
- Then develop the beam dimensions based on an objective shear stress limit (ϕv_s) on the order of $5\sqrt{f_c'}$.

2.1.2.4 Column Shear Demand The shear imposed on a column can reach undesirable levels, especially when the columns are short. Compressive loads tend to increase the shear capacity of members subjected to flexural loads, while tension reduces their ability to sustain shears. It is reasonable in the conceptual design phase to neglect axial loads and use the same objective shear limit state proposed for beams $(\phi v_n = \pm 5\sqrt{f_c'})$.

Short columns will be subjected to higher levels of shear, and the amount of shear will be a function of the beam reinforcement. A capacity-based approach has as its objective the creation of a column that is stronger than the beam. This will not always be the case insofar as the flexural capacity of the column is concerned (see Chapter 3), but the shear strength of the column should always exceed the demands imposed upon it regardless of the source. Because it is not possible to accurately predict the shear demand that the column might experience, a conservative approach is advised. The shear demand on a column will also be a function of the column's location in the building. Three regions need to be considered, the lowest level, intermediate levels, and the upper level.

The shear imposed on the midlevel (intermediate) column described in Figure 2.1.26 will reach a maximum value when plastic hinges form in the adjoining frame beams at the face of the column. The generalized relationship between beam moments and column shear is

$$V_c = \frac{\left[A_s'(\ell_1/\ell_{c1}) + A_s(\ell_2/\ell_{c2})\right](d - d')\lambda_o f_y}{(h_{x1}/2) + (h_{x2}/2)} \tag{2.1.22}$$

The reverse action must also be considered.

The center-to-center beam length divided by the clear span can be reasonably replaced by 1.1 (1/0.9). These simplifications produce the following relationship:

$$V_c = \frac{1.1(A_s' + A_s)(d - d')\lambda_o f_y}{h_x} \tag{2.1.23}$$

To account for dynamic actions and other uncertainties, Paulay and Priestley[2.5, Sec. 4.67] recommend an increase of 30%, and this is reasonable. Hence

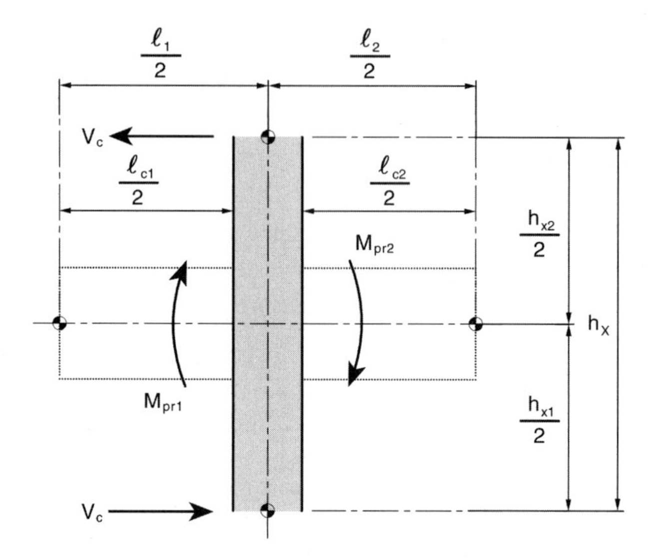

Figure 2.1.26 Actions imposed on an interior column.

$$V_{c,\mathrm{pr}} = 1.3V_c \tag{2.1.24}$$

Introducing our adopted value for λ_o (1.25), Eq. 2.1.23 can be reduced to

$$V_{c,\mathrm{pr}} = \frac{1.8(A_s + A_s')(d - d')f_y}{h_x} \tag{2.1.25}$$

which for Grade 60 steel becomes

$$V_{c,\mathrm{pr}} = \frac{108(A_s + A_s')(d - d')}{h_x} \tag{2.1.26}$$

Our example beam, reinforced with eight #11s top and bottom, would generate a column "design" shear of

$$V_{c,\mathrm{pr}} = \frac{108(2)(8)(1.56)(36 - 3)}{h_x}$$

$$= \frac{89,000}{h_x}$$

where h_x is, in this case, in inches. For a story height of 8 ft, 6 in., this amounts to

$$V_{c,\mathrm{pr}} = 872 \text{ kips}$$

This relationship can, as shown in the preceding sections, be reduced to one that suggests a minimum column area ($b_c h_c$) adopting a stress limit state (ϕv_m) of $5\sqrt{f'_c}$:

$$b_c h_c = \frac{22(A'_s + A_s)(d - d')}{h_x \sqrt{f'_c}} \tag{2.1.27}$$

For a concrete strength of 5000 psi, this becomes

$$b_c h_c \geq \frac{300(A'_s + A_s)(d - d')}{h_x} \tag{2.1.28}$$

which, for our example beam, becomes

$$b_c h_c = 2422 \text{ in.}^2 \qquad \text{(49 in. by 49 in.)}$$

Conclusion: Column shear stress levels need to be considered when beams are heavily reinforced and story heights are short. Otherwise, column shear is not likely to be a problem.

The shear demand imposed on columns in the upper and lower levels will not exclusively be a function of the beam capacity. Rather, the shear demand in the uppermost and lowermost columns must recognize that plastic hinges will probably form in the columns. If this column hinging is likely, the probable flexural strength of the column will define the shear demand, and this shear demand should be accommodated especially in the lowermost columns because if shear deterioration is allowed, the ductility of the column will be significantly reduced. This topic is addressed in Section 2.2 when we discuss column behavior.

2.1.2.5 Available Ductility The frame beam must be capable of deforming to the level of the story drift anticipated by the design without exceeding objective compressive strain levels. Ductility is reduced as flexural reinforcement is added, and this is because the neutral axis depth moves toward the center of the beam, thereby increasing the strain in the concrete ($\varepsilon_c = \phi c$). The fact that the beam is doubly reinforced, even to the extent that A'_s is equal to A_s, does not alter this shift in the neutral axis because the compressive steel cannot attain overstrength levels of stress in compression.

Maximum reinforcement ratios (ρ) of 2.5% and 1.5% are suggested by ACI[2.6] and Paulay and Priestley,[2.5] respectively.

Tests discussed in Section 2.3 (Beam-Column Joints) show that the compressive strain in the reinforcing near the face of the column is quite small (Figure 2.3.13 through 2.3.15). Accordingly, it is not unreasonably conservative to neglect the contribution of compression bars entirely.

Consider the available ductility given these maxima and presuming that $C_s = 0$. Following strength-based theory,[2.3]

$$c = \frac{a}{\beta_1} = \frac{\lambda_o A_s f_y}{0.85 f'_c b \beta_1} \tag{2.1.29}$$

If the probable level of rotation in the plastic hinge region is θ_p, the curvature is

$$\phi_p = \frac{\theta_p}{\ell_p} \qquad \text{(Eq. 2.1.9)}$$

and, as previously developed (Section 2.1.1.1), assuming that $\ell_p = 0.5h$,

$$\phi_p = \frac{\theta_p}{0.5h} \qquad \text{(see Figure 2.1.10)}$$

the concrete strain following Figure 2.1.5 is

$$\begin{aligned} \varepsilon_{cp} &= \phi_p c \\ &= \frac{\theta_p}{0.5h} \frac{\lambda_o A_s f_y}{0.85 f_c' b \beta_1} \end{aligned} \qquad (2.1.30)$$

and reducing this for design purposes, presuming that the concrete strength is 5000 psi and that $d \cong 0.9h$,

$$\varepsilon_{cp} = 40\theta_p \rho \qquad (2.1.31)$$

If an ultimate concrete strain objective (ε_{cu}) of 0.01 in./in. ($\varepsilon_{cp} = 0.009$ in./in.) is adopted, then

$$\rho_{\max} = \frac{0.000225}{\theta_p}$$

and for a postyield rotation (θ_p)of 1.5%,

$$\rho_{\max} = 0.015$$

The assumptions made in the development of this limit state for the reinforcing ratio (ρ) are obviously conservative, but they do suggest that the 1.5% suggested by Paulay and Priestley[2.5] is more rational than the 2.5% proposed by ACI.[2.6] The designer of frame beams will soon discover that reinforcement ratios in excess of 1.5% will be difficult to construct and will only be feasible if the column is very large so as to effectively control joint shear stresses.

If we treat the effectiveness of the compression reinforcement ($\lambda_e f_y$) as a variable, then

$$\varepsilon_{cp} = 32\theta_p(\lambda_o - \lambda_e)\rho \qquad (2.1.32)$$

and, for $\varepsilon_{cp} = 0.009$ in./in. and θ_p, a more realistic postyield deformation limit state of 2.5% is

$$\rho_{\max} = \frac{0.01125}{\lambda_o - \lambda_e} \qquad (2.1.33)$$

A more realistic assessment of λ_e may be derived from Reference 2.15. The effective stress range (λ_e) for outer layer reinforcing bars is between 0.33 and 0.5. Pessimistically, the maximum reinforcement ratio, as derived from Eq. 2.1.33, is 1.2% and optimistically 1.67%.

Conclusion: The maximum reinforcement ratio ($\rho_{\max}$) used in the design of frame beams should be 1.5% in order to maintain a reasonable level of ductility.

2.1.2.6 Design Process Summary

Step 1: Determine a Trial Value for $(A_s + A'_s)$ *and* d. It is recommended that $A_s = A'_s$ be presumed. For further discussion, see Section 2.1.3, which is devoted to analysis.

Step 2: Determine the Minimum Size of the Interior Beam-Column Joint.

$$A_j \geq 62\left(A_s + A'_s\right) \qquad \text{(Eq. 2.1.16)}$$

Step 3: Check the Feasibility of Placing the Reinforcement Suggested by the Trial Design. Use the reinforcement placing program developed in Section 2.1.2.2 to determine the minimum width of the beam (b).

Step 4: Check to Insure that the Reinforcement Ratio (ρ) *Does Not Exceed 1.5%.*

Step 5: Check the Shear Capacity/Demand Ratio to Insure That Objective Shear Stress Limit States Have Not Been Exceeded in the Beam.

$$V_{bu} = \frac{\left(A_s + A'_s\right)\lambda_o f_y (d - d')}{\ell_c} + \frac{P_D + P_L}{2} \qquad \text{(Eq. 2.1.20)}$$

$$bd \geq \frac{V_{bu}}{5\sqrt{f'_c}} \qquad \text{(see Eq. 2.1.21)}$$

Remember that P_D and P_L should, in this case, be realistically selected, not the factored loads used in a code compliance analysis.

Step 6: Check Column Shear If the Beam Is Deep and the Column Short.

$$b_c h_c \geq \frac{300\left(A_s + A'_s\right)(d - d')}{h_x} \qquad \text{(Eq. 2.1.28)}$$

This objective column size is based on a concrete strength of 5000 psi. Alternate concrete strengths will impact the column size according to the square root of the strength ratio.

2.1.2.7 Example Designs

Example 1

Beam length(ℓ)	30 ft
Story height(h_x)	12 ft

Column shear(V_{cE}) 250 kips (factored)
Maximum beam height(h) 36 in.

Step 1a: Determine Seismically Induced Shear in the Beam.

$$V_{bE} = \frac{V_{cE} h_x}{\ell}$$

$$= \frac{250(12)}{30}$$

$$= 100 \text{ kips}$$

Step 1b: Determine Required Flexural Strength of Beam.

$$M_{bE} = V_{bE} \frac{\ell_c}{2}$$

Assume a column dimension—probable is 3ft—to be conservative, use 2 ft.

$$M_{bE} = 100 \left(\frac{28}{2} \right)$$

$$= 1400 \text{ ft-kips}$$

Step 1c: Select Trial Beam Reinforcement.

$$A_s = \frac{M_{bE}}{\phi_b f_y (d - d')}$$

$$= \frac{1400}{0.9(60)2.5}$$

$$= 10.37 \text{ in.}^2$$

Step 1d: Select Trial Reinforcement. Propose seven #11 Top and Bottom $\left(A_s = A_s' = 10.92 \text{ in.}^2\right)$.

Comment: Observe that dead and live loads have not been considered. Assume that service level dead (D) and live (L) loads are 4.5 kips/ft and 3 kips/ft—these are large loads.

Factored dead and live loads are

$$U = 0.75(1.4D + 1.7L) \qquad \text{(see Ref. 2.6, Eq. 9-2)}$$

Recall that the seismic load was a factored load.

$$U = 1.05D + 1.28L$$
$$= 8.6 \text{ kips/ft}$$

$$M_{b_{D+L}} = \frac{w_{D+L}\ell_c^2}{12}$$
$$= \frac{8.6(28)^2}{12}$$
$$= 562 \text{ ft-kips}$$

The design moment for top steel (A_s) based on ACI[2.6] Eq. 9-2 would be

$$M_{bT} = 1400 + 562$$
$$= 1962 \text{ ft-kips}$$

The design moment for the bottom steel (A_s') requires the use of ACI[2.6] Eq. 9-3, where $U = -0.9\ D + \ E$:

$$M_{bD} = \frac{0.9(4.5)(28)^2}{12}$$
$$= 265 \text{ ft-kips}$$
$$M_{bB} = 1400 - 265$$
$$= 1135 \text{ ft-kips}$$

Moment redistribution is allowed by the ACI Code (ACI 318-99, Section 8.4.1),[2.6] and suggested by others.[2.5, Sec. 4.3.4] The maximum redistribution allowed by the ACI is 20%, while Paulay and Priestely[2.5] suggest a limit of 30%. Use a 20% redistribution to attain our reinforcing objective ($A_s = A_s'$).

$$\text{redistributed moment} = 0.2\ (1962)$$
$$= 392 \text{ ft-kips}$$

Design moments are

$$\text{top (T)} \Rightarrow 1962 - 392 = 1570 \text{ ft-kips}$$
$$\text{bottom (B)} \Rightarrow 1135 + 392 = 1527 \text{ ft-kips}$$

What Have We Learned?

- Design moments for top and bottom steel can be essentially the same.
- Dead and live loads even in this extreme case have little impact on the design of the reinforcement ($1570/1400 = 1.12$).

Conclusion: If dead and live loads imposed on the beam are significant when compared to the seismic moment ($562/1400 = 0.4$), include a 15% allowance in the design moment and provide the same flexural strength top and bottom $\left(A'_s = A_s\right)$. *Remember this is design.* The analysis section that follows discusses these issues and justifies their adoption. In the design of precast frames, these redistribution concepts will be especially important because differing flexural strengths will be difficult to attain.

Step 1e: Consider the Impact of Dead and Live Loads. Assume that dead and live loads are significant and select a reinforcing program that provides eight #11 T&B. Our objective is to select a beam size that will satisfy our "capacity-based" objectives. Reinforcing quantities may, and hopefully will, be reduced in the final design.

Step 2: Size the Beam-Column Joint.

$$\begin{aligned} A_j &\geq 62\left(A_s + A'_s\right) \qquad \text{(Eq. 2.1.16)} \\ &\geq 62(2)(8)(1.56) \\ &\geq 1547 \text{ in.}^2 \end{aligned}$$

Conclusion: Minimum column size is 40 in. by 40 in.

Comment: If this column size is not possible, the number of frames proposed in the initial system design should be revisited. A 36 in. by 42 in. column $\left(A_j = 1512 \text{ in.}^2\right)$ would, in the final analysis, work and should be proposed, the 42-in. dimension being parallel to the frame beam.

Step 3: Develop Beam and Column Reinforcing Placement Program. The minimum beam width should be the focus, for beam width may be increased but cannot fall below the so-developed minimum width. Stacked two-bar bundles will require the widest beam and produce a gap that allows the required space between column bars ($1.5b_d$).

Four #11 bars	4 (1.375)	=	5.5 in.
Three spaces @ 2 in.	3 (2)	=	6.0 in.
#5 hoop ties	2 (0.625)	=	1.25 in.
Fire cover	2 (1.5)	=	3.0 in.
			15.75 in.

Conclusion: The absolute minimum width is 16 in.

Comment: Observe that this will not allow the column bars to be bundled; the proposed column width is 36 in. so this should not become a problem.

Step 4: Check Beam Reinforcement Ratio.

$$\rho = \frac{A_s}{bd}$$
$$= \frac{8(1.56)}{16(33)}$$
$$= 0.024$$

Conclusion: A 16-in. wide beam should not be recommended unless beam ductility demands are very low. Propose a 24- or 26-in. wide beam realizing that the quantity of steel proposed is high. For a 24-inch wide beam,

$$\rho = \frac{8(1.56)}{24(33)}$$
$$= 0.016$$

Step 5: Check Beam Shear Capacity for Code $\left(8\phi\sqrt{f_c'}\right)$ *and Objective Limit* $\left(5\sqrt{f_c'}\right)$ *Compliance.*

$$V_b = \frac{2A_s\lambda_o f_y(d-d')}{\ell_c} + \frac{w_D\ell_c + w_L\ell_c}{2} \qquad \text{(see Eq. 2.1.20 and Design Step 1d)}$$
$$= \frac{2(8)(1.56)(1.25)(60)(33-3)}{(30-3.5)(12)} + \frac{8.6(26.5)}{2}$$
$$= 177 + 114$$
$$= 291 \text{ kips}$$

$$v_{bu} = \frac{291}{24(33)} = 0.367 \text{ ksi} < 8\phi\sqrt{f_c'} \qquad \text{(Complies with code)}$$

Probable dead and live loads are

$$V_D = \frac{4.5(26.5)}{2}$$
$$= 60 \text{ kips}$$
$$V_L = \frac{0.3(26.5)}{2} \qquad \text{(10 psf)}$$
$$= 4 \text{ kips}$$
$$V_{b,\text{pr}} = 177 + 64$$
$$= 241 \text{ kips}$$

$$v_{b,\text{pr}} = \frac{241}{24(33)}$$

$$= 0.304 \text{ ksi} < 5\sqrt{f'_c} \qquad \text{(Complies with objective shear stress levels)}$$

Conclusions: A 24 in. by 36 in. beam should be proposed. The column size should be at least 36 in. by 42 in. The maximum reinforcement program should not exceed eight #11 bars.

Example 2: Residential Building, Unequal Spans The frame geometry described in Figure 2.1.27 is fairly typical in a multistory apartment or condominium building. The objective factored interior column shear at this story is 200 kips. The architectural program limits the beam height to 24 in. above the floor and the slab thickness is 8 in. Hence, a beam depth of 32 in. becomes an economic objective because a beam that extends below the slab will increase the cost of forming. Function demands that the beam width be a minimum. Dead and live, unfactored loads, are 1.5 kips/ft and 0.5 kips/ft, respectively.

The design procedure advocated (Section 2.1.2.6) can be adapted to this geometry quite easily if the design approach is practical.

Step 1a: Select a Trial Beam Reinforcement Program. The optimum reinforcing program, from a design and construction perspective, will provide the same steel in the top and bottom of the beam. This is especially important here because one objective is to minimize the width of the beam. The most effective solution form will convert the moment capacity of the beam to a resistive moment at the centerline of the column ($M_{bu,c\ell}$) and thereby develop the objective column shear (V_c):

$$\sum M_{bu,c\ell} = A_s \left(\frac{\ell_1}{\ell_{c1}} + \frac{\ell_2}{\ell_{c2}} \right) (d - d')\phi f_y \tag{2.1.34}$$

$$V_c h_x = \sum M_{bu,c\ell}$$

$$V_c = \frac{A_s}{h_x} \left(\frac{\ell_1}{\ell_{c1}} + \frac{\ell_2}{\ell_{c2}} \right) (d - d')\phi f_y \tag{2.1.35}$$

$$200 = \frac{A_s}{9(12)} \left(\frac{24}{21} + \frac{16}{13} \right) (29 - 3)(0.9)(60)$$

$$A_s = 6.5 \text{ in.}^2$$

Because of the critical nature of this all important design step, a simple check should be made to confirm the conclusions reached in solving Eq. 2.1.35:

$$V_c h_x \cong 2M_{bu} \tag{2.1.36}$$

$$M_{bu} = \frac{200(9)}{2}$$

$$= 900 \text{ ft-kips}$$

$$A_s = \frac{M_{bu}}{\phi f_y(d - d')}$$

$$= \frac{900(12)}{0.9(60)(29 - 3)}$$

$$= 7.7 \text{ in.}^2$$

Comment: This 18% variation is the result of reducing the moment demand to the face of the column in Eq. 2.1.35. The order of magnitude is confirmed, however, so the design may proceed.

Step 1b: Assess the Impact of Dead and Live Loads on the Design. The worst case situation will impose a full factored load (U) on the longer span beam and a minimum load on the shorter beam:

$$M_{b1} = \frac{(1.05(1.5) + 1.28(0.5))(21)^2}{12} \quad \text{(see Ref. 2.6, Eq. 9-2)}$$

$$= 81 \text{ ft-kips}$$

$$M_{b2} = \frac{0.9(1.5)(13)^2}{12} \quad \text{(see Ref. 2.6, Eq. 9-3)}$$

$$= 19 \text{ ft-kips}$$

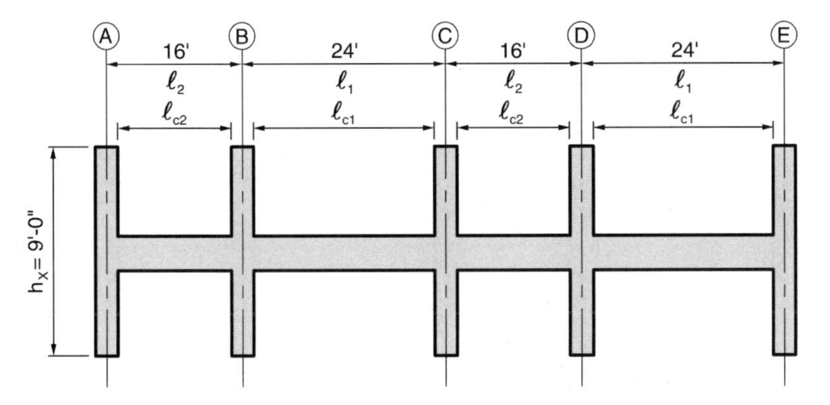

Figure 2.1.27 Frame geometry, unequal spans.

A first yield analysis would require a flexural capacity in the top of beam 1 of 982 ft-kips (+9%) and in the bottom of beam 2 of 881 ft-kips. Alternatively, we could redistribute the moments so as to attain a balanced increase.

The adjusted added increment of factored moment is

$$\Delta M_b = \frac{M_{b1} - M_{b2}}{2}$$
$$= \frac{82 - 19}{2}$$
$$= 31.5 \text{ ft-kips} \qquad (3.5\%)$$

The adjusted area of steel becomes

$$A_s = A_s' = 1.035(6.5)$$
$$= 6.73 \text{ in.}^2$$

Six #10 bars ($A_s = 7.62$ in.^{2}) should comfortably satisfy strength objectives. Remember a lesser amount of reinforcing may and probably will be used in the final analysis. The objective here is to decide how big the frame components must be—a further reduction in steel area will not adversely impact these decisions.

Step 2: Size the Beam-Column Joint.

$$A_j \geq 62\left(A_s + A_s'\right) \qquad \text{(Eq. 2.1.16)}$$
$$= 62(2)(6)(1.27)$$
$$= 945 \text{ in.}^2$$

Conclusion: The column size should be 30 in. by 32 in.

Comment: Observe that the 32-in. dimension matches the beam height and, as a consequence, should be aesthetically pleasing. Design engineers should be aware of aesthetic objectives.

Step 3: Develop Beam and Column Reinforcing Placement Program. Three rows of bundled or stacked #10 bars produce a minimum beam width.

Three #10 bars	3 (1.27)	=	3.81 in.
Two spaces @ 3 in.	2 (3)	=	6.0 in.
#5 hoop ties	2 (0.625)	=	1.25 in.
Fire cover	2 (1.5)	=	3.0 in.
			14.0 in.

Conclusion: The minimum beam width is 14 in.

Step 4: Check the Reinforcement Ratio (ρ).

$$\rho = \frac{6(1.27)}{14(29)}$$

$$= 0.019$$

Conclusion: Beam width should be at least 16 and preferably 18 in.

Step 5: Check the Shear Capacity of the Shorter Span (ℓ_2) *Using a Beam Width of 16 in.*

$$M_{b,\text{pr}} = A_s \lambda_o f_y (d - d')$$

$$= 6(1.27)(1.25)(60)(29 - 3)$$

$$= 14,860 \text{ in.-kips}$$

$$V_{bu} = \frac{2M_{b,\text{pr}}}{\ell_{c2}} + \frac{(w_D + w_L)\ell_c}{2}$$

$$= \frac{29,700}{192 - 32} + \frac{2.22(160)}{2(12)}$$

$$= 201 \text{ kips}$$

$$v_{b,\text{pr}} = \frac{201}{16(29)}$$

$$= 0.43 \text{ ksi} < 8\phi\sqrt{f_c'} \qquad \text{(Complies with code)}$$

$$V_{b,\text{pr}} = \frac{2M_{b,\text{pr}}}{\ell_{c2}} + V_{bD,\text{pr}}$$

$$= \frac{29,700}{192 - 32} + \frac{1.5(160)}{2(12)}$$

$$= 195.7 \text{ kips}$$

$$v_{b,\text{pr}} = \frac{195.7}{16(29)}$$

$$= 0.42 \text{ ksi} > 5\sqrt{f_c'} \qquad \text{(Does not meet our objective shear stress limit state)}$$

Conclusion: It is advisable to increase the beam width to 20 in.

Comment: Observe that this reduces our reinforcement ratio to 1.32%.

Step 6: Check Column Shear Capacity.

$$b_c h_c \geq \frac{300(A_s' + A_s)(d - d')}{h_x} \qquad \text{(Eq. 2.1.28)}$$

$$\geq \frac{300(2)(6)(1.27)(29 - 3)}{9(12)}$$

$$\geq 1101 \text{ in.}^2$$

$$30 \times 32 = 960 \text{ in.}^2 < 1101 \text{ in.}^2$$

Conclusion: It is advisable to increase the column size to at least 32 in. by 32 in. (1024 in.^{2}). This lower-than-objective column area can be rationalized based on the conservative assumptions used in the development of Eq. 2.1.28, specifically, the fact that the shear strength used did not consider the level of axial load and incorporated an uncertainty factor of 1.3. The analytical effort should insure that these factors are in fact conservative.

Proposed Design

Beam Size: 20 in. by 32 in. ($M_n = 1080$ ft-kips)
Beam Reinforcing: Six #10 bars top and bottom
Column Size: 32 in. by 32 in.

2.1.3 Analyzing the Frame Beam

Knowing how to conceptually design a frame beam so as to optimize its behavior is necessary but not sufficient. The conceived design must today be subjected to a careful and thorough analysis, the conclusions of which must be communicated to the builder. This section demonstrates how to avoid the many pitfalls that can convert a good conceptual design into one that is seriously flawed.

Today, the computer is king and its word the law of the land. Unfortunately, if the computer is used in a manner that is inconsistent with the design approach, the objectives of the design will not be attained. The analysis team must learn how to create a frame beam that closely resembles the one created in the conceptual design process. If, in the analysis phase, conclusions suggested are not consistent with the design, the designer and analyst must reconcile the differences and make sure that the final design produces the desired results.

The best way to demonstrate how far afield an analysis may go is by example. This also allows us to compare the behavior of the designed beam to one that might be created by an elastic-based analysis. The frame and loading program described in Figure 2.1.28 is the one that created the strength objectives used in Example 2 of the preceding Section 2.1.2.7. The designer who conceived the configuration described in Figure 2.1.28*a* did not know the exact beam and column sizes but had come to the conclusion that a four-bay frame would do the job—the procedures used to arrive

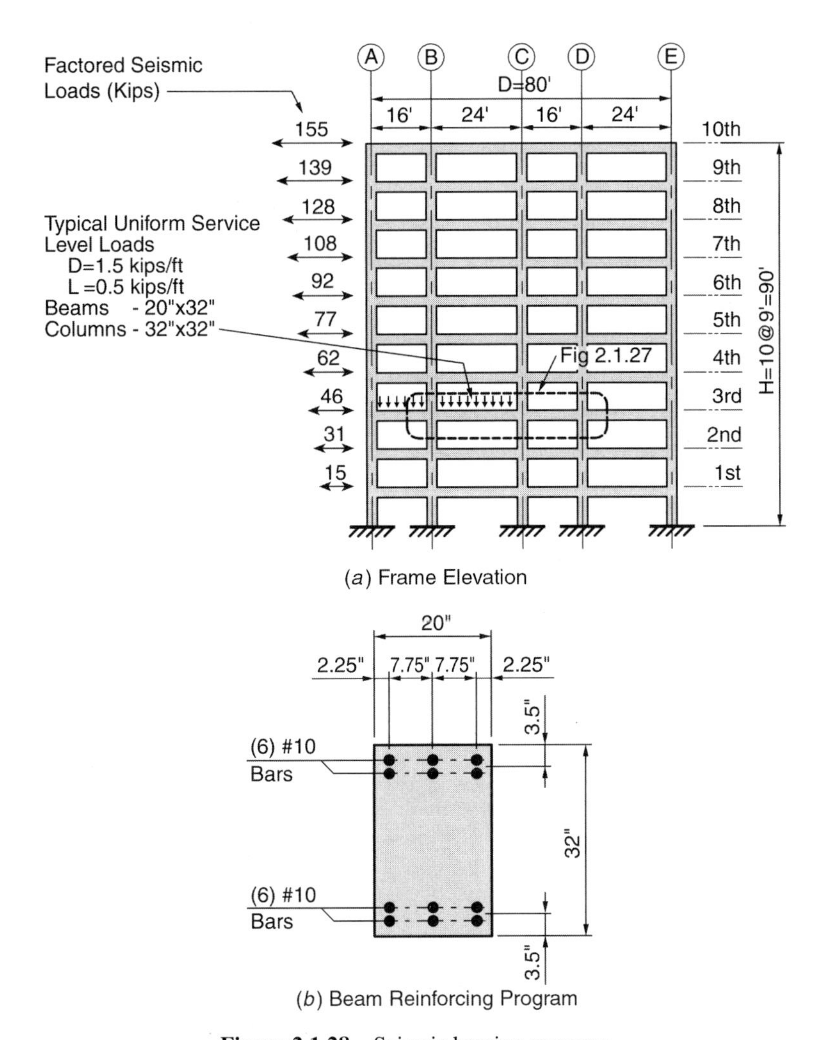

Figure 2.1.28 Seismic bracing program.

at this conclusion will be developed in Chapter 3. As we have seen in the preceding sections, the beam size and reinforcing program described in Figure 2.1.28*b* were the result of a conscientious effort to attain certain performance objectives, not the least of which was economy. A similar design process will be developed in Section 2.2 for the conceptual design of the beam columns, and this conceptual design step clearly must be completed before one proceeds with the analytical phase.

In the analysis phase, the first task will be to model the frame using the idealized stiffnesses previously developed (Section 2.1.1.1) or those believed to be appropriate for the system. Computer programs today take the drudgery out of the analysis and produce directly the critical moment demand for the frame beams. Essentially they select the critical combination of dead (D), live (L), and seismic (E) loads. Armed with this information, the analyst can use a software package or hand calculations to arrive at a flexural reinforcement program for each beam. A computer-generated factored design moment envelope is shown in Figure 2.1.29. Before we proceed, let us first identify the logical steps involved with the analysis of a frame beam.

2.1.3.1 Analysis Process Summary

Step 1: Check the Computer Program Used to Insure That the Output Is Appropriate.

- Figure 2.1.29 describes the output for the beam–column interface being considered. Figure 2.1.29*a* assumes that lateral loads are arriving from the left. Is that sufficient? No. For an unsymmetrical frame the lateral load must be assumed to

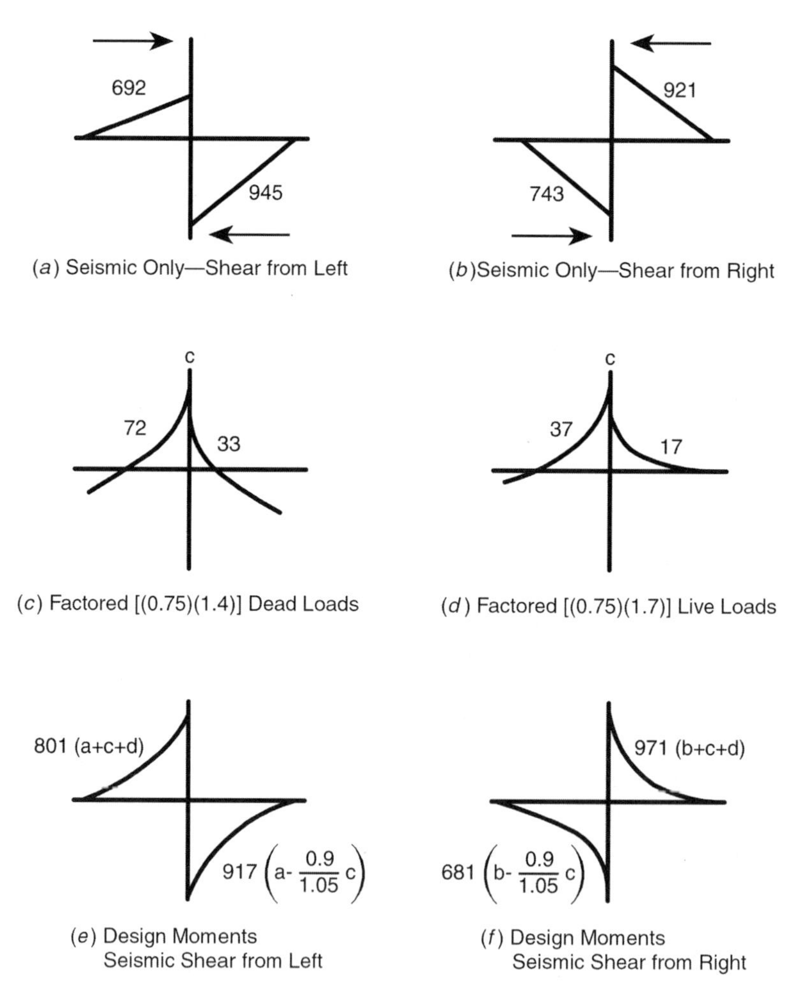

Figure 2.1.29 Moment (ft-kips) demands imposed on the beams at the center of column C, Level 3 (see Figure 2.1.28).

come from either direction. In this case observe that lateral loads arriving from the right (Figure 2.1.29*b*) create a seismic moment demand that is 33% higher than that generated by loads arriving from the left (Figure 2.1.29*a*).

- Are the moments generated at the centerline of the column or the face of the column? If they are developed at the column centerline, they may be reduced to the face of the column. Alternatively, they may be used in the design of the frame beam. The resultant design would be conservative, but this is acceptable provided every analysis step does not add to this conservatism.
- Were the appropriate load factors used to generate the design moments? We have been in a transitional state insofar as load factors are concerned for many years, and it does not appear that we will soon arrive at a consensus. Earthquake loads are now typically factored loads in most criteria. The ACI[2.6] load factors are

$$U = 1.05D + 1.28L + 1.4E \qquad \text{(see Ref. 2.6, Eq. 9-2)}$$

while the UBC[2.9] requires that their basic load factors,

$$U = 1.2D + 0.5L + E$$

be amplified by 10% for concrete buildings.[2.9, Sec. 1612.2.1] The ACI assumes that E, the earthquake load, is a service level load, while the UBC recognizes it as a factored load. Carefully check the load development process before proceeding. Errors undetected here will be fatal because the analytical efforts that follow will *all* be based on the flexural reinforcing program developed for the frame beam.

Comment: The likelihood is that the next generation of all codes will follow the UBC approach to load factors, and that the current 10% amplification factor[2.97] will be deleted. Concrete load factors will probably be revised also.

Step 2: Identify Your Objective.

- Do you wish to confirm the design or follow the dictates of the computer analysis? The objective should be to create a beam that can be simply constructed, one that will perform optimally during a seismic event. Recall that if we exceed the strength of the "designed" beam, we will need to repeat the design steps (Section 2.1.2), and this is counterproductive.
- What degree of accuracy is appropriate? Building codes must be complied with, however, seismic loads define an objective level of strength (Chapter 1) in a fairly arbitrary manner, and this level of strength is not as important as maximizing available ductility.
- How conservative should you be? The tendency is to say that stronger is better, but this is not necessarily the case in concrete frame beam design, for an increase in steel quantities not accompanied by a proportionate increase in beam width will result in a reduction in beam ductility. Further, it will increase the load

imposed on other components of the bracing program, potentially increasing the risk or likelihood of a failure in these less ductile elements.

Conclusion: Our objective should be to confirm the frame beam design, in this case, a 20-in. by 32-in. beam reinforced with six #10 bars top and bottom (Figure 2.1.28*b*).

Step 3: Develop the Design Moment. The designer should utilize moment redistribution to create as balanced a moment demand as possible. The result should be frame beams that do not contain heavily reinforced areas, for such areas would be more brittle and would define the deformation limit state for the building. Observe also that detailing and construction will be easier when reinforcing is not varied from point to point. How simple it is to instruct the contractor to, say, use six #10 bars top and bottom over the entire length of the frame beam. Would it not be great if this simple reinforcing program produced a bracing program whose behavior exceeded that of a more complex alternative? This turns out to be the case, so don't be afraid to be creative. Creativity is the essence of "design."

The concepts of moment redistribution are differently applied to a seismically dominated frame. The objective should be to balance the flexural reinforcement so that the top steel (A_s) is the same as the bottom steel (A_s') and yet the column shear remains equal to or greater than that imposed by seismic actions. The inclusion of dead and live loads somewhat complicates the process. Both procedures are most easily explained by example (see Section 2.1.3.2, Step 3).

Step 4: Develop a Flexural Reinforcing Program for the Entire Beam Length. Here the designer must be logical. Pick a program that is consistent throughout and satisfies minimum reinforcement requirements. Generally, minimum reinforcement ($A_{s,\min}$) requirements are 200 $b_w d/f_y$.[2.6, Sec. 21.3.2.1] This minimum may be reduced when the provided flexural reinforcement exceeds the analytical demand by 33%.[2.6, Sec. 10.5.3] Regardless, continuous reinforcing consisting of at least two bars top and bottom must be provided. The prescribed minimum ($A_{s,\min}$) should not override an intelligent bar selection because the exception to the rule (33%) can be dealt with in the detailing of the frame beam.

Comment: Prescriptive provisions that identify reinforcing minimums appear and disappear. The objective here is not to identify the current code requirements, but rather to foster the development of rational design and detailing procedures. The designer should at all times be rational. The rational designer will not, for example, terminate most of the flexural reinforcing at a point simply because a code allows it.

Step 5: Develop a Transverse Reinforcing Program. From this point forward in the analysis, the computer analysis that generated the moments based on the critical combinations of loading is of no use, and this is a fundamental tenet of capacity-based design. Critical shears used in the design of frame beams are functions of the probable moments (Section 2.1.2.3).

Key elements are

- The plastic hinge region (Figure 2.1.25)
- Splices in the flexural reinforcement
- Lateral support of compression bars
- Detailing for constructibility

Step 6: Review Strain States. This step need only be performed on critical members. The objective will be to determine the probable strain states given an identified frame deformation state (Chapter 3 and 4). This will be an important consideration in performance-based design wherein the client would be apprised of the probable condition of the building should it experience the design level earthquake. Strain states will also help the designer develop the transverse reinforcing program for bars located inside the outer bar layout. (See, for example, Figure 2.1.28*b*).

2.1.3.2 Example Analysis The frame described in Figure 2.1.28*a* is now analyzed. The beam described in Figure 2.1.28*b* has been developed by the conceptual design team. Our objective is to determine the effectiveness of this frame beam size and reinforcing program.

Step 1: Check Computer Program.

- Both loading conditions (direction of applied shear) appear to have been considered.
- Appropriate member stiffnesses have been used.
- The appropriate load factors have been used.
- Computer-generated moments are of the same order of magnitude of those generated during the conceptual design phase. If not, the discrepancies should be resolved before proceeding with the analysis.
- Moments generated are those at the centerline of the column.

Step 2: Determine the Objective. Confirm the conceptual design.

Step 3: Develop the Design Moment. Taken at face value the seismic actions described in Figures 2.1.29*a* and *b* suggest that both the top and bottom reinforcement should be essentially the same, for the demands are similar (945 and 921). Observe, however, that this would create a significant overcapacity in the beam on the opposite side of the column during the same seismic action. Were we to design to a beam capacity of 945 ft-kips, the resultant column shear at the design strength ($M_{bu} = 945$ ft-kips) would be

$$V_{cu} = \frac{2M_{bu}}{h_x}$$

$$= \frac{2(945)}{9}$$

$$= 210 \text{ kips}$$

The column shear required to satisfy our strength criterion is only

$$V_{cu} = \frac{M_{buL} + M_{buR}}{h}$$

$$= \frac{743 + 921}{9}$$

$$= 185 \text{ kips}$$

The thus proposed strength (210 kips) is 14% greater than the objective strength.

Comment: Recall that the moments used are those at the centerline of the column. The provided column or subassembly shear will be at least 10% higher for the strength of the beam at the face of the column will define the strength of the subassembly.

We could maintain a resistance that matches the demand by redistributing the beam moments:

$$M_{bu,\text{design}} = \frac{M_{buL} + M_{buR}}{2} \tag{2.1.37}$$

$$= \frac{743 + 921}{2}$$

$$= 832 \text{ ft-kips}$$

This reduction would do two things that can be viewed as positive: it would reduce the demand on the other elements along the load path while it increased the level of ductility in the beam.

Now consider the impact of the factored dead and live loads on the required strength (Figures 2.1.29*e* and *f*). The required design moments, taken at face value, are 971 ft-kips top and 917 ft-kips bottom. Moment redistribution changes these design moments to

$\frac{971 + 681}{2} = 826$ ft-kips Shear from left (see Eq. 2.1.37)

$\frac{801 + 917}{2} = 859$ ft-kips Shear from right (see Eq. 2.1.37)

Consider the approximate capacity of the proposed beam (Figure 2.1.28*b*):

$$M_u = \phi A_s f_y (d - d') \tag{2.1.38}$$

$$= \frac{0.9(6)(1.27)(60)(28.5 - 3.5)}{12}$$

$$= 857 \text{ ft-kips}$$

This frame beam capacity of 857 ft-kips does not satisfy the moment demand of 971 ft-kips (Figure 2.1.29*f*), but it does satisfy the redistributed moment demand of 859 ft-kips. The analyst must consider alternatives and conservatisms built into the design before making a final decision.

What are the alternatives and conservatisms?

(a) Increase the amount of reinforcement. If six #11 bars were used, the capacity would be 1053 ft-kips, and all flexural strength objectives would be clearly satisfied. A simple design decision, but it is likely to overstress the beam-column joint and thereby require a larger column as well as a regeneration of the frame analysis. The reinforcing programs for the other frame components, as previously pointed out, are all functions of the beam reinforcing program adopted, so the reinforcing costs will be significantly impacted. Further, the ductility of the beam will be decreased unless the beam width is proportionately increased. A 24-in. wide beam would be required if comparable ductility is desired.

Conclusion: This alternative is not attractive.

(b) Reduce the moment to the face of the column. This is possible; however, care must be taken because the moment envelopes described in Figure 2.1.29 come from a variety of loading conditions. An approximate impact can be developed from the design seismic moment of 945 ft-kips. The seismic moment at the opposite end is 920 ft-kips. This creates a seismic shear along the beam of

$$V_{bE} = \frac{M_{bEL} + M_{bER}}{\ell_c} \tag{2.1.39}$$

$$= \frac{(945 + 920)}{16}$$

$$= 117 \text{ kips}$$

Since this shear is essentially constant over the length of the beam, the resultant change in moment (ΔM_u) from the centerline of the column to the face of the column will be

$$\Delta M_u = \frac{V_u h}{2} \tag{2.1.40}$$

$$= \frac{117(32)}{12(2)}$$

$$= 156 \text{ ft-kips}$$

where h is the dimension of the column in the direction of the frame.

The resultant undistributed demand is now less than the capacity, and it appears as though the conceptual solution comfortably satisfies our strength objective.

(c) A strain-based analysis that considers the probable strength of the reinforcing ($f_y = 66$ ksi) may be performed on the beam section. This will define another conservatism. Computer programs are available that describe flexural capacities as a function of strain states and curvatures. Table 2.1.1 describes the

TABLE 2.1.1 Analysis of a 20-in. by 32-in. Beam Section (See Figure 2.1.28*b*)

Concrete Strain (in./in.)	Neutral Axis Depth (in.)	Steel Strain (in./in.)	Moment Capacity M_n (ft-kips)	Curvature (rad/in.)
0.0001	16.0	–0.00007	148.4	6.25e-06
0.0002	10.1	–0.00035	183.8	1.99e-05
0.0003	9.3	–0.00060	281.4	3.21e-05
0.0004	9.2	–0.00082	378.6	4.37e-05
0.0005	9.0	–0.00104	479.1	5.54e-05
0.0006	9.0	–0.00125	572.9	6.64e-05
0.0007	9.1	–0.00145	663.2	7.73e-05
0.0008	9.0	–0.00168	763.5	8.90e-05
0.0009	9.0	–0.00188	854.0	9.98e-05
0.0010	9.1	–0.00208	942.8	1.10e-04
0.0011	9.1	–0.00228	1027.2	1.21e-04
0.0012	8.4	–0.00279	1033.3	1.43e-04
0.0013	7.9	–0.00327	1043.2	1.64e-04
0.0014	7.4	–0.00387	1042.8	1.89e-04
0.0015	7.1	–0.00439	1046.6	2.11e-04
0.0016	6.7	–0.00505	1050.0	2.39e-04
0.0017	6.5	–0.00562	1051.6	2.63e-04
0.0018	6.3	–0.00617	1053.1	2.86e-04
0.0019	6.0	–0.00687	1058.8	3.15e-04
0.0020	5.9	–0.00746	1058.7	3.40e-04
0.0025	5.4	–0.01045	1058.1	4.65e-04
0.0030	5.1	–0.01332	1056.8	5.86e-04
0.0035	5.0	–0.01606	1079.0	7.02e-04
0.0040	4.9	–0.01870	1077.1	8.15e-04
0.0045	4.9	–0.02128	1075.0	9.25e-04
0.0050	4.9	–0.02324	1067.3	1.01e-04
0.0060	5.0	–0.02755	1056.7	1.20e-03
0.0070	5.0	–0.03185	1049.4	1.39e-03
0.0080	5.1	–0.03602	1130.9	1.58e-03
0.0090	5.1	–0.04028	1144.8	1.77e-03
0.0100	5.1	–0.04431	1190.1	1.95e-03
0.0120	5.1	–0.05412	1234.4	2.37e-03
0.0140	5.1	–0.06281	1291.4	2.76e-03
0.0160	5.3	–0.06758	1358.1	3.00e-03
0.0180	5.8	–0.06909	1378.6	3.13e-03
0.0200	5.8	–0.07611	1409.1	3.45e-03
0.0220	5.8	–0.08301	1433.1	3.77e-03
0.0240	5.9	–0.08991	1452.7	4.09e-03
0.0260	5.9	–0.09685	1468.8	4.41e-03

output for the beam of Figure 2.1.28*b*. This program uses the Mander[2.5, 2.16] concrete model wherein the peak unconfined concrete stress occurs at a strain of 0.0025 in./in. A concrete strain state of 0.0025 in./in. suggests a nominal moment capacity of 1058 ft-kips, and this is consistent ($M_u = 952$ ft-kips) with our elastic-based demand of 971 ft-kips.

Conclusion: The beam flexural reinforcement program described in Figure 2.1.28*b* attains our strength objectives.

Step 4: Develop a Flexural Reinforcement Program for the Entire Beam Length. Minimum flexural reinforcement is

$$A_{s,\min} = \frac{200 b_w d}{f_y} \tag{2.1.41}$$

$$= \frac{200(20)(28.5)}{60,000}$$

$$- 1.9 \text{ in.}^2$$

Conclusion: At least two #10 bars ($A_s = 2.54$ in.^{2}) should be continuous. The corner bars are appropriate.

Top and bottom flexural reinforcement must provide sufficient strength to satisfy dead and live load moment demands. The strength demand at the supports is clearly not a problem, and midspan demands are easily checked.

$$w_{u(D+L)} = 1.4D + 1.7L \qquad \text{(see Ref. 2.6, Eq. 9-1)}$$

$$= 1.4(1.5) + 1.7(0.5)$$

$$= 2.95 \text{ kips/ft}$$

$$M_{u(D+L)} = \frac{w_{u(D+L)}\ell_c^2}{16} \qquad \text{(Maximum condition)}$$

$$= \frac{2.95(21.33)^2}{16}$$

$$= 84 \text{ ft-kips}$$

Conclusion: This condition is clearly satisfied by the minimum flexural reinforcement requirement.

Splicing and bar cut-offs are critical to any flexural reinforcing program. Splices should not be located in the vicinity of a plastic hinge. The ACI code does not permit splices within 2*d* of the face of the column. All reinforcement must be developed beyond the point at which peak stress, in this case yielding, occurs. Critical to these determinations is the bar development length (ℓ_d). A discussion of development length is contained in the ACI code,[2.6] and in Paulay and Priestley.[2.5]

For our purposes we assume that the conditions allow the use of a modified version of ACI[2.6] Eq. 12–1:

$$\frac{\ell_d}{d_b} = \frac{f_y \alpha \beta \lambda}{20\sqrt{f'_c}} \qquad \text{(ACI,[2.6] Sec. 12.2.2)}$$

where

$$\begin{aligned} \alpha &= 1.3 \quad \text{For top bars} \\ &= 1.0 \quad \text{For bottom bars} \\ \beta &= \text{Coating factor} \\ &= 1.0 \quad \text{(Uncoated bars)} \\ \lambda &= \text{Lightweight aggregate concrete factor} \\ &= 1.0 \quad \text{(Normal weight concrete)} \end{aligned}$$

Hence

$$\begin{aligned} \frac{\ell_d}{d_b} &= \frac{60,000(1.0)(1.0)(1.0)}{20\sqrt{5000}} \\ &= 42.4 \quad \text{For bottom bars} \\ &= 55.2 \quad \text{For top bars} \end{aligned}$$

For #10 bars ($d_b = 1.27$ in.), the development lengths become

$$\begin{aligned} \ell_d &= 54 \text{ in.} \quad \text{Bottom bars} \\ &= 70 \text{ in.} \quad \text{Top bars} \end{aligned}$$

The required development length starts at the face of the column given the ACI[2.6] criterion. Paulay and Priestley[2.5] recommend that the development length start at the inboard edge of the plastic hinge, and this seems more logical (see Figure 2.1.9). The first cut-off point should be located, given the Paulay and Priestley formulation, at a distance

$$\begin{aligned} d + \ell_d &= 28.5 + 54 \\ &= 82.5 \text{ in.} \quad \text{Bottom bars} \\ &= 98.5 \text{ in.} \quad \text{Top bars} \end{aligned}$$

from the face of the column; this is somewhat conservative because the presumed hinge length is on the order of $0.5h$.

Splice lengths are also a function of the development length. Class A[2.6] splices (ℓ_d) are allowed when only 50% of the bars are spliced at one location. Otherwise Class B splices are required ($1.3\ell_d$). For #10 bars these development lengths become

$$
\begin{aligned}
\ell_d &= 54 \text{ in.} \qquad \text{Bottom bar—Class A} \\
&= 70 \text{ in.} \qquad \text{Top bar—Class A} \\
&= 70 \text{ in.} \qquad \text{Bottom bar—Class B} \\
&= 91 \text{ in.} \qquad \text{Top bar—Class B}
\end{aligned}
$$

The clear spans are

$$
\begin{aligned}
\ell_c &= 160 \text{ in.} \qquad \text{(16-ft bay)} \\
\ell_c &= 256 \text{ in.} \qquad \text{(24-ft bay)}
\end{aligned}
$$

This suggests that lap splices should be restricted to the 24-ft bays. Bar placement programs vary according to the designer's attitude or disposition and office detailing practice. As more rules are developed impossible situations will arise and, to a certain extent, compromises will have to be made. For example, suppose that the frame bays were all 16 ft. Then lap splices, if used, would have to be placed in this short a span. Detailing should then endeavor to compensate for an inability to comply with the prescriptive requirements in favor of a behavior-based approach; one that understands the objectives of the rules.

The development described in Figure 2.1.30 might reasonably be adopted. Two bottom bars could be terminated 83 in. (82.5 in.) from the face of column B. Two more bars could be terminated 96 in. from the face of column B because one might reasonably assume that the required strength one-third of the way to the point of inflection, presumed to be at midspan, would be only two-thirds of that required at the face of the column:

$$
\begin{aligned}
\frac{\ell_c}{6} + \ell_d &= 42 + 54 \\
&= 96 \text{ in.} \qquad \text{Bottom bars} \\
&= 112 \text{ in.} \qquad \text{Top bars}
\end{aligned}
$$

Splices for the continuous bars should occur between bays B–C and D–E. Class A splices are not possible since this would require a bar length in excess of 60 ft (mill order maximum length). Further, long bars with hooks on the end are very difficult to place. It is advisable to stagger the splice locations by 24 in., each splice being centered 12 in. from the centerline of the beam.

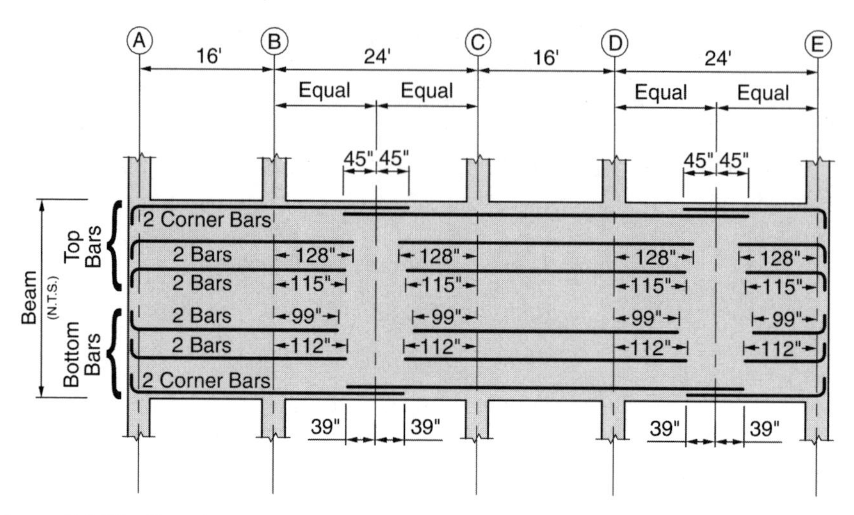

Figure 2.1.30 Bar placement program for the example beam of Figure 2.1.27. (For bar placement within the beam see Figure 2.1.28*b*.)

Many designers would opt to make four bars continuous, while others might require Type 1 mechanical splices.[2.6, Sec. 21.2.6.1] When the number of frame beams is large, a carefully detailed reinforcement program can reduce the cost of reinforcement significantly while it makes the shop drawing review process simpler.

Comment: The objective here is to identify detailing considerations only. A comprehensive treatment of the subject is left to others.

Step 5: Develop a Transverse Reinforcing Program. Now we switch to a capacity-based approach. Shear strength must exceed the shear demand created by the probable moment demand. An overstrength factor of 1.25 is normally used to generate the probable moment capacity (M_{pr}, see Figure 2.1.2):

$$M_{\text{pr}} = \lambda_o M_n \tag{2.1.42a}$$

The nominal moment capacity is 952 ft-kips (see Step 3, Eq. 2.1.38).

$$\begin{aligned} M_{\text{pr}} &= 1.25\,(952) \\ &= 1190 \text{ ft-kips} \end{aligned}$$

Observe (Table 2.1.1) that this probable moment corresponds to a concrete strain of 0.01 in/in. and that the corresponding postyield hinge rotation ($\phi \ell_p$) is 2.8%, and this is consistent with what might reasonably be expected (Chapter 3) and with the experimental data reviewed in Section 2.1.1.

The seismic shear induced in the 16-ft span is

$$V_{uE} = \frac{\lambda_o 2M_n}{\ell_c} \tag{2.1.42b}$$

$$= \frac{2(1190)}{16 - 2.67}$$

$$= 178.5 \text{ kips}$$

to which must be added the factored dead and live loads. Following the ACI,

$$U = 1.05D + 1.28L \qquad \text{(see Ref. 2.6, Eq. 9-2)}$$

$$= 1.05(1.5) + 1.28(0.5)$$

$$= 2.2 \text{ kips/ft}$$

Comment: A live load factor of 1.28 is unrealistically high, especially since live loads to frame beam members cannot, following code mandates, be reduced. The 0.5 live load factor and the 1.2 dead load factor used by the UBC[2.9, Section 1612.2.1] is more realistic, and will probably soon be adopted by the ACI.

The example that follows uses ACI Eq. 9-2.

$$V_u = V_{uE} + \frac{w_{u(D+L)}\ell_c}{2} \tag{2.1.42c}$$

$$= 178.5 + 14.6$$

$$= 193 \text{ kips}$$

It is most convenient to design transverse reinforcing using nominal stresses.

$$V_n = \frac{V_u}{\phi}$$

$$= \frac{193}{0.85}$$

$$= 227 \text{ kips}$$

$$v_n = \frac{V_n}{bd}$$

$$= \frac{227}{20(28.5)}$$

$$= 0.398 \text{ ksi}$$

The design shear strength required in the code defined[2.9] hinge region ($2d$) must be provided entirely by the transverse reinforcing.

$$A_v = \frac{v_s bs}{f_y} \tag{2.1.42d}$$

$$= \frac{0.398(20)s}{(60)}$$

$$= 0.133(s) \text{ in.}^2$$

Detailing considerations were discussed in Section 2.1.1.4 and examples developed in Sections 2.1.2.2 and 2.1.2.7. A practical approach is followed in this example because the objective is to assure optimal behavior in the beam, comply with the current code (ACI), and not expend construction dollars needlessly.

Hinge Region This is assumed here conservatively to be $0.75h$ (24 in.).

- Objective confining pressure is 450 psi (see Eq. 1.2.7, $f'_c = 5$ ksi). The maximum longitudinal tie spacing (s) should be 6 in. (Ref. Section 1.2.1):

$$A_v = \frac{v_s b_h s}{f_y}$$

$$= \frac{0.45(16.8)s}{60}$$

$$= 0.13(s) \text{ in.}^2$$

  where b_h is the out-to-out width of the confining hoop (see Figure 2.1.31).
- Objective hoop spacing is 4 to 6 in. on center.
- A stacked or bundled three-bar arrangement seems most practical to develop the adopted six-bar reinforcing program described in Figure 2.1.30. The interior bar should be laterally supported in the hinge region, and this suggests three vertical stirrups or hoop legs.
- Desired hoop configuration is a single hoop plus one added vertical leg:

$$s = \frac{A_{v,\text{provided}}}{A_{v,\text{required}}} \tag{2.1.43}$$

$$s = \frac{3(0.31)}{0.13} \qquad \text{\#5 ties, three legs}$$

$$= 7.2 \text{ in. on center} > 6 \text{ in.}$$

Code Defined Hinge Region The required hinge length is $2d$ (57 in.):

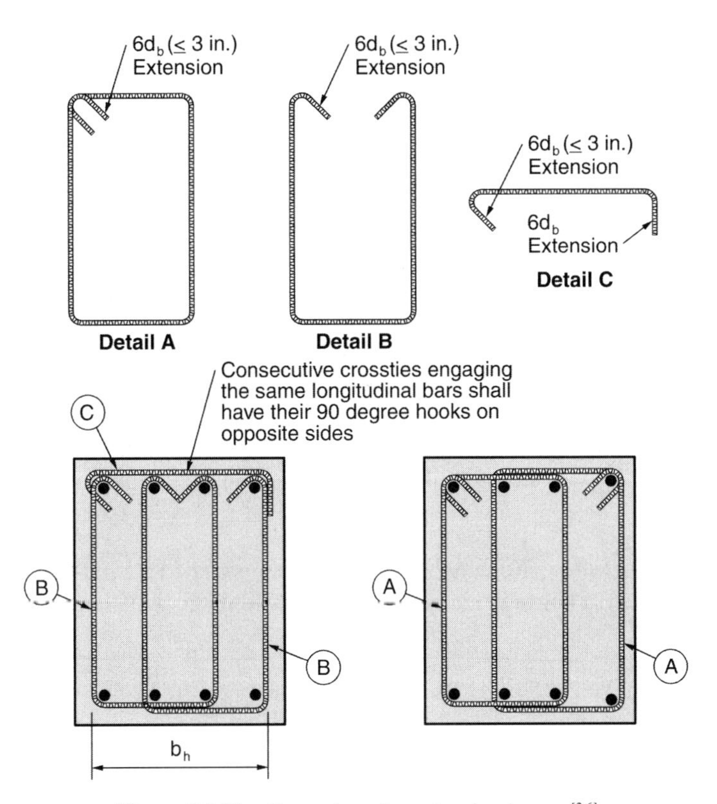

Figure 2.1.31 Examples of overlapping hoops.[2.6]

$$s = \frac{3(0.31)}{0.133} \qquad \text{(see Eq. 2.1.42d)}$$

$$= 7 \text{ in. on center}$$

Accordingly, one hoop at 2 in. from the face of the support and 12 spaces at 7 in. on center for #5 hoops having three legs would comply with code mandates.

Reinforcing outside the plastic hinge region would be developed as follows:

$$v_s = v_n - 2\sqrt{f_c'}$$

$$= 0.398 - 2\sqrt{5000}$$

$$= 0.257 \text{ ksi} < 4\sqrt{f_c'}$$

Comment: Stirrup spacing must be less than $d/2$ or 14 in.

$$A_{v,\text{required}} = \frac{v_s bs}{f_y}$$

$$= \frac{0.257(20)(s)}{60}$$

$$= 0.086(s) \text{ in.}^2$$

$$s = \frac{2(0.31)}{0.086}$$

$$= 7.2 \text{ in. on center} \qquad \# 5 \text{ stirrup}$$

Conclusion: Use a #5 hoop with three legs in the code defined hinge region ($2d$) and a #5 hoop elsewhere. Spacing of #5 hoops from the face of each column should be: three bar hoops—1 at 2 in., 4 at 6 in., 4 at 7in., remainder full hoops at 7 in. Since hoop ties are required in a lap splice region,[2.6, Sec. 21.3.2.3] the closed tie configuration that should be used is as described in Figure 2.1.31, tie types shown in details B and C. This will allow constructibility given the reinforcing program described in Figure 2.1.30.

Observe that each reinforcing bar is laterally supported in the plastic hinge region and that the spacing between longitudinal bars is within the limits suggested (Section 2.1.1.4).

Step 6: Review Strain States. Postyield rotations imposed on the hinge regions of frame beams will be evaluated in Chapters 3 and 4. For now we can approximate the level of imposed strain and confirm the selection of our reinforcing program by studying a story drift and assuming that the column and the beam-column joint are rigid. Let us assume that the component of probable story drift imposed on the beam will be on the order of 2%. This means that a total seismic rotation on the order of 0.02 radian will be required of each plastic hinge (see Figure 2.1.32).

Table 2.1.1 will assist us in making an estimate of the compressive strain state in the concrete. We can also evaluate the impact of our moment redistribution program on the ultimate strain state in the concrete.

Consider first the probable dead and live load moment imposed on the beam:

$$M_{D+L} = \frac{w\ell_c^2}{12} \tag{2.1.44}$$

where w is, in this case, the unfactored dead load plus the likely unfactored live load. Conservatively this imposed load would be on the order of 1.6 kips/ft.

$$M_{D+L} = \frac{1.6(21.33)^2}{12} \qquad \text{(see Eq. 2.1.44)}$$

$$= 61 \text{ ft-kips}$$

This moment, taken at the face of the column, would create a compressive strain on the bottom of the beam and a tensile strain in the reinforcing in the top of the beam:

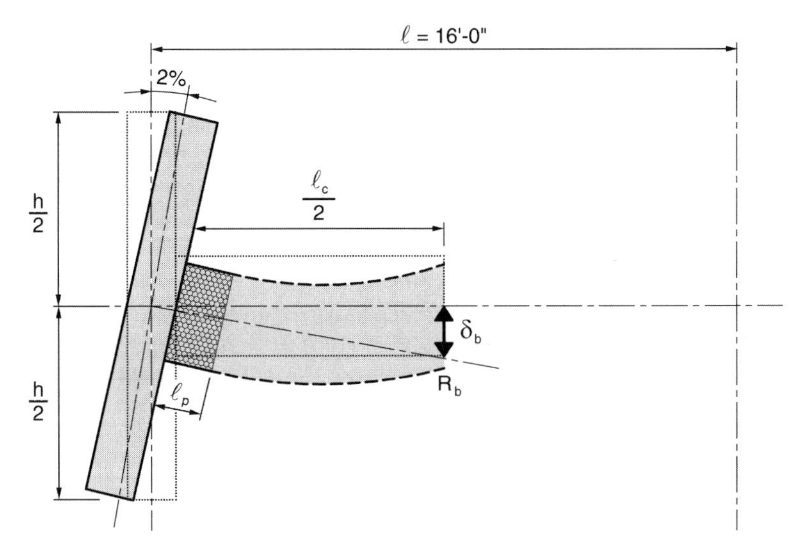

Figure 2.1.32 Subassembly subjected to a 2% story drift.

$$\varepsilon_s = 0.00013 \text{ in./in.} \qquad \left(\varepsilon_s = \frac{M}{kdA_sE_{so}}, \text{ see Figure 2.1.5}\right)$$

$$\varepsilon_c = 0.000065 \text{ in./in.}$$

These strains are very small in comparison with the strain states anticipated during the design seismic event. Observe that the vertical load induced strain state in the concrete is less than one-hundredth of the strain state associated with shell spalling. The seismic moments will also arrive at differing rates, but since the extant strain differential is small, the impact on behavior will be of no consequence (see Figure 2.1.29*b*). The probable moment differential, on the order of 180 ft-kips, is small when compared to the nominal strength ($\varepsilon_c = 0.0025$ in./in.) of 1058 ft-kips. Clearly these differences can be neglected. Recall that Paulay and Priestley[2.5, Sec. 4.3.4] suggest the use of a moment redistribution of up to 30% to achieve an efficient design.

The curvature in the hinge region is

$$\phi_p = \frac{\theta_p}{\ell_p} \qquad \text{(Eq. 2.1.9)}$$

If we conservatively assume that ℓ_p is $0.5h$ (Section 2.1.1.2), then

$$\phi_p = \frac{0.02}{16}$$

$$= 0.00125 \text{ rad/in.}$$

and this corresponds to a strain state in the concrete on the order of 0.0065 (see Table 2.1.1).

Conclusion: Shell spalling is not anticipated because the induced concrete strain is less than the threshold of spalling (0.007 in./in.). The indicated level of flexural overstrength (1050 ft-kips) does not approach the level assumed (1190 ft-kips) in developing the probable shear demand on the beam (see Eq. 2.1.42b).

Comment: A beam reinforcement program consisting of six #11 bars could reasonably have been adopted (Section 2.1.3.2, Step 3, Alternative a). If we were to predict the story drift associated with a concrete strain of 0.01 in./in. (upper range of shell spalling), we would conclude that the beam with six #10 bars would reach a story drift of

$$\theta \cong \phi \ell_p \qquad \text{(Eq. 2.1.9)}$$

$$\theta = 0.00195(16) \qquad \text{(see Table 2.1.1)}$$

$$= 0.031 \text{ radian} \qquad (3.1\%)$$

while that of the beam reinforced with six #11 bars could based on a similar analysis, only attain a story drift of about

$$\theta = 0.0016(16) \qquad \text{(Eq. 2.1.9)}$$

$$= 0.028 \text{ radian} \qquad (2.8\%)$$

In Chapter 4 we will investigate the dynamic response of alternative reinforcing programs and show that there is probably no advantage associated with using the stronger system.

Step 7: Estimate the Level of Provided Overstrength. As a final exercise we investigate the mechanism shear. The objective is to determine the level of system overstrength provided.

In order for the desired story mechanism to develop, plastic hinges must form in the end of each frame beam. Dead and live loads may be neglected since they do not impact sidesway or lateral mechanism loads (see Figure 2.1.45 and associated discussion in Section 2.1.4.3). The mechanism moment may be converted to a restoring moment (M_R) acting on the center of the column; hence

$$\sum M_R = \sum M_{\text{pr}} \left(4\frac{\ell_1}{\ell_{c1}} + 4\frac{\ell_2}{\ell_{c2}} \right) \qquad \text{(see Figures 2.1.27 and 2.1.45}c\text{)}$$

$$= 1190 \left(4 \left(\frac{288}{256} \right) + 4 \left(\frac{192}{160} \right) \right)$$

$$= 11{,}070 \text{ ft-kips}$$

The mechanism story shear is

$$V_M = \frac{\sum M_R}{h_x} \qquad \text{(see Figure 2.1.43}c\text{)}$$

$$= \frac{11{,}070}{9}$$

$$= 1230 \text{ kips}$$

The system overstrength based on the design base shear (Figure 2.1.28) would be

$$\Omega_o = \frac{1230}{853}$$

$$= 1.44$$

Had we elected to use the six #11 bar alternative, the system overstrength would be 1.77, and this is considerably less than the 2.8 suggested by the UBC.[2.9] The Blue Book[2.17] suggests a range of overstrengths from 1.25 to 2.0.

Comment: The approach to quantifying Ω_o described in this example is allowed by the UBC.[2.9] The difference (1.77 versus 1.44) clearly demonstrates the impact that a small increase (23%) in beam strength (idealized yield moment—M_{yi}) can have on the members located along the seismic load path.

2.1.3.3 Postyield Behavior Take a closer look at rotations imposed on the plastic hinge. Consider the hinge region in the 16-ft beam as described in Figure 2.1.32. A frame story drift of 2% will require a beam deflection (δ_b) of

$$\delta_b = 0.02\frac{\ell}{2}$$

$$= 1.92 \text{ in.}$$

The resultant rotation imposed on the hinge region can be developed as described in Section 2.1.1.2:

$$\phi_y = \frac{\varepsilon_{sy}}{0.67d} \qquad \text{(Eq. 2.1.7a)}$$

$$= \frac{0.002}{0.67(28.5)}$$

$$= 0.0001 \text{ rad/in.}$$

$$\delta_{by} = \frac{\phi_y}{3}\left(\frac{\ell_c}{2}\right)^2 \qquad \text{(see Eq. 2.1.7b)}$$

$$= \frac{0.0001(80)^2}{3}$$

$$= 0.21 \text{ in.}$$

The rotation required in the plastic hinge region ($\ell_p = 0.5h$) following the curvature model of Figure 2.1.11 is then

$$\theta_p = \frac{\delta_b - \delta_{by}}{\ell_c/2 - \ell_p/2} \qquad \text{(see Eq. 2.1.8)}$$

$$= \frac{1.92 - 0.21}{72}$$

$$= 0.0238 \text{ radian}$$

The average postyield curvature in the plastic hinge region is

$$\phi_p = \frac{\theta_p}{\ell_p} \qquad \text{(Eq. 2.1.9)}$$

$$= \frac{0.0238}{16}$$

$$= 0.00148 \text{ rad/in.}$$

$$\varepsilon_{cp} = \phi_p c \qquad \text{(Eq. 2.1.10a)}$$

$$= 0.00148(5)$$

$$\varepsilon_{cu} = \varepsilon_{cp} + \varepsilon_{cy} \qquad \text{(Eq. 2.1.10b)}$$

$$= 0.0074 + 0.001$$

$$= 0.0084 \text{ in./in.}$$

We should be convinced that this frame beam will perform quite well to story drifts of 2% because beam-column joint and column deformations have not been considered in the development of θ_p.

Comment: Observe that the procedures used here are the same as those used in Section 2.1.1 to predict the strain states in the test beam. As long as the designer uses a consistent methodology for the design and the evaluation of test data, the resultant design conclusion should be reasonable.

The relationships between curvature and strain states described in Table 2.1.1 were based on specific material stress strain relationships. Suspect is the manner in which the compressive load on the concrete was developed. The computer algorithm is based

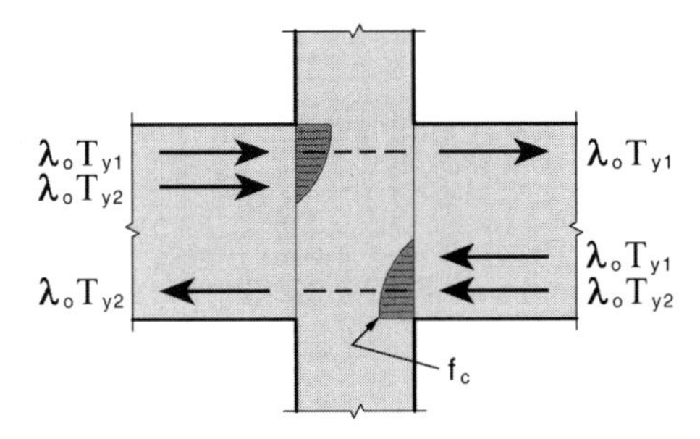

Figure 2.1.33 Forces delivered to the beam-column joint at the deformation limit state.

on a first principles approach $\left(C_s = A'_s \varepsilon'_s E_s\right)$. Steel strain states in the beam-column joint will be explored in Section 2.3. Tensile yielding will extend into the joint from both sides as the load is reversed. Bond deterioration within the joint will make it less likely that a transition from strain hardened steel in tension on one side of the column to steel yield in compression on the other side of the column will occur as is presumed in a first principles analysis. The level of resistance assumed by the algorithm is not likely to be attained by the compression reinforcing.

The most conservative assumption, and one that probably occurs as the deformation limit state is approached, is that described in Figure 2.1.33. Here, the tensile strength of the reinforcing ($\lambda_o T_{y1}$) is transferred directly to the opposite side of the column. Thus the force transfer at the beam-column interface is one that does not involve the compressive steel.

The compressive force transferred given this worst case scenario is

$$C_c = \lambda_o (T_{y1} + T_{y2})$$

Consider the implications of this transfer mechanism, recognizing the fact that the concrete outside the core will be ineffective while that inside the core is well confined:

$$a = \frac{C_c}{0.85 f'_{cc} b_h}$$

where b_h is the confined concrete beam width or effective width:

$$\begin{aligned} a &= \frac{1.25(12)(1.27)(60)}{0.85(7.5)(18)} \\ &= 10 \text{ in.} \end{aligned}$$

which suggests a neutral axis depth of 11 to 12 in.

Now the strain state associated with a curvature in the hinge region of 0.00148 rad/in. becomes

$$\varepsilon_c = \phi_p c$$
$$= 0.00148(12)$$
$$= 0.0178 \text{ in./in.}$$

Observe that, had we selected the six #11 bar design, the associated strain state in the concrete would be considerably higher.

Conclusion: Clearly the analysis performed does not produce an "exact" answer. The procedures explored do, however, suggest probable strain levels and define deformation-based boundaries. They can and should be used only to make a design decision. In this case my decision would have been to use the beam described in Figure 2.1.28*b*. Confinement of the hinge region is warranted anticipating a probable story drift on the order of 2%. The tie configuration within the beam hinge region should provide lateral support for the middle set of beam bars (Section 2.1.1.4).

2.1.4 Precast Concrete Beams

The precasting of concrete offers a wide variety of fabrication and assembly options. Economical solutions are, to a large extent, dependent on fabricator capabilities and the contractor's comfort with the manner in which a particular precast component or system is integrated into the building. As a consequence, innovation is the key to creating a successful solution because the options are many.

From a design perspective, the frame beam options can be placed in two categories: those that emulate cast-in-place concrete construction and those that provide connections between components that are capable of sustaining postyield deformations. We refer to these design alternatives as emulative and yielding. The term "jointed precast" is also used to identify precast concrete elements designed to yield at the precast interface.[2.18] These two approaches are shown in Figure 2.1.34. Systems *a*, *b*, and *d* of Figure 2.1.34 are emulative, for postyield rotations are expected to occur in the concrete beam away from the point at which precast members are connected.

The behavior described in Figure 2.1.34*b* may be attained in a variety of ways. Figure 2.1.35 shows how post-tensioning and conventional reinforcing can be combined so as to force the plastic hinge region away from the point at which the precast members are joined. The behavior described in Figure 2.1.34*d* may, for example, be attained as conceptually shown in Figure 2.1.36. Here, the beams may be cast to any practical length and spliced, away from the plastic hinge region, by a mechanical connector or a wet cast joint. The columns are one story in height and connected with grouted sleeves or proprietary connections that are capable of developing the fracture strength of the reinforcing bar.

The precast tree system described in Figure 2.1.34*a* was effectively used by Al Yee in Hawaii during the 1970s. In the early 1980s Rockwin Corporation (now a part of Coreslab) developed an H-shaped variation of the tree column (Figure 2.1.37).

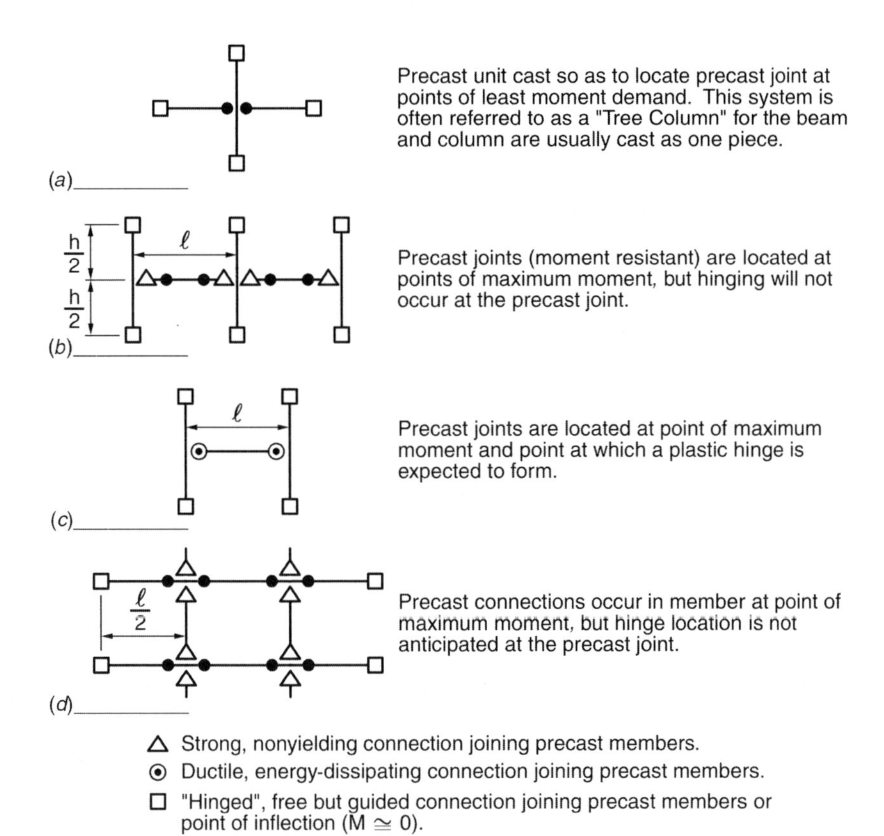

Figure 2.1.34 Classification of precast ductile frames according to component connector location.

This system was used to construct their headquarters, which survived the Whittier Earthquake without damage.

Unfortunately, emulative systems seldom proceed beyond the prototype, and this is typically because the cost of the prototype, including its development, exceeds expectations. These types of systems require close coordination between the design team and the builder. They are only possible when some form of the design/build delivery system is adopted by the owner.

Regardless of the type of emulative system adopted, the basic design follows the procedures developed in Section 2.1.2. The design of the connections follows strength-based concepts and must be capacity-based. This means that the connector must develop the probable strength of the plastic hinge. Because the number of alternative connection types is basically unlimited, and because the load-transfer mechanisms are understood and well described by others, our focus is on yielding connections.

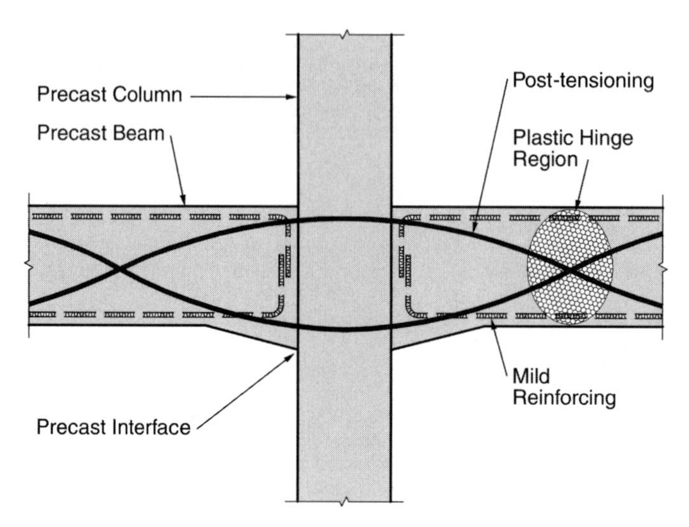

Figure 2.1.35 Strong nonyielding precast connection.

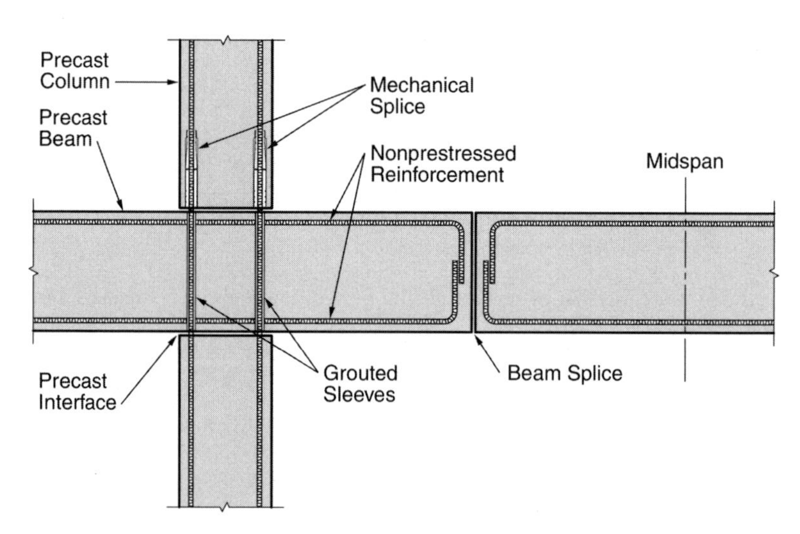

Figure 2.1.36 Continuous beam precast frame system.

The cost-effectiveness of a yielding connector requires a simple yet effective mechanism for transferring both shear and moment; accordingly, these basic load transfer mechanisms are discussed before exploring the design of beam systems.

2.1.4.1 Moment Transfer What makes a yielding connection different from a plastic hinge in a cast-in-place beam? The answer lies in the strain distribution in the region where the postyield rotation will take place. In Section 2.1.1 we hypothesized

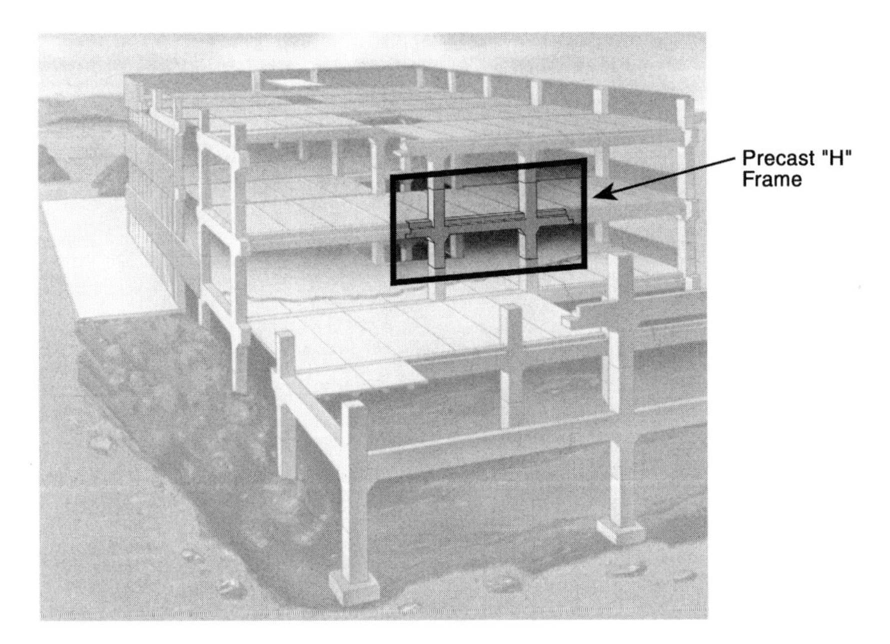

Figure 2.1.37 Precast building system developed by Rockwin Corporation.

and demonstrated that the postyield rotation in a cast-in-place beam would occur over a plastic hinge region (Figure 2.1.25). The strain in the concrete and reinforcing steel in this plastic hinge region was treated as though it were constant. When a precast beam and column are joined at the face of the column, a weakened plane is created. This causes a large portion of the rotation to occur at this discontinuity, and this condition will impact the strain state in the flexural connection as well as the shear transfer mechanism.

The strain imposed on the reinforcement in the cast-in-place system (Figure 2.1.38*a*) will tend to distribute itself over a region that extends some distance past the plastic hinge region (ℓ_p) and often through the beam-column joint (Section 2.3). The weakened plane created by the joining of precast elements causes this deformation to occur almost entirely at the interface. Debonding will occur over a much smaller region adjacent to the gap created between the precast beam and column (θ_p, Figure 2.1.38*b*). The length of debond appears to be on the order of d_b in precast assemblies because the reinforcing bars are usually grouted in tubes that provide confinement and promote a wedging action. This will be discussed in more detail in Section 2.1.4.4 when we talk about the hybrid system. Regardless of the extent of this added debond length, it should be clear that the ultimate strain imposed on the tension reinforcement will be significantly greater in the precast system than that imposed on the tension reinforcement in the cast-in-place system.

Consider, for example, the beam and system described in Figure 2.1.27 were it created of precast components. The debond region in a precast assembly, according to tests performed at the University of Washington that have not as yet been published,

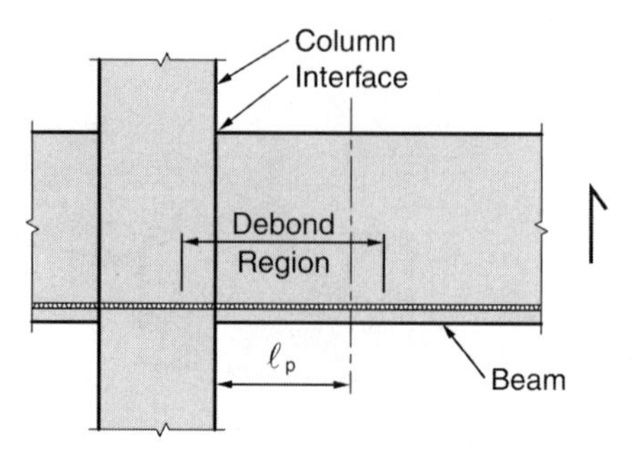

(*a*) Cast-in-Place Beam and Column

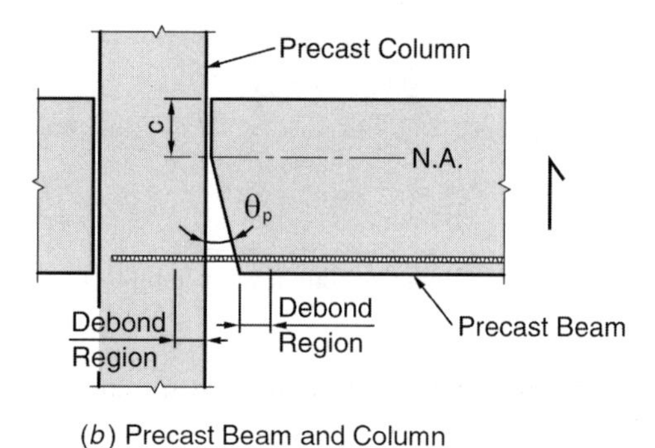

(*b*) Precast Beam and Column

Figure 2.1.38 Plastic hinge region—cast-in-place and precast systems.

might be as little as d_b on each side of the created gap. Postyield strain imposed on the #10 bars must occur over a length of

$$2(\pm 1d_b) + \theta_p(d - c) = 2.5 + \theta_p(d - c) \tag{2.1.45}$$

The neutral axis depth is about 5.5 in. at the strength limit state (Table 2.1.1). The created gap (δ_b) associated with a rotation (θ) of 3% is approximately

$$\begin{aligned} \delta_b &= \theta_p(d - c) \\ &= 0.03(23) \\ &= 0.69 \text{ in.} \end{aligned} \tag{2.1.46}$$

and the induced postyield steel strain is

$$\varepsilon_s = \frac{\delta_b}{2.54 + 0.69}$$

$$= \frac{0.69}{3.23}$$

$$= 0.214 \text{ in./in.}$$

Comment: This exceeds the probable rupture strain of mild steel (Figure 2.1.54). Observe that the strain state in a smaller bar would be significantly greater, and this has caused bars (#3 and #6) to rupture in the testing of scale models by NIST[2.8] of the hybrid beam system—see Figure 2.1.47*b*.

Conclusion: Special attention must be paid to the level of strain imposed on the flexural reinforcement connecting two precast members when this connection is expected to yield. It is advisable to intentionally debond the flexural reinforcing at the beam/column interface when this type of yielding connection (Figure 2.1.38*b*) is used.

For comparative purposes consider the strain imposed on the tension reinforcement of our cast-in-place example beam (Figure 2.1.27) when subjected to a rotation of 3 radians. An effective debond region for steel strain calculation purposes is conservatively assumed to be

$$\ell_p + 2(d_b) = 16 + 2.5 \tag{2.1.47}$$

$$= 18.5 \text{ in.}$$

More realistically, as discussed in Section 2.1.3.3 and Section 2.3, the debonding will probably extend through the joint if the subassembly is subjected to several cycles of large postyield deformations. Accordingly, the strain projection that follows must be viewed as quite conservative.

The neutral axis depth is approximately (see Table 2.1.1)

$$c \cong 5.5 \text{ in.}$$

Hence

$$d - c = 28.5 - 5.5$$

$$= 23 \text{ in.}$$

and the average curvature becomes

$$\phi = \frac{\theta}{\ell_p + 2d_b} \qquad \text{(see Eqs. 2.1.9 and 2.1.47)}$$

$$= \frac{0.03}{18.5}$$

$$= 0.00162 \text{ rad/in.}$$

Conservatively, the analytically based strain state becomes

$$\begin{aligned}\varepsilon_s &= \phi_p(d - c)\\ &= 0.00162(23)\\ &= 0.037 \text{ in./in.} \qquad (\cong 18\varepsilon_y)\end{aligned}$$

Clearly this explains why we never see beam bars fracture in cast-in-place ductile frames.

2.1.4.2 Shear Transfer Shear transfer across joints in concrete has been a focal research topic of Mattock[2.19] at the University of Washington since the 1960s (see Section 1.3.2). The major concern then was the transfer of shear across cracked planes, construction joints, and the steel-to-concrete interfaces that often are relied on in connections. This shear transfer mechanism is referred to as "shear friction," and much importance was given to the texture of the interface and the quantity of reinforcement crossing the joint. Essentially, the yielding of the reinforcing crossing the joint is assumed to create a frictional resistance by its equilibrating compressive counterpart.[2.6, Sec. 11.7.4]The Commentary to the ACI[2.6, Sec. R11.7.4.3] acknowledges the need to use "artificially high values of the coefficient of friction . . . so that the calculated shear will be in reasonable agreement with test results."

The basic shear transfer strength implies a shear friction factor (μ) of 0.6 to 1.4.

$$V_n = A_{vf} f_y \mu \qquad \text{(see Ref. 2.6, Eq. 11-25)} \qquad (2.1.48)$$

$A_{vf} f_y$ is in essence the clamping load, and it can be replaced or supplemented by a permanent reliable compressive force acting across the joint (see Eq. 1.3.19).

μ, the Greek letter used in codes to quantify the friction factor, is a function of the condition of the joint that, for an essentially smooth interface, is 0.6.

Studies of pure friction associated shear transfer across joints[2.20] suggest that the friction factors implied in the codification of shear friction theory are attainable and should not cause slippage along the interface when the axial load is reliable. Figure 1.3.14 describes the results of some of the shear transfer tests reported in Reference 2.20.

Figure 2.1.38*b* describes a yielded precast beam-column interface crossed by tensile reinforcing. The tensile force ($\lambda_o A_s f_y$) must be equilibrated by a compressive force of equal magnitude and opposite sense acting on the compression side of the neutral axis. An identical mechanism will be created in the comparable cast-in-place interface (Figure 2.1.38*a*). Given the prescriptive rules established for shear friction,[2.6]the friction coefficient for the cast-in-place system would be 1.4, while that for the precast assemblage would be 0.6. It is highly unlikely that the two friction factors would be significantly different after many postyield cycles of load; accordingly, a common friction factor seems reasonable and a friction factor of one (1) not unreasonably high. Typically, shear transfers will impose much lower friction factor demands.

Consider the example beam of Figure 2.1.28*b*. The developed tension force is

$$\lambda_o T_y = 1.25(6)(1.27)60$$
$$= 571.5 \text{ kips}$$

while the shear acting on the beam-column interface could be as high as 201 kips (see Section 2.1.2.7, Example 2). Accordingly, a friction factor of only 35% is required to effect a shear transfer, and one need not expect shear slip in the region of the plastic hinge, at least not until the concrete in the toe deteriorates.

Steel-to-steel and steel-to-concrete friction factors are also capable of transferring shear forces when the frictional demand is on this order of magnitude. The steel-to-concrete friction factor[2.6] is 0.7, while the AISC[2.4, Secs. 1.7(d) and 2.17] allows a friction factor of 0.33 for clean mill-surfaced finishes, and this includes a factor of safety of between 1.4 and 1.5.[2.4]

Conclusion: The use of a nominal friction coefficient of 0.5 for steel-to-steel and 1.0 for concrete-to-concrete, regardless of the type of surface finish, seems reasonable. A strength reduction factor of 0.85 is recommended by the ACI.[2.6]

2.1.4.3 Composite Systems Traditionally, structural engineers think of "composite" as referring to a structural system that contains components that are structural steel and reinforced concrete. In precast concrete systems the term "composite" refers to the combining of precast concrete components with cast-in-place elements as, for example, a precast column with a cast-in-place beam. In the development of frames, composite systems are often an attractive alternative. Typically the so-developed system tends to be emulative in its design approach, but yielding connectors can also be used and will often be more economical. Further, yielding connectors will usually allow larger deformation limit states and produce less postearthquake damage. The basic difference between the emulative connector and the yielding connector is the point toward which the yielding is directed. The emulative connector develops the strength of the reinforcing at the beam column interface, and this forces the yielding to occur in the reinforcing bar, for the connector must be designed to develop the rupture strength of the bar. The yielding connector will yield at loads that are below the yield strength of the reinforcing bar, thereby absorbing all postyield deformation.

Emulative Connectors The columns of the frame described in Figure 2.1.28*a* can be precast as multistory elements and the frame beams can be cast-in-place. Figure 2.1.39 describes a six-story parking structure that utilizes a precast post and beam vertical load-carrying system and is braced by a composite frame (see Photo 2.3). This structure was constructed in thirteen weeks from the start of foundation to occupancy. The frame beam design used in the building described in Figure 2.1.39 did not follow the emulative approach, but that alternative was considered. Where emulation is the adopted design approach the prescriptions contained in the ACI[2.6] must be followed and, given the reinforcing bar layout described in Figure 2.1.28, this would be difficult. Long bars would need to be threaded through sleeves in the columns and then grouted or otherwise mechanically connected to the column so as

Figure 2.1.39 Precast framing for a composite system.

to develop the rupture strength of the bar. Mechanical splices will add to the placing constraints discussed in Section 2.1.3.2. In either case beam cages would have to be assembled in place. It is not likely that this emulative type of system could be constructed in thirteen weeks.

Yielding Connectors Yielding connectors may be used to develop a composite system, and this is the approach used in the building described in Figure 2.1.39. One way to ensure yielding in the connector is to cast a ductile insert in the column. A high-strength (Grade 160) reinforcing bar is then threaded into the ductile insert. The assembly used is described in Figure 2.1.40, it is manufactured and distributed by Dywidag Systems International. The stress–strain curve for this assembly is shown in Figure 2.1.41. All of the yielding takes place in the rod of the yielding connector (ductile rod). Construction speed is attained by prefabricating the frame beam cages. Flexural reinforcing bars are coupled at midspan. This allows the bars to be turned out of the couplers and into the yielding connectors (Photo 2.3).

The yielding connector described in Figure 2.1.40 has a minimum yield strength of 141 kips. Were it to be proposed as an alternative to the beam designed in Example 2 (Section 2.1.2.7), Step 1 of the design procedure (Section 2.1.2.6) would be repeated.

Step 1: Determine a Trial Value for $(A_s + A_s')$. Equation 2.1.35 could be modified to reflect a design force (T_y) at the beam column interface:

$$V_c = \left(\frac{\ell_1}{\ell_{c1}} + \frac{\ell_2}{\ell_{c2}}\right)\left(\frac{d - d'}{h_x}\right)\phi T_y \qquad \text{(see Eq. 2.1.35)}$$

$$= \left(\frac{24}{21} + \frac{16}{13}\right)\left(\frac{29-3}{9(12)}\right)0.9T_y$$

$$= 0.51T_y$$

Solving for T_y, given an objective column shear of 200 kips, we get

$$T_y = \frac{200}{0.51}$$

$$= 392 \text{ kips}$$

Conclusion: Three ductile rod assemblies are required to resist the seismic load.

In Step 1b of Example 2 (Section 2.1.2.7), the impact of dead and live loads was estimated for the beam and found for a first yield analysis to be on the order of 9%. Observe that an increase in moment demand of this magnitude would suggest that four ductile rod assemblies be used:

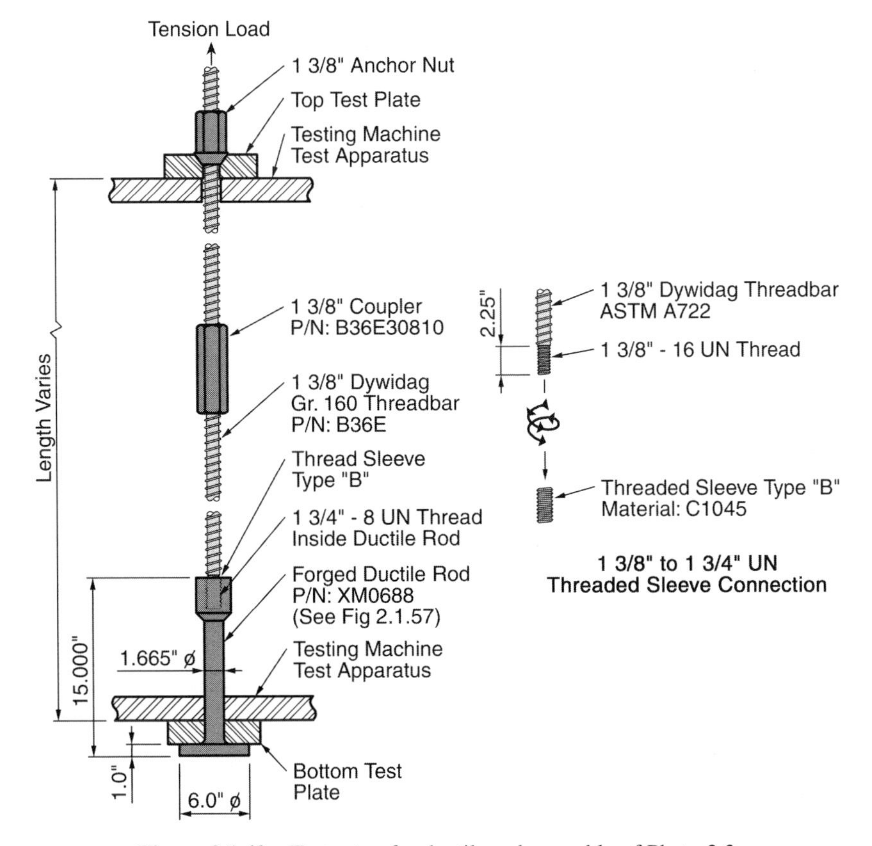

Figure 2.1.40 Test setup for ductile rod assembly of Photo 2.3.

Photo 2.3 Composite frame beam, DDC connection to precast column, Staples Parking Structure, Los Angeles, CA. (Courtesy of Englekirk Partners, Inc.)

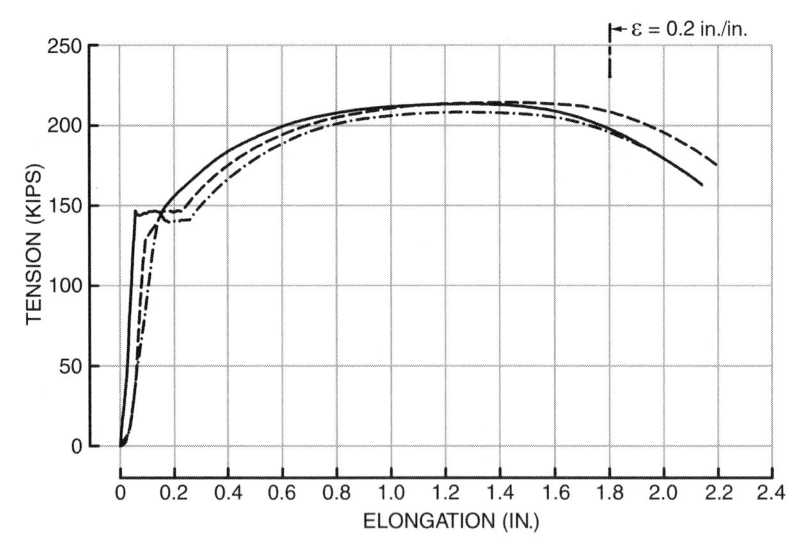

Figure 2.1.41 Stress–strain curves developed from tests of the forged ductile rod assembly described in Figure 2.1.40.

$$T_y = 1.09(392)$$
$$= 427 \text{ kips} \qquad \text{(Demand)}$$
$$T_y = 3(141)$$
$$= 423 \text{ kips} \qquad \text{(Capacity)}$$

This 1% exceedance is not in itself of any consequence. But we also understand that an elastic analysis will suggest somewhat higher demands and we do not want, in subsequent stages of the design of this building, to have to increase the size and capacity of the section. If we were to provide four ductile rods, the overcapacity would be significant. Both column and beam sizes would need to be increased, and the yielding alternative quite possibly abandoned. The introduction of another ductile rod is not warranted in this case, but we would certainly want the provideable capacity to exceed the demand in the conceptual design phase by 5 to 10%.

Adjusting the level of provided strength is much more difficult in precast assemblies where the transfer mechanisms have established capacities that are large and do not allow modest changes in provided strength. A mechanism approach[2.4, Sec. 2.3] can be used (ACI[2.6, Appendix B]) to define the strength limit state for a system. The mechanism approach would entirely discount the impact of dead and live loads on most seismic bracing programs. In essence the mechanism approach is the basis for the moment redistribution discussed in Section 2.1.3.2 and in Reference 2.6, Appendix B. The mechanism approach was introduced as "plastic design" in the 1950s and then was exclusively applied to indeterminate steel systems. The mechanism approach recognizes that system strength is based on the load required to produce a mechanism in the system and that a first yield criterion, as a limit state, does not produce indeterminate systems that possess consistent factors of safety.

If we consider the frame and extracted subassembly described in Figure 2.1.42, the strength limit state would be defined by one of the mechanisms shown in Figure 2.1.43. Observe that dead and live loads only impact the mechanisms of Figures

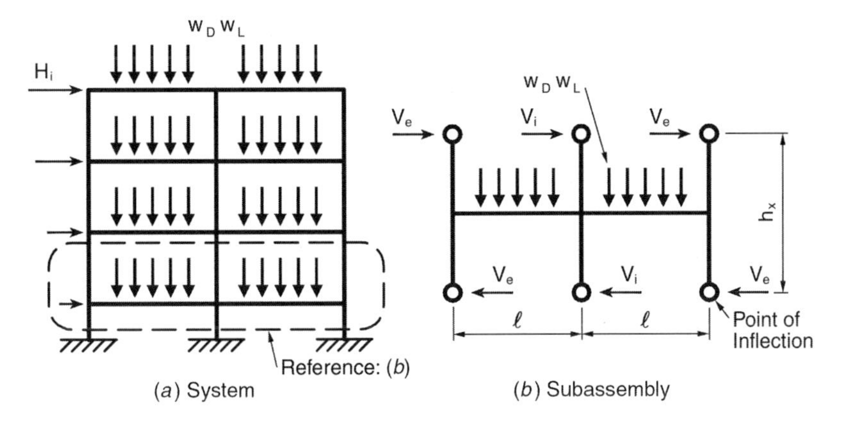

Figure 2.1.42 Frame elevation.

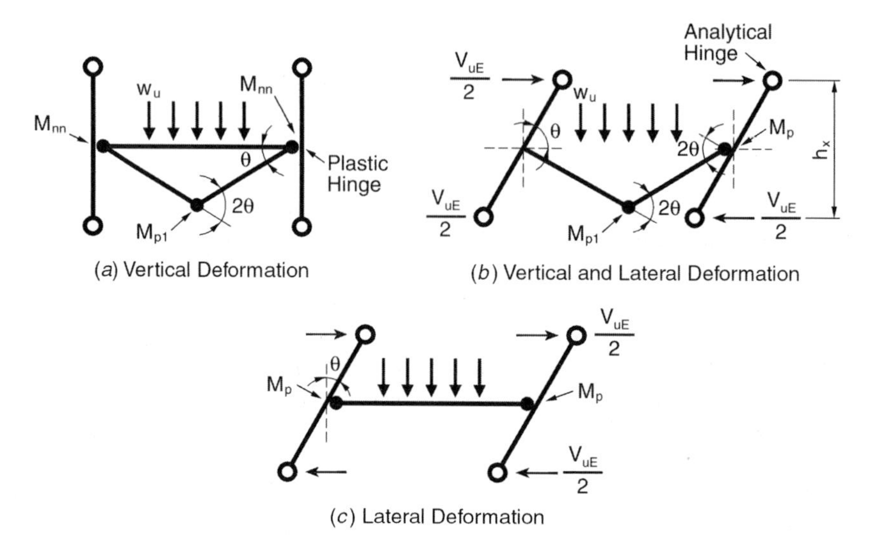

Figure 2.1.43 Subassembly mechanisms.

2.1.43*a* and *b*, because in the mechanism of Figure 2.1.43*c* dead and live loads create no external work. It should be clear from the design example (Section 2.1.2.7) that the mechanism of Figure 2.1.43*a* will not govern, so we need only evaluate the likelihood of the mechanism of Figure 2.1.43*b* defining the strength limit state.

Before rushing off into an elaborate analysis, consider the essence of the problem.

- If the mechanism of Figure 2.1.43*c* defines the strength limit demand, our three ductile rod system should prove to be acceptable.
- The mechanism identified in Figure 2.1.43*b* will include the moment capacity of the beam away from the support; hence the beam can be internally strengthened to avoid the formation of this mechanism.

Accordingly, let us compare the lateral loads required to create the two mechanisms described in Figures 2.1.43*b* and 2.1.43*c*.

Mechanism of Figure 2.1.43c

$$\text{External work} = \text{Internal work}$$

$$V_{uE} h_x \theta = 2M_p \theta \tag{2.1.49}$$

where M_p is assumed to be the nominal strength of the ductile rod assembly.

Mechanism of Figure 2.1.43b

$$\text{External work} = \text{Internal work}$$

$$V_{uE} h_x \theta + 2\left(w_{uD} + w_{uL}\right)\left(\frac{\ell}{2}\right)\left(\frac{\ell}{4}\right)\theta = 2M_p\theta + 2M_{p1}\theta \tag{2.1.50a}$$

where M_{p1} is the internal strength of the beam, assumed to be a minimum at midspan.

What we would like to know is the strength relationship between the minimum internally provided flexural strength (M_{p1}) and that provided at the ends of the beam (M_p) that would cause the mechanism of Figure 2.1.43*c* to be critical.

If we define A as

$$A = \frac{M_{p1}}{M_p} \tag{2.1.50b}$$

then M_{p1}, the strength provided in the interior of the beam, must be greater than AM_p to attain our objective—the dominance of the sidesway mechanism. Accordingly,

$$(A + 1)M_p > (w_{uD} + w_{uL})\frac{\ell^2}{8} + \frac{V_{uE}h_x}{2} \tag{2.1.51}$$

Thus if the strength of the internal plastic hinge (AM_p) exceeds $(w_{uD} + w_{uL})\ell^2/8$ and the provided moment capacity at the support is equivalent to or larger than M_p as developed by the mechanism of Figure 2.1.43*c* ($V_{uE}h_x/2$), then the impact of vertical loads on the design strength provided in the yielding connector may be neglected.

An example analysis should increase designer confidence. The 24-ft internal bay of the frame of Figure 2.1.28 is shown in Figure 2.1.44.

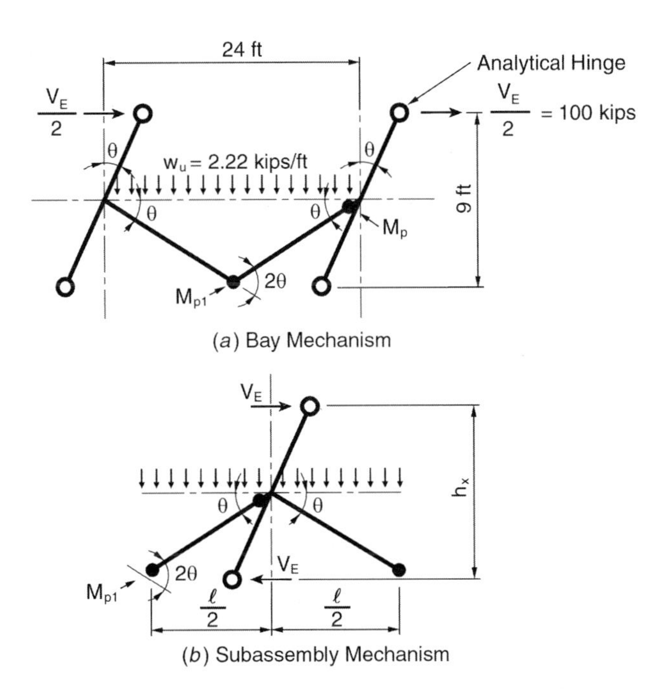

Figure 2.1.44 Internal mechanism of Example 2.

External work = Internal work

$$2\frac{V_E}{2}h_x\theta + 2w_u\left(\frac{\ell}{2}\right)\frac{\ell}{4}\theta = 2M_p\theta + 2M_{p1}\theta \qquad \text{(Eq. 2.1.50a)}$$

Solve for the minimum value of M_{p1} if

$$\begin{aligned}\phi M_p &= \phi T_y(d - d') \\ &= (0.9)423\left(\frac{26}{12}\right) \\ &= 825 \text{ ft-kips}\end{aligned}$$

$$200(9) + \frac{2.22(24)^2}{4} = 2(825) + 2\phi M_{p1} \qquad \text{(Eq. 2.1.50a)}$$

$$M_{p1} > 261 \text{ ft-kips}$$

$$A = \frac{261}{825} = 0.32$$

Observe that this is the same result obtained by the application of Eq. 2.1.51.

$$(0.32 + 1)825 \geq 160 + 900$$

$$1089 > 1060$$

Comment: Observe that the bay mechanism of Figure 2.1.44*a* and the subassembly mechanism of Figure 2.1.44*b* are, of course, the same—the subassembly mechanism may be easier to understand.

Conclusion: A three-ductile-rod beam may be substituted for the beam described in Figure 2.1.28*b*, provided that the internal strength (M_n) is at least 261 ft-kips. The factored mechanism shear will be

$$\begin{aligned}V_{cM} &= \frac{\phi M_p}{h_x}\left(\frac{\ell_1}{\ell_{c1}} + \frac{\ell_2}{\ell_{c2}}\right) \qquad \text{(see Eqs. 2.1.35 and 2.1.49)} \\ &= \frac{825}{9}\left(\frac{24}{21} + \frac{16}{13}\right) \\ &= 217 \text{ kips}\end{aligned}$$

and this suggests a 9% exceedance or overcapacity.

Step 2: Size the Beam-Column Joint. The force delivered to the joint is less than that delivered by the conventionally reinforced beam (six #10 bars, Figure 2.1.28*b*).

Accordingly, the demand imposed on the joint will be less than its capacity. This step is satisfied provided we use only three ductile rods.

Alternatively, the required joint size may be developed from Eq. 2.1.16:

$$A_j = 62\left(A_s + A_s'\right) \qquad \text{(Eq. 2.1.16)}$$

$$T_y = A_s f_y$$

$$A_j = \frac{2T_y}{f_y}(62)$$

$$= 2.07T_y \qquad (2.1.52)$$

where T_y is the yield strength provided by the yielding assemblage, in this case three ductile rods.

$$T_y = 3(141)$$

$$= 423 \text{ kips}$$

$$A_j = 2.07(423)$$

$$= 876 \text{ in.}^2$$

Step 3: Develop the Beam and Column Reinforcing Program. The ductile rod (Figure 2.1.40) has a head diameter of 6 in. Accordingly, the center-to-center spacing of the beam bars must be 6 in., and the minimum beam width becomes:

Center-to-center of flexural bars	=	12.0 in.
d_b $\left(1\frac{3}{8}\text{-in. bar}\right)$	=	1.38 in.
Two #5 tie diameters	=	1.25 in.
Cover (2 at 1.5 in.)	=	3.0 in.
		17.63 in.

Conclusion: The beam should be 18 in. wide.

Comment: d' need not be 3.5 in. as it was in the beam of Example 2 (Figure 2.1.28*b*). d' is:

$\frac{1}{2}$ bar diameter	=	0.69 in.
1 tie diameter	=	0.625 in.
Fire cover	=	1.5 in.
d'	=	2.82 in.

Step 4: Check the Reinforcement Ratio (ρ). The area of steel that will yield is the throat of the ductile rod (Figure 2.1.40).

$$A_B = \frac{T_y}{f_y}$$

$$= \frac{141}{60}$$

$$= 2.35 \text{ in.}^2$$

$$\rho = \frac{3(2.35)}{18(29.18)}$$

$$= 0.0134$$

and this is below our objective maximum of 1.5%.

Design Steps 5 and 6 are clearly satisfied, and the analysis procedures need not be repeated, with the exception of a part of Step 5, because the criterion for shear reinforcing is different. The hinge region is now in the column section and no postyield rotation will be required of the beam; accordingly, the shear carried by the concrete (V_c) may be included in the shear capacity (ϕV_n) of the beam.

Step 5 (Analysis, Section 2.1.3.1): Develop a Transverse Reinforcing Program. Following the capacity-based approach, determine the probable flexural strength of the beam:

$$M_{\text{pr}} = \lambda_o M_n \qquad \text{(see Eq. 2.1.42a)}$$

$$= \frac{1.25(3)(141)(26.4)}{12}$$

$$= 1163 \text{ ft-kips}$$

The seismically induced shear in the 16-ft beam ($\ell_c = 13.33$ ft) is

$$V_{uE} = 2\frac{M_{pr}}{\ell_c} \qquad \text{(see Eq. 2.1.42b)}$$

$$= \frac{2(1163)}{13.33}$$

$$= 174.5 \text{ kips}$$

The factored dead and live load shear (Section 2.1.3.2, Step 5) is 14.7 kips, and

$$V_u = 174.5 + 14.7$$

$$= 189.2 \text{ kips}$$

$$v_u = \frac{189.2}{18(29.18)}$$

$$= 0.36 \text{ ksi} < 4\sqrt{f'_c}$$

$$v_s = 0.36 - 2\phi_v\sqrt{f'_c}$$

$$= 0.24 \text{ ksi}$$

Maximum spacing is $d/4$, or 7.3 in. on center.

Comment: The designer should be aware that the high-strength longitudinal bars will be highly stressed in the region of the beam-column interface. The adopted overstrength factor of 1.25 may not be sufficient (see Figure 2.1.41) so a conservative stirrup spacing should be adopted. Further, the critical buckling length (ℓ_{cr}) for a $1\frac{3}{8}$-in. diameter bar is on the order of 23 in. so a longitudinal support spacing of $6d_b$ (see Section 2.1.1.4) does not represent a rational criterion.

Paulay and Priestley[2.5, Sec. 4.5.4] suggest that the capacity of ties in tension should be at least one-sixteenth of the yield force in the flexural reinforcing.

$$\frac{A_{te}}{s} = \frac{\sum A_b f_y}{64 f_{yh}} \qquad \text{(see Eq. 2.1.11b)}$$

where A_{te} is the area of one tie leg and $\Sigma A_b f_y$ is the force in the longitudinal bars being restrained by the tie leg (A_{te}).

$$\frac{A_{te}}{s} = \frac{141(1.5)}{64(60)}$$

$$= 0.055$$

$$s = \frac{0.20}{0.055}$$

$$= 3.6 \text{ in. on center} \qquad \#4 \text{ bars}$$

$$s = \frac{0.31}{0.055}$$

$$= 5.6 \text{ in. on center} \qquad \#5 \text{ bars}$$

A tie reinforcing program for this beam that would satisfy Eq. 2.1.11b would place one #5 3-leg hoop at 2 in. from the face of the column and 4 at $5\frac{1}{2}$ in. on center for a total distance of $0.75h$ from the face of the column. The interior portion of the beam would require the use of #5 ties at 7 in. on center ($\phi v_s = 250$ psi).

Analysis Step 6 (Section 2.1.3.1): Review Strain States. The critical strain state for this beam is that imposed on the ductile rod, for we have already established that the strain state in the concrete will be within acceptable limits (Step 6, Section 2.1.3.1). Figure 2.1.32 describes the behavior objective except that now the hinge region (strain imposed on the steel) is within the column and ℓ_p is (see Figure 2.1.57) that portion of the rod that will yield. The depth to the neutral axis will start

Photo 2.4 Paramount Apartments, San Francisco, CA. (Courtesy of Charles Pankow Builders.)

at $(d - d')/2$, but ultimately, as the concrete carries the bulk of the compressive load, will conservatively approach the neutral axis depth of the conventionally reinforced beam, or about 6 in. (see Table 2.1.1).

The elongation required of the ductile rod (δ_b) to attain a beam rotation of 0.02 radian is

$$\begin{aligned}\delta_b &= \theta_p(d - c)\\ &= 0.02(23.25)\\ &= 0.46 \text{ in.}\end{aligned}$$

The strain imposed on the ductile rod is

$$\begin{aligned}\varepsilon_s &= \frac{\delta_b}{\ell_p}\\ &= \frac{0.46}{9}\\ &= 0.051 \text{ in.} \qquad (37\varepsilon_y)\end{aligned}$$

This is certainly an acceptable level of strain, for the ultimate strain of the ductile material used to create the ductile rod is at least 20% (see Figure 2.1.41). Observe also that the use of an overstrength factor (λ_o) of 1.25 is unconservative for an elongation of 0.46 in.

2.1.4.4 Post-Tensioned Assemblages Bob Park and his associates in New Zealand studied and tested the post-tensioned assemblage of precast beams and columns in the early 1970s. In the late 1980s H. S. Lew at the National Institute of Standards and Technology (NIST)[2.8] continued the development and testing of this concept. In the 1990s several parking structures were built using versions of what is now referred to as the hybrid system[2.21] The system was then used to construct the Paramount apartment building (Photo 2.4). A comprehensive test program in support of the design of this building was conducted at the University of Washington. The exterior subassembly is shown in Figure 2.1.45. Beam and column reinforcing programs are shown in Figure 2.1.46. The capacity of the column and the joint in the subassembly significantly exceed the demand imposed on them by the beam. The fact that postyield deformation occurred almost exclusively in the beam was confirmed by the test program.

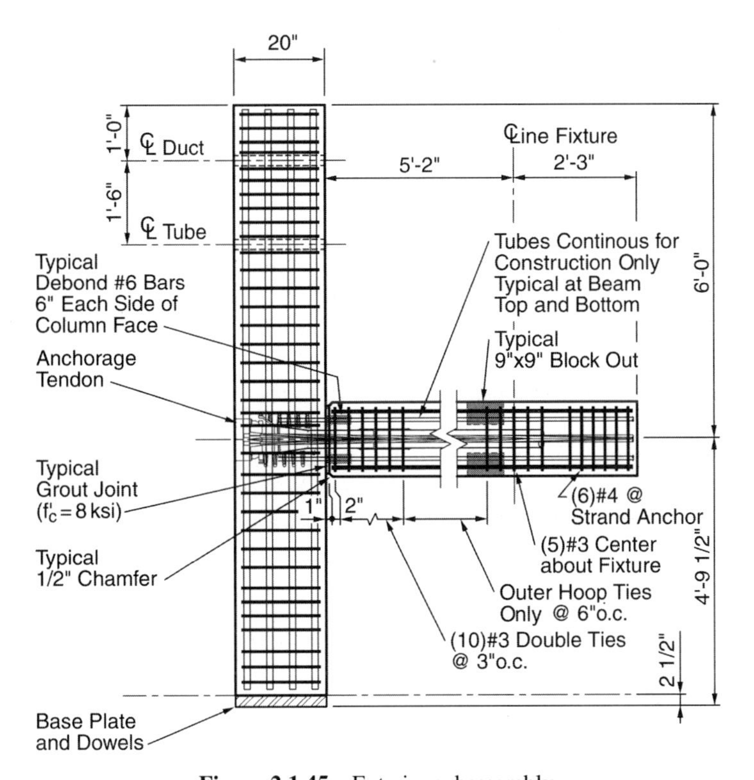

Figure 2.1.45 Exterior subassembly.

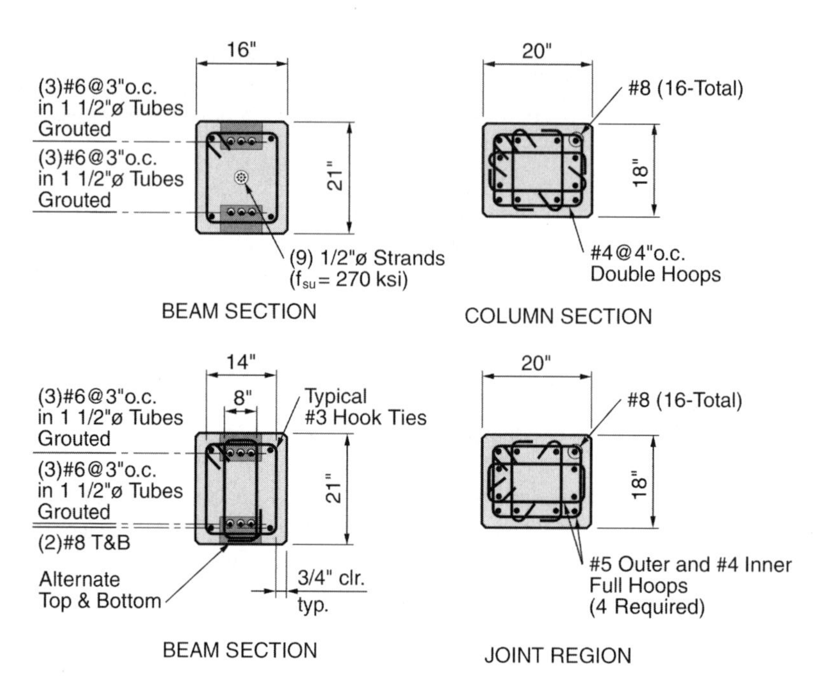

Figure 2.1.46 Beam and column cross sections—hybrid subassembly test program.

Let us start by explaining the design methodology and then compare the design conclusions with the test results. Flexural strength in the hybrid beam is provided by a combination of unbonded post-tensioning strands and bonded mild steel. Nine $\frac{1}{2}$-in. ϕ, 270-ksi strands, if stressed to 162 ksi as specified, would have provided an effective concentric post-tensioning force of 223.1 kips. Three #6 (Gr. 60) reinforcing bars were placed in the top and bottom of the beam in tubes that were subsequently grouted with high-strength grout. These bars provide energy dissipation—an attribute not provided by the unbonded post-tensioning. The strength provided by the 16 in. by 21 in. deep beam is developed from the size of the grout pad, which for this test was 16 in. by 20 in.

In the design of the exterior subassembly of Figure 2.1.45, it was assumed that the stress in the mild steel compression reinforcement reached yield. Accordingly, the flexural strength provided by the mild steel (M_{ns}) was

$$M_{ns} = T_{ns}(d - d') \tag{2.1.53}$$

$$= 3(0.44)(60)(16.5)$$

$$= 1307 \text{ in.-kips}$$

The flexural strength provided by the unbonded post-tensioning (M_{nps}) is developed as follows:

$$T_{nps} = A_{ps} f_{pse} \tag{2.1.54}$$

$$T_{nps} = 9(0.153)(162)$$

$$= 223.1 \text{ kips}$$

$$a = \frac{T_{nps}}{0.85 f'_c b}$$

$$= \frac{223.1}{0.85(5)(16)}$$

$$= 3.3 \text{ in.}$$

$$M_{nps} = T_{nps}\left(\frac{h}{2} - \frac{a}{2}\right) \qquad (h = 20 \text{ in., grout pad dimension})$$

(see Figure 2.1.46) (2.1.55)

$$= 223.1(10 - 1.65)$$

$$= 1865 \text{ in.-kips}$$

The nominal moment capacity of the hybrid frame beam of Figure 2.1.46 is

$$M_n = M_{ns} + M_{nps} \tag{2.1.56}$$

$$= 1307 + 1865$$

$$= 3172 \text{ in.-kips}$$

This corresponds to a beam load or shear of

$$V_{nb} = \frac{M_n}{\ell_c}$$

$$= \frac{3172}{62} \qquad \text{(see Figure 2.1.45)}$$

$$= 51.2 \text{ kips}$$

The associated column shear or applied test frame force (F_{col}) is

$$V_c = F_{\text{col}} = V_{nb}\left(\frac{\ell}{h_x}\right)$$

$$= 51.2\left(\frac{72}{117.5}\right)$$

$$= 31.4 \text{ kips}$$

Figure 2.1.47*a* describes the behavior of the test specimen. Figure 2.1.47*b* identifies critical behavior milestones. The stresses imposed on the post-tensioning strands are shown in Figure 2.1.48.

Observe that the predicted nominal strength of 31.4 kips (V_c) is not reached until a drift ratio of 2% is attained (Figure 21.1.47*b*). This is explained, at least in part, by the fact that the initially delivered post-tensioning force was only 216 kips, or 4% less than the specified 223.1 kips. A story drift of 1% was required to develop the assumed design force in the post-tensioning ($T_{nps} = 223.1$ kips). Figure 2.1.48 provides the relationship between post-tensioning force and drift.

Subassembly stiffness, as predicted by idealized member stiffnesses, is developed as follows:

$$\Delta_x = \frac{V_e h_x^2}{6E}\left(\frac{\ell}{I_b} + \frac{h_x}{2I_c}\right) \qquad \text{(see Eq. 3.2.2)}$$

$$h_x = 117.5 \text{ in.}$$

$$\ell = 72 \text{ in.}$$

$$I_{ce} = 0.7 I_g$$

$$= \frac{0.7(18)(20)^3}{12}$$

$$= 8400 \text{ in.}^4$$

$$I_{be} = 0.35 I_g$$

$$= \frac{0.35(16)(21)^3}{12}$$

$$= 4322 \text{ in.}^4$$

$$\Delta_x = \frac{31.4(117.5)^2}{6(4000)}\left[\frac{72}{4322} + \frac{117.5}{8400}\right]$$

$$= 0.55 \text{ in.}$$

The associated drift ratio (Δ / h_x) is 0.47%. This drift ratio might be accepted as an idealized representation of stiffness to a column shear of about 50% of the nominal column shear (31.4 kips) but not a good idealization for the behavior described in Figure 2.1.47. The hybrid subassembly appears to be considerably softer than the cast-in-place system whose behavior is described in Figure 2.1.2.

Probable flexural strength and ultimate strain states should be predicted during the analytical phase of the design's development. The process starts by estimating the strain states in the reinforcing at a selected level of drift. Consider a 4% postyield drift ratio (θ_p) as our objective drift and realize that any elastic component of story drift

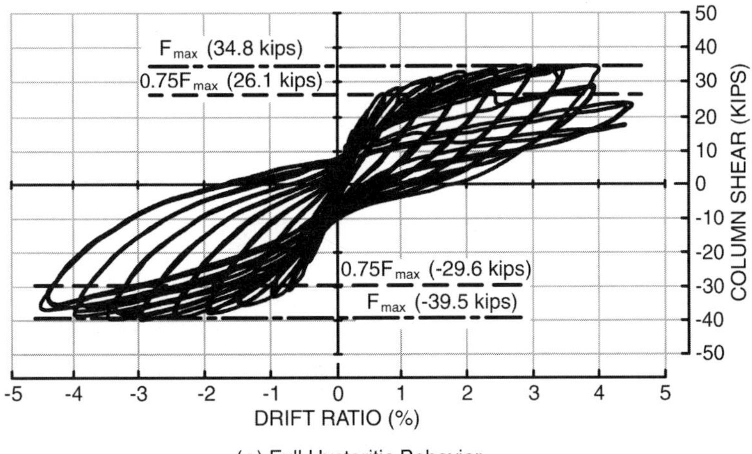

(*a*) Full Hysteritic Behavior

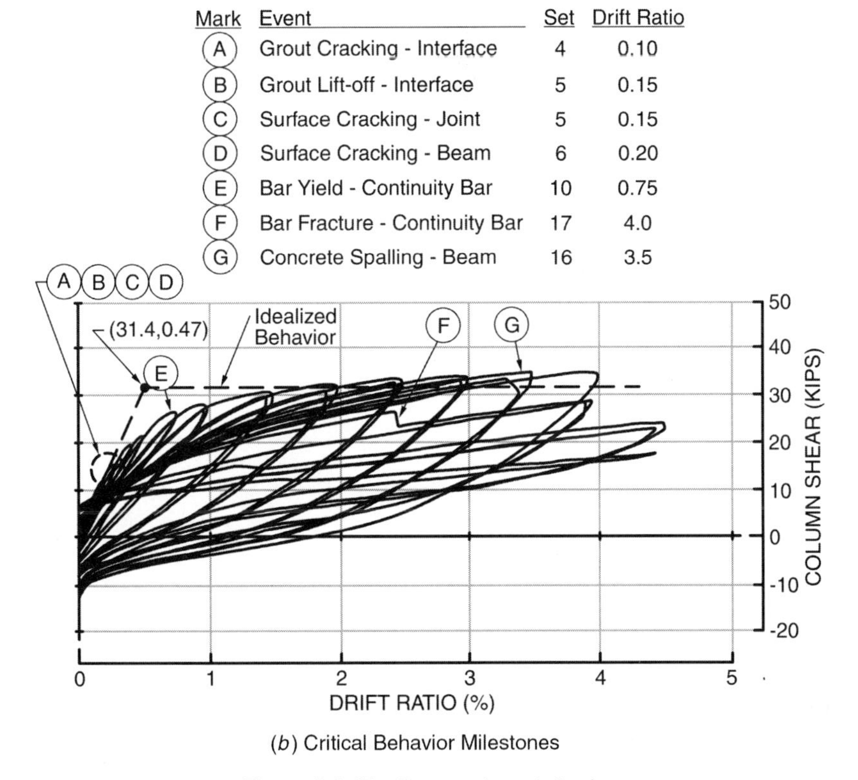

(*b*) Critical Behavior Milestones

Figure 2.1.47 Test specimen behavior.

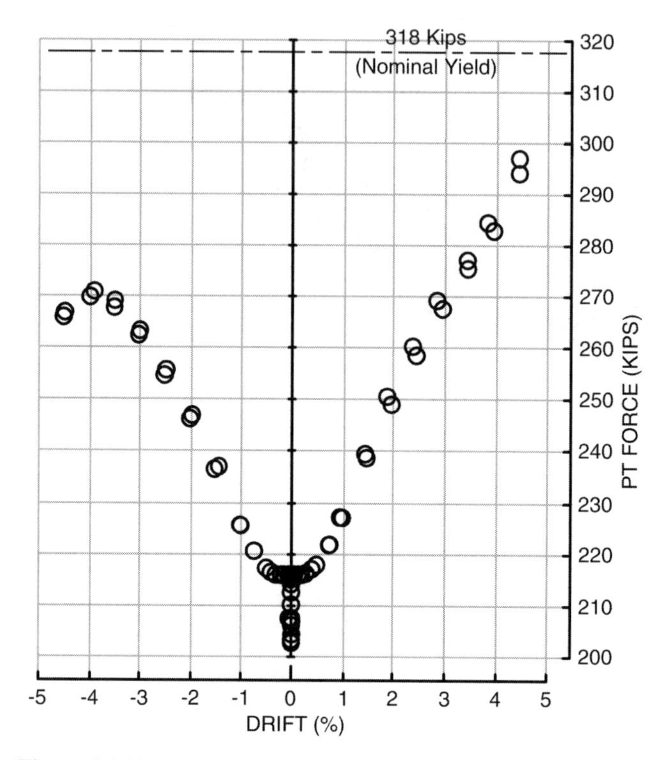

Figure 2.1.48 Post-tensioning (PT) force versus drift relationship.

will be small (Figure 2.1.47). The plastic hinge length initially will be the debonded length of the mild steel—in this case 6 in.

$$
\begin{aligned}
\theta_y &= \phi_y \ell_p \\
&= \frac{\varepsilon_{sy}}{h - c}(6) \\
&= \frac{0.002}{20 - 4}(6) \\
&\cong 0.00075 \text{ radian}
\end{aligned}
$$

Further, assume that all of this postyield rotation occurs at the beam-column interface (Figure 2.1.49). The elongations of the tensile reinforcement components (δ_{ps} andδ_s) are best made using an iterative process, for the stress levels in the reinforcement will dictate the location of the neutral axis. The adopted stress/strain relationships are those shown in Figure 2.1.50. If the compression load imposed on the concrete is increased by 50% (Figure 2.1.48), it is reasonable to start the iterative process with a neutral axis depth of 6 in.

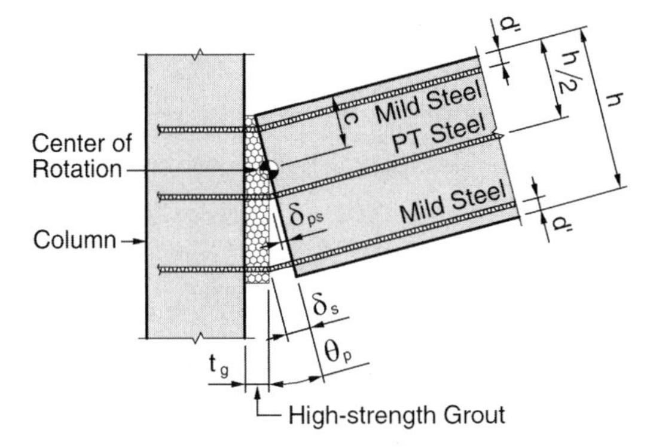

Figure 2.1.49 Rotation at beam-column interface—hybrid system.

Assume $c = 6$ in. Then

$$\begin{aligned}\delta_s &= (d - c)\theta_p \\ &= (18.25 - 6)(0.04) \\ &= 0.49 \text{ in.}\end{aligned}$$

The intentional mild steel debond length (Figure 2.1.45) was 6 in. but can be reasonably expanded to include some adjacent debonding. Hence the effective debond length (ℓ_d) is on the order of

$$\begin{aligned}\ell_d &= 6 + 2(d_b) \\ &= 7.5 \text{ in.}\end{aligned} \qquad \text{(Eq. 2.1.47)}$$

and the postyield strain state in the debond region (ε_{sp}) is

$$\begin{aligned}\varepsilon_{sp} &= \frac{\delta_s}{\ell_d} \\ &= \frac{0.49}{7.5} \\ &= 0.065 \text{ in./in.}\end{aligned}$$

This corresponds to a stress in the bar of

$$\lambda_o f_y \cong 86 \text{ ksi} \qquad \text{(see Figure 2.1.50}a\text{)}$$

The elongation of the post-tensioning strand (δ_{psp}) is

$$\delta_{psp} = \left(\frac{h}{2} - c\right)\theta_p$$

$$= (10 - 6)0.04$$

$$= 0.16 \text{ in.}$$

The overall length of the strand (Figure 2.1.45) is on the order of 100 in.; hence

$$\Delta\varepsilon_{psp} = \frac{0.16}{100}$$

$$= 0.0016 \text{ in./in.}$$

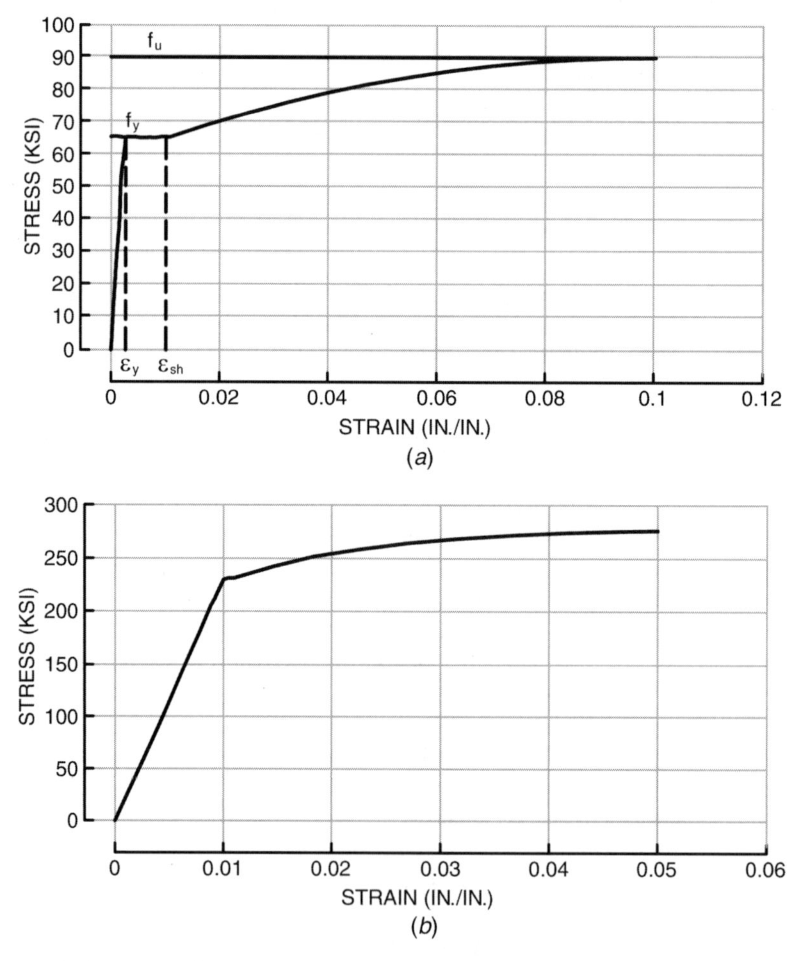

Figure 2.1.50 (*a*) Idealized behavior of Grade 60 reinforcing steel assumed in the design of the test specimen of Figure 2.1.45. (*b*) Idealized stress-strain relationship for 270-ksi stress-relieved strand assumed in the design of the test specimen of Figure 2.1.45.

$$
\begin{aligned}
\Delta f_{ps} &= \Delta \varepsilon_{psp} E_{ps} \\
&= 0.0016(28{,}000) \\
&= 45 \text{ ksi} \\
f_{ps} &= f_{se} + \Delta f_{psp} \\
&= 162 + 45 \\
&= 207 \text{ ksi} < f_{py} \cong 230 \text{ ksi} \qquad \text{(see Figure 2.1.50}b\text{)}
\end{aligned}
$$

The total tensile force in the post-tensioning steel at a joint rotation (θ_p) of 0.04 radian is

$$
\begin{aligned}
\Delta T_{ps} &= A_{sp} \Delta f_{ps} \\
&= 9(0.153)45 \\
&= 61.7 \text{ kips} \\
T_{ps} &= 223.1 + 61.7 \\
&= 285 \text{ kips} \qquad \text{(Theoretical)}
\end{aligned}
$$

Comment: Observe that the projected force in post-tensioning force is consistent with that measured (on the positive cycle) at a drift angle of 4% (Figure 2.1.48).

The tensile force provided by the mild reinforcing is

$$
\begin{aligned}
T_s &= 3(0.44)86 \\
&= 114 \text{ kips}
\end{aligned}
$$

The depth to the neutral axis may now be estimated:

$$
\begin{aligned}
T_{ps} + T_s - C_{sy} &= 285 + 114 - 79 \\
&= 320 \text{ kips} \\
a &= \frac{320}{0.85(5)(16)} \\
&= 4.7 \text{ in.} \\
c &= \frac{a}{0.8} \\
&= 5.9 \text{ in.}
\end{aligned}
$$

Conclusion: A neutral axis depth of 6 in. is reasonably presumed.

Strain levels in the concrete are quite high. If we assume a plastic hinge length (ℓ_p) equivalent to the effective debond length ($\ell_d + 2d_b$), then

$$\phi_p = \frac{\theta_p - \theta_y}{\ell_p} \tag{2.1.57}$$

$$= \frac{0.0392}{7.5}$$

$$= 0.0052 \text{ rad/in.}$$

$$\varepsilon_c = \phi_p c$$

$$= 0.0052(6)$$

$$= 0.031 \text{ in./in.}$$

$$\varepsilon_s = \phi_p(d - c)$$

$$= 0.0052(12.25)$$

$$= 0.064 \text{ in./in.}$$

It was for this reason that the developers of the hybrid system armored beam corners with angles during the NIST tests.[2.8] The University of Washington test specimens were not armored. Beam corners did exhibit surface cracking early, but beam strength did not begin to deteriorate until drifts exceeded 4% (see Figure 2.1.47). The plasticity of high-strength grout probably absorbs a disproportionate amount of the postyield concrete strain, for these grouts are usually quite ductile.

Comment: The fact that one of the mild steel bars fractured at a drift angle of 2.5% is disconcerting because the apparent strain in this bar is below that normally associated with fracture. Seven identical beams were tested, and this bar was the only one that fractured. A conservative selection of the debond length based on a fracture strain of 5% is suggested.

The probable moment capacity (M_{pr}) at a drift angle of 4%, discounting the ruptured #6 bar, is

$$M_{pr} = T_{ns}(d - d') + T_{ps}\left(\frac{h}{2} - \frac{a}{2}\right) + (T_s - T_{ns})\left(d - \frac{a}{2}\right) \tag{2.1.58}$$

$$= 52.8(16.5) + 285(10 - 2.35) + 23(18.25 - 2.35)$$

$$= 871 + 2180 + 366$$

$$= 3417 \text{ in.-kips}$$

This corresponds to a column shear force of

$$V_{c,\mathrm{pr}} = M_{\mathrm{pr}}\left(\frac{\ell}{\ell_c}\right)\left(\frac{1}{h_x}\right)$$

$$= 3417\left(\frac{72}{62}\right)\left(\frac{1}{117.5}\right)$$

$$= 33.8 \text{ kips}$$

This predicted column shear force is consistent with the test results (Figure 2.1.47).

Design Process (See Section 2.1.2.6.)

Step 1: Determine a Trial Reinforcing Program. It is advisable to maintain a reasonable level of restoring force, for this is clearly a very positive attribute of post-tensioning. Accordingly, a design objective should be to provide at least 50% of the moment capacity with the post-tensioning. Start by conservatively selecting the appropriate level of mild steel reinforcing:

$$M_{us} \cong 0.4M_u \tag{2.1.59}$$

$$A_s = \frac{0.4M_u}{\phi f_y(d - d')}$$

Then determine the amount of post-tensioning steel required to satisfy strength objectives:

$$M_{ups} = M_u - 0.9f_y(d - d')A_s \tag{2.1.60}$$

$$A_{ps} = \frac{M_{ups}}{\phi f_{ps}(h/2 - a/2)} \tag{2.1.61}$$

where f_{ps} may conservatively be assumed to be the effective level of prestress (f_{pse}), usually 162 ksi.

Comment: A nominal strength projection based on an effective prestress of 162 ksi will result in a strength equivalent to about 95% of the nominal strength commonly assumed to exist in continuous post-tensioned beams. The use of f_{ps} equal to f_{pse} seem reasonable based on the performance described in Figure 2.1.47.

Step 2: Determine the Minimum Size of the Beam-Column Joint. In Section 2.3 we will develop, for the hybrid beam system, a relationship between the area of the beam-column joint and the amount of beam reinforcing, a relationship that is similar in form to Eq. 2.1.16:

$$A_j = 62\left(A_s + A_s'\right) + 210A_{ps} \tag{2.1.62}$$

Step 3: Check the Feasibility of Placing the Reinforcement Suggested by the Trial Design.

Step 4: Check to Insure That the Provided Level of Post-Tensioning Is Reasonable—On the Order of 1000 psi.

Comment: The hybrid beam will become an integral part of the floor system. Large stress differentials may cause undesirable cracking in unstressed floors. Accordingly, it is best to use 1000 psi as an objective prestress limit.

Step 5: Check the Beam Shear Capacity/Demand Ratio to Insure That Objective Shear Stress Limit States Have Not Been Exceeded. For an exterior subassembly (Figure 2.1.45) this is

$$V_b = \frac{\lambda_o(M_{ns} + M_{nps})}{\ell_c/2} + \frac{P_D + P_L}{2} \qquad \text{(see Eq. 2.1.20)}$$

Comment: Sufficient accuracy for design purposes may be attained through the use of an overstrength factor (λ_o) of 1.25. Equation 2.1.58 is more appropriately used for analysis.

$$bd \geq \frac{V_b}{5\sqrt{f_c'}} \qquad \text{(Eq. 2.1.21)}$$

Remember that P_D and P_L in this case should be realistically selected, not the factored loads used in a code compliance analysis. This limit state $\left(v < 5\sqrt{f_c'}\right)$ is perhaps somewhat conservatively applied to the hybrid system because the hinge deformation tends to accumulate at the beam-column interface (see Figure 2.1.49), and the hinge region is prestressed.

Step 6: Check Column Shear If the Beam Is Deep and the Column Short. For an exterior subassembly (Figure 2.1.45) this is

$$b_c h_c \geq \frac{\lambda_o(M_{ns} + M_{nps})}{5\sqrt{f_c'}h_x} \qquad \left(f_c' = 5000 \text{ psi}\right)$$

$$\geq \frac{3.6(M_{ns} + M_{nps})}{h_x} \qquad \text{(see Eq. 2.1.28)}$$

Design Example: Design a hybrid beam for the residential building of Example 2 in Section 2.1.2.7. See Figure 2.1.28.

Step 1: Select a Trial Beam Reinforcement Program.

$$V_c h_x = M_{bu}\left(\frac{\ell_1}{\ell_{c1}} + \frac{\ell_2}{\ell_{c2}}\right) \qquad \text{(see Eq. 2.1.35)}$$

For an interior column and an objective column shear of 200 kips,

$$\begin{aligned} M_{bu} &= \frac{V_c h_x}{\ell_1/\ell_{c1} + \ell_2/\ell_{c2}} \qquad (2.1.63) \\ &= \frac{200(9)}{1.12 + 1.2} \\ &= 776 \text{ ft-kips} \qquad (9310 \text{ in.-kips}) \end{aligned}$$

Select a mild steel reinforcing program:

$$\begin{aligned} M_{us} &= 0.4M_{bu} \qquad \text{(Eq. 2.1.59)} \\ &= 0.4(776) \\ &= 310 \text{ ft-kips} \end{aligned}$$

$$\begin{aligned} A_s &= \frac{M_{us}}{\phi f_y(d - d')} \\ &= \frac{310(12)}{0.9(60)(29 - 3)} \\ &= 2.65 \text{ in.}^3 \end{aligned}$$

Try three #9 bars.

Select the post-tensioning reinforcement:

$$\begin{aligned} M_{us} &= A_s \phi f_y(d - d') \\ &= 3.0(0.9)(60)(29 - 3) \\ &= 4212 \text{ in.-kips} \qquad (M_{ns} = 4680 \text{ in.-kips}) \end{aligned}$$

$$\begin{aligned} \phi M_{nps} &= M_{bu} - M_{us} \\ &= 776(12) - 4212 \\ &= 5100 \text{ in.-kips} > 4212 \text{ in.-kips} \qquad \text{(OK)} \end{aligned}$$

Accordingly, the flexural strength provided by the post-tensioning exceeds that provided by the mild steel, and this was identified as a design objective.

For design purposes the ACI code and most texts[2.6, 2.21] suggest that the nominal strength of unbonded strands (f_{ps}) may be assumed to be at least

$$f_{ps} = f_{pse} + 10{,}000 \qquad \text{(see Ref. 2.6, Eq. 18-4)}$$

Comment: This development is based on a correlation between test data and analysis. Observe that a similar relationship between effective and nominal strength may not be appropriate for the hybrid system. This may be a consequence of the straight (undraped) post-tensioning program used in the hybrid system. Recall that the tensile force in the tendon corresponding to a strand stress of 172 ksi in the test described in Figure 2.1.45 would be 237 kips and that this force is not developed until the drift reaches 1.5% (Ref. Figure 2.1.48). Accordingly, the use of f_{pse} to define the nominal strength of a hybrid beam is viewed as being more appropriate.

$$\begin{aligned} A_{ps} &= \frac{\phi M_{nps}}{\phi f_{pse}(h/2 - a/2)} \\ &= \frac{5100}{0.9(162)(16-3)} \end{aligned} \tag{2.1.64}$$

where the depth of the compressive stress block (a) is presumed to be about 6 in.

$$A_{ps} = 2.69 \text{ in.}^2$$

Use 0.6-in. ϕ strands ($A = 0.217$ in.^{2}).

Comment: 0.6-in. ϕ strand hardware is most common in the United States, but local availability should be confirmed.

$$\begin{aligned} \text{Number of strands} &= \frac{2.69}{0.217} \\ &= 12.4 \text{ strands required} \end{aligned}$$

Check the level of prestress provided by 12 strands:

$$\begin{aligned} f_c &= \frac{12(0.217)(162)}{20(32)} \\ &= 0.660 \text{ ksi} \qquad \text{(OK)} \end{aligned}$$

Conclusion: The trial reinforcing program will consist of twelve 0.6-in. ϕ strands and three #9 bars top and bottom.

$$M_n = M_{nps} + M_{ns} \tag{2.1.65}$$

$$a = \frac{T_{ps}}{0.85 f_c' b}$$

$$= \frac{12(0.217)(162)}{0.85(5)(20)}$$

$$= 5 \text{ in.}$$

$$M_{nps} = T_{ps}\left(\frac{h}{2} - \frac{a}{2}\right)$$

$$= 422(16 - 2.5)$$

$$= 5697 \text{ in.-kips}$$

$$M_n = 5987 + 4680$$

$$= 10{,}380 \text{ in.-kips}$$

$$\phi M_u = 9340 \text{ in.-kips} > 9310 \text{ in.-kips} \qquad \text{(OK)}$$

Step 2: Determine the Minimum Size of the Beam-Column Joint.

$$A_j = 62\left(A_s + A_s'\right) + 210A_{ps} \qquad \text{(Eq. 2.1.62)}$$

$$= 62(6) + 210(12)(0.217)$$

$$= 919 \text{ in.}^2$$

Conclusion: Column size should be 30 in. by 32 in.

Step 3: Develop Beam and Column Reinforcing Program. The mild steel must be placed in ducts and grouted. The post-tensioning strands will also be placed in ducts and stressed using a multistrand jack. Mild steel tubes should have a diameter of at least $2d_b$ and be spaced so as to provide a gap of no less than d_b in width (Figure 2.1.51).

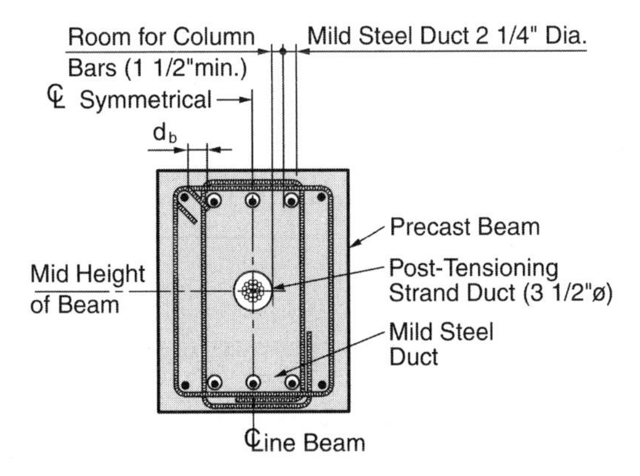

Figure 2.1.51 Hybrid beam reinforcement program.

The post-tensioning duct will be of the size suggested by the supplier. Column bars will need to be placed in this central region, and they must pass the post-tensioning duct and be inside the outer mild steel duct.

Beam bars cast in the precast beam are also required if for no other reason than to reinforce an otherwise unreinforced region. The size of these bars will probably depend to some extent on the adopted mild steel splicing program. Minimum flexural reinforcing requirements are satisfied by the post-tensioning, so the basis for the sizing of these bars depends on the designer's convictions relative to splicing concerns. The ACI[2.6, Sec. 18.9] requires a minimum amount of bonded reinforcement in all flexural members with unbonded prestressing tendons:

$$A_{s,\min} = 0.004A \qquad \text{(see Ref. 2.6, Eq. 18-6)}$$

where A is the area between the tension face and the center of gravity of the section.

The beam sizing process must then allow for the flexural reinforcing provided in the precast beam as well as the space that may be required to pass the column bars (see Figure 2.1.51). Minimum beam width becomes

One post-tensioning duct	=	3.50 in.
Two #11 column bars	=	2.75 in.
Two mild steel ducts	=	4.50 in.
Two corner bar diameters	=	2.00 in.
Two corner bars (#8)	=	2.00 in.
#5 hoop ties	=	1.25 in.
Fire cover	=	3.00 in.
		19.00 in.

Conclusion: The minimum beam width should be 20 in. The minimum bonded reinforcement[2.6, Eq. 18-6] is

$$\begin{aligned} A_{s,\min} &= 0.004b\frac{h}{2} \\ &= 0.004(20)16 \\ &= 1.28 \text{ in.}^2 \end{aligned}$$

Conclusion: The beam should be 20 in. wide. At least one #8 bar should be provided in each corner.

Comment: This mild steel may need to be increased to reduce concrete cracking during transportation, but this will be the concern of the fabricator.

Step 4: Check the Reinforcement Ratio. A maximum prestress reinforcement ratio is identified by the ACI. The appropriateness of this requirement is discussed when we analyze the beam below (Analysis, Step 4).

$$\omega_p < 0.36\beta_1 \qquad \text{(see Ref. 2.6, Sec. 18.8.1)}$$

$$\frac{A_{ps}}{bd_p}\frac{f_{ps}}{f'_c} < 0.36\beta_1$$

$$\frac{12(0.217)}{20(16)}\left(\frac{162}{5}\right) < 0.36(0.8)$$

$$0.26 < 0.288 \qquad \text{(Complies with Ref. 2.6, Sec. 18.8.1)}$$

Comment: The limitation imposed on ω_p is based on a bonded system. The objective is to insure some level of ductility. The imposed limit $(0.36\beta_1)$ is not appropriate for the hybrid system because a significant amount of ductility is available. Observe that had A_{ps} been 50% greater than the quantity proposed $(P/A \cong 1000$ psi) this code provision would not have been satisfied.

Step 5: Check Shear Capacity of the Shorter Beam.

$$\begin{aligned} M_{ns} &= T_{ns}(d - d') \\ &= 3(60)(29 - 3) \\ &= 4680 \text{ in.-kips} \end{aligned}$$

The overstrength attained in the mild steel will be equilibrated by a compressive force in the concrete. The imposed compressive force on the concrete (C_c) is

$$\begin{aligned} C_c &= (\lambda_o - 1)T_{ns} + \lambda_o T_{nps} \\ &= 0.25(3)(60) + 1.25(12)(0.217)(162) \\ &= 572 \text{ kips} \end{aligned}$$

$$\begin{aligned} a &= \frac{C_c}{0.85f'_c b} \\ &= \frac{572}{0.85(5)(20)} \\ &= 6.73 \text{ in.} \end{aligned}$$

These forces $\left(C_c, (\lambda_o - 1)T_{ns}, \text{and} \lambda_o T_{nps}\right)$ generate a probable moment component $\left(M'_{b,\text{pr}}\right)$ of

$$\begin{aligned} M'_{b,\text{pr}} &= (\lambda_o - 1)T_{ns}\left(d - \frac{a}{2}\right) + \lambda_o T_{nps}\left(\frac{h}{2} - \frac{a}{2}\right) \\ &= 45(29 - 3.36) + 527(16 - 3.36) \\ &= 7815 \text{ in.-kips} \end{aligned}$$

The probable flexural strength is

$$\begin{aligned} M_{\text{pr}} &= M_{ns} + M'_{b,\text{pr}} \\ &= 4680 + 7815 \\ &= 12{,}500 \text{ in.-kips} \end{aligned}$$

The level of provided overstrength is

$$\begin{aligned} \lambda_{o,\text{prov}} &= \frac{M_{\text{pr}}}{M_{nb}} \\ &= \frac{12{,}500}{10{,}380} \\ &= 1.20 \end{aligned}$$

Comment: The preceding development is too long and detailed to be a part of the design process. Clearly the use of $\lambda_o M_n$ as the probable driving seismic moment is more appropriate for capacity-based component designs. The use of an overstrength factor of 1.25 in design is typically conservative for the hybrid beam, and this conclusion can also be reached by reviewing the test discussed at the beginning of this section.

$$\begin{aligned} V_{b,\text{pr}} &= \frac{2\lambda_o M_n}{\ell_c} + \frac{(w_D + w_L)\ell_c}{2} \qquad \text{(see Eq. 2.1.20)} \\ &= \frac{2(12,500)}{192 - 32} + \frac{2.22(160)}{2(12)} \\ &= 171.0 \text{ kips} \end{aligned}$$

Code compliance is the objective—hence w_D and w_L are factored loads (see Ref. 2.6, Eq. 9-2).

$$\begin{aligned} v_u &= \frac{V_{b,\text{pr}}}{bd} \\ &= \frac{171.0}{20(29)} \\ &= 0.3 \text{ ksi} < 8\phi\sqrt{f'_c} \qquad \text{(Complies with code)} \end{aligned}$$

More realistically, the probable level of shear imposed on the beam will be

$$V_{b,\text{pr}} = \frac{2\lambda_o M_n}{\ell_c} + V_{bD,\text{pr}}$$

$$= 156.3 + 10.5$$
$$= 166.8 \text{ kips}$$

$$v_{b,\text{pr}} = \frac{V_{b,pr}}{bd}$$
$$= \frac{166.8}{20(29)}$$
$$= 0.288 \text{ ksi} < 5\sqrt{f_c'} \qquad \text{(Complies with objective shear stress levels)}$$

(See Section 2.1.1.3).

Step 6: Check Column Shear Capacity. The column shear is developed from the probable moment capacities of the beams. For column C in Figure 2.1.28 this is

$$V_{c,\text{pr}} = \frac{2M_{b,\text{pr}}}{h_x}\left(\frac{\ell}{\ell_c}\right)(1.3) \qquad \text{(see Eq. 2.1.24)}$$

$$= \frac{2(12{,}500)}{9(12)}\left(\frac{240}{208}\right)(1.3)$$

$$= 347 \text{ kips}$$

This corresponds to an imposed shear stress of

$$v_{cu} = \frac{V_{c,\text{pr}}}{bd}$$
$$= \frac{347}{32(29)}$$
$$= 0.374 \text{ ksi} \qquad \text{(use 6 ksi concrete.)}$$

Proposed design:

Beam size:	20 in. by 32 in.
Beam reinforcing:	Twelve 0.6-in ϕ, 270-ksi strands
	Three #9 bars top and bottom
Column size:	32 in. by 32 in., $f_c' = 6$ ksi

Analysis: An analysis procedure was developed in Section 2.1.3.1. The first two steps were discussed in the development of the conventionally reinforced frame beam (Section 2.1.3.2). Accordingly, we start with Step 3.

Step 3: Develop the Design Moment. The computer-generated moments for the frame of Figure 2.1.28 were presented in Figure 2.1.29. Hybrid beam moment capacities can be different on the top and bottom of the beam. This would be accomplished by providing more mild steel in the top than the bottom. Energy dissipation and ductility would be unbalanced. Further, the objective relationship between the restoring forces provided by the post-tensioning and mild steel would probably be more difficult to attain. The author's preference is to create a beam that has the same moment capacity on each face and on both sides of an interior column. Two methodologies for attaining this objective, a balanced flexural strength, have previously been presented. They are not repeated here, but rather shortened as they would be in a design situation.

The nominal moment capacity of the hybrid beam is developed as follows:

$$f_{ps} = f_{pse} \qquad \text{(See hybrid beam design process, step 1)}$$

$$= 162 \text{ ksi}$$

$$T_{nps} = 12(0.217)162$$

$$= 422 \text{ kips}$$

$$a = \frac{T_{nps}}{0.85 f_c' b}$$

$$= \frac{422}{0.85(5)20}$$

$$= 4.96 \text{ in.}$$

$$M_{nps} = T_{nps}\left(\frac{h}{2} - \frac{a}{2}\right)$$

$$= 422(16 - 2.48)$$

$$= 5705 \text{ in.-kips}$$

$$M_{ns} = 4680 \text{ in.-kips}$$

$$M_n = M_{nps} + M_{ns}$$

$$= 10{,}385 \text{ in.-kips}$$

$$= 865 \text{ ft-kips}$$

$$M_u = \phi M_n$$

$$= 779 \text{ ft-kips}$$

Moment redistribution was used in Section 2.1.3.2 to establish a demand of 859 ft-kips. In Section 2.1.4.3 a mechanism approach was used to establish the provided strength limit state. It was demonstrated that, by adjusting the positive moment capacity of the frame beam, we could cause the sidesway mechanism (Figure 2.1.43*c*) to be critical; hence

$$\phi M_n > \frac{V_{uE} h_x}{2} \qquad \text{(see Eq. 2.1.49)}$$

The shear imposed on the column can be developed from the seismic moment demands of Figure 2.1.29*a* and *b*. It is more convenient, however, to use the generated seismic only moment demands directly:

$$\begin{aligned} 2\phi M_n &> M_{uL} + M_{uR} \\ &> 743 + 921 \\ &> 1664 \text{ ft-kips} \\ \phi M_n &> 832 \text{ ft-kips} \end{aligned}$$

It seems advisable since even the mechanism demand is less than the proposed capacity (779 ft-kips), to increase the strength of the beam in spite of the fact that the moment demands were generated at the column centerline. The addition of two more strands should accomplish this objective:

$$\begin{aligned} T_{nps} &= 14(0.217)(162) \\ &= 492 \text{ kips} \\ a &= \frac{492}{0.85(5)(20)} \\ &= 5.8 \text{ in.} \\ M_{nps} &= 492(16 - 2.9) \\ &= 6445 \text{ in.-kips} \\ M_n &= 11{,}125 \text{ in.-kips} \\ \phi M_n &= 834 \text{ ft-kips} > 832 \text{ ft-kips} \end{aligned}$$

Step 4: Develop a Flexural Reinforcement Program for the Hybrid Beam. The development of a reinforcing program for the hybrid beam is simple. The only element that needs to be considered is the length of the energy dissipating bonded mild steel bars. It is most convenient that the length of these bars not exceed the length of the precast beam. This is a constructibility consideration, for the bars are most easily installed by sliding them through the column after the precast members have been assembled. Shop drawings will reflect the location of access ports for assembly and grouting. Access port locations need be considered from an aesthetic perspective only where the precast member is to serve as the architectural finish.

The minimum length of the bar (ℓ_b) must include the debond regions (ℓ_{db}) and a bar development length, as well as the width of the column and grout spaces (see Figure 2.1.52):

$$\ell_b = 2\ell_d + 2\ell_{db} + h_{col} + 2 \text{ in.} \leq \ell_{pre}$$

The development length is usually established or confirmed by test, for it will be a function of the duct type and the strength of the grout. For plan production purposes the length available for mild steel development is all that need be specified along with a requirement that the contractor demonstrate that the proposed tube and grout can develop the rupture strength of the mild steel bar.

Comment: From a practical perspective little is gained by reducing the length of these bars from the length of the precast beam (ℓ_{pre}) because the location of the debond region is critical and bar placement must be visually monitored. Shorter bars will require two access points where one access point could accomplish the quality control objective.

In order to determine the debond length and the probable moment (M_{pr}), the strain states in the concrete and steels must be estimated. Accordingly, Step 6, Review Strain States, must precede the completion of Steps 4 and 5.

The hybrid beam does not fit into our classic strain-based limit states. Accordingly, it is appropriate here to discuss strain states for conventionally reinforced beams as well as prestressed and post-tensioned reinforced members. Strain limit states are subliminally identified in codes and design rules. The objective of these rules is to insure that at least a minimum amount of ductility is provided in all

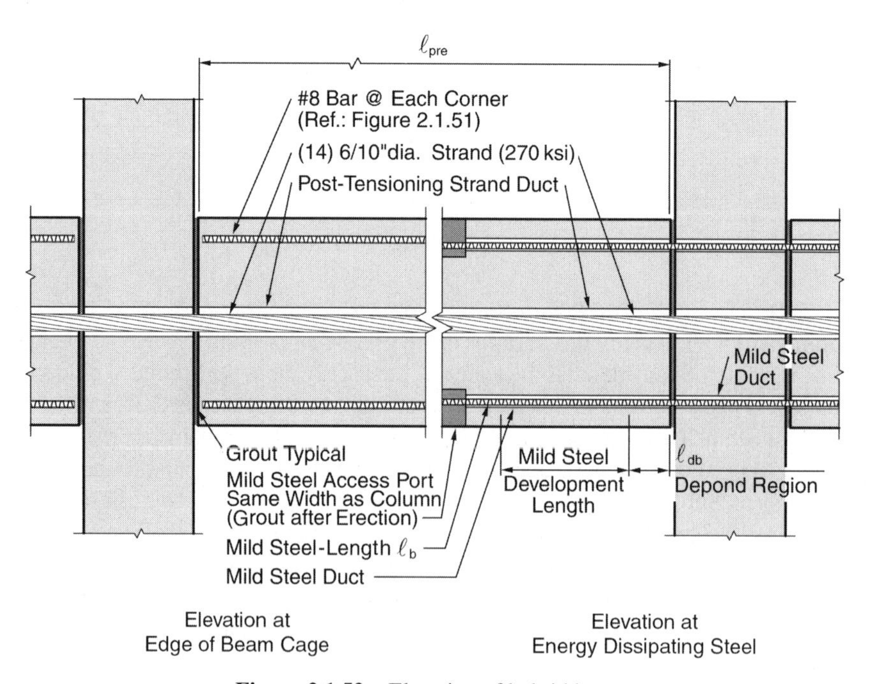

Figure 2.1.52 Elevation of hybrid beam.

beams and that reinforcing steel will yield but not fracture before the strength limit state of the member is reached.

Consider minimum reinforcing objectives. Classically, the minimum requirement for conventionally reinforced beams is

$$A_{s,\min} = \frac{200bd}{f_y} \tag{2.1.66}$$

The objective of Eq. 2.1.66 is to insure that the flexural reinforcing does not rupture should the beam be subjected to a moment that causes the section to crack. This minimum steel requirement is derived from the analysis of a rectangular section.

$$\begin{aligned} M_{\text{cr}} &= f_r S_x \\ &= 7.5\sqrt{f_c'}\,\frac{bd^2}{6} \\ A_s &= \frac{M_{\text{cr}}}{f_y\,(d - a/2)} \end{aligned}$$

The concrete strength used in the development of this relationship many years ago was on the order of 4000 psi, and since the moment was understood to be small, the effect of the compressive stress block $(a/2)$ was deleted for it too would be very small. This allowed the required amount of reinforcement (A_s) to be described in terms of stress limit states and the member size:

$$\begin{aligned} A_s &= \frac{7.5\sqrt{f_c'}bd^2}{6f_y d} \\ &= \frac{7.5\sqrt{f_c'}}{6}\,\frac{bd}{f_y} \\ &= 1.25\sqrt{f_c'}\,\frac{bd}{f_y} \end{aligned} \tag{2.1.67a}$$

The factor of safety or overcapacity $(1/\phi)$ inherent in Eq. 2.1.66 was in excess of 2 for 4000-psi concrete:

$$\begin{aligned} \frac{1}{\phi} &= \frac{200}{1.25\sqrt{4000}} \\ &= 2.53 \\ \phi &= 0.4 \end{aligned}$$

We have significantly increased the strength of concrete over the years, and the factor of safety inferred by the traditional limit $(200\ bd/f_y)$ continues to be reduced.

For 12,000-psi concrete the factor of safety, given the limit proposed by Eq. 2.1.66, would be less than 1.5. This apparent factor of safety is further impacted by the level of cracking stress (f_{cr}) typically attained in high performance concrete, one that is significant above that traditionally assumed to be the mean.[2.3]

The minimum reinforcing requirement as now defined by the ACI[2.6] is a function of f_c':

$$A_{s,\min} = \frac{3\sqrt{f_c'}b_w d}{f_y} \qquad \text{(see Ref. 2.6, Eq. 10-3)} \quad (2.1.67b)$$

and this corresponds to an overstrength factor of only 2.4 (3(6)/7.5), which based on the tensile overstrengths characteristic of high-strength concrete[2.3] is minimal. Where the flange of a T-section is in tension, the limit state must be altered so as to attain the same objective (see Ref. 2.6, Sec. 10.5).

From the standpoint of seismic design, minimum steel will provide a very inefficient member, but it is a criterion away from the plastic hinge region as discussed in Section 2.1.3.2, Step 4:

$$\rho_{\min} = \frac{3\sqrt{f_c'}}{f_y} \qquad \text{(see Ref. 2.6, Eq. 10-3)}$$

For 5000-psi concrete and 60-ksi steel, $\rho_{\min}$ is only

$$\rho_{\min} = \frac{3\sqrt{5000}}{60{,}000} = 0.35\%$$

This is a reasonable objective prior to the activation of the post-tensioning force.

A similar concept applies to minimum prestressed and nonprestressed reinforcement in a prestressed member. The objective is to avoid a sudden failure. ACI[2.6, Sec. 18.8.3] requires that the provided reinforcement, both conventional and prestressing, supply a moment strength that is 20% higher than the cracking moment for the section. This provision is not a factor in the design of hybrid beams because it results in uneconomical sections and is inappropriate for unbonded tendons; this latter because the tendon cannot be fractured or even yielded absent the attainment of high levels of deformation—the desired warning. At midspan the strength provided by the post-tensioning (M_{nps}) and minimum mild steel (Eq. 2.1.67b) should always satisfy this requirement.

Maximum reinforcement ratios have as an objective the avoidance of a premature compression failure in the concrete. In other words the steel should yield before the concrete reaches its strain limit state in compression. The reinforcement limit is 75% of the balanced moment capacity of the section. Balanced conditions are commonly used as a reference point in the design of concrete members. The reference is to balanced conditions of strain or when the strain in the concrete and steel simultaneously reach yield (Figure 2.1.53).

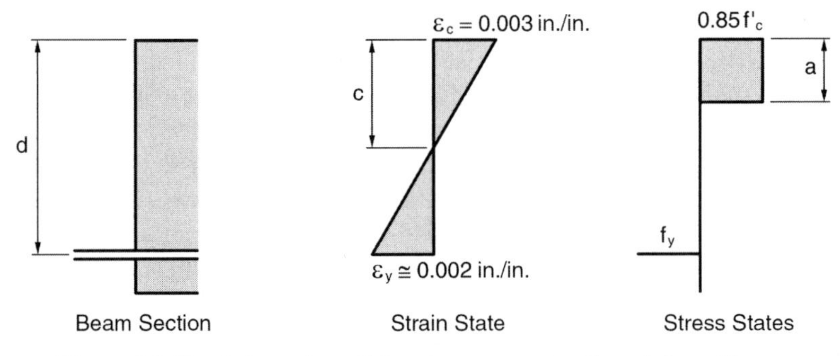

Figure 2.1.53 Balanced conditions in a conventionally reinforced beam.

The balanced moment (M_b) is usually developed from the Whitney stress block and a concrete strain limit state of 0.003 in./in., though other procedures and concrete strain limit states may be used.

$$\begin{aligned}
c &= 0.6d \\
a &= \beta_1 c \\
&= 0.8c \qquad \left(f_c' = 5000 \text{ psi}\right) \\
&= 0.48d \\
T &\equiv C \\
A_s f_y &= 0.85 f_c' b(0.48d) \qquad (2.1.68) \\
&= 0.408 f_c' bd
\end{aligned}$$

The balanced reinforcement ratio (ρ_b) is then

$$\begin{aligned}
\rho_b &= \frac{A_{sb}}{bd} \\
&= 0.408 \frac{f_c'}{f_y} \qquad (2.1.69)
\end{aligned}$$

and the maximum allowed reinforcement ratio ($\rho_{\max}$) is

$$\begin{aligned}
\rho_{\max} &= 0.75\rho_b \\
&= 0.306 \frac{f_c'}{f_y}
\end{aligned}$$

and this, for a concrete strength of 5000 psi and Grade 60 reinforcing steel, becomes

$$\rho_{\max} = 0.0255$$

Hence the limit of 2.5% identified in many seismic design rules. As previously pointed out, this limit state will seldom be constructible and will, if attained, produce minimal ductility, as should be evident from its proximity to balanced strain states. A more realistic maximum is on the order of 1.5%.

The addition of compression reinforcement will alter the development of balanced conditions and, since in beams of any reasonable depth the compression steel will yield provided $((c - d')/c) > 0.67$, limiting reinforcing ratios often include (deduct) the compression reinforcement ratio ρ'. This is not appropriate in seismic design (see Section 2.1.3.3).

Prestressing maximums are also developed from balanced conditions. The development of a limit state is affected by the fact that prestressing strands do not have a well-defined yield point (Figure 2.1.54). The accepted yield strength of stress relieved strand (f_{py}) is defined by the specified minimum load at a 2% elongation, and this is 85% of its minimum breaking strength f_{pu} (Figure 2.1.54).

In the case of 270-ksi strand, f_{py} is 230 ksi $(0.85 f_{pu})$, and the effective prestress (f_{pse}) is assumed to be 162 ksi $(0.6 f_{pu})$ for design purposes. Balanced conditions are developed from Figure 2.1.55.

The strain required to yield the strand is the adopted yield stress for a bonded strand, f_{py} less the effective stress (f_{pse}) divided by the elastic modulus:

$$\begin{aligned} \varepsilon_{sy} &= \frac{f_{py} - f_{pse}}{E} \\ &= \frac{230 - 162}{28{,}000} \\ &= 0.0024 \text{ in./in.} \end{aligned}$$

Balanced conditions are developed in the standard manner:

$$\begin{aligned} c &= \frac{0.003}{0.0054} d_p \\ &= 0.56 d_p \\ a &= \beta_1 c \qquad \left(f_c' = 5000 \text{ psi}\right) \\ &= 0.56 \beta_1 d_p \\ T &= C \\ A_{ps} f_{py} &= 0.85 f_c' \left(0.56 \beta_1 d_p\right) b \\ &= 0.48 \beta_1 d_p b f_c' \end{aligned}$$

Since ρ_p is by definition $d_p b$, the percentage of reinforcement that produces balanced conditions is

$$\rho_b = 0.48\beta_1 \frac{f'_c}{f_{py}} \tag{2.1.70}$$

The prestress reinforcement limit state is 75% of that which creates balanced conditions:

$$\rho_{p,\max} = 0.36\beta_1 \frac{f'_c}{f_{py}} \tag{2.1.71}$$

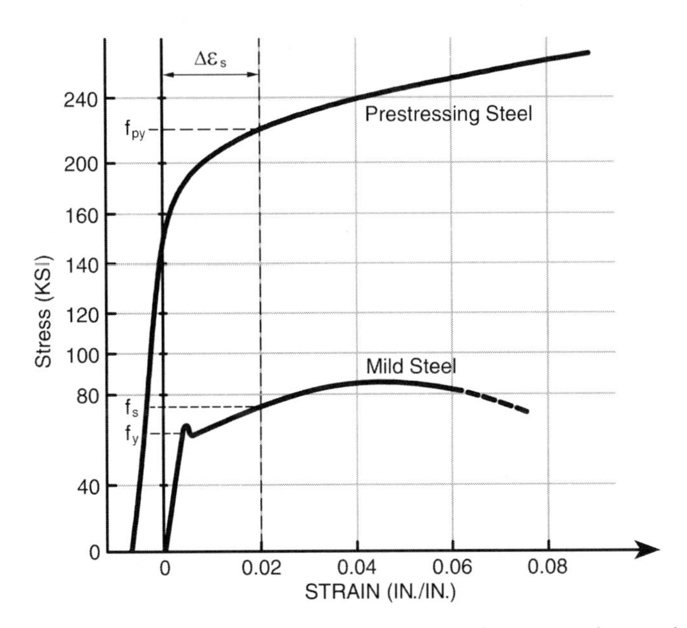

Figure 2.1.54 Generally accepted stress–strain diagram for prestressing steel strands in comparison with mild steel bar reinforcement.[2.21]

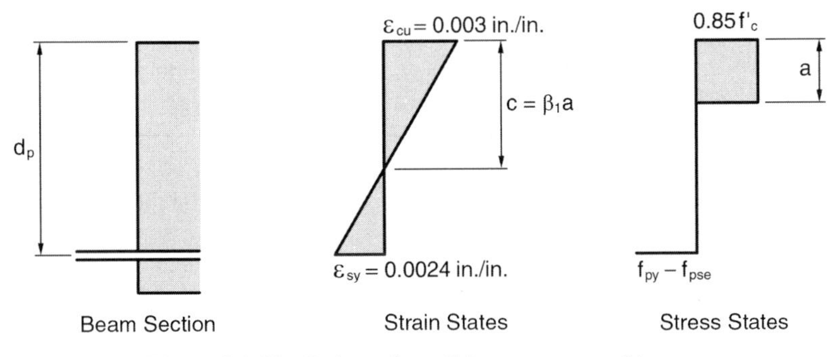

Figure 2.1.55 Balanced condition—prestressed beam.

For 5000-psi concrete,

$$\rho_{p,\max} = 0.63\%$$

Prestressed concrete design procedures identify a reinforcement index (ω_p):

$$\omega_p = \rho_p \frac{f_{py}}{f'_c} \tag{2.1.72}$$

and place a limit on the reinforcement index of

$$\omega_{p,\max} = 0.36\beta_1$$

Compressive reinforcement, mild steel, and variations in the section are accounted for directly from the stress and strain states of Figure 2.1.55. The interested reader is referred to Reference 2.21.

The unbonded tendons used in hybrid beams are not strained locally; rather, postyield deformations elongate the tendon from anchorage to anchorage. When we analyzed the test assembly described in Figure 2.1.45 (see Figure 2.1.48) we found that the stress induced in the tendon at a drift angle (θ_p) of 4% was less than the adopted yield strength (230 ksi). It is not desirable to exceed 90% of the ultimate strength of a strand because the reduced area of strand in the anchorage region caused by the anchoring wedges could result in a rupture of the strand.[2.21, Fig. 2.24]

The prestress reinforcement index for the test module was

$$\omega_p = \rho_p \left(\frac{f_{py}}{f'_c}\right) \qquad \text{(Eq. 2.1.72)}$$

$$\cong \frac{9(0.153)}{10(16)} \left(\frac{213}{6}\right) \qquad \text{(Figure 2.1.48)}$$

$$= 0.0086(35.5)$$

$$= 0.305 > 0.36\beta_1 = 0.288$$

The use of ω_p as a limit state is not appropriate in unbonded post-tensioned construction because the objectives insofar as strand yielding are diametrically opposite. Given the strain between yield and the strength limit state when extended over the distances between anchorages, it is not likely that strands will rupture during the severest of earthquakes, and this will be a condition that we will guard against (Analysis, Step 6).

Conclusion: Hybrid reinforcement limit states must be developed from concrete strain limit states.

Step 5: Develop a Transverse Reinforcing Program. This step is identical in form to the procedure developed for cast-in-place beams (Steps 5 of both Section 2.1.3.1 and Section 2.1.3.2).

Step 6: Review Strain States.

Objective story drift	4%
Tendon length	$\ell_T = 80$ ft (see Figure 2.1.27)
Number of plastic hinges	Eight (see Figure 2.1.27)
Assumed depth to neutral axis	8 in.

This assumption as to the neutral axis depth, though somewhat of a guess, should at least be an educated guess. The depth of the compressive stress block (a) is easily estimated:

$$0.85 f_c' ab = \lambda_o T_{pse} + (\lambda_o - 1) T_{ns}$$

$$0.85(5)(20)a = 1.25(14)(0.217)(162) + 0.25(3)60$$

$$85a = 615 + 45$$

$$a \cong 7.76 \text{ in.}$$

$$c = 9.13 \text{ in.} \qquad \text{(Apparent)}$$

This presumption is, however, probably large and, from a steel strain projection perspective, somewhat unconservative. This is because the actual strength of the concrete will be greater than that specified and higher yet because of the speed with which the concrete is loaded, and this increase can be on the order of 30%.[2.2, Sec. 3.2.2(d)]

Were we to assume that the effective stress in the concrete was 25% greater than that specified, a would become

$$a = 6.2 \text{ in.}$$

$$\beta_1 = 0.75$$

$$c = 8.3 \text{ in.}$$

Conclusion: 8 in. is reasonable and somewhat conservative in terms of predicting the strains imposed on the steel.

The change in tendon length ($\Delta\ell_T$) is now

$$\Delta\ell_T = N\theta_p \left(\frac{h}{2} - c\right)$$

where N is the number of plastic hinges, so

$$\Delta\ell_T = 8(0.04)(16 - 8)$$
$$= 2.56 \text{ in.}$$

The change in tendon strain ($\Delta\varepsilon_p$) is

$$\Delta\varepsilon_p = \frac{\Delta\ell_T}{\ell_T}$$
$$= \frac{2.56}{80(12)}$$
$$= 0.00267 \text{ in./in.}$$

The effective tendon stress (f_{pse}) is 162 ksi.
The change in tendon stress (Δf_{ps}) is

$$\Delta f_{ps} = \Delta\varepsilon_{ps} E_{ps}$$
$$= 0.00267(28{,}000)$$
$$= 75 \text{ ksi}$$
$$f_{ps} = 162 + 75$$
$$= 237 \text{ ksi} \qquad \text{(Essentially elastic)}$$

A similar procedure may be used to estimate the strain states in the mild steel and the concrete. We start by conservatively assuming that the total story drift will be imposed on the beam. The elastic component of the story drift (beam component) may be estimated using the methods developed in Section 2.1.1 or alternatively presumed to be on the order of 0.5%. Observe that this is consistent with the idealization of the experimental behavior reported in Figure 2.1.47*b*.

The remaining rotation (3.5%) will manifest itself in beam liftoff or the opening of a gap at the beam-column interface (see Figure 2.1.38). We can now solve for the debond length associated with a given stress limit in the mild steel or select a debond length and then predict the associated stress level in the mild steel. The identification of a debond length creates a debond region (see Section 2.1.4.1), which we will presume to be the extent of the plastic hinge ℓ_p in spite of the fact that it extends into the column.

$$\ell_p = \text{Intentional debond length} + 2d_b \qquad \text{(Eq. 2.1.47)}$$
$$= 6 + 2(1.125)$$
$$= 8.25 \text{ in.}$$

The postyield strain (ε_{sp}) imposed on the mild steel is

$$\phi_p = \frac{\theta_p}{\ell_p}$$

$$= \frac{0.035}{8.25}$$

$$= 0.0042 \text{ rad/in.}$$

$$\varepsilon_{sp} = \phi_p(d - c)$$

$$= 0.0042(29 - 8)$$

$$= 0.088 \text{ in./in.}$$

This appears quite high given the fact that the strain limit state described in most representations (i.e., Figure 2.1.54) stops at about 10%. ASTM A615, which defines limit states for mild steel, requires that mild steel reinforcing be capable of attaining a 7% elongation. Accordingly, a larger debond length appears to be warranted. We might, however, accept the 6-in. debond region ($5.33d_b$) because we are more concerned about buckling of the bar during the compression cycle than about the possibility of reaching a rotation of 4%. Accepting the 6-in. debond region, we would proceed as follows:

$$C_s' = (f_s - f_y)A_s$$

$$= (90 - 66)3.0 \qquad \text{(see Figure 2.1.50}a\text{)}$$

$$= 72 \text{ kips}$$

$$C_{ps} = f_{ps}A_{ps}$$

$$= 237(14)(0.217)$$

$$= 720 \text{ kips}$$

$$a = \frac{C_s' + C_p}{0.85 f_c' b}$$

$$= \frac{792}{0.85(5)(20)}$$

$$= 9.32 \text{ in.}$$

$$c = 11.6 \text{ in.}$$

Accordingly, steel strains using conventional procedures suggest that induced strains are within accepted limits (7.4%) especially since a beam rotation of 4% is quite conservative.

Concrete strains will be high and depend on the adopted plastic hinge length and neutral axis depth. For a plastic hinge length of 8.25 in. and neutral axis depth of 8 in., the suggested concrete strain is

$$\begin{aligned}\varepsilon_c &= c\phi \\ &= 8(0.0042) \\ &= 0.0336 \text{ in./in.}\end{aligned}$$

Observe that this is consistent with the strain levels predicted for the test beam (0.031 in./in.) using a similar analytical approach (see calculations following Eq. 2.1.57, Section 2.1.4.4).

Clearly, confinement should be provided in the hinge region and shell spalling should be anticipated if the building analysis suggests that rotations of this magnitude (4%) are to be expected. Observe that 4% was the experimentally observed limit state (Figure 2.1.49).

2.1.4.5 Bolted Assemblages The development of the assemblage described in Figure 2.1.56 commonly referred to as the Dywidag Ductile Connector (DDC®), was motivated by a desire to improve the postyield behavior of concrete ductile frames. The basic ideas was to introduce a ductile rod (Figure 2.1.57) or fuse into the load path away from the toe of the beam. The adaptation of the ductile connection concept to precast concrete is logical, because it allows postyield deformations to be accommodated where members are joined (see Figure 2.1.34*c*).

Photo 2.5 Preparing to stress a hybrid frame beam, Paramount Apartments, San Francisco, CA. (Courtesy of Charles Pankow Builders.)

Postyield Behavior The desired behavior is accomplished through a merging of steel technology with the basic objectives of seismic load limiting principles essential to the development of ductile behavior in structural systems that must survive earthquakes.

The Achilles heel of a properly conceived concrete ductile frame beam has always been the toe (no pun intended) of the frame beam where large compressive and shear stresses combine (see Figures 2.1.9 and 2.1.16). This condition of high stress is further aggravated by the preceding cycle tensile overstraining of the flexural reinforcement in the toe region. During reverse cycles of load, the now permanently elongated bar tends to buckle when it is subjected to the entire compression load.

The logical mitigation alternative is to relocate the causative actions. To this end, the ductile connector described in Figure 2.1.56 was proposed and developed as a prototype (see Figure 2.1.61). Observe that it

- Relocates the yielding element to within the column where the confined concrete can provide it with nondeteriorating lateral support,
- Allows the strain in the toe region of the beam to be controlled,
- Transfers shear forces by friction from steel to steel.

As a consequence, the frame beam may be designed to behave elastically because the yielding experienced by the ductile rods will limit the shear and flexural load imposed

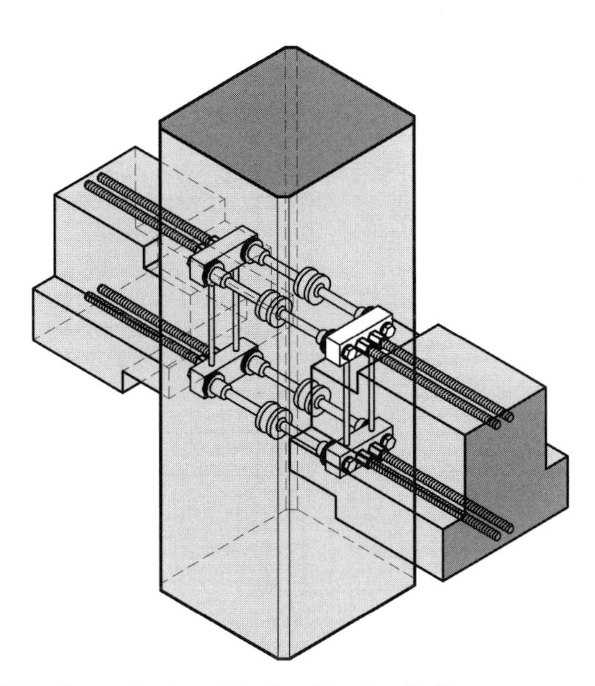

Figure 2.1.56 Isometric view of the Dywidag Ductile Connector system—DDC®.

on the frame beam. Span to depth ratios commonly invoked on cast-in-place concrete frame beams are not appropriately applied to precast assemblies of this type.

System capacity is developed directly from accepted load transfer mechanisms and conditions of equilibrium. The strength reduction factors and overstrength factors are consistent with values used in the design of concrete ductile frames (see Section 2.1.2).

The key element in a ductile frame that contains the ductile connector described in Figure 2.1.56 is a ductile rod (see Figure 2.1.57). This ductile rod is the yielding element. The function of the ductile rod is to accommodate postyield system deformations.

Our analytic understanding of system behavior, then, logically starts from the ductile rod and moves first to the beam and then into the column.

When a moment couple is developed between two sets of N rods separated by a distance $d - d'$ (Figure 2.1.58), the ideal or nominal moment capacity (M_n) developed is

$$M_n = NT_y(d - d') \tag{2.1.73}$$

where T_y is the nominal tensile strength of one ductile rod.

The nominal capacity of the set of ductile rods must be developed in the beam. Because the adopted design objective for the rest of the system is elastic behavior, an overstrength factor (λ_o) must be introduced.

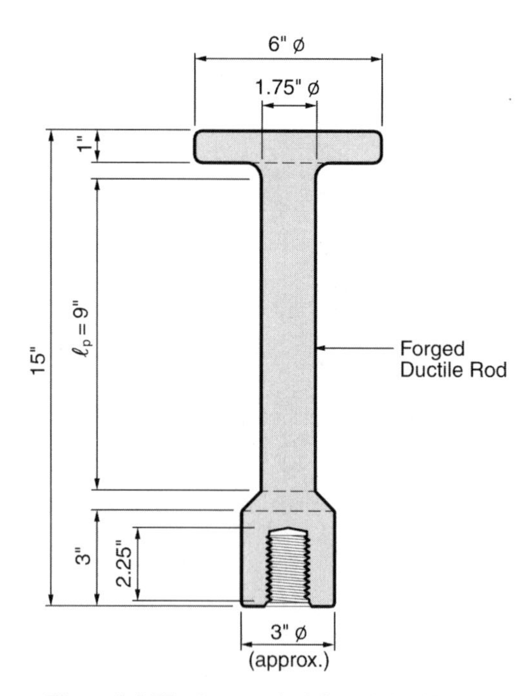

Figure 2.1.57 Prototypical forged ductile rod.

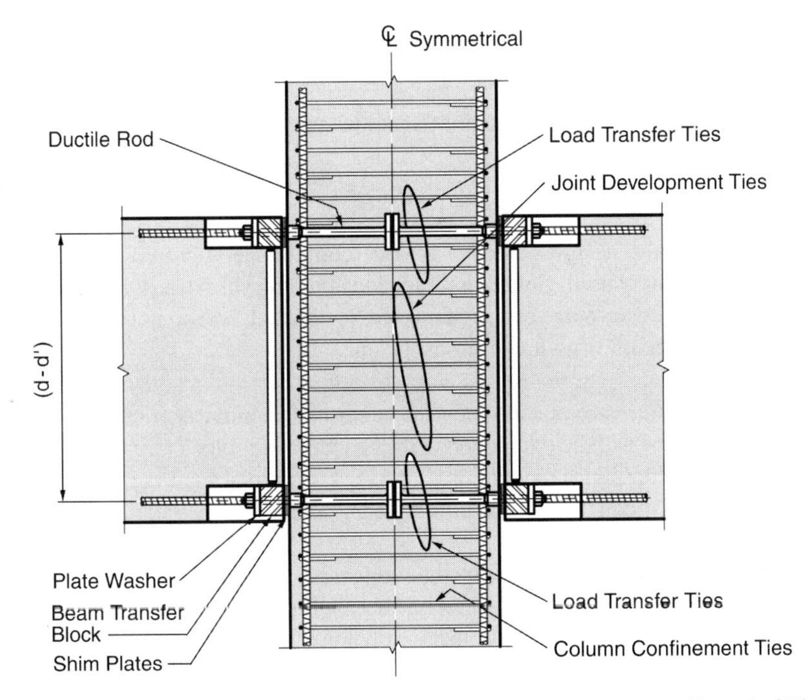

Figure 2.1.58 Frame beam-to-column connection incorporating the ductile rod of Figure 2.1.57.

The first load transfer point proceeding toward the beam is the beam-column interface where the appropriate level of shear and moment must be transferred. High-strength (A490SC) bolts are used to accomplish this transfer. The nominal area (NA_B) required of the bolt group is developed from LRFD Specifications[2.17]:

$$T_{bn} = \lambda_o \frac{M_n}{(d - d')} \tag{2.1.74}$$

$$NA_B = \frac{T_{bn}}{\phi F_t} \tag{2.1.75}$$

where

N is the number of bolts in one transfer block.
A_B is the nominal area of a single bolt.
F_t is the nominal yield strength of the bolt in tension.

Observe that the bolts will not yield provided λ_o/ϕ is appropriately chosen. The shear load (V_{nE}) induced by the ductile rod at mechanism on the beam-column interface is

$$V_{nE} = \frac{2\lambda_o M_n}{\ell_c}$$

The nominal shear capacity required of the connector is

$$V_n = V_{nE} + V_D + V_L \qquad \text{(see Eq. 2.1.20)} \qquad (2.1.76)$$

Comment: Since the design objective is code compliance, factored dead and live loads are appropriately used. Concerns relative to attaining shear transfer in the plastic hinge region of a cast-in-place frame beam (see Section 2.1.2.3) do not apply because the plastic hinge region is no longer in the beam.

The shear transfer mechanism between beam and column is friction. The load proceeds from the face of the ductile rod to the beam transfer block (see Figure 2.1.58) through a set of shim plates (see Figure 2.1.59). The normal load that activates this friction load path is the larger of the bolt pretension (NT_p) or the flexurally induced compression $[M/(d - d')]$.

The ability of the connector described in Figure 2.1.59 to transfer load will depend on the level of pretensioning ($2NT_p$) and the applied moment (M). The level of applied moment (M) must at some instant be zero. At this instant both the upper and lower connections will participate in the transfer of shear. Accordingly,

$$V_D + V_L < 2NT_p f \qquad (2.1.77)$$

where f is the friction factor allowed by the LRFD Specifications.[2.17]

As moment is applied to the connection, the effective level of pretensioning on the tensile bolt group will be relieved. Observe that the force applied to the ductile rods is unaffected by the level of bolt preload, for the bolts serve only to clamp the beam transfer block to the ductile rod.

However, once the preload (NT_p) has been relieved, the ability of the compression face connector to transfer shear will continue to increase, for the compression (C) crossing the surface described in Figure 2.1.59 will now be entirely a function of the level of moment imposed on the connection. Hence, C will be the larger of NT_p or $M/(d - d')$.

Accordingly, the nominal capacity of the shear transfer mechanism is the larger value generated from Eq. 2.1.78:

$$V_n = \left(\frac{M}{d - d'} \quad \text{or} \quad NT_p\right) f \qquad (2.1.78)$$

Beam component design should logically proceed based on the adoption of a variable overstrength factor (λ_o). The required capacity of each element should be modified (λ_o/ϕ) to appropriately account for uncertainties associated with each of the considered load transfer mechanisms. The required area of beam flexural reinforcement (A_s) is

$$A_s = \frac{\lambda_o N T_y}{\phi f_y} \tag{2.1.79}$$

The yield strength (f_y) of the beam reinforcement, because it need not yield, may significantly exceed Grade 60. The yield strength of Threadbars (high-strength threaded bars manufactured by Dywidag) is guaranteed; accordingly, λ_o/ϕ need not be overconservatively adopted. The assembly described in Figure 2.1.56 has been tested to assure that yielding does in fact occur exclusively in the ductile rod. The behavior is essentially that developed from the test described in Figure 2.1.40 (see Figure 2.1.41).

Shear reinforcement is developed from Eq. 2.1.76 and high-strength shear reinforcement ($f_y = 75$ ksi) may also be used here because postyield behavior in the stirrups is guarded against through the use of capacity-based design.

The load path from the ductile rod to the column is by bearing (Figure 2.1.60). Shear loads are equilibrated by bearing stresses under the compression side rod ends at the face of the column. The bearing stress allowed for confined concrete may appropriately be used because the shear load is only transferred through the compressed zone of the frame beam, and the shim plates and grout provide a significant normal or confining pressure in this part of the column.

The internal bearing at the rod end when two rods abut (see Figure 2.1.60) is subjected to a tensile load from the ductile rod on one side and a compressive load

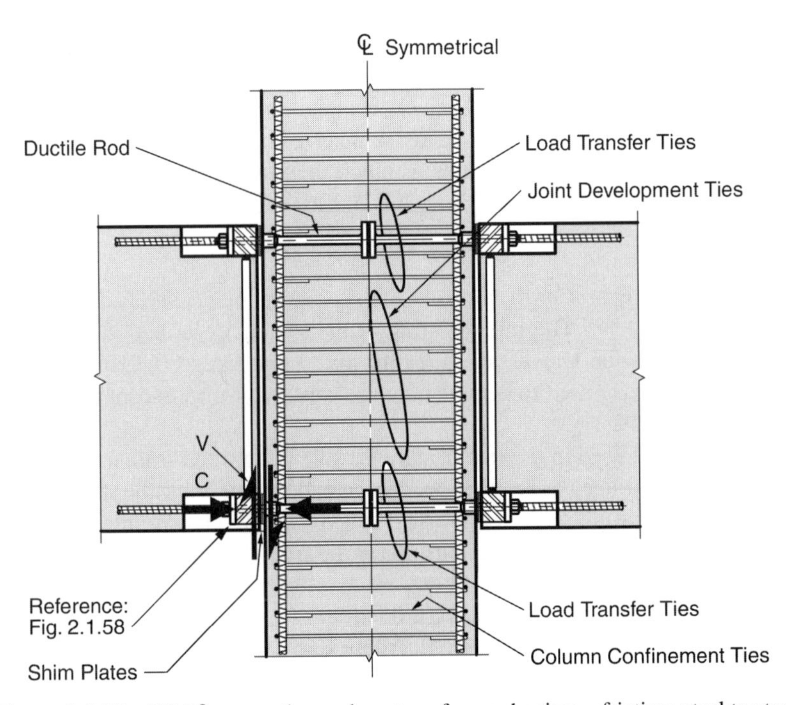

Figure 2.1.59 DDC® connection—shear transfer mechanism—friction: steel to steel.

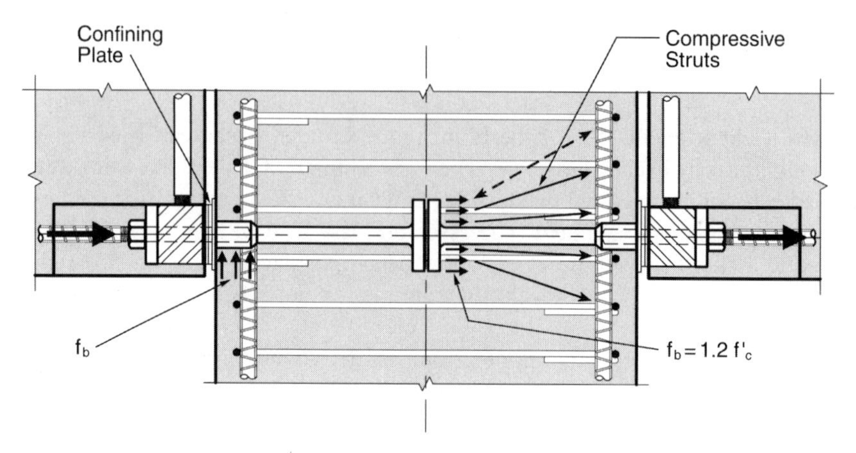

Figure 2.1.60 DDC® connection—shear transfer mechanism (concrete bearing: confined region).

from the rod on the opposite side. The tensile load will at some point exceed T_{yi} and is accordingly factored to account for probable overstrength. The worst case bearing load imposed on the anchored end of a ductile rod is $2\lambda_o T_{yi}$, but this will not be realized because any overstrength compression side demand will be resisted by bearing on the face of the column. The concrete that resists this load is well confined, and the supporting surface is wider than the bearing area on all sides; thus the design bearing stress may conservatively be presumed to be $0.85\phi(2)f'_c$,[2.6, Sec. 10.17.1] and this is $1.2f'_c$.

A set of compressive struts distribute bearing stresses imposed on the rod ends to joint reinforcement located above, below, and alongside the ductile rod assembly (Figure 2.1.60). The internal load transfer mechanism within the joint itself, with the exception of the load transfer ties, is much the same as that which occurs in the panel zone of a concrete ductile frame. It is discussed in considerably more detail in Section 2.3.

2.1.4.6 Experimental Confirmation A test program was developed to confirm the behavior predicted.[2.23] The initial testing of the system described in Figure 2.1.60 was undertaken at the University of California at San Diego (UCSD) in the early 1990s. The connector used to demonstrate the validity of the concept is described in Figures 2.1.61 and 2.1.62.

The ductile rod used in the test (Figure 2.1.62) was milled from a steel alloy bar (AISI 1045). This alloy was selected because it tends to minimize the strain hardening characteristically produced in steels when subjected to cyclic straining well beyond yield strain. The stress–strain relationship for the forged ductile rod of Figure 2.1.62 is described in Figure 2.1.63.

The focus of the test program was the ductility of the subassembly (Figure 2.1.64) provided by the incorporation of the ductile rods. Accordingly, the elements along the load path were conservatively proportioned so as to insure that their strength limit states would not be exceeded when the probable ultimate capacity of the ductile rod

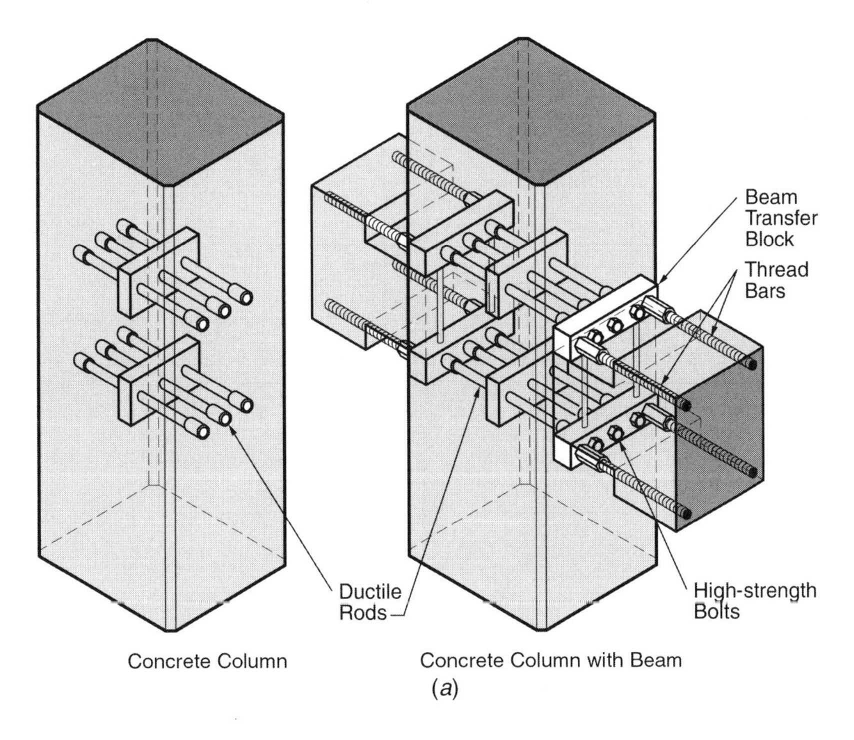

(*a*)

(*b*)

Figure 2.1.61 Ductile rod assembly—test program.

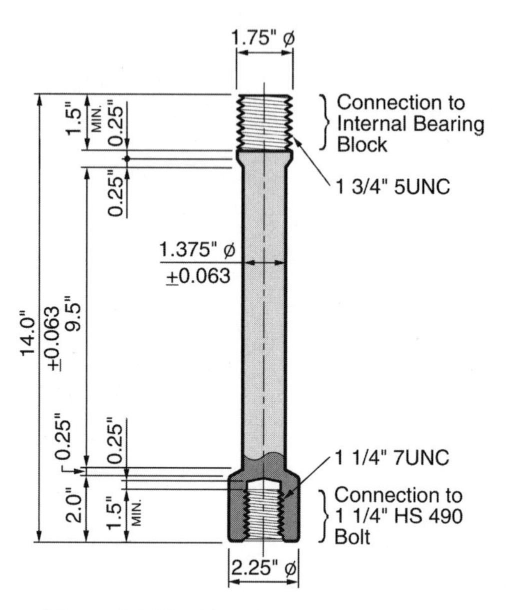

Figure 2.1.62 Shop drawing—ductile rod.

was realized (Figure 2.1.63). The test module was created by anchoring the milled rods (Figure 2.1.62) to a bearing block cast in the column (Figure 2.1.61).

The force required to yield the subassembly described in Figure 2.1.64 is developed from the yield strength of the milled ductile rods. These rods had a nominal yield strength of 60 ksi. The actual yield strength of the rod material was 62 ksi.

The tensile strength at nominal yield (T_y) of the three $1\frac{3}{8}$-in. diameter ductile rods was

$$NT_y = NA_{\text{rod}}f_y$$

where A_{rod} is the area of the provided rod, and thus

$$\begin{aligned} NT_y &= 3(1.48)60 \\ &= 266 \text{ kips} \end{aligned}$$

The nominal moment capacity of the connector was

$$\begin{aligned} M_n &= NT_y(d - d') \qquad \text{(Eq. 2.1.73)} \\ &= 266(2.25) \\ &= 599 \text{ ft-kips} \end{aligned}$$

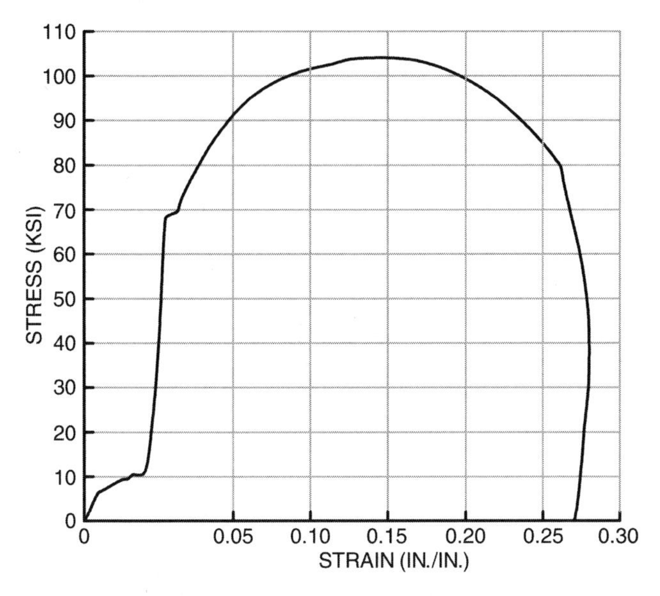

Figure 2.1.63 Relationship between stress and strain for the ductile rod of Figure 2.1.62.

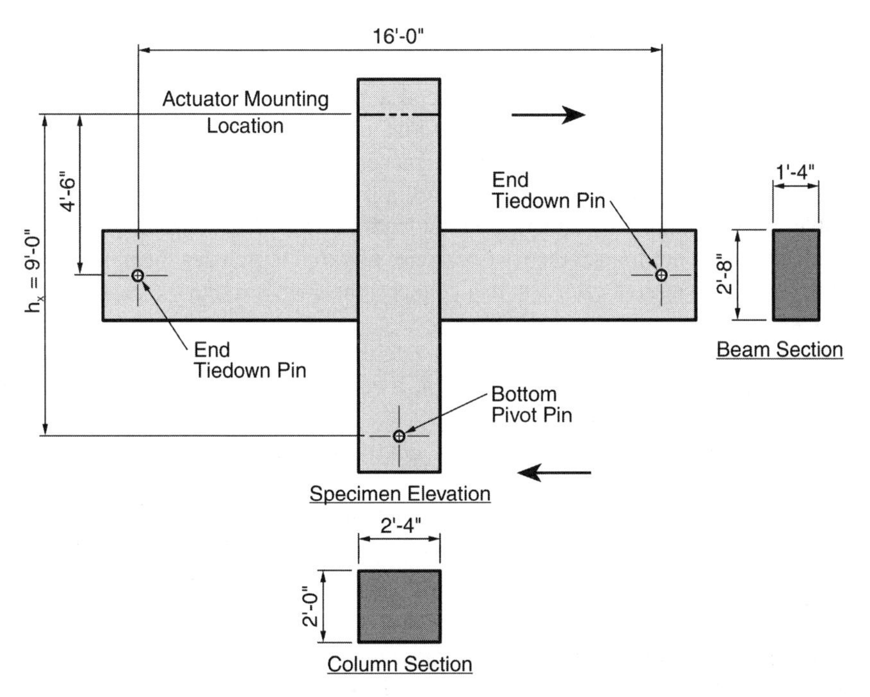

Figure 2.1.64 Test specimen dimensions.

The column shear required to activate this moment is

$$V_c = \frac{M_n}{\ell_c}\left(\frac{2\ell}{h_x}\right)$$

$$= \frac{599 \times 12}{82}\left(\frac{16}{9}\right) \qquad \text{(see Figure 2.1.64)}$$

$$= 156 \text{ kips}$$

$$V_b = V_c\left(\frac{h_x}{2\ell}\right)$$

$$= 156\left(\frac{108}{192}\right)$$

$$= 87.8 \text{ kips}$$

The level of overstrength provided along the load path is developed from the beam shear (87.8 kips). The compression imposed on the bolt group that is being compressed is 266 kips (NT_y). The friction coefficient (f) required to effect a shear transfer is

$$f = \frac{87.8}{266} = 0.33$$

The LRFD Specifications[2.17] recommend a service level friction coefficient of 33% for surfaces that are clean and free of mill scale (Class A). Slipping should not be expected,[2.4] however, at friction coefficients of up to 0.45. The plate surfaces provided were cleaned with a wire brush. Accordingly, they were presumed to have been capable of developing the requisite friction load transfer. Observe that overstrength considerations do not impact the friction factor required to transfer shear.

The tensile strength of each bolt that connects the ductile rod to the beam transfer block is

$$T_{by} = A_b F_b$$

where A_b and F_b are the nominal area and yield strength of the bolt. Thus

$$T_{by} = 1.23(112.5) \qquad \left(1\tfrac{1}{4}\text{-in. } \phi \text{ bolt, see Figure 2.1.62}\right)$$

$$= 138 \text{ kips}$$

The three bolts provided an overstrength of 1.56 (414/266).

Two #11 threadbars were used as flexural reinforcement in the beam. The provided overstrength in the flexural reinforcement of the beam, based on a minimum yield strength of 120 ksi, was

$$\lambda_o = \frac{374}{266} = 1.4$$

The nominal shear strength provided by the concrete of the frame beam, was

$$V_c = 2\sqrt{f_c'}bd$$

$$= \frac{2\sqrt{4500}(16)(29.5)}{1000}$$

$$= 63.3 \text{ kips}$$

For stirrups (#4 at 6 in. on center),

$$V_s = v_s bd$$

$$= 0.25(16)(29.5)$$

$$= 118 \text{ kips}$$

Shear strength is

$$V_c + V_s = 181.3 \text{ kips}$$

$$V_u = \frac{M_n}{\ell_c}$$

$$= \frac{(599)(12)}{82}$$

$$= 87.6 \text{ kips}$$

Provided overstrength is

$$\lambda_o = \frac{V_n}{V_u}$$

$$= \frac{181.3}{87.6}$$

$$= 2.07$$

The probable bearing stress under the ductile rod ends (see Figure 2.1.60) at nominal demand ($V = 87.6$ kips), based on a bearing area external to the shank of the rod of 13.5 in.^{2}, was 6.48 ksi. This is 30% greater than the nominal strength of the concrete ($1.3f_c'$), and more than that allowed by ACI[2.6, Sec. 10.17.1] ($1.20f_c'$). The bearing stress allowed by ACI is based on $0.85f_c'$, which assumes a permanent load and this, of course, is not appropriate for this condition. The provided level of overstrength is 1.32, and this is slightly below our objective margin (λ_o/ϕ) of 1.47 (1.25/0.85).

Comment: The development of the forged rod (Figure 2.1.57) allowed for a higher overstrength factor in bearing.

The internal bearing stress imposed on the 8 in. by 12 in. anchor block (see Figure 2.1.61) at nominal yield could include the force delivered by the compression side rods or $2NT_y$ ($NT_y = 266$ kips). The maximum compressive force imposed on the concrete could be

$$f_b = \frac{2(266)}{8(12) - 3(1.48)}$$

$$= 5.8 \text{ ksi}$$

For short-term bearing[2.3] the sustainable stress in the concrete $f_{b,\text{pr}}$, is

$$f_{b,\text{pr}} = 2\phi_p f_c'$$

$$= 2(1.25)(4.5)$$

$$= 11.3 \text{ ksi}$$

$$\lambda_o = \frac{11.3}{5.8}$$

$$= 1.9$$

Two tie groups surrounded the anchor block (eight legs of #6 bar) and a triple tie group was located immediately above and below the anchor block (six legs of #5 and six legs of #6 bar). The ties would deliver the flexurally induced force to the joint much as beam reinforcement delivers this load to the panel zone (see Figure 2.1.60).

The nominal tensile strength that could be mobilized by these tie groups to resist the bearing load is

$$T_n = (14 \times 0.44 + 6 \times 0.31)60$$

$$= 481 \text{ kips}$$

The nominal force imposed on the proximate tie group is

$$2NT_{yi} - V_{\text{col}} = 532 - 156$$

$$= 376 \text{ kips}$$

This was considered acceptable in spite of the fact that provided overstrength of 1.28 (481/376) was less than the objective ($\lambda_o/\phi = 1.25/0.9 = 1.39$) because yielding of these ties is not critical and a further load sharing from farther removed tie groups is available.

The shear stress at nominal yield induced on the joint was

$$V_j = 2NT_{yi} - V_{\text{col}}$$
$$= 2(266) - 156$$
$$= 376 \text{ kips}$$
$$v_j = \frac{376}{24(28)}$$
$$= 0.56 \text{ ksi}$$

The provided overstrength within the joint, based on the ACI identified nominal capacity of $15\sqrt{f_c'}$, was 1.89, and this significantly exceeds our objective ($\lambda_o/\phi = 1.47$).

The hysteretic behavior of the subassembly of Figure 2.1.64 is described in Figure 2.1.65. The nominal strength is reached at a drift angle of about 1%, and the level of overstrength is 1.33. The strength does not degrade through a drift angle of 4.5%. The limit of 4.5% was established by the capacity of the test frame. Observe that the strength loss between postyield deformation cycles 16, 17, and 18 was minimal.

The stiffness of the subassembly is reasonably predicted using ACI recommendations.[2.6, Sec. 10.11.1]

$$\Delta = \frac{Vh_x^2}{12E}\left(\frac{\ell}{I_b} + \frac{h_x}{I_c}\right) \qquad \text{(see Ref. 2.4, Eq. 4.2.17)}$$

$$I_{ce} = \frac{0.7bd^3}{12}$$
$$= \frac{0.7(24)(28)^3}{12}$$
$$= 30{,}700 \text{ in.}^4$$
$$I_{be} = \frac{0.35(16)(32)^3}{12}$$
$$= 15{,}300 \text{ in.}^4$$
$$h_x = 108 \text{ in.}$$
$$\ell = 192 \text{ in.}$$
$$\frac{\Delta}{h_x} = \frac{156(108)}{12(3800)}\left(\frac{192}{15{,}300} + \frac{108}{30{,}700}\right)$$
$$= 0.0059 \qquad (0.59\%)$$

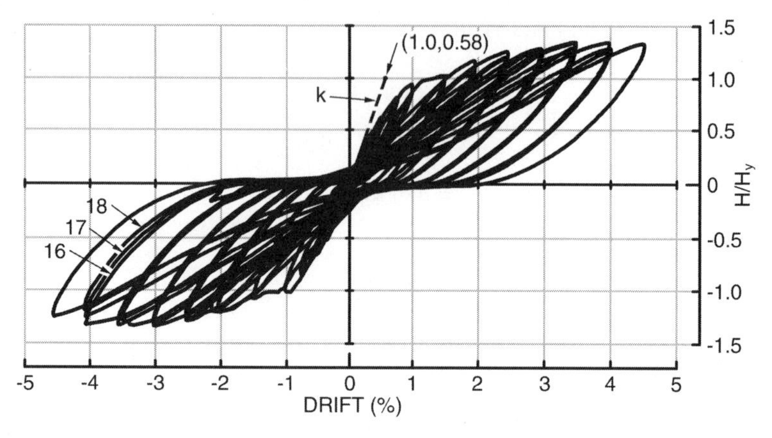

Figure 2.1.65 Hysteretic behavior of test subassembly.

Observe that this is consistent with the drifts described in Figure 2.1.65 and represents a good idealization, one that is consistent with the idealization used to characterize the behavior of a cast-in-place beam. (See Section 2.1.1.)

Comment: The condition of the subassembly at a drift of 3.5% is shown in Figure 2.1.66. Observe that beam cracks are minimal, and this undoubtedly impacts the stiffness idealization of the beam.

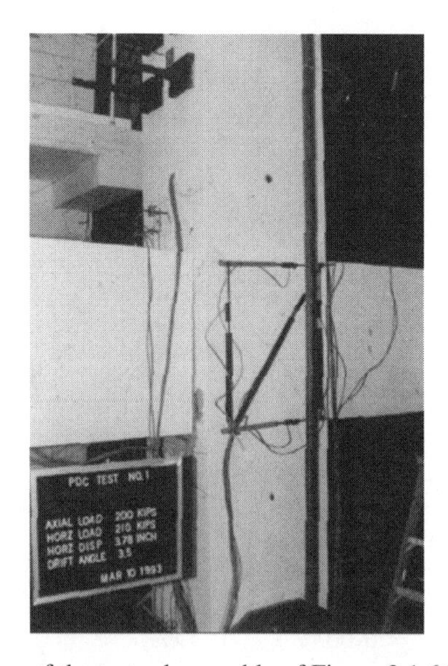

Figure 2.1.66 Condition of the test subassembly of Figure 2.1.64 at a drift angle of 3.5%.

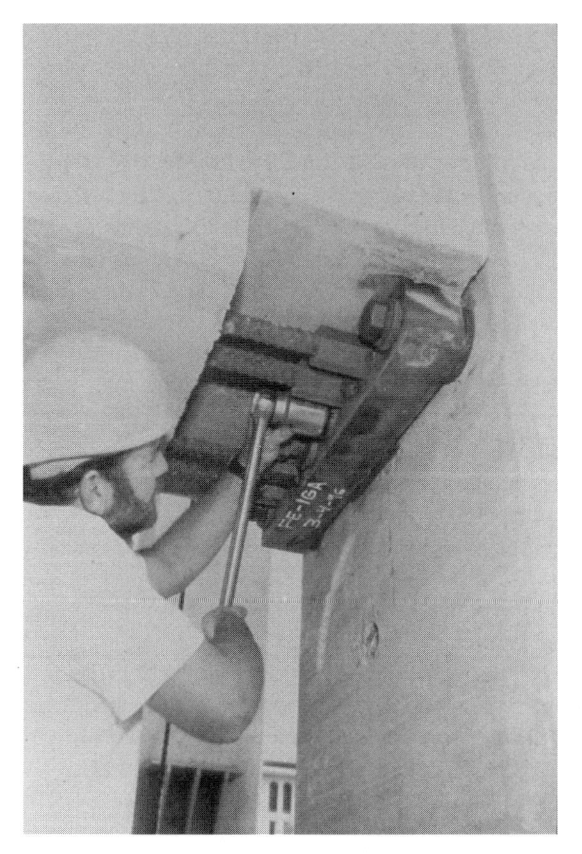

Photo 2.6 Assembling a DDC frame beam, Wiltern Parking Structure, Los Angeles, CA. (Courtesy of Englekirk Partners, Inc.)

The level of postyield strain imposed on the ductile rod may be estimated by assuming a lever arm of $d - d'$. At a drift angle of 4.5% and assuming that this rotation is experienced entirely at the beam-column interface,

$$\begin{aligned}\delta_{\text{rod}} &= \theta(d - d') \\ &= 0.045(27) \\ &= 1.2 \text{ in.} \\ \varepsilon_{\text{rod}} &= \frac{1.2}{9.5} \\ &= 0.127 \text{ in./in.}\end{aligned}$$

or about $58\varepsilon_y$. Admittedly, this estimate of postyield strain is quite conservative but the indicated strain is still well within the capabilities of the ductile rod (see Figure 2.1.63).

Design Procedure: (See Section 2.1.2.7.) The design process could, at least initially, follow that summarized in Section 2.1.2.6. The major system constraint lies in the fact that one Dywidag Ductile Connector—DDC®—assembly contains two ductile rods and as a consequence has a relatively high yield strength, which cannot be fine-tuned. This constraint must be considered very early in the design process and the bracing program modified to fit system and functional objectives.

The assembly described in Figure 2.1.56 contains two ductile rods (Figure 2.1.57) that establish its yield strength at 282 kips (T_n), and this sets the strength increments. When we studied the composite system (Section 2.1.4.3) three ductile rods were required, and this translates into one and a half DDCs®. The consequences of providing two DDCs® per beam if say 1.5 were required would be significant not only from the stand point of connector costs but also from a doubling of the size of the beam and column. The increase in cost would extend to the entire frame as capacity-based design is implemented. For example, in Section 2.1.3.2 we analyzed a conventionally reinforced beam containing six #10 bars (Figure 2.1.28*b*, $T_n = 457$ kips). This reinforcing program appeared to require a 30 in. by 32 in. column section in order to satisfy joint shear strength objectives. Accordingly, if we provided two DDC® assemblies ($T_n = 564$ kips), we would exceed the design strength provided by the selected conventionally reinforced beam, and this means that we would need to increase the size of the column by at least 23% (564/457).

Alternative DDC® design solutions need to be explored in the design of the system described in Figure 2.1.28, or the DDC® approach abandoned. One alternative is to create a nominal flexural frame beam strength on the order of that provided by the beam of Figure 2.1.28*b*. This corresponds to a nominal flexural strength (M_n) of about 1080 ft-kips (see Table 2.1.1). Two DDC assemblies (Figure 2.1.56) would provide a tensile strength NT_n of 564 kips. The distance between these two assemblies, given our flexural strength objective of 1080 ft-kips, should be

$$d - d' = \frac{M_n}{NT_n} \tag{2.1.80}$$

$$= \frac{1080(12)}{564}$$

$$= 23 \text{ in.}$$

In order to attain the stated functional objective—minimum beam width (Section 2.1.2.7, Example 2), the two connectors would need to be oriented vertically, thereby providing a maximum lever arm ($d - d'$) in a 32-in. deep beam of 17 in. (see Figure 2.1.67), and this does not satisfy the flexural demand; further, it is not a logical design approach. Discard this approach immediately.

Placing the two connectors (Figure 2.1.56) in a horizontal position provides an overstrength. The lever arm allowing for a $1\frac{1}{2}$-in. fire cover over the transfer block would now be

$$d - d' = 32 - 5 - 2(1.5)$$

$$= 24 \text{ in.}$$

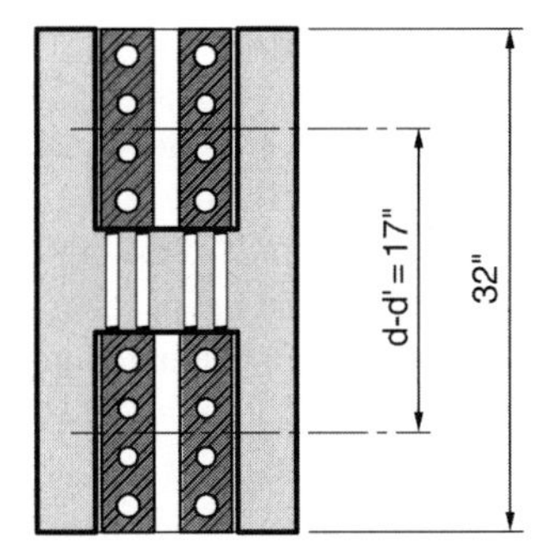

Figure 2.1.67 Two DDC connectors—vertical orientation.

Hence M_n is

$$\begin{aligned} M_n &= NT_y(d - d') \\ &= 2(282)(24) \\ &= 13{,}563 \text{ in.-kips} \qquad (1128 \text{ ft-kips}) \end{aligned}$$

The beam width would need to be at least

$$\begin{aligned} b &= 2(15) + \text{Fire cover} \\ &= 34 \text{ in.} \end{aligned}$$

The design of the DDC® system must return to the preceding design step wherein the frame geometry of Figure 2.1.28 was developed. A mechanism-based approach is the most design-efficient way to create as cost-effective a program as possible.

The mechanism approach was developed in Section 2.1.4.3, wherein it was demonstrated that a lateral deformation mechanism (Figure 2.1.43*c*) could be assured by providing a sufficient positive moment capacity. A relationship between external and internal work was established for the lateral mechanism:

$$V_{uE} h_x \theta = 2NM_p\theta \qquad \text{(see Eq. 2.1.49)}$$

$$M_p = \frac{V_{uE} h_x}{2N} \qquad (2.1.81)$$

where N is, in this case, the number of frame beams provided.

Example 2—DDC® system

Step 1: Question the Constraints and Their Applicability to the Development of the DDC®. Several architectural constraints were imposed on the design of Example 2 (Section 2.1.2.7). One constraint was that imposed on the depth of the beam. A beam depth that extended below the slab soffit would have an impact on forming costs in a cast-in-place system. A deeper beam would not be as much of a constraint on a precast spandrel system because an interior form could easily be dropped and removed. It might, in fact, be advantageous to have the beam soffit extend below the bottom of the slab, for the straight line created by the precast element would not only be architecturally pleasing, but would create a slab edge form. Further, the inclusion of drips and the acceptance of glazing would be more reliably accomplished. This increase in beam depth would also conceal the drapery valance. The beam depth required of five- and six-bay DDC® frames is easily determined, as are other logical options.

Step 2: Consider Any Project Enhancements That Might Be Derived from the Use of the DDC® System. The use of a precast concrete façade will enhance the project aesthetically, and this advantage should be exploited. This means that the DDC® assembly should not require an exposed patch. Accordingly, field grout pockets should be recessed. Allow at least a 3-in. precast element to project on the outer side of the spandrel panel. The minimum width for a horizontal DDC orientation (Figure 2.1.68) becomes:

Exterior cover and tolerance	=	3.5 in.
DDC® assembly	=	15.0 in.
Fire protection	=	1.5 in.
Minimum width		20.0 in.

The minimum width for a vertical DDC® orientation becomes:

Exterior cover and tolerance	=	3.5 in.
DDC® assembly	=	5.0 in.
Fire protection	=	1.5 in.
		10.0 in.

Step 3: Determine the Flexural Capacities Associated with Potential Alternatives. The mechanism strength provided by the beam described in Figure 2.1.28 is

$$V_{uE} h_x = 2NM_n \quad \text{(see Eq. 2.1.81)}$$

$$= 8(1080) \quad \text{(see Example 2—Proposed Design Section 2.1.2.7)}$$

$$= 8640 \text{ ft-kips}$$

A horizontal orientation of the DDC® results in a lever arm $(d - d')$ of $(h - 8)$. See Figure 2.1.69.

Cover (× 2)	=	3.0 in.
Connector depth	=	5.0 in.
		8.0 in.

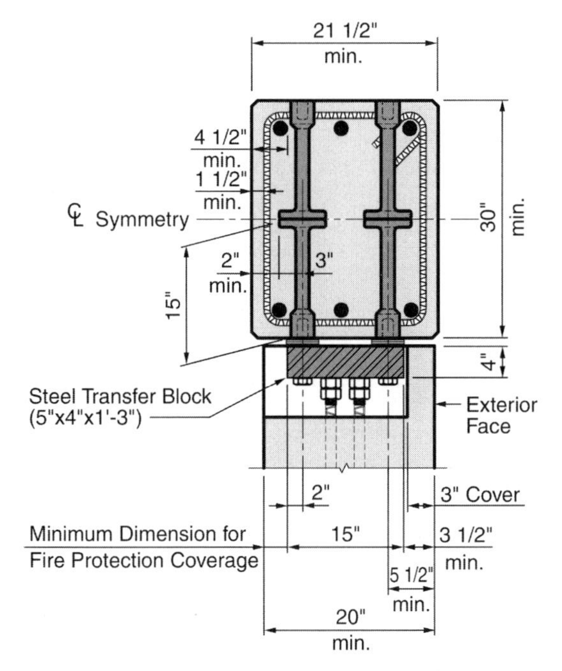

Figure 2.1.68 Plan view of column at DDC® connection.

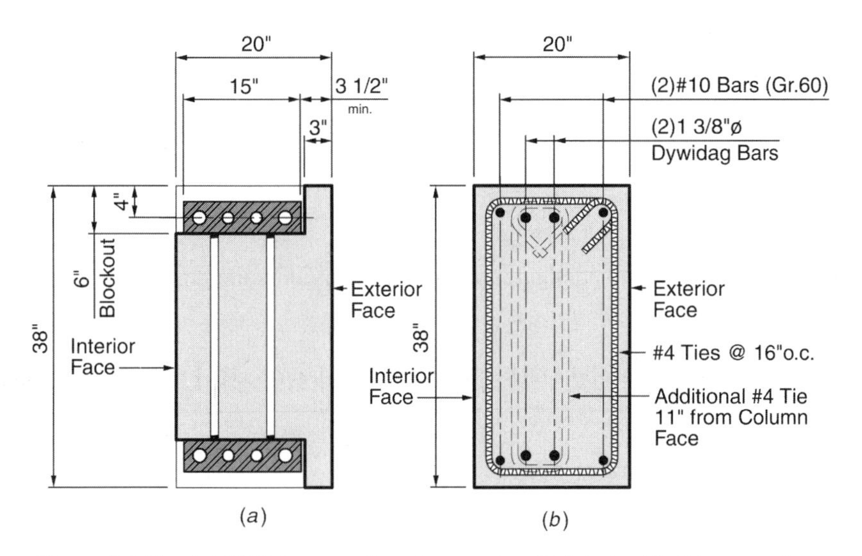

Figure 2.1.69 (*a*) DDC beam section at connector. (*b*) Interior DDC beam section.

The depth required of a five-bay frame would be

$$M_{n5} = \frac{V_{uE}h_x}{2N} \qquad \text{(see Eq. 2.1.81)}$$

$$= \frac{8640}{10}$$

$$= 864 \text{ ft-kips}$$

$$h_5 - 8 = \frac{M_{n5}}{T_n} \qquad \text{(see Eq. 2.1.80)}$$

$$= \frac{864}{282}(12)$$

$$= 37 \text{ in.}$$

$$h_5 = 45 \text{ in.}$$

The depth required of a six-bay frame would be

$$M_{n6} = \frac{V_{uE}h_x}{2N} \qquad \text{(see Eq. 2.1.81)}$$

$$= \frac{8640}{12}$$

$$= 720 \text{ ft-kips}$$

$$h_6 = \frac{M_{n6}}{T_n}(12) + 8 \qquad \text{(see Eq. 2.1.80)}$$

$$= \frac{720(12)}{282} + 8$$

$$= 38.6 \text{ in.}$$

Comment: This six-bay configuration should satisfy our strength objective given a 38-in. deep beam.

A vertical orientation of the DDC® similar to that described in Figure 2.1.67 should also be considered.

Cover (× 2)	=	3.0 in.
Connector depth	=	15.0 in.
		18.0 in.

$$d - d' = h - 18$$

The nominal moment capacity is

$$M_n = T_n(h - 18)$$

A seven-bay system with one vertical DDC connector/beam end would require the beam depth to be developed as follows:

$$M_{n7} = \frac{V_{uE}h_x}{2N} \qquad \text{(see Eq. 2.1.81)}$$

$$= \frac{8460}{2(7)}$$

$$= 604 \text{ ft-kips}$$

$$h = \frac{M_{n7}(12)}{T_n} + 18$$

$$= \frac{604(12)}{282} + 18$$

$$= 44 \text{ in.}$$

A nine-bay system with one vertical DDC connector/beam end would require the beam to be

$$M_{n9} = \frac{V_{uE}h_x}{2N} \qquad \text{(see Eq. 2.1.81)}$$

$$= \frac{8460}{2(9)}$$

$$= 470 \text{ ft-kips}$$

$$h = \frac{M_{n9}(12)}{T_n} + 18$$

$$= \frac{470(12)}{282} + 18$$

$$= 38 \text{ in.}$$

The designer can now suggest a six- or nine-bay program with 38-in. deep beams, understanding that some conservatism exists through the use of column centerline dimensions.

Comment: The decision as to which of the alternatives should be used will normally require the input of the architect, because it involves aesthetics and function, and the owner, because it involves cost. The basic question is whether the use of a 10-in. fascia panel is worth the extra cost, and this will need to be evaluated on a case-by-case basis. Some architects may, for example, choose to express the six-bay frame and change the cladding materials used on the other bays. Alternatively, the objective may be to create a uniform exterior, in which case the added three bays of frame will replace architectural panels. Local precasters can and should assist the design team by providing comparative cost information. The important thing is that a decision be made as to how many frames are to be used and what their component sizes will be during the development of the conceptual design. If we are to be consistent with our dynamic design process, this important decision cannot be reversed.

Conclusion: Let us presume that the design team has elected to proceed with the six-bay frame and the 20-in. by 38-in. beam with one DDC assembly oriented horizontally.

Step 4: Determine Minimum Column Size. The minimum size of the beam-column joint (Step 2) should be checked.

$$\begin{aligned} A_j &> 2.07NT_n \qquad \text{(see Eq. 2.1.52)} \\ &> 2.07(282) \\ &> 584 \text{ in.}^2 \end{aligned}$$

The column size must provide the area required to place the ductile rods. Figure 2.1.68 shows a plan view at the DDC® transfer block. It is presumed that the exterior face of the column and the face of the beam are aligned, though this, of course, is not a constraint. Given this alignment, the required width of the column may be determined. The minimum column width is 21.5 in. The required depth of the column (h) must accommodate the placement of back-to-back ductile rods (30 in.). The minimum column size is then 21.5 in. by 30 in. (645 in.^{2}), and this exceeds that required to satisfy joint shear demands (584 in.^{2}).

Designers may be concerned about the impact of dead and real live loads on the strain induced in the ductile rods. An alternative that does not prestrain the rods is to assemble the system in stages. The bottom bolts are installed during the erection of the precast beams and prestressed so as to provide a slip critical shear transfer. The upper set of bolts can then be placed to provide torsional stability but not stressed. Once most of the dead load is in place, the upper bolts can then be prestressed. The level of required prestress is a function of the service level shear that is likely to be experienced at the connector when no moment is imposed. The basic test program was conducted without any real prestress, because the bolts were tightened by the "turn of the nut" method to a probable preload of only 10 kips.

The sequential bolt-tightening program, if applied to our Example 2 frame now extended to six bays, would proceed as follows for the 24-ft bay:

$$w_D = 1.5 \text{ kips/ft} \qquad \text{(Service load)}$$

$$R_D = w_D\left(\frac{\ell_c}{2}\right)$$

$$= 1.5\left(\frac{24 - 2.5}{2}\right)$$

$$= 16.1 \text{ kips}$$

The load per bolt is 8.05 kips, and the effective preload required becomes

$$T_{\text{pre}} = \frac{R_D/\text{bolt}}{f}$$

$$= \frac{8.05}{0.33}$$

$$= 24.4 \text{ kips}$$

The specified preload for a slip critical bolt is 185 kips given the required $1\frac{1}{2}$-in. ϕ A490 bolt.

Comment: A sequential prestsressing program is not required, for it does not reduce the level of mechanism strength (see Section 2.1.4.3) and only results in the premature yielding of the tension side ductile rod, a concept recognized by moment redistribution.

Step 5: Beam Flexural Reinforcement Ratios Need Not Be Checked Because Elastic Behavior in the Beam Is Assured by the Criterion Used to Develop the Strength of the Components along the Load Path. The load path within the connector was reviewed during the analysis of the experimental program. It need not be reviewed in either the design or analysis phase because it is always a function of the strength of the ductile rod, which is standard ($T_y = 141$ kips/rod—282 kips/connector).

Step 6: Shear Capacity Development in the Beam Need Not Follow the Constraints Imposed on Hybrid Frame Beams or the Objective Levels Previously Suggested for Cast-in-Place Beams ($5\sqrt{f_c'}$). This is because postyield behavior will not occur in the beam. Code compliance following capacity-based concepts is required. The short span beam ($\ell_c = 13.5$ ft) will have a shear demand of

$$V_{uE} = \frac{2\lambda_o M_n}{\ell_c}$$

$$= \frac{2(1.25)(705)}{13.5}$$

$$= 131 \text{ kips}$$

$$V_u = V_{uE} + 1.05V_D + 1.28V_L \qquad \text{(see Ref. 2.6, Eq. 9-2)}$$

$$
\begin{aligned}
&= 131 + \frac{2.2\ell_c}{2} \\
&= 146 \text{ kips} \\
v_u &= 0.215 \text{ ksi} < 4\sqrt{f_c'} \\
v_c &= 2\sqrt{f_c'} \\
&= 2\sqrt{5000} \\
&= 141 \text{ psi} \\
V_{cn} &= (0.141)(20)(34) \\
&= 95.9 \text{ kips} \\
v_s bd &= \frac{V_u}{\phi} - V_{cn} \\
&= 75.9 \text{ kips} \\
v_s &= \frac{75.9}{20(34)} \\
&= 0.11 \text{ ksi}
\end{aligned}
$$

The required spacing for #5 stirrups is

$$
\begin{aligned}
v_s bs &= A_{sh} F_y \\
s &= \frac{A_{sh} f_y}{v_s b} \\
&= \frac{0.62(60)}{0.11(20)} \\
&= 16.9 \text{ in. on center}
\end{aligned}
$$

Conclusion: Provide #5 stirrups at 16 in. on center.

Step 7: Develop a Flexural Reinforcing Program for the Entire Beam. The criteria that need to be considered are:

- *Minimum Steel:*

$$
\begin{aligned}
\frac{200 b_w d}{f_y} &= \frac{200(20)(34)}{60{,}000} \\
&= 2.27 \text{ in.}^2
\end{aligned}
$$

- *Flexural Strength—Positive Moment Demand:*

$$
\begin{aligned}
M_u &\cong (1.4W_D + 1.7W_L)\,\frac{\ell^2}{16} \\
&= (1.4(1.5) + 1.7(0.5))\frac{(24)^2}{16} \\
&= 106 \text{ ft-kips} \\
A_s &= \frac{M_u}{\phi f_y (d - d')} \\
&= \frac{106(12)}{0.9(60)(30)} \\
&= 0.79 \text{ in.}^2
\end{aligned}
$$

- *Unbalanced Seismic Moment Attack and Splicing Requirements:* In Section 2.1.3.2 we discussed splicing considerations and bar development requirements. We have elected to use a high-strength bar that is capable of developing the ductile rod to an overstrength of 1.25. The development and splicing program for the high-strength bar must be based on the adopted approach. First, let us define the bond stress implied in Ref. 2.6, Sec. 12.2.2:

$$\frac{\ell_d}{d_b} = \frac{f_y \alpha \beta \lambda}{20\sqrt{f_c'}}$$

Rearranged and converted so as to express a force, this becomes, given α (1.0 for a bottom bar), β (1.0 for uncoated), and λ (1.0 for regular weight concrete) values of 1.0:

$$f_y A_b = T_y = 5\sqrt{f_c'}\pi d_b \ell_d \tag{2.1.82}$$

and μ is the implied bond stress:

$$\mu = 5\sqrt{f_c'} \tag{2.1.83}$$

In the case of the DDC® assembly, the comparable force that must be developed is 141 kips (T_y). The development length is then

$$\ell_d = \frac{T_y}{5\pi d_b \sqrt{f_c'}} \tag{2.1.84}$$

For the #11 high-strength ($f_y = 120$ ksi) bar connected to the transfer block in 5000-psi concrete, this becomes

$$\ell_d = \frac{141}{5(3.14)(1.41)\sqrt{5000}}$$
$$= 90 \text{ in.}$$

and for a top bar,

$$\ell_d = \alpha(90)$$
$$= 1.3(90)$$
$$= 117 \text{ in.}$$

The DDC® bars can be developed in the longer beams ($\ell_c = 21.5$ ft), but couplers are required in the shorter beams. A Class B splice ($1.3\ell_d$) will be required at midspan in the long beams unless couplers are used here also.

The splice force required is that force that must be considered at midspan of the long-span beam. This is the minimum required steel (2.26 in.^{2}). Observe that this provides a moment capacity of 48% of that provided by the DDC®. Accordingly, this would allow the point of inflection to shift from the middle of the beam to the end of the beam, and this is not likely.

Minimum steel would be two #10 bars (2.54 in.^{2}). The basic development length for these bars is

$$\ell_d = \frac{f_y d_b}{20\sqrt{f_c'}}$$
$$= \frac{60{,}000(1.25)}{20\sqrt{5000}}$$
$$= 53 \text{ in.}$$

The splice length (Class B) for a bottom bar would be 70 in. and for a top bar 91 in. The logical reinforcement program would be as described in Figure 2.1.68 for the long beam. The coupled high-strength bars in the short beam would satisfy minimum strength requirements, and the corner bars for the short beam need be only #6 bars.

Step 8: Develop the Transverse Reinforcing Program for the Long Beam. This was accomplished for the short beam in the design phase and need not be repeated. For the long beam,

$$V_{uE} = \frac{2\lambda_o M_n}{\ell_c}$$
$$= \frac{2(1.25)(705)}{21.5}$$
$$= 82 \text{ kips} \qquad \text{(see design phase)}$$

$$V_u = V_{uE} + V_D + V_L \qquad \text{(Step 6)}$$

$$= 97 \text{ kips}$$

$$V_c = 95.9 \text{ kips} \qquad \text{(Step 6)}$$

$$v_s bd = \frac{V_u}{\phi} - V_c \qquad \text{(see design phase)}$$

$$= \frac{96}{0.85} - 95.9$$

Conclusion: Minimum shear reinforcement is required (v_s > 50 psi). Maximum spacing is $d/2$.

$$A_v = \frac{0.05b(d/2)}{f_{yh}}$$

$$= \frac{0.05(20)(17)}{60}$$

$$= 0.28 \text{ in.}^2$$

Provide #4 stirrups at 16 in. on center. An elevation of the propsed beam is shown in Figure 2.1.70.

Step 9: Review Strain States. The DDC assembly does not require that the interface between the precast beam and column be grouted. The designer may elect to grout

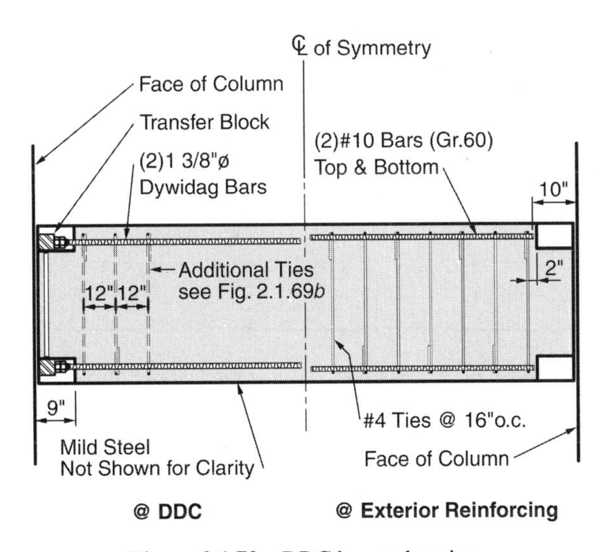

Figure 2.1.70 DDC beam elevation.

this gap in which case the strain states would be developed as they were for the composite system (see Section 2.1.4.3). Ungrouted the neutral axis will remain on the centerline of the beam ($c = h/2$). For a rotation of 4%, the elongation required of the ductile rod is

$$\begin{aligned}\delta_{\text{rod}} &= \theta\left(\frac{d-d'}{2}\right)\\ &= 0.04(15)\\ &= 0.6 \text{ in.}\end{aligned}$$

This corresponds to a strain of

$$\begin{aligned}\varepsilon_{\text{rod}} &= \frac{0.60}{9}\\ &= 0.067 \text{ in./in.}\end{aligned}$$

and this is well within the capacity of the ductile rod.

2.2 THE BEAM COLUMN

The objective of capacity-based design is to create a load path, the strength of which is based on the strength that can be developed in the member or members toward which postyield deformations are directed. In a ductile frame postyield deformations are directed toward the frame beam. The frame beam is designed so as to restrict the postyield deformation to one that is flexure, and this process was the topic of Section 2.1. Shear, for example, in the frame beam was constrained to remain in the elastic range by creating a capacity that exceeded the demand that would be imposed on the frame beam when it reached its maximum probable flexural strength ($\lambda_o M_n$). A similar design objective is rational for the frame column. Unfortunately, the attainment of this elastic or nonyielding objective, were it in fact possible, would produce an overly conservative member design in the vast majority of the columns. Accordingly, ductility in columns must also be encouraged.

Three conditions are likely to create a flexural strength demand that exceeds the capacity of a beam column:

- The column may behave like a cantilever. This condition is not uncommon, and virtually unavoidable at the base of the building when a foundation cannot be created in a manner that allows it to absorb energy at an imposed strength level that is less than the strength of the beam column. This situation also can occur at the roof, especially when beam spans are long, which causes beam flexural strength demands to be impacted by large dead and live load moments.

- The dynamic response of a building is not exclusively in the fundamental or first mode. Often, second- and third-mode responses will create distortions that cause the point of inflection to move away from the midheight of the column, and this can result in an imposed moment demand on the column that reaches a level equivalent to the probable moments that the frame beams can deliver $[\lambda_o f_y(A_s + A'_s)(d - d')]$. This unbalanced moment attack is well described by Paulay and Priestley.[2.5]
- Frame boundary columns are usually subjected to large tensile loads. At frame capacity these tension loads will often exceed the strength of the column in pure tension. Given this level of axial tension, the flexural strength virtually disappears. As a consequence, the absence of shear in these columns can reduce the restoring force of the frame. Clearly, ductility must be available in these members.

The overriding objective in the design of frame columns is the avoidance of a story mechanism. The most critical story mechanism is one created through the formation of plastic hinges in the top and bottom of all columns on a story. The approaches to attaining this objective are many and still in what might best be described as evolving form. The approach used herein is capacity-based, and it attempts to provide a reasonable guard against the formation of plastic hinges in columns, as well as a methodology for avoiding story mechanisms. Code alternatives are presented only in the context of demonstrating that the current code approaches have as their basis capacity design objectives.

In this section we describe how ductility may be provided and demonstrate that it exists in columns that are appropriately designed. The impact of system behavior on column design will be discussed in more detail in Chapters 3 and 4.

2.2.1 Strength Limit States

The strength of a concrete member subjected to a compressive load and bending is defined by the strain state in the concrete (see Figure 2.2.1). Traditionally, the ACI[2.6] has identified the concrete strain limit as 0.003 in./in., but this is probably conservative.

An interaction diagram (see Figure 2.2.2) is developed directly from assumed strain states. The degree of available ductility will depend on the strain states that define the strength limit state and accordingly these strain states merit some review prior to analyzing the behavior of test specimens.

Consider first the balanced strain state, where the strain in the concrete and the strain in the steel reach their respective limit states simultaneously (0.003 in./in. in the concrete and ε_y in the reinforcing steel). These strain states, when imposed on a column, define the maximum moment a section can sustain, for the maximum moment occurs when the imposed axial load reaches the balanced axial load. This balanced axial load (P_b) will define, at least in theory, the boundary between a section that is, or can be, ductile and a column that is brittle. If, for example, the imposed axial load

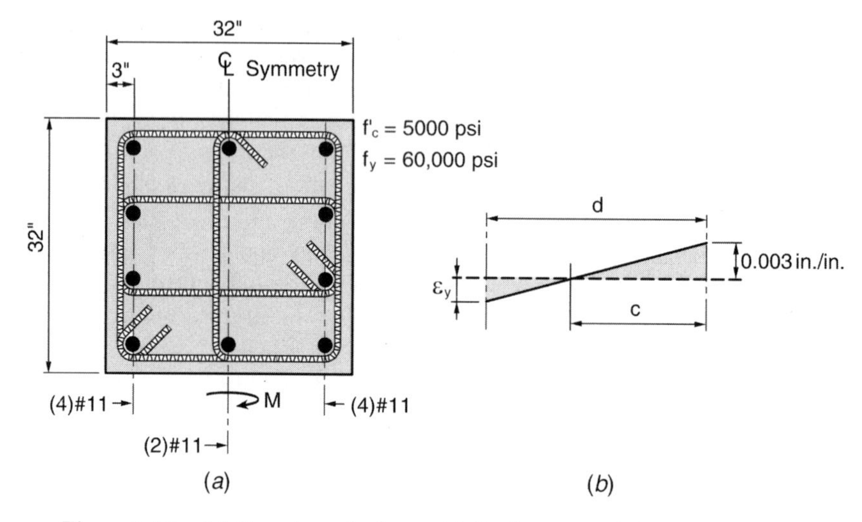

Figure 2.2.1 (*a*) Plan view of column. (*b*) Strain states at balanced conditions.

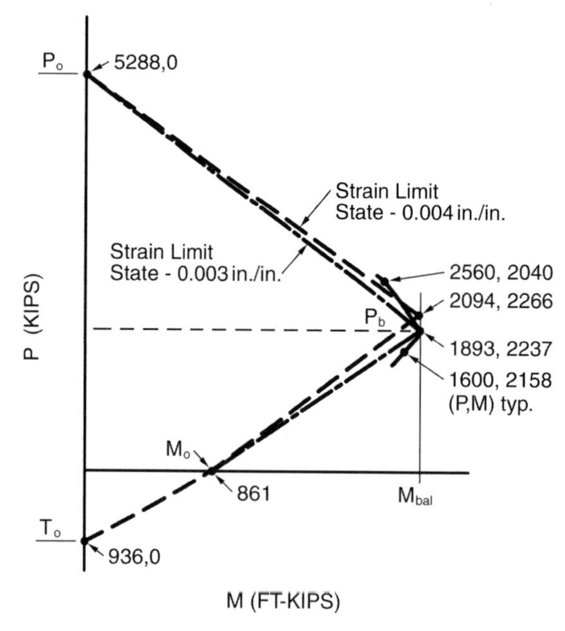

Figure 2.2.2 Interaction diagram for the 32-in. by 32-in. column described in Figure 2.2.1*a*.

is larger than the balanced axial load (P_b), the strain in the steel will be less than ε_y when a flexural load equivalent to the flexural limit state is applied to the beam column. In addition, if the flexural deformation exceeds that associated with the flexural strength limit state, the established strain limit state (0.003 in./in.) in the concrete will be exceeded. The consequences of this exceedance are presumed to be a brittle failure. If, on the other hand, the imposed level of axial load is less than the balanced axial load (P_b), the moment imposed curvature can increase beyond first yield of the reinforcing steel; hence the column possesses at least some degree of ductility.

When the axial load is eliminated, a beam is created and the flexural capacity of the column is the nominal flexural capacity of the section (M_n). This behavior point in the design of a column is M_o, and the subscript identifies the axial load. An equally important limit state is that associated with pure axial load (P_o). In this case the strain limit state in compression has been reached across the entire section. This load, from a design perspective, has little meaning because the continuity inherent in concrete construction makes it impossible to reach a condition of pure axial load. Hence the nominal axial load limit state is defined by the ACI as 80% of P_o.

2.2.1.1 Developing an Interaction Diagram Let us start by considering the balanced loads P_b and M_{bal}. If we assume that the steel yield strain is reasonably approximated by 0.002 in./in., the depth to the neutral axis becomes

$$c = \frac{\varepsilon_c}{\varepsilon_y + \varepsilon_c} d \qquad \text{(see Figure 2.2.1}b\text{)} \qquad (2.2.1)$$

$$= \frac{0.003}{0.005} d$$

$$= 0.6d$$

The depth of the compressive stress block (a) is

$$a = \beta_1 c$$

If we assume that the tension and compression steel located on the flexural extremities will yield, and further, that steel between these extremities will be subjected to inconsequentially low levels of strain. The balanced axial load (P_b) is approximated by

$$P_b = 0.85 f'_c ab \qquad (2.2.2a)$$

$$= 0.85\beta_1 f'_c cb$$

$$= 0.85(0.6)\beta_1 f'_c db$$

When we review tests in the subsection that follows, we find that axial load levels of interest are described in terms of $f'_c A_g$. Further, we see that ductility will noticeably decrease as load levels reach $0.3 f'_c A_g$. Observe that if $d \cong 0.9h$ and β_1 is 0.8, as it would be for 5000-psi concrete, then

$$P_b = 0.37 f'_c A_g \qquad (2.2.2b)$$

As higher strength concretes are used, β_1 becomes smaller, and for $\beta_1 = 0.6$,

$$P_b = 0.28 f'_c A_g$$

Accordingly it is reasonable to assume that loads on the order of $0.3 f_c' A_g$ will result in a significant decrease in available ductility.

In order to better understand the impact of various adopted parameters, consider the column described in Figure 2.2.1*a*.

$$A_s = 6.24 \text{ in.}^2 \qquad \text{Four \#11}$$

$$A_{st} = 15.6 \text{ in.}^2 \qquad \text{Ten \#11} \qquad (\rho_g = 1.5\%)$$

$$f_c' = 5000 \text{ psi}$$

$$P_b = 0.85(0.6)(0.8)(5)(29)(32) \qquad \text{(see Eq. 2.2.2a)}$$

$$= 1893 \text{ kips}$$

$$a = \beta_1 c$$

$$= 0.8(0.6)(29)$$

$$= 13.92 \text{ in.}$$

$$M_{\text{bal}} = A_s f_y (d - d') + P_b \left(\frac{h}{2} - \frac{a}{2} \right) \tag{2.2.3}$$

$$= 6.24(60)(26) + 1893(16 - 6.96)$$

$$= 26{,}850 \text{ in.-kips} \qquad (2237 \text{ ft-kips})$$

The consequences of selecting a concrete strain limit state of 0.004 in./in. are

$$c = \frac{0.004}{0.006} d$$

$$= 0.667(29)$$

$$= 19.3 \text{ in.}$$

$$a = 15.4 \text{ in.}$$

$$P_b = 0.85 f_c' ab$$

$$= 0.85(5)(15.4)(32)$$

$$= 2094 \text{ kips}$$

and this corresponds to an increase of 10% in the balanced load.

The flexural moment capacities in the absence of axial load are on the order of

$$a = \frac{A_s f_y}{0.85 f_c' b}$$

$$= \frac{6.24(60)}{0.85(5)(32)}$$

$$= 2.75 \text{ in.}$$

$$M_o = M_n = 6.24(60)(29 - 1.375) \qquad \text{(see Eq. 2.2.3)}$$
$$= 10{,}330 \text{ in.-kips} \qquad \text{(861 ft-kips)}$$

Observe that the impact of the centrally placed bars and the compression steel have been neglected for simplicity.

In order to consider the impact of both lesser and greater axial loads on the interactive limit states and strain states, assume that the imposed axial load is 1600 kips ($< P_b$).

$$a = \frac{P}{0.85 f'_c b}$$
$$= \frac{1600}{0.85(5)(32)}$$
$$= 11.8 \text{ in.}$$
$$c = 14.7 \text{ in.}$$
$$\varepsilon_s = \left(\frac{14.7}{14.3}\right)(0.003)$$
$$= 0.0031 \text{ in./in.} \qquad (1.5\varepsilon_y)$$
$$M_{1600} = 6.24(60)(26) + 1600(16 - 5.9) \qquad \text{(see Eq. 2.2.3)}$$
$$= 25{,}900 \text{ in.-kips} \qquad \text{(2158 ft-kips)}$$

while if P is greater than P_b, it becomes more convenient to identify a strain state and solve for P and M. If $\varepsilon_c = 0.003$ in./in. and $\varepsilon_s = 0.001$ in./in., then

$$c = \frac{0.003}{0.004} d$$
$$= 0.75d$$
$$= 21.8 \text{ in.}$$
$$T_s = \varepsilon_s E_s A_s$$
$$= 0.001(29{,}000)(6.24)$$
$$= 181 \text{ kips}$$
$$C_c = 0.85 f'_c \beta_1 c b$$
$$= 0.85(5)(17.4)32$$
$$= 2366 \text{ kips}$$
$$C_s = A_s f_y \qquad \text{(Compression steel has yielded)}$$
$$= 6.24(60)$$
$$= 374 \text{ kips}$$

$$P = C_c + C_s - T_s$$
$$= 2560 \text{ kips}$$

Taking moments about T_s,

$$M = C_c\left(d - \frac{a}{2}\right) + C_s(d - d') - P\left(\frac{h}{2} - d'\right)$$
$$= 2366(29 - 8.7) + 374(26) - 2560(16 - 3)$$
$$= 24{,}474 \text{ in.-kips} \qquad (2040 \text{ ft-kips})$$

Additional key points on the interaction diagram are

$$P_o = 0.85 f'_c A_g + f_y A_{st} \tag{2.2.4}$$
$$= 0.85(5)(32)^2 + 60(15.6)$$
$$= 5288 \text{ kips}$$
$$T_o = f_y A_{st} \tag{2.2.5}$$
$$= 60(15.6)$$
$$= 936 \text{ kips}$$

Figure 2.2.2 plots the various points in an interactive format.

2.2.1.2 Design Relationships From a design perspective the material developed in support of the interaction diagram (Figure 2.2.2) is extremely important. Consider the following points.

- *Sizing the Column for Axial Loads:* The balanced axial load should serve as a limit state in a column expected to sustain significant postyield deformations. The balanced axial load can be developed from Eq. 2.2.2a or, more simply, from Eq. 2.2.2b. For design purposes the objective area of a frame column should be at least

$$A_g = \frac{3P}{f'_c} \tag{2.2.6}$$

  The axial load (P) in this development must be a reasonable estimate of the probable load. Real, unfactored dead loads and realistically reduced live loads should be used to define the axial load. Earthquake loads should be capacity based.
- *Sizing the Column for Flexural Loads:* If the axial load is in the maximum objective design range(P_b), the reinforcing steel required can be developed from

a simplification of Eq. 2.2.3. The depth of the stress block is set by β_1 and the adopted concrete strain limit state (0.003 in./in.).

$$\begin{aligned} a &= 0.6\beta_1 d \\ &\cong 0.54\beta_1 h \end{aligned} \tag{2.2.7}$$

For a concrete strength of 5000 psi,

$$a \cong 0.43h$$

Observe that the depth of the compressive stress block (a) will diminish as the strength of the concrete is increased. Accordingly, the moment capacity provided by the shifting of the axial load may be quickly and conservatively estimated provided that the strength of the concrete (f_c') is at least 5000 psi:

$$P_b\left(\frac{h}{2} - \frac{a}{2}\right) \cong 0.3P_b h \qquad \text{(see Eq. 2.2.3)}$$

Now the required level of reinforcement can easily be ascertained.

$$A_s = \frac{(M_u/\phi) - 0.3P_b h}{f_y(d - d')} \tag{2.2.8}$$

Equation 2.2.8 seeks to determine an objective capacity. Accordingly, factored loads are appropriately used. Earthquake loads should be capacity based, and this subject will be discussed in more detail in Section 2.2.3. As axial loads decrease, the amount of reinforcing required will increase but so will the available level of ductility.

2.2.2 Experimentally Based Conclusions

2.2.2.1 Strength A series of columns were tested at the University of Southern California under the direction of Professor Yan Xiao[2.24] in the late 1990s. The objective of these tests was to study the behavior of high-strength concrete and the effectiveness of confining reinforcing. Several specimens were transversely reinforced to ACI standards and the concrete strength was 9 ksi. Because the specimen size was large (20 in. by 20 in.) in comparison with previous efforts and the axial loads high (up to $0.34f_c'A_g$), these tests allow an excellent opportunity to compare design procedures with behavior.

The constructed specimen is described in Figure 2.2.3. The axial load and shear were applied at the top of the cantilever column shown. A computer generated interaction diagram for the column of Figure 2.2.3 is shown in Figure 2.2.4, as are the experimentally determined strength limit states for the specimens whose behavior is discussed below. Start by comparing both analysis and behavior with the relations developed in Section 2.2.1:

$$P_b = 0.85(0.6)\beta_1 f_c' bd \qquad \text{(see Eq. 2.2.2a)}$$
$$= 0.51(0.65)(9)(20)(17.25)$$
$$= 1030 \text{ kips}$$

This is reasonably consistent with the analytical prediction and midway between the applied loads (744 kips and 1200 kips).

$$a = \frac{P_b}{0.85 f_c' b}$$
$$= \frac{1030}{0.85(9)20}$$
$$= 6.73 \text{ in.}$$

$$M_{\text{bal}} = A_s f_y (d - d') + P_b \left(\frac{h}{2} - \frac{a}{2} \right) \qquad \text{(Eq. 2.2.3)}$$
$$= 4.12(60)(14.5) + 1030(10 - 3.36)$$
$$= 10{,}423 \text{ in.-kips} \qquad (869 \text{ ft-kips})$$

And this too is reasonably consistent with the computer-generated estimate (see Figure 2.2.4).

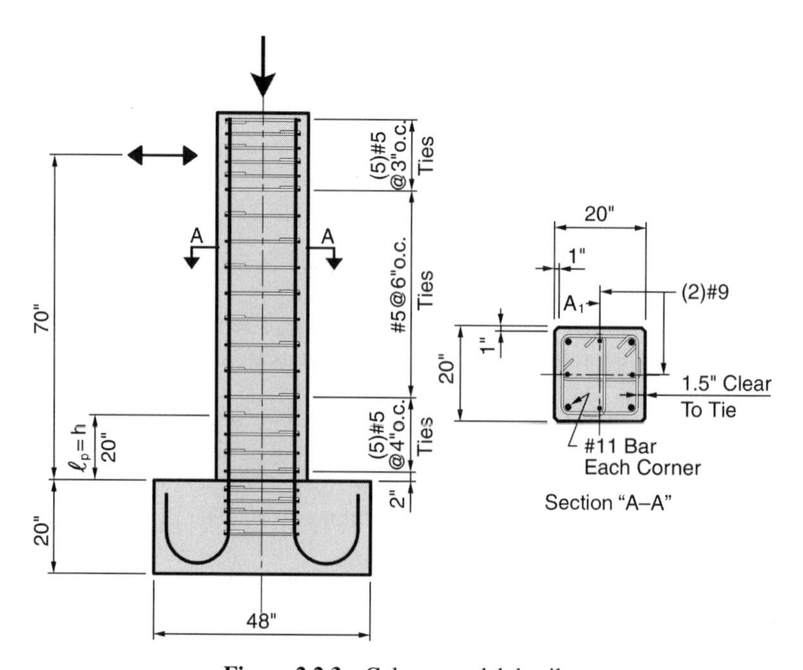

Figure 2.2.3 Column model details.

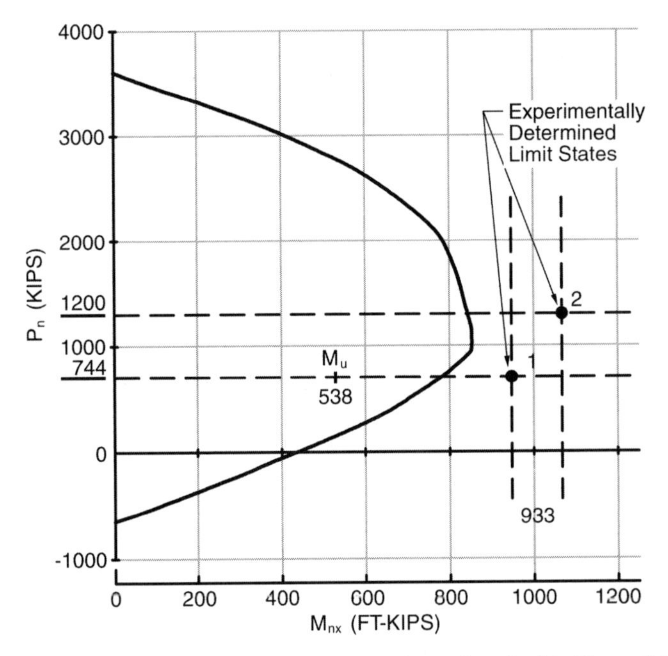

Figure 2.2.4 Interaction diagram for the column described in Figure 2.2.3.

Following the development of Eq. 2.2.3 for the test specimens:

Specimen 1—P = 744 kips

$$a = \frac{P}{0.85 f_c' b}$$

$$= \frac{744}{0.85(9)(20)}$$

$$= 4.86 \text{ in.}$$

$$c = \frac{a}{\beta_1}$$

$$= \frac{4.86}{0.65}$$

$$= 7.5 \text{ in.}$$

$$M_{744} = A_s f_y (d - d') + P\left(\frac{h}{2} - \frac{a}{2}\right) \qquad \text{(see Eq. 2.2.3)}$$

$$= 4.12(60)(14.5) + 744(10 - 2.43)$$

$$= 9217 \text{ in.-kips} \qquad (768 \text{ ft-kips})$$

Specimen 2—P = 1200 kips

$$
\begin{aligned}
a &= \frac{P}{0.85 f_c' b} \\
&= \frac{1200}{0.85(9)(20)} \\
&= 7.84 \text{ in.} \\
c &= \frac{a}{\beta_1} \\
&= \frac{7.84}{0.65} \\
&= 12 \text{ in.} \\
M_{1200} &= A_s f_y (d - d') + P\left(\frac{h}{2} - \frac{a}{2}\right) \qquad \text{(see Eq. 2.2.3)} \\
&= 4.12(60)(14.5) + 1200(10 - 3.92) \\
&= 10{,}811 \text{ in.-kips} \qquad (907 \text{ ft-kips})
\end{aligned}
$$

The peak shear load was 180 kips (see Figure 2.2.6). This corresponds to a shear stress of 514 psi or $5.4\sqrt{f_c'}$. Shear reinforcement (v_s) significantly exceeded the imposed shear stress in the plastic hinge region and was 466 psi outside the plastic hing region (see Figure 2.2.3). Accordingly, it is difficult to draw any conclusions relative to the impact of shear on behavior in the beam column. Based on experimental evidence developed from beams and walls it seems reasonable and prudent to design to shear limits in the $\pm 5\sqrt{f_c'}$ range.

The following observations can be made:

- The computer-generated analysis of Specimen 1 (Figure 2.2.4) is reasonably consistent with the results predicted by the application of Eq. 2.2.3. This is because the tension steel has yielded as it will for axial loads below P_b. Equation 2.2.3 may be reasonably used for all column designs—provided the axial load is less than P_b
- The computer-generated analysis for Specimen 2 predicts a flexural strength that is considerably below that projected by Eq. 2.2.3, and this is because the tension steel has not yielded; a condition assumed in the development of Eq. 2.2.3. The strain in the tension steel is

$$
\begin{aligned}
\varepsilon_s &= \frac{d - c}{c}\varepsilon_c \\
&= \frac{17.25 - 12}{12}(0.003) \\
&= 0.0013 \text{ in./in.} < \varepsilon_y
\end{aligned}
$$

Equation 2.2.3 and the design procedure proposed by Eq. 2.2.8 are not valid if the applied axial load exceeds P_b. These relationships are appropriate for design, however, because our objective will be to provide a column section whose balanced axial strength (P_b) will be greater than the real axial demand.

- The experimentally determined strength of the column exceeds that predicted by interaction methodologies, and this is especially true for Specimen 2 (see Figure 2.2.4). The reasons for this exceedance are discussed when we discuss strain states.
- If significant postyield deformation is anticipated the shear stress should be limited to $\pm 5\sqrt{f_c'}$. As was the case for the frame beam, v_s should exceed the probable shear stress in the plastic hinge region.

2.2.2.2 Strain States The test specimens were subjected to displacement controlled story drifts with three cycles at each level of displacement interest. The hysteretic behavior recorded is shown in Figures 2.2.5 and 2.2.6. Of most interest is the fact that Specimen 1 is capable of sustaining a story drift of 6% for three cycles and Specimen 2 is capable of sustaining a story drift of 3% through three cycles. In both cases the strength degradation is small.

Figure 2.2.7 shows the damage to the column of Specimen 1 at various levels of drift, while Figure 2.2.8 describes the damage to Specimen 2.

The axial load imposed on a beam-column affects the length of the plastic hinge (ℓ_p). Compare the observed behavior in Figures 2.2.7*c* and 2.2.8*c* with that of the

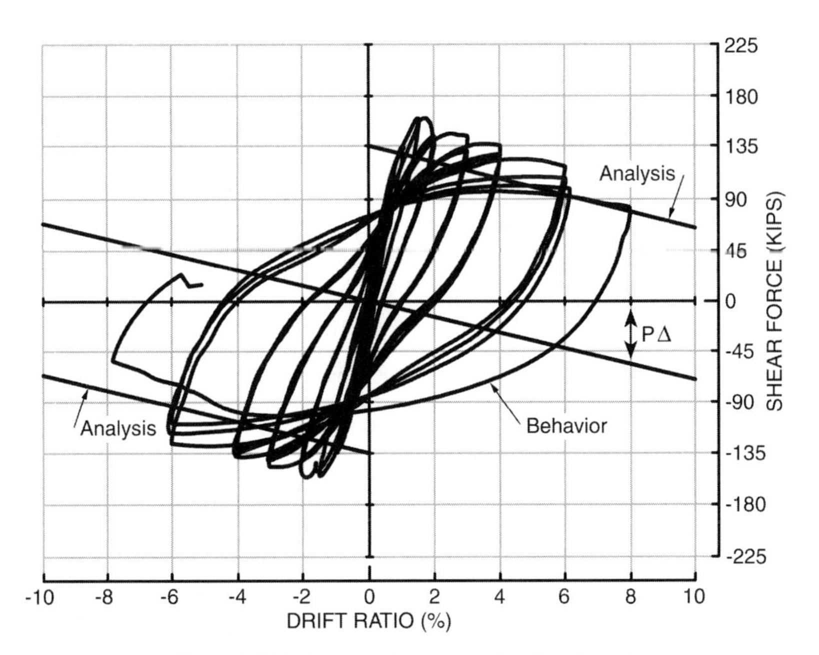

Figure 2.2.5 Hysteretic response for Specimen 1.

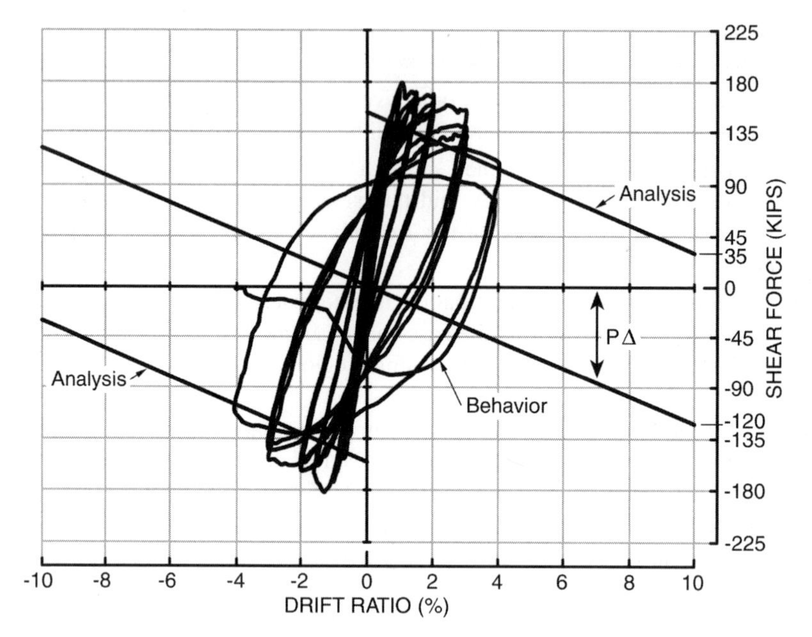

Figure 2.2.6 Hysteretic response for Specimen 2.

frame beam (Figure 2.1.9). Clearly the plastic hinge region in the column is significantly greater than that of the beam. Paulay and Priestley[2.5, Sec. 3.5.5(c)] suggest that the plastic hinge length (ℓ_p) in a column is typically 0.2ℓ. This is more than twice the value they suggest for a frame beam ($0.08\ell + 0.15 d_b f_y$). Accordingly, the analysis that follows presumes that the plastic hinge region in a column is twice what we assumed for the beam ($\ell_{pcol} = h$).

The yield strength of the reinforcing used in the test specimen was 69 ksi. The associated moment for Specimen 1, given a tension strain hardening of 10% ($1.1 f_y$), is developed as follows:

$$\begin{aligned}
\lambda_o T_y - C_s &= A_s(1.1) f_y - A_s' f_y \\
&= A_s f_y (0.1) \\
&= 4.12(69)(0.1) \\
&= 28.8 \text{ kips} \\
P &= 744 \text{ kips}
\end{aligned}$$

Assume $c = 8$ in. Then

$$C_c = (\lambda_o T_y - C_s) + P + A_1 \left(\frac{h/2 - c}{c} \right) E_s \varepsilon_c$$

and this now accounts for A_1, the contribution of the two centrally located #9 bars (see Figure 2.2.3).

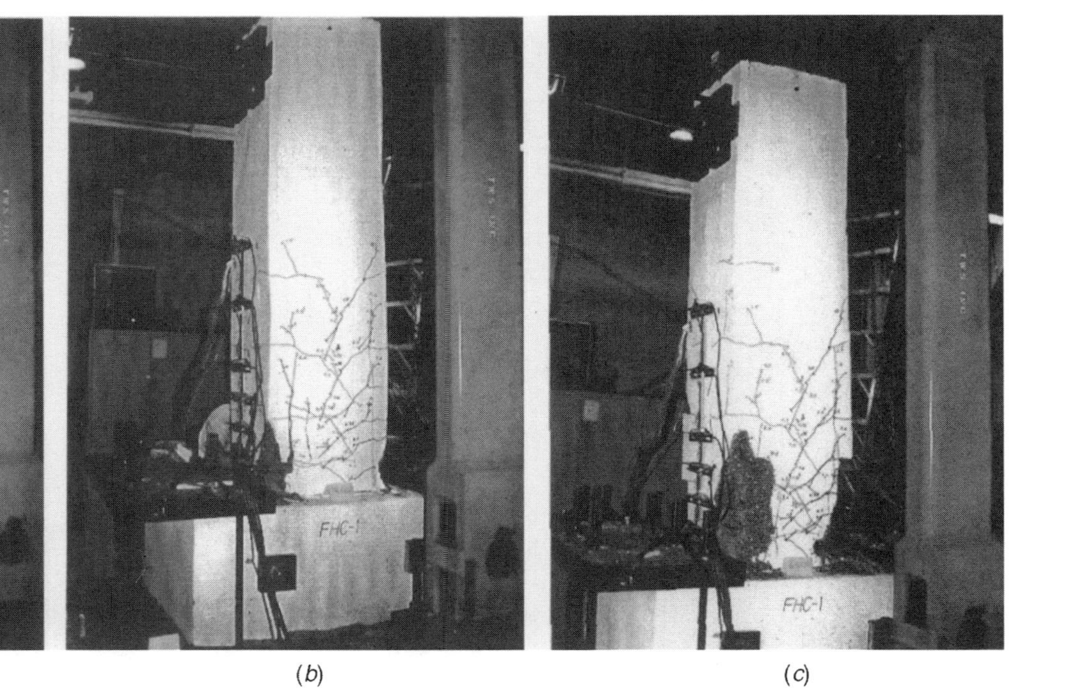

Figure 2.2.7 Crack patterns for Specimen 1 at various loading stages. (*a*) At $\delta/h_x = 2.0\%$. (*b*) At drift ratio $\delta/h_x = 3.0\%$. (*c*) At $\delta/h_x = 6.0\%$.

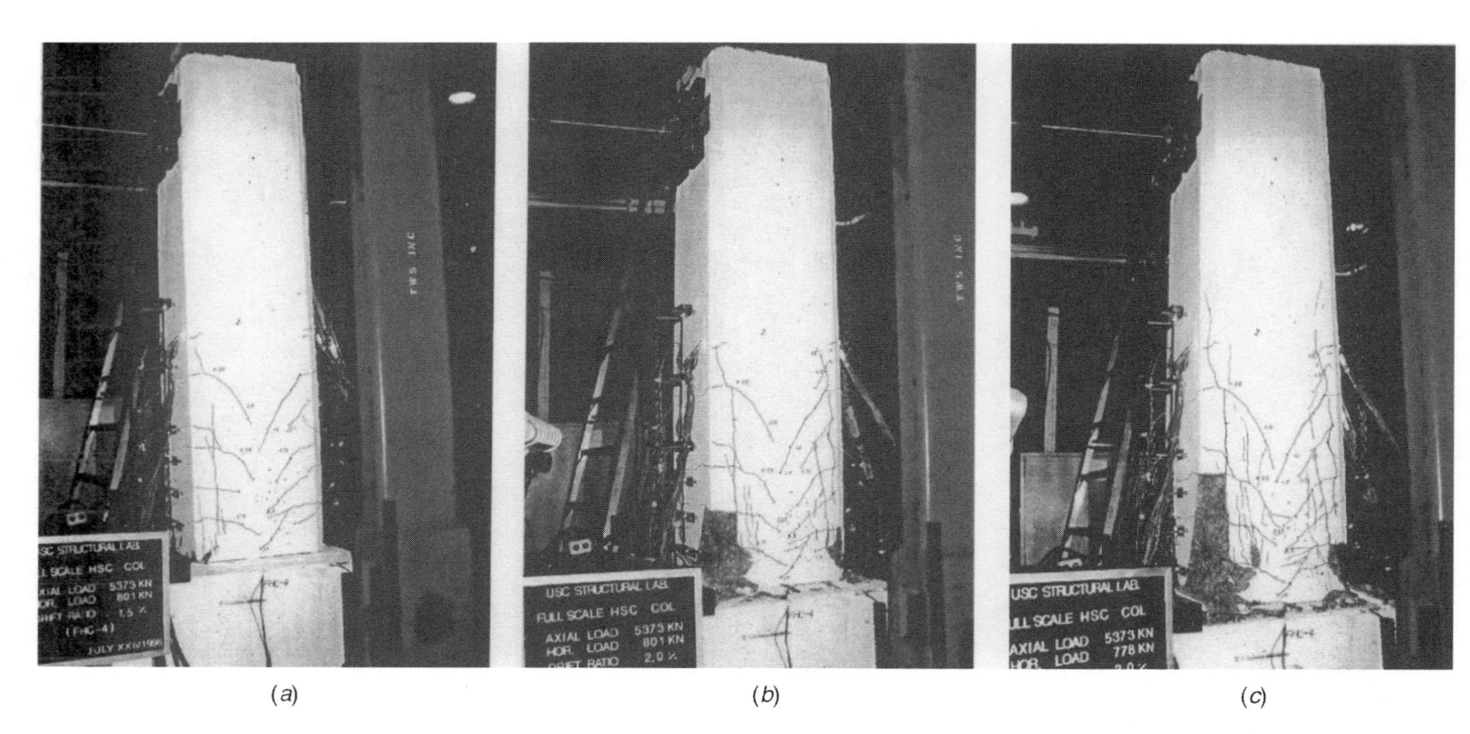

(*a*) (*b*) (*c*)

Figure 2.2.8 Crack patterns for Specimen 2 at various loading stages. (*a*) At $\delta/h_x = 1.5\%$. (*b*) At drift ratio $\delta/h_x = 2.0\%$. (*c*) At $\delta/h_x = 3.0\%$.

$$
\begin{aligned}
C_c &= 28.8 + 744 + 21.8 \\
&= 795 \text{ kips} \\
a &= \frac{C_c}{0.85 f_c b} \\
&= \frac{795}{0.85(9)(20)} \\
&= 5.2 \text{ in.} \\
c &= \frac{a}{\beta_1} \\
&= \frac{5.2}{0.65} \\
&= 8.0 \text{ in.}
\end{aligned}
$$

$$
\begin{aligned}
M_{744} &= A_s f_y(d - d') + (P + A_1 \varepsilon_s E)\left(\frac{h}{2} - \frac{a}{2}\right) + A_s\left(1.1 f_y - f_y\right)\left(d - \frac{a}{2}\right) \\
&= 4.12(69)(14.5) + (744 + 21.8)(10 - 2.6) + 4.12(7)(17.25 - 2.6) \\
&= 10{,}200 \text{ in.-kips} \qquad (851 \text{ ft-kips})
\end{aligned}
$$

The projected flexural strength is significantly less than the attained flexural capacity (Figure 2.2.4 and 2.2.5) and this implies that the depth of the compressive stress block is less than predicted. Accordingly, the assumption of a neutral axis depth (c) of 8 in. appears to be conservative.

Comment: The appropriateness of the Whitney stress block for concrete strengths of 9 ksi is reasonably questioned. This is because the stress-strain characteristic of higher strength concretes tends to be more linear (see Figure 1.2.4) than those of normal strength concretes. Alternative stress blocks are proposed by Englekirk and Pourzanjani.[2.25] They tend to provide better estimates of the capacity of a beam column.

Now examine the suggested strain states. The elastic deflection of the portion of the column above the plastic hinge is developed in the standard way:

$$
\begin{aligned}
I_g &= \frac{bd^3}{12} \\
&= \frac{(20)^4}{12} \\
&= 13{,}300 \text{ in.}^4 \\
I_e &= 0.7 I_g
\end{aligned}
$$

$$= 0.7(13,300)$$
$$= 9300 \text{ in.}^4$$

The moment at the top of the plastic hinge (M_{+20}) in Specimen 1 is on the order of

$$M_{+20} = \left(\frac{\ell - \ell_p}{\ell}\right) M_n$$
$$= \frac{50}{70}(10{,}200)$$
$$= 7286 \text{ in.-kips}$$
$$E = 40\sqrt{f_c'} + 1000$$
$$= 40\sqrt{9000} + 1000$$
$$= 4800 \text{ ksi}$$
$$\phi = \frac{M}{EI_e}$$
$$= \frac{7286}{4800(9300)}$$
$$= 0.000163 \text{ rad/in.}$$

following the conjugate beam model of Figure 2.1.10.

$$\delta_e = \frac{\phi(\ell - \ell_p)^2}{3} \qquad \text{Above plastic hinge}$$
$$= \frac{0.000163(50)^2}{3}$$
$$= 0.14 \text{ in.}$$

The component associated with curvature in the hinge region to an average concrete strain of 0.003 in./in., and accordingly its nominal strength, is

$$\phi_p = \frac{\varepsilon_c}{c}$$
$$= \frac{0.003}{8}$$
$$= 0.000375 \text{ rad/in.}$$
$$\theta_p = \phi_p \ell_p$$
$$= 0.000375(20)$$
$$= 0.0075 \text{ radian}$$

$$\delta_p = \theta_p \left(\ell - \frac{\ell_p}{2}\right)$$

$$= 0.0075(60)$$

$$= 0.45 \text{ in.}$$

The deflection at nominal moment capacity is

$$\delta_e + \delta_p = 0.59 \text{ in.}$$

This corresponds to a drift ratio of 0.84%, which is consistent with experimentally based conclusions (Figure 2.2.5). At a story drift of 2%, the induced concrete strain in the hinge region is

$$\Delta\theta_p = 0.02 - 0.0084$$

$$= 0.0116 \text{ radian}$$

$$\Delta\phi_p = \frac{\Delta\theta_p}{\ell_p}$$

$$= \frac{0.0116}{20}$$

$$= 0.00058 \text{ rad/in.}$$

$$\Delta\varepsilon_c = \Delta\phi_p c$$

$$= 0.00058(8)$$

$$= 0.0046 \text{ in./in.}$$

and the probable concrete, strain is

$$\varepsilon_{c,2\%} \cong 0.003 + 0.005$$

$$= 0.008 \text{ in./in.}$$

This suggests a state of incipient spalling and, in fact, that is the condition described by the test (Figure 2.2.7*a*).

Extending this to a drift ratio of 3%, we find

$$\Delta\theta_p = 0.01 \text{ radian}$$

$$\Delta\phi_{\mathrm{p}} = \frac{\Delta\theta_p}{\ell_p}$$

$$= \frac{0.01}{20}$$

$$= 0.0005 \text{ rad/in.}$$

$$
\begin{aligned}
\Delta\varepsilon_c &= \Delta\phi_p c \\
&= 0.0005(8) \\
&= 0.004 \text{ in./in.} \\
\varepsilon_{c,3\%} &= 0.012 \text{ in./in.}
\end{aligned}
$$

and this too is confirmed by spalling of the test specimen (see Figure 2.2.7*b*).

The contribution of the shell to flexural strength beyond a drift of 3% is significantly reduced (see Figure 2.2.7), yet the level of flexural strength remains essentially constant to a drift ratio of 6% (see Figure 2.2.5). The added strength provided by the confined core makes this possible.

The confining pressure is provided by a #5 hoop and a cross leg at 4 in. on center (see Figure 2.2.3).

$$
\begin{aligned}
f_\ell &= \frac{3(0.31)(60)}{4(14.5)} \\
&= 960 \text{ psi}
\end{aligned}
$$

The strength of the concrete in the confined core (f'_{cc}) is on the order of

$$
\begin{aligned}
f'_{cc} &= f'_c + 4.1 f_\ell \qquad \text{(Eq. 1.2.1)} \\
&= 9000 + 4.1(960) \\
&= 12{,}900 \text{ psi} \qquad (1.44 f'_c)
\end{aligned}
$$

The depth of the compressive stress block is approximately

$$
\begin{aligned}
C_c &= P + A_1 f_y + (\lambda_o - 1) A_s f_y \\
a &= \frac{C_c}{h_c f'_{cc}} \\
&= \frac{953}{17(13.5)} \\
&= 4.2 \text{ in.}
\end{aligned}
$$

$$
\begin{aligned}
M_{6\%} &= (P + A_1 f_y)\left(\frac{h}{2} - d' - \frac{a}{2}\right) + \lambda_o A_s f_y (d - d') \\
&= 882(10 - 2.75 - 2.1) + 1.25(4.12)(69)(14.5) \\
&= 4540 + 5150 \\
&= 9690 \text{ in.-kips} \qquad (807 \text{ ft-kips})
\end{aligned}
$$

This corresponds to a shear force of 94 kips which is consistent with the experimental conclusions (see Figure 2.2.5). Accordingly, it is reasonable to assume that strength degradation will be minimal even at a drift ratio of 6% provided axial loads are limited (see Eq. 2.2.6). It remains to estimate the strain state in the confined core. At a 3% story drift the strain state at the edge of the confined core was

$$\begin{aligned}\varepsilon_{c,d'} &= \varepsilon_c\left(\frac{c-d'}{c}\right)\\ &= 0.012\left(\frac{8-2.75}{8}\right)\\ &= 0.0079 \text{ in./in.}\end{aligned}$$

The curvature associated with an additional plastic hinge rotation of 3% is

$$\begin{aligned}\Delta\phi &= \frac{\Delta\theta}{\ell_p}\\ &= \frac{0.03}{20}\\ &= 0.0015 \text{ rad/in.}\end{aligned}$$

and the increase in strain is

$$\begin{aligned}\Delta\varepsilon_c &= \Delta\phi c\\ &= 0.0015(5.0) \qquad (c \cong 5 \text{ in.})\\ &= 0.0075 \text{ in./in.}\end{aligned}$$

and the concrete strain sustained is on the order of

$$\begin{aligned}\varepsilon_c &= \varepsilon_{d'} + \Delta\varepsilon_c\\ &\cong 0.0079 + 0.0075\\ &\cong 0.0164 \text{ in./in.}\end{aligned}$$

2.2.2.3 Stiffness The gross moment of inertia of the column described in Figure 2.2.3 is

$$\begin{aligned}I_g &= \frac{bh^3}{12}\\ &= \frac{20(20)^3}{12}\\ &= 13{,}333 \text{ in.}^4\end{aligned}$$

The elastic deflection of this column, given a shear load of 135 kips, would be

$$\delta_c = \frac{Vh^3}{3EI_g}$$

$$= \frac{135(70)^3}{3(5400)(13{,}333)}$$

$$= 0.21 \text{ in.} \qquad (0.3\%)$$

The stiffness of a column will depend to a large extent on the axial load imposed on it. This can be seen by comparing the initial stiffnesses of Specimen 1 (Figure 2.2.5), which was subjected to an axial load of $0.25f'_c$ with those of Specimen 2 (Figure 2.2.6), whose axial load was $0.34f'_c$. The use of I_g as a measure of stiffness predicts a behavior that is stiffer than that suggested for either column. The ACI recommendation of $0.7I_g$ for the effective moment of inertia seems reasonably consistent with the test results. Accordingly, for design purposes,

$$I_e = 0.7I_g$$

2.2.3 Conceptual Design of the Beam Column

The objective is to develop the appropriate column size in an expeditious manner. Concern should focus on:

- Attaining an appropriate level of ductility
- Constructibility
- The need to reconsider prior decisions

2.2.3.1 Estimating Probable Levels of Demand Demands imposed on a beam-column are difficult to identify, for they involve quite a few variables. Seismic axial loads can be large and on taller buildings are likely to produce net tension in some columns. Moment demands can also be large, and the capacity-based demand is not easily established because of the potential for higher modes to alter elastic based distributions (see Section 2.2). Exceeding the flexural capacity of the column must be anticipated. The satisfaction of performance objectives and code criterion requires the designer to consider real loads as well as factored loads. The basic strength criterion should be capacity-based and developed from a system mechanism (see Section 2.1.2.4 and Figure 2.1.43*c*).

From a strength perspective the flexural demands on an interior column are developed from capacity-based estimates of the column shear as developed in Eq. 2.1.22 (see Figure 2.1.26):

$$M_{cu} = V_c \frac{\ell_c}{2} \qquad (2.2.9)$$

Since V_c in Eq. 2.1.22 is developed from the probable strength of the beams $(\lambda_o M_n)$, it is a factored load. The probable strength of the frame beam should also include the strength provided by slab reinforcing (see Ref. 2.5, Sec. 4.5.1(b)). The so-developed probable moment demand, which includes an overstrength factor, will typically exceed ACI requirements,

$$\sum M_c \geq \frac{6}{5} \sum M_g \qquad \text{(Ref. 2.6, Eq. 21-1)}$$

because M_g, given the ACI definition, is the nominal flexural strength of the girder and does not include provisions for overstrength. In the development of a conceptual design procedure (Section 2.1.2.4), an additional overstrength factor of 1.3 (Eq. 2.1.24) was included, and this was intended to insure a conservative estimate of shear demand, primarily because ductility in shear is not reasonably anticipated. The use of $M_{b,\text{pr}}(\lambda_o M_{bn})$ as a driving force in the development of a flexural design criterion for the column is sufficient because flexural ductility will exist in a column provided axial loads do not exceed P_b (Eq. 2.2.2a).

The axial load from a code perspective is the factored dead and live loads plus a prescriptive estimate of seismic demand. A realistic mechanism-based (Section 3.2.6), unfactored axial load should be used to evaluate performance. Exterior column strength *must* consider capacity-based axial loads. The frame of Figure 2.1.28 will be subjected to a lateral load that, in terms of defining the demand on exterior columns, can be presumed to arrive in the form of a primary or first mode distribution. The frame shear at the strength limit state is reasonably estimated from the story mechanism described in Figure 2.2.9. The story mechanism shear is

$$V_M = \frac{\sum M_{b,\text{pr}}\,(\ell/\ell_c)}{h_x} \qquad (2.2.10)$$

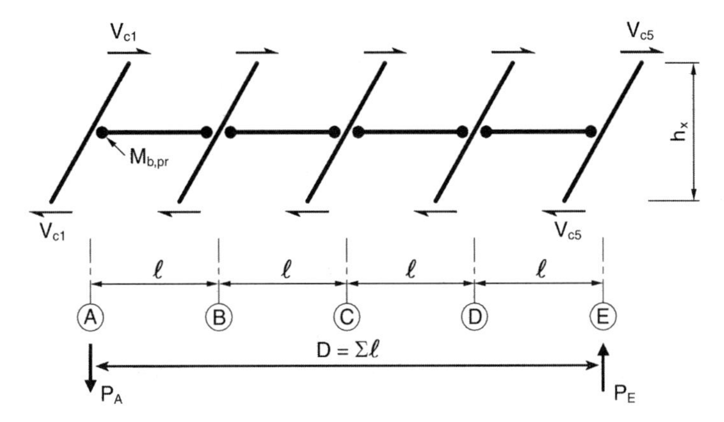

Figure 2.2.9 Story mechanism.

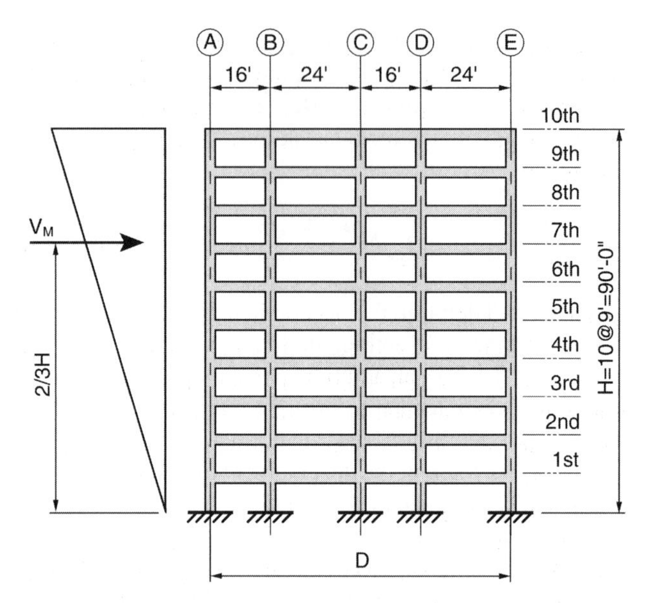

Figure 2.2.10 Mechanism loads applied to a frame.

This shear should be distributed in a triangular form varying from a maximum at the top to a minimum value at the base (Figure 2.2.10). In this form it reasonably represents a primary mode response and, as a consequence, becomes a good estimate of the capacity-based axial demand imposed on the exterior columns. In Chapter 3 we study more precise methods of estimating the mechanism shear. For now it is important to understand that the mechanism shear quantified by Eq. 2.2.10 will be nonconservative, for once the capacities of all of the beams at the lowest or most critical level have been exceeded, the columns will tend to bridge that story. The designer can consider this subliminally in the column sizing process or factor it directly into design equations. We choose the latter approach in the conceptual design process and adopt frame or system overstrength factor (Ω_f) of 1.15.

The axial load imposed on the exterior columns when a frame mechanism is attained (see Figure 2.2.9) is

$$P_E = -P_A = \frac{\Omega_f V_M}{D}\left(\frac{2}{3}H\right) \tag{2.2.11}$$

where H is the height of the building and D is the overall length of the frame (see Figure 2.2.10).

The sequential application of Eq. 2.2.10 and 2.2.11 is easily accomplished, but they may be combined with sufficient accuracy for conceptual design purposes as follows:

$$P_E = \frac{\Omega_f \sum M_{b,\text{pr}}(\ell/\ell_c)}{Dh_x}\left(\frac{2}{3}H\right)$$

If the probable moment capacities and bay widths are the same, ℓ/ℓ_c is assumed to be 1.1, and H, the total height of the building, is replaced by nh_x, where n is the number of floors. In addition, N is the number of bays. Thus we get

$$P_E = \frac{(1.15)2NM_{b,\text{pr}}(1.1)(0.67)n}{D}$$

$$= \frac{1.7nNM_{b,\text{pr}}}{D} \tag{2.2.12}$$

If Eq. 2.2.12 is not used to develop the axial load, it should be used to confirm the conclusion reached using Eq. 2.2.10 and 2.2.11.

The preceding development assumes that bay widths (ℓ) and beam strengths ($M_{b,\text{pr}}$) are the same or essentially the same, because having unequal bays in a frame or unequal beam strengths will produce a different distribution of axial load; this is a result of the total overturning load at mechanism not being entirely resisted by the exterior columns. When frame bays are not equal the shears delivered to interior columns will differ even when the probable moment capacities of the beams framing into them are the same. For example, the column at D (Figure 2.2.10) will, at mechanism for the direction of lateral load shown, receive a net increment of axial load equivalent to

$$R_D = 2\left(\frac{M_{b,\text{pr}}}{\ell_{c2}} - \frac{M_{b,\text{pr}}}{\ell_{c1}}\right) \qquad \text{(see Figure 2.1.27)}$$

Regardless of the lengths of the frame beams and their probable moment capacities, a story mechanism will form and equilibrium ($\sum R$) will be maintained. A capacity-based estimate of the axial demand on each column is still possible and should be developed in the conceptual design phase. Capacity-based concepts, coupled with a mechanism-based approach, are also effectively used to improve a conceptual design.

Consider the frame geometry of Figure 2.1.27 for example. If the probable moment capacities of the frame beams are the same throughout, the shear developed in span $AB(\ell_2 = 16$ ft) will be about 50% greater than those developed in span $BC(\ell_1 = 24$ ft). If we refer to the shear in the longer beam as V_{b1}, we can develop a normalized distribution of axial loads. The shears imposed and normalized reactions (V^*) are shown in Figure 2.2.11*a*. Observe that the column reactions are in equilibrium.

The mechanism shear is not affected by the span lengths and for the case of equal beam moments is still defined by Eq. 2.2.10 and the overturning moment (OTM) by

$$\text{OTM} = V_M\left(\frac{2}{3}H\right) \tag{2.2.13}$$

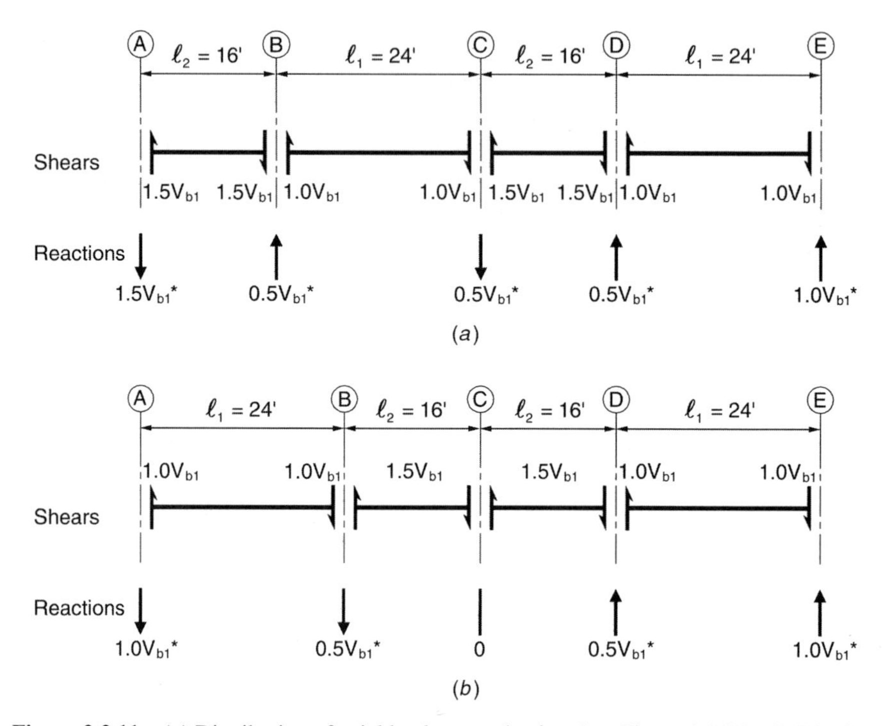

Figure 2.2.11 (*a*) Distribution of axial loads at mechanism (see Figure 2.1.27). (*b*) Distribution of axial loads—rearranged bays.

The resisting moment may be developed directly from mechanism-induced (normalized) reactions (Figure 2.2.11*a*). The resisting moment may be developed by taking moments about A.

$$\text{OTM} = 0.5V_{b1}^* (\ell_2) + V_{b1}^* (2\ell_1 + 2\ell_2) + 0.5V_{b1}^* (\ell_2) \qquad \text{(see Figure 2.2.11)}$$

$$= V_{b1}^* (2\ell_1 + 3\ell_2) \tag{2.2.14}$$

The reaction at A will be higher for the unequal bay system than that developed by Eq. 2.2.11 based on the equal bay presumption:

Equal Bays

$$P_A = \frac{\text{OTM}}{D} \qquad \text{(see Eq. 2.2.11)}$$

Unequal Bays

$$P_A = 1.5V_{b1}^*$$

$$= (1.5)\frac{\text{OTM}}{D + \ell_2} \tag{2.2.15}$$

$$P_E = (1.0)\frac{\text{OTM}}{D + \ell_2}$$

For a frame depth D, equal to 80 ft, the difference becomes

Equal bays	– 0.0125 OTM
Unequal bays	– 0.0156 OTM

This amounts to a 25% increase. A similar impact would occur on reverse loading when the column at A is subjected to compression.

Design control over structures can significantly reduce the seismic impact on key columns. For example, if the bays described in Figure 2.2.11a are rearranged so that the exterior bay spans are 24 ft and the interior bays 16 ft, the axial load redistribution becomes that described in Figure 2.2.11b. Now the column loads on columns D and E are $0.5V_{b1}^*$ and $1.0V_{b1}^*$, respectively. The overturning moment is

$$\text{OTM} = 0.5V_{b1}^*(2\ell_2) + V_{b1}^*(D) \tag{2.2.16}$$

and the seismic load imposed on columns A and E becomes

$$V_{b1}^* = \frac{\text{OTM}}{D + 2\ell_2} \tag{2.2.17}$$

For a frame depth (D) equal to 80 ft, the load on columns A and E for the rearranged bays (24-ft exterior span) is

Equal bays	– 0.0125 OTM	(see Eq. 2.2.11)
Rearranged bays	– 0.0089 OTM	(see Eq. 2.2.17)

This amounts to a reduction of almost 30% in the axial load that must be sustained by the exterior columns. The corresponding seismic load on the columns at B and D (0.0045 OTM) are half of those imposed on the columns at A and E.

The important point is that the designer is in control. The designer can easily apply capacity-based concepts to determine and control the probable axial demand on frame columns by adjusting bay widths or beam strengths. The use of capacity-based procedures allows an insight into behavior not afforded by elastic-based procedures.

Conclusions:

- Unequal bays will cause the axial load distribution to vary.
- Columns near short span beams will be subjected to higher loads than those estimated through the use of Eq. 2.2.11.

- Column loads can be altered through a considered choice of span length and probable frame beam moment capacities.

2.2.3.2 *Sizing the Beam Column* Our performance based objectives are, as previously stated, three (Introduction to Section 2.2.3). We want to promote:

- *Ductility:* The anticipated load should produce a stress on the gross column area (concrete only) that is optimally less than $0.3f'_c$.
- *Constructibility:* The total amount of reinforcing should not exceed 3% of the gross area of the section.
- *Compatibility with Frame Beam:* The selected column size must exceed that required to accommodate beam reinforcing and satisfy both column and joint shear stress limit states (see Sections 2.1.2.4 and 2.1.2.1). The appropriate transverse reinforcing program must be developed.

First Consideration: Ductility The sizing process should start with the attainment of ductility objectives. The required input is the probable axial load. The dead and live loads should be based on a reasonable estimate of those loads that are likely to be experienced; hence, not only are they not factored, but they may be significantly less than service level loads. Let us refer to these loads as probable D_{pr} and L_{pr}. The imposed earthquake load should be capacity-based (E_{pr}) and developed as described in the preceding section. Accordingly,

$$0.3f'_cA_g > D_{\text{pr}} + L_{\text{pr}} + E_{\text{pr}} = P_{\text{pr}} \tag{2.2.18}$$

Alternatively, Eq. 2.2.2a may be used, the objective being to limit the probable axial load to P_b. Hence,

$$0.5\beta_1 f'_c bd > D_{\text{pr}} + L_{\text{pr}} + E_{\text{pr}} = P_{\text{pr}} \qquad \text{(see Eq. 2.2.2a)}$$

Second Consideration: Constructability Once ductility objectives have been attained, flexural strength objectives need to be satisfied and reinforcing quantities determined to fall within constructable limits.

From a purely capacity-based perspective, the moment demand may be those moments imposed on the subassembly (Figure 2.1.23) by the beams when they reach their probable flexural capacities. Some account should be made of the potential for dynamic redistribution, and a system overstrength factor (Ω_f) of 1.15 is suggested. Hence, the design column moment at the face of the beam is

$$M_c = \frac{\lambda_o\Omega_f \sum M_{bn}}{h_x}\left(\frac{\ell}{\ell_c}\right)\left(\frac{h_c}{2}\right) \tag{2.2.19a}$$

$$= \frac{1.25(1.15)}{2}\sum M_{bn}\left(\frac{\ell}{\ell_c}\right)\left(\frac{h_c}{h_x}\right)$$

$$= 0.72 \sum M_{bn} \left(\frac{\ell}{\ell_c}\right) \left(\frac{h_c}{h_x}\right) \tag{2.2.19b}$$

We now can determine the appropriate amount of column reinforcing. The process is somewhat complicated by the introduction of strength reduction factors. The strength reduction factors (ϕ) are intended philosophically to account for the difference between design strength and provided member strength as well as uncertainties in the analysis. For axial loads in excess of $0.1 f_c' A_g$, a strength reduction factor of 0.7 is recommended by ACI,[2.6] but this appears somewhat excessive when applied to a confincd column whose loads are capacity based. Consider the development of Eq. 2.2.3. The capacity reduction factor would be applied in the classical manner. Regenerating this relationship

$$M_u = \phi M_n$$

$$= \phi P_{\mathrm{pr}} \left(\frac{h}{2} - \frac{a}{2}\right) + \phi A_s f_y (d - d') \tag{2.2.20a}$$

The effective safety factor, when imposed on the test specimen discussed in Section 2.2.2 at an axial load of 744 kips, is

$$M_u = 0.7(744)(10 - 2.43) + 0.7(4.12)(60)(14.5)$$

$$= 538 \text{ ft-kips}$$

and this is 58% of its real strength (see Figure 2.2.4).

Thus the provided safety factor is in excess of 1.5. This level of conservatism should be reserved for cases where the flexural and axial load imposed on the column is far less certain, and this is how such safety factors evolved. Modern codes impose a conservative overstrength factor on strength-based axial loads and the conclusions can be misleading.

The designer must be careful in assigning the appropriate level of axial load. Let us suppose that a capacity-based load (real) is 300 kips but that the code strength-based load amplified by a prescriptive overstrength factor is 1000 kips (hypothetical). The nominal flexural strength of the column based on the real load (300 kips) as described in Figure 2.2.4 is on the order of 600 ft-kips, while the flexural strength based on the hypothetical axial load (1000 kips) is 850 ft-kips.

Conclusion: An artificially high axial load assumption can produce a column that will not satisfy design objectives.

Consider also the decision-making difficulties created by factoring the resistance function (interaction diagram). The apparent balanced load is now 70% of the real balanced load. This can be misleading and result in significant conservatism. Suppose, for example, your design requires that the column support a real axial load of 1200 kips. The implication developed from the factored interaction diagram would be that the column was brittle yet, as we have seen, the column has a displacement ductility of about 3 (Figure 2.2.6). The nominal moment capacity of the column (Figure 2.2.4) is 840 ft-kips. If we plot a factored interaction curve, the factored balanced load (P_{bu}), would be 840 kips and the balanced moment (M_{bu}) would be 560 ft-kips. The flexural capacity at an axial load of 1200 kips would identify M_u as about 430 ft-kips, and this would produce a column whose factor of safety is at least 1.95 (840/430). Observe that the experimental flexural capacity is in excess of 1000 ft-kips (Figure 2.2.4).

Conclusion: Alternative means of assigning column capacity/demand ratios should be developed especially if the design objective is to optimize performance. My preference is to work with unfactored beam column interaction diagrams. This allows me to more appropriately assign the capacity/demand ratio.

A more consistent factor of safety would be attained by imposing the strength reduction factor on the level of provided flexural strength. Rearranging Eq. 2.2.20a, we arrive at a flexural steel component demand of

$$A_s = \frac{M_u - P_{\text{pr}}(h/2 - a/2)}{\phi f_y (d - d')} \tag{2.2.20b}$$

where P_{pr} is as developed in Eq. 2.2.18.

It is reasonable, when our objective is to optimize performance, to question the universal use of a strength reduction factor of 0.7 when capacity-based concepts are used in the load development process, and the designer realizes that significant ductility exists. A strength reduction factor of 0.9 is used in the designs that follow because capacity-based design procedures will have been used to develop the axial load and the (real) level of applied axial load is by design less than $0.75P_b (\cong 0.25 f_c' A_g)$. Observe that this ($0.75P_b$) is consistent with the approach used to define flexural reinforcement maxima. Where high axial loads ($P_{\text{pr}} > 0.25 f_c' A_g$) are expected, a capacity reduction factor of 0.7 is suggested.

The introduction of Eq. 2.2.19b into Eq. 2.2.20b produces our basic flexural strength objective:

$$A_s = \frac{0.72 \sum M_{bn}(\ell/\ell_c)(h_c/h_x) - P_{\text{pr}}(h/2 - a/2)}{\phi f_y (d - d')} \tag{2.2.21a}$$

Rearranging Eq. 2.2.21a so as to adjust the impact of the capacity reduction factor to be consistent with our proposed design basis produces

$$A_s = \frac{0.72 \sum M_{bn}(\ell/\ell_c)(h_c/h_x)}{\phi f_y (d - d')} - \frac{P_{\text{pr}}(h/2 - a/2)}{f_y (d - d')} \tag{2.2.21b}$$

We can reduce this for conceptual design purposes, assuming that

$$d - d' \cong 0.8h$$

$$\frac{\ell}{\ell_c} \cong 1.1$$

$$\frac{h}{2} - \frac{a}{2} \cong 0.3h$$

$$\phi = 0.9$$

$$\frac{h_c}{h_x} = 0.75$$

The area of tensile steel (A_s) required to satisfy our performance flexural capacity objectives becomes

$$A_s = \frac{0.83 \sum M_{bn}}{hf_y} - \frac{0.38 P_{\text{pr}}}{f_y} \tag{2.2.21c}$$

Contrast this procedure with the referenced code approach.[2.6] The design axial load becomes

$$P_u = 1.05D + 1.28L + \Omega_o E \tag{2.2.21d}$$

The intent is to emulate capacity-based concepts. Observe that each component may considerably overestimate the probable demand.

From a code perspective the column design moment need not be more than 20% greater than the nominal demand of the frame beams that drive it.

The code area of tension steel (A_s) is now developed from Eq. 2.2.20a:

$$A_s = \frac{1.2 \sum M_{bn}(\ell/\ell_c)(h_c/h_x) - P_u(h/2 - a/2)}{2\phi(f_y)(d - d')} \tag{2.2.22a}$$

and, using the previously defined approximations and a ϕ of 0.7 for axial loads as developed from Eq. 2.2.21d that are in excess of $0.1 f_c' A_g$, we get

$$A_s = \frac{1.2 \sum M_{bn}(1.1)(0.75) - P_u(0.3h)}{2(0.7)0.8hf_y}$$

$$= \frac{0.88 \sum M_{bn} - 0.27 P_u h}{f_y h} \tag{2.2.22b}$$

For columns whose axial loads are less than $0.1 f_c' A_g$, ϕ need only be 0.9 and Eq. 2.2.22b becomes

$$A_s = \frac{0.69 \sum M_{bn} - 0.21 P_u h}{f_y h} \tag{2.2.22c}$$

We must not, however, forget that both Eq. 2.2.21a and 2.2.22a assume that the tension steel will yield, and this will only be the case when the axial load is less than P_b. Observe that the use of factored loads makes it difficult to understand if the tension side steel will in fact yield. We consider both design procedures, capacity-based and code compliant, in our example conceptual designs and analysis. In either case the reinforcing ratio (ρ) should not exceed 3%.

Third Consideration: Compatibility with Frame Beam The proposed column size should be checked and a transverse reinforcing program must be developed. Three conditions must be considered in the development of a transverse reinforcement program—shear, confinement, and lateral support of the column bars.

Shear in the column is limited to the moments that can be delivered by the beams. Exceptions include the uppermost and lowermost columns. Shifts in the point of inflection will have little or no impact on column shear.[2.5, Sec. 4.6.4, Fig. 4.21, and Sec. 4.6.7] Shear forces developed using capacity-based concept need not be magnified. A conservative design approach is justified, however, because of the consequence of a shear failure. In any case, the probable magnitude of column shear will be developed during design confirmation. For conceptual design purposes we assume that the use of a capacity-based load and the introduction of a strength reduction factor will provide a section that will prove, in the design verification phase, to be acceptable. If we assume that frame beam reinforcement above and below the column is the same or essentially the same, then

$$V_{cu} = \frac{\sum M_{b,\text{pr}}}{h_x}\left(\frac{\ell}{\ell_c}\right) \tag{2.2.23}$$

where $\sum M_{b,\text{pr}}$ is the sum of the probable beam moments at the lower joint assuming that it is equal to or larger than $\sum M_{b,\text{pr}}$ at the upper joint.

Concrete in columns is allowed to carry shear by the ACI[2.6] but this acceptance carries conditions, and those conditions reflect a concern with the potential for a tension force to be imposed on the column. Performance objectives can only be attained if shear failures in the plastic hinge region are avoided and this, as we saw in Section 2.1, requires that the shear strength provided by the concrete (V_c) be considered ineffective ($V_c = 0$):

$$v_s = \frac{V_{cu}}{\phi bd} \qquad (\text{if } v_c = 0) \tag{2.2.24a}$$

$$= \frac{\sum M_{b,\text{pr}}(\ell/\ell_c)}{\phi h_x bd} \tag{2.2.24b}$$

Using the approximations previously developed, this reduces to

$$v_s = \frac{\sum M_{b,\text{pr}}(1.1)}{0.85 h_x bd} \tag{2.2.24c}$$

$$= \frac{1.3 \sum M_{b,\text{pr}}}{h_x bd}$$

Confinement is, at least for columns larger than 24 in. by 24 in. (the usual case), quantified by

$$A_{sh} = \frac{0.09 s h_c f'_c}{f_{yh}} \qquad \text{(Ref. 2.6, Eq. 21-4)}$$

This may be reduced to a confining pressure (f_ℓ), for direct comparison with the required shear strength (v_s) required by the transverse steel:

$$\frac{A_{sh} f_{yh}}{b_c s} = 0.09 f'_c \qquad (2.2.25)$$

This represents a stress acting over the core width (b_c) while v_s acts over the beam width (b). Since b_c/b is conservatively 0.85, a direct comparison is possible. Accordingly, $v_{s,\text{min}}$ in the hinge region is $0.077 f'_c$.

Our conceptual design rule is

$$v_s = \frac{\sum M_{b,\text{pr}}(\ell/\ell_c)}{\phi h_x bd} \qquad \text{(see Eq. 2.2.24b)}$$

or

$$v_s = \frac{1.3 \sum M_{b,\text{pr}}}{h_x bd}$$

$$v_{s,\text{min}} = 0.077 f'_c \qquad \text{(Hinge region)} \qquad (2.2.26)$$

The hinge region is the depth of the member but not less than one-sixth of the clear column height (h_c). (See Ref. 2.6, Sec. 21.4.4.4.)

The maximum spacing between crossties of the transverse confining steel (see Figure 2.2.1*a*), given the code[2.6] as a criterion, is 14 in. on center, and the bars must be anchored around a vertical bar. Accordingly, the spacing between vertical bars in a column should not exceed 14 in.

When significant postyield rotations are expected to occur in a column, every longitudinal bar subjected to strain hardening should be supported adequately (135° hook).

2.2.3.3 Story Mechanism Considerations In the introduction to this section, the importance of avoiding a story mechanism was emphasized. In Section 2.2.3.1, mechanism load levels were quantified (Eq. 2.2.10). The desired mechanism story shear, (V_M) is a function of the probable strength of the frame beams. Capacity-based design requires that a column-based mechanism strength conservatively exceed the frame beam induced mechanism shear (V_M) as quantified by Eq. 2.2.10.

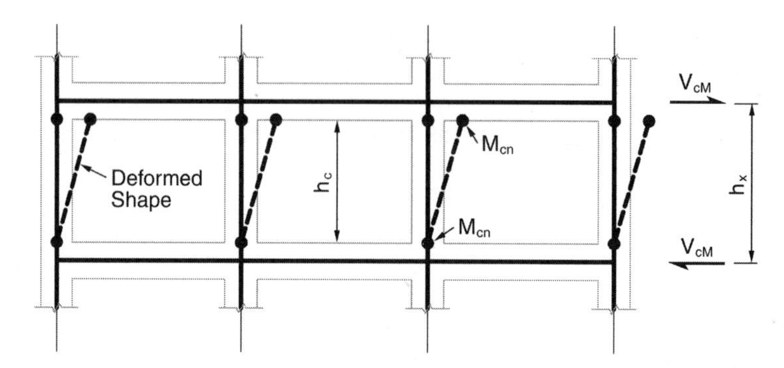

Figure 2.2.12 Column mechanism.

The column mechanism shear (V_{cM}) is (Figure 2.2.12)

$$V_{cM} = \frac{\sum M_{cn}}{h_c} \tag{2.2.27}$$

where ΣM_{cn} represents the *available* flexural capacity of the columns both top and bottom.

The available flexural capacity must take into consideration the probable level of axial load, and this will significantly discount the shear strength of columns subjected to large tensile loads.

One advantage associated with the use of mechanism loads is the inclusion of $P\Delta$ effects directly into analysis.[2.4, Sec. 3.7(b)] The frame beam mechanism load provides the shear necessary to balance $P\Delta$ loads as well as the restoring force that provides dynamic stability. Accordingly, the column mechanism (V_{cM}) need only be comfortably larger than the beam mechanism (V_M).

Our objective in the conceptual design phase should be to create a column-based mechanism that satisfies

$$V_{cM} > 1.5V_M \tag{2.2.28}$$

The attainment of this objective is demonstrated in Example 2, which follows. Frame stability is discussed in more detail in Chapter 4.

2.2.3.4 *Design Process Summary*

Step 1: Determine the Objective Capacities. During the beam design phase, approximate loads were reasonably used to size frame columns. These loads should be reviewed and refined prior to the design of the column.

(a) Capacity-based loads are probable loads.

(b) Code-compliant loads should include axial earthquake loads that include overstrength amplification factors (Ω_o) and factored service loads.

Step 2: Determine the Required Minimum Area of the Column.

$$A_g > \frac{D_{\text{pr}} + L_{\text{pr}} + E_{\text{pr}}}{0.3 f_c'} \qquad \text{(Eq. 2.2.18)}$$

Step 3: Review Minimum Column Size Required to Satisfy Column Shear and Beam-Column Joint Objectives. See Section 2.1.2.6, Steps 2 and 6.

Step 4: Determine the Amount of Reinforcing Required for the Following.

(a) To satisfy design objectives:

$$A_s = \frac{0.83 \sum M_{bn}}{h f_y} - \frac{0.38 P_{\text{pr}}}{f_y} \qquad \text{(Eq. 2.2.21c)}$$

(b) To comply with code criterion when P_u exceeds $0.1 f_c' A_g$:

$$A_s = \frac{0.88 \sum M_{bn} - 0.27 P_u h}{f_y h} \qquad \text{(Eq. 2.2.22b)}$$

Step 5: Reinforce Column. Check to insure that total reinforcing (A_{st}) is 3% or less. Propose a reinforcing program, both longitudinal and transverse.

Shear reinforcement must exceed that required by Eq. 2.2.24b (see Figure 2.1.26):

$$v_s = \frac{\sum M_{b,\text{pr}} (\ell / \ell_c)}{\phi h_x b d} \qquad \text{(Eq. 2.2.24b)}$$

In the potential column hinge region v_s may not be less than that required to satisfy confinement objectives:

$$v_{s,\text{min}} = 0.077 f_c' \qquad \text{(Eq. 2.2.26)}$$

Additionally, in the lowermost level column, shear reinforcement should not be less than that required to sustain a shear of

$$V_{cu,1} = \frac{0.5 \sum M_{b,\text{pr}} (\ell / \ell_c) + M_{c,\text{pr}}}{h_x} \qquad (2.2.29)$$

Step 6: Check Story Column Mechanism Strength. Determine story mechanism shear demand (see Figure 2.2.12):

$$V_M = \frac{\sum M_{b,\text{pr}} (\ell / \ell_c)}{h_x} \qquad \text{(Eq. 2.2.10)}$$

Determine column mechanism shear capacity:

$$V_{cM} = \frac{\sum M_{cn}}{h_c} \qquad \text{(Eq. 2.2.27)}$$

Strengthen columns if

$$V_{cM} < 1.5V_M \qquad \text{(see Eq. 2.2.28)}$$

2.2.3.5 Example Designs

Example 1 See Section 2.1.2.7. The beams of an interior subassembly are typically designed so as to minimize seismic axial loads imposed on the interior columns. In this example we will conclude that earthquake-induced axial loads are small (see Figure 2.2.9) but that dead and live loads are high. For a 30-ft by 30-ft bay, the service dead and live loads are 150 lb/ft^2 and 100 lb/ft^2. These translate to story service loads of 135 kips and 90 kips, respectively.

Step 1: Determine the Objective Capacities. The loads that must be considered for a twenty-story building are:

$$\begin{aligned} \text{Probable} \quad D_{\text{pr}} &= 135\ (20) &&= 2700 \text{ kips} \\ L_{\text{pr}} &= 18\ (20) &&= \underline{\ 360 \text{ kips}} \\ P_{\text{pr}} &= &&\ \ 3060 \text{ kips} \end{aligned}$$

Note that 18 for L_{pr} corresponds to a live load of 20 lb/ft^2.

$$\begin{aligned} \text{Factored} \quad D &= 1.05\ (2700) &&= 2835 \text{ kips} \\ L &= 1.28\ (20)(900)(0.04) &&= \underline{\ 922 \text{ kips}} \\ P_u &= &&\ \ 3757 \text{ kips} \end{aligned}$$

Here, the 0.04 in the equation for L assumes that a 60% reduction is allowed for the design of columns.

Step 2: Determine Required Area of the Column.

$$A_g > \frac{D_{\text{pr}} + L_{\text{pr}} + E_{\text{pr}}}{0.3 f_c'} \qquad \text{(see Eq. 2.2.18)}$$

$$> \frac{3060}{0.3(5)} \qquad f_c' = 5 \text{ ksi}$$

$$> 2040 \text{ in.}^2 \qquad \text{(45 in. by 45 in.)}$$

Step 3: Review Minimum Size Required to Satisfy Beam and Beam-Column Joint Objectives. Selected minimum column size—36 in. by 42 in. $A_g = 1512$ in.^{2} (see Sec. 2.1.2.7, Example 1).

Conclusion: Use a higher strength concrete.

$$f'_{c,\min} = \frac{D_{\text{pr}} + L_{\text{pr}} + E_{\text{pr}}}{0.3(A_g)}$$

$$= \frac{3060}{0.3(1512)}$$

$$= 6.7 \text{ ksi}$$

Conclusion: Propose a 36 in. by 42 in. column; $f'_c = 7$ ksi.

Step 4: Determine the Amount of Column Reinforcing Required.

(a) To satisfy design objectives:

$$A_s = \frac{0.83 \sum M_{bn}}{f_y h} - \frac{0.38 P_{\text{pr}}}{f_y} \qquad \text{(Eq. 2.2.21)}$$

$$\sum M_{bn} = 2A_s f_y (d - d')$$

$$= 2(12.48)(60)(30) \qquad \text{(Eight \#11 bars)}$$

$$= 44{,}930 \text{ in.-kips}$$

$$P_{\text{pr}} = 3060 \text{ kips}$$

$$A_s = \frac{0.83(44,930)}{60(36)} - \frac{0.38(3060)}{60} \qquad \text{(see Eq. 2.2.21c)}$$

$$= 17.3 - 19.4$$

$$= 0$$

Conclusion: Minimum steel is required.

Comment: Observe that the axial load need only develop an eccentricity of

$$e = \frac{44{,}930}{3060} = 14.7 \text{ in.}$$

in order to balance the design moment. Accordingly, the provided flexural capacity will significantly exceed the demand

(b) To comply with code criterion:

$$A_s = \frac{0.88 \sum M_{bn} - 0.27 P_u h}{f_y h} \qquad \text{(Eq. 2.2.22b)}$$

$$= \frac{0.88(44{,}930) - 0.27(3757)(36)}{60(36)}$$

$$= 1.40 \text{ in.}^2$$

Conclusion: Minimum steel is required.

Step 5: Reinforce Column.

$$\begin{aligned} \text{Minimum reinforcing} &= 0.01A_g \\ &= 0.01(36)\ (42) \\ &= 15.12 \text{ in.}^2 \end{aligned}$$

Since column bars should be spaced no further apart than 14 in., four bars are required on each face. Accordingly, at least twelve bars are required. Observe that the flexural overstrength of the column is significant. Accordingly, a lesser spacing of transverse ties does not, in this case, seem warranted:

$$\begin{aligned} A_B &= \frac{A_{st}}{n} \\ &= \frac{15.12}{12} \\ &= 1.26 \text{ in.}^2 \end{aligned}$$

Conclusion: Provide twelve #10 bars.

Transverse Reinforcing Program—Intermediate Level Column

Since the axial load is reliable and significant the concrete may be presumed to transfer shear $\left(v_c = 2\sqrt{f'_c}\right)$

$$v_s = \frac{\sum M_{b,\text{pr}}(\ell/\ell_c)}{\phi h_x bd} \qquad \text{(Eq. 2.2.24b)}$$

$$= \frac{1.25(44,930)(30/26.5)}{0.85(12)(12)(36)(39)} \qquad \text{(see Step 4)}$$

$$= 0.371 \text{ ksi} \qquad (4.5\sqrt{f'_c})$$

$$v_{s,\min} = 0.077 f'_c \qquad \text{(Eq. 2.2.26)}$$

$$= 0.077(7000)$$

$$= 0.54 \text{ psi} \qquad \text{(Hinge region)}$$

Comment: Given the high shear demand and the short deep column ($h_c = 9$ ft, $d = 3.25$ ft), confinement would be provided throughout. Four stirrup legs are required to satisfy detailing considerations; accordingly, one might proceed to determine the spacing of #5 hoops:

$$A_{sh} = \frac{v_{s,\min} bs}{f_{yh}}$$

$$4(0.31) = \frac{0.54(36)(s)}{60}$$

$$s = 3.83 \text{ in. on center}$$

Conclusion: Provide two #5 hoop ties at 3.75 in. on center in each direction.

Transverse Reinforcing Program—Base Level First determine the probable moment likely to be experienced at the column foundation interface. Recall that the maximum attainable moment is that associated with balanced conditions (M_{bal}).

$$P_b = 0.85(0.6)\beta_1 f_c' bd \qquad \text{(see Eq. 2.2.2a)}$$

$$= 0.85(0.6)(0.7)(7)(36)(39)$$

$$= 3509 \text{ kips} \qquad (f_c' = 7000 \text{ psi})$$

$$M_{\text{bal}} = A_s f_y (d - d') + P_b \left(\frac{h}{2} - \frac{a}{2}\right) \qquad \text{(Eq. 2.2.3)}$$

$$= 4(1.56)(60)(36) + 4(1.27)(60)(12) + 3509(21 - 8.19)$$

$$= 62{,}086 \text{ in.-kips} \qquad (5167 \text{ ft-kips})$$

The inclusion of the interior layer of column reinforcing (12 in the second term of the second equation) must be viewed as an approximation at this point. The inner steel layer may yield; if it doesn't, the shear developed with its inclusion will be slightly conservative. In this case, given a conservative approach, it contributes about 5% to the column shear. Observe that this degree of accuracy is consistent with our estimate of probable moment in the column:

$$V_{cu,1} = \frac{0.5 \sum M_{b,\text{pr}}(\ell/\ell_c) + M_{c,\text{pr}}}{h_x} \qquad \text{(Eq. 2.2.29)}$$

The column height used must be the distance between the centerline of the beam and the face of the foundation or point of base fixity. Should the fixity be assumed to be 18 in. below the floor, the use of h_x would be correct. If, however, it were at the floor line, then h_x in this case would be 10.5 ft, as opposed to 12 ft. Let us assume (conservatively) that fixity will occur at the floor line; hence $h_x = 10.5$ ft.

$$V_{cu,1} = \frac{0.5(1.25)(44{,}930)(30/26.5) + (1.25)62{,}086}{(10.5)(12)}$$

$$= 868 \text{ kips}$$

$$v_s = \frac{V_{cu,1}}{\phi bd}$$

$$= \frac{868}{0.85(36)(39)}$$

$$= 0.727 \text{ ksi} \quad (8.65\sqrt{f'_c})$$

and this slightly exceeds the transverse reinforcement required to produce confinement ($0.077 f'_c$).

Conclusion: Provide two hoops at 2.75 in. on center over the entire height of the first floor in the direction of shear demand at least.

Comment: Observe that the shear demand (v) is $8.65\sqrt{f'_c}$ and this is higher than we might like (see Section 2.1.2.3). The use of a higher strength concrete should be considered. The use of 12,000-psi concrete would result in a shear demand of $6.6\sqrt{f'_c}$, and its use is advisable. The conservative nature of the shear development process ($M_{c,\text{pr}} = \lambda_o M_{\text{bal}}$) should be considered before making a design decision.

Step 6: Insure That a Story Mechanism Will Not Develop. This condition must be demonstrated on a complete frame. See Example 2 for its application.

Example 2 See Section 2.1.2.7 and Figure 2.1.28. Both exterior and interior columns need to be designed, and to this end we assume that the alternating span pattern continues beyond the frame.

Step 1: Develop the Objective Capacities. The live load of 0.5 kip/ft was based on a tributary width of 12.5 ft and a uniform load of 40 lb/ft^2. The dead load of 120 lb/ft^2 included an allowance for the beam weight but not the column. Before proceeding, these loads should be reviewed. A $6\frac{1}{2}$-in. post-tensioned slab is proposed. The dead load tributary to a frame column is

Slab	82 lb/ft^2 (12.5)	= 1025 lb/ft^2
Beam	$\frac{(20)(32-6.5)(150)}{144}$	= 530 lb/ft^2
Additional dead load	10 (12.5)	= 125 lb/ft^2
		1680 lb/ft^2

Design dead load	$= 1.7$ kips/ft	
Column load		
Beam and slab	1.7 (20)	$= 34$ kips/floor
Column self weight	$\dfrac{9(32)^2(150)}{144}$	$= 9.6$ kips/floor
Design load/floor	P_D	$= 44$ kips/floor

Live loads may be reduced by 60% in the design of columns. Code-based service load (P_L) is

$$\begin{aligned} P_L &= 20(12.5)(0.016) \\ &= 4 \text{ kips/floor} \end{aligned}$$

(a) Capacity-based loads are

$$\begin{aligned} P_{D,\text{pr}} &= 10(44) = 440 \text{ kips} \\ P_{L,\text{pr}} &= 0.006(20)(12.5)(10) = 15 \text{ kips} \\ P_{D,\text{min}} &= 408 \text{ kips} \end{aligned}$$

6 lb/ft^2 ($\pm$), as used in the second equation, is a generous allocation for the weight of people and furniture.

(b) Code-compliant loads are

$$\begin{aligned} P_{uD} &= 1.05 P_D \\ &= 1.05(440) \\ &= 462 \text{ kips} \\ P_{uL} &= 1.28 P_L \\ &= 1.28(4)(10) \\ &= 51 \text{ kips} \\ P_{uD,\text{min}} &= 0.9 P_D \\ &= 0.9(440) \\ &= 396 \text{ kips} \end{aligned}$$

Earthquake loads are capacity based and the procedure for developing the loads for this frame was developed in Section 2.2.3.1. Specifically, the axial load imposed on column A was summarized as

$$E_A = \frac{\text{OTM}}{D + \ell_2}(1.5) \qquad \text{(see Eq. 2.2.15 and Figures 2.2.9 and 2.2.10)}$$

The overturning moment was estimated by

$$\text{OTM} = V_M\left(\frac{2}{3}H\right) \qquad \text{(Eq. 2.2.13)}$$

and the mechanism shear by

$$V_M = \frac{\sum M_{b,\text{pr}}(\ell/\ell_c)}{h_x} \qquad \text{(Eq. 2.2.10)}$$

The probable moment capacity of the beam described in Figure 2.1.28*b* was established in the beam design process (Example 2, Step 5, Section 2.1.2) and confirmed in the analysis (Table 2.1.1).

$$M_{b,\text{pr}} = 14{,}860 \text{ in.-kips} \qquad (1233 \text{ ft-kips})$$

It follows that

$$V_M = \frac{\sum M_{b,\text{pr}}(\ell/\ell_c)}{h_x} \qquad \text{(Eq. 2.2.10)}$$

$$V_M = \frac{8(1233)}{9}(1.1 + 1.17/2)$$

$$= 1277 \text{ kips}$$

where $(1.1 + 1.17/2)$ represents an average value, and

$$\text{OTM} = V_M\left(\frac{2}{3}H\right) \qquad \text{(Eq. 2.2.13)}$$

$$= 1277(0.67)(90)$$

$$= 77{,}000 \text{ ft-kips}$$

Accordingly, the earthquake-induced axial load on the column at A is

$$E_A = \frac{\text{OTM}}{D + \ell_2}(1.5) \qquad \text{(see Eq. 2.2.15)}$$

$$= \frac{77{,}000}{(80 + 16)}(1.5)$$

$$= 1200 \text{ kips}$$

and that imposed on interior columns one-third of that imposed on A (see Figure 2.2.11) and that imposed on the column at E two-thirds of that imposed on A.

$$E_A = 1200 \text{ kips}$$
$$E_B = E_C = E_D = 400 \text{ kips}$$
$$E_E = 0.67E_A = 800 \text{ kips}$$

The maximum axial load imposed on the column at A is

$$\begin{aligned} P_A &= P_{D,\text{pr}} + P_{L,\text{pr}} + E_A \\ &= 440 + 15 + 1200 \\ &= 1655 \text{ kips} \end{aligned}$$

while the maximum tensile load will be

$$\begin{aligned} -P_A &= P_{D,\text{pr}} - E_{A,\text{pr}} \\ &= 440 - 1200 \\ &= -760 \text{ kips} \end{aligned}$$

A similar development for an interior column is

$$\begin{aligned} P_B &= P_D + P_L + E_B \\ &= 440 + 15 + 400 \\ &= 855 \text{ kips} \end{aligned}$$

The minimum axial load will be

$$\begin{aligned} P_D &= 440 - 400 \\ &= 40 \text{ kips} \end{aligned}$$

Step 2: Determine the Required Area of the Columns. Start with the most heavily loaded column.

$$A_{gA} > \frac{P_{A,\max}}{0.3f'_c} \qquad \text{(see Eq. 2.2.18)}$$

$$> \frac{1655}{0.3(5)} \qquad f'_c = 5 \text{ ksi}$$

$$> 1100 \text{ in.}^2$$

And this is reasonably satisfied by our proposed 32-in. by 32-in. column ($A_g =$ 1024 in.^{2}). The concrete strength should probably be increased to 6 ksi.

Conclusion: The column size proposed is 32 in. by 32 in., and the concrete strength is 6 ksi. This column should possess a reasonable level of ductility.

Step 3: Review Minimum Column Size Required to Satisfy Column Shear and Beam-Column Joint Objectives.

Conclusion: All columns should be 32 in. by 32 in. (see Section 2.1.2.7, Example 2).

Step 4: Determine the Amount of Reinforcing Required. Start with an *interior column (Grid B)* and consider the minimum axial load.

Design Objectives

$$A_s = \frac{0.83\sum M_{bn}}{hf_y} - \frac{0.38P_{\text{pr}}}{f_y} \qquad \text{(see Eq. 2.2.21c)}$$

$$= \frac{0.83(2)(986)(12)}{(32)(60)} - \frac{0.38(40)}{60} \qquad (M_{nb} = 986 \text{ ft-kips})$$

$$= 10.0 \text{ in.}^2$$

$$A_{st,\min} > 0.01(32)^2$$

$$> 10.24 \text{ in.}^2$$

Conclusion: At least 20.0 in.^{2} of steel is required ($2A_s$). This steel must be placed so as to provide maximum flexural resistance (see Eq. 2.2.20a). Tension reinforcement of at least 10 in.^{2} and a four-bar module blend well with the three-bar beam module. #11 bars are probably desirable so as to limit the number of bars.

$$A_s = 12.48 \text{ in.}^2 > 10.0 \text{ in.}^2 \qquad \text{(Eight \#11 bars)}$$

These bars can be bundled or arranged in two lines of four bars (see Figure 2.2.13*a*).

The proposed design is described in Figure 2.2.13*a*, where 26.54 in.^{2} of reinforcing steel are shown (sixteen #11s and two #8s). This 2.6% of reinforcing may be reduced during the analysis phase.

Now consider the design of the *exterior column at A*. The tensile load is

$$P_{A,\min} = P_D - E_A$$

$$= 440 - 1200$$

$$= -760 \text{ kips}$$

The design moment is

$$M_{cu} = \frac{M_{b,\mathrm{pr}}}{h_x}\left(\frac{\ell}{\ell_c}\right)\frac{h_c}{2} \tag{2.2.30}$$

$$= \frac{1233}{(9)}\left(\frac{16}{13.33}\right)\left(\frac{6.33}{2}\right)$$

$$= 520 \text{ ft-kips}$$

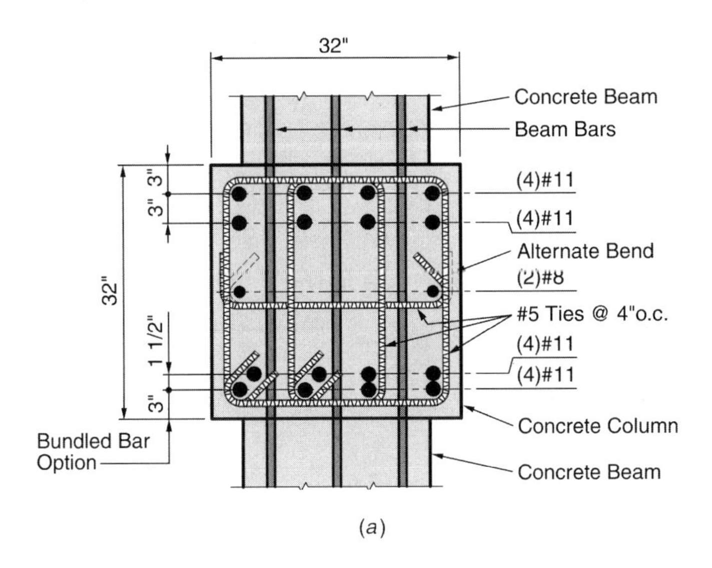

(a)

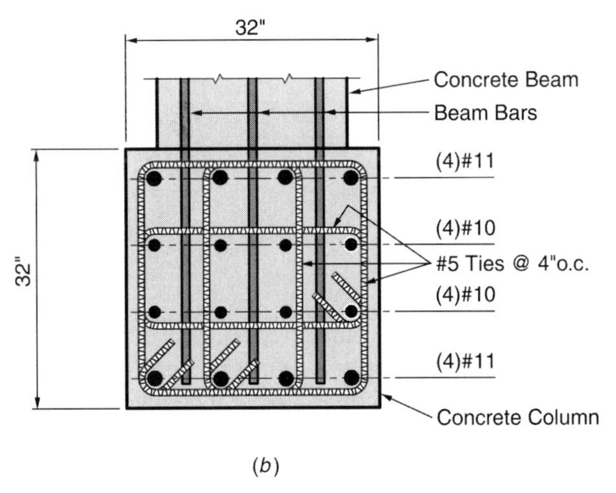

(b)

Figure 2.2.13 (*a*) Interior column detail. (*b*) Exterior column detail.

Adopting a strength reduction factor of 0.9, we find that the required area of steel becomes

$$A_{st,\text{tension}} = \frac{P_{A,\min}}{f_y} \qquad \text{(Tension only)}$$
$$= \frac{760}{60}$$
$$= 12.67 \text{ in.}^2$$

The increment of reinforcing steel required to develop the objective flexural capacity is

$$\Delta A_s = \frac{M_{cu}}{f_y(d - d')}$$
$$= \frac{520(12)}{60(32 - 6)}$$
$$= 4.0 \text{ in.}^2$$

The total area of steel required is

$$A_{st} > \frac{A_{st,\text{tension}} + 2\Delta A_s}{\phi}$$
$$> \frac{12.67 + 2(4)}{0.9}$$
$$> 23.0 \text{ in.}^2$$

and A_s must be at least 4.44 in.^{2} $(\Delta A_s/\phi)$.

Step 5: Check Total Reinforcing Percentage.

$$\rho = \frac{23.0}{(32)^2}$$
$$= 0.022 < 0.03 \qquad \text{(O.K.)}$$

Propose a reinforcing program.

The adopted reinforcing program for the exterior column would probably use four #11 bars on the tension face ($A_s = 6.24$ in.^{2}) and the interior bars should be aligned within these bars. Accordingly, eight more interior bars are appropriate. The total area of interior bars should exceed

$$A_{s,\text{int}} = 23.0 - 8(1.56)$$
$$= 10.5 \text{ in.}^2$$

Conclusion: The column at A should be detailed as described in Figure 2.2.13*b*:

Column size	32 in. by 32 in.
Concrete strength	6 ksi
Reinforcing	Eight #11s and eight #10 bars (Gr. 60).

Observe that the alignment of column bars for both the interior and exterior columns is compatible and allows the beam reinforcing to pass freely between them. It remains to develop a transverse reinforcing program before we adopt the bar placement programs shown in Figure 2.2.13.

Transverse Reinforcing Program For interior column B above Level 2:

$$v_s = \frac{\sum M_{b,\text{pr}}(\ell/\ell_c)}{\phi h_x bd} \qquad \text{(Eq. 2.2.24a)}$$

$$= \frac{2(14{,}860)(288/256)}{0.85(9)(12)(32)(29)}$$

$$= 0.393 \text{ ksi}$$

$$v_{s,\min} = 0.077(6) \qquad \text{(see Eq. 2.2.26)}$$

$$= 0.462 \text{ ksi}$$

Provided shear reinforcing (Figure 2.2.13a) produces a shear strength of

$$v_n = \frac{A_{sh} f_{yh}}{bs}$$

$$= \frac{4(0.31)60}{32(4)} \qquad \text{(\# 5 at 4 in. on center)}$$

$$= 0.58 \text{ ksi} \qquad \text{(OK)}$$

Conclusion: Provide #5 hoops at 4 in. on center (see Figure 2.2.13*a*).

For exterior column A at Level 1:

$$P_b = 0.85(0.6)\beta_1 f_c' bd \qquad \text{(see Eq. 2.2.2a)}$$

$$= 0.85(0.6)(0.75)(6)(32)(29)$$

$$= 2130 \text{ kips}$$

$$M_{\text{bal}} = A_s f_y(d - d') + P_b\left(\frac{h}{2} - \frac{a}{2}\right) \qquad \text{(Eq. 2.2.3)}$$

$$a = 0.6\beta_1 d \qquad \text{(Balanced conditions)}$$

$$= 0.6(0.75)(29)$$

$$= 13.0 \text{ in.}$$

$$M_{\text{bal}} = 4(1.56)(60)(26) + 2130(16 - 6.5)$$

$$= 9734 + 20{,}236$$

$$= 30{,}000 \text{ in.-kips} \qquad (2237 \text{ ft-kips})$$

$$V_{cu,1} = \frac{\sum (M_{b,\text{pr}}/2)(\ell/\ell_c) + M_{c,\text{pr}}}{h_x} \qquad \text{(see Eq. 2.2.29)}$$

Assume that h_x, the distance between the point of column fixity and the centerline of the beam, is 92 in. (9(12) – 32/2).

$$V_{cu,1} = \frac{14{,}860(288/256)}{2(92)} + \frac{1.25(30{,}000)}{92}$$

$$= 90.5 + 407.5$$

$$= 498 \text{ kips}$$

$$v_s = \frac{V_{cu,1}}{\phi b d}$$

$$= \frac{498}{0.85(32)(29)}$$

$$= 0.63 \text{ ksi}$$

Conclusion: Two #5 hoops (four legs) at 3.5 in. on center should be provided in the direction of shear (see Figure 2.2.13*b*).

Step 6: Insure That a Story Mechanism Will Not Develop: The story mechanism shear was developed in Step 1:

$$V_M = 1277 \text{ kips}$$

A detailed analysis is not warranted unless a quick check suggests that a column mechanism is likely to occur.

Start by considering the column mechanism shear strength provided by the interior columns at grid lines *B*, *C*, and *D* of Figure 2.2.10. These columns are detailed in Figure 2.2.13*a*.

Generate first M_o, their flexural strength in the absence of axial load:

$$M_o = A_s f_y (d - d')$$

$$= 8(1.56)(60)(2)$$

$$= 1498 \text{ ft-kips}$$

At the very least ($P = 0$) the mechanism shear (see Figure 2.2.12) provided by these three columns is

$$V_{cM,BCD} = \frac{\sum M_{cn}}{h_c} \qquad \text{(Eq. 2.2.27)}$$

$$= \frac{6(1498)(12)}{(9(12) - 32)}$$

$$= 1419 \text{ kips}$$

$$\frac{V_{cM,BCD}}{V_M} = \frac{1419}{1225}$$

$$= 1.16$$

Observe that the flexural strength provided by the axial load is estimated easily because we know the total dead load supported by the frame ($5\sum P_D$). The total story shear resistance provided by this axial load is the sum of the added flexural capacities divided by the clear story height. Accordingly,

$$\Delta \sum M_{cn} = 5 \sum P_D \left(\frac{h}{2} - \frac{a}{2}\right) \qquad \text{(see Eq. 2.2.3)}$$

$$= 5(440)(16 - 1.5)$$

$$= 31{,}900 \text{ in.-kips}$$

$$\Delta V_{cM} = \frac{\Delta \sum M_{cn}}{h_c}$$

$$= \frac{31,900}{76}$$

$$= 420 \text{ kips}$$

The exterior columns will also contribute to the mechanism strength based on their flexural strength, which is approximately half of the interior columns (see Figure 2.2.13*a*).

$$V_{cM,AE} = \frac{2}{3}\left(\frac{1}{2}\right) V_{cM,BCD}$$

$$= 0.33(1419)$$

$$= 468 \text{ kips}$$

$$V_{cM} = V_{cM,AE} + V_{cM,BCD} + \Delta V_{cM}$$

$$= 468 + 1419 + 420$$

$$= 2307 \text{ kips}$$

$$\frac{V_{cM}}{V_M} = \frac{2307}{1277}$$

$$= 1.81 > 1.5 \qquad \text{(OK)}$$

Conclusion: A column story mechanism (soft story) will not be a concern in this frame.

2.2.4 Analyzing the Beam Column

The analysis of a column must proceed based on the capacity requirements generated from the analysis of the frame. One should first evaluate the computer output to ensure that the results are consistent with design expectations. The imposed axial load now must be compared with the objectives identified in Section 2.2.3. The strain level induced in columns where plastic hinges are likely to occur should also be carefully reviewed and confinement levels increased when and where appropriate.

Step 1: Review Analysis. Capacity-based loads on columns are checked in several ways. The frame of Figure 2.2.10 was analyzed using the procedures described in Chapters 3 and 4. An elastic time history analysis, inelastic time history analysis, and a sequential yield analysis was performed.

Our concern here is with the axial loads and postyield rotations generated for the lower level columns. Figure 2.2.14 summarizes the distribution of peak seismically induced axial loads.

Clearly the impact of frame ductility on behavior is significant and desirable (see Section 1.1.1). When compared to the distribution suggested in the design phase (Figure 2.2.11), it is clear that the analytical results are reasonable.

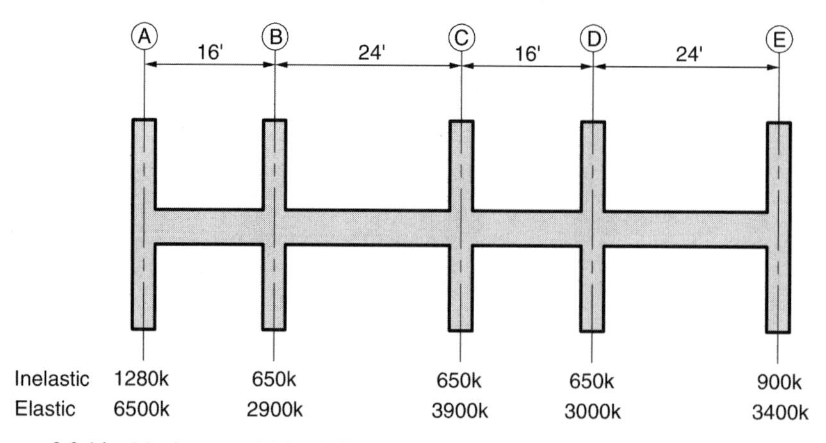

Figure 2.2.14 Maximum axial loads induced on the columns of the frame of Figure 2.2.10 when subjected to an inelastic time history analysis.

The magnitude of the seismic load was predicted in Section 2.2.3.5, Example 2, step 1. The seismic axial load predicted for the columns was

$$E_A = 1200 \text{ kips}$$

$$E_B = E_C = E_D = 400 \text{ kips}$$

$$E_E = 800 \text{ kips}$$

Observe that this distribution is consistent with that described in Figure 2.2.11*a*.

A review of the computer analyses suggests that hinging has extended to roof girders—a condition not anticipated by the design but clearly a possibility on intermediate height buildings. In spite of this fact, the predicted axial load on the column at A is only 7% greater than that estimated in the design phase.

Conclusion: The analytical results appear to be reasonable.

Step 2: Check the Axial Stress (P/A) Imposed on the Columns.

$$\begin{aligned} P_A &= P_{D,\text{pr}} + P_{L,\text{pr}} + E_A \\ &= 440 + 15 + 1200 \\ &= 1655 \text{ kips} \end{aligned}$$

$$\begin{aligned} \frac{P_A}{A_g} &= \frac{1655}{(32)^2} \\ &= 1.6 \text{ ksi} \qquad (0.27 f_c' \text{ for 6000 psi concrete}) \end{aligned}$$

Conclusion: Consider increasing the concrete strength in the lower level to improve its ductility.

Step 3: Check the Concrete Strain in the Shell of the Column. The inelastic time history suggests that the postyield hinge rotation imposed on column B at the base is 0.0104 radian.

$$\theta_p = 0.0104 \text{ radian}$$

$$\begin{aligned} \ell_p &= h \qquad \text{(see Section 2.2.2.2)} \\ &= 32 \text{ in.} \end{aligned}$$

$$\begin{aligned} \phi_p &= \frac{\theta_p}{\ell_p} \\ &= \frac{0.0104}{32} \\ &= 0.000325 \text{ rad/in.} \end{aligned}$$

The bundled bar arrangement described in Figure 2.2.13*a* was used in the inelastic time history analysis. The neutral axis depth is

$$
\begin{aligned}
P &= P_{D,\text{pr}} + P_{L,\text{pr}} + E_B \\
&= 440 + 15 + 650 \\
&= 1105 \text{ kips} \\
T_y &= A_s f_y \\
&= 8(1.56)(60) \\
&= 749 \text{ kips} \\
\lambda_o T_y &= 1.25(749) \\
&= 936 \text{ kips} \\
C_s &= A_s f_y \\
&= 749 \text{ kips} \\
C_c &= P + (\lambda_o - 1)T_y \\
&= 1105 + 187 \\
&= 1292 \text{ kips} \\
a &= \frac{C_c}{0.85 f'_c b} \\
&= \frac{1292}{0.85(6)(32)} \\
&= 7.9 \text{ in.} \\
c &= \frac{a}{\beta_1} \\
&= \frac{7.9}{0.75} \\
&= 10.5 \text{ in.} \\
\varepsilon_{cp} &= \phi_p c \\
&= 0.000325(10.5) \\
&= 0.00341 \text{ in./in.} \\
\varepsilon_c &= \varepsilon_{cn} + \varepsilon_{cp}
\end{aligned}
$$

where ϕ_{cn} is the strain in the concrete at idealized yield M_n. Recall from the development of Figure 2.1.9 that ε_{cy} may be as low as 0.001 in./in. but this does not represent our idealization of yield rotation as developed in Section 2.2.2.2.

$$\begin{aligned}\varepsilon_c &= 0.003 + 0.0034\\ &= 0.0064 \text{ in./in.}\end{aligned}$$

Conclusion: This type of analysis tends to be conservative, and shell spalling should not be anticipated.

Step 4: Check the Probable Shear Stress Imposed on the Column. This can be done in several ways, the easiest of which is from the sequential yield computer program analysis. A shear of 400 kips was predicted for an interior column. A quick confirmation can be attained from the probable moment at peak load imposed on an interior column. Proceeding from the data developed in Step 3, we get

$$\begin{aligned}M_{c,\mathrm{pr}} &= P\left(\frac{h}{2} - \frac{a}{2}\right) + T_y(d - d') + (\lambda_o - 1)T_y\left(d - \frac{a}{2}\right)\\ &= 1105(16 - 3.95) + 749(28.25 - 3.75) + 0.25(749)(28.25 - 3.92)\\ &= 13{,}360 + 18{,}350 + 4550\\ &= 36{,}000 \text{ in.-kips} \qquad (3000 \text{ ft-kips})\end{aligned}$$

The maximum probable moment at the top of an interior column is 1233 ft-kips (Step 1, Section 2.2.3.5). The induced level of column shear would be

$$\begin{aligned}V_{c,\max} &= \frac{M_{cT} + M_{cB}}{h_c}\\ &= \frac{1233 + 3000}{9.67} \qquad (\text{foundation is at } -12 \text{ in.})\\ &= 438 \text{ kips}\end{aligned}$$

$$\begin{aligned}v_{c,\max} &= \frac{V_{c,\max}}{bd}\\ &= \frac{438}{32(28.75)}\\ &= 485 \text{ psi} \qquad \left(6.3\sqrt{f_c'} - f_c' = 6000 \text{ psi}\right)\end{aligned}$$

Consider the shear suggested by the inelastic time history (probable shear demand):

$$V_{c,\mathrm{pr}} = 400 \text{ kips}$$

$$\begin{aligned}v_{c,\mathrm{pr}} &= \frac{V_{c,\mathrm{pr}}}{bd}\\ &= \frac{400}{32(28.25)}\end{aligned}$$

$$= 442 \text{ psi} \qquad \left(5.7\sqrt{f'_c} - f'_c = 6000 \text{ psi}\right)$$

Conclusion: In step 2 it seemed advisable to increase the strength of the concrete in the lower level and this tends to be confirmed by the induced level of shear stress. For a concrete strength of 8000 psi the induced stress would correspond to $4.9\sqrt{f'_c}$.

The provided level of shear reinforcement is

$$v_s = 663 \text{ psi}$$

Conclusion: The columns should be expected to perform well, given the predicted force and deformation demands.

2.3 BEAM-COLUMN JOINTS

The seismic load path in a ductile frame flows through the beam-column joint. The traditional design objective for the beam-column joint has been to treat it as though it were brittle. In addition, the objective of the capacity-based approach is to create a beam-column joint that is stronger than the frame beams that drive it. An overstrength factor is used to create the probable demand $(\lambda_o M_n)$ and thereby attain a conservative design criterion for the beam-column joint. Further, the strength limit state for the joint is conservatively approximated. Much of this apparent conservatism is reasonably attributed to the complexity of the load transfer mechanism within the joint and the uncertainty of the impact of column axial loads, bond deterioration within the joint, and the reinforcing program adopted for the joint.

It has been demonstrated that a significant amount of ductility can be developed in a well-designed beam-column joint. This topic is discussed in Section 2.3.2. The fact that ductility exists in the joint should not, however, alter our adopted capacity-based design approach. This is because the most logical approach, that of shared ductility between beam and joint, is virtually unattainable. Since the postyield behavior of the beam is very reliable and not impacted by as many variables, our design objective should remain to develop the beam mechanism. However, it should be comforting to know that the joint will not fail in a brittle manner if subjected to an overstrength demand.

Before we examine experimental behavior, let us endeavor to understand how researchers have incorporated conditions that seem to impact behavior into beam-column joint design procedures.

2.3.1 Behavior Mechanisms

Forces flow through a joint following a logical path that can be visualized using strut and tie modeling. Initially, the load path is along the principal diagonal strut that

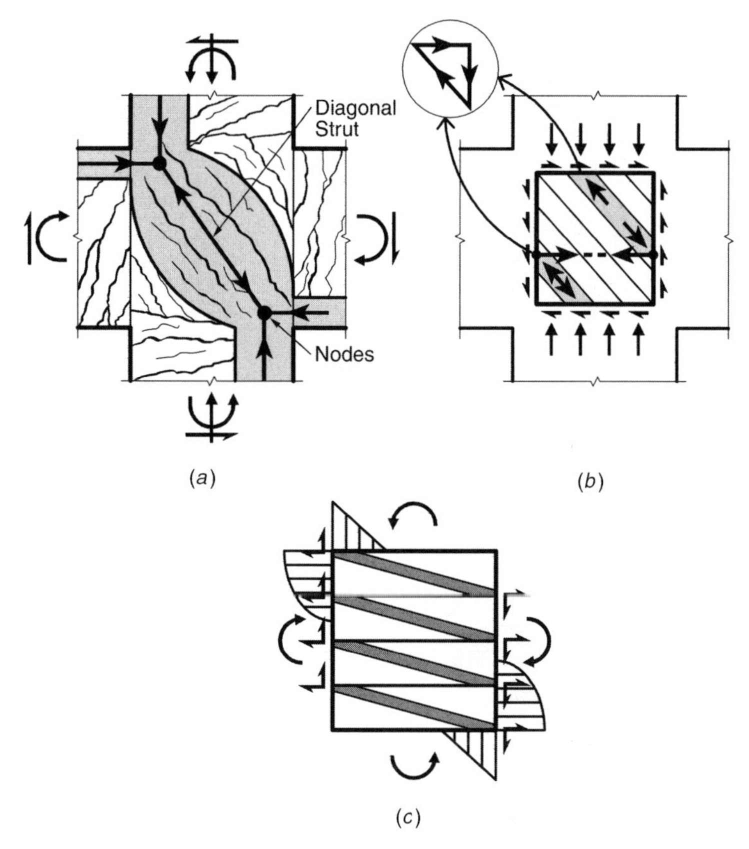

Figure 2.3.1 (*a*) Mechanisms of shear transfer at an interior joint. (*b*) Substrut mechanism. (*c*) Truss mechanism.

links the load nodes (Figure 2.3.1*a*). Also effective is a secondary load path that is generated by a bond transfer mechanism and a strut and tie load path (Figure 2.3.1*b*) usually referred to as a truss mechanism. As the principal diagonal strut reaches its strength limit and bond stresses deteriorate, the strain in the concrete that creates the strut increases and load is shed to a truss mechanism which is similar to the substrut mechanism described in Figure 2.3.1*b* and a broader compression field is created (Figure 2.3.1*c*). Ultimately this compression field will also break down and diagonal cracking will become severe (Figure 2.3.2). Now, a (pure) truss mechanism forms (Figure 2.3.1*c*) and the role of the transverse reinforcement becomes dominant.

The ACI,[2.6] recognizing the complex nature of this varying load path, has utilized a fairly simple strength defining limit state that is a function of the square root of the strength of concrete ($\sqrt{f'_c}$). The use of a tension-based limit state ($\sqrt{f'_c}$) is probably as rational as any, for it would tend to identify the point at which diagonal cracking causes the compression field to break down (Figure 2.3.2). Initially, the ACI design procedures had required that the transverse reinforcing program develop the

Figure 2.3.2 Joint shear cracks are widening and extending. The cover has begun to spall.

shear imposed on the joint; thereby recognizing the importance of the truss mechanism (Figure 2.3.1*c*). This requirement has been relaxed to one based on providing confinement, and this increases the required level of transverse reinforcement in direct proportion to the compressive strength of the concrete (Section 1.2). Established strength limit states have been experimentally confirmed.

Conceptually the procedures used in the development of an interior beam-column joint are summarized as follows. The driving force is developed in the beam (see Figure 2.3.3). The demand imposed on the joint is developed from the probable strength of the beam:

$$\lambda_o M_{n1} = \lambda_o f_y A_s \left(d - \frac{a}{2}\right) \tag{2.3.1}$$

The shear in the column is consistent with the driving mechanism:

$$V_c = \frac{\lambda_o \left(M_{n1}(\ell_1/\ell_{c1}) + M_{n2}(\ell_2/\ell_{c2})\right)}{h_x} \tag{2.3.2}$$

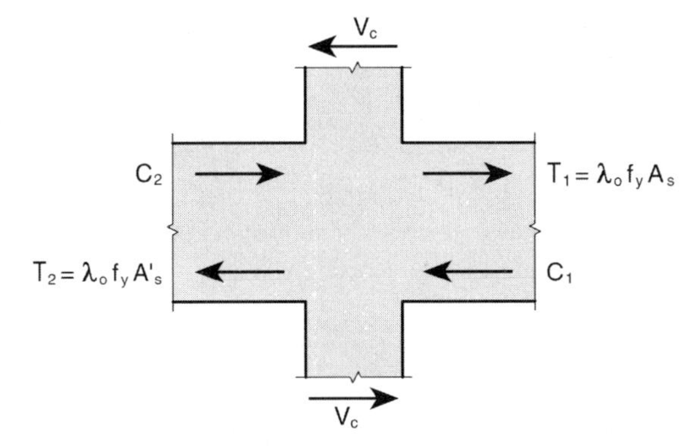

Figure 2.3.3 Forces imposed on an interior beam-column joint.

The shear imposed on an interior beam-column joint is

$$V_{jh} = T_1 + C_2 - V_c \qquad \text{(see Figure 2.3.3)} \qquad (2.3.3)$$

The associated joint shear stress is

$$v_{jh} = \frac{V_{jh}}{bh} \qquad (2.3.4)$$

The ACI-established[2.6] strength limit state of an interior joint is $15\phi\sqrt{f_c'}$, and the joint must be confined. This strength limit state applies to all concrete strengths.

Paulay and Priestley[2.5, Sec. 4.8.7] develop joint strength from the principal diagonal strut (Figure 2.3.1*a*) and the substrut mechanism described in Figure 2.3.1*b*; hence they describe the capacity of the beam-column joint in terms of its two constitutive components.

$$V_{jh} = V_{ch} + V_{sh} \qquad (2.3.5)$$

where

V_{ch} is the capacity of the principal concrete strut.
V_{sh} is the strength of the reinforced substrut mechanism.

The suggested maximum stress level (V_{jh}/bh) is $0.25f_c'$ and this is 29% higher than the ACI limit state for 6000-psi concrete. An absolute maximum joint shear stress of 1300 psi is also suggested. Bond, axial load, and reinforcement ratios are factors that define the amount of horizontal reinforcement required to attain the desired level of strength in their opinion. An elastic behavior limit (V_{sh}) may also be attained through the introduction of large amounts of transverse reinforcing.

Lin, Restrepo, and Park[2.22] suggest that the strength of a beam-column joint is a function of the concrete strength (V_c) the transverse reinforcement (V_{sh}), and the

Photo 2.7 Beam-column joint—test specimen, University of California San Diego.

axial load (V_N). They conclude that the axial load will improve the strength of the joint once it exceeds $0.1 f'_c bh$. Clearly the presence of a reasonable axial load (0.1 to $0.4 f'_c bh$) should, when combined with transverse confining reinforcing, provide improved biaxial confinement and thereby increase the capacity of the joint.

2.3.1.1 Bond Stresses The depth of the joint and its relationship to the diameter of the beam bars (h/d_b) have been studied. Ciampi and coauthors[2.26] feel that this ratio must be between 35 to 40 in order to prevent slippage. This is virtually impossible, so we must conclude that at least some slippage will occur in a beam-column joint subjected to seismic load reversals in the postyield range. Leon[2.27] concluded that h/d_b must be 28 in order to withstand cyclic loading without significant bond deterioration. The ACI requires that h/d_b be at least 20. Paulay and Priestley describe an analytical procedure[2.5, Sec. 4.8.6] and suggest that bond stresses as high as $16\sqrt{f'_c}$ may be attained.

Conditions that impact bond transfer are many. They include:

- The load that must be transferred—should it be the overstrength of the tension bar ($\lambda_o f_y$) combined with the yield strength of the compression bar ($A'_s f_y$)?
- How effective is the bond within an overstrained tensile portion of the column?
- The extent to which yield penetration in the joint will reduce the effective bond transfer mechanism.
- The amount of axial load imposed on the joint.

Accordingly, it appears unlikely that research efforts will soon develop a comprehensive treatment of the impact of bond on the behavior of a beam-column joint. The ACI[2.6] bond limit state ($h/d_b > 20$) requires that a column be 28 in. deep in order to use a #11 bar and 40 in. deep to accommodate bundled #11 bars (two-bar bundle).

The effective treatment of bond transfer within a beam-column joint is beyond the scope of this book. Rather, we examine the results of tests that use lesser development lengths and still produce ductile behavior within the joint and subassembly.

2.3.1.2 Biaxially Loaded Joints Typically, framing programs that require frame beams to intersect along both axes should be avoided because of the constructibility problems their union creates. Controversy exists as to limit states and joint reinforcement on interior joints where induced levels of joint shear are bound to be quite high. The ACI[2.6] raises the nominal joint shear limit state 33% and this applies to either direction of the imposed shear stresses.

$$v_{jx} \le 20\phi\sqrt{f_c'} \qquad \text{and} \qquad v_{jy} \le 20\phi\sqrt{f_c'} \tag{2.3.6}$$

Paulay and Priestley[2.5] recommend limits for both axes of $0.2f_c'$, or 80% of what they recommend for joints loaded on only one axis. Hence the established limit for biaxially loaded joints as defined by the ACI[2.6] exceeds that suggested by Paulay and Priestley by 29% for 6000-psi concrete. Interested readers are referred to the work of Japanese researchers, for biaxial loading is common in Japanese construction.

Reinforcing requirements for biaxially loaded joints vary considerably. The Japanese do not require any joint reinforcement, while the ACI requires one-half of that required for a planar joint, and Paulay and Priestley, develop the reinforcement required (V_{sh}) based on the excess shear demand.

2.3.1.3 Exterior Joints Load transfer mechanisms in exterior joints deserve special attention. The development of nodes and bond transfer is essential. Figure 2.3.4 describes the forces imposed on an exterior beam-column joint. In contrast to the interior beam-column joint (Figure 2.3.1), no beam compression force acts to develop the strut mechanism. A node must be developed, and this can be created by a developed 90° hook on the beam bars or an anchor plate. The hook must be turned down in order to create the desired node, and the tail must be developed within the joint.

Beam bars on reverse loading are subjected to compressive loads prior to closing the cracks in the beam created by overstaining the concrete in tension. If this load is not resisted by bond stresses within the joint, the beam bars will push through the back face of the joint (Figure 2.3.5). Bond deterioration on the tensile load cycle at the edge of the column will aggravate this condition, and Paulay and Priestley[2.5, Sec. 4.8.11] recommend that the $10d_b$ adjacent to the beam-column interface not be relied upon to develop the strength of the bars and that ties be located so as to restrain the hooks when the bars are subjected to compression (Figure 2.3.6).

2.3.1.4 Eccentric Beams Beams are quite often located on the face of the column —usually to attain aesthetic or functional objectives. The current approach is to reduce the effective area of the joint to that which is obviously capable of developing the

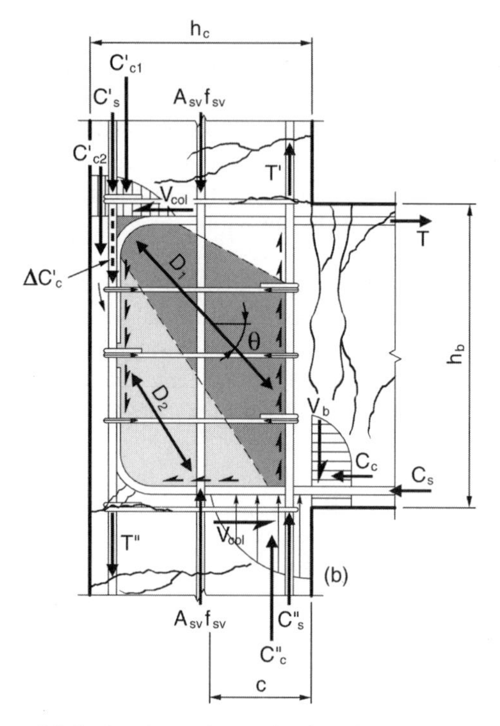

Figure 2.3.4 Load transfer mechanisms in an exterior joint.

desired force transfer mechanisms (Figure 2.3.1). This is clearly a conservative approach, and a more realistic approach must be developed analytically and confirmed by test. The alternative offered is to consider torsion on the joint. This is rationally accomplished in much the same manner as one combines vertical and torsional shear in the design of a beam.

2.3.2 Experimentally Based Conclusions

A series of beam-column subassemblies were tested at the University of California at San Diego in the fall of 1997 and spring of 1998.[2.15, 2.28, 2.29] The objective of these tests was to evaluate the strength of beam-column joints constructed using high-strength concrete (HSJ). One of the parameters being studied was the strength of the concrete. The baseline test was constructed using a concrete whose nominal strength was 6000 psi (HSJ6). The subassembly tested is described in Figure 2.3.7. The subassembly was driven by beam actuators. The column ends were free to rotate but not displace. The strength of the beam was developed so as to cause the joint to absorb the bulk of the postyield deformation to which the subassembly would be subjected.

Figure 2.3.8 describes the resultant force-displacement relationship. The peak load applied to the beam was 142 kips. This corresponds to a column shear of 175 kips.

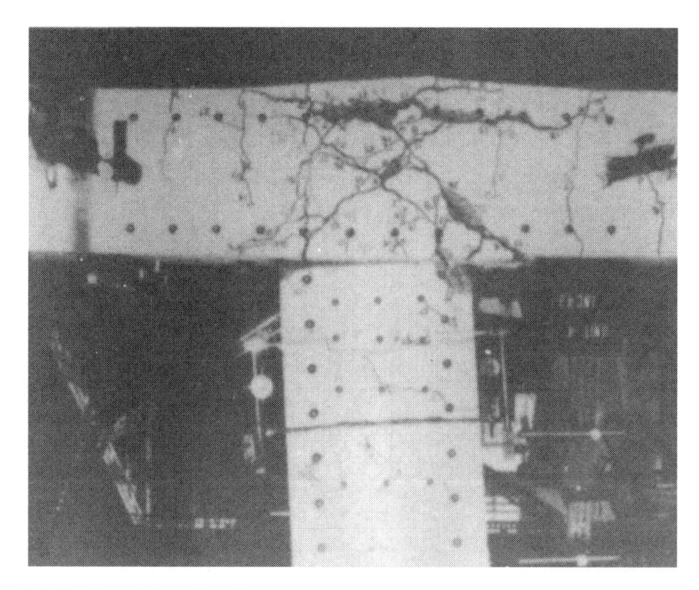

Figure 2.3.5 Bond deterioration along outer column bars passing through a joint [2.5]

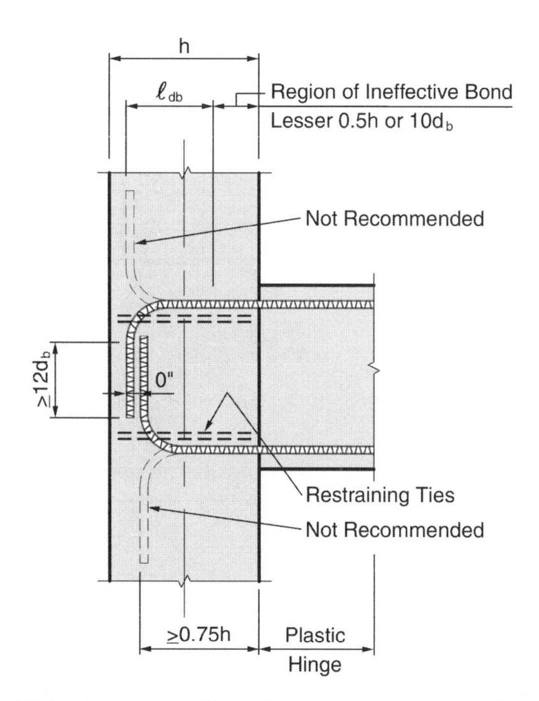

Figure 2.3.6 Anchorage of beam bars at exterior beam-column joints.

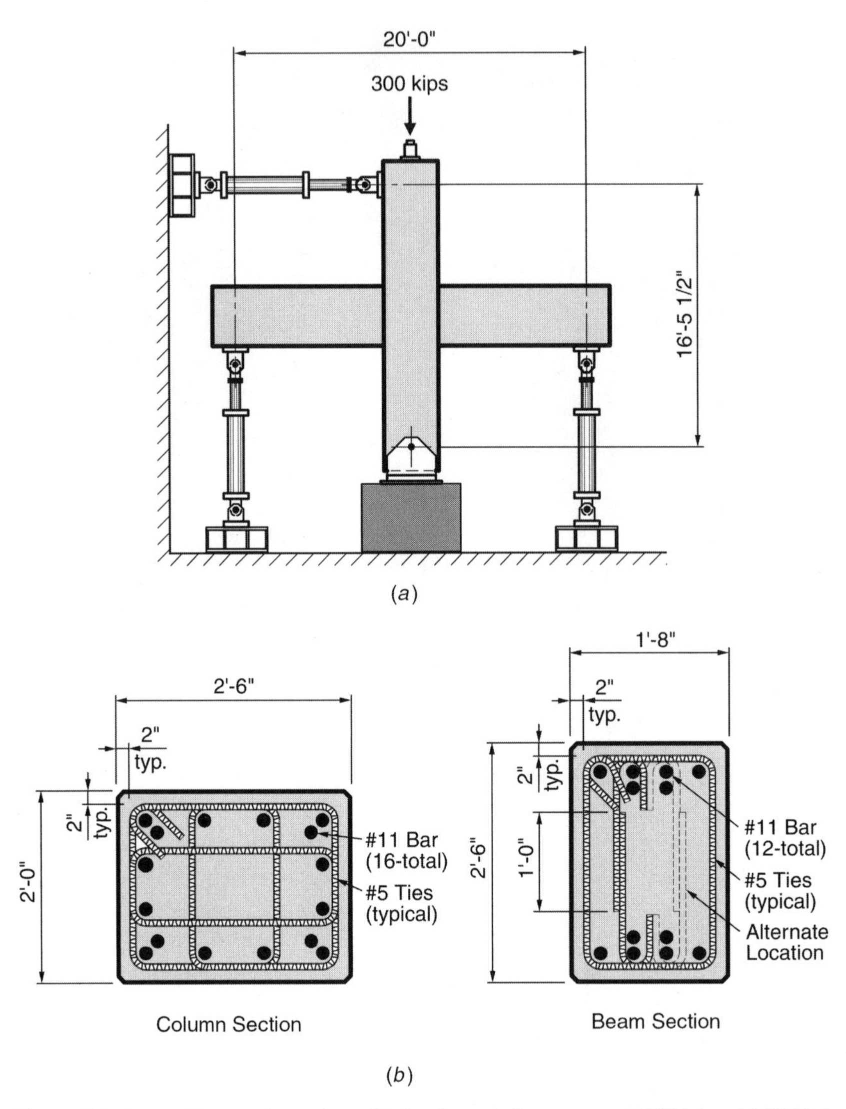

Figure 2.3.7 (*a*) Test configuration. (*b*) Section reinforcement of HSJ6-1 and HSJ12-1.

The associated level of joint shear, following the procedures described in Section 2.3.1, is

$$T_s = \frac{P_b \ell_c}{d - a/2} \tag{2.3.7}$$

$$= \frac{142(105)}{26}$$

$$= 573 \text{ kips} \qquad (f_s = 61.2 \text{ ksi})$$

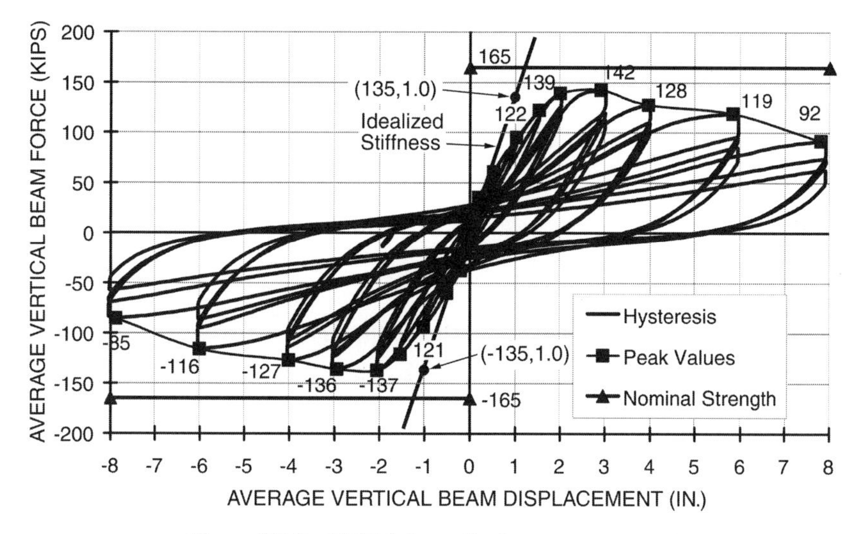

Figure 2.3.8 HSJ6-1 force-displacement response.

The shear imposed on the joint is

$$V_{jh} = 2T_s - V_c \qquad \text{(see Eq. 2.3.3)}$$

$$= 2(573) - 175$$

$$= 971 \text{ kips}$$

and the empirical shear stress imposed on the joint is

$$v_{jh} = \frac{V_j}{bh} \qquad \text{(Eq. 2.3.4)}$$

$$= \frac{971}{24(30)}$$

$$= 1.35 \text{ ksi}$$

Expressed as a function of measured concrete strength ($f_c' = 7200$ psi), the stress is

$$v_{jh} = 15.9\sqrt{f_c'}$$

and, as a function of the nominal concrete strength (6000 psi),

$$v_{jh} = 17.4\sqrt{f_c'}$$

The axial load imposed on the column expressed as a function of $f_c' A_g$ was quite low at 0.058. Joint reinforcement was

$$v_s = \frac{\sum A_v f_y}{bs} \tag{2.3.8a}$$

$$= \frac{4(0.31)(60)}{24(3)}$$

$$= 1.03 \text{ ksi}$$

and the confining pressures were

$$f_\ell = \frac{\sum A_v f_y}{b_c s} \quad \text{(Direction of load)} \tag{2.3.8b}$$

$$= \frac{4(0.31)(60)}{20(3)}$$

$$= 1.24 \text{ ksi}$$

$$f_\ell = \frac{\sum A_v f_y}{h_c s} \quad \text{(Perpendicular to the beam)} \tag{2.3.8c}$$

$$= \frac{4(0.31)(60)}{26(3)}$$

$$= 0.95 \text{ ksi}$$

Minimum joint reinforcement as required by ACI is

$$f_\ell = 0.09 f_c' \tag{2.3.9}$$

$$= 0.09(6000)$$

$$= 0.54 \text{ ksi}$$

Observe that the shear reinforcement provided was only capable of supporting a joint shear force of

$$V_s = v_s bd \quad \text{(see Eq. 2.3.4)}$$

$$= 1.03(24)(27)$$

$$= 667 \text{ kips}$$

and this corresponds to a beam load (R_b) of

$$R_b = \frac{V_s}{V_{jh}}(142) \tag{2.3.10}$$

$$= \frac{667}{971}(142)$$

$$= 97.5 \text{ kips}$$

This is only slightly above the sustained beam load (92 kips) at a displacement of 8 in. (see Figure 2.3.8). Accordingly, one might conclude that the principal strut mechanism (Figure 2.3.1*a*) had lost its capacity to carry load and that the truss mechanism (Figure 2.3.1*c*) was almost entirely responsible for the strength available in the joint. Figure 2.3.9 shows the condition of the joint at the conclusion of the test. This behavior shift is demonstrated in Section 2.3.4 when the amount of joint reinforcement is increased.

The criterion contained in the ACI Code is largely based on the work of Meinhelt and Jirsa,[2.30] and the UCSD test seems to be consistent with these experiments if a proper account is made of the impact of axial load on performance.

Several of Meinhelt and Jirsa's specimens (II and XIII)[2.30] are quite comparable to the UCSD subassembly. The concrete strengths are nominally 6 ksi and the geometry of the joints is square. The imposed level of axial stress is $0.25 f_c'$ and the member sizes are considerably smaller. The peak joint shear developed in the Jirsa specimen was in excess of $22\sqrt{f_c'}$, and this makes them 26% stronger than the UCSD specimen. Lin and coauthors[2.22] suggest that higher joint shears are to be expected when the column is subjected to a significant axial load ($> 0.1 f_c' A_g$). The increase suggested by Lin et al. is

$$\frac{V_N}{V_{jh}} = 2\left[\frac{N^*}{A_g f_c'} - 0.1\right] \tag{2.3.11}$$

Figure 2.3.9 HSJ6-1: third cycle of 8.0 in.

where V_N is the additional joint shear strength provided by the axial load imposed on the column (N^*).

$$\frac{V_N}{V_{jh}} = 2[0.25 - 0.1]$$
$$= 0.3$$

The 30% increase is consistent with the relationship between the strength of the Jirsa specimen and the UCSD specimen. The loss of strength on subsequent cycles of postyield rotation experienced in the Jirsa specimens (Figure 2.3.10) reduces the capacity to a level that is lower than that attained by the UCSD specimen. It appears

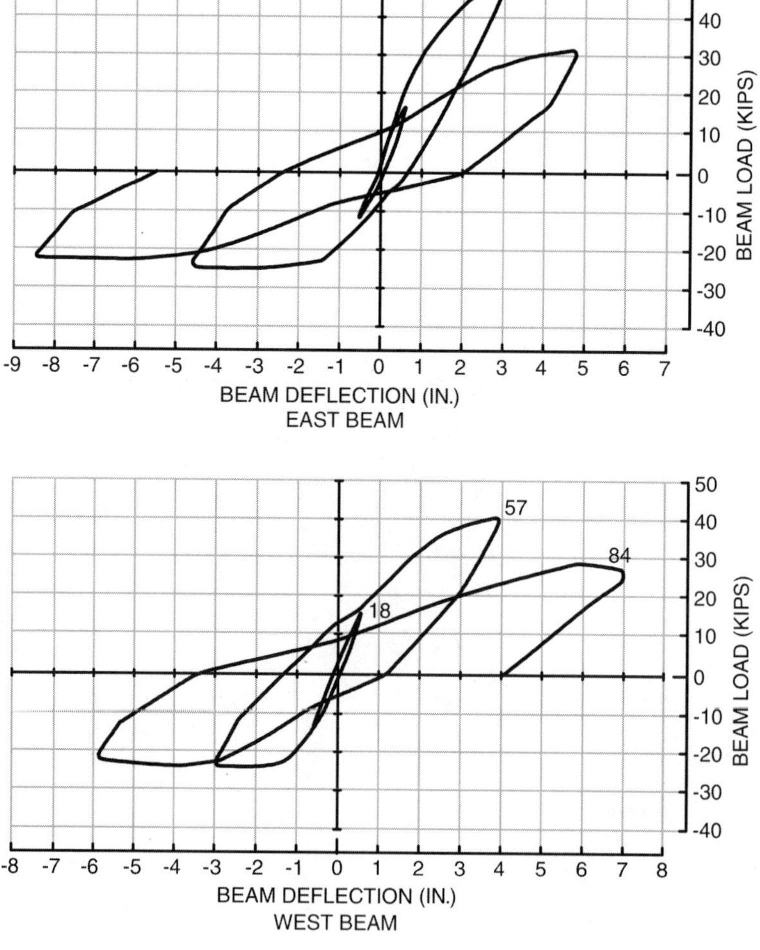

Figure 2.3.10 Load deflections for Meinhelt and Jirsa[2.30] Specimen XII.

as though the axial load improves the effectiveness of the compression strut, but this increased capacity is lost as the compression strut deteriorates. The extent of the loss experienced by the Jirsa specimens is probably attributable to the small size of the specimens and the amount of reinforcement provided in the beam.

Conclusion: It appears reasonable to follow current ACI limit states in the design of beam-column joints, provided the strength of the concrete is not presumed to exceed 6000 psi, a limit state that is developed in Section 2.3.3.

Comment: In Chapter 3 we will rely on estimates of subassembly stiffness to predict the response of systems. Equation 3.2.2a will be used for this purpose. The behavior of the subassembly described in Figure 2.3.7*a* when subjected to cyclic loads provides an opportunity to calibrate commonly used effective member stiffnesses.

$$I_{be} = 0.35 I_g$$

$$= 15{,}750 \text{ in.}^4$$

$$I_{ce} = 0.7 I_g$$

$$= 37{,}800 \text{ in.}^4$$

$$\ell = 240 \text{ in.}; \qquad \ell_c = 210 \text{ in.}; \qquad (\ell_c/\ell)^2 = 0.77$$

$$h_x = 197 \text{ in.}; \qquad h_c = 167 \text{ in.}; \qquad (h_c/h_x)^2 = 0.72$$

$$V_i = V_b(\ell/h_x); \qquad \Delta_i = 2\delta_b h_x/\ell$$

$$\Delta_i = \frac{V_i h_x^2}{12E}\left[\frac{\ell_c}{I_{be}}\left(\frac{\ell_c}{\ell}\right)^2 + \frac{h_c}{I_{ce}}\left(\frac{h_c}{h_x}\right)^2\right] \qquad \text{(Eq. 3.2.2a)}$$

$$= \frac{165(197)^2}{12(4400)}\left[\frac{210}{15{,}750}(0.77) + \frac{167}{37{,}800}(0.72)\right] \qquad (V_h = 135 \text{ kips})$$

$$= 121.3(0.01 + 0.0032)$$

$$= 1.6 \text{ in.} \qquad \text{(Subassembly drift)}$$

$$\frac{\Delta_i}{h_x} = \frac{1.6}{197}$$

$$= 0.008 \text{ radian}$$

$$\delta_b = 1.0 \text{ in.} \qquad \text{(Corresponding beam displacement)}$$

This reasonably represents subassembly stiffness in the idealized elastic behavior range, but once the subassembly is cycled through a number of reverse cycles of loading at loads in this range (± 120 kips), the stiffness of the subassembly decreases (see Figure 2.3.8).

Computer programs allow the user to adopt an effective panel zone dimension. The designs and the computer analyses undertaken in Chapter 3 will consider the consequences associated with the use of an effective joint dimension that is 60% of the actual joint dimension. The consequences of this modeling assumption produce the following estimates of subassembly behavior:

$$\ell = 240 \text{ in.}; \qquad \ell_c = 222 \text{ in.}; \qquad \left(\frac{\ell_c}{\ell}\right)^2 = 0.86$$

$$h_x = 197 \text{ in.}; \qquad h_c = 179 \text{ in.}; \qquad \left(\frac{h_c}{h_x}\right)^2 = 0.83$$

$$\Delta_i = \frac{V_i h_x^2}{12E}\left[\frac{\ell_c}{I_{be}}\left(\frac{\ell_c}{\ell}\right)^2 + \frac{h_c}{I_{ce}}\left(\frac{h_c}{h_x}\right)^2\right] \qquad \text{(see Eq. 3.2.2a)}$$

$$= \frac{165(197)^2}{12(4400)}\left[\frac{222}{15{,}750}(0.86) + \frac{179}{37{,}800}(0.83)\right]$$

$$= 121.3(0.0121 + 0.0039)$$

$$= 1.94 \text{ in.} \qquad \text{(Subassembly drift)}$$

$$\delta_b = \frac{1.92}{197}(120)$$

$$= 1.18 \text{ in.} \qquad \text{(Corresponsing beam displacement)}$$

Observe that this represents an 18% reduction in stiffness, but this loss of stiffness will not be realized unless the joint is subjected to many cycles at high levels of joint stress.

2.3.3 Impact of High-Strength Concrete

The subassembly described in Figure 2.3.7 was constructed using high-strength concrete (HSC—$f_c' = 12$ ksi) and the loading cycle repeated. Beam force–displacement relationships are compared for the 6-ksi (HSJ6-#1 and the 12-ksi (HSJ12-#1) subassemblies in Figure 2.3.11. The peak beam load for the HSC specimen is only 8% higher than that attained by the 6-ksi subassembly. This corresponds to an attained joint shear stress of

$$v_{jh} = 1460 \text{ psi}$$

or $13.3\sqrt{f_c'}$. The attained strength is 11% below the nominal shear strength ($15\sqrt{f_c'}$) currently proposed by the ACI code (1643 psi).

A review of the axial load component and the amount of provided confinement is interesting. The imposed axial load of 300 kips, though the same as that imposed on

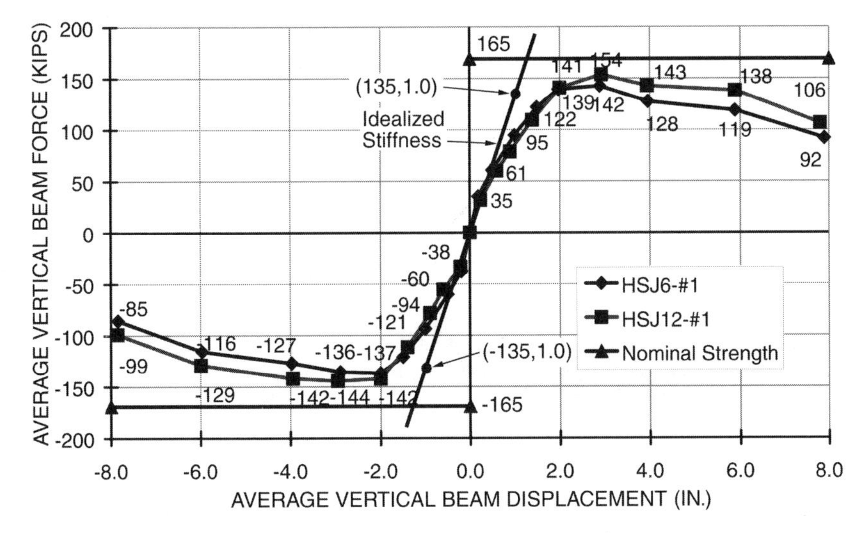

Figure 2.3.11 Force–displacement envelopes.

the 6-ksi specimen (HSJ6-#1), is a much smaller percentage (9.4% or $0.035 f_c' A_g$) of the balanced load.

$$P_b \cong 0.37 f_c' A_g \qquad \text{(Eq. 2.2.2b)}$$
$$= 0.37(12)(24)(30)$$
$$= 3197 \text{ kips}$$

Also, the level of confinement, when expressed in terms of the strength of the concrete, is reduced. Recall that the objective confining pressure as defined by the ACI is

$$f_\ell = 0.09 f_c' \qquad \text{(Eq. 2.3.9)}$$
$$= 0.09(12{,}000)$$
$$= 1080 \text{ psi}$$

while that provided in the direction of the load was 1240 psi (15% more than required). The ratio of provided confinement to that required by the code for the 6-ksi specimen was 2.3. Hence, the confining pressure provided in the 6-ksi specimen was conservatively developed and could in fact have been much less.

The capacity of the HSC joint at higher drift levels is 16% greater than that of the 6-ksi joint ($\delta_b = 6$ in., Figure 2.3.11). The strength of the HSC subassembly at a displacement of 8 in. (drift ratio 6.7%) is also 15% higher. Accordingly, it appears that the concrete strut does not deteriorate as rapidly in the 12-ksi specimen as it did in the 6-ksi specimen. The attainment of a drift ratio of 6.7% is remarkable for either material especially given the fact that most of the postyield rotation occurs within the joint.

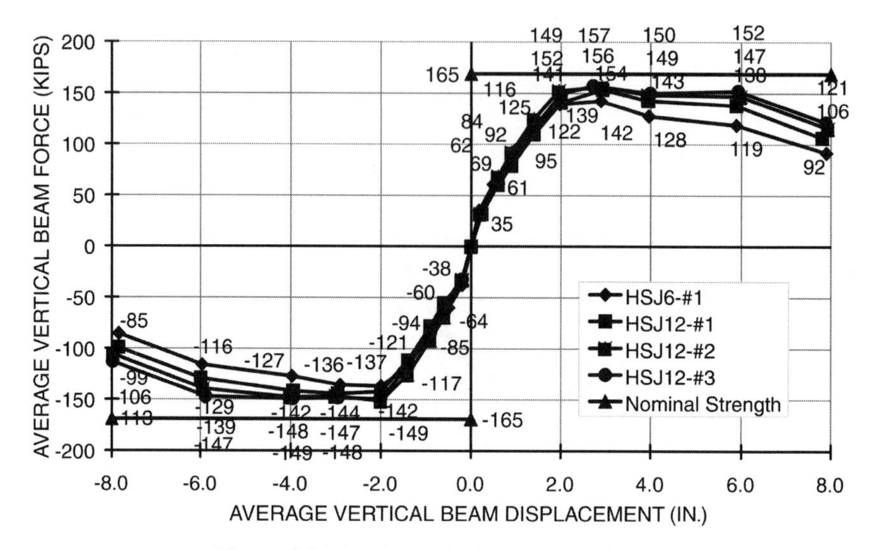

Figure 2.3.12 Force–displacement envelopes.

Imposing high axial loads on test specimens containing large columns constructed of HSC is difficult. The load imposed on the 12-ksi specimen was doubled so as to duplicate the load level imposed on the 6-ksi specimen in terms of the percentage of the balanced load. This test is identified as HSJ12-#2. The peak load attained in HSJ12-#1 was repeated, and the only impact observed occurred at higher drift angles (Figure 2.3.12). The extended participation of the concrete strut mechanism is again clear. The increased capacity at a drift angle of 6.7% is now 26%.

Conclusion: It is clear that HSC promotes better behavior in the beam-column joint. The nominal shear strength limit for HSC should not, however, exceed 1500 psi. Observe that this absolute maximum joint shear stress is reasonably consistent with the 1300 psi proposed by Paulay and Priestley.[2.5]

2.3.4 Impact of Joint Reinforcing

The impact of the amount of beam-column joint reinforcement on behavior has been much debated. The UCSD test program endeavored to shed some light on the subject by comparing the behavior of otherwise identical HSC specimens. The baseline specimen was HSJ12-#1. The confining pressure provided was developed in Section 2.3.3 (1240 psi in the direction of load). This level of confinement or shear reinforcement (v_s = 1030 psi) was increased by 50% in specimen HSJ12-#3. No increase in subassembly strength was observed (see Figure 2.3.12), essentially confirming the initial dominance of the concrete strut mechanism (Figure 2.3.1*a*). The sustained strength (R_b = 121 kips) increased by 14% and, more importantly, the amount of strength degradation was only 23%, as opposed to 32% in the baseline specimen. Accordingly, there is some advantage associated with providing a higher level of joint reinforcing but it will not be effective unless high levels of postyield deformations are imposed on the beam-column joint. Observe that the confining pressure provided in specimen HSJ12-#3 exceeded that required by the ACI ($0.09f_c'$) by 72%.

Strain levels measured in the hoop reinforcing within the joints of these specimens were reported. The strain levels reported for HSJ12-#2 ($v_s = 1030$ psi) and HSJ12-#3 ($v_s = 1545$ psi) were high, as might be expected since the joint shear stress (V_{jh}/bh) reached 1460 psi. At a beam displacement of 2 in., the inner ties were at incipient yield (HSJ12-#2); in fact, they had yielded near the centerline of the joint but not at points near the nodes. The outer ties did not yield until the beam displacement reached 3 in., while the inner ties experienced strains of 3 to $4\varepsilon_{yh}$. At a beam displacement of 3 in. in specimen HSJ12-#3 ($v_s = 1545$ psi) the outer ties were at incipient yield while the inner ties had reached a strain of about $1.5\varepsilon_{yh}$. The column tie strains reported for HSJ12-#1 were higher ($> 7\varepsilon_{yh}$) at beam displacements of 3 in. Recall that this specimen had only half of the axial load (300 kips) imposed on subassembly HSJ12-#2.

Reported strain levels in all ties increased by orders of magnitude at beam drifts of 2 in. or more. Clearly the truss mechanism described in Figure 2.3.1*c* is activated. After multiple cycles of postyield deformation a permanent irreversible straining (additive) of the ties takes place, and this contributes significantly to the pinching recorded (Figure 2.3.8). Observe that this pinching becomes extreme for repeated displacement demands.

Photo 2.8 Beam-column joint at 2.5% drift, University of California San Diego.

2.3.5 Bond Deterioration within the Beam-Column Joint

Beam bar strains were measured for all tests. Figure 2.3.13 describes the measured strains at the beam bars of the baseline 12-ksi joint (HSJ12-#1). Measured strains are keyed to beam displacement ductilities (μ) assuming that $\delta_b = 2$ in. corresponds to subassembly yield ($\mu = 1$). Figure 2.3.13*a* is for an interior (bundled) #11 bar ($h/d_b = 15$). Bond transfer at low levels of subassembly drift ($\mu = 0.4$) is uniformly distributed along the length of the bar. As, however, the subassembly drift reaches a ductility of 0.7 ($\delta_b = 1.4$ in.), less than half of the bar force is transferred by bond. At subassembly ductilities of 1 and 1.5, very little force is transferred to the joint through bond, and load transfer ($\varepsilon_s \cong 0$) does not appear to occur until the bar is embedded in the beam on the opposite side of the joint.

Figure 2.3.13*b* describes the strains recorded in one of the exterior #11 bars ($h/d_b = 22$). A similar pattern is apparent. Observe that these bars are not bundled—hence the impact of bundling appears to be minimal.

Bond transfer within the beam-column joint does appear to completely transfer the tensile load to the joint when axial load levels are higher (HSJ12-#2) or transverse reinforcement within the joint is increased (HSJ12-#3). Measured strains are described in Figure 2.3.14 for HSJ12-#2 and in Figure 2.3.15 for HSJ12-#3. Observe that in all cases the strain in the beam bars is essentially zero at the far side of the joint regardless of whether or not the bars are bundled. It is interesting that the compressive strains recorded in the exterior beam bars (Figure 2.3.13*b*) on the far side of the column suggest a compressive stress that is approaching $f_y/2$ at beam yield ($\mu = 1$), and this is contrary to our usual assumptions. We can conclude, however, that the existence of axial compression and the inclusion of an increased level of transverse reinforcing within the beam column joint significantly improve bond transfer within the joint. This improved bond transfer mechanism undoubtedly contributes to the observed improvement in postyield behavior of these specimens (see Figure 2.3.12).

Conclusion: The factors that impact bond transfer within the joint appear to be related more to the level of axial load and the amount of transverse reinforcing provided than to the h/d_b ratio. Clearly, as Leon[2.27] points out, bond transfer deteriorates rapidly when a subassembly is subjected to cyclic excitation absent a significant axial load. In these cases, the designer should consider increasing the level of transverse reinforcing.

2.3.6 Design Procedure

Though noble in their intent, design procedures that attempt to include too many variables are not appropriately used to design beam-column joints. Rather, they should be reexamined with each new experimental program and considered subliminally by the designer in the detailing of a beam-column joint.

The development of the guidelines that follow are generally consistent with the ACI[2.6] approach. The important features are for a planar frame.

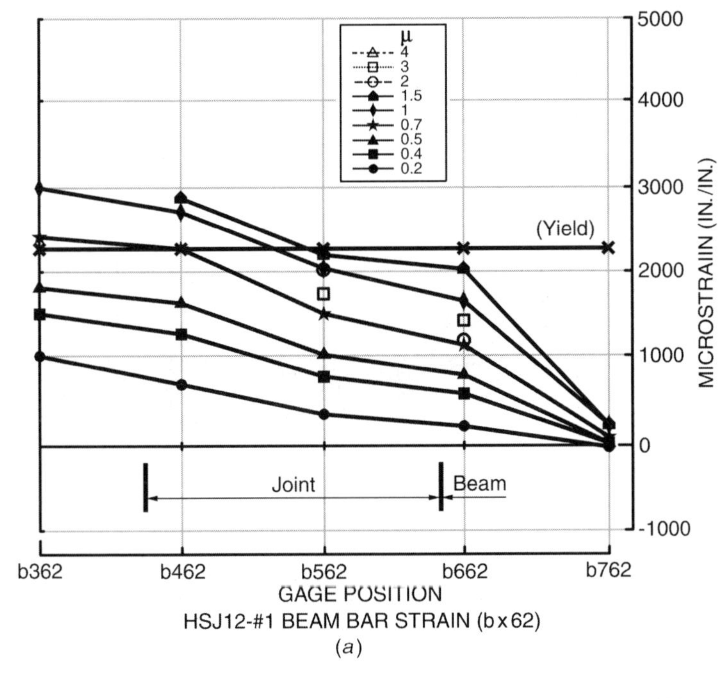

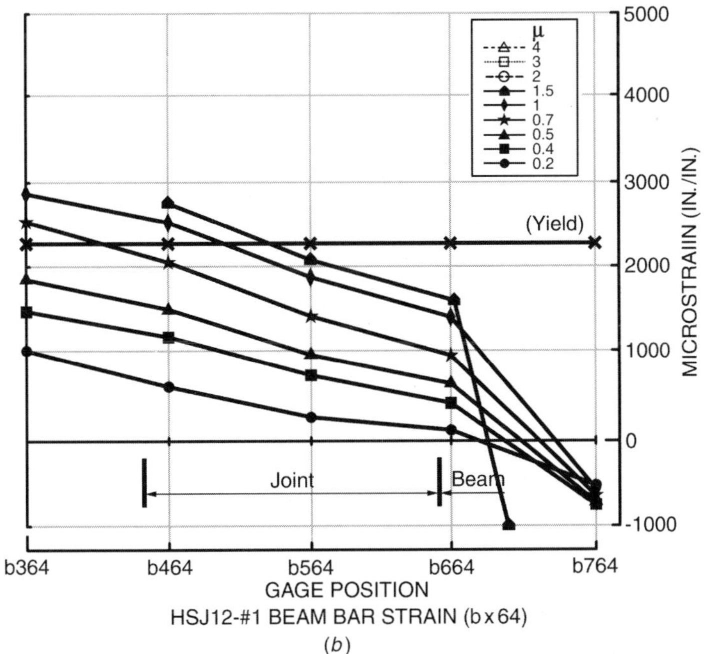

Figure 2.3.13 (*a*) HSJ12-#1 beam bar strains, interior (bundled) #11 bar. (*b*) HSJ12-#1 beam bar strains, exterior #11 bar.

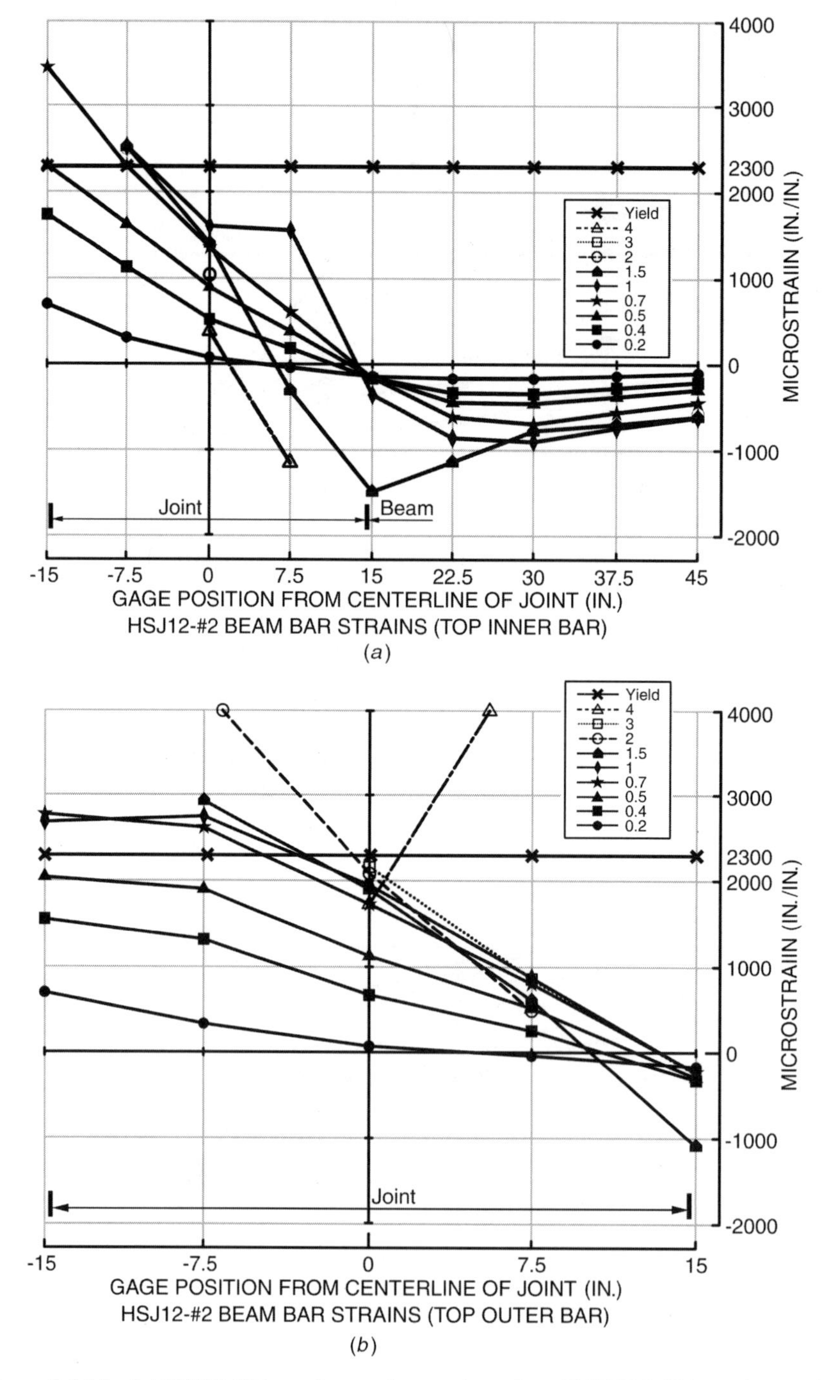

Figure 2.3.14 (*a*) HSJ12-#2 beam bar strains, top inner bar. (*b*) HSJ12-#2 beam bar strains, top outer bar.

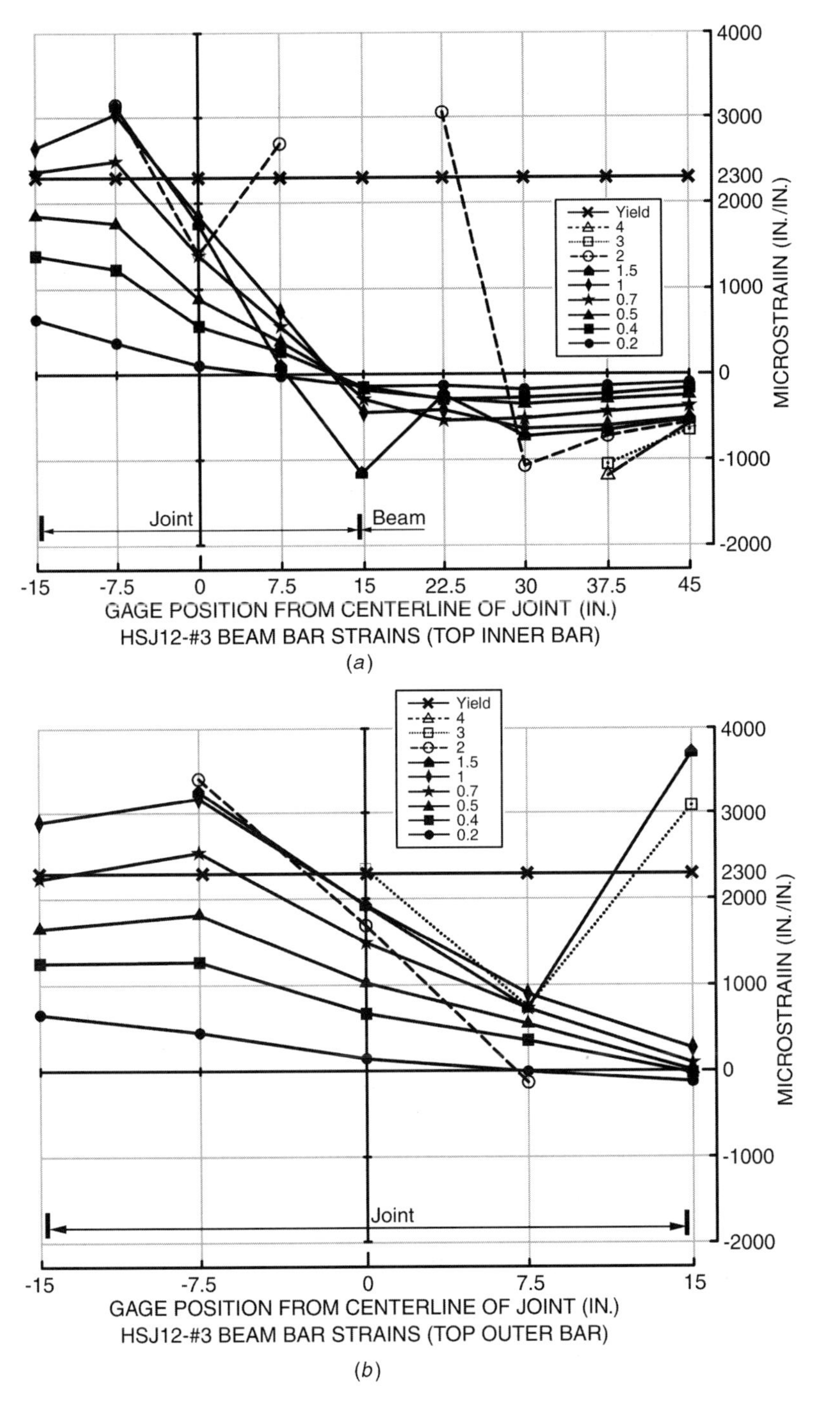

Figure 2.3.15 (*a*) HSJ12-#3 beam bar strains, top inner bar. (*b*) HSJ12-#3 beam bar strains, top outer bar.

(1) Postyield behavior should be directed toward the beam in spite of the fact that considerable ductility is available within the beam-column joint.
(2) Nominal joint shear stresses (v_{jh}) for an interior beam-column joint should be limited to $15\sqrt{f'_c}$. They should not, however, be allowed to exceed 1500 psi (Figure 2.3.12).
(3) Transverse reinforcement within the joint should be that required to satisfy ACI[2.6] objective confining pressures within the core ($0.09 f'_c$).
(4) Development lengths (h/d_b) of 20 should be a design objective but it need not be the overriding consideration. Ease of construction, especially as it relates to the effective placement of concrete within the joint, is equally, if not more, important.

The development of a design procedure for an interior joint starts with Eq. 2.3.3, which is developed from Figure 2.3.3:

$$V_{jh} = T_1 + C_2 - V_c \qquad \text{(Eq. 2.3.3)}$$

The shear in the column (V_c) can be developed directly from the amount of provided beam reinforcing, assuming that the point of inflection is at the midspan of the beam ($\ell_c/2$).

$$M_{b1} = \lambda_o f_y A_s (d - d') \qquad (2.3.12)$$

$$V_{b1} = \frac{\lambda_o f_y A_s (d - d')}{\ell_{c1}/2} \qquad (2.3.13a)$$

Similarly,

$$V_{b2} = \frac{\lambda_o f_y A'_s (d - d')}{\ell_{c2}/2} \qquad (2.3.13b)$$

When these shears are imposed on a subassembly (Figure 2.1.23), the column shear is

$$V_c = \frac{V_{b1}\ell_1 + V_{b2}\ell_2}{2h_x} \qquad (2.3.13c)$$

The joint shear (V_{jh}) may now be restated as

$$\begin{aligned} V_{jh} &= T_1 + C_2 - V_c \qquad \text{(Eq. 2.3.3)} \\ &= \lambda_o f_y A_s + \lambda_o f_y A'_s - \frac{V_{b1}\ell_1}{2h_x} - \frac{V_{b2}\ell_2}{2h_x} \\ &= \lambda_o f_y \left[A_s \left(1 - \frac{\ell_1 (d - d')}{\ell_{c1} h_x} \right) + A'_s \left(1 - \frac{\ell_2 (d - d')}{\ell_{c2} h_x} \right) \right] \qquad (2.3.14) \end{aligned}$$

The required area of the joint may be developed by introducing the objective strength limit state,

$$V_{jh,\text{allow}} = 15\phi\sqrt{f'_c}bh \tag{2.3.15}$$

and

$$A_j = bh = \frac{\lambda_o f_y}{15\phi\sqrt{f'_c}}\left[A_s\left(1 - \frac{\ell_1(d-d')}{\ell_{c1}h_x}\right) + A'_s\left(1 - \frac{\ell_2(d-d')}{\ell_{c2}h_x}\right)\right] \tag{2.3.16a}$$

This relationship is appropriate for analysis but should be simplified for design purposes. The following assumptions allow its reduction:

- The areas of steel provided, top and bottom $\left(A_s \text{ and} A'_s\right)$, are typically equal or, if not equal, may be combined such that $A_s + A'_s = 2A_{s,\text{effective}}$.
- Alternatively, if the beam spans are equal, the reduction need not combine A_s and A'_s.
- $d - d'$ is usually on the order of $0.9d$.
- d/h_x and ℓ/ℓ_c are also typically reducible to a constant.

Hence

$$\frac{\ell(d-d')}{\ell_c h_x} \cong 0.25$$

Equation 2.3.16a may now be reduced to

$$A_j = \frac{\lambda_o f_y}{15\phi\sqrt{f'_c}}\left(A_s + A'_s\right)(0.75) \tag{2.3.16b}$$

If $\lambda_o = 1.25$, $\phi = 0.85$, and $f_y = 60$ ksi, Eq. 2.3.16b is further reduced to

$$A_j = \frac{4.41}{\sqrt{f'_c}}\left(A_s + A'_s\right)$$

which for 5000-psi concrete further reduces to

$$A_j = 62\left(A_s + A'_s\right) \tag{2.1.16}$$

Now, if equal steel areas are provided on each face of the beam $\left(A_s = A'_s\right)$, as I recommend, the area of the joint or maximum amount of beam steel, both expressed in square inches, that can be tolerated in a given joint becomes

$$\boxed{A_j = 124A_s} \tag{2.3.17}$$

The usual design problem involves balancing the shear strength of an interior subassembly and the area of the beam-column joint.

The shear in the column can be developed directly from Eq. 2.3.13c and, for conceptual design purposes, Eq. 2.3.13c must be reduced much as we did in the development of Eq. 2.3.16c:

$$V_c = \frac{V_{b1}\ell_1 + V_{b2}\ell_2}{2h_x} \qquad \text{(Eq. 2.3.13c)}$$

Since the objective subassembly shear is usually stated in terms of its strength objective (V_{cu}), the following representation is appropriate:

$$V_{b1} = \frac{\phi f_y A_s(d - d')}{\ell_{c1}/2} \qquad (2.3.18a)$$

$$V_{b2} = \frac{\phi f_y A'_s(d - d')}{\ell_{c2}/2} \qquad (2.3.18b)$$

$$V_{cu} = \frac{\phi f_y}{h_x}\left[\frac{A_s\ell_1}{\ell_{c1}}(d - d') + \frac{A'_s\ell_2}{\ell_{c2}}(d - d')\right] \qquad (2.3.18c)$$

which reasonably reduces to

$$V_{cu} = \frac{\phi f_y}{h_x}\left[A_s d + A'_s d\right] \qquad (2.3.19)$$

and then for equal amounts of reinforcing steel $(A_s = A'_s)$ to

$$V_{cu} = \frac{2\phi f_y A_s d}{h_x} \qquad (2.3.20)$$

If the ratio of h_x to d is a factor of $4(h_x = 4d)$, as it might reasonably be in an office building, $\phi = 0.9$, and f_y is 60 ksi, then

$$V_{cu} = 27A_s \qquad (2.3.21a)$$

In a residential building h_x/d is often as low as 3.2(8.5/2.67), in which case

$$V_{cu} = 22A_s \qquad (2.3.21b)$$

and now a direct relationship can be created between the area of an interior beam-column joint (A_j), expressed in square inches, and the ultimate subassembly shear, expressed in kips:

$$\boxed{V_{cu} = 0.22A_j} \qquad \text{Office buildings } (h_x/d \cong 4) \qquad (2.3.22a)$$

$$\boxed{V_{cu} = 0.18A_j} \qquad \text{Residential buildings } (h_x/d \cong 3.2) \qquad (2.3.22b)$$

Design Summary: The conceptual design process reduces to the satisfaction of Eq. 2.3.22a or b if the reductions identified are reasonably appropriate. In any event, after the conceptual design has been completed, the acceptability of the selected joint size may be confirmed by checking Eq. 2.3.16a.

2.3.7 Design Example

Example 2 of Section 2.1 (Figure 2.1.27) required that an interior subassembly be developed so as to resist an ultimate shear of 200 kips. Following Eq. 2.3.22a, the required interior beam-column joint area is

$$\frac{h_x}{d} = \frac{108}{28.5} = 3.8$$

$$A_j = \frac{V_{cu}}{0.18} \qquad \text{(Eq. 2.3.22b)}$$

$$= \frac{200}{0.18}$$

$$= 1111 \text{ in.}^2$$

Ultimately six #11 bars were selected, and the adopted column size was 32 in. by 32 in. The frame geometry is described in Figure 2.1.28.

The interior subassembly described in Figure 2.1.28 has the following characteristics:

$$\begin{aligned} h_x &= 108 \text{ in.} & & \\ \ell_1 &= 24 \text{ ft} & \ell_{c1} &= 21.33 \text{ ft} \\ \ell_2 &= 16 \text{ ft} & \ell_{c2} &= 13.33 \text{ ft} \\ d &= 28.5 \text{ in.} & d' &= 3.5 \text{ in.} \end{aligned}$$

Equation 2.3.16a is now checked:

$$A_j = \frac{\lambda_o f_y}{15\phi\sqrt{f_c'}}\left[A_s\left(1 - \frac{\ell_1(d-d')}{\ell_{c1}h_x}\right) + A_s'\left(1 - \frac{\ell_2(d-d')}{\ell_{c2}h_x}\right)\right] \qquad \text{(Eq. 2.1.16a)}$$

$$= \frac{1.25(60)}{15(0.85)\sqrt{5000}}\left[9.36\left(1 - \frac{24(25)}{21.33(108)}\right) + 9.36\left(1 - \frac{16(25)}{13.33(108)}\right)\right]$$

$$= 82.8[9.36(0.74) + 9.36(0.72)]$$

$$= 1131 \text{ in.}^2$$

The final design of Example 2 proposed a 32-in. by 32-in. column ($A_g = 1024$ in.^{2}). The concrete strength could be increased to accommodate this exceedance.

$$f'_{c,\min} = \left(\frac{A_{j,\text{req'd}}}{A_{j,\text{provided}}}\right)^2 5000 \qquad (2.3.23)$$

$$f'_{c,\min} = \left(\frac{1131}{1024}\right)^2 5000$$

$$= 6100 \text{ psi}$$

2.3.8 Precast Concrete Beam-Column Joints—DDC Applications

The composite system (Section 2.1.4.3) and the bolted system (Section 2.1.4.5) rely on the introduction of a ductile rod (Figures 2.1.60 and 2.1.61) into the column. Otherwise, the load path within the beam-column joint is essentially the same as that developed in Section 2.3.1

2.3.8.1 Experimentally Based Conclusions The focus of the design of the test specimen discussed in Section 2.1.4.5 (Figure 2.1.68) was the behavior of the beam and the deformability of the subassembly. The beam-column joint was designed so as to preclude, or at least minimize, yielding within the beam-column joint (Figure 2.1.61). The produced joint was considerably stronger than it needed to be.

To confirm the strength of the beam-column joint, the subassembly described in Figure 2.3.7*a* was redesigned to include ductile rods in the columns (Figure 2.1.57) connected to high-strength thread bars in the beam (see Figure 2.1.40). The arrangement of column ties is described in Figure 2.3.16. The level of shear reinforcing (v_s) was the same as that used in specimen HSJ12-#3 ($v_s = 1545$ psi). The confining pressure parallel to the direction of load was 1860 psi, and this exceeds that currently required by the ACI ($f_{\ell x} = 1080$ psi) by 72%.

The load path through the beam-column joint is a function of the load delivered by the ductile rods. The yield strength of one rod is 141 kips; hence the load delivered to the internal head of one ductile rod in the case of abutting heads could be as much as

$$2\lambda_o T_{yi} = 2(1.25)(141)$$

$$= 352.5 \text{ kips}$$

This is probably conservative, for at tensile strains associated with $\lambda_o T_y$ and drift angles of 3%, much of the compressive load will be transferred directly to the column by bearing (see Figure 2.3.18) once the created gap has been closed. Nevertheless, the compressive force imposed on the head of the ductile rod could be as high as

$$f_{c,\text{bearing}} = \frac{2\lambda_o T_{yi}}{A_{\text{bearing}} - A_{\text{rod}}} \qquad (2.3.24)$$

$$= \frac{352.5}{28.26 - 2.4}$$

$$= 13.62 \text{ ksi}$$

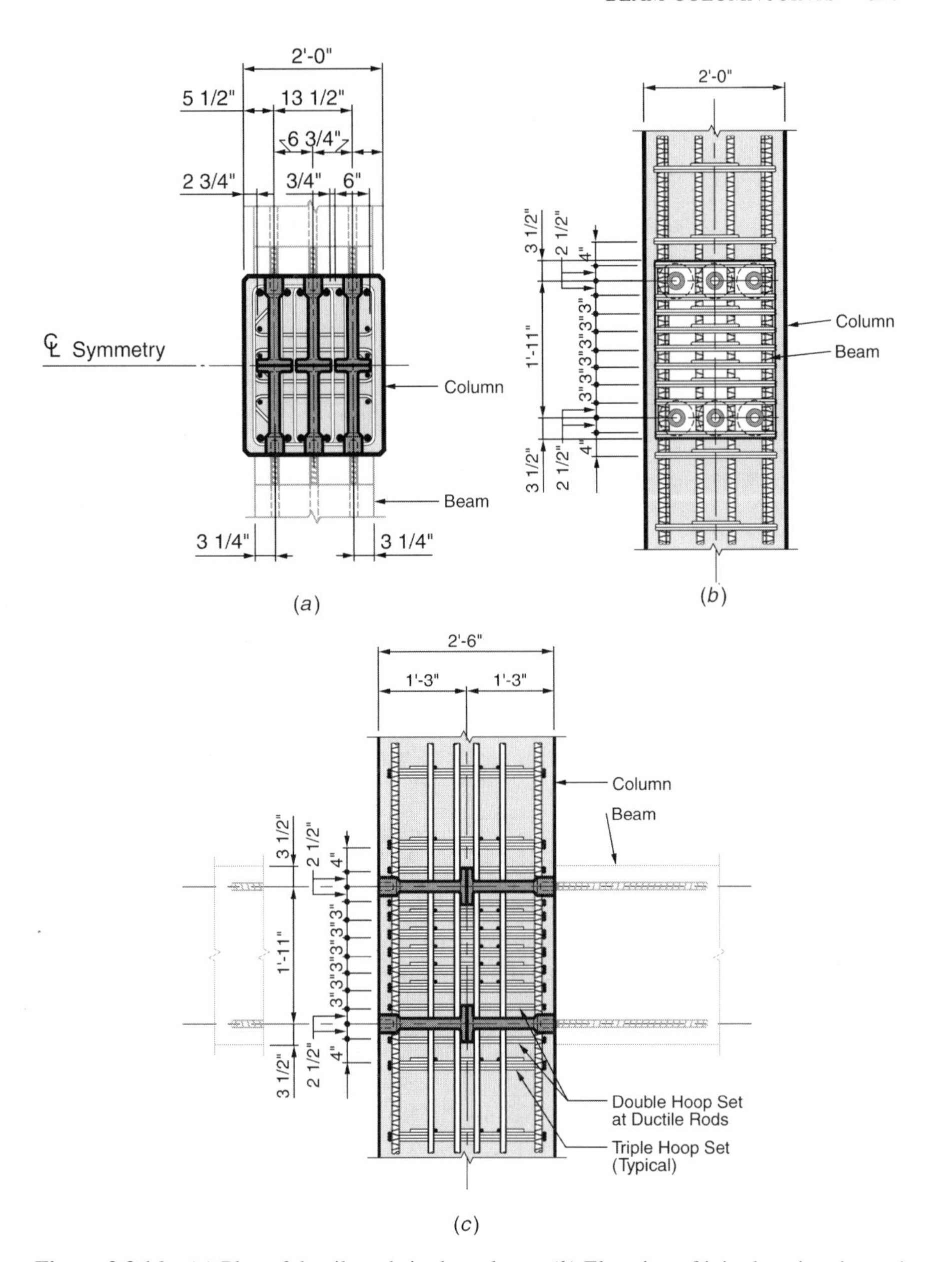

Figure 2.3.16 (*a*) Plan of ductile rods in the column. (*b*) Elevation of joint locating ties and ductile rods. (*c*) Elevation of joint locating ties and ductile rods.

The compressive strength of the concrete in this particular subassembly (HSJ12-#4) was 9.27 ksi, and the resulting stress ratio was

$$\frac{f_{c,\text{bearing}}}{f_c'} = \frac{13.62}{9.27}$$
$$= 1.47$$

and this is consistent with ACI[2.6] allowables $(1.7\phi f_c')$.

Following ACI joint design procedures, the stress imposed on the joint of the test assembly is developed as follows:

$$\lambda_o M_{bn} = \lambda_o N T_y \left(d - \frac{a}{2}\right) \tag{2.3.25}$$
$$= 1.25(3)(141)(24.8)$$
$$= 13{,}130 \text{ in.-kips}$$

This corresponds to a beam load (R_b) of

$$\lambda_o R_b = \frac{\lambda_o M_{bn}}{\ell_c} \tag{2.3.26}$$
$$= \frac{13{,}130}{105}$$
$$= 125 \text{ kips}$$

Figure 2.3.17 describes the hysteretic behavior of the subassembly. The peak beam load (129 kips) is consistent with this estimate.

The column shear following the ACI procedure is

$$V_c = \frac{\lambda_o R_b \ell}{h_x} \tag{2.3.27}$$
$$= \frac{125(20)}{16.46}$$
$$= 152 \text{ kips}$$

$$V_{jh} = 2\lambda_o N T_{yi} - V_c \tag{2.3.28}$$
$$= 2(1.25)(3)(141) - 152$$
$$= 905.5 \text{ kips}$$

$$v_{jh} = \frac{V_{jh}}{bh} \tag{2.3.29}$$

$$= \frac{905.5}{24(30)}$$
$$= 1.26 \text{ ksi}$$

The joint shear stress allowed by the ACI is

$$\begin{aligned} v_{jh,\text{allow}} &= 15\phi\sqrt{f_c'} \\ &= 15(0.85)\sqrt{9300} \\ &= 1.23 \text{ ksi} \end{aligned}$$

The discontinuity of the flexural reinforcing within the joint has concerned some reviewers, but a reasonable load path was provided and it has been proven (Figure 2.3.17) to be reliable. Consider the load flow described in Figure 2.3.18. Strut and tie modeling used to develop a plastic truss[2.3] suggest that an angle of load distribution of 65° is attainable. Accordingly, all of the interior ties could be activated.

A reasonable load flow must start by estimating the compression load on the compression side ductile rods (Figure 2.3.18). Figure 2.3.19 describes the strains mea sured in the ductile rods during the preyield behavior cycles. Observe that the compression load imposed on ductile rods is low and soon transferred to the concrete. Figure 2.3.20 shows the crack pattern at a drift angle of 1.7% ($\delta_b = 2$ in.). The load transfer mechanism within the beam-column joint is the same as one might expect of any beam-column joint. Tensile straining in the middle of the column at the ductile rod is evident, but the strut mechanism is clearly activated. At low levels of ductility demand on the beam-column joint, an effort should be made to activate the concrete strut

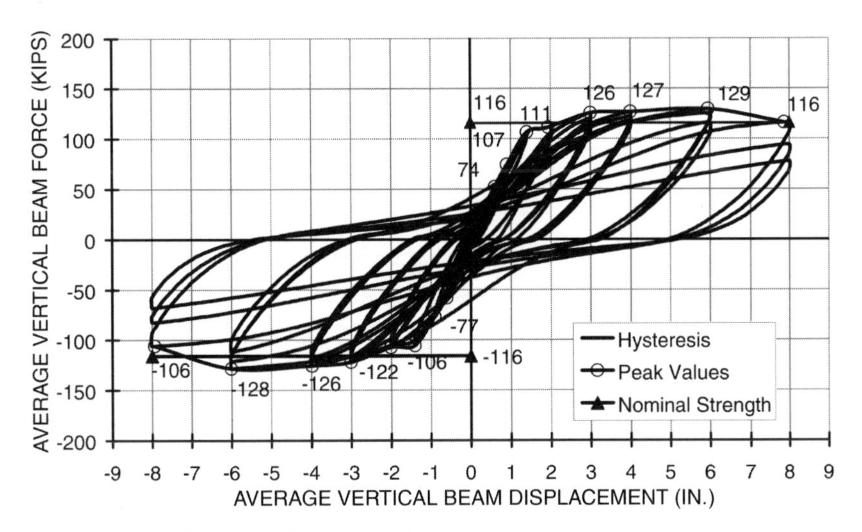

Figure 2.3.17 HSJ12-#4 Force–displacement response.

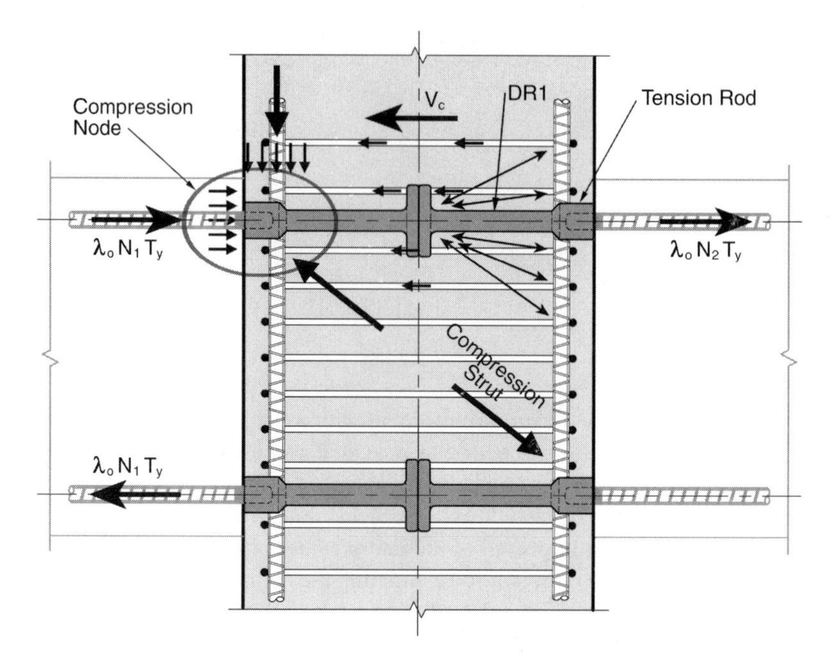

Figure 2.3.18 Load flow within a DDC beam-column joint.

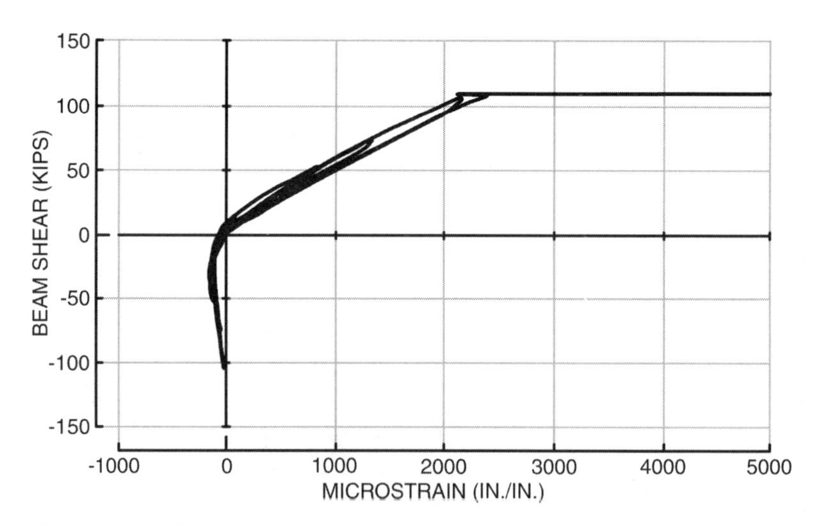

Figure 2.3.19 HSJ12-#4 hysteresis of bottom inner DDC connector.

mechanism. This can be accomplished by transferring the load from the tensile rods to the compression node (Figure 2.3.18). Four sets of proximate ties accomplish this objective in the test specimen—two double and three triple tie sets (see Figure 2.3.16*c*).

$$NT_y < NA_{sh} f_{yh}$$

$$3(141) < 26(0.31)(60)$$

$$423 < 483 \text{ kips}$$

Postyield strain levels experienced in the ductile rods were off scale (Figure 2.3.19), as one might expect since the reported gap, reasonably confirmed by analysis, was on the order of $\frac{1}{4}$ in. ($\varepsilon_{DR} = 0.03$ in./in.). After three cycles of load at a drift angle of 5%, the concrete interface has deteriorated and the entire shell has spalled (Figure 2.3.21). Now all of the force imposed on the beam-column joint by the tensile rod must be transferred through the truss mechanism. Accordingly, equilibrium requires that the load imposed on the compression side ductile rod group be $\lambda_o N T_y$. Observe that the peak beam load is recorded (Figure 2.3.17) at a drift angle of 5% ($\delta_b = 6$ in.).

Strain levels recorded in the ties confirm both pre- and postyield mechanisms. Figure 2.3.22 describes the strain gage locations. The vertical strain profiles of the inner and outer ties are shown in Figure 2.3.23. Yielding of the inner tie group first occurs in the ties located at the midheight of the joint (Figure 2.3.23*a*) at a ductility of one ($\delta_b = 2$ in.). Ultimately ($\mu > 1$) high strains develop in the vicinity of the ductile rods (first triple hoop set, Figure 2.3.16*c*). The strains in outer ties (Figure 2.3.23*b*) suggest a uniform distribution of stress more characteristic of the truss mechanisms described in Figure 2.3.1*c*. At a drift angle of 2.5% ($\delta_b = 3$ in.), the ties at the ductile rod first yield. Observe that this is consistant with the strain pattern developed in the inner ties but less severe at ductilities of two or more.

At peak load and a drift angle of 5% ($\delta_b = 6$ in.), all of the ties have yielded, and this includes the two tie sets above and below the ductile rods.

Figure 2.3.20 HSJ12-#4 first full cycle of 2.0-in. vertical beam displacements.

Figure 2.3.21 HSJ12-#4 third full cycle of 6.0-in. displacements.

At peak load ($R_b = 129$ kips) the load imposed on the tensile ductile rod set ($N = 3$) is

$$T_{5\%} = \frac{R_b \ell_c}{d - d'} \tag{2.3.30}$$

$$= \frac{129(105)}{23}$$

$$= 589 \text{ kips}$$

The capacity of each tie set (three hoops) is

$$V_{sh} = A_{sh} f_y \tag{2.3.31}$$

$$= 6(0.31)(60)$$

$$= 111.6 \text{ kips}$$

The total load delivered to the compression node (C_1) is

$$C_1 = T_{5\%} - V_c \tag{2.3.32}$$

$$= 589 - 157$$

$$= 432 \text{ kips}$$

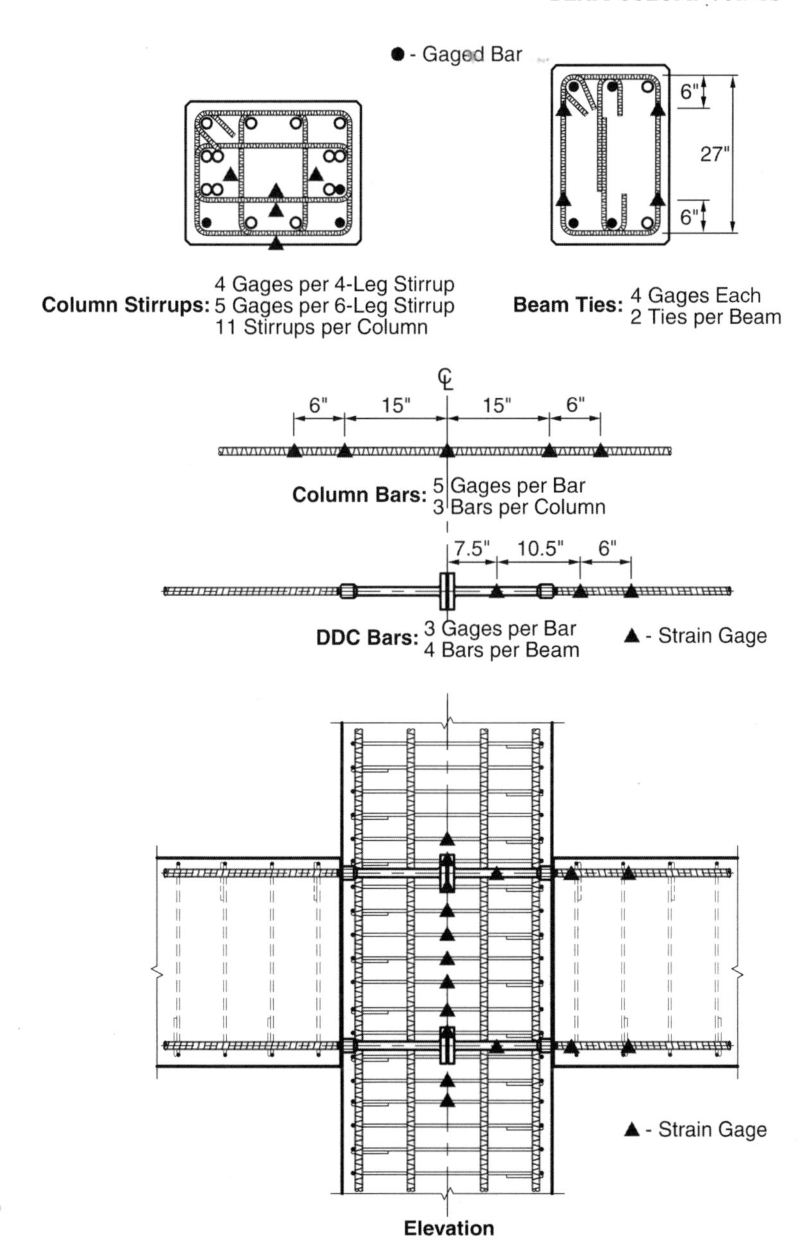

Figure 2.3.22 Tie strain gage locations.

and that delivered to the head of the tensile rod (DR1) is 589 kips ($T_{5\%}$). If the internal truss mechanism (Figure 2.3.1*c*) were to be developed entirely from the compression node (Figure 2.3.18), more than nine (9.17) tie sets would be required and only five were available (plus two doubles). The load path within the joint and the load transfer mechanism are further complicated by the observed behavior at peak load (Figure

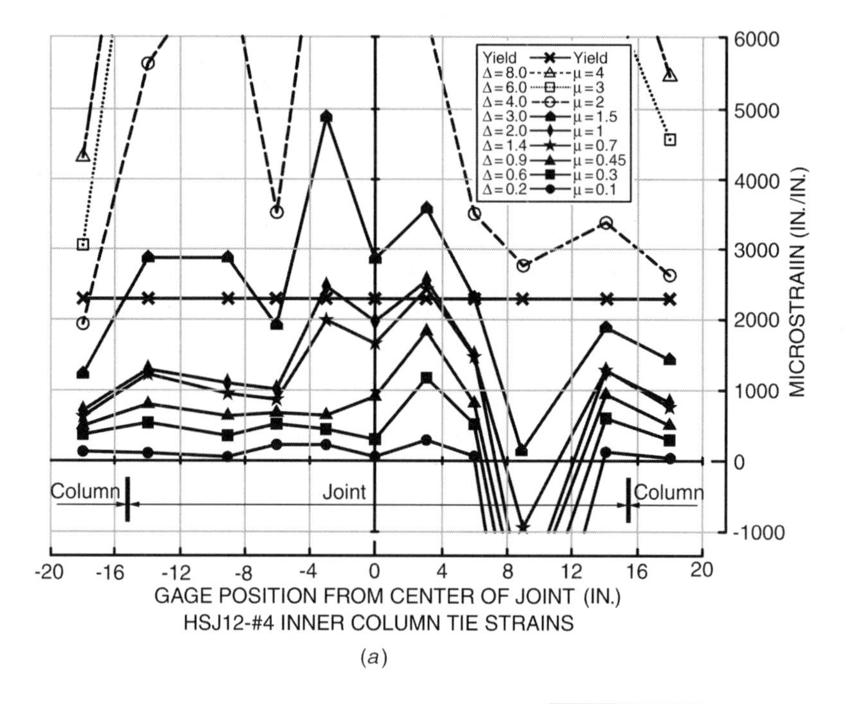

(*a*)

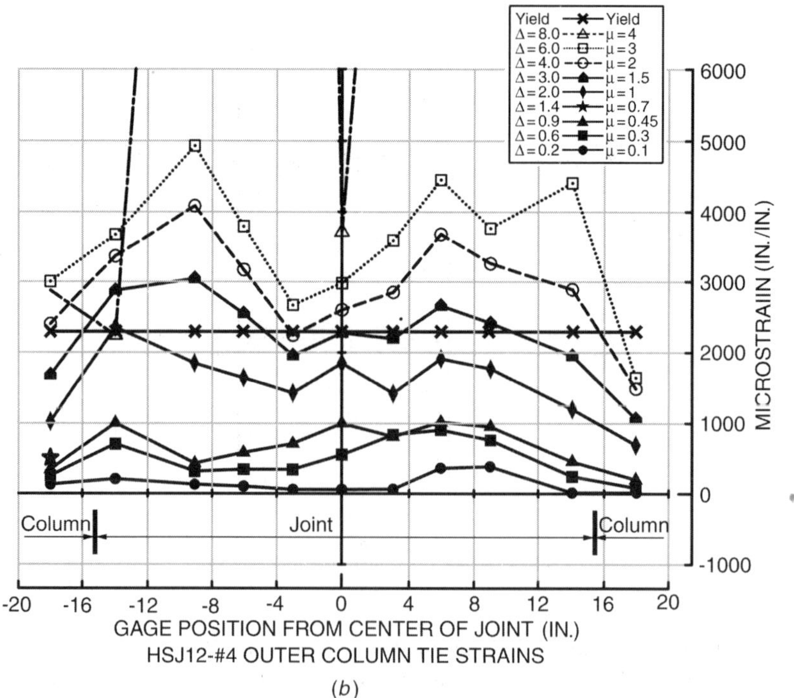

(*b*)

Figure 2.3.23 (*a*) HSJ12-#4 inner column tie strains-vertical strain profile along column center line. (*b*) HSJ12-#4 outer column tie strains-vertical strain profile along column centerline.

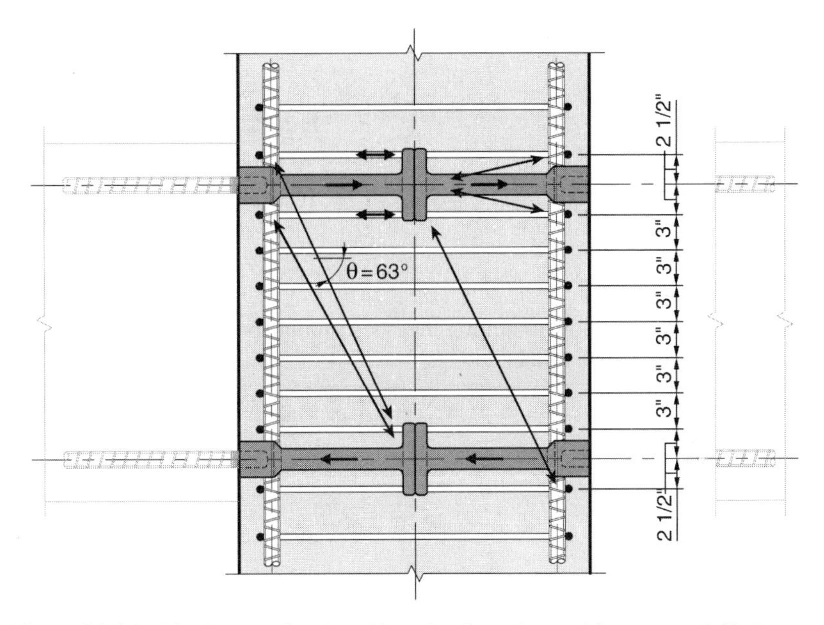

Figure 2.3.24 Plastic truss developed in a ductile rod assembly at strength limit state.

2.3.21), where it is clear that a significant gap has been created between the beam and column as a result of spalling of the concrete shell. Shear transfer obviously occurs at the edge of the core in the column. The location of the compresison load transfer is probably in this vicinity also, for a secondary strut and tie mechanism similar to that described in Figure 2.3.24 must be developed. This load transfer mechanism is consistent with the plastic truss analogy. Observe also (Figure 2.3.17) that, even after three displacement cycles at a drift angle of 5% ($\delta_b = 6$ in.), the subassembly is still capable of sustaining its design strength.

Shear transfer at the beam-column interface is somewhat different than that which occurs in a traditional cast-in-place beam column subassembly. This is because the strain in the DDC assemblage is concentrated at the interface (see Figure 2.3.21) as opposed to distributed over the plastic hinge region. The elongation of the ductile rod is developed for a drift angle of 5% as follows:

$$\theta = \frac{\delta_b}{\ell_c} \tag{2.3.33}$$

This assumes that the deflection in the beam, column, and beam-column joint are small. This is probably not true for the beam-column joint but reasonable for the beam and the column. In any case it represents a conservative assessment.

$$\theta = \frac{6}{105} \qquad \text{(Eq. 2.3.33)}$$

$$= 0.057 \text{ radian}$$

This creates a rod elongation of

$$\delta_{\text{rod}} = \theta\left(\frac{d - d'}{2}\right) \tag{2.3.34}$$

$$= 0.057\left(\frac{23}{2}\right)$$

$$= 0.65 \text{ in.}$$

On the compression side, assuming that the column shell is ineffective, the effective bearing length, ℓ_{be}, is reduced to

$$\ell_{be} = \ell_b - \text{cover} + \delta_{\text{rod}} \qquad \text{(see Figure 2.1.61)}$$

$$= 3 - 1.5 + 0.65$$

$$= 2.15 \text{ in.}$$

In the beam-column joint test program, assuming that all of the beam shear (R_b) is transferred through the compression side ductile rods, the bearing pressure under the exterior head of each ductile rod is

$$P_{\text{rod}} = \frac{R_b}{N}$$

$$= \frac{129}{3}$$

$$= 43 \text{ kips}$$

$$f_{c,\text{bearing}} = \frac{P_{\text{rod}}}{A_{\text{bearing}}}$$

$$= \frac{43}{3(2.15)}$$

$$= 6.67 \text{ ksi}$$

and this is reasonably attainable in confined concrete, especially given the conservative nature of the assumptions used in the development coupled with the fact that a shear deformation was not observed in this test.

2.3.8.2 Beam-Column Joint Design Procedures The design objective should be to create a serviceable, strong beam-column joint that will allow the ductile rods to yield and create the desired strong joint–weak beam objective. The test specimen studied in Section 2.1.4.5 (Figure 2.1.64) attained these objectives to drift ratios in excess of 4% (Figure 2.1.65). The beam-column joint examined in this section was capable of attaining a drift ratio of 6.7%, but only after a significant amount of spalling was experienced.

The nominal shear stresses imposed on the DDC beam-column joint of the test specimen described in Figure 2.1.64 was $10\sqrt{f_c'}$ while that imposed on the DDC reinforced beam-column joint described in Figure 2.3.16 was $15\sqrt{f_c'}$.

For the reasons discussed in Section 2.3.6, it seems most logical to follow the procedures currently advocated by the ACI to design beam-column joints that include ductile rods. The design process must, however, be extended so as to include elements particular to the ductile rod.

The design procedure, insofar as beam-column joint strength is concerned, is essentially that developed in Section 2.3.6.

The shear imposed on the beam-column joint is

$$V_{jh} = 2\lambda_o N T_y - V_c \qquad \text{(Eq. 2.3.28)}$$

If we adopt the proposed allowable shear stress for an internal joint of $15\phi\sqrt{f_c'}$, we find that it follows from Section 2.3.6, and specifically the development starting with Equation 2.3.12, that the relationship appropriate for use in the conceptual design of beam-column joints that contain ductile rods is

$$A_j = \frac{\lambda_o f_y}{15\phi\sqrt{f_c'}}\left(A_s + A_s'\right)(0.75) \qquad \text{(Eq. 2.3.16b)}$$

Converted to N ductile rods, top and bottom, this becomes

$$A_j = \frac{\lambda_o T_{yi}(2N)0.75}{15\phi\sqrt{f_c'}}$$

and this reduces to

$$\boxed{A_j = 292N \text{ in.}^2} \qquad \text{(Eq. 2.3.17)}$$

for $f_c' = 5000$ psi and the reductions proposed in Section 2.3.6.

It also follows that a convenient relationship may be developed between the number of ductile rods and the design shear (V_u) for the column.

For office buildings,

$$V_{cu} = 0.22A_j \qquad \text{(Eq. 2.3.22a)}$$

$$= 0.22(292)N$$

$$\boxed{V_{cu} = 64N \text{ kips}} \qquad (2.3.35)$$

For residential buildings,

$$V_{cu} = 0.18A_j \qquad \text{(Eq. 2.3.22b)}$$

$$= 0.18(292)N$$

$$\boxed{V_{cu} = 52N \text{ kips}} \tag{2.3.36}$$

The proposed joint reinforcement program should be capable of developing the maximum load imposed upon it to drift ratios on the order of 5%.

$$\begin{aligned} V_{sh} &= V_{jh} \\ &= 2\lambda_o N T_{yi} - V_c \qquad \text{(Eq. 2.3.28)} \end{aligned}$$

and this can be reduced as it was for Eq. 2.3.14:

$$V_{sh} = \lambda_o T_{yi}\left[N_1\left(1 - \frac{\ell_1(d - d')}{\ell_{c1}h_x}\right) + N_2\left(1 - \frac{\ell_2(d - d')}{\ell_{c2}h_x}\right)\right] \tag{2.3.37}$$

and for the usual case where $\ell_1/\ell_{c1} = 1.1$, $d - d' = 0.8h$, and $h/h_x \cong 0.33$,

$$V_{sh} = \lambda_o T_{yi}(1.4)N \tag{2.3.38a}$$

$$= 1.25(141)(1.4)N$$

$$\boxed{V_{sh} = 247N \text{ kips}} \tag{2.3.38b}$$

The total area of internal shear ties should be

$$A_{sh} = \frac{V_{sh}}{f_{yh}} \tag{2.3.39a}$$

$$= \frac{247N}{60}$$

$$\boxed{A_{sh} = 4.11N \text{ in.}^2} \tag{2.3.39b}$$

Observe that this is essentially the reinforcing provided in the test specimen (Figure 2.3.16c):

$$\begin{aligned} A_{sh} &= 4.11(3) \qquad \text{(Eq. 2.3.39b)} \\ &= 12.33 \text{ in.}^2 \qquad \text{(Required)} \\ A_{sh} &= \big((5 \times 6) + (2 \times 4)\big)(0.31) \\ &= 11.78 \text{ in.}^2 \qquad \text{(Provided)} \end{aligned}$$

The bearing stress imposed on the internal bearing plate need not be checked since it is capable of transferring $2\lambda_o T_{yi}$ for concrete strengths of 4000 psi. However, the bearing stress under the external ductile rod head must be checked because the imposed load will be a function of member and subassembly geometry.

2.3.9 Precast Concrete Beam-Column Joints—Hybrid System

The load paths developed in the beam-column joints of a hybrid subassembly are only slightly different than those created in a cast-in-place concrete beam-column joint. The design of the beam-column joints used in the 39-story building shown in Photo 2.4 was based on a series of tests performed at the University of Washington. The adopted design criterion was that contained in the ACI Code.[2.6]

2.3.9.1 Experimentally Based Conclusions—Interior Beam-Column Joint The subassembly described in Figure 2.3.25 was designed with the express intent of identifying the strength limit state of the beam-column joint.

The design of the test module followed the concepts developed in Section 2.1.4.4. The dependable strength of the subassembly was developed as follows:

$$T_{nps} = A_{ps} f_{pse} \qquad \text{(Eq. 2.1.54)}$$

$$= 9(0.153)(162)$$

$$= 223.1 \text{ kips}$$

$$T_{ns} = A_s f_y$$

$$= 3(0.44)(60)$$

$$= 79.2 \text{ kips}$$

$$a = \frac{T_{nps}}{0.85 f_c' b}$$

$$= \frac{223.1}{0.85(5)(16)}$$

$$= 3.3 \text{ in.}$$

$$M_{nps} = T_{nps}\left(\frac{h}{2} - \frac{a}{2}\right) \qquad \text{(Eq. 2.1.55)}$$

$$= 223.1(10 - 1.65)$$

$$= 1865 \text{ in.-kips}$$

$$M_{ns} = T_{ns}(d - d') \qquad \text{(Eq. 2.1.53)}$$

$$= 79.2(16.5)$$

$$= 1307 \text{ in.-kips}$$

$$M_n = M_{ns} + M_{nps} \qquad \text{(Eq. 2.1.56)}$$

$$= 1307 + 1865$$

$$= 3172 \text{ in.-kips}$$

$$V_{nb} = \frac{M_n}{\ell_c}$$

$$= \frac{3172}{62}$$

$$= 51.2 \text{ kips}$$

$$V_{nc} = V_{nb}\left(\frac{\ell}{h_x}\right)$$

$$= 51.2\left(\frac{144}{117.5}\right) \qquad \text{(see Figure 2.3.25)}$$

$$= 62.8 \text{ kips}$$

The probable strength of the subassembly (Figure 2.3.25) at a drift of 3.5% may also be developed from the relationships described in Figure 2.1.54.

The stress levels in the reinforcing are developed from imposed strain states. Mild steel strain is based on an intentional debond length of 6 in. plus bond penetrations of d_b:

$$\ell_d = 6 + 2d_b \qquad \text{(Eq. 2.1.47)}$$

$$= 6 + 2(0.75)$$

$$= 7.5 \text{ in.}$$

Some iteration combined with speculation is required to determine the neutral axis depth (c). Assume that

$$c \cong 6 \text{ in.}$$

The elongation of the mild steel at a postyield drift of 3% (θ_p) is

$$\delta_s = (d - c)\theta_p \qquad \text{(see Figure 2.1.52)}$$

$$= (18.25 - 6)(0.03)$$

$$= 0.37 \text{ in.}$$

and the postyield strain in the mild steel is

$$\varepsilon_{sp} = \frac{\delta_s}{\ell_d}$$

$$= \frac{0.37}{7.5}$$

$$= 0.049 \text{ in./in.}$$

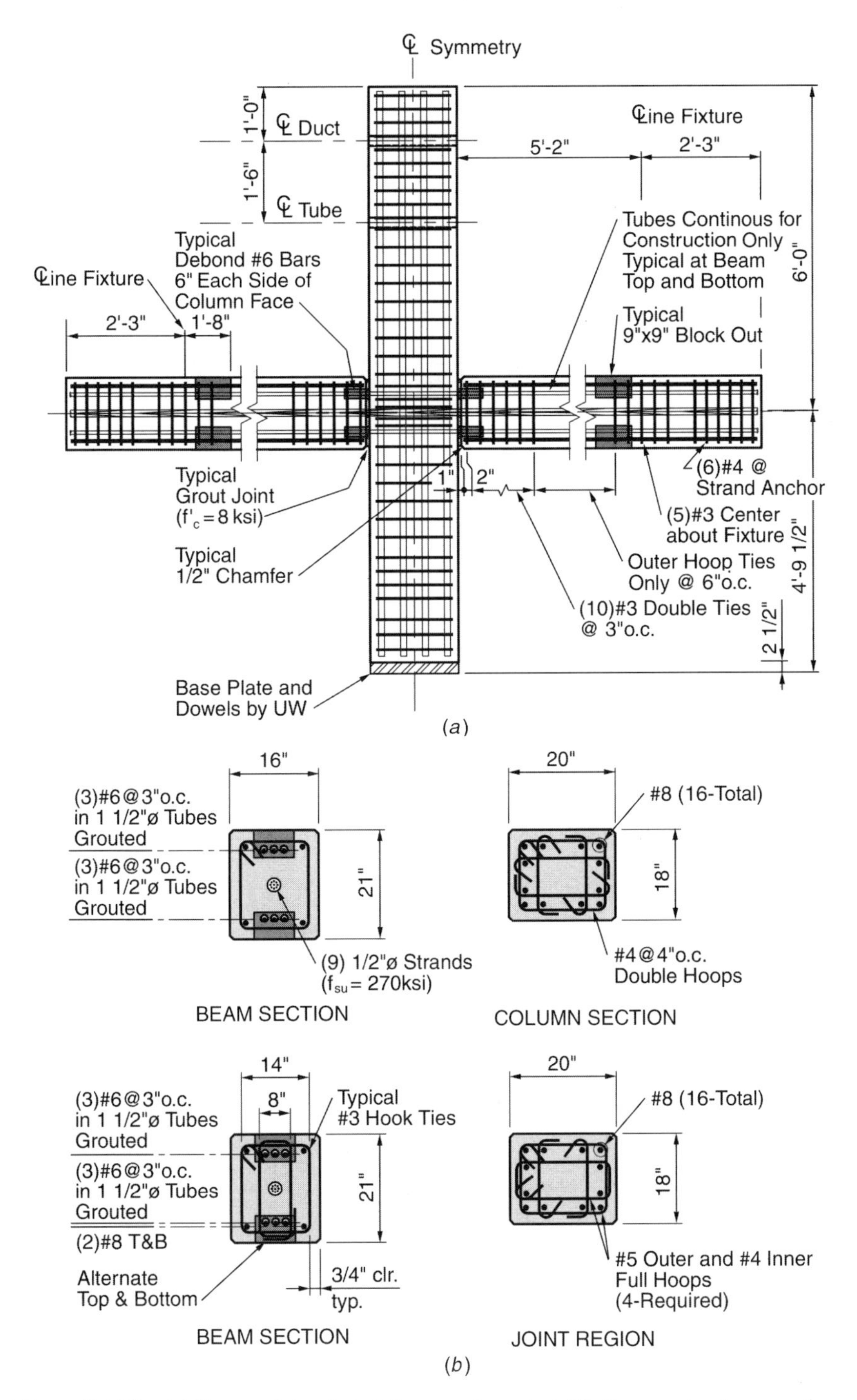

Figure 2.3.25 (*a*) Test module—hybrid beam-column test. (*b*) Beam and column sections.

This corresponds to an induced stress in the mild steel of about 80 ksi (see Figure 2.1.54). Thus

$$\begin{aligned}\lambda_o T_s &= A_s f_{su}\\ &= 3(0.44)(80)\\ &= 106 \text{ kips}\end{aligned}$$

The elongation imposed on the post-tensioning tendons (δ_{ps}) is developed as follows:

$$\begin{aligned}\delta_{ps} &= \left(\frac{h}{2} - c\right)\theta_p\\ &= (10 - 6)(0.03)\\ &= 0.12 \text{ in.}\end{aligned}$$

The postyield rotation induced strain imposed on the tendon is

$$\begin{aligned}\varepsilon_{psp} &= \frac{2\delta_{ps}}{\ell}\\ &= \frac{2(0.12)}{144}\\ &= 0.00167 \text{ in./in.}\end{aligned}$$

Since the stress imposed on the tendon will be, by design, less than 230 ksi, elastic behavior may be presumed. Hence

$$\begin{aligned}f_{psp} &= \varepsilon_{psp} E_{ps}\\ &= 0.00167(28,000)\\ &= 47 \text{ ksi}\end{aligned}$$

and the force on the tendon becomes

$$\begin{aligned}\lambda_o T_{ps} &= \left(f_{pse} + f_{psp}\right) A_{ps}\\ &= (162 + 47)(9)(0.153)\\ &= (209)(1.38)\\ &= 288 \text{ kips}\end{aligned}$$

The depth of the compressive stress block is based on the assumption that the mild steel subjected to compression (C_s) is stressed to yield. Accordingly,

$$C_c = \lambda_o T_{ps} + \lambda_o T_{sy} - T_{sy}$$

$$= 288 + 106 - 79.2$$

$$= 314.8 \text{ kips}$$

$$a = \frac{C_c}{0.85 f_c' b}$$

$$= \frac{314.8}{0.85(5)(16)}$$

$$= 4.63 \text{ in.}$$

Observe that this suggests a neutral axis depth of 5.8 in., which is reasonably consistent with the 6-in. estimate. The probable subassembly shear becomes

$$M_{b,\text{pr}} = \lambda_o T_{psp}\left(\frac{h}{2} - \frac{a}{2}\right) + T_{sy}(d - d') + \left(\lambda_o T_{sy} - T_{sy}\right)\left(d - \frac{a}{2}\right)$$

$$= 288(10 - 2.32) + 79.2(16.5) + (106 - 79.2)(18.25 - 2.32)$$

$$= 2212 + 1307 + 481$$

$$= 4000 \text{ in.-kips}$$

Observe that this corresponds to a member overstrength of

$$\lambda_o = \frac{M_{b,\text{pr}}}{M_n}$$

$$= \frac{4000}{3182}$$

$$= 1.26$$

The column shear that can be sustained by the subassembly at a story drift of about 3.5% ($\theta_p = 0.03$ radian) is

$$V_{b,\text{pr}} - \frac{M_{b,\text{pr}}}{\ell_c}$$

$$= \frac{4000}{62}$$

$$= 64.5 \text{ kips}$$

$$V_{c,\text{pr}} = V_{b,\text{pr}}\left(\frac{\ell}{h_x}\right)$$

$$= 64.5\left(\frac{144}{117.5}\right)$$

$$= 79 \text{ kips}$$

The force (V_c) displacement relationship record of the test of this subassembly (Figure 2.3.25) is described in Figure 2.3.26. Observe that both V_{cn} and V_{cp} are confirmed by the experiment.

The subassembly stiffness proposed for cast-in-place subassemblies is plotted on Figure 2.3.26 and developed as follows:

$$\begin{aligned} V_{cn} &= \frac{V_{c,\text{pr}}}{\lambda_o} \\ &= 63.2 \text{ kips} \\ I_{be} &= 0.35 I_{bg} \\ &= 4320 \text{ in.}^4 \\ I_{ce} &= 0.7 I_{cg} \\ &= 8400 \text{ in.}^4 \\ \ell &= 144 \text{ in.}; \qquad \ell_c = 124 \text{ in.}; \qquad (\ell_c/\ell)^2 = 0.74 \\ h_x &= 117.5 \text{ in.}; \qquad h_c = 97 \text{ in.}; \qquad (h_c/h_x)^2 = 0.68 \end{aligned}$$

$$\Delta_i = \frac{V_i h_x^2}{12E}\left(\frac{\ell_c}{I_{be}}\left(\frac{\ell_c}{\ell}\right)^2 + \frac{h_c}{I_{ce}}\left(\frac{h_c}{h_x}\right)^2\right) \qquad \text{(Eq. 3.2.2a)}$$

where

h_x is the story height.
E is the modulus of elasticity.
ℓ is the beam length—node to node.
ℓ_c is the clear span of the beam.
h_c is the clear span of the column.
I_{be} is the effective moment of inertia of the beam.
I_{ce} is the effective moment of inertia of the column:

$$\begin{aligned} \Delta_i &= \frac{63.2(117.5)^2}{12(4000)}\left(\frac{124}{4320}(0.74) + \frac{97}{8400}(0.68)\right) \\ &= (18.18)(0.0212 + 0.0078) \\ &= 0.53 \text{ in.} \end{aligned}$$

This corresponds to a drift ratio of 0.0045, and though this reasonably reflects the elastic stiffness, a significant loss in stiffness occurs at subassembly yield (Figure 2.3.26). This stiffness degradation is essentially the same as that observed in cast-in-place subassemblies (Figure 2.3.8).

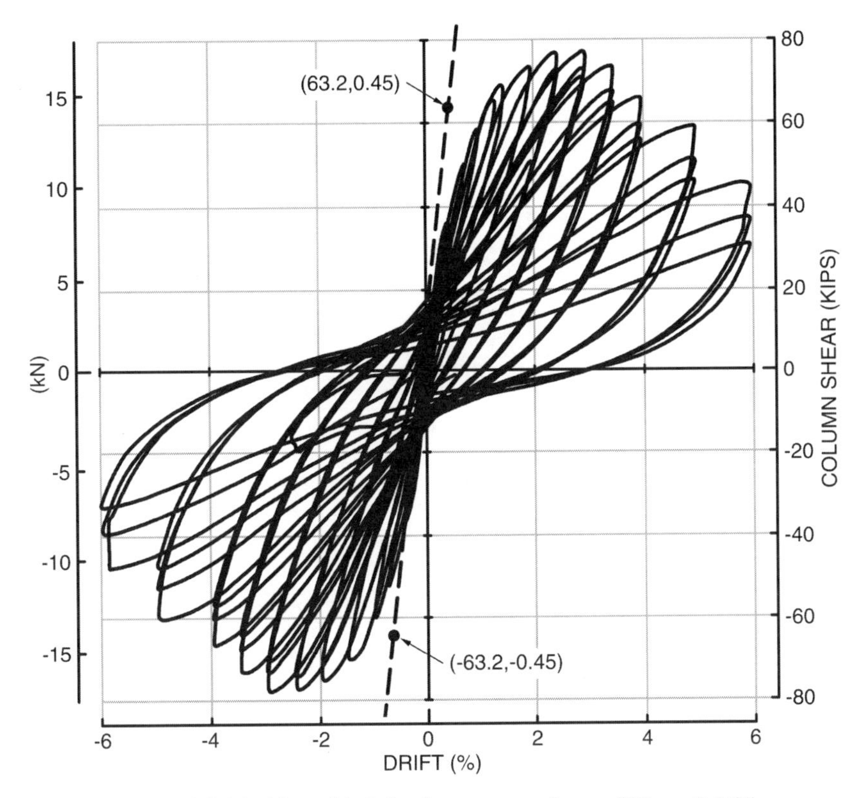

Figure 2.3.26 Hysteritic behavior—test specimen of Figure 2.3.25.

Before proceeding to analyze the stresses in the joint, it is interesting to compare measured force in the post-tensioning tendon with the analytical projection ($\lambda_o T_{ps} =$ 288 kips). The measured force in the post-tensioning only reached 252 kips, or 88% of the predicted force. Reinforcing bar strains were between 2 and 2.5%, and this too is considerably less than the predicted 4.4%. From this we might reasonably conclude that prediction methodologies are conservative both in terms of predicting overstrength and peak strain states.

2.3.9.2 Design Procedures—Interior Beam-Column Joints Joint shear stress analysis procedures are developed as they were in Section 2.3.1 for conventionally reinforced cast-in-place concrete beam-column joints. From the forces and strains recorded during the test the following induced shear stress may be developed at peak load:

$$V_{c,\max} = 79 \text{ kips} \qquad \text{(see Figure 2.3.26)}$$

$$T_{ps,\max} = 252 \text{ kips}$$

$$T_{s,\max} = A_s f_s$$

$$= 3(0.44)80$$
$$= 105 \text{ kips}$$

The shear imposed on the beam-column joint is shown in Figure 2.3.27:

$$\begin{aligned} V_{jh,\max} &= T_{ps,\max} + 2T_{s,\max} - V_{c,\max} \\ &= 252 + 210 - 79 \\ &= 383 \text{ kips} \end{aligned}$$

And the shear stress imposed on the joint is

$$\begin{aligned} v_{jh,\max} &= \frac{V_{jh,\max}}{A_j} \\ &= \frac{383}{18(20)} \\ &= 1.06 \text{ ksi} \qquad \left(13.7\sqrt{6000}\right) \end{aligned}$$

If, however, ACI[2.6] prescriptive procedures were used to develop the forces imposed on the beam-column joint, the level of induced shear stress would be developed as follows:

$$\begin{aligned} \lambda_o T_{nps} &= 1.25 A_{ps} f_{pse} \\ &= 1.25(9)(0.153)(162) \\ &= 280 \text{ kips} \end{aligned} \tag{2.3.40}$$

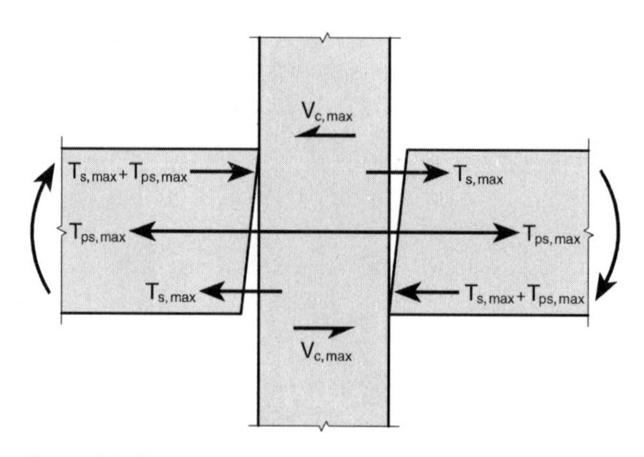

Figure 2.3.27 Forces acting on a hybrid beam-column joint.

$$\lambda_o T_{ns} = 1.25(3)(0.44)(60) \tag{2.3.41}$$
$$= 99 \text{ kips}$$
$$\lambda_o V_{cn} = 1.25(62.8)$$
$$= 78.5 \text{ kips}$$
$$\lambda_o V_{jh} = \lambda_o T_{nps} + 2\lambda_o T_{ns} - \lambda_o V_{cn} \tag{2.3.42}$$
$$= 280 + 2(99) - 78.5$$
$$= 399.5 \text{ kips}$$
$$v_{jh} = \frac{\lambda_o V_{jh}}{A_j}$$
$$= \frac{399.4}{18(20)}$$
$$= 1.11 \text{ ksi}$$

The allowable stress following ACI procedures is

$$v_{jh,\text{allow}} = 15\phi\sqrt{f_c'}$$
$$= 15(0.85)\sqrt{6000}$$
$$= 0.988 \text{ ksi}$$

from which one might conclude that joint shear controlled the strength of the subassembly, and this was confirmed by the deterioration of the joint in the test program.

The emerging criterion seems to include the net area of the joint as a gage of reliable shear strength. Based on the net size of the column joint of the test specimen, one would conclude that the beam-column joint was capable of reaching its nominal strength:

$$A_{j,\text{net}} = bh - t_d h \tag{2.3.43}$$
$$= 18(20) - 20(2.75)$$
$$= 305 \text{ in.}^2$$

where t_d is the out to out diameter of the tube.

The attained level of beam-column joint shear stress in the test specimen was

$$v_{jh,\text{net}} = \frac{V_{jh,\max}}{A_{j,\text{net}}}$$
$$= \frac{383}{305}$$
$$= 1.256 \text{ ksi} \qquad \left(16.2\sqrt{f_c'}\right)$$

Design Summary—Interior Hybrid Beam-Column Joint

- Calculate the overstrength shear load imposed on the beam-column joint using Eq. 2.3.42 (see Figure 2.3.27).
- Determine the shear stress level imposed on the net area of the beam-column joint (Eq. 2.3.43).
- Adjust the reinforcing and/or area of the beam-column joint until the ultimate stress imposed on the net area of the beam-column joint is less than $15\phi\sqrt{f_c'}$.
- Confine the beam-column joint in accordance with Eq. 1.2.3b.

2.3.9.3 Design Procedures—Exterior Beam-Column Joints An exterior hybrid beam-column subassembly test was described and analyzed in Section 2.1.4.4 (see Figure 2.1.45). The strength limit state of this subassembly was a result of beam spalling (see Figure 2.1.47*b*), and this will in general be the case because the loads imposed on an exterior beam-column joint are significantly less than those imposed on an interior beam-column joint. The loads imposed on an exterior beam-column joint are shown in Figure 2.3.28 and, as can be seen, the developed flow of compressive stresses is well distributed over the joint. The forces imposed on the beam-column joint at a drift angle of 4% are

$$V_{c,\max} = 34.8 \text{ kips} \qquad \text{(see Figure 2.1.47)}$$

$$T_{ps,\max} = 282 \text{ kips} \qquad \text{(see Figure 2.1.48)}$$

$$T_{s,\max} \cong 105 \text{ kips}$$

The shear imposed on the joint was

$$\begin{aligned} V_{jh,\max} &= T_{ps,\max} + T_{s,\max} - V_{c,\max} \\ &= 282 + 105 - 34.8 \\ &= 352.2 \text{ kips} \end{aligned}$$

The shear stress associated with the gross area would have been on the order of

$$\begin{aligned} v_{jh,\max} &= \frac{V_{jh,\max}}{A_j} \\ &= \frac{352.2}{18(20)} \\ &= 0.978 \text{ ksi} \qquad \left(12.6\sqrt{f_c'}\right) \end{aligned}$$

and this is more than the ACI[2.6] nominal allowable of $12\sqrt{f_c'}$.

The stressing/anchorage assembly is typically cast iron and the maximum shear occurs between the stressing head and the compression face of the beam. Accordingly,

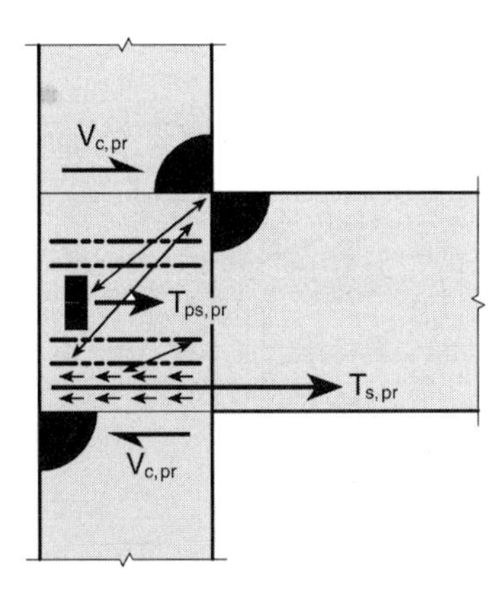

Figure 2.3.28 Forces imposed on an exterior hybrid beam-column joint.

Figure 2.3.29 Exterior beam-column joint assembly.

the use of the gross area of an exterior beam-column joint and ACI allowables seems reasonable and conservative. Figure 2.3.29 shows an exterior beam column assembly prior to the casting of the column.

2.3.9.4 Corner Hybrid Beam-Column Joints Corner hybrid beam-column joints were tested at the University of Washington and, in general, they performed quite

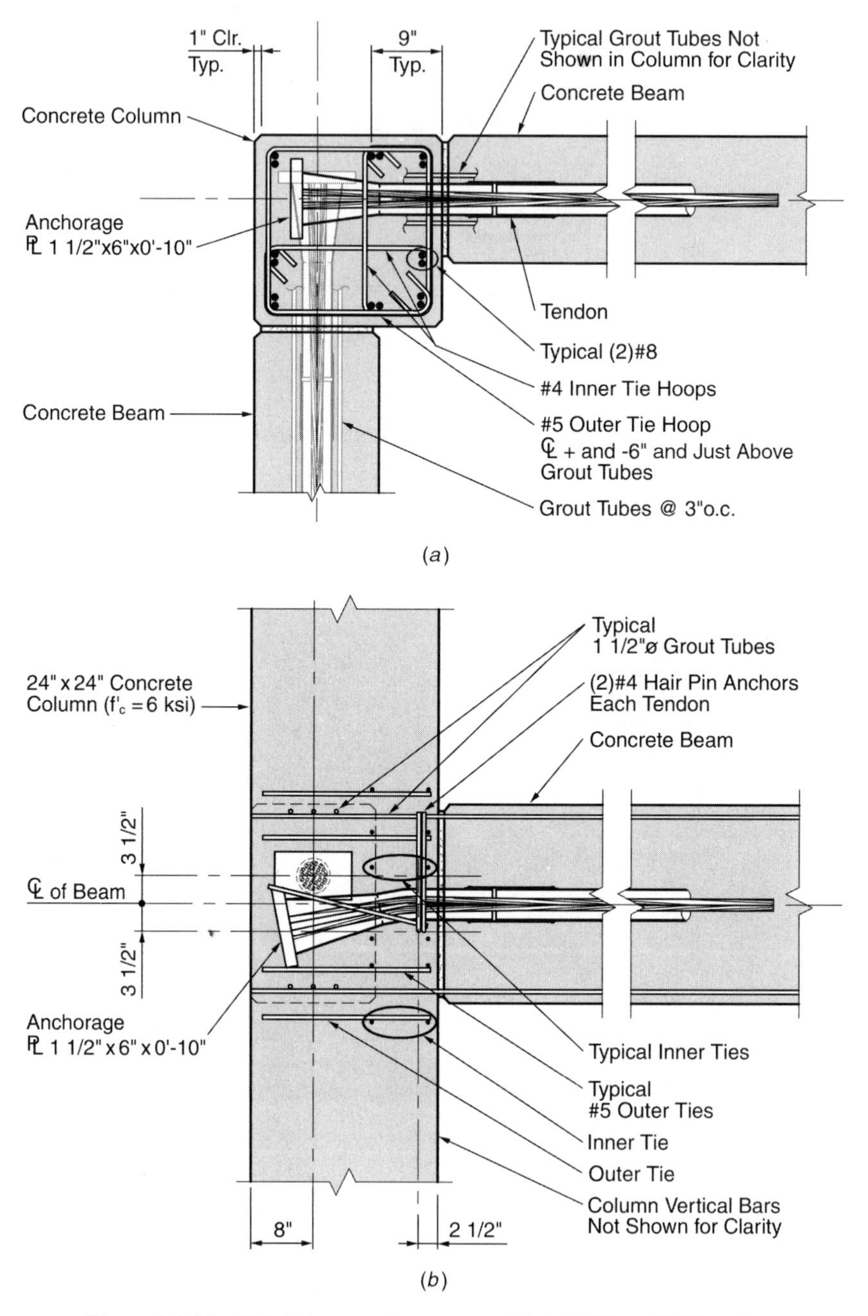

Figure 2.3.30 Hybrid beam-column corner joint. (*a*) Plan. (*b*) Elevation.

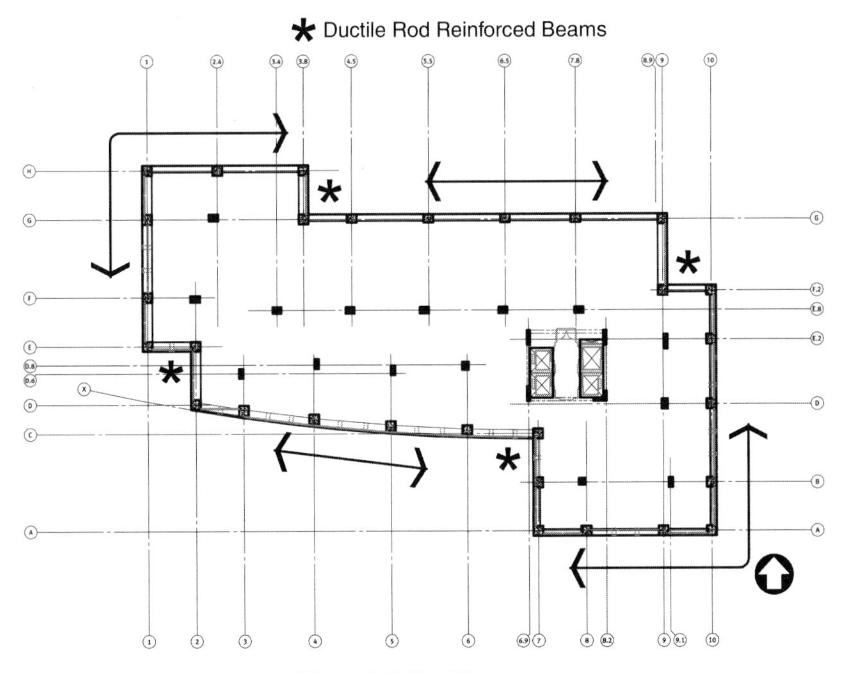

Figure 2.3.31 Floor plan.

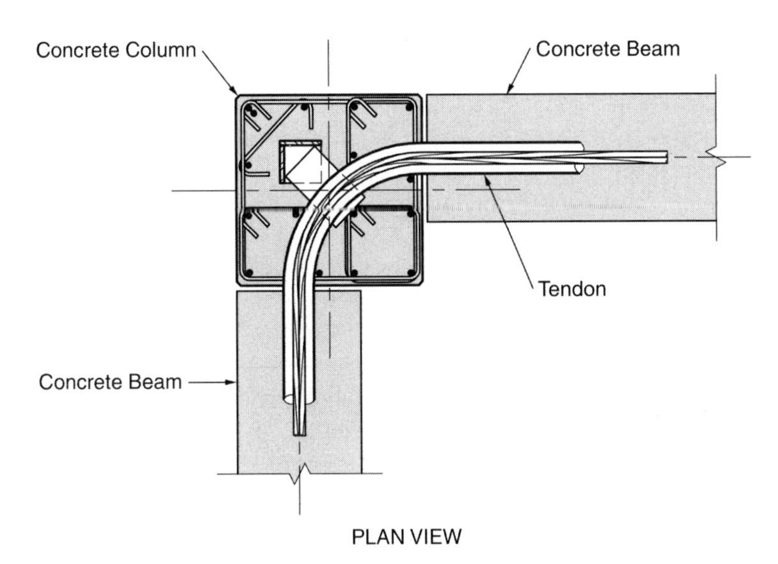

Figure 2.3.32 Around-the-corner test hardware.

well. They are extremely difficult to build, however, as can readily be seen in Figures 2.3.30*a* and 2.3.30*b*. Tie and reinforcing placement must be carefully laid out. In the development of the Paramount apartment building (Photo 2.4), this cross-over stressing program (Figure 2.3.30) was abandoned in favor of an around-the-corner post-tensioning program.

The plan of a typical floor (Figure 2.3.31) allowed for around-the-corner stressing because composite beams (Photo 2.3) were used on the short reentrant corners. A significant amount of care in the form of alternating and sequential stressing was required to create the objective levels of post-tensioning without damaging strands. The hardware installed in the corner column that allowed the around the corner stressing is described in Figure 2.3.32. A strap was placed around the curved pipe and anchored to an angle that allowed the resolution of internal forces shown in Figure 2.3.28. Once the forces are resolved, the design follows that described in this section for a typical exterior beam-column joint.

2.4 SHEAR DOMINATED SYSTEMS

Shear walls are often used to provide lateral support for buildings. In spite of their usual strength and stiffness, they will in most cases be expected to deform beyond their elastic limit. Should the strength of the shear wall exceed the seismic load imposed on it, either the foundation or the diaphragm will be called upon to absorb earthquake-induced energy. Concrete diaphragms behave in much the same way as shear walls. Accordingly their design and postyield deformability will generally follow the procedures developed for shear walls.

The design of ductile shear walls or diaphragms will follow one of two paths: postyield deformability will be directed toward either the flanges or the web. The design of tall walls ($h_w/\ell_w > 2$) should endeavor to create a wall that will yield in flexure. When the flexural capacity is less than the shear strength, the behavior of the shear wall or diaphragm will closely follow that of a beam or beam-column. The principal deviation will be the stability of thin wall sections when subjected to high axial loads, and this will be one of the focal issues discussed in this section.

When the provided shear strength is less than the flexural strength of the wall, a ductile or deformable load path must be created in shear, and this too will become a focal topic. Quite often openings occur in shear walls, and the seismic load path must flow around the created discontinuity. When openings are aligned the so-created discontinuity can impose large deformations on the "coupling" beams that link two much stiffer elements. Since these coupling beams cannot be expected to be stronger than the elements they connect, postyield deformability becomes critical to the success of the bracing program. Thus the design procedure for these critical elements is also explored in this section. Precast concrete shear walls will be increasingly used if they can be effectively joined. Experimental efforts have demonstrated that properly connected precast elements will not only survive earthquake actions, but also experience less damage in the process. This emerging construction alternative is examined and a design approach is developed.

Photo 2.9 Six 6-story precast concrete wall panels, Hilton Hotel, Oxnard, CA, 1968. (Courtesy of Englekirk Partners, Inc.)

2.4.1 Tall Thin Walls

Bearing wall construction is well suited to residential type buildings. In densely populated areas this type of construction is viewed as essential because of its inherent economy and functional attributes. This need has promoted some recent experimental work that should adequately support the use of thin bearing walls in seismically active regions.

2.4.1.1 Experimentally Based Conclusions Experimental work on thin bearing walls in the United States has been, for the most part, performed by Wallace and his associates.[2.31–2.34] In New Zealand the work of Yañez, Park, and Paulay[2.35] has confirmed the work of Wallace. Recent experimental work of Yunfeng Zhang and Zhihao Wang[2.36] on walls that support high axial loads ($0.24 f'_c A_g$ and $0.35 f'_c A_g$)

extend the conclusions of others. The experimental work of Zhang and that of Wallace are reviewed in this section. This experimental work clearly supports Wallace's conclusion[2.33] that "commonly used code equations for solid walls do not provide adequate design guidance."

Zhang and Wang[2.36] studied the postyield behavior of a series of tall thin walls subjected to two levels of axial loading (0.24 and $0.35 f_c' A_g$). Our focus is on the behavior of shear wall #9 (SW9), because the axial load ($0.24 f_c' A_g$) did not result in buckling, while those walls subjected to an axial load of $0.35 f_c' A_g$ reportedly buckled. The test specimen and setup are described in Figure 2.4.1. The lateral load was applied 59 in. above the base, and this produced an aspect ratio (h_w/ℓ_w) of 2.15.

The following are pertinent parameters and material strengths:

$$f_c' = 5000 \text{ psi} \qquad (\text{Measured } f_c' = 5750 \text{ psi})$$

$$f_y = 54{,}000 \text{ psi}$$

$$\frac{h_w}{t_w} = \frac{59}{3.9} = 15$$

Lateral support was not provided at the top of the wall. Accordingly, the effective aspect ratio (h_w/t_w) of the wall is 30 and

$$\frac{kh_w}{r} = \frac{2.0(59)}{0.3(3.9)} = 100$$

The axial load imposed on the wall (P) was

$$P = 0.24 f_c' A_g = 130 \text{ kips}$$

For comparative purposes the current empirical ACI[2.6] axial strength limit state for this wall would be

$$\phi P = 0.55\phi f_c' A_g \left[1 - \left(\frac{kh_w}{32t_w}\right)^2\right] \qquad \text{(see Ref. 2.6, Eq. 14-1)}$$

$$= 0.55(0.7) f_c' A_g \left[1 - \left(\frac{2(59)}{32(4)}\right)^2\right]$$

$$= 0.041 f_c' A_g$$

The hysteretic behavior of the specimen described in Figure 2.4.1 is that shown in Figure 2.4.2*a*. The wall was capable of sustaining a horizontal load (H) of 68 kips

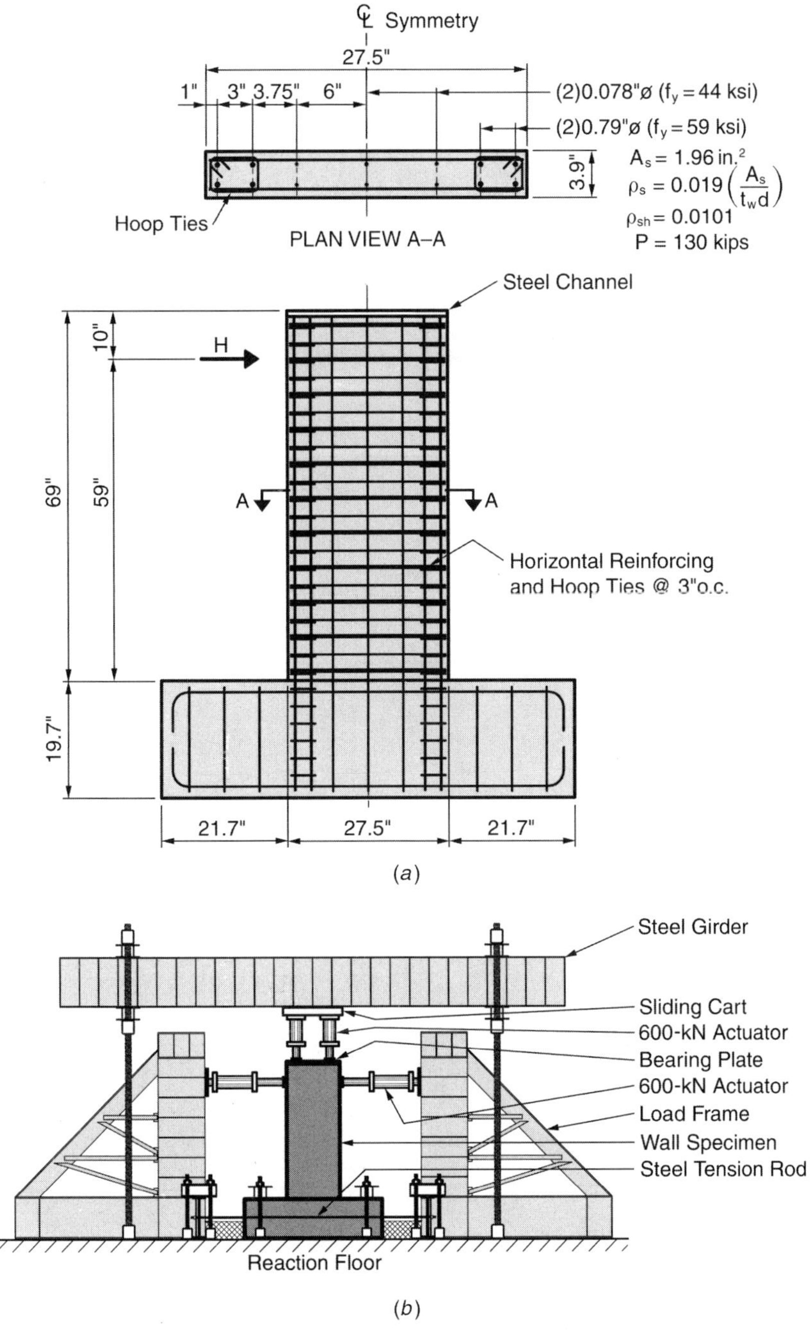

Figure 2.4.1 Geometry and reinforcement details of wall specimen SW9.[2.36]

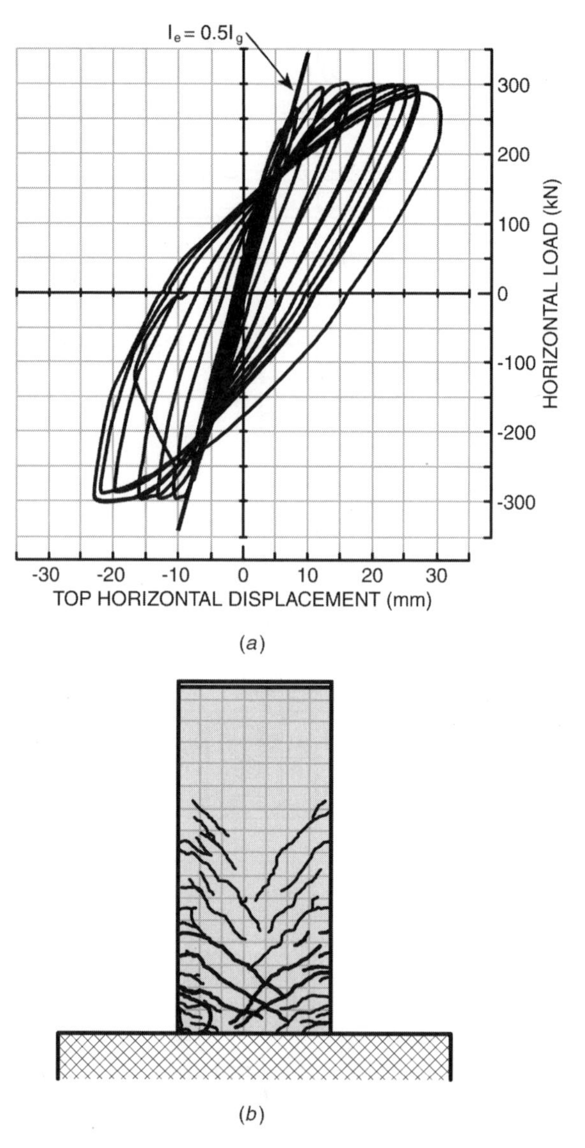

Figure 2.4.2 (*a*) SW9: Lateral load versus top horizontal displacement (1.5 m from base). (*b*) Crack pattern at failure—test specimen of Figure 2.4.1.[2.36]

(300 kN) to a story drift of slightly more than 1 in. (27 mm). This corresponds to a drift ratio of 1.8%. Observe that the peak lateral load (H) was maintained throughout, and that strength hardening was not experienced. Observe also that the analytically predicted strength (Table 2.4.1) describes a perfectly plastic system (M = constant). This is partially explained by the fact that the strain in the steel remains relatively low ($2.6\varepsilon_y$) at the nominal strength of the wall ($\varepsilon_c = 0.003$ in./in., Table 2.4.1). If

we assume that shear deformation is negligible, which it clearly is not, the estimated strain in the steel only reaches 0.02 in./in. at Δ_u. Shear deformation is undoubtedly high, and this, combined with the perfectly plastic behavior of the steel, produces the uncharacteristic behavior of described in Figure 2.4.2a. P/Δ effects impact the horizontal load by 3%.

The stiffness of this wall may be estimated following the usual procedures and assuming that the contribution of shear to the displacement of the wall will be minimal in the elastic range.

$$\begin{aligned} I_g &= \frac{t_w \ell_w^3}{12} \\ &= \frac{3.9(27.5)^3}{12} \\ &= 6759 \text{ in.}^3 \\ \Delta &= \frac{Hh_w^3}{3EI_g} \\ &= \frac{68(59)^3}{3(4000)(6759)} \\ &= 0.17 \text{ in.} \qquad (4.4 \text{ mm}) \end{aligned}$$

It is generally accepted[2.6] that the effective moment of inertia (I_e) for a shear wall is 50% of the gross moment of inertia (I_g), and this is consistent with the experimental evidence provided behavior is essentially elastic. SW9 was subjected to a significant axial load ($0.24 f_c'$) and this causes the initial stiffness (Figure 2.4.2a) to significantly exceed the generally accepted model ($I_e = 0.5 I_g$).

The associated shear stress imposed on the wall (SW9) was

$$\begin{aligned} v &= \frac{H}{dt_w} \\ &= \frac{68{,}000}{(26.5)(3.9)} \\ &= 658 \text{ psi} \qquad \left(9.3\sqrt{f_c'}\right) \end{aligned}$$

Comment: The fact that this wall was capable of sustaining this high a shear stress is clearly attributed to the level of axial load imposed upon it. Observe that the "shear friction" factor (H/C_c) is on the order of 50% and this prevents the occurrence of a sliding shear failure at the base of the wall.

Wall construction and reinforcing are described in Figure 2.4.1a. The nominal shear stress limit state defined by the ACI for this wall would be

$$v_n = 2\sqrt{f_c'} + \rho_{sh} f_{yh} \qquad \text{(see Ref. 2.6, Eq. 21-7)}$$

$$= 2\sqrt{5000} + 0.0101(44{,}000)$$

$$= 144 + 444$$

$$= 588 \text{ psi}$$

where ρ_{sh} is as tabulated by the authors, and ACI Eq. 11-5 would allow the term 144 in the third equation to be 159 psi ($v_n = 600$ psi).

From these comparisons it is clear that the applied levels of load clearly exceed the limit states defined by current codes. Observe also that shear reinforcing provided in the hinge region is not capable of supporting the imposed level of shear ($v_s = 444$ psi < 658 psi).

The mechanism associated with shear transfer appears to follow the plastic truss analogy (see Section 1.3). A shear fan emanating from the toe of the test specimen (Figure 2.4.1) should pick up at least 16 stirrup sets ($\theta_{\max} \cong 65°$). The development of this shear fan is confirmed by the crack pattern at failure (Figure 2.4.2*b*)

The shear strength limit state, based on the reported shear reinforcement percentage (ρ_{sh}) and an effective node point at the base of the wall removed a distance $a/2$ from the compressed end of the wall, becomes

$$V_s = (\tan 65°)t_w \left(d - \frac{a}{2}\right) \rho_{sh} f_{yh}$$

$$= 2.14(3.9)(19.5)(0.01)(44)$$

$$= 71.6 \text{ kips}$$

and this exceeds the peak shear demand of 68 kips. Observe that all of the horizontal reinforcing in this region appears to have reached yield (Fig. 2.4.2*b*).

An estimate of the strain states imposed on the concrete and reinforcing steel suggest reasonable strain limit states. The neutral axis may be located through the use of approximate techniques or by one of many biaxial load computer programs (see Table 4.2.1).

As the tension steel yields, one might reasonably assume that the load sustained by the compression reinforcing (A_s') is subyield and, as a consequence, only partially effective. Thus the differential between T_y and C_s might conservatively be assumed to be $0.33 f_y A_s$. The corresponding depth of the compression block is

$$a = \frac{P + (0.33)A_s f_y}{0.85 f_c' t_w}$$

$$= \frac{130 + 0.33(4)(0.49)(59)}{0.85(5)3.9}$$

$$= 8.9 \text{ in.}$$

$$c = \frac{a}{\beta_1}$$

$$= \frac{8.9}{0.8}$$

$$= 11.0 \text{ in.}$$

And this is essentially the same result arrived at by a computer that considers the impact of all of the vertical wall reinforcing as well as that in the boundary (Table 2.4.1).

At an ultimate displacement of 27 mm or 1.06 in. (Figure 2.4.2a), the strain in the concrete is estimated following the procedures developed in Section 2.1.1.2 as

TABLE 2.4.1 Biaxial Behavior of the Wall Section Described in Figure 2.4.1 (P = 130 kips, f'_c = 5 ksi, f_y = 59 ksi)

ε_c (in./in.)	ε_s (in./in.)	c (in.)	M_u (ft-kips)	ϕ (rad/in.)
0.00050	0.00006	23.63	76.4	0.0000212
0.00070	0.00032	18.26	116.2	0.0000383
0.00080	0.00047	16.76	134.4	0.0000477
0.00090	0.00062	15.68	152.2	0.0000574
0.00100	0.00077	14.93	169.5	0.0000670
0.00110	0.00094	14.29	186.7	0.0000770
0.00120	0.00110	13.80	203.6	0.0000869
0.00130	0.00127	13.43	220.1	0.0000968
0.00140	0.00142	13.16	236.0	0.0001064
0.00150	0.00158	12.89	252.0	0.0001164
0.00160	0.00174	12.70	267.3	0.0001260
0.00170	(f_y) 0.00189	12.54	282.3	0.0001355
0.00180	0.00204	12.41	296.7	0.0001451
0.00190	0.00225	12.14	306.5	0.0001565
0.00200	0.00246	11.87	316.0	0.0001685
0.00210	0.00274	11.49	321.5	0.0001827
0.00220	0.00307	11.06	323.8	0.0001988
0.00230	0.00337	10.74	324.7	0.0002141
0.00240	0.00364	10.53	324.8	0.0002280
0.00250	0.00389	10.37	324.9	0.0002412
0.00260	0.00415	10.21	324.7	0.0002548
0.00270	0.00439	10.10	324.6	0.0002674
0.00280	0.00463	9.99	324.3	0.0002803
0.00290	0.00483	9.94	324.2	0.0002919
0.00300	0.00504	9.88	324.0	0.0003036
0.00310	0.00530	9.78	323.1	0.0003171
0.00320	0.00547	9.78	323.1	0.0003274
0.00330	0.00570	9.72	322.4	0.0003394

$$\phi_y = \frac{\varepsilon_{sy}}{d - c} \qquad \text{(Eq. 2.1.7a)}$$

$$= \frac{0.00186}{26.5 - 11.8}$$

$$= 0.000126 \text{ rad/in.}$$

$$\delta_y = \frac{\phi_y (h_w)^2}{3} \qquad \text{(Eq. 2.1.7b)}$$

$$= \frac{0.000126(59)^2}{3}$$

$$= 0.15 \text{ in.} \qquad (3.7 \text{ mm})$$

Comment: Observe that this is consistent with the projected elastic deflection of 0.17 in. or 4.4 mm and only half of the idealized deflection ($I_e = 0.5 I_g$). See Figure 2.4.2*a*.

$$\varepsilon_{cy} = \phi c$$

$$= 0.000126(11.8)$$

$$= 0.0015 \text{ in./in.}$$

and this is slightly below that predicted in Table 2.4.1.

An ultimate displacement of 1.06 in. corresponds to a displacement ductility factor (Δ_u/Δ_y) of 7 based on a predicted yield displacement of 0.15 in. The probable curvature at a displacement of 1.06 in. is

$$\delta_p = \delta_u - \delta_y$$

$$= 1.06 - 0.15$$

$$= 0.91 \text{ in.}$$

$$\ell_p = \frac{\ell_w}{2}$$

$$= \frac{27.5}{2}$$

$$= 13.75 \text{ in.}$$

Comment: A plastic hinge length (ℓ_p) of $0.5\ell_w$ is becoming increasingly accepted though it is conservative. Paulay[2.37] suggests a range of 0.5 to 1.0.

$$\theta_p = \frac{\delta_p}{h_w - \ell_p/2}$$

$$= \frac{0.91}{59 - 6.88}$$

$$= 0.017 \text{ radian}$$

$$\phi_p = \frac{\theta_p}{\ell_p} \qquad \text{(Eq. 2.1.9)}$$

$$= \frac{0.017}{13.75}$$

$$= 0.00127 \text{ rad/in.}$$

$$\phi_u = \phi_p + \phi_y$$

$$= 0.00127 + 0.000126$$

$$= 0.0014 \text{ rad/in.}$$

Observe that this corresponds to a curvature ductility factor (ϕ_u/ϕ_y) of 11. Tests performed at the University of California at Berkeley in the 1970s[2.38] developed curvature ductilities of 13. Concrete strains were reportedly quite a bit less (0.004 in./in.) than the strains attained in either the Zhang[2.36] or Wallace[2.39, 2.40] tests a discussion of which follows (Figure 2.4.4*b*). Accordingly, a universal quantification of curvature ductility is not rationally established. It seems far more rational to identify deformation limit states from estimates of induced concrete strains.

As displacements increase, the neutral axis depth (c) will become smaller as the tension steel continues to yield and the concrete approaches its strain limit state.

$$a = \frac{P}{0.85 f_c' b}$$

$$= \frac{130}{0.85(5)(3.9)}$$

$$= 7.8 \text{ in.}$$

$$c = \frac{a}{\beta_1}$$

$$= \frac{7.8}{0.8}$$

$$= 9.8 \text{ in.}$$

and the probable strain in the concrete is

$$\varepsilon_{cp} = \phi_p c$$

$$= 0.00127(9.8)$$

$$= 0.0124 \text{ in./in.}$$

$$\begin{aligned}\varepsilon_{cu} &= \varepsilon_{cy} + \varepsilon_{cp} \\ &= 0.0015 + 0.0124 \\ &= 0.0139 \text{ in./in.}\end{aligned}$$

Conclusion: The strain limit state associated with incipient shell spalling in thin walls subjected to significant levels of axial load can be expected to be reached at concrete strain states on the order of 0.01 in./in. Failure of this test specimen occurred when the shell spalled. Observe that the concrete strain limit state associated with shell spalling in thin walls is consistent with that observed in frame beams (Section 2.1.1.2).

Confinement was provided in the boundary elements but the confining pressure was somewhat low:

$$\begin{aligned}f_\ell &= \frac{A_{sh} f_{yh}}{s h_c} \\ &= \frac{2(0.043)(44)}{3(3.25)} \\ &= 0.38 \text{ ksi}\end{aligned}$$

This is 87% of the ACI objective, $0.09 f'_c$ (see Eq. 1.2.7):

$$\begin{aligned}\frac{f_\ell}{0.09 f'_c} &= \frac{0.39}{0.09(5)} \\ &= 0.87\end{aligned}$$

Confinement of the toe region is considered prudent when concrete strains are expected to exceed 0.003 in./in. It is not likely, however, that confinement specifically improves the strain limit state in the concrete, for it was the strain in the unconfined shell that established the strength limit state for the wall whose limit state is described in Figures 2.4.2*a* and 2.4.2*b*. The confinement provided undoubtedly served to extend the deformation limit state of the wall, but the primary contribution of the ties lies in preventing the overstrained (in tension) compression reinforcing from buckling. Observe (Figure 2.4.2*a*) that once the deformation limit state had been reached, the strength of the wall was entirely lost. In thin walls it is unlikely that a large enough region of confined concrete can be effectively activated, as it is in a beam or a beam column. This is because the increased strength in the confined core will not compensate for the strength lost as the shell spalls.

$$\begin{aligned}f'_{cc} &= f'_c + 4.1 f_\ell \qquad \text{(Eq. 1.2.1)} \\ &= 5 + 4.1(0.34) \\ &= 6.4 \text{ ksi}\end{aligned}$$

$$C_{\text{core}} = f'_{cc} A_{c,\text{core}}$$
$$= 6.4(3.5)(3.75)$$
$$= 84 \text{ kips} \ll 130 \text{ kips}$$

Thomsen and Wallace studied the behavior of tall thin bearing walls axially loaded to $0.1 f'_c A_g$.[2.39] The behavior of Wallace specimen RW2[2.40] confirms the preceding analytical work, albeit for specimens supporting lesser axial loads. Figure 2.4.3 describes the test specimen. The response of this wall to cyclic displacements is described in Figure 2.4.4*a*. A lateral drift ratio in excess of 2% was attained. The wall was braced laterally at the top and, as a consequence, had the following stability characteristics:

$$\frac{h_w}{t_w} = \frac{144}{4}$$
$$= 36$$
$$\frac{kh_w}{r} = \frac{1.0(144)}{0.3(4)}$$
$$= 120$$

The displacement of the wall using usual procedures is

$$I_g = \frac{t_w \ell_w^3}{12}$$
$$= \frac{4(48)^3}{12}$$
$$= 36{,}860 \text{ in.}^4$$
$$\Delta_w = \frac{H h_w^3}{3EI_g}$$
$$= \frac{35(144)^3}{3(3600)(36{,}860)}$$
$$= 0.26 \text{ in.}$$

As in the case of the Zhang wall, the use of an effective moment of inertia (I_g) of 50% of the gross moment of inertia reasonably idealizes system behavior. It is clear that higher levels of axial load (Figure 2.4.2*a*) maintain the stiffness to higher levels of drift, and this is a logical conclusion.

Strain profiles were also measured (Figure 2.4.4*c*), and reported limit states are consistent with those analytically developed for the Zhang[2.36] test specimen described in Figure 2.4.1 ($\varepsilon_{cu} \cong 0.014$ in./in.).

Moment curvature relationships were also developed for wall RW2, and they are shown in Figure 2.4.4*d*. Essential to the development of deformation-based design

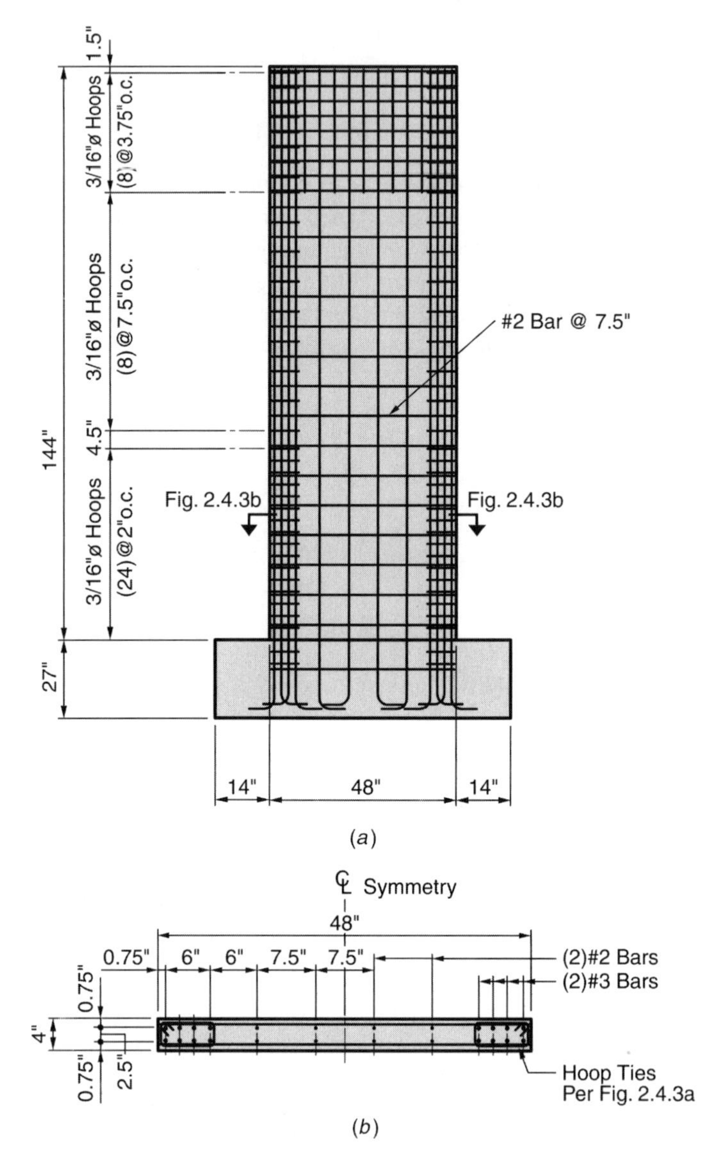

Figure 2.4.3 (*a*) Geometry and reinforcing details—RW2. (*b*) Plan view of section indicated.

procedures is a reasonably reliable moment curvature relationship, which must easily define idealized values of curvature (ϕ_{yi}) and moment (M_{yi}).

Figure 2.4.4*d* plots the analytical, as developed by Wallace, as well as the experimentally determined moment curvature relationships and compares them to the proposed idealization. Behavior of the test wall as analytically developed is presented in Table 2.4.2

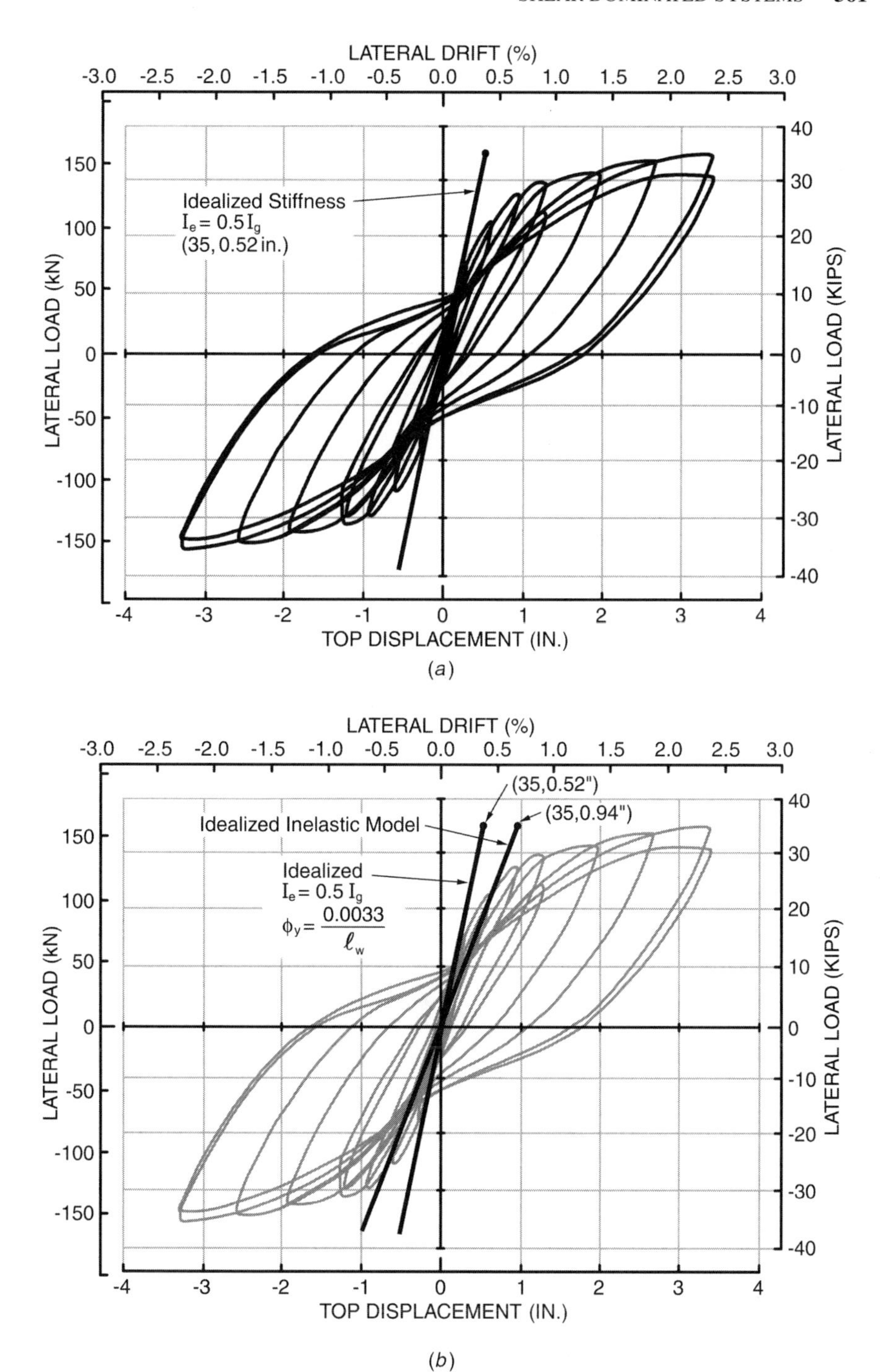

Figure 2.4.4 (*a*) RW2: Lateral load versus top displacement.[2.10] (*b*) RW2: Analytical versus measured force displacement relationships.[2.10]

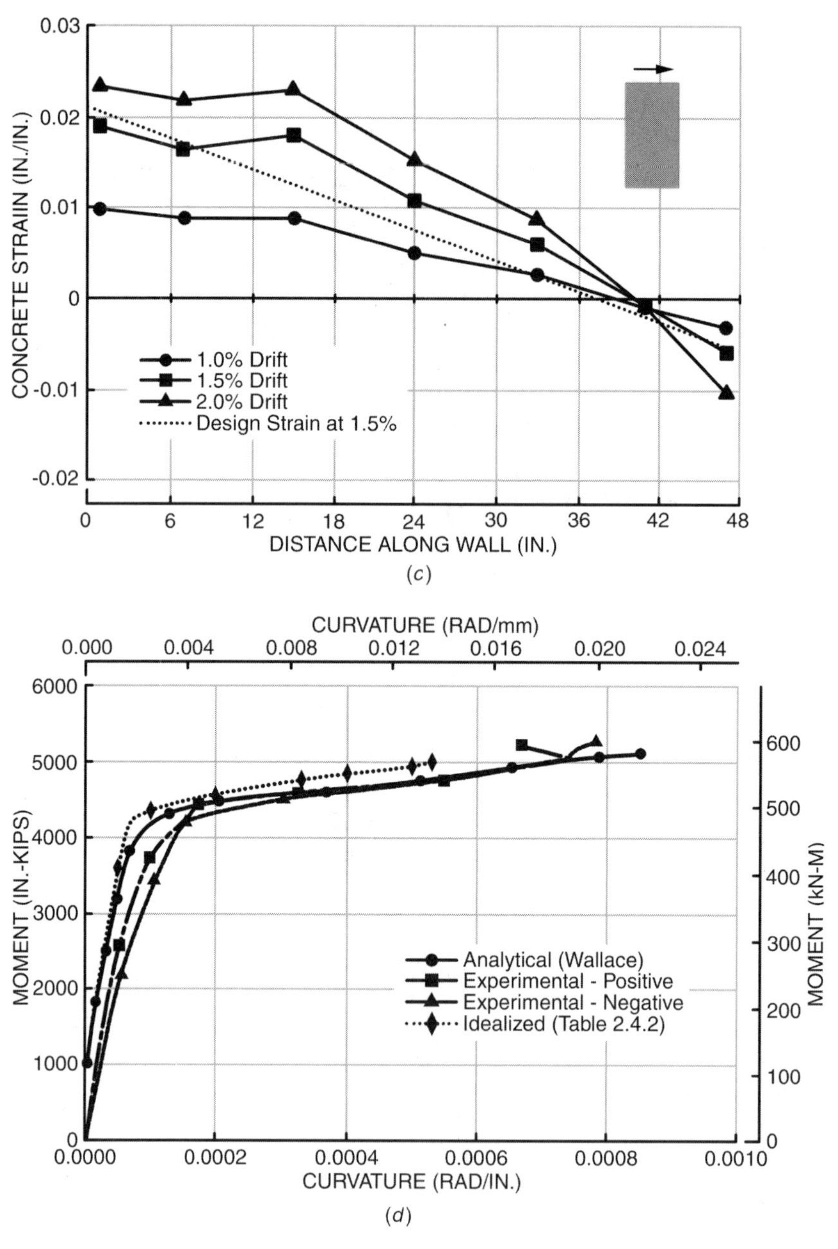

Figure 2.4.4 (*Continued*) (*c*) RW2: Analytical versus measured concrete strain profiles (positive displacement).[2.10] (*d*) Analytical and experimental versus idealized moment curvature response.

TABLE 2.4.2 Biaxial Behavior of the Wall Section Described in Figure 2.4.3*b* ($P = 96$ kips, $f_c' = 4$ ksi, $f_y = 60$ ksi)

ε_c (in./in.)	ε_s (in./in.)	N.A. (in.)	M_u (ft-kips)	ϕ (rad/in.)
0.0003	–0.00003	40.6	97.8	0.0000074
0.0004	–0.00014	32.6	138.8	0.0000123
0.0005	–0.00033	26.5	160.4	0.0000189
0.0006	–0.00056	22.9	183.5	0.0000263
0.0007	–0.00081	20.5	207.1	0.0000342
0.0008	–0.00106	19.0	231.4	0.0000422
0.0009	–0.00133	17.8	255.9	0.0000505
0.0010	–0.00160	16.9	279.1	0.0000591
0.0011	–0.00185	16.4	303.5	0.0000670
0.0012	–0.00218	15.7	320.9	0.0000766
0.0013	–0.00256	14.8	327.2	0.0000877
0.0014	–0.00307	13.8	327.3	0.0001010
0.0015	–0.00359	13.0	331.3	0.0001150
0.0016	–0.00407	12.4	332.1	0.0001290
0.0017	–0.00455	12.0	333.3	0.0001420
0.0018	–0.00515	11.4	333.7	0.0001580
0.0019	–0.00570	11.0	337.4	0.0001730
0.0020	–0.00623	10.7	337.6	0.0001870
0.0025	–0.00898	9.6	346.3	0.0002610
0.0030	–0.01162	9.0	358.3	0.0003320
0.0035	–0.01435	8.6	364.3	0.0004050
0.0040	–0.01673	8.5	368.9	0.0004700
0.0045	–0.01791	8.8	370.7	0.0005090
0.0050	–0.01845	9.4	372.6	0.0005320
0.0060	–0.01710	11.4	346.9	0.0005240
0.0070	–0.01065	17.5	280.8	0.0004010

Design idealizations proposed are

$$\begin{aligned}\phi_{yi} &= \frac{\varepsilon_{sy}}{d - kd} \\ &= \frac{0.002}{0.625d} \qquad \left(\cong \frac{0.0033}{\ell_w}\right) \\ &= \frac{0.002}{29.4} \qquad \text{(Steel first yield)} \\ &= 0.00007 \text{ rad/in.}\end{aligned}$$

This type of idealization has also been proposed by Paulay.[2.37] A behavior idealization may be developed from this pseudoelastic beam model:

$$
\begin{aligned}
M_{yi} &= \phi_{yi} E I_e \\
&= 0.00007(3600)(18{,}432) \qquad (I_e = 0.5 I_g) \\
&= 4644 \text{ in.-kips}
\end{aligned}
$$

This reasonably idealizes stiffness characteristics but does not predict strength because strength will depend on the level of axial load and the amount of reinforcing in the wall.

The strength of a wall is most readily estimated using procedures similar to those developed in Sections 2.1.1.1 and Section 2.2.1.1:

$$
M_n = A_s f_y (d - d') + P\left(\frac{\ell_w}{2} - \frac{a}{2}\right) \tag{2.4.1}
$$

where

$$
\begin{aligned}
a &= \frac{P}{0.85 f_c' t_w} \\
&= \frac{96}{0.85(4)4} \\
&= 7 \text{ in.} \\
M_n &= 8(0.11)(60)(44.25 - 3.75) + 96(24 - 3.5) \\
&= 4100 \text{ in.-kips}
\end{aligned}
$$

which, when idealized as proposed by Eq. 2.1.2c, becomes

$$
\begin{aligned}
M_{yi} &= \lambda_o M_n \qquad \text{(see Eq. 2.1.2c)} \\
&= 1.25(4100) \\
&= 5125 \text{ in.-kips} \qquad (P_{yi} = 35.6 \text{ kips})
\end{aligned}
$$

This reasonably predicts an elastic/perfectly plastic behavior model (see Figure 2.4.4*a*).

Comment: Observe that the use of an overstrength factor of somewhat less than 1.25 might be more appropriate. Clearly the use of a universal overstrength factor is questionable (see Section 2.1.1.1).

The deflections described in Figure 2.4.4*a* are easily predicted. The curvature diagram of Figure 2.4.5*a* is developed from Table 2.4.2, as is the idealization described on Figure 2.4.4*d*. The curvatures described in Figure 2.4.5*b* are developed from the beam design model (Figure 2.1.10). Both suggest a wall displacement of 1 in., and this is consistent with the experimental and analytical predictions (see Figure 2.4.4*a*).

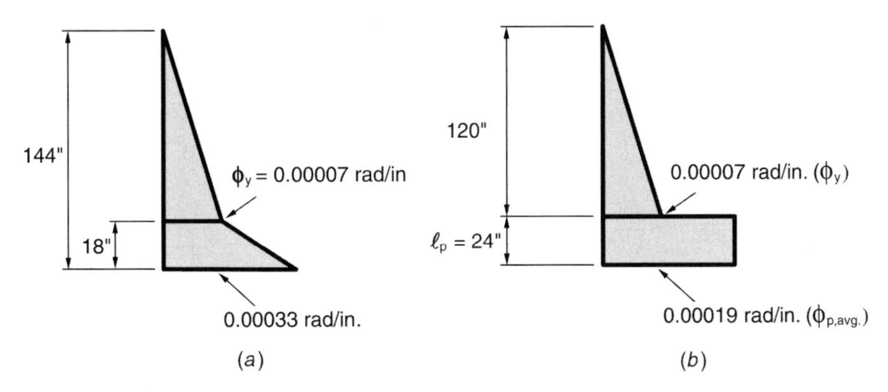

Figure 2.4.5 (*a*) Curvature model from Table 2.4.2 at $\varepsilon_c = 0.003$. (*b*) Curvature diagram—design model ($\ell_p = \ell_w/2$).

If we adopt the following as being sufficiently accurate for our purposes:

$$\phi_y = 0.00007 \text{ rad/in.} \qquad (\varepsilon_s \cong \varepsilon_y)$$

$$\phi_u = 0.00038 \text{ rad/in.} \qquad (\varepsilon_c = 0.003 \text{ in./in., Table 4.2})$$

along with the curvature diagrams of Figure 2.4.5, then we conclude that, given the curvature model,

$$\delta_y = \phi_y \frac{(h_w - \ell_p)^2}{3} + \phi_y \ell_p \left(h_w - \frac{\ell_p}{2}\right) + \frac{\phi_p \ell_p}{2}\left(h_w - \frac{\ell_p}{3}\right) \qquad \text{(see Figure 2.4.5}a\text{)}$$

$$= \frac{0.00007(144-18)^2}{3} + 0.00007(18)(144-9) + \frac{0.00026}{2}(18)(144-6)$$

$$= 0.37 + 0.17 + 0.32$$

$$= 0.86 \text{ in.}$$

And, based on the design model,

$$\Delta_y = \frac{\phi_y (h_w - \ell_p)^2}{3} + \phi_{p,\text{avg}} \ell_p \left(h_w - \frac{\ell_p}{2}\right) \qquad \text{(see Figure 2.4.5}b\text{)}$$

$$= \frac{0.00007(120)^2}{3} + 0.00019(24)(132)$$

$$= 0.34 + 0.6$$

$$= 0.94 \text{ in.}$$

Observe (Figure 2.4.4*b*) that a yield deflection of 0.94 in. provides a better idealization of system stiffness than $I_e = 0.5 I_g$.

Shear reinforcing provided in RW2 consists of #2 bars at 7.5 in. on center on each face.

$$v_s = \frac{A_{sh} f_y h}{bs}$$

$$= \frac{2(0.05)(60)}{4(7.5)}$$

$$= 0.2 \text{ ksi}$$

The shear sustained by the wall was

$$v = \frac{H}{bd}$$

$$= \frac{35}{4(44.25)}$$

$$= 0.197 \text{ ksi}$$

Accordingly, the shear strength provided by the shear reinforcing acting alone exceeded the demand.

Stability limit states for the entire wall will not be approached in most wall designs. Buckling failures of concern are those that are likely to occur in the toe region where high axial strains and significant shear loads are anticipated. The primary causative action will, however, be the buckling of the flexural reinforcement as postyield tensile strains are overcome on reverse load cycles. This topic was discussed by Paulay,[2.37] who proposed stability limit states. Paulay assumes that lateral support for the toe region, if the wall has been significantly cracked in tension during reverse cycle loading, must be provided by the uncompressed region located at the neutral axis, a distance c from the outermost compression fiber.

Paulay[2.37] intuitively defines the minimum wall thickness in terms of the neutral axis depth c. The recommended critical wall thickness ($t_{w,\mathrm{cr}}$) proposed by Paulay[2.37] is

$$t_{w,\mathrm{cr}} = \frac{c}{4}$$

Paulay and Priestley[2.5] (Section 5.4.3(c)) suggest a conservative critical wall thickness for doubly reinforced walls in the plastic hinge region of

$$t_{w,\mathrm{cr}} = 0.017\ell_w \sqrt{\mu_\phi}$$

where ℓ_w need not be assumed to be larger than 1.6 times the story height.

The absence of the axial load in this relationship acknowledges the fact that the limit state is tensile strain based. Paulay and Priestley[2.5] also recommend that the thickness of the wall in the plastic hinge region be not less than the story height in the hinge region divided by 16. These limit states appear to be overly conservative, and this is demonstrated by Wallace in tests performed on T sections (Figure 2.4.6).

The treatment of c/t_w as a critical stability limit state, when stability along the longitudinal axis is not reasonably considered, is consistent with elastic plate theory as applied to a simply supported plate element. The critical buckling ratio in steel plate elements that are expected to be stable in the postyield range is developed from elastic theory.[2.4, Sec. 3.4 (b)] Were we to apply these stability parameters to concrete, we would conclude that the critical relationship between c and t_w would be eight ($c/t_w \leq 8$). Tests performed on steel beams[2.14, Sec. 2.2(d)] suggest, however, that this relationship is nonconservative.

Zhang's tests[2.36] allow some insight into this critical condition. The previously described test specimen SW9 (Figure 2.4.1) had a neutral axis depth (c) of 11.8 in. Hence c/t_w was 3 (11.8/3.9). Specimen SW9 did not buckle. Buckling was reported as the limit state for Zhang specimen CW8, which was axially loaded to $0.35 f_c' A_g$. The neutral axis depth for Zhang specimen CW8 would have been about 15 in.

$$
\begin{aligned}
C_c &= P + (\lambda_o - 1) A_s f_y \\
&= 0.35(5)(3.9)(27.5) + 0.25(1.96)(59) \\
&= 216 \text{ kips} \\
a &= \frac{C_c}{0.85 f_c' b} \\
&= \frac{216}{0.85(5)3.9} \\
&= 13 \text{ in.} \\
c &= \frac{a}{\beta_1} \\
&= \frac{13}{0.8} \\
&= 16.3 \text{ in.} \\
\frac{c}{t_w} &= \frac{16.3}{3.9} \\
&= 4.2
\end{aligned}
$$

This apparent confirmation of the stability limit state proposed by Paulay[2.37] is clearly refuted by Wallace's tests on flanged walls.[2.40]

The presence of intersecting walls can significantly alter the behavior of a tall thin wall. This is because the neutral axis of the wall will shift as the intersecting wall is subjected alternately to tension and compression. Two conditions need to be considered carefully: the stability of the thin wall and the potential level of overstrength.

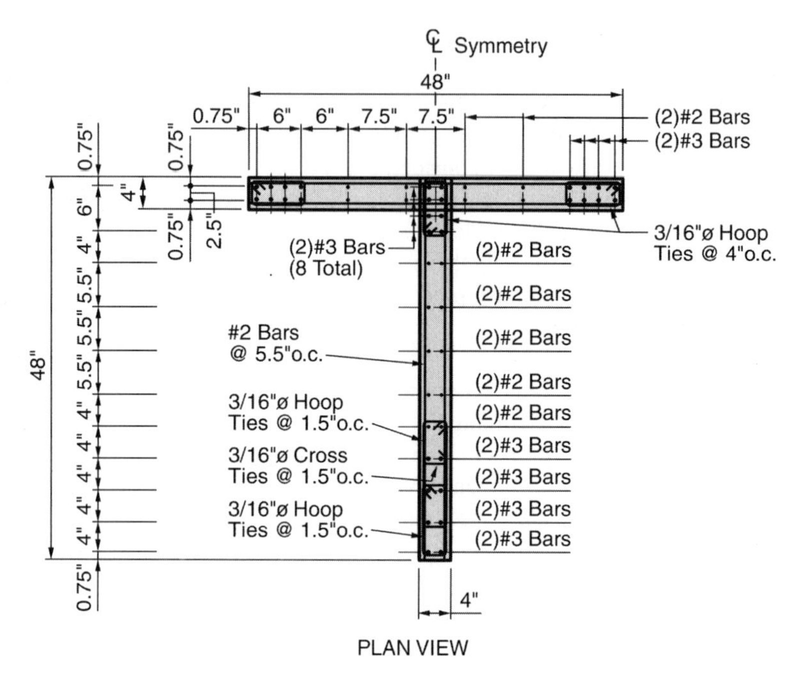

Figure 2.4.6 Geometry and reinforcing details—TW2.[2.40]

The T section described in Figure 2.4.6 was studied by Wallace. This wall was modeled after the rectangular wall RW2 (Figure 2.4.3). The level of applied axial stress was the same as that applied to RW2, and it was placed at the middle of the stem and the flange of the T. The behavior of the test specimen is described in Figure 2.4.7. The neutral axis depth required to attain the tensile strength in the flange is developed as follows:

$$P = 150 \text{ kips} \qquad (0.1 f_c' A_c)$$

$$T_y = A_s f_y$$

$$= 3.0(60) \qquad \text{(Twenty-four \#3 bars + eight \#2 bars)}$$

$$= 180 \text{ kips}$$

$$C_s = A_s' f_y$$

$$= 0.88(60) \qquad \text{(Eight \#3 bars)}$$

$$= 53 \text{ kips}$$

$$C_c = P + T_y - C_s$$

$$= 150 + 180 - 53$$

$$= 277 \text{ kips}$$

$$a = \frac{C_c}{0.85 f_c' t_w}$$

$$= \frac{277}{0.85(4)(4)}$$

$$= 20.4 \text{ in.}$$

$$c = \frac{a}{\beta_1}$$

$$= \frac{20.4}{0.85}$$

$$= 24.0 \text{ in.}$$

The shear imposed on the wall is 80 kips (Figure 2.4.7). This brings into question several limit states suggested by the testing of rectangular sections.

The critical wall thickness was

$$t_{w,\mathrm{cr}} = \frac{c}{4}$$

$$= \frac{24.0}{4}$$

$$= 6.0 \text{ in.} > t_w \qquad \left(\frac{c}{t_w} = 6.0\right)$$

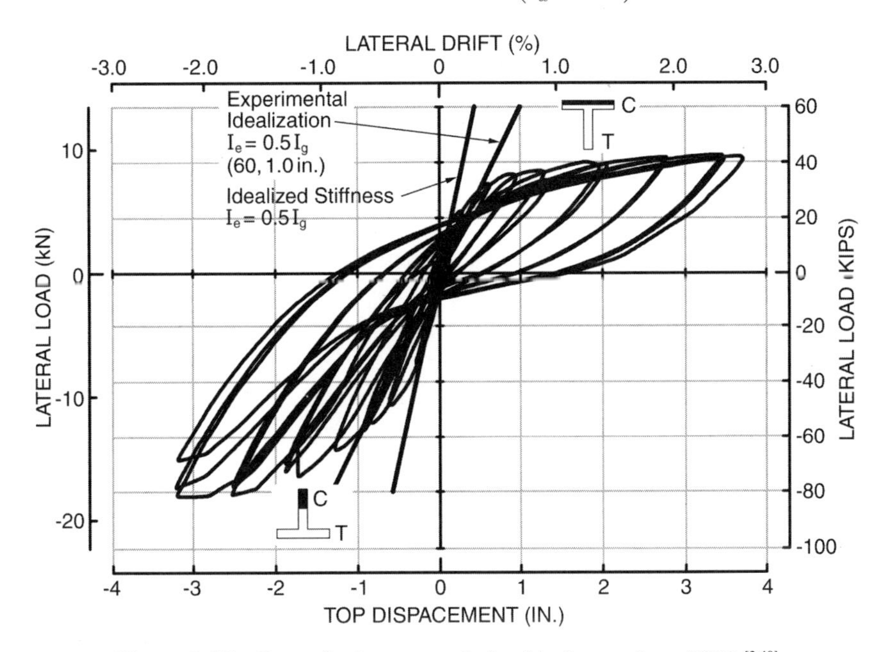

Figure 2.4.7 Force displacement relationship for specimen TW2.[2.40]

This thickness of the stem was significantly less than the Paulay-suggested stability limit state.

The strain imposed on the steel in the stem was on the order of $40\varepsilon_y$, and this suggests that buckling of the reinforcing bars in the stem on reverse cycle loading should have promoted instability in this region.

The shear stress imposed on the web was

$$v = \frac{80{,}000}{44(4)}$$

$$= 454 \text{ psi} \qquad 6.5\sqrt{f_c'}$$

and this is consistent with what in most cases is proposed by me as an objective limit state. It should be noted that the compressive force imposed on the concrete in the stem was on the order of 260 kips. The friction force required to transfer the shear would accordingly be quite low at 30%. Further, the depth of the activated compressive stress block would have allowed this shear transfer to occur at a significant distance from the toe.

The confinement provided was

$$f_\ell = \frac{A_{sh} F_y}{h_c s}$$

$$= \frac{5(0.0275)(60)}{16(1.5)}$$

$$= 0.35 \text{ ksi}$$

and this would have activated a confined compressive strength in the concrete (f_{cc}') of

$$f_{cc}' = f_c' + 4.1 f_\ell \qquad \text{(Eq. 1.2.1)}$$

$$= 4 + 4.1(0.35)$$

$$= 5.4 \text{ ksi}$$

The strength of the confined core was

$$C_{\text{core}} = f_{cc}' h_c b_{\text{core}}$$

$$= 5.4(16)(3.25)$$

$$= 260.8 \text{ kips}$$

The strength provided by the core (C_{core}) is essentially equivalent to the compression force that would have been imposed on it (± 277 kips), and this begins to explain how a shear load of 80 kips could be sustained by the specimen to drifts of 2%. Figure 2.4.7 describes curvatures projected for various wall configurations and loading directions.

Clearly confinement in deep compression zones ($c > 4t_w$) improves behavior. It is also apparent that curvature ductilities of 10 are not overly optimistic even when c/t_w ratios exceed 4.

The ultimate curvature and concrete strain states that the specimen of Figure 2.4.6 could have sustained would be estimated using previously developed relationships.

$$\delta_u = 3.25 \text{ in.} \qquad \text{(see Figure 2.4.7)}$$

$$\phi_y = 0.00009 \text{ rad/in.} \qquad \text{(Computer analysis)}$$

$$\delta_y = \frac{\phi_y h_w^2}{3} \qquad \text{(see Eq. 2.1.7b)}$$

$$= \frac{0.00009(144)^2}{3}$$

$$= 0.62 \text{ in.}$$

This corresponds to a displacement ductility factor μ_Δ of 5.25 (3.25/0.62).

$$\delta_p = \delta_u - \delta_y$$

$$= 3.25 - 0.62$$

$$= 2.63 \text{ in.}$$

$$\theta_p = \frac{\delta_p}{h_w - \ell_p/2}$$

$$= \frac{2.63}{144 - 12}$$

$$= 0.02 \text{ radian}$$

$$\phi_p = \frac{\theta_p}{\ell_p}$$

$$= \frac{0.02}{24}$$

$$= 0.00083 \text{ rad/in.}$$

$$\phi_u = \phi_p + \phi_y$$

$$= 0.00083 + 0.00009$$

$$= 0.00092 \text{ rad/in.}$$

$$\mu_\phi = \frac{\phi_u}{\phi_y}$$

$$= \frac{0.00092}{0.00009}$$

$$= 10.2$$

$$\varepsilon_{cu} = \phi_u c \qquad (c \cong 16.3 \text{ in.})$$
$$= 0.00092(16.3)$$
$$= 0.0115 \text{ in./in.}$$

Observe that these limit states are consistent with those developed from rectangular test specimens.

Shear reinforcing provided in TW2 (Figure 2.4.6) consisted of #2 bars at 5.5 in. on center on each face.

$$v_s = \frac{A_{sh} f_y}{bs}$$
$$= \frac{2(0.05)(60)}{4(5.5)}$$
$$= 0.273 \text{ ksi}$$

The shear load sustained by this wall when the flange was in tension was 80 kips (Figure 2.4.7).

$$v = \frac{H}{bd}$$
$$= \frac{80}{4(46.0)}$$
$$= 0.435 \text{ ksi}$$

and this exceeds $v_s + v_c$ (400 psi).

A shear fan radiating from the confined core at an angle of 65° would engage seventeen pairs of #2 bars. Thus

$$V_p = 17(0.1)(60)$$
$$= 102 \text{ kips}$$

Again, as was the case for Zhang's wall, it appears that the plastic strength of the web reinforcing provided the shear load path.

The idealization of the stiffness of a T wall section is complicated by the fact that it is a function of the direction of the applied load (see Figure 2.4.7). The moment of inertia of the T wall section described in Figure 2.4.6 is

$$I_g = 81{,}500 \text{ in.}^4$$
$$I_{\text{cr}} \cong 33{,}000 \text{ in.}^4 \qquad \text{(Stem in compression)}$$
$$I_{\text{cr}} = 10{,}000 \text{ in.}^4 \qquad \text{(Stem in tension)}$$

For a shear load of 60 kips the deflection of the T wall (Figure 2.4.6) would be

$$\delta_{w,60} = 0.41 \text{ in.} \qquad (I_e = 0.5I_g)$$

$$\delta_{w,60} = 0.5 \text{ in.} \qquad (I_e = 33{,}000 \text{ in.}^4)$$

$$\delta_{w,60} = 1.0 \text{ in.} \qquad (I_e = 0.2I_g)$$

These behavior idealizations are shown on Figure 2.4.7. Clearly the designer must adopt the idealization that provides the best possible insight into the design decision at hand.

Conclusions: Recent code changes have encouraged the use of strain-based design procedures. In spite of this positive direction, little guidance is provided in critical areas dealing with element stability and shear strength, required or provided. Capacity-based design is not required, and shear limit states appear to be too liberal in most cases. Limits are not placed on the amount of shear reinforcing, and this can lead to compression field failures in shear walls. Much work remains if a design procedures is to be created that will cost-effectively and safely create shear wall braced structures. Conclusions supported by the experiments reviewed herein and those of other scholars that can assist in the development of a rational design procedure include:

(1) Strain distribution in tall walls is reasonably approximated by the plane section model used in concrete beam theory.
(2) Curvature models can be used to predict deflections provided beam behavior represents a rational analytical model (see Section 3.3, Diaphragms).
(3) Commonly used analytical tools can be used in conjunction with estimates of effective stiffness to predict strain states in tall walls.
(4) Strain limit states can be used to predict deformation limit states of the wall. Concrete strains of 0.01 in./in. are reasonably used to describe the deformation limit state in a concrete wall.
(5) Ductility will be available in a shear wall provided the real axial load in the plastic hinge region is limited. This is consistent with the theory developed for a beam column (Section 2.2). An objective real axial design stress of $0.25f'_c$ is suggested but higher levels of load may satisfy strain based design objectives.
(6) Curvature ductilities (μ_ϕ) of at least 10 are attainable in tall walls.
(7) Displacement ductilities of 5 are attainable but curvature ductilities and concrete strain states should be used to confirm the attainability of displacement ductilities.
(8) Shear strength limit states should be developed from shear load paths. Conventional procedures used to reinforce beams in the plastic hinge region ($v_c = 0$) are not appropriate for the design of shear walls provided the wall is not in tension. Shear reinforcing provided should exceed the capacity-based demand, but this can be accomplished using the plastic truss analogy discussed in Chapter 1.

Shear reinforcement in excess of $8\sqrt{f_c'}$ should not be provided unless the stress induced in the concrete is carefully reviewed and found to be within acceptable limits.

(9) Confinement should be provided when concrete strains become large, but its effectiveness must be evaluated.

(10) Flexural reinforcement in the plastic hinge regions of walls must be supported laterally so as to inhibit its loss of stability on reverse cycle loading. $8d_b$ is the suggested bracing interval.

(11) Wall stability in the hinge region should be carefully evaluated. Conservative limit states are

$$t_{w,\text{cr}} = \frac{c}{4}$$

$$t_{w,\text{min}} = \frac{h_x}{16}$$

where h_x is the clear story height. When the c/t_w limit of 4 is exceeded, the compression region should be confined.

(12) Outside the plastic hinge region the critical thickness of the wall may be as high as $h_x/36$.

(13) The plastic hinge region may conservatively be assumed to be $0.5\ell_w$.

(14) Lateral support for primary flexural reinforcing should be provided in the plastic hinge region. Longitudinal spacing should be at least $8d_b$. It does not seem necessary to support every bar (Figure 2.4.3*b*) unless tensile strains are very large (Figure 2.4.6).

(15) Elastic behavior may be idealized through the use of an effective moment of inertia (I_e) equal to 50% of the gross moment of inertia (I_g).

(16) Orthogonal walls built integrally with a shear wall must be considered in the design. The effective extension of these appendages, especially those in tension, could be as much as 25% of the wall height unless shear lag becomes a factor.

2.4.1.2 Design Procedures In Section 1.1 the basis for modern seismic designs was briefly reviewed. The definition of a design forcing function evolved directly from an understanding of the response of a single-degree-of-freedom system to ground motion. Shear wall behavior is easily patterned after that of a single-degree-of-freedom system because postyield deformation is concentrated at the base of the wall. As a consequence, the design of the shear wall as an element can proceed directly from the establishment of an objective behavior in the plastic hinge region.

The design process starts with the adoption and appropriate development of a ground motion criterion. Once the objective ground motion has been quantified, the designer must elect to follow either a force- or displacement-based approach. The force- or strength-based approach places the region of design interest below an objective strength and displacement of F_o and Δ_o, as described in Figure 1.1.1. The displacement-based approach has as its focus the region described by F_{yi} and Δ_u in Figure 1.1.1.

Force-based design approaches usually convert the single-degree-of-freedom response spectrum (Figure 1.1.5) to what is, in effect, a base shear spectrum (Figure 1.1.8). This conversion is accomplished subtly, as discussed in the SEAOC Blue Book.[2.41] Tall shear walls are multi-degree-of-freedom systems that are usually modeled by uniformly distributing the mass, thereby creating a system that contains an infinite number of nodes or degrees of freedom. The direct conversion from a single-degree-of-freedom acceleration response spectrum to a base shear spectrum appropriate for the design of multi-degree-of-freedom systems requires the inclusion of various modes, and this is accomplished by considering the modal participation factors (Γ) and mode shape effects. This is discussed in more detail in Reference 2.4, Sec. 4.7.3(c). Suffice it for now to realize that the conversion from single-degree-of-freedom spectral acceleration to base shear coefficient (V/W) is for a frame-braced or lumped mass multi-degree-of-freedom system asymptotic to 0.75.[2.4, Eq. 4.7.14] For systems of distributed mass and stiffness, the participation factor and mode shape are slightly different, but a conversion factor of about 0.7, as currently used,[2.9, 2.41] is reasonable enough for design purposes, especially given the many other uncertainties involved in quantifying objective levels of system strength. This convenient correlation between the single-degree-of-freedom response spectrum and the base shear spectrum allows the designer to work with either a force- or displacement-based design approach, provided we properly account for the conversion factor.

Force-based approaches recognize this distinction between a single-degree-of-freedom spectrum and a base shear response spectrum. The base shear spectrum for a multi-degree-of-freedom system could be developed from a single-degree-of-freedom spectrum by multiplying the single-degree-of-freedom spectral acceleration ordinates by a factor of 0.7. Hence the correlation

$$\frac{V}{W} = 0.7S_a \tag{2.4.2a}$$

The 70% factor is in current developments[2.9, 2.41] then dropped to account for the participation of higher modes, and this is appropriate if the interest is in base shear only. The resultant most common conversion ($V/W = S_a$) is also convenient, for it accurately describes the base shear for a single-story structure. Unfortunately, the inclusion of the multimode effect (1.0/0.7) introduces a level of conservatism in the design of a shear wall braced structure, for the strength parameter of interest in shear wall design is the moment at the base of the wall, and the moment is not sensitive to the inclusion of higher modes. This multimode conservatism inherent in the base shear spectrum is inappropriately applied to displacement calculations developed from design base shears because displacement response is dominated by a response in the fundamental or first mode. The increase in displacement (1.0/0.7) is currently compensated for by a reversal of the 0.7 factor in the development of the drift limit state from the design level displacement (Δ_S), which corresponds to Δ_o in Figure 1.1.1 ($\Delta_M = 0.7R\Delta_S$).[2.9]

Displacement-based approaches can be developed from the single-degree-of-freedom response spectrum or from the base shear spectrum, since they are equivalent. The system displacement developed from a base shear spectrum is appropriate for a single-degree-of-freedom system only. As the number of nodes increases, so does the ratio of peak drift (Δ_u) to spectral displacement (S_d). The amplification associated with the inclusion of multinode effects is limited, however, and for lumped mass systems asymptotic to 1.5. Thus the probable displacement of the roof will be up to 50% greater than the spectral displacement. For systems of distributed mass (shear walls), the ratio of peak displacement to spectral displacement can be as much as 1.6.

One thing should be clear from this brief overview—quite a bit of "creative accounting" must be utilized if a single-degree-of-freedom spectrum is to be the design basis for structures of all types. This creative accounting is usually conservatively done, but unfortunately, this introduced conservatism does not tend to produce optimal designs either in terms of cost or performance, for shear wall braced buildings.

Seismic ground motions fall into three basic behavior regions as discussed in Section 1.1.3: constant spectral acceleration, velocity, and displacement.[2.4, Sec. 4.7 & 2.42, Sec. 6.9] The velocity defining coefficient (C_v) used by building codes is directly related to a constant spectral velocity (S_v) by

$$S_v = \frac{386.4C_v}{2\pi} \text{ (in./sec)}$$

$$= 61.5C_v \text{ in./sec} \tag{2.4.2b}$$

In seismic zone 4, for example, the velocity defining coefficient (C_v) for a building founded on a very dense soil (S_C) within 2 km of a B fault, would be 0.896.[2.9] The corresponding spectral velocity would be 55 in./sec.

The base shear in the acceleration constant range is defined in the UBC[2.9] as

$$V = \frac{2.5C_a I}{R} W \qquad \text{(see Ref. 2.9, Eq. 30-5)}$$

The boundary between the velocity constant and acceleration constant region can be determined by equating the UBC's Eq. 30-4,

$$V = \frac{C_v I}{RT} W \qquad \text{(see Ref. 2.9, Eq. 30-4)}$$

to its Eq. 30-5. The boundary defining period is

$$T = \frac{C_v}{2.5C_a} \tag{2.4.3}$$

For the site conditions described the boundary period is

$$T = \frac{0.896}{2.5(0.52)}$$

$$= 0.69 \text{ second}$$

and this is a little high for most ground motions, the usual generic boundary being 0.5 to 0.6 seconds.

The boundary between the velocity constant region and the displacement constant region could be developed from

$$V = 0.11 C_a I W \qquad \text{(see Ref. 2.9, Eq. 30-6)}$$

Accordingly, the boundary defining period is

$$T = \frac{C_v}{0.11 C_a R} \tag{2.4.4}$$

which for the site described and a system based reduction factor (R) of 4.5 for (bearing) shear wall structures becomes

$$T = \frac{0.896}{0.11(0.52)(4.5)}$$

$$= 3.48 \text{ seconds}$$

and this too is consistent with the generic boundary of 3.5 seconds (or more).

Comment: C_a is used in code[2.9]based design procedures to identify the spectral acceleration in the constant acceleration range (Fig. 1.1.5). The development of Eq. 30-6 and 30-7 of Ref. 2.9 strive to establish a minimum base shear for use in force based design procedures. Accordingly, the boundary between velocity and displacement control regions is more realistically set at 4 seconds.

This information is easily used in the design of a tall thin bearing wall because most tall bearing wall braced buildings will fall into the spectral velocity constant region. Further, a maximum building displacement is easily identified by inserting the upperbound period ($T = 3$ seconds) into a design that is based on S_v, the objective spectral velocity (Eq. 2.4.6).

$$\omega_n = \frac{2\pi}{T} \tag{2.4.5}$$

$$= \frac{2(3.14)}{4.0}$$

$$= 1.57 \text{ rad/sec}$$

$$S_v = 61.5 C_v \qquad \text{(Eq. 2.4.2b)}$$

$$= 61.5(0.896)$$

$$= 55 \text{ in./sec}$$

$$S_d = \frac{S_v}{\omega_n} \tag{2.4.6}$$

$$= \frac{55}{1.57}$$

$$= 35 \text{ in.}$$

$$\Delta_{\max} = \Gamma S_d \tag{2.4.7}$$

$$= 1.6(35)$$

$$= 56 \text{ in.}$$

At a building period of about 4 seconds, we have reached the peak displacement, and any softening of the structure would not lead to higher levels of displacement demand.

The identification of this displacement limit state is especially relevant to the design of high-rise shear wall braced buildings. In essence it means that the postyield rotation imposed on the plastic hinge region will be reduced with increasing building height since θ_p is inversely proportional to h_w (see Eq. 2.1.8).

The maximum code design base shear ($V_{\max}$) is developed directly from UBC's Eq. 30-5. For our previously defined site and structure, this becomes

$$V_{\max} = \frac{2.5C_a I}{R} W \qquad \text{(see Ref. 2.9, Eq. 30-5)} \tag{2.4.8}$$

$$= \frac{2.5(0.52)(1.0)}{4.5} W$$

$$= 0.29W$$

Comments

- The importance factor (I) is assumed to be one in this case. The importance factor is common to all code-based equations and will not affect the velocity response region boundaries.
- The use of representative participation factors, as well as the suggested relationship between single-degree-of-freedom spectral values (S_a, S_v, S_d) and spectral-based constants used to define base shear (C_v) will not be as accurate for buildings that fall into the acceleration constant region. Accordingly, displacement-based design procedures are not suggested for buildings whose probable periods are less than 0.6 second unless energy equivalence is considered in the development of spectral displacement estimates.[2.42, Sec. 7.11] See also Chapter 1.

Proceed now to formulate a design approach for a tall shear wall based on the presumption that the proposed structure will fall into the velocity constant response region.

The period of the structure must first be estimated. Should the designer opt to use a force- or strength-based approach, the estimated period should be less than the probable period, thereby creating a conservative estimate of the design force.[2.9, Eq. 30-4] If a displacement-based approach is adopted, the estimated period should exceed the probable period in order to identify a conservative prediction of building drift.

The usual design case for tall thin walls is one where the designer knows where the shear walls are to be located and the question becomes one of determining how strong the wall should be. When quantity and length are an issue, a balance between strength and stiffness should be sought, and this is discussed in Section 3.1.1. Start by projecting a building period that is neither high nor low. This may be done by determining the period based on the flexural stiffness of the proposed shear wall. In order to do this we must identify the mass tributary to the wall and reduce this to a uniform load (w) applied to a cantilever beam—the shear wall. This procedure is discussed in more detail in Section 2.4.1.4.

$$w = \frac{W}{h_w} \tag{2.4.9}$$

where W quantifies the tributary weight (mass) laterally supported by the shear wall expressed in kips, and h_w is the height of the wall expressed in feet. The deflection of the shear wall subjected to this uniform load (w) is

$$\Delta_f = \frac{wh_w^4}{8EI} \tag{2.4.10}$$

It is most convenient to use units of feet because the desired quantifications are typically large enough in feet; hence E, for say 5000-psi concrete, becomes 576,000 kips/ft^2.

Example 1 If, for example, we were considering using a 12-in. thick wall (t_w) 20 ft deep (ℓ_w) to brace a ten-story building whose overall height (h_w) was 100 ft and whose tributary weight (mass) was 150 kips/floor, then the hypothetical deflection Δ_f of this wall loaded by its tributary weight (mass) (W) would be calculated as follows:

$$\begin{aligned} w &= \frac{W}{h_w} \\ &= \frac{10(150)}{100} \\ &= 15 \text{ kips/ft} \end{aligned}$$

$$I_g = \frac{bd^3}{12}$$

$$= \frac{1(20)^3}{12}$$

$$= 667 \text{ ft}^4$$

and presuming an effective stiffness of 50%,

$$I_e = 333 \text{ ft}^4$$

The deflection of this wall subjected to its tributary mass (W) would be

$$\Delta_f = \frac{wh_w^4}{8EI_e} \qquad \text{(Eq. 2.4.10)}$$

$$= \frac{15(100)^4}{8(576{,}000)333}$$

$$= 0.98 \text{ ft}$$

and the period associated with flexural stiffness only may be determined from

$$\boxed{T_f = 0.89(\Delta_f)^{0.5}} \qquad \text{(see Ref. 2.4, Sec. 4.7.4(b))} \qquad (2.4.11)$$

where Δ_f is expressed in feet.

Comment: Shear deformation could also contribute to the period of the structure. This is demonstrated in Section 2.4.1.4 (Eq. 2.4.25).

$$T_f = 0.89(0.98)^{0.5} \qquad \text{(see Eq. 2.4.11)}$$

$$= 0.88 \text{ second}$$

Compare this with the relationship proposed by the UBC:

$$T = C_t(h_n)^{3/4} \qquad (2.4.12)$$

where $C_t = 0.02$

Comment: Following the UBC, C_t may also be developed from the effective area (square feet) of the shear wall at the base of the structure (A_c) in the following manner:

$$A_c = \sum A_e \left[0.2 + \left(\frac{D_e}{h_n}\right)^2\right] \qquad (2.4.13)$$

where D_e is the length of the shear wall (ℓ_w) at the base in feet, and h_n is the height of the building. It is assumed that four such walls brace the proposed building.

$$A_c = 4(1)(20)\left[0.2 + \left(\frac{20}{100}\right)^2\right]$$

$$= 19.2 \text{ ft}^2$$

and

$$C_t = \frac{0.1}{\sqrt{A_c}} \tag{2.4.14a}$$

$$= \frac{0.1}{\sqrt{19.2}}$$

$$= 0.023$$

Accordingly,

$$T = C_t h_w^{3/4} \qquad \text{(Eq. 2.4.12)}$$

$$T = 0.023(100)^{3/4}$$

$$= 0.73 \text{ second}$$

In this case the period as developed through the use of an emulation of a Rayleigh-based approach (Eq. 2.4.11), extended to include the shear deformation component, may be used for either the force-based or deformation-based approach, because a 30% amplification of the prescriptive period (Eq. 2.4.12) is allowed provided it does not exceed the period determined using a Rayleigh procedure.

Proceed to calculate the strength required of the structure as determined using the strength- or force-based approach and the deflection-based approach, assuming a fundamental period of 0.9 second and a design spectral velocity (S_v) of 55 in./sec^2.

The strength-based approach following the UBC[2.9] proceeds as follows:

$$V = \frac{C_v I}{RT} W \qquad \text{(see Ref. 2.9, Eq. 30-4)}$$

$$= \frac{0.896}{4.5(0.9)} W$$

$$= 0.22W$$

The displacement- or strength-based approach as developed directly from the response spectrum proceeds as follows:

$$\omega_n = \frac{2\pi}{T} \qquad \text{(Eq. 2.4.5)}$$

$$= \frac{2(3.14)}{0.9}$$

$$= 6.98 \text{ rad/sec}$$

$$\Delta_{\max} = \frac{S_v}{\omega_n}\Gamma \tag{2.4.14b}$$

$$= \frac{55}{6.98}(1.6)$$

$$= 12.6 \text{ in.}$$

$$S_a = \omega_n S_v \tag{2.4.14c}$$

$$= 6.98(55)$$

$$= 384 \text{ in./sec}^2$$

$$V = \frac{S_a}{Rg}W \tag{2.4.14d}$$

$$= \frac{0.7(384)}{4.5(386.4)}$$

$$= 0.154W$$

The spectral-based shear is, as expected, 70% of that derived through the use of UBC,[2.9] Eq. 30-4. This, as previously discussed, is because it does not include higher mode contributions, a subject that is discussed in more detail in Chapter 3. For now, accept the fact that, in terms of identifying a flexural strength objective, higher modes have virtually no impact (see Eq. 3.1.8).

The ultimate design moment, developed from the strength-based criterion assuming a triangular distribution of code lateral loads, is

$$M_u = V(0.67)h_w \tag{2.4.14e}$$

$$= 0.22W(0.67)h_w$$

$$= 0.22(1500)(0.67)(100)$$

$$= 22{,}100 \text{ ft-kips}$$

$$M_n = \frac{M_u}{\phi} \tag{2.4.14f}$$

$$= \frac{22{,}000}{0.9}$$

$$= 24{,}600 \text{ ft-kips}$$

If the tributary dead load from a vertical load perspective is 50 kips/floor, the flexural strength of the wall may be determined from

$$M_n = P_D\left(\frac{\ell_w}{2} - \frac{a}{2}\right) + A_s f_y(d - d') \qquad \text{(Eq. 2.4.1)}$$

If P_D is factored ($0.9P_D$), or otherwise altered to conservatively reflect the probable dead load ($P_{D,\text{eff}}$) so as to determine the maximum quantity of required steel, then

$$\begin{aligned} a &= \frac{0.9P_D}{0.85f_c'b} \\ &= \frac{0.9(500)}{0.85(5)(12)} \\ &= 8.8\text{ in.} \end{aligned}$$

Comment: Observe that c (11 in.) is less than $4t_w$. Accordingly, the stability of the compression flange should not be a problem.

$$\begin{aligned} A_s f_y(d - d') &= M_n - 0.9P_D\left(\frac{\ell_w}{2} - \frac{a}{2}\right) \qquad \text{(see Eq. 2.4.1)} \quad (2.4.15) \\ &= 24{,}600 - \frac{0.9(500)}{12}\left(\frac{240}{2} - \frac{8.8}{2}\right) \\ &= 24{,}600 - 4335 \\ &= 20{,}265\text{ ft-kips} \\ A_s &= \frac{20{,}265}{60(17)} \\ &= 19.86\text{ in.}^2 \qquad (\rho = 0.72\%) \end{aligned}$$

The displacement-based approach proposed by Wallace[2.31] assumes a curvature loosely described in Figure 2.4.8 (see also Figures 2.1.10 and 2.1.11). Wallace[2.31] suggests that the objective displacement may be expressed in terms of the yield curvature (ϕ_y) and the ultimate curvature (ϕ_u) as

$$\Delta_u = \Delta_y + \Delta_p \qquad (2.4.16)$$

$$= \frac{11}{40}\phi_y h_w^2 + \frac{1}{2}\left(\phi_u - \phi_y\right) h_w \ell_w \qquad (2.4.17)$$

I prefer to use a relationship similar to those developed in Section 2.1.1.2 (see Figure 2.1.11) that, for $\ell_p = 0.5\ell_w$ and $\phi_p = (\mu_\phi - 1)\phi_y$, becomes

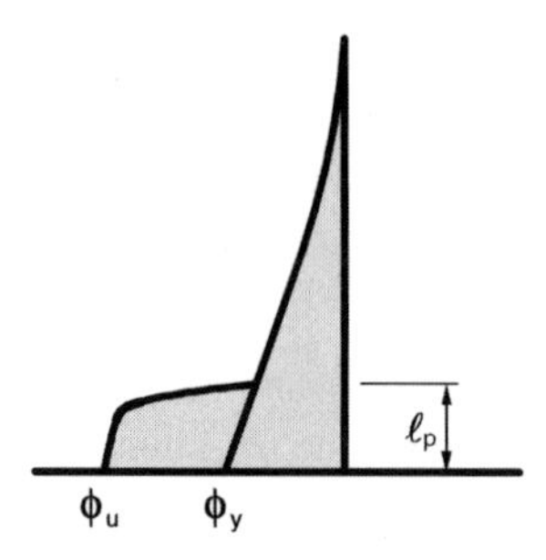

Figure 2.4.8 Relationship between global and local deformations: curvature.[2.31]

$$\Delta_u = \frac{11}{40}\phi_y h_w^2 + (\mu_\phi - 1)\,\phi_y 0.5\ell_w\left(h_w - \frac{\ell_p}{2}\right) \qquad (2.4.18)$$

This allows a direct solution for ϕ_y:

$$\phi_y = \frac{\Delta_u}{0.275h_w^2 + (\mu_\phi - 1)\ell_w(0.5h_w - 0.125\ell_w)} \qquad (2.4.19)$$

For our example structure, if μ_ϕ, the curvature ductility, is taken as 9 (see Section 2.4.1.1), then

$$\phi_y = \frac{12.6}{0.275(1200)^2 + 8(240)(600 - 30)}$$
$$= 0.0000084 \text{ rad/in.}$$

The curvature of the wall at steel first yield is reasonably estimated from our knowledge of the strain profile and the strain in the steel.

$$\phi_y = \frac{0.0033}{\ell_w} \qquad \text{(see Eq. 2.1.7a)}$$

For this example the yield curvature would be

$$\phi_y = \frac{0.0033}{240}$$
$$= 0.0000135 \text{ rad/in.}$$

Conclusion: This wall should meet our curvature objectives.

Comment: Methodologies for predicting the displacement of a concrete beam, column, or wall are many. They all involve the introduction of simplifying assumptions.

Accordingly, accuracy is not a word reasonably used in describing any approach. The best one can hope for is reasonable agreement between analysis and testing. The approach proposed by Wallace, which is included in Eq. 2.4.17, is undoubtedly based on an assumed curvature in the wall that is not linear (see Figure 2.4.8), and this is certainly reasonable. Curvature in a beam or beam column is far from constant. This characteristic is discussed by MacGregor[2.3, Sec. 9–3] and also by Paulay and Priestley.[2.5, Sec. 3.5.3] In Section 2.1.1.2 an idealization of yield curvature was extrapolated from flexural test data (Eq. 2.1.6). An effective moment of inertia (I_e) was developed so as to allow the use of elastic-based relationships. In effect measured curvatures and associated moments were used to develop idealized moments of inertia (see Figure 2.1.3).

Establishing a strength objective based on a displacement objective coupled with system displacement ductilities can be accomplished in a variety of ways. Alternatives are explored in Chapter 3. For now, realize that a strength objective can be developed from a displacement objective provided the system displacement ductility factor is consistent with performance based objectives related to strain states.

In our review and analysis of the (experimental) wall described in Figure 2.4.3*a*, an overstrength factor was developed. The attempt was to differentiate between real strength (experimentally based) and design strength (M_n). This overstrength factor is appropriately used in displacement-based design procedures since Δ_u is the objective ultimate displacement (refer also to Figure 2.1.2). Accordingly, the design flexural strength is developed as follows:

$$V = 0.154W \qquad \text{(Eq. 2.4.14d)}$$

$$\begin{aligned} M_{yi} &= 0.154W(0.67h_w) \\ &= 0.103Wh_w \\ &= 0.103(1500)(100) \\ &= 15{,}450 \text{ ft-kips} \end{aligned}$$

$$M_n = \frac{M_{yi}}{\lambda_o} \qquad \text{(Eq. 2.1.2c)}$$

$$\frac{15{,}450}{1.25}$$

$$= 12{,}360 \text{ ft-kips}$$

and this is about half of the nominal strength suggested by strength- or force-based design procedures (see Eq. 2.4.14f).

A curvature ductility objective may also be used to develop a flexural strength objective. This is accomplished through the use of Eq. 2.4.16 and 2.4.17 in conjunction with Eq. 2.4.19. In the example,

$$\phi_y = 0.0000084 \text{ rad/in.}$$

$$\Delta_y = \frac{11}{40}\phi_y h_w^2 \qquad \text{(Eq. 2.4.17)}$$

$$= 0.275(0.0000084)(1200)^2$$

$$= 3.3 \text{ in.}$$

$$\mu_\phi = \frac{\Delta_u}{\Delta_y}$$

$$= \frac{12.6}{3.3}$$

$$= 3.82$$

and the determination of V, M_u, and M_n follows most simply from Eq. 2.4.14d, using a curvature reduction or ductility factor of R_ϕ equal to 3.82.

If the designer elects to follow a curvature-based ductility factor of 9 (Eq. 2.4.18), the nominal strength objective becomes

$$M_{n,\mu_\phi} = M_{n,\mu_\Delta}\left(\frac{R_\Delta}{R_\phi}\right)$$

$$= 12{,}360\left(\frac{4.5}{3.82}\right)$$

$$= 14{,}640 \text{ ft-kips}$$

Comment: The overstrength factor must be developed with care. In the analysis of the experimental wall of Figure 2.4.3 one could develop an overstrength factor of 1.23 based on the nominal strength as developed from Eq. 2.4.1 ($F_n = 33$ kips). However, the nominal strength as predicted by the computer analysis (Table 2.4.2—$\varepsilon_c = 0.003$ in./in.) would suggest the use of an overstrength factor of 1.03.

Concluding Remarks: The fact that the displacement-based approach requires only 50% (12,360/24,600) of the flexural capacity required by the strength-based alternative will trouble many raised on code-based procedures, but, as previously discussed, this is attributable to built-in conservatisms in the code. The postyield behavior of systems designed to various strength levels are studied in Chapter 4. Inelastic time history analyses will demonstrate that the increase in postyield demand in most cases will be adequately compensated for by the associated increase in system ductility.

2.4.1.3 Design Summary

Step 1: Determine the Design Approach or Basis. Alternative design approaches are strength- or deflection-based. See Section 1.1.3 for a discussion of these alternatives and Section 2.4.1.2 for their application to shear walls.

Step 2: Determine the Required Thickness of the Shear Wall. Outside the plastic hinge region:

$$t_{w,\min} = \frac{P_D}{0.25 f_c' \ell_w} \tag{2.4.20}$$

or

$$t_{w,\min} < \frac{h_x}{25} \tag{2.4.21a}$$

where h_x in this case is the distance vertically between points of lateral support.

Comment: These should not typically govern.

Within the plastic hinge region ($\ell_p = 0.5\ell_w$):

$$t_{w,\min} = \sqrt{\frac{P_D}{3.4 f_c' \beta_1}} \tag{2.4.21b}$$

Comment: This will limit the neutral axis depth (c) to $4t_w$. The inclusion of confined core will alter this relationship. The selected wall thickness need not be greater than $h_x/16$ provided strain objectives are attained.

Step 3: Determine the Probable Fundamental Period of the Building.

$$T_f = 0.13 \left(\frac{h_w}{\ell_w}\right)^2 \sqrt{\frac{w\ell_w}{E t_w}} \qquad \text{(see Eq. 2.4.10, 2.4.11)} \tag{2.4.22}$$

where E is expressed in kips per square inch, t_w, ℓ_w, and h_w in feet and w is the tributary weight (mass), expressed in kips per foot, that must be supported laterally by the shear wall (W/h_w).

Comment: If a strength-based approach is to be used, underestimate the period; if a displacement-based approach is to be used, overestimate the period. A good estimate for design purposes may be developed from the flexural stiffness of the wall, assuming that the effective stiffness is half of the uncracked stiffness ($I_e = 0.5I_g$).

Step 4:

(a) *Determine the Spectral Velocity.* If spectral velocity is defined by the building code (C_v), convert it to the spectral velocity for a single-degree-of-freedom system:

$$S_v = 61.5C_v \qquad \text{(Eq. 2.4.2b)}$$

(b) *Insure That the Structure Falls within the Velocity Constant Region:*

$$T > \frac{C_v}{2.5C_a} \qquad \text{(Eq. 2.4.3)}$$

$$T_{\max} < 4 \text{ seconds} \qquad \text{(see Eq. 2.4.4)}$$

(c) *Quantify the Angular Frequency* (ω_n)*:*

$$\omega_n = \frac{2\pi}{T} \qquad \text{(Eq. 2.4.5)}$$

(d) *Determine the Spectral Displacement* (S_d)*:*

$$S_d = \frac{S_v}{\omega_n} \qquad \text{(Eq. 2.4.6)}$$

(e) *Determine the Probable Ultimate Displacement of the Roof* (Δ_u)*:*

$$\begin{aligned} \Delta_u &= \Gamma S_d \\ &\cong 1.6S_d \qquad \text{(conservatively)} \qquad \text{(Eq. 2.4.7)} \end{aligned}$$

(f) *Determine the Spectral Acceleration:*

$$S_a = \omega_n S_v \qquad \text{(Eq. 2.4.14c)}$$

Step 5: Determine the Yield Moment Required to Attain the Objective Displacement Given an Objective Displacement Ductility Factor (μ_Δ) *or an Objective Curvature Ductility Factor* (μ_ϕ).

(a) *Establish a Reduction Factor* (R). When a displacement ductility objective is adopted,

$$R = \mu_\Delta$$

or, when a curvature ductility is proposed, proceed as follows

$$\phi_y = \frac{0.0033}{\ell_w} \qquad \text{(see Eq. 2.1.7a)}$$

and

$$R = \frac{\Delta_u}{0.275\phi_y h_w^2} \qquad \text{(see Eq. 2.4.17)}$$

(b) Determine the Base Shear (V)

$$V = \frac{0.7 S_a W}{Rg} \qquad \text{(see Eq. 2.4.14d)}$$

(c) Quantify the Idealized Moment (M_{yi})

$$M_{yi} = 0.67 V h_w$$

(d) Quantify the Objective Nominal Flexural Strength (M_n)

$$M_n = \frac{M_{yi}}{\lambda_o} \qquad \text{(see Eq. 2.1.2c)}$$

Step 6: Determine the Flexural Reinforcement Required to Attain This Objective Strength.

$$M_n = (P_D)\left(\frac{h}{2} - \frac{a}{2}\right) + A_s f_y (d - d') \qquad \text{(see Eq. 2.4.1)}$$

$$A_s \cong \frac{(M_n/\phi) - P_{D,\text{eff}}(h/2 - a/2)}{f_y(d - d')} \qquad (2.4.23)$$

Step 7: Check to Insure That Wall Thickness Objectives Have Been Attained.

$$t_w > 0.25c$$

Step 8: Select a Wall Reinforcing Program.

Step 9: Estimate the Ultimate Strain in the Concrete.

Step 10: Reinforce the Wall for Shear.

Step 11: Detail the Wall for Shear and Confine the Concrete in the Hinge Region, If Appropriate.

2.4.1.4 Design Example

Example 2 A residential building is to be braced by thin shear walls. The wall shown in Figure 2.4.9 is presumed to be a functional necessity. The tributary axial load/floor including the wall weight is 60 kips. The tributary weight (mass) per floor is 140 kips/floor. Use a displacement-based approach and assume that the wall is capable of attaining a curvature ductility factor of 9. Use the displacement based

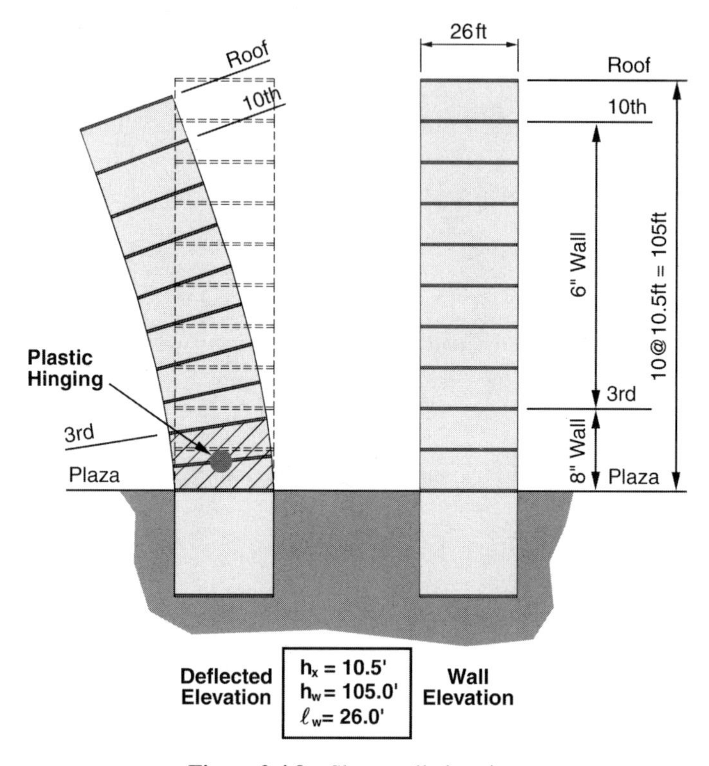

Figure 2.4.9 Shear wall elevation.

procedures developed in Section 2.4.1.2. Use an effective stiffness of 50% of the provided stiffness. The spectral velocity (S_v) for the site is 44 in./sec.

Step 1: Determine the Design Basis. A displacement design basis has been selected.

Step 2: Determine the Thickness of the Wall. Outside the hinge region,

$$\frac{P_D}{f'_c A_g} < 0.25 \qquad \text{or} \qquad \frac{h_x}{t_w} < 25 \qquad \text{(Eq. 2.4.20, 2.4.21a)}$$

$$\frac{P_D}{f'_c A_g} = \frac{10(60)}{5(26)(t_w)(12)} \leq 0.25$$

$$t_{w,\min} = \frac{600}{5(26)(12)(0.25)}$$

$$= 1.53 \text{ in.}$$

Conclusion: Outside the hinge region a 6-in. wall may be used ($h_x/t_w = 20$).

Hinge region—$0.5\ell_w$ (13 ft):

$$t_{w,\min} = \sqrt{\frac{P_D}{3.4 f'_c \beta_1}} \qquad \text{(Eq. 2.4.21b)}$$

$$= \sqrt{\frac{600}{3.4(5)(0.8)}}$$

$$= 6.64 \text{ in.}$$

Conclusion: An 8-in. thick wall should be used in the plastic hinge region. The proposed wall should be 8-in. thick in Levels 1 and 2 and 6-in. thick above Level 2.

Step 3: Determine the Period of the Structure. The procedure used was developed in Section 2.4.1.2 as well as several pre-computer treatments of structural dynamics (see also Ref. 2.4, Sec. 4.7.4(b)). The period of the building is developed from estimates of the building deflection were it laterally loaded with the weight of its tributary mass.

The tributary weight per foot (w) is

$$w = \frac{W}{h_w}$$

$$= \frac{1400}{105}$$

$$= 13.3 \text{ kips/ft}$$

$$I_g = \frac{0.5(26)^3}{12}$$

$$= 732 \text{ ft}^4$$

$$I_e = 0.5 I_g$$

$$= 366 \text{ ft}^4$$

$$EI = 4{,}000 I_e$$

$$= 3600(144)(366)$$

$$= 190{,}000{,}000 \text{ kip-ft}^2$$

Comment: The effective stiffness of the wall is somewhat subjectively assigned in this case ($0.5I_g$). First principles or the conjugate beam may be used to better gauge the fundamental period. Several considerations suggest the adopted analytical simplification. First, since we have adopted a displacement-based approach, the use of a constant minimum thickness will be conservative. Second, the wall will be stiffer in the upper levels where cracking is not expected to be as severe as it is in the plastic hinge region. Accordingly, the use of I_g based on 6 in. throughout is not unreasonable. Recall also that this is conceptual design.

The deflection of the wall under its self-weight is

$$\Delta_f = \frac{wh_w^4}{8EI_e} \qquad \text{(Eq. 2.4.10)}$$

$$= \frac{13.3(105)^4}{8(190{,}000{,}000)}$$

$$= 1.06 \text{ ft}$$

$$T_f = 0.89(\Delta_f)^{0.5} \qquad \text{(Eq. 2.4.11)}$$

where Δ_f is the flexural component of roof deflection expressed in feet. Thus

$$T_f = 0.89(1.06)^{0.5}$$

$$= 0.91 \text{ second}$$

The contribution of shear deformation based on an effective shear stiffness of 50% may also be included.

$$\Delta_s = \frac{wh_w^2}{1.67A_cG} \qquad (2.4.24)$$

$$= \frac{13.3(105)^2}{1.67(6)(26)(12)720}$$

$$= 0.07 \text{ ft}$$

$$T_s = (0.288)(\Delta_s)^{0.5} \qquad (2.4.25)$$

where Δ_s is the shear component of roof deflection expressed in feet. Thus

$$T_s = (0.288)(0.06)^{0.5}$$

$$= 0.07 \text{ second}$$

$$T^2 = T_f^2 + T_s^2 \qquad (2.4.26)$$

$$= (0.91)^2 + (0.07)^2$$

$$T = 0.91 \text{ second}$$

This methodology slightly underestimates the period of the building because, in the averaging process, one-half of the mass at the roof has been neglected. It may be compensated for easily:

$$\Delta'_f = \frac{W'(h_w)^3}{3(EI_e)}$$

$$= \frac{70(105)^3}{3(190{,}000{,}000)}$$

$$= 0.14 \text{ ft}$$

$$T_f = 0.89\left(\Delta_f + \Delta'_f\right)^{0.5}$$

$$= 0.89(1.20)^{0.5}$$

$$= 0.96 \text{ second}$$

$$T^2 = T_f^2 + T_s^2$$

$$= (0.96)^2 + (0.07)^2$$

$$T = 0.96 \text{ second}$$

Step 4: Determine the Spectral Displacement and Acceleration. The spectral velocity (S_v) for the site is 44 in./sec. Thus

$$\omega_n = \frac{2\pi}{T}$$

$$= \frac{2(3.14)}{0.96}$$

$$= 6.54 \text{ rad/sec}$$

$$S_d = \frac{S_v}{\omega_n} \qquad \text{(Eq. 2.4.6)}$$

$$= \frac{44}{6.54}$$

$$= 6.7 \text{ in.}$$

$$S_a = \omega_n S_v$$

$$= 6.54(44)$$

$$= 288 \text{ in./sec}^2$$

and the expected ultimate deflection at the roof is

$$\Delta_{10} = \Gamma S_d \qquad \text{(Eq. 2.4.7)}$$

$$= 1.6(6.7)$$

$$= 10.8 \text{ in.}$$

Step 5: Determine the Yield Moment Required to Attain This Displacement with a Curvature Ductility in the Hinge Region of 9. (See Section 2.4.1.1).

$$\phi_y = \frac{\Delta_u}{0.275h_w^2 + (\mu_\phi - 1)\,\ell_w\,(0.5h_w - 0.125\ell_w)} \qquad \text{(Eq. 2.4.19)}$$

$$= \frac{10.8}{0.275(1260)^2 + (9-1)(312)(630-39)}$$

$$= 0.0000058 \text{ rad/in.}$$

$$R = \frac{\Delta_u}{0.275\phi_y h_w^2} \qquad \text{(see Eq. 2.4.17)}$$

$$= \frac{10.8}{0.275(0.0000058)(1260)^2}$$

$$= 4.27$$

$$V = \frac{0.7S_a W}{Rg} \qquad \text{(see Eq. 2.4.14d)}$$

$$= \frac{0.7(288)(1400)}{4.27(386.4)}$$

$$= 171 \text{ kips}$$

Comment: 0.7 is a spectral reconciliation factor (see Eq. 2.4.2a) appropriately used in the design of tall walls when the forcing function is presented in the form of a base shear spectrum (see Eq. 3.1.8).

$$M_{yi} = 0.67Vh_w$$

$$= 0.67(171)(105)$$

$$= 12{,}034 \text{ ft-kips}$$

$$M_n = \frac{M_{yi}}{\lambda_o} \qquad \text{(see Eq. 2.1.2c)}$$

$$= \frac{12{,}034}{1.25}$$

$$= 9630 \text{ ft-kips} \qquad (0.066Wh_w)$$

Step 6: Determine the Flexural Reinforcing Required to Attain This Objective Strength.

$$a = \frac{P_{D,\text{eff}}}{0.85f_c' t_w}$$

$$= \frac{600}{0.85(5)(8)}$$

$$= 17.6 \text{ in.}$$

$$A_s \cong \frac{(M_n/\phi) - P_{D,\text{eff}}(\ell_w/2 - a/2)}{f_y(d - d')} \qquad \text{(Eq. 2.4.23)}$$

$$= \frac{10{,}700 - 600(13 - 0.66)}{(60)(26 - 3)}$$

$$\cong 2.4 \text{ in.}^2$$

Step 7: Check to Insure That Thickness Requirements Have Been Met.

$$c = \frac{a}{\beta_1}$$

$$= \frac{17.6}{0.8}$$

$$= 22.0 \text{ in.}$$

$$\frac{c}{t_w} = \frac{22.0}{8}$$

$$= 2.75 < 4.0 \qquad \text{(OK)}$$

Step 8: Select a Wall Reinforcing Program. The steel area suggested by the preceding analysis falls below minimums intended to provide a strength greater than the cracking moment. A minimum reinforcement ratio of 0.25% is suggested. Eight #7 bars, arranged in the pattern described in Figure 2.4.10*a*, should be evaluated to insure that the program meets our strength and strain objectives.

The flexural strength (M_u) at a concrete strain of 0.003 in./in. is 17,467 ft-kips (see Table 2.4.3), and this corresponds to a provided overstrength of about 100%. The estimated depth to the neutral axis at M_n is 30.16 in. ($c/t_w = 3.77$).

Comment: Clearly it is important to not overreinforce a tall thin wall.

Step 9: Estimate the Strain in the Concrete. The elastic strain can be approximated from the probable depth to the neutral axis (kd) and the strain in the steel at yield.

$$\varepsilon_{sy} = 0.00207 \text{ in./in.}$$

$$kd = 0.375d$$

$$= 108 \text{ in.}$$

$$\varepsilon_{cy} = \left(\frac{kd}{d - kd}\right)\varepsilon_{sy}$$

$$= \frac{108}{(288 - 108)}(0.00207)$$

$$= 0.0012 \text{ in./in.}$$

or, alternatively, from a biaxial computer run (Table 2.4.3) as 0.0007 in./in. Observe that the yield moment at first yield is in excess of 14,840 ft-kips; hence our objective strength has been exceeded.

An estimate of the yield curvature may be developed from the elastic strain states $(d - kd = 0.67d)$:

$$\phi_y = \frac{\varepsilon_{sy}}{0.67\ell_w} \qquad \text{(Eq. 2.1.7)}$$

$$= \frac{0.00207}{0.67\ell_w}$$

$$= \frac{0.0033}{\ell_w} \qquad (2.4.27)$$

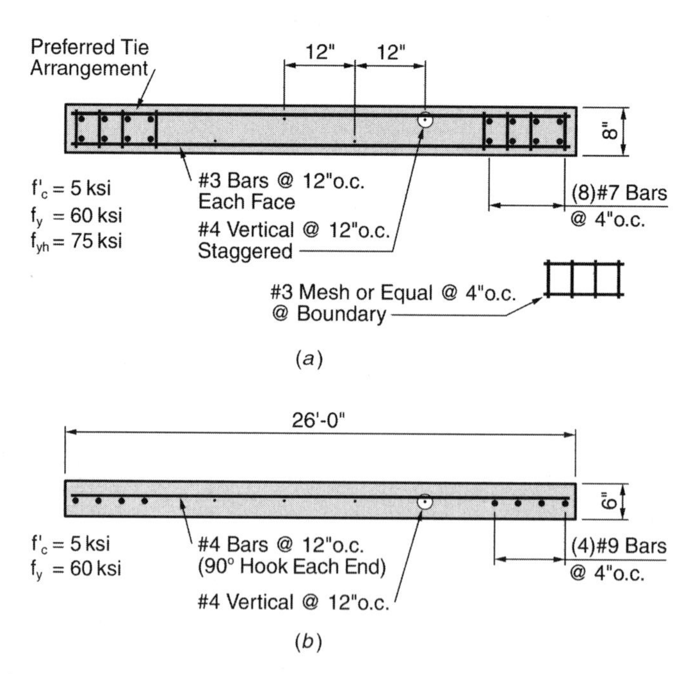

Figure 2.4.10 (*a*) Wall reinforcement for Levels 1 and 2 (hinge region). (*b*) Wall reinforcing program Levels 3 through 10.

$$= \frac{0.0033}{312}$$

$$= 0.0000105 \text{ rad/in.} > 0.0000058 \text{ rad/in.}$$

Observe that the curvature associated with first yield of the steel is 0.000009 rad/in. (Table 2.4.3). Accordingly our curvature ductility demand will be much less than the design objective of 9.

TABLE 2.4.3 Moment-Curvature Diagram of the Rectangular Section Described in Figure 2.4.10*a*

ε_c (in./in.)	ε_s (in./in.)	N.A. (in.)	M_u (ft-kips)	ϕ (rad/in.)
0.00010	0.00001	294.94	2965.0	0.0000003
0.00020	0.00020	154.78	5934.0	0.0000013
0.00030	0.00052	113.34	7873.4	0.0000026
0.00040	0.00091	94.45	9741.6	0.0000042
0.00050	0.00134	84.40	11,639.3	0.0000059
0.00060	0.00178	78.30	13,548.5	0.0000077
0.00070	0.00227	73.13	15,157.7	0.0000096
0.00080	0.00294	66.42	15,732.5	0.0000120
0.00090	0.00369	60.94	16,140.6	0.0000148
0.00100	0.00451	56.37	16,444.9	0.0000177
0.00110	0.00542	52.41	16,660.2	0.0000210
0.00120	0.00641	49.05	16,831.4	0.0000245
0.00130	0.00742	46.31	16,999.6	0.0000281
0.00140	0.00854	43.72	17,088.4	0.0000320
0.00150	0.00974	41.64	17,149.6	0.0000362
0.00160	0.01094	39.61	17,244.8	0.0000404
0.00170	0.01222	37.93	17,305.1	0.0000448
0.00180	0.01349	36.56	17,390.6	0.0000492
0.00190	0.01486	35.19	17,412.5	0.0000540
0.00200	0.01620	34.13	17,483.4	0.0000586
0.00210	0.01762	33.06	17,488.5	0.0000685
0.00220	0.01895	32.30	17,515.3	0.0000681
0.00230	0.02024	31.69	17,516.7	0.0000726
0.00240	0.02146	31.23	17,531.5	0.0000768
0.00250	0.02272	30.77	17,510.2	0.0000812
0.00260	0.02376	30.62	17,540.6	0.0000849
0.00270	0.02482	30.47	17,539.2	0.0000886
0.00280	0.02588	30.32	17,515.7	0.0000924
0.00290	0.02695	30.16	17,472.7	0.0000961
0.00300	0.02788	30.16	17,467.8	0.0000995
0.00310	0.02865	30.32	17,476.9	0.0001023
0.00320	0.02941	30.47	17,465.6	0.0001050
0.00330	0.03016	30.62	17,441.0	0.0001078

$$\Delta_y = \frac{11}{40}\phi_y h_w^2 \qquad \text{(see Eq. 2.4.17)}$$

$$= \frac{11}{40}(0.0000105)(1260)^2$$

$$= 4.6 \text{ in.} \qquad (0.37\%)$$

$$\mu_\Delta = \frac{\Delta_u}{\Delta_y}$$

$$= \frac{10.8}{4.6}$$

$$= 2.35$$

Curvature ductility demands are

$$\Delta_p = \Delta_u - \Delta_y$$

$$= 10.8 - 4.6$$

$$\phi_p = \frac{\Delta_p}{\left(h_w - \ell_p/2\right)\ell_p}$$

$$= \frac{6.2}{(105 - 13/2)\,(13)(144)}$$

$$= 0.000034 \text{ rad/in.}$$

$$\phi_u = \phi_y + \phi_p$$

$$= 0.0000105 + 0.000034$$

$$= 0.000045 \text{ rad/in.} \qquad (\mu_\phi = 4.2)$$

An estimate of the concrete strain state is developed as follows:

$$\varepsilon_{cy} = \varepsilon_{sy}\left(\frac{kd}{d - kd}\right)$$

$$= 0.00207(0.5)$$

$$= 0.001 \text{ in./in.}$$

$$\varepsilon_{cp} = \phi_p c$$

$$= 0.000034(22.0)$$

$$= 0.00075 \text{ in./in.}$$

$$\begin{aligned}\varepsilon_c &= \varepsilon_{cy} + \varepsilon_{cp}\\ &= 0.001 + 0.00075\\ &= 0.00175 \text{ in./in.}\end{aligned}$$

Table 2.4.3 suggests that the neutral axis depth associated with a curvature of 0.000045 rad/in. is 37.93 in., and the associated strain in the concrete 0.0017 in./in.

Conclusion: Boundary area confinement is not required by current codes because the concrete strain is less than 0.003 in./in. Given the projected level of steel strain ($5.9\varepsilon_y$), confinement might reasonably be provided in this case to stabilize the tension flange reinforcing on reverse cycle loading.

Step 10: Reinforce Wall for Shear. The required ultimate moment demand (M_n) was estimated at 9630 ft-kips (Step 5). The vertical wall reinforcing provided (Figure 2.4.10*a*) will contribute to this strength, and overstrength must be included in the capacity-based development of the wall. In the preceding example, for conceptual design purposes, we used an overstrength factor λ_o of 1.25 while the UBC[2.9] suggests an overstrength factor of 2.8. Observe that the UBC factor is unconservative when applied to the development of an objective strength and that the design value proposed (1.25) is inappropriate for capacity-based design purposes. The nominal capacity suggested by Table 2.4.3 is 19,400 ft-kips (M_u/ϕ). An increase in provided steel yield stress and strain hardening will add another 4 to 5000 ft-kips of strength. This suggests an overstrength factor of about 2.5 for a wall whose design was governed by minimum reinforcing.

Comment: Observe that the adoption of a universal overstrength factor is not realistic. The adopted overstrength factor must be based on the design task at hand.

A more realistic estimate of the shear capacity required to meet capacity-based design objectives is developed as follows:

$$M_n = 19{,}400 \text{ ft-kips} \qquad \text{(see Table 2.4.3)}$$

$$\begin{aligned}\lambda_o M_n - P_D\left(\frac{\ell_w}{2} - \frac{a}{2}\right) &= 1.5(19{,}400) - 600(13 - 0.66)\\ &= 21{,}700 \text{ ft-kips}\end{aligned}$$

$$\begin{aligned}V_u &= \frac{21{,}700}{0.67h_w}\\ &= \frac{21{,}700}{0.67(105)}\\ &= 308 \text{ kips}\end{aligned}$$

The associated shear stress is

$$v_u = \frac{V_u}{t_w d} = \frac{308}{8(25)12} = 130 \text{ psi} \qquad \left(1.9\sqrt{f_c'}\right)$$

$$v_n = \frac{v_u}{\phi} = \frac{130}{0.85} = 153 \text{ psi}$$

Should we elect to provide shear reinforcement in sufficient quantities to carry all of the shear in the plastic hinge region, in spite of the fact that this requirement is not supported by the experiments reviewed in Section 2.4.1.1, we would find

$$A_{vh} = \frac{v_n t_w s}{f_y} = \frac{0.153(8)(12)}{60} = 0.25 \text{ in.}^2/\text{ft}$$

Conclusion: Provide #3 bars at 10 in. on center ($\rho_h = 0.0026$).

Vertical reinforcement is required to restrain crack growth. The suggested quantity (ACI[2.6], $\rho_v = 0.0025$) is reasonably provided and, acting in conjunction with the reliable axial load, is more than sufficient to carry shear across construction joints.

$$V_n = \left(A_{vf} f_y + P\right) \mu \qquad \text{(see Eq. 1.3.19)}$$
$$= [0.0025(26)(12)(8)(60) + 600](1)$$
$$= 375 + 600 \gg 366 \text{ kips}$$

Comment: The relationship between the ultimate shear capacity (V_u) of this wall and the tributary weight (mass) of the system is less than that required by current codes.

$$W = 140(10) = 1400 \text{ kips}$$

$$V_u = \frac{\phi M_n}{0.67 h_w}$$

$$= \frac{8670}{0.67(105)} \qquad \text{(see Step 5)}$$

$$= 123 \text{ kips}$$

$$\frac{V_u}{W} = \frac{123}{1400}$$

$$= 0.088$$

The base shear coefficient[2.9] for a bearing wall structure located at this site ($S_v = 44$ in./sec) would be

$$C_v = \frac{S_v}{61.5} \qquad \text{(see Eq. 2.4.2b)}$$

$$= \frac{44}{61.5}$$

$$= 0.72$$

$$V = \frac{C_v}{RT} W \qquad \text{(see Ref. 2.9, Eq. 30-5)}$$

$$= \frac{0.72}{4.5(0.96)} W$$

$$= 0.167W$$

Comment: Flexural strengths that induce high shear stresses in the plastic hinge region should be avoided whenever possible; in this case the capacity based demand of $2\sqrt{f_c'}$ is well below my objective limit state. The attainment of an appropriate balance between restoring force (strength), concrete strain, and shear stress is the key to a good design.

Conclusion: Calculations involving the prediction of strain states in concrete should be used to help the engineer make design decisions. Based on the analyses performed, the engineer is now in a better position to detail the shear wall.

Step 11: Detail the Shear Wall. Alternative wall reinforcing programs are proposed. Outside the hinge region the program described in Figure 2.4.10*b* seems appropriate.

Within the hinge region, which should extend at least $0.5\ell_w$ above the base (13 ft), the program described in Figure 2.4.10*a* seems appropriate.

The confining pressure provided is

$$f_\ell = \frac{0.11(75)}{4(4)}$$
$$= 0.51 \text{ ksi}$$

This meets ACI objective levels of confinement $(0.09 f_c')$. Observe that it exceeds that provided in the tests described in Section 2.4.1.1.

2.4.2 Shear Walls with Openings

Shear walls often contain openings. The most troublesome are those that are stacked, for they tend to decouple walls (Figure 2.4.11), and this can reduce building stiffness and attainable strength. Randomly placed openings interrupt the otherwise normal beam shear load path (Figure 2.4.12). Procedures for dealing with the change of shear flow around these openings must be understood if the strength and performance of the wall are not to be compromised. These procedures are the focal topics of this section.

2.4.2.1 Coupling Beams Two types of reinforcement programs are commonly used in coupling beams today—one is the pseudoflexural reinforcing program described in Figure 2.4.13*a*; the other is the diagonally reinforced coupling beam of Figure 2.4.13*b*. Recently proposed alternatives include the reinforcing program described in Figures 2.4.13*c* and the steel or steel composite coupling beam described in Figure 2.4.13*d*. The three concrete options were studied experimentally by Galano and Vignoli,[2.43] and this experimental effort will help us propose design guidelines and limit states for these systems. Further, the behavior reported by Galano and Vignoli is consistent with that reported by others. The Galano studies are reviewed in detail because they endeavor to compare the behavior of coupling beams with various reinforcing programs (Figure 2.4.13*a*, *b*, and *c*). The work of Park and Paulay[2.2] is also reviewed to aid in establishing deformation guidelines.

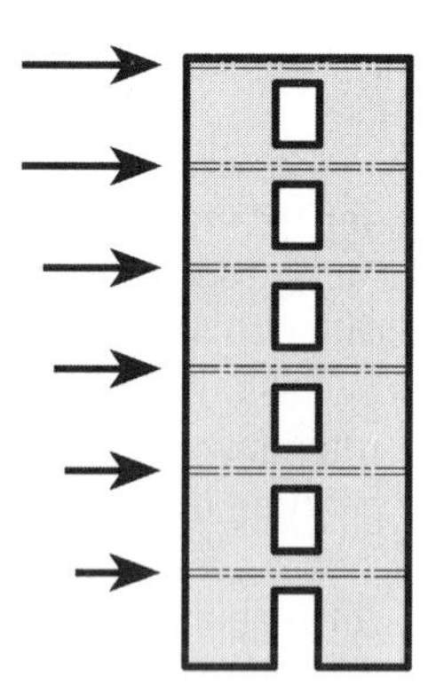

Figure 2.4.11 Coupled shear wall system[2.5] with stacked openings.

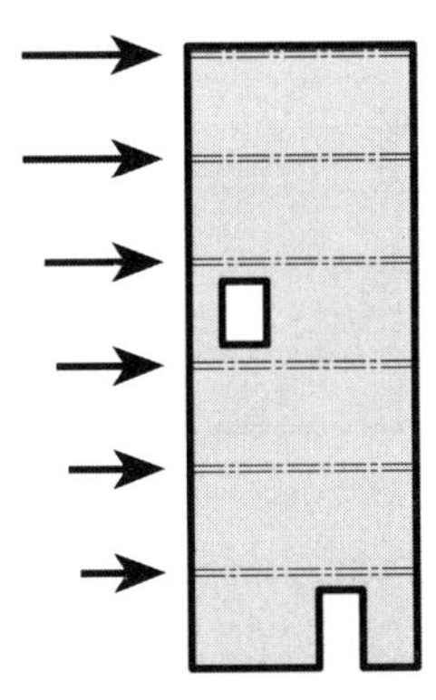

Figure 2.4.12 Shear wall with randomly placed openings.

The steel coupling beam is analytically modeled after the link beam developed for a steel eccentrically braced frame. This system has been studied by Harries and his associates.[2.44] The postyield deformability of this type of coupling beam makes it especially attractive where coupling beam spans are short and coupling beam strength and deformability are critical.

Flexural Behavior Model The reinforcing program described in Figure 2.4.13*a* is usually designed as though it were a beam (Figure 2.4.14*a*). The shear transferred is usually developed from the probable flexural capacity (M_{pr}) of the beam:

$$V = \frac{2M_{pr}}{\ell_c} \tag{2.4.28a}$$

Analytically this correctly predicts the capability of the beam to transfer shear, but it misrepresents the behavior experienced in the coupling beam. This is because beams, whose shear span to depth ratio (a/d or M/Vd) is short, transfer shear by an arching action and, as a consequence, a compressive strut connects the nodes, as described in Figure 2.4.14*b*. The top and bottom reinforcing bars are actually tension ties. Experimentally measured stresses in short coupling beams confirm that a tensile stress exists in these bars along the entire length of the beam. Beam longitudinal reinforcing bars are more appropriately referred to as ties and the coupling beam is analytically modeled using a strut and tie model as described in Figure 2.4.14*b*. The so-developed capacity is the same as that predicted by beam theory (Eq. 2.4.28a).

The shear resisted by the test beam of Figure 2.4.14*a*, as developed from the flexural model or the equivalent two-truss strut and tie model of Figure 2.4.14*b*, is

$$V = \frac{2M_n}{\ell_c} \quad \text{(Beam model)} \tag{2.4.28b}$$

$$= \frac{2T_y(d - a/2)}{\ell_c}$$

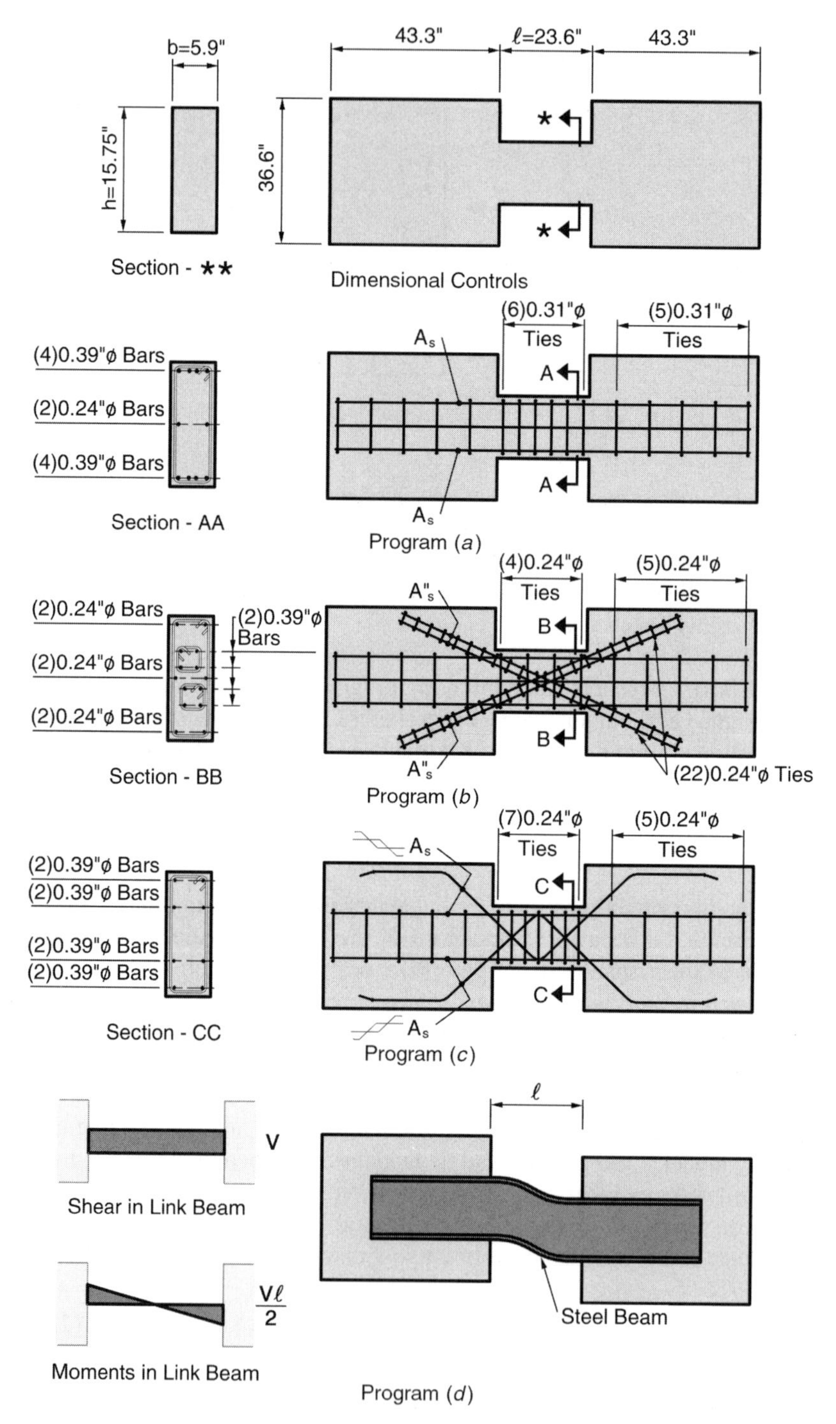

Figure 2.4.13 Alternative coupling beam reinforcement programs.[2.43, 2.44]

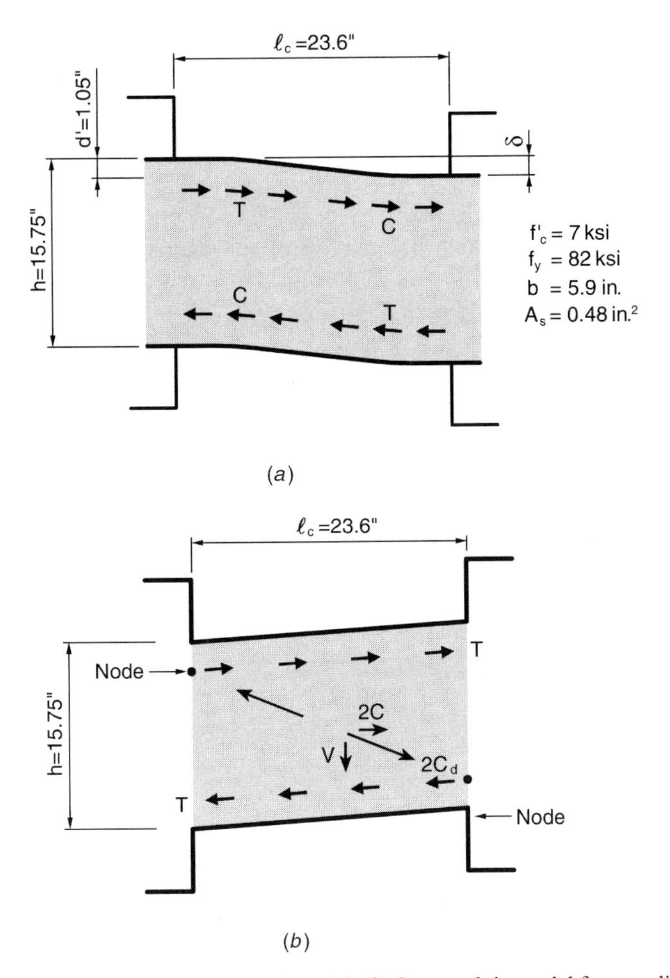

Figure 2.4.14 (*a*) Traditional flexural model. (*b*) Strut and tie model for coupling beam.

$$\cong \frac{2T_y(d - d')}{\ell_c}$$

$$= \frac{2A_s f_y(d - d')}{\ell_c} \qquad \text{(Strut and tie model)} \qquad (2.4.28c)$$

$$= \frac{2(0.48)(82)(13.7)}{23.6}$$

$$= 45.7 \text{ kips} \quad (203 \text{ kN})$$

The associated shear stress is

$$v = \frac{V}{bd}$$

$$= \frac{45.7}{5.9(14.7)}$$

$$= 0.527 \text{ ksi} \qquad \left(6.3\sqrt{f'_c}\right)$$

and this strength was experimentally confirmed (see Figure 2.4.15*a*).

Park and Paulay[2.2] studied the behavior of conventionally reinforced coupling beams (Flexural Behavior Model). The beam characteristics of the conventionally reinforced coupling beam of Figure 2.4.16 were

$$h = 39 \text{ in.} \qquad (h/\ell_c) = 0.975$$

$$b = 12 \text{ in.} \qquad f'_c = 4000 \text{ psi}$$

$$\ell_c = 40 \text{ in.}$$

The strength limit state of this beam is defined by the level of compressive stress imposed on the compression diagonal (see Figure 2.4.14*b*) since the flexurally based load projection of 184 kips was not attained (Figure 2.4.16). The axial load imposed on a compression diagonal (Figure 2.4.14*b*) based on the attained level of shear (170 kips) would have been

$$F_{cd} = V\left(\frac{\ell_d}{d - d'}\right) \tag{2.4.29a}$$

$$= 170\left(\frac{52}{33}\right)$$

$$= 268 \text{ kips}$$

Given an assumed compressive strut width of 13 in. ($0.25\ell_d$), the compressive stress in the strut becomes

$$f_{cd} = \frac{F_{cd}}{bw} \tag{2.4.29b}$$

$$= \frac{268}{12(13)}$$

$$= 1.72 \text{ ksi} \qquad (0.43 f'_c)$$

and this clearly exceeds its strut and tie suggested capacity for a severely cracked web (see Section 1.3.1 and Section 2.3) of 1.4 ksi $\left(0.35 f'_c\right)$.

The level of shear stress imposed on the coupling beam of Figure 2.4.16 was also high:

$$v = \frac{V}{bd}$$

$$= \frac{170}{12(36)}$$

$$= 0.394 \text{ ksi} \qquad 6.2\sqrt{f'_c}$$

and this tends to confirm our objective peak shear in beams of $5\sqrt{f'_c}$ (see Section 2.1.2.3), which in the coupling beam should be used as an alert.

Park and Paulay[2.2] also reported the behavior of coupling beams whose shear span to depth ratio was 1 as opposed to the 0.5 of Figure 2.4.16. Figure 12.7*b* of Reference 2.2, repeated here as Figure 2.4.17, is developed from this conventionally reinforced coupling beam. The developed shear stress in beams *a* and *b* of Figure 12.7 of Reference 2.2 was on the order of $6.5\sqrt{f'_c}$, and the stress imposed on the diagonal was $0.37 f'_c$; both are consistent with previously identified limit states. Shear reinforcement was $2\sqrt{f'_c}$ in beam *b* and even less in beam *a*, clearly refuting the beam model.

Diagonally Reinforced Coupling Beam Paulay[2.2, 2.37] and his associates introduced the diagonal reinforcing program described in Figure 2.4.13*b* many years ago. It is now a part of current building code requirements (ACI[2.6, Sec. 21.6.7.3]) when the ultimate shear demand imposed on the coupling beam exceeds $4\sqrt{f'_c}bd$.

The strength of the reinforcing program developed in Figure 2.4.13*b* is best modeled using a pair of struts and their companion ties (Figure 2.4.18). The shear force V_d[2.2, Sec. 5.6.2] is

$$V_d = 2T \sin\alpha \tag{2.4.30a}$$

where

$$\alpha = \tan^{-1}\left(\frac{h/2 - d'}{\ell_c/2}\right)$$

For the test beam of Figure 2.4.13*b*,

$$V_d = 2T \sin\alpha \qquad \text{(Eq. 2.4.30a)}$$

$$= 2A''_s f_y \left(\frac{2(h/2 - d')}{\ell_d}\right) \tag{2.4.30b}$$

$$= 2(0.48)(82)\left(\frac{13.7}{27.3}\right)$$

$$= 39.5 \text{ kips}$$

where A''_s is the reinforcing steel in one diagonal.

Two 6-mm bars located at the top and bottom of the tested coupling beam (Figure 2.4.13*b*) add some shear strength that we refer to as V_f.

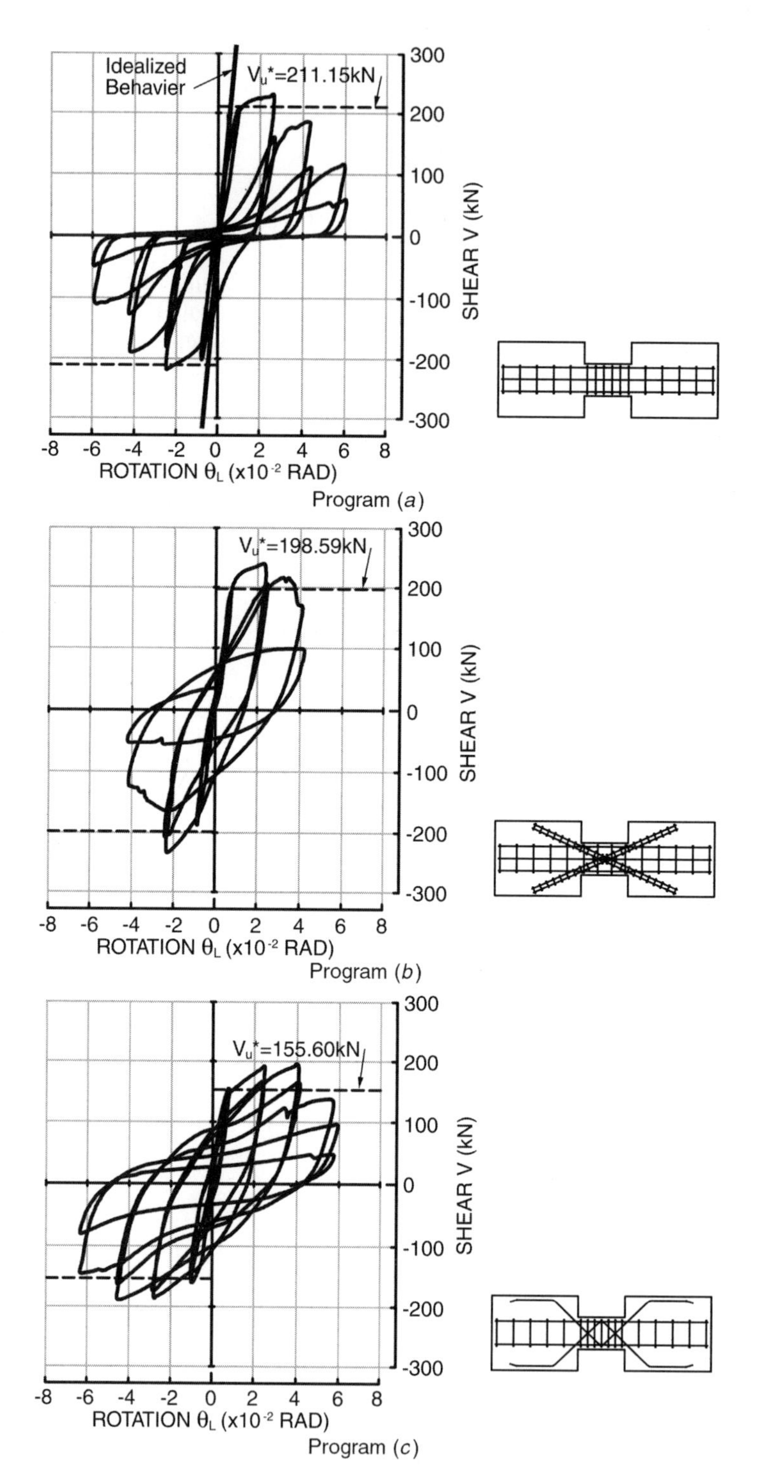

Figure 2.4.15 Shear rotation relationships for reinforcement programs *a* to *c* of Figure 2.4.13.[2.43]

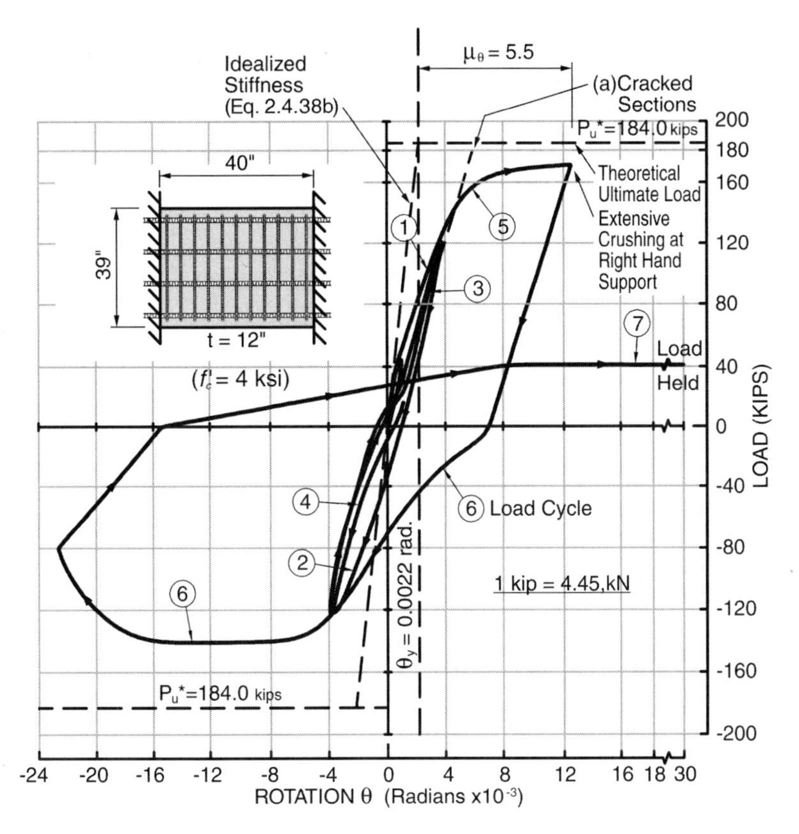

Figure 2.4.16 Details and behavior of a typical coupling beam.

$$V_f = \frac{2A_s f_y (d - d')}{\ell_c} \qquad \text{(Eq. 2.4.28a)}$$

$$= \frac{2(0.088)(82)(13.7)}{23.6}$$

$$= 8.4 \text{ kips}$$

$$V_{\text{total}} = V_d + V_f$$

$$= 39.5 + 8.4$$

$$= 47.9 \text{ kips} \qquad (213 \text{ kN})$$

and this matches the experimental results (Figure 2.4.15*b*).

Park and Paulay[2.2] studied the behavior of a fairly robust diagonally reinforced coupling beam (see Figure 2.4.19). This beam was ultimately capable of sustaining

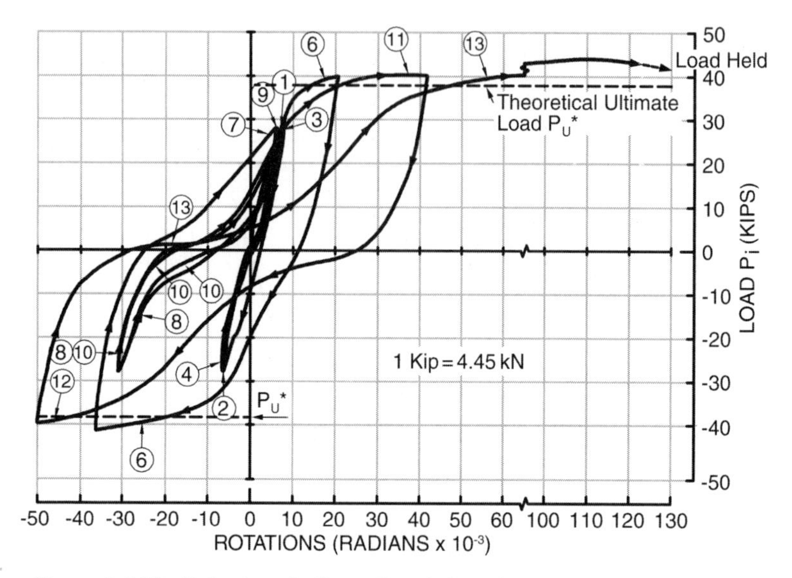

Figure 2.4.17 Behavior of a flexurally reinforced coupling beam ($a/d = 1$).

a shear (V_d) of 151.5 kips. This corresponds to a compression stress along the diagonal of

$$f_{cd} = V_d \left[\frac{\ell_d}{b(0.25)\ell_d(d - d')} \right] \qquad \text{(see Eq. 2.4.29a and Eq. 2.4.29b)}$$

$$= \frac{4V_d}{b(d - d')} \qquad (2.4.30c)$$

$$= \frac{4(151.5)}{12(24)}$$

$$= 2.1 \text{ ksi} \qquad (0.52 f_c')$$

Clearly, the presence of diagonal reinforcement increases the ability of the compression strut described in Figure 2.4.14*b* to carry load.

Comment: The analytical model of the diagonally reinforced coupling beam (Figure 2.4.18), which presumes that both diagonals share the load, does not accurately describe behavior. Observe that the strut and tie model of Figure 2.4.14*b* is confirmed by the post-test crack pattern of Figure 2.4.19.

Truss Reinforcing Program The reinforcing program described in Figure 2.4.13*c* represents a significant and promising departure from traditional coupling beam reinforcing programs.

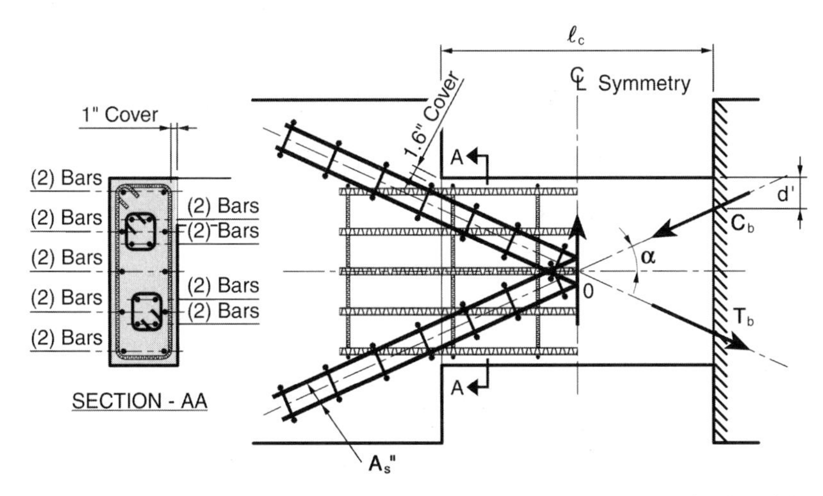

Figure 2.4.18 Primary truss describing a component of the load path developed by the reinforcing program of Figure 2.4.13*b*.

The primary load transfer mechanism of the system described in Figure 2.4.13*c* is best represented by the truss shown in Figure 2.4.20 taken to its yield capacity. A secondary load path is created by the "global" strut and tie model shown in Figure 2.4.14*b*. The load transfer limit state will coincide with yielding of all of the tension diagonals, provided the so-produced compression loads do not exceed the capacity of the concrete compression strut.

The yield strength of the primary truss (Figure 2.4.20) is governed by the tensile strength (T_d) of its diagonal, which in this case is

$$\begin{aligned} T_d &= A_s'' f_y \\ &= 0.24(82) \\ &= 19.7 \text{ kips} \end{aligned} \tag{2.4.31}$$

The shear or vertical component of one diagonal is

$$\begin{aligned} V_1 &= T_d \left(\frac{d - d'}{\ell_d} \right) \\ &= 19.7 \left(\frac{12.0}{16.8} \right) \\ &= 14.0 \text{ kips} \end{aligned} \tag{2.4.32}$$

The primary truss transfer mechanism (Figure 2.4.20) must include the shear traveling along the compression diagonal (Figure 2.4.14*b*); hence the yield strength of the primary truss mechanism is $2V_1$. The bottom chord at *a* of Figure 2.4.20 will not have yielded given the yielding of the primary truss. The residual chord strength (T') is

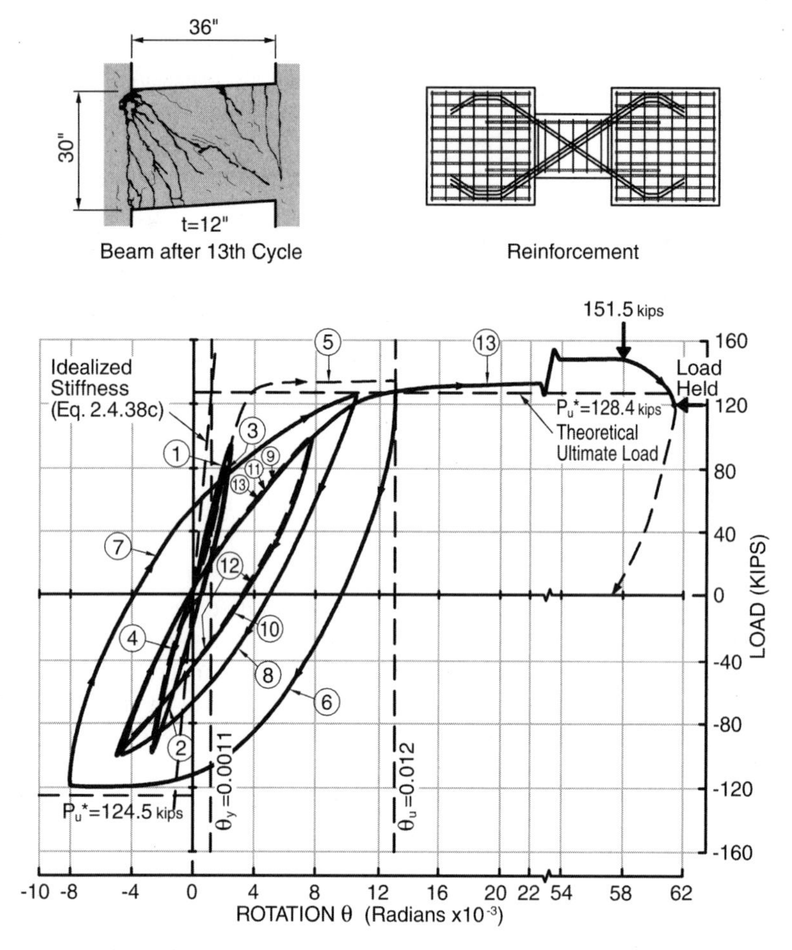

Figure 2.4.19 Robust diagonally reinforced coupling beam.[2.2]

$$T' = \left(A_s - A_s''\right) f_y \tag{2.4.33}$$

where A_s is the total steel at point a of Figure 2.4.20 (0.48 in.^{2}) and A_s'' is the area of the truss diagonal reinforcing in Figure 2.4.20.

$$\begin{aligned} T' &= (0.48 - 0.24)(82) \qquad \text{(Eq. 2.4.33)} \\ &= 19.7 \text{ kips} \end{aligned}$$

This residual tension force activates the load transfer mechanisms of Figure 2.4.14*b* and in effect produces a secondary truss:

$$V_2 = 2T'\left(\frac{d - d'}{\ell_{ab}}\right) \qquad \text{(see Eq. 2.4.28a)}$$

$$= 2(19.7)\left(\frac{12}{26.5}\right)$$

$$= 17.8 \text{ kips}$$

and

$$V_n = 2V_1 + V_2$$
$$= 2(14.0) + 17.8$$
$$= 45.8 \text{ kips} \qquad (204 \text{ kN})$$

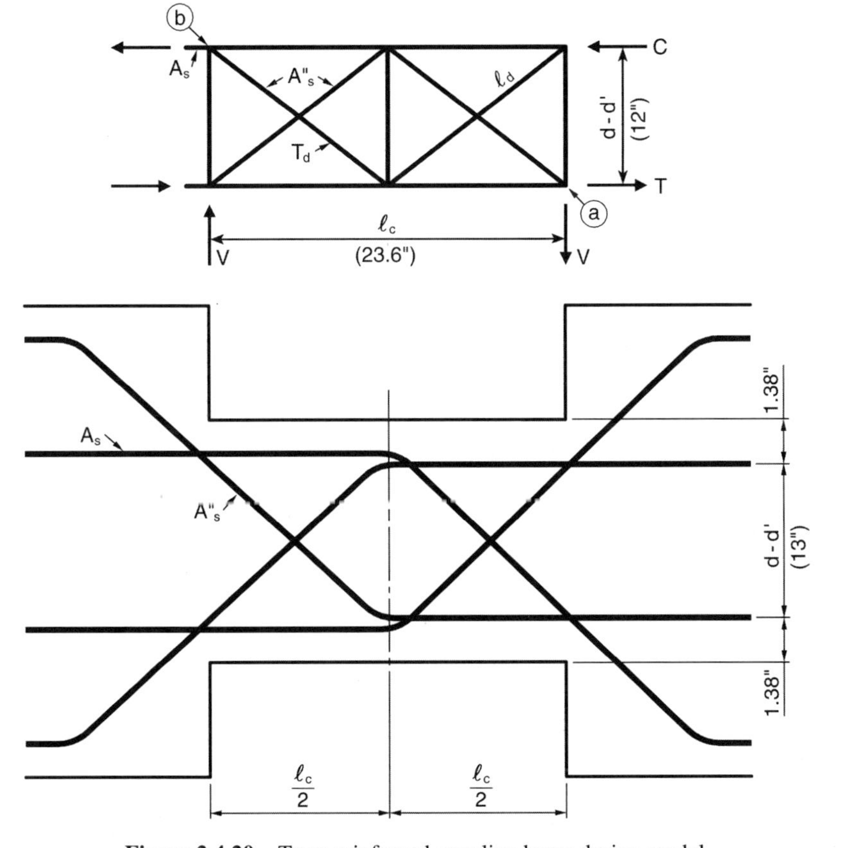

Figure 2.4.20 Truss reinforced coupling beam design model.

and this design process is supported by the experimentally based conclusion (see Figure 2.4.15*c*). The existence of these analytical load paths is supported by the "after failure" photographs shown in Figure 2.4.21.

Comment: Observe that the design procedure developed for the flexural behavior model is equally effective in predicting the strength if a proper accounting of d' is made.

The stress imposed on the Galano coupling beam diagonals was $0.35f_c'$ for beam types *a* and *c* as described in Figure 2.4.13 ($V = 200$ kN) and slightly higher for the diagonally reinforced beam ($0.4f_c'$) of Figure 2.4.13*b*.

Steel Coupling Beams The two steel coupling beams studied experimentally by Harries and associates[2.44] in the early 1990s are described in Figure 2.4.22*a*. Specimen 2, which performed quite well, is now discussed.

The coupling or link beam strength is evaluated based on the limit states developed for link beams used in structural steel eccentrically braced frames.[2.45 & 2.4, Sec. 6.2] The "plastic" shear strength limit (V_{sp}) is defined in the Seismic Provisions for Structural Steel Buildings (SPSSB)[2.45] as

$$\begin{aligned} V_{sp} &= 0.6F_y(d - 2t_f)t_w \\ &= 0.6(43.5)(12.1)(0.19) \\ &= 60 \text{ kips} \qquad (266 \text{ kN}) \end{aligned} \tag{2.4.34a}$$

Comment: Earlier codified versions[2.4] of shear strength were based on the depth of the beam as opposed to $(d - 2t_f)$. In this case, V_{sp} would be

$$\begin{aligned} V_{sp} &= 0.6F_y d t_w \\ &= 0.6(43.5)(13.58)(0.19) \\ &= 67 \text{ kips} \qquad (297 \text{ kN}) \end{aligned} \tag{2.4.34b}$$

The flexural strength of the coupling beam is

$$\begin{aligned} M_p &= F_y Z_x \\ &= 43.5(58) \\ &= 2523 \text{ in.-kips} \end{aligned} \tag{2.4.34c}$$

The deformed shape of the coupling beam will cause the point of maximum moment to occur ultimately at the face of the reinforcing bar in the pier because the concrete that covers these bars will become ineffective. The total length of the steel link beam (e) becomes

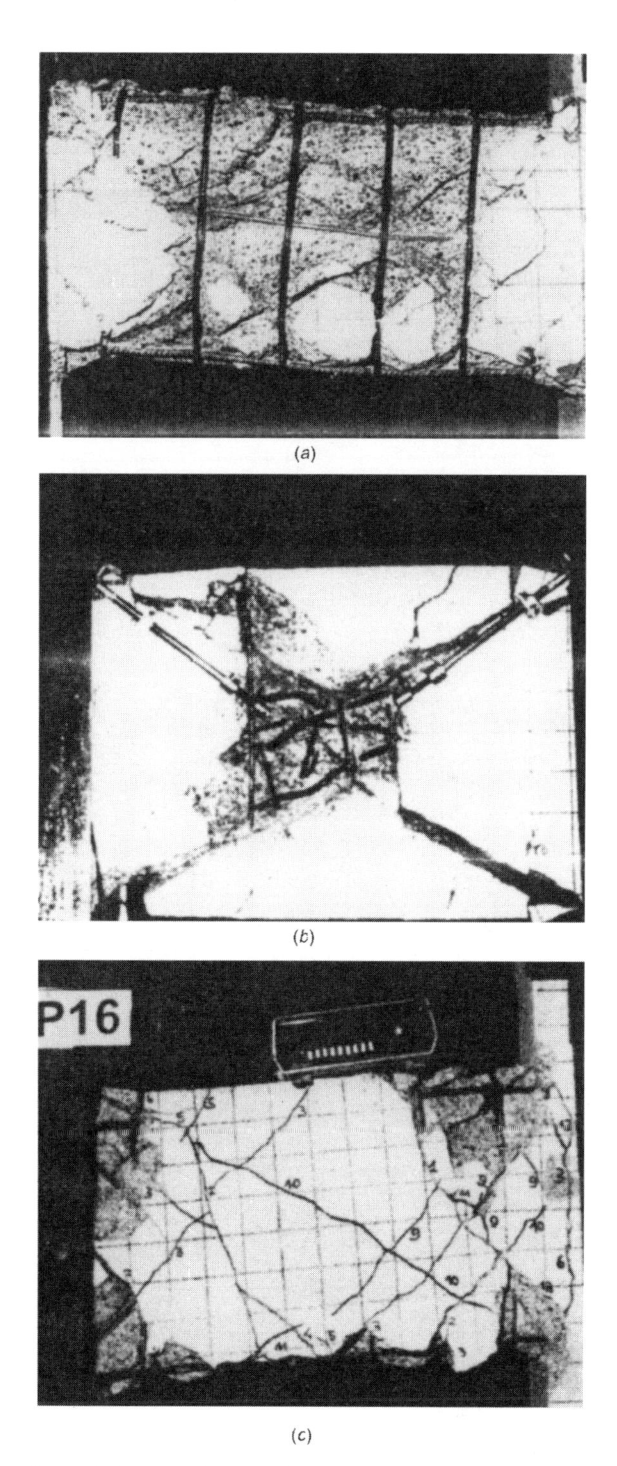

Figure 2.4.21 State of three specimens (Figure 2.4.13*a*, *b*, and *c*) after failure stage (cyclic tests). (From Ref. 2.43.)

$$e = \ell_c + 2(\text{cover}) \tag{2.4.35a}$$
$$= 47.2 + 1.6$$
$$= 48.8 \text{ in.}$$

The shear force required to yield test beam 2 in flexure (V_f) is

$$V_{fp} = \frac{M_p}{e/2} \tag{2.4.35b}$$
$$= \frac{2523}{24.4}$$
$$= 103 \text{ kips}$$

The SPSSB criterion would describe the behavior of this link beam as being dominated by shear since V_{sp} is less than $0.8V_{fp}$. Further, it would suggest that the plastic or

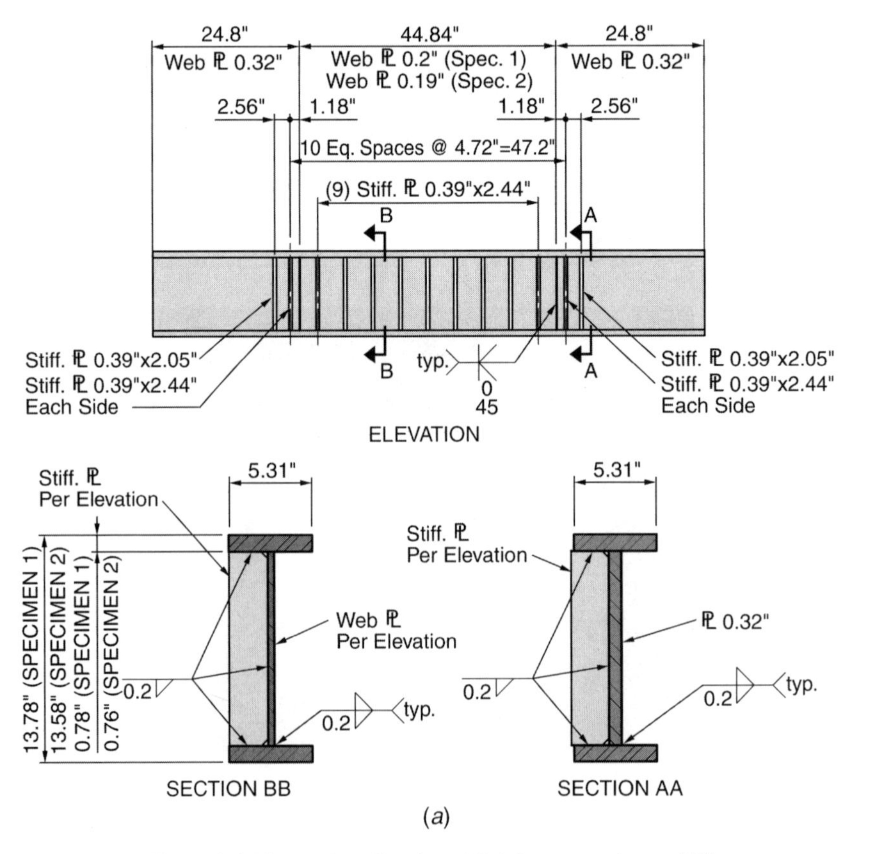

Figure 2.4.22 (*a*) Details of steel link beam specimens.[2.44]

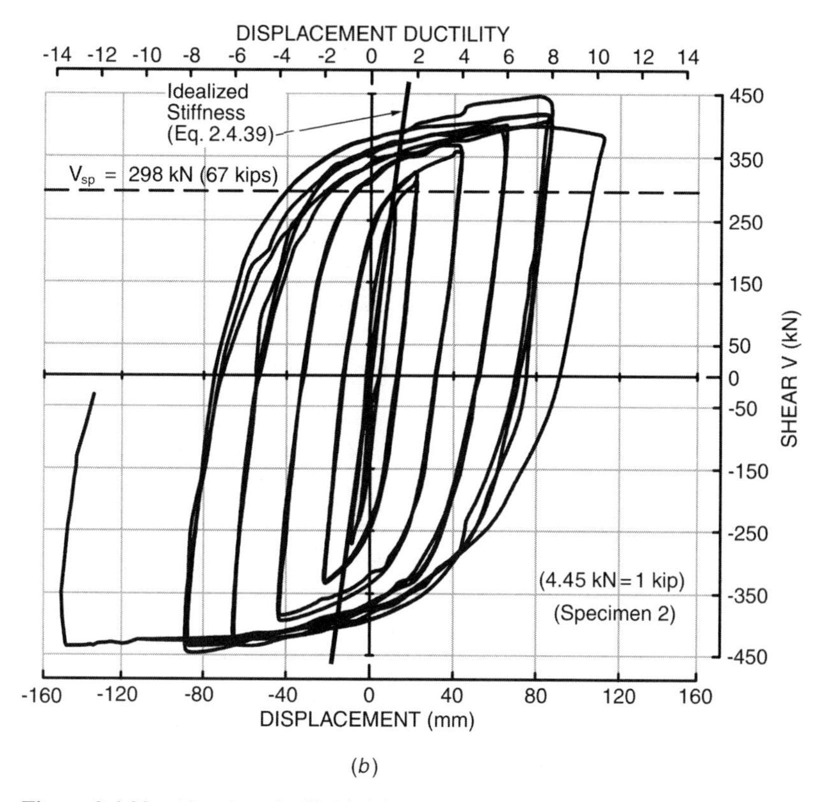

Figure 2.4.22 (*Continued*) (*b*) Link beam shear versus relative displacement.[2.44]

postyield rotation of this link beam could reach 0.08 radian. Figure 2.4.22*b* describes the response of this coupling beam when subjected to cyclic deformations. Observe that a displacement of 84 mm (3.3 in.) was repeatedly attained. This displacement corresponds to a rotation of 0.07 radian, and this is somewhat lower than the postyield rotation (θ_p) limit of 0.08 radian proposed by the SPSSB. It should, however, be noted that a rotation of 0.07 radian is not the limit state because the system was pushed to a displacement of 155 mm (6 in.) or 0.12 radian. Accordingly, the limit states proposed by SPSSB appear to be appropriate. Harries and coauthors report that rotation within the embedded region, described in Figure 2.4.23, accounted for only 3% of the total displacement.

The attained level of beam shear was considerably higher ($V_{b,\max} = 100$ kips) than the proposed (Eq. 2.4.34a) SPSSB limit (V_{sp}) of 60 kips. This suggests that the capacity-based strength be developed using Eq. 2.4.34b and an overstrength factor (λ_o) of about 1.5 (100/67). V_{sp} does, however, reasonably describe the yield strength of the system (67 kips).

2.4.2.2 Analytical Modeling of the Coupling Beam Coupled walls are usually modeled as frames (Figure 2.4.24). The column or pier element is typically centered

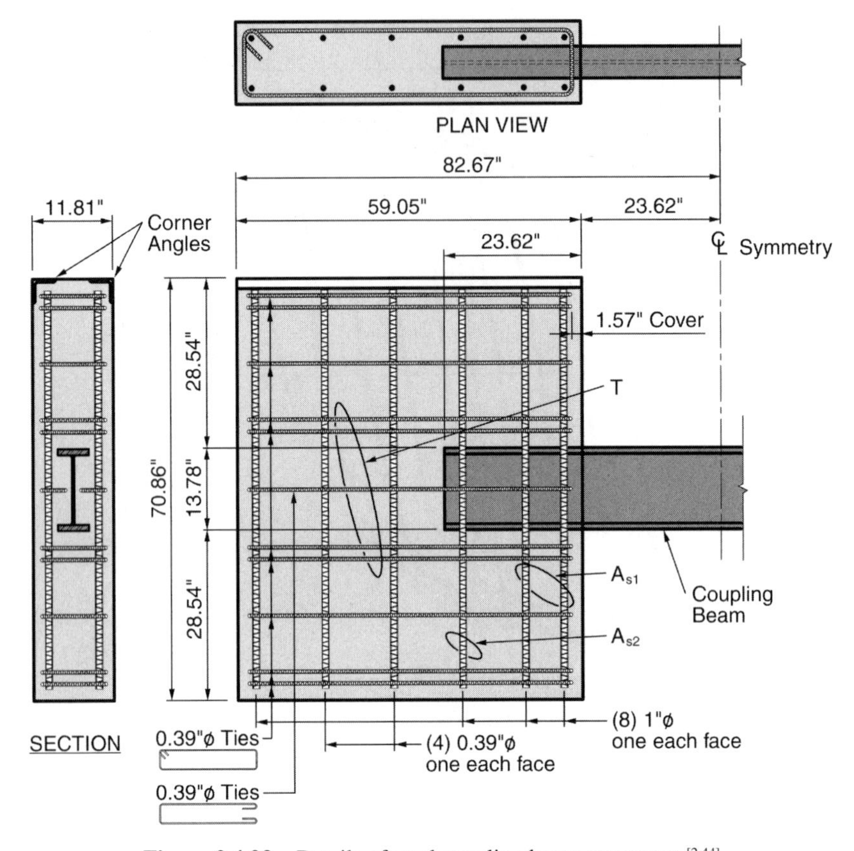

Figure 2.4.23 Details of steel coupling beam at support.[2.44]

on the midlength of the wall ($\ell_w/2$) and the distance between the centerline of the wall and the face of the wall modeled as an infinitely rigid beam, while the coupling beam is assigned a specific stiffness that is intended to idealize its behavior. A rigorous static analysis might identify the location of the neutral axis as the center of rotation of each pier (neutral axis), and this will be discussed in more detail when design examples are studied in Section 2.4.2.3. For now our focus is on developing analytical models for coupling beams.

Flexural Behavior Model The shear stresses imposed on the concrete coupling beams that were studied by Galano and Vignoli[2.43] and discussed in the preceding sections were high (Figure 2.4.15*a*):

$$V \cong 50 \text{ kips} \qquad (222 \text{ kN})$$

$$v = \frac{V}{bd}$$

$$= \frac{50}{5.9(14.7)}$$

$$= 580 \text{ psi} \qquad \left(7\sqrt{f_c'}\right)$$

The compressive load imposed on the diagonal strut (Figure 2.4.14*b*) of the coupling beam of Figure 2.4.13*a* was also high:

$$F_{cd} = V\left(\frac{\ell_d}{d - d'}\right) \qquad \text{(Eq. 2.4.29a)}$$

$$= 50\left(\frac{27.3}{13.7}\right)$$

$$= 99.6 \text{ kips}$$

This corresponds to a stress along the compression diagonal, assuming that the width of the strut (w) is 25% of the length of the diagonal of

$$f_{cd} = \frac{F_{cd}}{0.25\ell_d b} \qquad (2.4.36)$$

$$= \frac{99.6}{0.25(27.3)(5.9)}$$

$$= 2.47 \text{ ksi} \qquad \left(0.35 f_c'\right)$$

Accordingly, the damage described in Figure 2.4.21*a* might reasonably be predicted.

The design objective for a flexural concrete coupling beam (Figure 2.4.13*a*) should be to control the compressive stress imposed on the diagonal strut to $0.35 f_c'$, and this is consistent with generally accepted strut and tie limit states (see Section 1.3.1 and

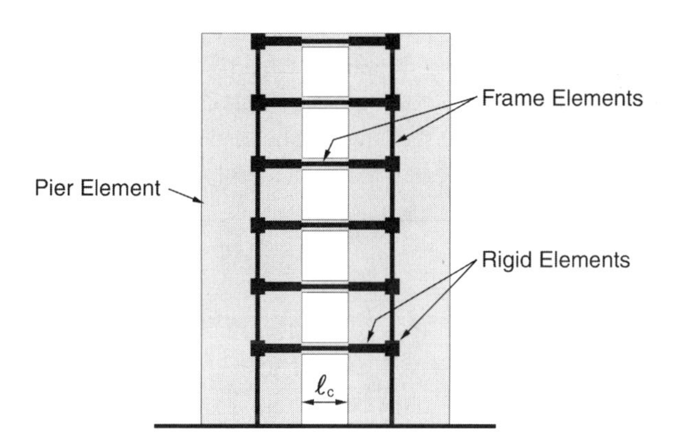

Figure 2.4.24 Two-dimensional equivalent frame model of coupled wall system.[2.46]

Section 2.3). The maximum shear carried by a flexural coupling beam, given this $0.35f_c'$ criterion and Eq. 2.4.30c, would be

$$\lambda_o V_n \leq 0.08 f_c' bd \tag{2.4.37a}$$

This corresponds to limiting the shear stress imposed on a coupling beam to

$$v_{n,\max} = \frac{0.08 f_c'}{\lambda_o} \tag{2.4.37b}$$

An upper limit is likely to be developed for v_n as was the case for the beam-column joint (Figure 2.3.11). Accordingly, based on the work of Galano and Vignoli[2.43] (f_c' = 7 ksi), a maximum nominal shear stress of 450 psi ($\lambda_o = 1.25$) seems appropriate.

Modeling the stiffness of the coupling beam is difficult. Stiffness predictions tend to be conservative and depend on the depth or, more typically, on the overall height of the beam (h), to clear span ratio (h/ℓ_c). The Canadian Standard Association (CSA) and the New Zealand Standard Association (NZS) propose procedures for identifying the effective stiffness of coupling beams.[2.46] For conventionally reinforced coupling beams (Figure 2.4.13a), the effective moment of inertia (I_e) is described by each as

$$\text{CSA} \qquad I_e = \frac{0.2 I_g}{1 + 3(h/\ell_c)^2} \tag{2.4.38a}$$

$$\text{NZS} \qquad I_e = \frac{0.4 I_g}{1 + 8(h/\ell_c)^2} \tag{2.4.38b}$$

The effective moment of inertia for the beam of Figure 2.4.13a becomes

$$\text{CSA} \qquad I_e = 0.086 I_g \qquad \text{(Eq. 2.4.38a)}$$

$$\text{NZS} \qquad I_e = 0.088 I_g \qquad \text{(Eq. 2.4.38b)}$$

The development of an effective moment of inertia implies that a flexural beam model is appropriately used to predict more complex system behavior. The flexural component of deflection (δ_{bf}) for the coupling beam described in Figure 2.4.14a may be developed for the segment of beam between the support and the point of inflection at midspan:

$$\begin{aligned}\delta_{bf} &= \frac{V\,(\ell_c/2)^3}{3EI_e} \\ &= \frac{49(11.8)^3}{3(4350)(0.088)1920} \\ &= 0.036 \text{ in.}\end{aligned}$$

Recall that Eq. 2.4.28b endeavored to convert a strut and tie model (Eq. 2.4.28c) to a simple flexural behavior base. In the stiffness modeling process Eq. 2.4.38a was undoubtedly developed based on Grade 60 ($f_y = 60$ ksi) reinforcing steel. Since the idealized deflection of a coupling beam is a function of the yield strength of the steel tension tie (see Eq. 2.4.28c), a deflection calculated using the flexural model (Eq. 2.4.38a and Eq. 2.4.38b) should reasonably be increased by 33% (82/60) if our objective is to predict the behavior described in Figure 2.4.15*a*:

$$\delta_{bf} = 1.33(0.036)$$
$$= 0.048 \text{ in.}$$

The coupling beam rotation at yield would be

$$\theta_{by} = \frac{\delta_{bf}}{\ell_c/2}$$
$$= \frac{0.048}{11.8}$$
$$= 0.004 \text{ radian}$$

Galano and Vignoli[2.43] report that the experimentally determined rotation at yield θ_y was 0.00842 radian. Clearly, Eq. 2.4.38b describes an idealized stiffness of the coupling beam that is significantly greater than that developed experimentally (Figure 2.4.15*a*). Observe that the relationship between idealized stiffness and experimental stiffness in the coupling beam is not too dissimilar to the same relationship developed for the frame beam (Figure 2.1.3) and the shear wall (Figure 2.4.4*a*). If the ultimate rotation from a design perspective is 0.04 radian (see Figure 2.4.15*a*), the rotational ductility when Eq. 2.4.38b is used to define the rotation at yield is 10.

$$\mu_\theta = \frac{0.04}{0.004}$$
$$= 10$$

The effective stiffness of the Park and Paulay[2.2] flexural coupling beam, following the Canadian Standard (CSA) is,

$$I_e = \frac{0.2I_g}{1 + 3(h/\ell_c)^2} \qquad \text{(Eq. 2.4.38a)}$$
$$= \frac{0.2}{1 + 3(0.975)^2} I_g$$
$$= 0.052I_g$$
$$= 3090 \text{ in.}^4$$

The yield rotation based on this estimate of flexural stiffness is

$$\theta_y = \frac{V(\ell_c/2)^2}{3EI_e}$$

$$= \frac{184(20)^2}{3(3600)(3090)}$$

$$= 0.0022 \text{ radian}$$

The ultimate rotation attained during the test (Figure 2.4.16*a*) was slightly more than 0.012 radian. The ductility (μ_θ) of this beam is then about 5.5. Recall that the stress imposed on the compression diagonal of this beam ($0.43 f_c'$) was higher than our objective limit state of $0.35 f_c'$. The behavior of the flexural coupling beam described in Figure 2.4.17 was considerably better than that of the beam whose behavior was described in Figure 2.4.16, and this is due in part to a reduction in the stress imposed on the concrete diagonal. Figure 2.4.17 suggests that a rotational limit state (θ_u) of 0.04 radian is conservative.

It appears reasonable to conclude that if a capacity-based shear stress limit of $5\sqrt{f_c'}$ is maintained in a coupling beam, then a rotational ductility of up to 10 may be attained. This ductility must be developed from the idealized rotation at yield (θ_{yi}) which, as is demonstrated in Figures 2.4.15*a* and 2.4.16, is much less than the experimental stiffness typically observed during tests. The appropriateness of the suggested shear limit should be verified by evaluating the capacity-based stress imposed on the compression diagonal (see Eqs. 2.4.29a and 2.4.36) using a compression stress limit of $0.35 f_c'$.

Diagonally Reinforced Coupling Beams Diagonally reinforced coupling beams (Figure 2.4.13*b*) are generally presumed to be somewhat stiffer than those that are conventionally reinforced (Figure 2.4.13*a*).

The Canadian Standard (CSA) suggests that diagonally reinforced coupling beams are twice as stiff as conventionally reinforced coupling beams:

$$\text{CSA} \qquad I_e = \frac{0.4 I_g}{1 + 3(h/\ell_c)^2} \tag{2.4.38c}$$

while the New Zealand Standard suggests that they are generally less stiff than predicted by the Canadian Standard:

$$\text{NZS} \qquad I_e = \frac{0.4 I_g}{1.7 + 2.7(h/\ell)^2} \tag{2.4.38d}$$

For the diagonally reinforced coupling beam of Figure 2.4.13*b*, the effective moment of inertia predicted by the two proposals is

CSA $$I_e = \frac{0.4I_g}{1 + 3(0.67)^2} \qquad \text{(Eq. 2.4.38c)}$$

$$= 0.172I_g \qquad (2 \times I_e \text{ for flexural behavior model; Eq. 2.4.38a})$$

NZS $$I_e = \frac{0.4I_g}{1.7 + 2.7(0.67)^2} \qquad \text{(Eq. 2.4.38d)}$$

$$= 0.137I_g \qquad (1.56 \times I_e \text{ for flexural behavior model; Eq. 2.4.38b})$$

The Galano and Vignoli[2.43] tests do not confirm this increase in stiffness (Figure 2.4.15), and the tests reported on by Park and Paulay in Reference 2.2 (Figure 2.4.19) suggest that these stiffness projections are too high also.

The diagonally reinforced beam tested by Park and Paulay[2.2] (Figure 2.4.19) had a theoretical ultimate load of 128.4 kips. This beam was 12 in. by 30 in. and had a clear span of 36 in. The yield rotation (θ_y), based on the Canadian Standard, is developed as follows:

$$I_e = \frac{0.4I_g}{1 + 3(h/\ell_c)^2} \qquad \text{(Eq. 2.4.38c)}$$

$$= 0.13I_g$$

and

$$\theta_y = \frac{V(\ell_c/2)^2}{3EI_e}$$

$$= \frac{128.4(18)^2}{3(3600)(3510)}$$

$$= 0.0011 \text{ radian}$$

and this describes a much stiffer system than that which is suggested by the tests. The available level of rotational ductility is in excess of 11 (Figure 2.4.19).

Steel Coupling Beams The effective stiffness proposed by Harries and coauthors[2.46] for steel coupling beams is 60% of that developed from first principles. When shear deformations are considered, as they must be in short coupling beams, an effective stiffness for use in "flexure only" system modeling is conveniently developed:

$$I_e = \frac{0.6I_x}{1 + 36I_x/A_w e^2} \qquad (2.4.39)$$

Now the deflection of the coupling beam of Figure 2.4.22 at the idealized shear yield (V_{sp}) is

$$\delta_y = \frac{V_{sp}e^3}{12EI_e}$$

where e is the clear span plus the cover on each side, 48.8 in., and

$$I_x = 350 \text{ in.}^4$$

$$A_w = 2.4 \text{ in.}^2$$

$$I_e = \frac{0.6I_x}{3.2} \qquad \text{(Eq. 2.4.39)}$$

$$= \frac{0.6(350)}{3.2}$$

$$= 65.6 \text{ in.}^4$$

$$\delta_y = \frac{V_{sp}e^3}{12EI_e}$$

$$= \frac{67(48.8)^3}{12(29{,}000)(65.6)} \qquad (V = 290 \text{ kN})$$

$$= 0.34 \text{ in.} \qquad (9 \text{ mm})$$

and this is reasonably consistent with the test (Figure 2.4.22*b*). It should, however, be recalled that the idealizations proposed for the concrete coupling beams were also stiffer than those measured experimentally.

Conclusions: From a design perspective it is more convenient to use a single estimate of coupling beam stiffness for the concrete alternatives because the choice of coupling beam reinforcing will be made later in the design process and it will be based on the shear imposed on the coupling beam and the rotation to which it is likely to be subjected.

It appears reasonable to conclude that

- The idealized capacities (M_{yi}, θ_{yi}, and θ_u) of each coupling beam type are predictable, and a nonstrength-degrading load path can be developed through careful detailing, providing rotation limit states are not exceeded.
- The load path created by the system described in Figure 2.4.13*c* seems to have the least amount of strength degradation of the concrete coupling beam alternatives. Further, the attained level of rotation exceeds that of the other concrete coupling beams by 50%.
- Concrete coupling beams of the flexural type (Figure 2.4.13*a*) can be designed and detailed so as to reliably attain rotations of 4 and 5%, but the load imposed on the compression diagonal must be controlled through the use of capacity-based design. A shear stress alert level on the order of $4\sqrt{f_c'}$ for short shear spans and

$5\sqrt{f'_c}$ for shear spans of one or more can be used for conceptual design. The stress imposed on the compression diagonal should be limited to $0.35 f'_c$.

- Diagonally reinforced coupling beams can sustain higher loads, but ultimate deformations will be less than those attained in well-designed flexural coupling beams.
- The use of a steel or composite coupling beam provides a reliable load path to rotation levels not attainable in concrete coupling beams.

2.4.2.3 Design Procedures—Coupling Beams Coupling beam design procedures will have as an objective the identification of member strength and deformation limit states. The usual system design process will be one in which the configuration of the wall system and its penetrations are established based upon the functional requirements of the building. The system conceptual design process should identify the magnitude of the required strength and the ultimate rotation the coupling beam will probably experience. The development of these system design objectives (M_n and θ_u) is the focus of this section. We start by considering component deformation limit states and how to optimize them.

Flexural Behavior Model (Figure 2.4.13 a)

DEFORMATION LIMIT: A deformation limit for this type of coupling beam has not as yet been proposed. The tests performed by Park and Paulay[2.2] suggest a probable limit of 0.02 radian (Figure 2.4.16) for an overreinforced coupling beam that is 39 in. deep and has a span-to-depth ratio of 1 (shear span $= 0.5d$). This was also the deformation level at which strength degradation was observed by Galano and Vignoli[2.43] (Figure 2.4.15*a*). The Galano beam had a span-to-depth ratio of 1.5 (shear span $= 0.75d$). The diagonal compressive stress (f_{cd}) in the Paulay beam (Figure 2.4.16) was, however, higher than we would suggest. The deformation limit for an overreinforced coupling beam ($\ell_c/d \leq 1$) should be about 0.02 radian. This deformation limit state (0.02 radian) should be increased for longer shear spans or when induced stresses in diagonal compression strut are controlled. A rotation limit state of 0.04 radian seems appropriate for coupling beams whose span-to-depth ratio (ℓ_c/d) exceeds one (1) based on experimental evidence currently available (Figure 2.4.17).

A conceptual design, given a shear span of at least $0.5d$, may be based on the idealized yield rotation and a rotational ductility factor (μ_θ) of 10:

$$\theta_{u,\max} = \mu_\theta \theta_{by} < 0.04 \text{ radian}$$

$$= 10\frac{V\ell_c^2}{12EI_e} \tag{2.4.40a}$$

$$\boxed{\theta_{u,\max} = \frac{0.83V\ell_c^2}{EI_e} \leq 0.04 \text{ radian}} \tag{2.4.40b}$$

The *idealized stiffness* of a conventionally reinforced coupling beam should be developed from its effective moment of inertia as defined by Eq. 2.4.38a:

$$I_e = \frac{0.2I_g}{1 + 3\,(h/\ell_c)^2} \qquad \text{(Eq. 2.4.38a)}$$

STRENGTH LIMIT STATE: Stress imposed on the compressive diagonal should be controlled. This may be accomplished by limiting the shear imposed on the coupling beam:

$$\lambda_o V_n \le 0.08 f_c' bd \qquad \text{(Eq. 2.4.37a)}$$

FLEXURAL REINFORCING: Flexural reinforcing should be limited so as to encourage yielding of the longitudinal reinforcement. The nominal flexural strength of the section should include the strength provided by all longitudinal reinforcing:

$$M_n \le \frac{0.08 f_c' bd\ell_c}{2\lambda_o} \qquad (2.4.41)$$

where λ_o is 1.25.

DETAILING CONSIDERATIONS: Horizontal bars should be placed at 12 in. on center (see Figure 2.4.16) and confinement reinforcement in the form of closed hoops should be provided such that v_s exceeds $0.09 f_c'$ or 450 psi, whichever is greater.

Diagonally Reinforced Coupling Beams (Figure 2.4.13b) The current perception is that diagonally reinforced coupling beams are capable of sustaining design shears to much higher levels of rotation than those that are conventionally reinforced. This "current perception" is based on the work of Paulay and Priestley[2.5] (Figure 2.4.19), but this conclusion is obtained from essentially monotonic loading (see load cycle 13 of Figure 2.4.19). Even higher levels of deformability are possible in conventionally reinforced coupling beams (Figure 2.4.17) provided the shear span and the level of induced shear stress are controlled.

DEFORMATION LIMIT: Figure 2.4.15*b* suggests that a rotation of 0.04 radian can be attained for a diagonally reinforced coupling beam, while Figure 2.4.19 suggests that the deformation of a diagonally reinforced coupling beam subjected to monotonic loading can reach 0.06 radian and still maintain its load. Given the available data there seems reason to assume that a properly designed conventionally reinforced coupling beam (Figure 2.4.17) will be capable of sustaining larger ultimate deformations than a properly designed diagonally reinforced coupling beam. It is also clear that the diagonally reinforced coupling beam is at least somewhat stiffer than its conventionally reinforced counterpart. This suggests that a rotational ductility factor (μ_θ) of about 20 be used in the conceptual design phase to analytically predict attainable deformation limit states.

For conceptual design purposes it is reasonable to presume that

$$\boxed{\theta_{u,\max} = \frac{1.65V\ell^2}{EI_e} \leq 0.04 \text{ radian}} \tag{2.4.42a}$$

and that the idealized stiffness is

$$\boxed{I_e = \frac{0.4I_g}{1+3(h/\ell)^2}} \quad \text{(Eq. 2.4.38c)}$$

STRENGTH LIMIT STATE: The strength of a diagonally reinforced coupling beam is

$$V_n = V_d + V_f \tag{2.4.42b}$$

$$= \frac{4A_s'' f_y(h/2 - d')}{\ell_d} + \frac{2A_s f_y(d - d')}{\ell_c} \quad \text{(see Eq. 2.4.30b and Eq. 2.4.28c)}$$

where A_s'' is the reinforcing in the diagonal strut and A_s is the flexural reinforcing (see Figure 2.4.18).

Photo 2.10 Diagonally reinforced coupling beam. (Courtesy of Morley Construction Co.)

The so-provided strength should not exceed the capacity of the diagonal strut, and this will depend to a large extent on the additional capacity provided by the diagonal strut. A limit state of $0.5 f_c'$ is reasonably used for conceptual design purposes. Accordingly,

$$\frac{4(V_d + V_f)}{b(d - d')} \leq 0.5 f_c' \tag{2.4.42c}$$

DETAILING CONSIDERATIONS: Dealing with congestion will be the main problem. Beam size must be sufficient to allow multiple levels of reinforcement (Figure 2.4.18) and the effective placement of concrete. Compression struts must be confined if their effectiveness is to be developed. Hoop reinforcing should also be provided.

Comment: Diagonal reinforcement should not be attempted in walls that are less than 16 in. thick. The ineffectiveness of the diagonal reinforcing in the thin Galano walls is clearly demonstrated (Figure 2.4.15). In my opinion, unless strength is an overriding consideration, diagonally reinforced coupled beams should not be used.

Truss Reinforced Coupling Beam (See Figure 2.4.13c) The deformation limit state for this system seems to be higher than that available in either the conventionally reinforced or diagonally reinforced coupling beam. Clearly, its behavior follows that of the conventionally reinforced system and, accordingly, so should its design procedure. However, the rotational ductility could be increased to 20 and the deformation limit state to 0.06 radian.

DEFORMATION LIMIT STATE:

$$\boxed{\theta_{u,\max} = \frac{1.65 V \ell^2}{E I_e} \leq 0.06 \text{ radian}} \tag{2.4.42d}$$

$$\boxed{I_e = \frac{0.2 I_g}{1 + 3(h/\ell)^2}} \qquad \text{(Eq. 2.4.38a)}$$

STRENGTH LIMIT STATE: Follow conventionally reinforced coupling beam (Flexure Behavior Model) design process.

$$\boxed{V_n = \frac{2 A_s f_y (d - d')}{\ell_c} \leq \frac{0.08 f_c' b d \ell_c}{2\lambda_o}} \quad \text{(see Eq. 2.4.28b and Eq. 2.4.41)} \tag{2.4.43}$$

where A_s is the sum of the longitudinal and diagonal steel provided at ends (node of Figure 2.4.20) of the coupling beam.

Comment: This type of reinforcing program has not been used. Detailing and placement problems must be carefully studied if its use is contemplated. Clearly, additional experimentation is called for because the system appears to have merit, especially in thin walls (see Figure 2.4.15).

Steel Coupling Beams The design procedures used for steel coupling beams follow those developed for link beams used in steel eccentric braced frames (EBF). Certain modifications are essential to accommodate the embedment of the link or coupling beam in the concrete piers that constrain the rotation of the ends of the steel beams. Link beams in EBFs are intended to accommodate postyield deformations while the supporting structure behaves elastically (Figure 2.4.25). This postyield deformation may be accommodated by either a flexural or shear deformation in the link. Obviously there will be a behavior region that involves both flexural and shear postyield deformations. These behavior regions are defined from the idealized plastic capacities of the member in flexure and shear. The theoretical balanced link length (e_b) is

$$e_b = \frac{2M_p}{V_{sp}} \tag{2.4.44}$$

The link beam length for link beams embedded in concrete should include the depth of the concrete covering the reinforcing in the supporting pier:

$$e = \ell_c + 2c \tag{2.4.45}$$

where c is the depth of concrete cover.

DEFORMATION LIMIT STATES: The link rotation angle allowed by codes[2.45] is based on the nature of the yield mechanism. For link beam (coupling beam) lengths that are shear dominant, $\theta_{\max}$ (λ_p in Figure 2.4.25) is 0.08 radian. The link beam length (e)

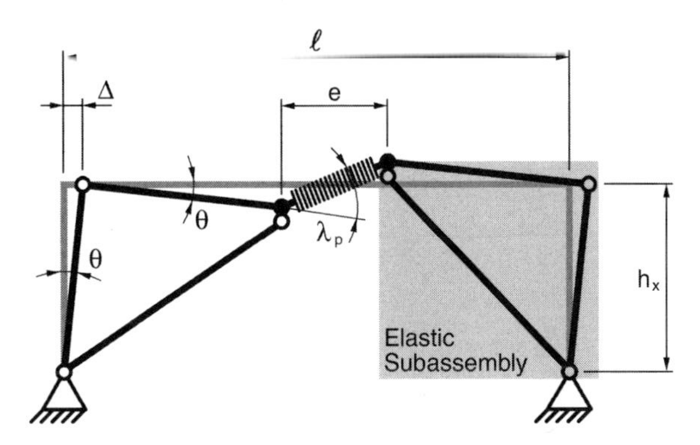

Figure 2.4.25 Postyield deformation of an EBF.

must be less than 80% of the balanced link beam length (e_b in Eq. 2.4.44) in order to qualify as shear dominant.

$$\theta_{\max} = 0.08 \text{ radian} \qquad \text{(Shear dominant)}$$

Shear dominance is presumed when

$$e \leq \frac{1.6M_p}{V_{sp}} \tag{2.4.46}$$

where

$$M_p = Z_x F_y \tag{2.4.47a}$$

and

$$V_{sp} = 0.6F_y t_w (d - 2t_f) \tag{2.4.47b}$$

For link beams that are considered to be flexure dominant, the link beam length must be 30% greater than the balanced link beam length (e_b) defined by Eq. 2.4.44:

$$\theta_{\max} = 0.02 \text{ radian} \qquad \text{(Flexure dominant)}$$

Flexure dominance is presumed when

$$e \geq \frac{2.6M_p}{V_{sp}} \tag{2.4.48}$$

A straight line interpolation is allowed between these postyield rotational limit states.

STRENGTH LIMIT STATE: The transferable shear force (V) is the lesser of

$$V_{nf} = \frac{2M_p}{e} \tag{2.4.49a}$$

and

$$V_{ns} = V_{sp} \qquad \text{(see Eq. 2.4.47b)} \tag{2.4.49b}$$

EFFECTIVE STIFFNESS: The impact of embedment length on coupling beam stiffness is accounted for by presuming that the analytical stiffness is 60% effective. When the link beam deformation is dominated by shear, shear deformations must be considered. First principles are used to determine the stiffness and the effective stiffness may be developed directly from

$$I_e = \frac{0.6I_x}{1 + 36I_x/A_w e^2} \qquad \text{(Eq. 2.4.39)}$$

DETAILING CONSIDERATIONS: Bearing in the embedded region is created through the formation of a stabilizing couple as described in Figure 2.4.26. The design shear (V) should be developed using a capacity-based approach ($\lambda_o V$). Bearing stress allowables are those prescribed by ACI[2.6] ($\phi 0.85 f_c' A_{be}$). The effective bearing area (A_{be}) is larger than the contact area (A_1) when the supporting area (A_2) is wider on all sides. The area increase is $\sqrt{A_2/A_1} \leq 2$ and ϕ is 0.7.

The number of variables involved make a direct solution impractical. At this point in the design, the designer should have developed the geometry required of the supporting shear wall and should be in a position to simplify the design process. The wall, for example must be at least 8 in. wider than the link beam. The cover concrete will spall so the effective face of the concrete should coincide with the location of the reinforcing bars; accordingly, any increase in area will be severely limited according to ACI provisions. The available bearing stress will, however, be conservative because the reinforcing bars placed in the ends of the supporting piers, if effectively tied, will confine the concrete that supports the steel coupling beam. It is realistic to assume that the confined compressive strength (f_{cc}') of the concrete will be activated (Section 1.2). Accordingly, the use of a bearing stress of $\phi 0.85 f_{cc}'$ is not unreasonable.

$$f_b = \phi 0.85 f_{cc}' \qquad (2.4.50)$$

$$= 0.7(0.85)1.5 f_c' \qquad \text{(see Eq. 1.2.1)}$$

$$= 0.9 f_c' \qquad (2.4.51)$$

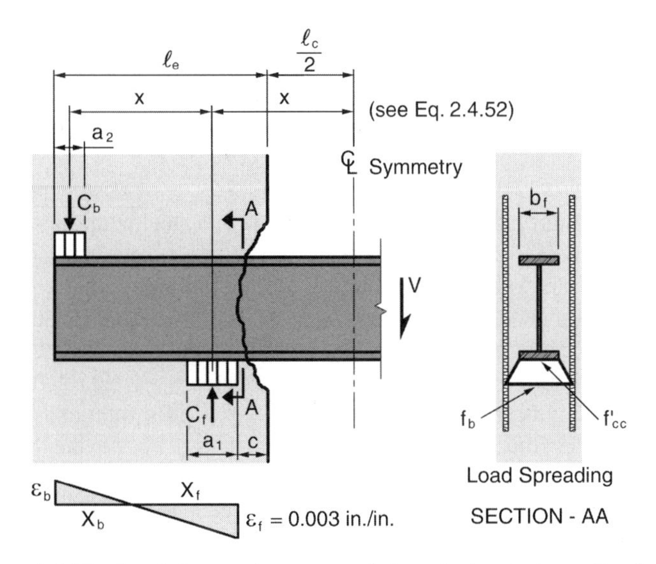

Figure 2.4.26 Loads imposed on encapsulating pier by steel coupling beam.

For design purposes it is reasonable to assume that the compressive force (C_f, see Figure 2.4.26), will be approximately equal to twice the shear force in the coupling beam. Since a bearing failure is not desired, the probable overstrength of the link beam should be considered.

$$C_f \cong 2\lambda_o V \tag{2.4.52}$$

The length of bearing (a_1) will be

$$a_1 = \frac{C_f}{0.9 f_c' b_f} \tag{2.4.53}$$

We have assumed, in the development of the embedment geometry, a relationship between C_b, C_f, and $V_n (C_f = 2\lambda_o V_n)$, and this requires that

$$C_b = \lambda_o V \tag{2.4.54}$$

From this we can solve directly for the required length of the beam. The length of bearing required to resist C_b is

$$a_2 = \frac{\lambda_o V}{0.9 f_c' b_f} \tag{2.4.55}$$

and x (Figure 2.4.26) is then

$$x = \frac{a_1}{2} + c + \frac{\ell_c}{2} \tag{2.4.56}$$

The embedded length (ℓ_e) becomes

$$\begin{aligned} \ell_e &= x + \frac{a_1}{2} + c + \frac{a_2}{2} \\ &= 1.25a_1 + 2c + \frac{\ell_c}{2} \quad \left(a_2 \cong \frac{a_1}{2}\right) \end{aligned} \tag{2.4.57}$$

For conceptual design purposes the total length of the steel coupling beam may be presumed to be on the order of $2.5\ell_c$.

The coupling beam web in the link area of a shear dominant coupling beam must be stiffened. A stiffener need only be placed on one side of the web provided that the link depth is less than 25 in. The thickness of the stiffener should be the larger of t_w or $\frac{3}{8}$ in. The spacing of the stiffeners is a function of the probable postyield rotational demand. The required spacing between stiffeners is $(30t_w - d/5)$ for rotations of 0.08 radian and $(52t_w - d/5)$ for rotation angles of 0.02 radian or less. For flexure dominant links $(2.6M_p/V_{sp} > e > 5M_p/V_{sp})$, an intermediate stiffener is also required. This means that a full height stiffener must be placed a distance of $1.5b_f$ from the end of the link. These design rules are a part of the Seismic Provisions for Steel Buildings[2.45, Sec. 10.3] and they should be reviewed for currency. A full height stiffener, on each side of the web, should also be provided at the edge of the reinforcing bar cage.

Figure 2.4.26 describes the loads imposed on the encapsulating piers. Tensile stress will be imposed on the concrete and cracking should be controlled so as not to allow gaps to form on the sides of the beams opposite to that on which the bearing load is being delivered. Harries and coauthors[2.44] provided longitudinal reinforcing over the entire embedment length. The provided strength was three times the shear strength of the link beam, and this seems excessive. This reinforcing was distributed so that two-thirds was placed near the link. The after-failure condition of the test specimen (Figure 2.4.27) suggests that this was sufficient to control cracking. Figure 2.4.23 describes the pier reinforcing program used by Harries and coauthors.[2.44] The suggested reinforcing program transfers one-half of C_f in the boundary region. The process is described in the example that follows and detailed in Figure 2.4.28.

Quantities required are

$$A_{s1} = \frac{\lambda_o V}{f_y} \tag{2.4.58}$$

$$A_{s2} = \frac{A_{s1}}{2} \tag{2.4.59}$$

Tension bars should be provided in sufficient quantities to assure compatibility between the coupled piers—a subject that will be discussed in more detail in Chapter 3. These bars may extend into the concrete surrounding the coupling beam when composite coupling beams are used. A_{s1} bars (Figure 2.4.28) should be confined with hoop ties extending at least 16 in. above and below the steel coupling beam. The hoop

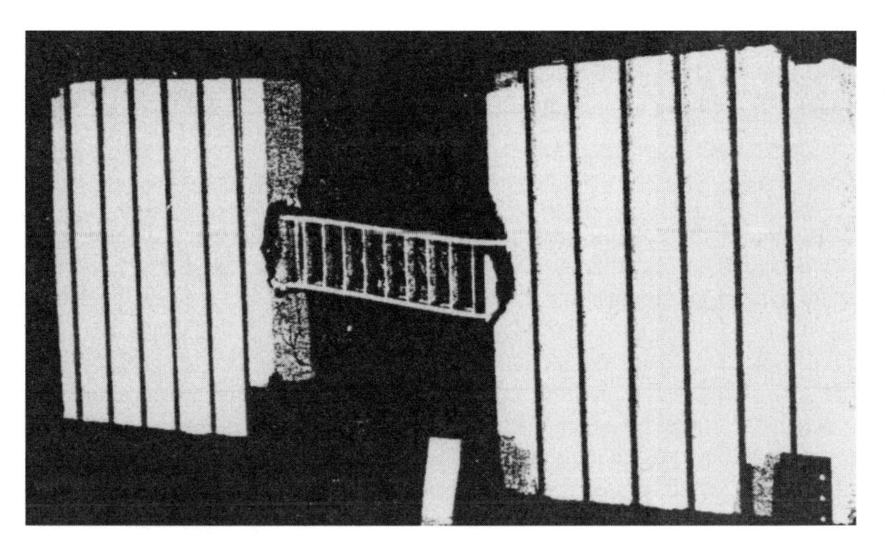

Figure 2.4.27 Specimen 2 after testing. (From K. A. Harries, D. Mitchell, W. D. Cook, and R. G. Redwood, "Structural Response of Steel Beams Coupling Concrete Walls," *Journal of Structural Engineering, ASCE*, Vol. 119, No. 12, December 1993. Reproduced by permission of the publisher, ASCE.)

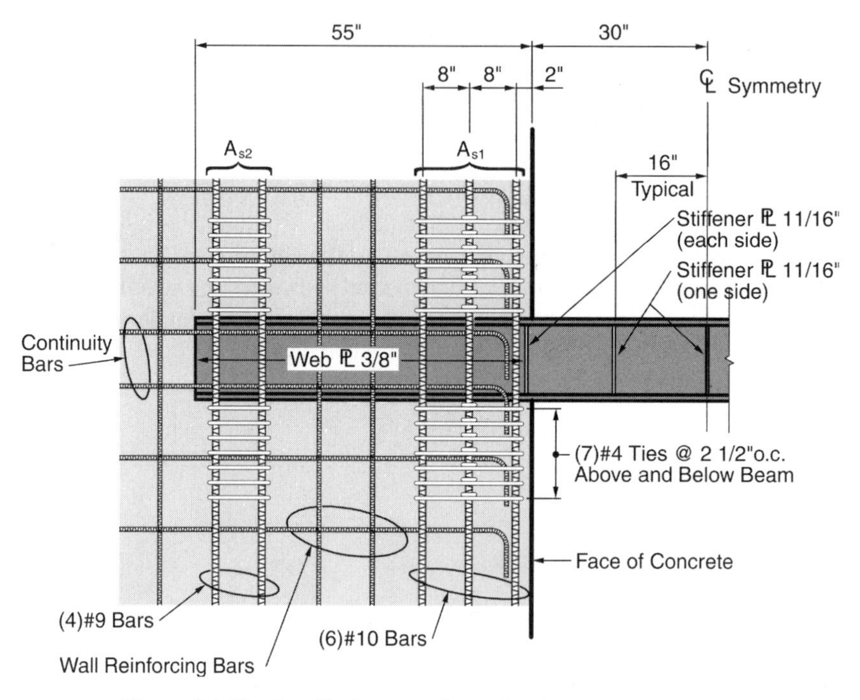

Figure 2.4.28 Detail of proposed coupling beam at support pier.

reinforcing should provide an "effective" confining pressure of about $0.09f_c'$ in order to support the attainment of a confined concrete strength (f_{cc}') on the order of $1.5f_c'$. Paulay and Priestley[2.5] argue that confinement is not as effective in a wall section as it is in a column. They propose that, in order to develop an effective confining pressure of $0.09f_c'$ in the wall, a confining pressure of $0.15f_c'$[2.5, Fig 3.6] should be provided.

$$f_\ell = 0.15f_c' \tag{2.4.60}$$

The portion of the coupling beam that extends into the pier (ℓ_e in Figure 2.4.26) must not yield in either flexure or shear beyond the link. SPSSB[2.45] provisions require a flexural strength that exceeds the link demand by 25% ($\lambda_o V$). Harries and coauthors[2.44] suggest that only the flanges of the coupling beam be used to establish the capacity of the section. In a flexure dominant link it suggests that the flanges be strengthened, but this tends to be dangerous because the introduction of welding in the plastic hinge region may cause the beam to fail prematurely. The shear strength provided outside the link region is also an important consideration. Harries and associates demonstrated that, when the shear strength outside the link region was the same as that provided in the link, both the postyield strength and the ductility of the coupling beam were reduced. In a shear dominant link this means that the embedded portion of the web must be reinforced. Quantitatively, once a decision has been made to reinforce the web, the thickness of the added web plate is not a major cost. Harries increased the thickness of the web outside the link region by 72% and shear yielding was clearly restricted to the link region. Based on the level of overstrength reported (Figure 2.4.22), it seems reasonable to provide a shear strength outside the link region that exceeds the nominal demand by 50%.

$$V_{p,\text{ext}} \geq \lambda_o V_{p,\text{link}} \tag{2.4.61a}$$

where λ_o in this case is 1.5 and $V_{p,\text{ext}}$ is that portion of the link beam embedded in the pier.

Example Design: W18 × 130 ($F_y = 50$ ksi) coupling beam.

$$\ell_c = 5 \text{ ft}, \qquad c = 2 \text{ in.} \qquad f_c' = 5 \text{ ksi}$$

$$\begin{aligned} V_{sp} &= 0.6F_y(d - 2t_f)t_w && \text{(Eq. 2.4.47b)} \\ &= 30(16.85)(0.67) \\ &= 339 \text{ kips} \\ M_p &= Z_x F_y && \text{(Eq. 2.4.47a)} \\ &= 291(50) \\ &= 14{,}550 \text{ in.-kips} \\ V_{pf} &= \frac{2M_p}{e} \\ &= \frac{2(14{,}550)}{64} \\ &= 455 \text{ kips} \\ \frac{1.6M_p}{V_p} &= \frac{1.6(14{,}550)}{339} && \text{(see Eq. 2.4.46)} \\ &= 68.6 \text{ in.} \qquad (e = 64 \text{ in.}) \end{aligned}$$

Conclusion: Link is shear dominant ($V = V_{sp} = 339$ kips).

$$\begin{aligned} a_1 &= \frac{2\lambda_o V_{sp}}{0.9(f_c')b_f} && \text{(see Eq. 2.4.53)} \\ &= \frac{2(1.25)(339)}{0.9(5)(11.16)} \qquad (b_f = 11.16 \text{ in.}) \\ &= 16.9 \text{ in.} \\ a_2 &= \frac{\lambda_o V_{sp}}{0.9 f_c' b_f} && \text{(Eq. 2.4.55)} \end{aligned}$$

$$= \frac{1.25(339)}{0.9(5)(11.16)}$$

$$= 8.43 \text{ in.}$$

$$x = \frac{a_1}{2} + \frac{\ell_c}{2} + c \qquad \text{(see Eq. 2.4.56)}$$

$$= 8.45 + 30 + 2$$

$$= 40.45 \text{ in.}$$

$$\ell_e = 1.25a_1 + 2c + \frac{\ell_c}{2} \qquad \text{(Eq. 2.4.57)}$$

$$= 1.25(16.9) + 2(2) + 30$$

$$= 55 \text{ in.}$$

STIFFENER DESIGN

Comment: The link is shear dominant—design the stiffeners so as to assure the maximum attainable rotation. This is appropriate because, as we shall see in the next subsection and in Chapter 3, coupling beam rotations will tend to be large and a realistic quantification of probable deformation demand difficult to make.

$$\theta_p \cong 0.08 \text{ radian}$$

Thickness of stiffener $\quad t_w = \frac{11}{16} = t_{\text{stiffener}}$

Spacing

$$s = \left(30t_w - \frac{d}{5}\right) \qquad (2.4.61\text{b})$$

$$= 30(0.67) - \frac{19.25}{5}$$

$$= 20.1 - 3.85$$

$$= 16.25 \text{ in.}$$

Conclusion: Provide an $\frac{11}{16}$-in. stiffener on each side of the web inside of the face of the pier (see Figures 2.4.26 and 2.4.28). Provide an $\frac{11}{16}$-in. thick stiffener at the quarter points of the beam.

$$e = \ell_c + 2c \qquad \text{(Eq. 2.4.45)}$$

$$= 60 + 2(2)$$

$$= 64 \text{ in.}$$

$$\frac{e}{4} = \frac{64}{4}$$
$$= 16 < 16.25 \text{ in.}$$

BOUNDARY REINFORCING

$$A_{s1} = \frac{\lambda_o V}{f_y} \qquad \text{(Eq. 2.4.58)}$$
$$= \frac{1.25(339)}{60}$$
$$= 7.1 \text{ in.}^2$$

Provide four #9 bars or six #8 bars for A_{s1} and two #9 bars or four #7 bars at A_{s2} (see Figure 2.4.28).

Confinement must be provided at the ends of the supporting piers (A_{s1} bars). The flange width of W18 by 130 is $11\frac{1}{8}$ in. and the length of bearing required, based on satisfying Eq. 2.4.53, was 16.9 in. The selected wall width should be 20 in. Accordingly, a 17 in. by 18-inch concrete column section should be provided. Three reinforcing bars must be provided on each face to develop a section of this size. The vertical spacing (s) between #4 bars is

$$s = \frac{A_{sh} f_y}{0.15 f_c' h} \qquad \text{(see Eq. 2.4.60)}$$
$$= \frac{3(0.2)(60)}{0.15(5)(18)}$$
$$= 2.67 \text{ in.}$$

The resultant design is conceptually detailed in Figure 2.4.28.

2.4.2.4 Coupled Shear Walls with Stacked Openings—Design Process and Example The design of a coupled shear wall must flow from a fundamental understanding of how the system will behave and the acceptance of a criterion that acknowledges this behavior. A first yield criterion is not appropriate and will probably produce a system that will fail. This is because the coupling beams that connect shear piers like those described in Figure 2.4.29 will be subjected to large rotations before the system approaches its deformation limit. Accordingly, a displacement-based design procedure is most appropriate. If, however, a force-based procedure is used, it is essential that the deformation level imposed on the coupling beam be understood in the conceptual design phase and that the appropriate wall system geometry and coupling beam type be selected. The conceptual design example that follows will be extended into a final design in Section 3.1.3, where a sequential yield analysis should confirm the validity of the conceptual design approach proposed.

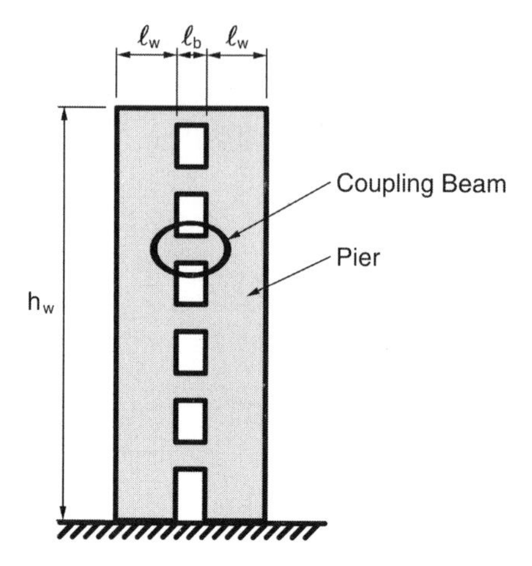

Figure 2.4.29 Symmetrical coupled shear walls.

The behavior of coupled shear walls is most easily understood by making some simplifying assumptions and reductions. First, the symmetrically organized coupled shear walls of Figure 2.4.29 can be reduced to a representation of a story as described in Figure 2.4.30. The rotation imposed on the coupling beam is clearly a function of the rotation that exists in each pier or individual wall and the geometry of the coupled wall system.

The rotated shear pier will cause the center of the coupling beam to raise or drop δ_1 where

$$\delta_1 = \theta_w \left(\frac{\ell_w}{2} + \frac{\ell_b}{2} \right) \tag{2.4.62a}$$

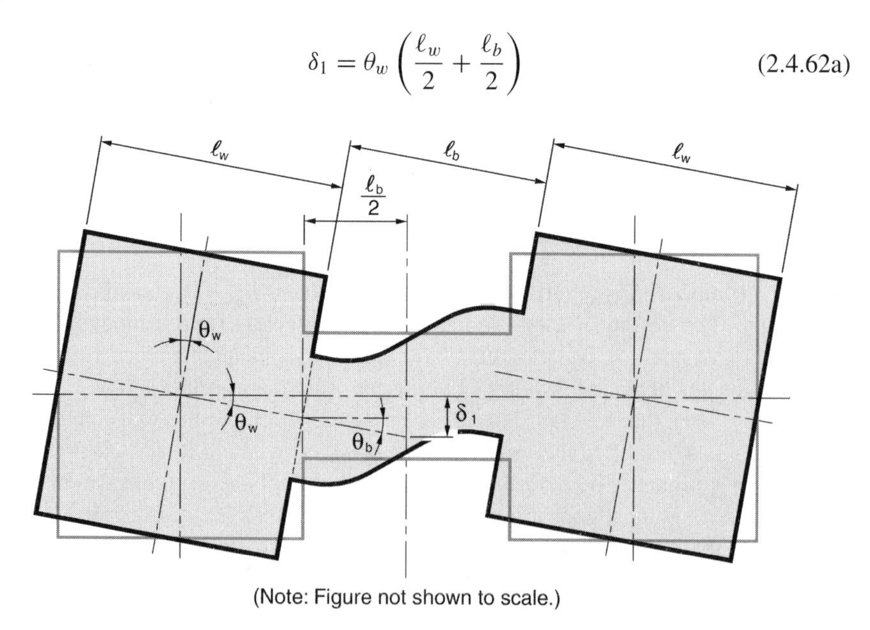

Figure 2.4.30 Deformed shape of a story-symmetrical coupled shear wall (see Figure 2.4.29).

The rotation imposed on the coupling beam will be

$$\theta_b = \frac{\delta_1}{\ell_b/2} \tag{2.4.62b}$$

Combining Eq. 2.4.62a and Eq. 2.4.62b allows the creation of a relationship between the rotation of a pier (θ_w) and the coupling beam (θ_b):

$$\theta_b = \theta_w\left(\frac{\ell_w}{\ell_b} + 1\right) \tag{2.4.63}$$

The rotation of the pier (θ_w) will be a function of the strength of the pier (elastic behavior) and the postyield rotation imposed on the base of the pier at system deformation limit states. These relationships were studied for a single shear wall in Section 2.4.1.2. Specifically, the idealized yield moment (M_{yi}) and curvature at yield (ϕ_{yi}) were quantified:

$$\phi_y = \frac{0.0033}{\ell_w} \qquad \text{(Eq. 2.4.27)}$$

$$M_{yi} = \lambda_o M_n \qquad \text{(see Eq. 2.1.2c)}$$

and an expression for the displacement of the shear wall at yield (Δ_{wy}) was proposed:

$$\Delta_{wy} = \frac{11}{40}\phi_y h_w^2 \qquad \text{(see Eq. 2.4.17)}$$

This allowed us to develop an elastic drift or, when combined with a postyield rotation (θ_p) at the base, an inelastic drift capability (see Eq. 2.4.17).

In order to select the appropriate coupling beam type or adjust the geometry of the system, an estimate of the probable pier rotation at the level of interest (θ_{wx}) must first be made. Equation 2.4.17 suggests that the elastic component of curvature in the wall is not quite linear (Figure 2.4.8), for a linear distribution of curvature would cause the deflection of the wall at yield (Δ_{wy}) to be described by $0.33\phi_y h_w^2$ (Eq. 2.4.17), as opposed to $0.275\phi_y h_w^2$. The adoption of a linear distribution of curvature is more convenient from a design perspective, even though it will tend to be conservative, for it allows us to easily identify the probable magnitude of pier rotation at a particular level. Subsequent design decisions should subliminally consider this conservatism.

The conjugate beam (Figure 2.1.10) developed from the elastic portion of the curvature diagram of Figure 2.4.8, which we now linearize, when combined with an estimate of the curvature in the wall at idealized yield, allows us to predict the elastic deformation of a pier, and this will provide significant design insight.

If we assume that the idealized curvature at first yield is

$$\phi_y = \frac{0.0033}{\ell_w} \qquad \text{(Eq. 2.4.27)}$$

then the rotation of the shear pier at a particular level (h_x) may be determined as

$$\theta_{wx} = \frac{\phi_y}{2}\left(\frac{h_x}{h_n}\right)(2h_n - h_x) \tag{2.4.64a}$$

Combining Eq. 2.4.27b and Eq. 2.4.64a, we can refine the expression for the rotation in the shear pier at the level of interest (θ_{wx}):

$$\theta_{wx} = \frac{0.00165}{\ell_w}\left(\frac{h_x}{h_n}\right)(2h_n - h_x) \tag{2.4.64b}$$

and from this the coupling beam rotation at that level can be developed using Eq. 2.4.63:

$$\theta_{bx} = \frac{0.00165}{\ell_w}\left(\frac{h_x}{h_n}\right)(2h_n - h_x)\left(\frac{\ell_w}{\ell_b} + 1\right) \tag{2.4.65}$$

Once the wall yields, the rotation at every level will be increased by the postyield rotation of the wall, θ_{wp}, and θ_{bx} becomes

$$\theta_{bx} = \left[\left(\frac{0.00165}{\ell_w}\right)\left(\frac{h_x}{h_n}\right)(2h_n - h_x) + \theta_{wp}\right]\left(\frac{\ell_w}{\ell_b} + 1\right) \tag{2.4.66}$$

The insight provided in the conceptual design phase by the relationships developed in this section are best demonstrated by example. We use the coupled shear wall described in Figure 2.4.31 to describe how one might proceed with a conceptual design.

First, consider how this wall system will behave as it approaches the anticipated ultimate roof deformation Δ_u. We would expect that most, if not all, of the coupling

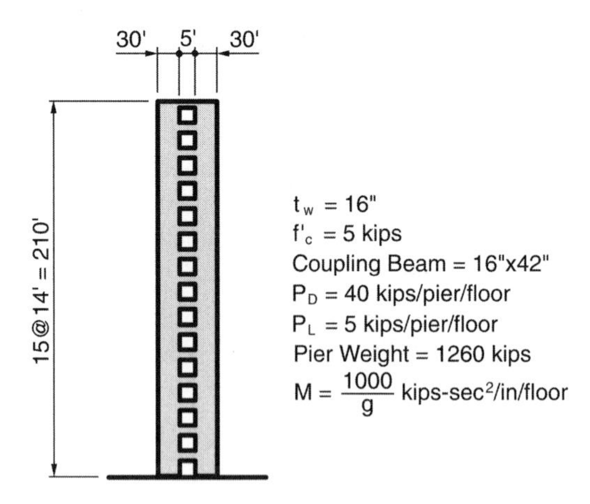

Figure 2.4.31 Coupled wall system.

beams would yield, and that yielding would also be expected at the base of each shear pier. Our primary concern in the conceptual design phase is with the extent of rotation imposed on the coupling beams because, if the imposed level of rotation is too high, the strength of the coupling beams will be reduced or disappear (see Figure 2.4.15), and the coupled system will "unzip," producing two essentially independent walls. Accordingly, our design objective is to select a coupling beam system that would effectively transfer shear at the anticipated level of system drift. The experimental evidence previously reviewed (Section 2.4.2.1) suggests that the rotation imposed on a concrete coupling beam be limited to somewhere between 0.012 and 0.04 radian, depending to a large extent on the design and detailing of the coupling beam. If we expect to exceed this level of coupling beam rotation, a steel or steel composite coupling beam should be considered, for nonstrength degrading rotations of 0.08 radian are attainable (see Figure 2.4.22*b*). Should higher rotations be suggested, the geometry of the system must be altered or the behavior of the uncoupled system accepted.

For the coupled wall of Figure 2.4.31, the rotation imposed on the uppermost coupling beam at h_n when the shear piers reach our adopted yield is developed as follows:

$$\ell_w = 30 \text{ ft}$$

$$\ell_b = 5 \text{ ft}$$

$$h_w = 210 \text{ ft}$$

$$\theta_{bn} = \frac{0.00165}{\ell_w} h_n \left(\frac{\ell_w}{\ell_b} + 1 \right) \qquad \text{(see Eq. 2.4.65)}$$

$$= \frac{0.00165(210)}{30} \left(\frac{30}{5} + 1 \right)$$

$$= 0.081 \text{ radian}$$

This conclusion will be conservative for it assumes that the pier is infinitely stiff longitudinally, but this will not be the case because the coupling beams will create an axial load differential in the piers.

It appears as though at least the uppermost coupling beams, if they are constructed of concrete, cannot be expected to survive a building drift that causes the piers of the coupled shear wall to reach our adopted yield curvature (Eq. 2.4.28a). The deflection at yield may be estimated through the use of Eq. 2.4.17 and Eq. 2.4.27b:

$$\phi_y = \frac{0.0033}{\ell_w} \qquad \text{(see Eq. 2.4.27)}$$

$$= \frac{0.0033}{30(12)}$$

$$= 0.0000092 \text{ rad/in.}$$

$$\Delta_{yw} = \frac{11}{40}\phi_y h_w^2 \qquad \text{(see Eq. 2.4.17)}$$

$$= 0.275(0.0000092)(210)^2(144)$$

$$= 16 \text{ in.} \qquad (\theta_{wy} = 0.064 \text{ radian})$$

The anticipated level of postyield building drift can be approximated using the procedures developed in Section 2.4.1.2. An estimate of the fundamental period of the coupled wall structure can be made through the use of Eq. 2.4.11 by starting with one of the walls (piers) and treating it as though it were uncoupled. The wall system described in Figure 2.4.31 has a tributary mass (weight) of 1000 kips/floor. Hence the mass tributary to one wall or pier corresponds to a weight (W_f) of 500 kips/floor, and this translates to a uniform load on the wall of

$$w = \frac{W_f}{h_x} \qquad \text{(see Eq. 2.4.9)}$$

$$= \frac{500}{14}$$

$$= 35.7 \text{ kips/ft}$$

The deflection of this wall, subjected to its own weight, assuming an effective stiffness of 50% of I_g, is

$$I_e = \frac{0.5(t_w)\ell_w^3}{12}$$

$$= \frac{0.5(1.33)(30)^3}{12}$$

$$= 1496 \text{ ft}^4$$

$$\Delta_f = \frac{wh_w^4}{8EI_e}$$

$$= \frac{35.7(210)^4}{8(4000)(144)(1496)}$$

$$= 10 \text{ ft}$$

and the probable period of the wall or pier, assuming that shear drift is negligible, is

$$T_w = 0.89(\Delta_f)^{0.5} \qquad \text{(Eq. 2.4.11)}$$

$$= 0.89(10)^{0.5}$$

$$= 2.81 \text{ seconds}$$

Now if the spectral velocity for the site is 48 in./sec, the drift anticipated is developed as follows:

$$\omega_n = \frac{2\pi}{T} \qquad \text{(Eq. 2.4.5)}$$

$$= \frac{2(3.14)}{2.81}$$

$$= 2.23 \text{ rad/sec}$$

$$\Delta_{\max} = \frac{S_v}{\omega_n}\Gamma \qquad \text{(Eq. 2.4.14b)}$$

$$= \frac{48}{2.23}(1.5)$$

$$= 32.3 \text{ in.}$$

Absent the coupling action the individual wall would surely yield. Accordingly, an estimate of the effect of the coupling beams on the dynamic characteristics of the system must be made. An elastic analysis is, of course, possible using our estimates of effective coupling beam stiffnesses, but we know the coupling beams and the tension side pier will have yielded long before the system reaches what we consider to be idealized system yield. Accordingly, an elastic analysis of the coupled wall system would be of little value, for the objective is to develop an expression for the idealized or secant stiffness of the system when the deflection of the coupled system reaches system yield, not coupling beam first yield. The effective increase in stiffness provided by the coupling beams can be developed in a relative sense.

When a wall is coupled to another wall, as it is in Figure 2.4.31, it is reasonable to presume that most, if not all, of the coupling beams will reach their strength limit state. The force imposed on a coupled shear wall (pier), once yielding occurs in all of the coupling beams, can be expressed as an increment of moment (M') at the base of each wall or pier (see Figure 2.4.32):

$$M' = \sum V_{by}\left(\frac{\ell_w}{2} + \frac{\ell_b}{2}\right) \qquad (2.4.67)$$

The hypothetical, in the sense that it is developed for comparative purposes only, stiffness of an independent pier (k_w) is then increased by k', and the relative stiffness of the coupled system (k^*) is

$$k^* = k_w + k' \qquad (2.4.68)$$

$$= M_{yi} + M'$$

$$= \phi_{yi} E I_e + \sum V_{by}\left(\frac{\ell_w}{2} + \frac{\ell_b}{2}\right) \qquad (2.4.69)$$

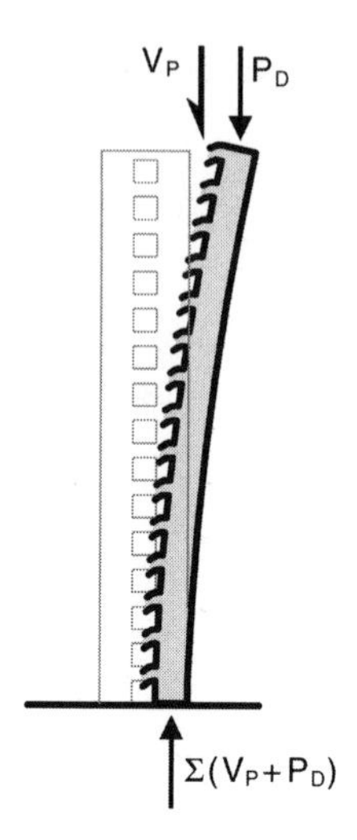

Figure 2.4.32 Elastic model for the compression side pier described in Figure 2.4.31.

Comment: M_{yi} may not reach its elastic idealization $(\phi_{yi} EI_e)$. This need not be a deterrent because stiffness is not a function of the level of applied load. The force-based relationship of Eq. 2.4.69 is accordingly used for comparative purpose.

Our interest is in the incremental increase in stiffness (A) where

$$A = \frac{k_w + k'}{k_w}$$

$$= 1 + \frac{k'}{k_w}$$

$$= 1 + \frac{M'}{M_{yi}} \tag{2.4.70}$$

for this will allow an estimate of the period of the coupled system (T_{cw}) to be made:

$$T_{cw} = \frac{T_w}{\sqrt{A}} \tag{2.4.71}$$

For the example coupled system (Figure 2.4.31), if in the design process we control the shear stress imposed on the coupling beams (Eq. 2.4.37b), the strength of the coupling beams can be estimated.

$$V_{by} = \frac{0.08 f_c' bd}{\lambda_o}$$

$$= 0.26(16)(39)$$

$$= 160 \text{ kips} \qquad (f_c' = 4 \text{ ksi}; \lambda_o = 1.25)$$

$$M' = \sum V_{by}\left(\frac{\ell_w}{2} + \frac{\ell_b}{2}\right) \qquad \text{(Eq. 2.4.67)}$$

$$= 15(160)(17.5)$$

$$= 42{,}000 \text{ ft-kips}$$

and the strength of a pier at yield is

$$\phi_{yi} = \frac{0.0033}{\ell_w}$$

$$M_{yi} = \phi_{yi} E I_e \qquad (2.4.72)$$

$$= \frac{0.0033}{30}(4000)(144)(1496)$$

$$= 94{,}787 \text{ ft-kips}$$

$$A = 1 + \frac{M'}{M_{yi}} \qquad \text{(see Eq. 2.4.70)}$$

$$= 1 + \frac{42{,}000}{94{,}787}$$

$$= 1 + 0.44$$

$$= 1.44$$

The period of the coupled system is

$$T_{cw} = \frac{T}{\sqrt{A}} \qquad \text{(Eq. 2.4.71)}$$

$$= \frac{2.81}{\sqrt{1.44}}$$

$$= 2.34 \text{ seconds}$$

The maximum anticipated displacement of the coupled wall system is developed as before:

$$\omega_n = \frac{2\pi}{T}$$

$$= \frac{2(3.14)}{2.34}$$

$$= 2.68 \text{ rad/sec}$$

$$\Delta_{\max,cw} = \frac{S_v}{\omega_n}\Gamma \qquad \text{(Eq. 2.4.14b)}$$

$$= \frac{48}{2.68}(1.5)$$

$$= 26.9 \text{ in.}$$

Since Δ_{wy} is 16 in., one might reasonably conclude that the piers will yield. The postyield rotation at the base of each pier should be

$$\theta_p = \frac{\Delta_{\max} - \Delta_y}{h_w - \frac{\ell_p}{2}} \qquad \text{(see Eq. 2.1.8)}$$

$$= \frac{26.9 - 16}{(210 - 7.5)12}$$

$$= 0.0045 \text{ radian}$$

Equation 2.4.64a may be used to determine the rotation of the shear pier at any level. Alternatively, the rotation likely to be imposed on the coupling beam may be determined directly by amplifying Eq. 2.4.64b (see Eq. 2.4.63).

$$\theta_{bx} = \left\{\frac{\phi_y}{2}\left[\left(\frac{h_x}{h_n}\right)(2h_n - h_x)\right] + \theta_{wp}\right\}\left(\frac{\ell_w}{\ell_b} + 1\right) \qquad \text{(see Eq. 2.4.66)}$$

For the design example this produces

$$\theta_{bx} = \left(\frac{0.00165}{\ell_w}\left\{\frac{h_x}{h_n}(2h_n - h_x)\right\} + 0.0045\right)(7)$$

For $h_x = h_n$,

$$\theta_b = (0.0000046(210)12 + 0.0045)7$$

$$= (0.0116 + 0.0045)7$$

$$= 0.113 \text{ radian}$$

For $h_x = 0.5h_n$,

$$\theta_b = 0.092 \text{ radian}$$

For $h_x = 0.25h_n$,

$$\theta_b = 0.067 \text{ radian}$$

Given our adopted strength limit state for the coupling beam, it is reasonable to presume a rotation limit state for a concrete coupling beam on the order of 0.04 radian. Anticipated levels of rotations are quite a bit higher than the deformation limit state for a concrete coupling beam. Accordingly, steel coupling beams should be proposed for the upper levels. The proposed conceptual design for the coupling beams of the wall system described in Figure 2.4.31 might reasonably start with

$$V_{bn} = 160 \text{ kips}$$

Conclusion: Coupling beam types should be

Levels 1–5: Concrete.

Levels 6–15: Steel.

Pier Strength: The flexural strength of the pier must be large enough to limit the amount of postyield rotation demand on the coupling beam. The displacement ductility demand proposed to this point is approaching 2 (26.9/16). This imposes a post wall yield rotation demand on the coupling beam of 0.032 radian (0.0045 (7)). The strength of the wall should be sufficient to be consistent with a yield displacement of about 16 in.

Proceed now to design the components of the system.

Step 1: Select Steel Coupling Beam. See Section 2.4.2.3, Subsection on Steel Coupling Beams.

$$V_p = 160 \text{ kips}$$

$$A_w = \frac{V_p}{f_{yv}} \qquad \text{(see Eq. 2.4.47b)}$$

$$\begin{aligned} f_{yv} &= 0.6F_y \\ &= 0.6(50) \\ &= 30 \text{ ksi} \end{aligned}$$

$$\begin{aligned} A_w &= \frac{160}{30} \\ &= 5.33 \text{ in.}^2 \end{aligned}$$

Evaluate a W18 × 50 beam:

$$\begin{aligned} A_w &= t_w(d - 2t_f) \qquad \text{(see Eq. 2.4.47b)} \\ &= 0.355(17.99 - 1.14) \\ &= 5.98 \text{ in.}^2 \end{aligned}$$

$$e = \ell_c + 2c \qquad \text{(Eq. 2.4.45)}$$
$$= 60 + 2(1.5)$$
$$= 63 \text{ in.}$$
$$M_p = Z_x F_y \qquad \text{(Eq. 2.4.47a)}$$
$$= 101(50)$$
$$= 5050 \text{ in.-kips}$$
$$V_{sp} = A_w(0.6)F_y \qquad \text{(Eq. 2.4.47b)}$$
$$= 5.98(30)$$
$$= 179.4 \text{ kips}$$

The maximum rotation for a W18 × 50 beam is developed from its postyield deformation behavior. Recall that a balance between shear and flexural yielding in a link beam is attained when

$$\frac{eV_p}{M_p} = 2 \qquad \text{(Eq. 2.4.44)}$$

for a W18 × 50 that has an effective (link) length (e) of 63 in. (Figure 2.4.25).

$$\frac{eV_p}{M_p} = \frac{63(179.4)}{5050}$$
$$= 2.24$$

This describes the behavior as tending toward flexure dominant and, based on the established behavior limits for link beams (Section 2.4.2.3, subsection on Steel Coupling Beams), produces a deformation limit state for a W18 × 50 of

$$\theta_{\text{max,W18}\times 50} = 0.08 - \frac{2.24 - 1.6}{1.0}(0.06)$$
$$= 0.042 \text{ radian}$$

The W18 × 50 may, in the final analysis, provide the rotational capacity required, but it seems advisable to develop as much rotational capacity in the link beam as possible. Accordingly, a shear yielding of the link beam must be promoted.

The beam selection process is made easier by converting Eq. 2.4.46 into one that directly relates section properties to the link beam length (e) required to produce a yielding in pure shear.

$$e \leq \frac{2.67Z_x}{t_w(d - 2t_f)} \qquad (2.4.73)$$

A W18 × 130 satisfies our objective provided

$$e \le \frac{2.67(291)}{(0.67)(19.25 - 2(1.2))}$$

$$\le 68.8 \text{ in.} \qquad \text{(OK)}$$

The shear strength of this beam is significantly higher than our objective of 160 kips, but this is acceptable and probably desirable, for we have increased the available ductility along with the strength and stiffness of the coupling beam. Observe also that this will further reduce the period and probable level of system drift.

$$V_{sp} = 0.6F_y t_w(d - 2t_f) \qquad \text{(Eq. 2.4.47b)}$$

$$= 339 \text{ kips} \qquad (\text{W18} \times 130)$$

Step 2: Select and Design the Concrete Coupling Beam. See Section 2.4.2.1. Select a flexurally reinforced beam so as to attain our design shear objective (160 kips). The reinforcing program may be changed in the final design if rotation levels warrant.

$$A_s = \frac{M_n}{(d - d')f_y}$$

$$= \frac{V(\ell_b/2)}{(d - d')f_y}$$

$$= \frac{160(30)}{36(60)}$$

$$= 2.22 \text{ in.}^2$$

Observe that this is minimum steel ($200/f_y$).

Conclusion: Reinforce the coupling beam with three #8 bars top and bottom.

$$M_n = A_s f_y(d - d')$$

$$= \frac{2.37(60)(36)}{12}$$

$$= 426.6 \text{ ft-kips}$$

$$V_n = \frac{2M_n}{\ell_b}$$

$$= \frac{2(426.6)}{5}$$

$$= 171 \text{ kips}$$

$$V_{n,\max} = \frac{0.08 f_c' bd}{\lambda_o} \qquad \text{(see Eq. 2.4.37b)}$$

$$= \frac{0.08(4)(16)(39)}{1.25}$$

$$= 160 \text{ kips}$$

Conclusion: Use 5000-psi concrete.

Step 3: Develop a Strength Criterion for the Shear Piers. A decision must first be made regarding the types of coupling beams proposed:

Levels 1–5: Concrete; $M_n = 426.6$ ft-kips, $V_p = 171$ kips.
Levels 6–15: Steel W18 × 130; $V_p = 339$ kips.

This will create an axial load on each pier of ΣV_p.

$$\sum Vp = 5(171) + 10(339)$$

$$= 4245 \text{ kips}$$

The tension side pier could be designed to remain in the pseudoelastic range. This would require that the pier be capable of sustaining a net tension load of 2385 kips (4245—P_D) and the moment associated with an idealized yield curvature of 0.0000092 rad/in. (0.0033/360). This, however, is not desirable, for it would require a significant amount of reinforcing and undoubtedly reduce pier ductility when the pier is subjected to compression.

The pier reinforcing required to cause the tension side pier to remain in the psuedoelastic range could be developed as follows:

$$M_{yw} = \phi_{yi} E I_e \qquad \text{(Eq. 2.4.72)}$$

$$= 0.0000092(4000)(1496)(1728)$$

$$= 95{,}000 \text{ ft-kips}$$

Comment: Some engineers might be uncomfortable with extending the pseudoelastic approach. The logical alternative is to use the relationship between tip displacement and moment for a cantilever beam loaded with a triangular (first mode) load distribution.

$$M_y = \frac{3.65 E I_e \Delta_y}{h_w^2}$$

$$= \frac{3.64(4000)(1496)(12)16}{(210)^2}$$

$$= 94{,}800 \text{ ft-kips}$$

Following the procedures developed in Section 2.1, we find

$$M_n = A_s f_y (d - d') \qquad \text{(Eq. 2.1.1)}$$

$$A_s = \frac{95{,}000}{60(28)}$$

$$\cong 56 \text{ in.}^2$$

and adding to this the tension balancing reinforcing, we get

$$A_s = \frac{T}{2f_y}$$

$$= \frac{2385}{2(60)}$$

$$\cong 20 \text{ in.}^2$$

The boundaries of each pier would require about 66 in.^{2} of reinforcing (1.2%). Intuitively, based on an understanding of frame beam design considerations, this is high and, when the pier is subjected to compression forces on the reverse loading cycle, likely to significantly reduce the level of available ductility.

It seems clear that the tension side pier should be allowed to yield, but this will change the system response and complicate the conceptual design process. Regardless of how one might elect to proceed, the postyield behavior of the system, when one or both piers yield, should be better understood. The behavior and how to arrive at a conceptual design are best described by continuing the design example.

An iterative design process converges rapidly, at least to the degree of accuracy possible at the conceptual stage in the design process. What would the consequences be if we elected to provide only about 0.5% reinforcement in the piers ($A_s = 27$ in.^{2})? The tension side pier would undoubtedly yield, as would the coupling beams. The stiffness of the system would decrease as the tension side pier loses its stiffness. In effect a system similar to the one described in Figure 2.4.32 would be created. The idealized yield strength of the compression side coupled single pier could be developed as follows:

$$\sum (P_D + V_p) = 1860 + 4245$$

$$= 6105 \text{ kips}$$

Following the methodology for the design of beam columns developed in Section 2.2,

$$a = \frac{\sum (V_p + P_D)}{0.85 f'_c b}$$

$$= \frac{6105}{0.85(5)(16)}$$

$$= 90 \text{ in.}$$

$$M_{yi} = A_s f_y(d - d') + \sum (V_p + P_D)\left(\frac{h}{2} - \frac{a}{2}\right) \quad \text{(see Eq. 2.2.3)}$$

$$= 27(60)(25) + 6105(15 - 3.75)$$

$$= 109{,}200 \text{ ft-kips}$$

From a stiffness perspective it would be conservative were we to neglect the stiffness provided by the tension side pier but yet include the additional strength provided by the coupling beam. Since the yield moment of the compression side pier is of the same order of magnitude as the moment developed by Eq. 2.4.72 (95,000 ft-kips) and subsequently extended to predict the probable period range of the coupled system, we may expand on the period previously developed (2.34 seconds). The effective mass has doubled because the tension side pier has lost its stiffness and the strength of the coupling beams (V_{by}) has also doubled.

$$T = \sqrt{2}\, T_{cw}$$

$$= 1.414(2.34)$$

$$= 3.3 \text{ seconds}$$

$$M' = \sum V_{bp}\left(\frac{\ell_w}{2} + \frac{\ell_b}{2}\right) \quad \text{(see Eq. 2.4.67)}$$

$$= 4245(17.5)$$

$$= 74{,}300 \text{ ft-kips}$$

$$A = 1 + \frac{M'}{M_{yi}} \quad \text{(Eq. 2.4.70)}$$

$$= 1 + \frac{74{,}300}{109{,}200}$$

$$= 1 + 0.68$$

$$= 1.68$$

$$T_{cw} = \frac{T_w}{\sqrt{A}} \quad \text{(Eq. 2.4.71)}$$

$$= \frac{3.3}{\sqrt{168}}$$

$$= 2.54 \text{ seconds}$$

The building drift for the coupled wall (cw) system now becomes

$$\Delta_{\max,cw} = \frac{T S_v}{2\pi} \cdot \Gamma$$

$$= \frac{2.54(48)(1.5)}{6.28}$$

$$= 29 \text{ in.}$$

The components of postyield deformation demand on the compression side pier are

$$\Delta_y = \frac{11}{40}\phi_y h_w^2 \qquad \text{(Eq. 2.4.17)}$$

$$= 0.275(0.0000092)(210)^2(144)$$

$$= 16 \text{ in.}$$

$$\theta_{wp} = \frac{\Delta_p}{h_w - \ell_w/4} \qquad \text{(see Eq. 2.1.8)}$$

$$= \frac{29 - 16}{(210 - 7.5)(12)}$$

$$= 0.0054 \text{ radian}$$

$$\phi_p = \frac{\theta_p}{\ell_p}$$

$$= \frac{0.0054}{0.5\ell_w}$$

$$= 0.00003 \text{ rad/in.}$$

This corresponds to a curvature ductility demand of

$$\mu_\phi = \frac{\phi_y + \phi_p}{\phi_y}$$

$$= \frac{0.000009 + 0.00003}{0.000009}$$

$$= 4.3$$

which is well within the limit of 10 suggested in Section 2.4.1.1.

Now the postyield rotational demand on the coupling beams must be estimated:

$$\theta_{bn} = \left(\frac{\phi_y}{2}h_n + \theta_{wp}\right)\left(\frac{\ell_w}{\ell_b} + 1\right) \qquad \text{(see Eq. 2.4.66)}$$

$$= \left(\frac{0.000009}{2}(210)(12) + 0.0054\right)(7)$$

$$= 0.12 \text{ radian}$$

This exceeds our established deformation limit state for the uppermost coupling beam. Recall that conservative assumptions were made in the development, so this should represent a good starting point for the development of the design. If need be, the strength of the piers may easily be increased. Reasonably, steel coupling beams throughout should be considered in the next iteration.

The capacity-based shear demand on a pier under compression would be

$$V_u = \frac{\lambda_o M_n}{0.67 h_w}$$

$$= \frac{1.25(109{,}200)}{0.67(210)}$$

$$= 970 \text{ kips}$$

$$v_u = \frac{V_u}{bd}$$

$$= \frac{970}{16(28.5)12}$$

$$= 0.178 \text{ ksi} \qquad \left(2.5\sqrt{f_c'}\right)$$

The proposed flexural reinforcing program is described in Figure 2.4.33 ($A_s = 28$ in.^{2}—eighteen #11 bars).

Step 4: Provide the Analysis Team with the Information Required to Confirm the Conceptual Design.

(a) Design sketch of coupled wall system and basic design data (Figure 2.4.31)

(b) Design sketch of proposed shear pier reinforcing program (Figure 2.4.33)

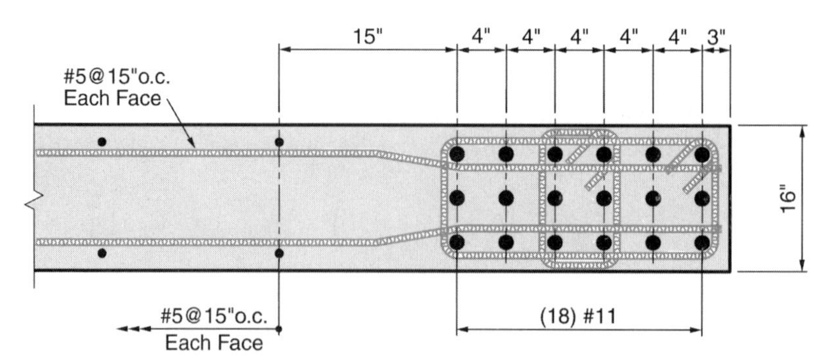

Figure 2.4.33 Conceptual pier reinforcing program (end of each pier of Figure 2.4.31).

(c) Determine a yield criterion for the coupling beams:
- Levels 1–15 (steel):

$$M_p = 850 \text{ ft-kips} \qquad (V_p e/2)$$

(d) Determine a stiffness criterion for the components of the coupled wall system:
- Wall: $I_e = 0.5 I_g$
- Coupling beams: $I_e = 500 \text{ in.}^4$ (see Eq. 2.4.50)

(e) Identify the design approach:
- Displacement-based analysis
- Spectral velocity of 48 in./sec

The preceding conceptual design process will be extended with the aid of a computer in Section 3.1.3.

2.4.2.5 Capped and Belted Shear Walls The preceding sections have clearly demonstrated that using coupling beams to join shear walls is difficult and will surely cause some postyield distress in the coupling beams were the system to approach its deformation limit state. Often it is functionally impossible to join shear walls, as for example in residential buildings where story heights are held to a minimum.

Let us revisit the thin wall system designed in Section 2.4.1.4. This wall (Figure 2.4.9) was 26 ft long (ℓ_w) and would conveniently be located between residential units or along an interior corridor, as described in Figure 2.4.34. At ten stories the aspect ratio of this wall (h_w/ℓ_w) was 4 and this, as we saw, presented no significant design problems (Section 2.4.1.4).

If, however, the height of this building were increased so as to accommodate twenty floors, the aspect ratio (h_w/ℓ_w) of a 26-ft long wall would be 8, and the design objectives for the walls would become much more difficult to attain. One means of attaining our behavior objectives would be to couple the walls with a cap beam, as

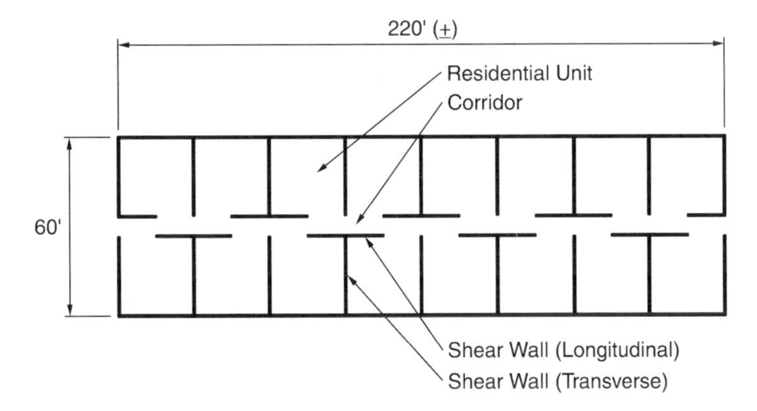

Figure 2.4.34 Typical residential floor plan.

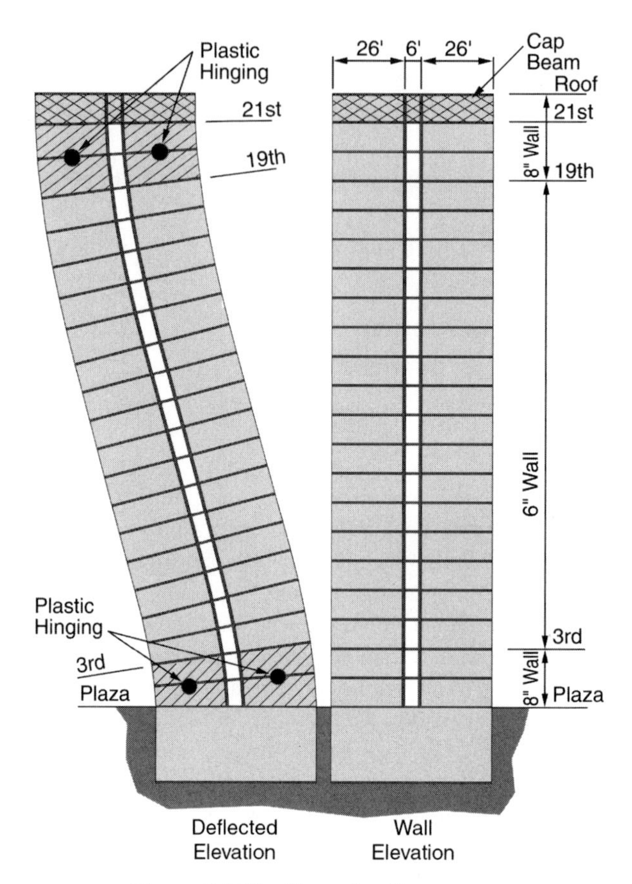

Figure 2.4.35 Capped wall system.

shown in Figure 2.4.35 or with a belt wall at a middle level. Observe how a cap wall alters the deflected shape, creating plastic hinges at the roof as well as the base of the walls or piers. Essentially four wall segments similar to that described in Figure 2.4.9 have been created.

Proceed now with the design of the coupled shear wall system shown in Figure 2.4.35.

A residential building is to be braced by thin shear walls. The tributary axial load/floor including the wall weight is 120 kips. The tributary weight (mass) per floor is 280 kips/floor. Use a displacement-based approach and assume that the wall is capable of attaining a curvature ductility factor of 9. Follow the displacement-based procedures summarized in Section 2.4.1.3. Use an effective wall (pier) stiffness of 50% of I_g. The spectral velocity (S_v) for the site is 44 in./sec.

Step 1: Determine the Design Basis. A displacement design basis has been selected.

Step 2: Determine the Thickness of the Wall. Outside the hinge region:

$$\frac{P_D}{f'_c A_g} < 0.25 \qquad \text{or} \qquad \frac{h}{t_w} < 25 \qquad \text{(Eq. 2.4.20 and Eq. 2.4.21a)}$$

$$\frac{P_D}{f'_c A_g} = \frac{20(60)}{5(26)(t_w)(12)} \le 0.25$$

$$t_{w,\min} = \frac{1200}{5(26)(12)(0.25)} \qquad \text{(Eq. 2.4.20)}$$

$$= 3.08 \text{ in.}$$

Conclusion: Outside the hinge region a 6-in. wall may be used ($h_x/t_w = 20$).

For the hinge region—$0.5\ell_w$ (13 ft):

$$t_{w,\min} = \sqrt{\frac{P_D}{3.4 f'_c \beta_1}} \qquad \text{(Eq. 2.4.21b)}$$

$$= \sqrt{\frac{1200}{3.4(5)(0.8)}}$$

$$= 9.4 \text{ in.}$$

$$t_{w,\min} = \frac{h_x}{16}$$

$$= \frac{10(12)}{16}$$

$$= 7.5 \text{ in.}$$

Conclusion: An 8-in. thick wall should be used in the upper plastic hinge regions (levels 19 thru 21). Stability in the lower plastic hinge region (below level 3) must be insured, and this will require a review of tensile strains, concrete strains, and probably the development of an effective confined region. The proposed conceptual wall should be 8-in. thick in Levels 1, 2, 19, and 20, otherwise 6 in. thick.

Step 3: Determine the Period of the Structure. The procedure used follows that developed in Section 2.4.1.2, modified to account for the revised boundary condition at Level 20. The period of the building is developed from estimates of the building deflection were it laterally loaded with the weight of its tributary mass (Eq. 2.4.11).

The tributary weight per foot (w) to each pier is

$$
\begin{aligned}
w &= \frac{W}{2h_x} \\
&= \frac{280}{2(10.5)} \\
&= 13.3 \text{ kips/ft}
\end{aligned}
$$

The effective stiffness of each pier is

$$
\begin{aligned}
I_g &= \frac{0.5(26)^3}{12} \\
&= 732 \text{ ft}^4 \\
I_e &= 0.5 I_g \\
&= 366 \text{ ft}^4 \\
E I_e &= 4000 I_e \\
&= 4000(144)(366) \\
&= 211{,}000{,}000 \text{ kips-ft}^2
\end{aligned}
$$

The flexural component of the tip deflection of a cantilever beam that is restrained against tip rotation at the end (Figure 2.4.36*b*) is

$$
\begin{aligned}
\Delta_{f,1-20} &= \frac{w h_w^4}{24 E I_e} \\
&= \frac{13.3(210)^4}{24(211{,}000{,}000)} \\
&= 5.12 \text{ ft}
\end{aligned}
$$

To this we must add the contribution of the weight of Level 21 and half of Level 20:

$$
\begin{aligned}
\Delta_{f,21} &= \frac{P h_w^3}{12 E I} \\
&= \frac{210(210)^3}{12(211{,}000{,}000)} \\
&= 0.77 \text{ ft}
\end{aligned}
$$

$$
T_f = 0.89(\Delta_f)^{0.5} \qquad \text{(Eq. 2.4.11)}
$$

where Δ_f is the flexural component of roof deflection expressed in feet.

$$T_f = 0.89(5.89)^{0.5}$$
$$= 2.16 \text{ seconds}$$

The contribution from shear deflection may be neglected as being inconsequentially small (see Eq. 2.4.24).

Step 4: Determine the Expected Displacement Δ_u. The spectral velocity (S_v) for the site is 44 in./sec.

$$\omega_n = \frac{2\pi}{T}$$
$$= \frac{2(3.14)}{2.16}$$
$$= 2.9 \text{ rad/sec}$$

$$S_d = \frac{S_v}{\omega_n} \qquad \text{(see Eq. 2.4.6)}$$
$$= \frac{44}{2.9}$$
$$= 15.2 \text{ in.}$$

and the expected ultimate deflection at the roof is

$$\Delta_{20} = \Gamma S_d \qquad \text{(see Eq. 2.4.7)}$$
$$= 1.6(15.2)$$
$$= 24.3 \text{ in.}$$

Accordingly, the deflection of each half pier (see Figure 2.4.9) would be 12.15 in.

Step 5: Reinforce the Wall Segments So That They Can Attain the Displacement Objective without Exceeding the Curvature Ductility Limit State. The design process must start by understanding the consequences associated with the design decisions. Consider the coupled upper two piers as described in Figure 2.4.36*a*. With a positive sense displacement the moments shown will be activated at the base of the cap beam, and these moments will soon reach idealized yield (M_{yi}). The free body will be equilibrated by axial loads on the piers (C and T). The yield strength of each pier will depend to a certain extent on the level of axial load imposed on the pier in the hinge region. The restoring force will soon reach the mechanism strength of the entire wall system (Figure 2.4.36b) and

$$V_M = \frac{M_{y,TT} + M_{y,CT} + M_{y,TB} + M_{y,CB}}{h_w} \tag{2.4.74}$$

A first yield approach is clearly not realistic because M_{TT} will reach yield quite early, but its available ductility will be large.

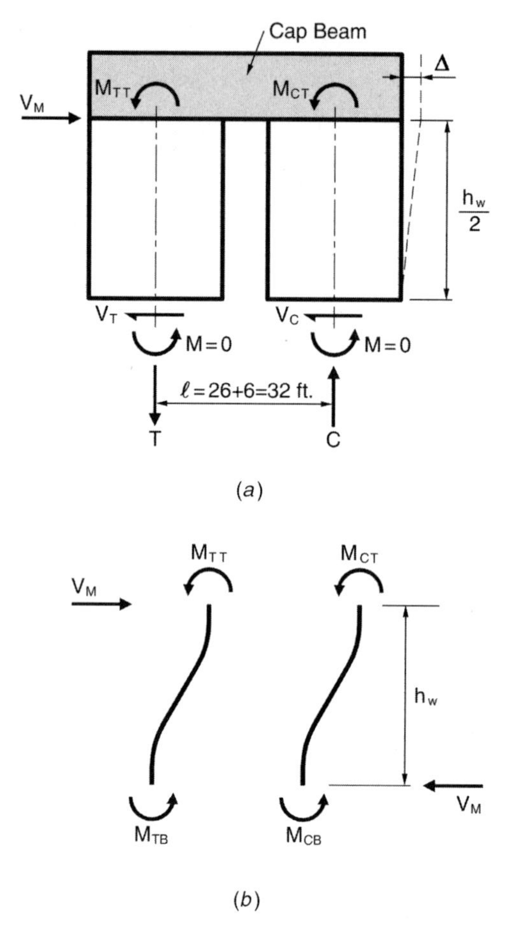

Figure 2.4.36 (*a*) Free body of the upper portion of the wall described in Figure 2.4.35. (*b*) Mechanism forces imposed on the wall system of Figure 2.4.35.

Given the complexity of the identified interrelationships, it is advisable to select a wall design and then determine if each pier can attain the probable level of displacement. To demonstrate the process adopt a pier design that satisfies minimum steel requirements and evaluate its behavior. The pier shown in Figure 2.4.10*a* accomplishes this objective.

Start by determining the mechanism shear, for this will establish the compression and tension load (C and T) imposed on each pier by a postyield displacement in the upper portion of the system (Figure 2.4.36*a*). Since C and T must be equal, the moment capacities added to the compression side pier will be the same as those relieved from the tension side pier. The dead load imposed on each pier will contribute to the yield moment and must be considered. The tributary dead load per floor is 120 kips or 60 kips/floor/pier. This produces the following loads:

$$P_{D,TT} = 120 \text{ kips} \qquad \text{(Levels 21 and roof)}$$

$$P_{D,CT} = 120 \text{ kips}$$

$$P_{D,TB} = 1260 \text{ kips} \qquad \text{(Levels 2 through roof)}$$

$$P_{D,CB} = 1260 \text{ kips}$$

The effective moment associated with pier yielding absent an axial load would be

$$\begin{aligned} M_{w,yf} &= A_s f_y(d - d') && \text{(see Eq. 2.4.1)} \\ &= 4.8(60)(25) && \text{(see Figure 2.4.10}a\text{)} \\ &= 7200 \text{ ft-kips} \end{aligned}$$

The additional yield moment caused by the effective dead load is

$$a_1 = \frac{120}{0.85(5)8} \qquad \text{(Level 20)}$$

$$= 3.52 \text{ in.}$$

$$a_2 = \frac{1260}{0.85(5)(8)} \qquad \text{(Level 1)}$$

$$= 37 \text{ in.}$$

$$M_{DT} = P_D\left(\frac{h}{2} - \frac{a}{2}\right) \qquad (h = \ell_w)$$

$$= 120(13 - 0.15)$$

$$= 1542 \text{ ft-kips}$$

$$M_{DB} = 1260(13 - 1.54)$$

$$= 14{,}440 \text{ ft-kips}$$

This results in a mechanism shear of

$$V_M = \frac{4M_{wy,f} + 2M_{DT} + 2M_{DB}}{h_w} \tag{2.4.75a}$$

$$= \frac{4(7200) + 2(1542) + 2(14{,}440)}{210}$$

$$= 288 \text{ kips} \tag{2.4.75b}$$

The C–T couple (Figure 2.4.36a) would be

$$C = T = \frac{2M_{wy,f} + 2M_{DT}}{\ell} \tag{2.4.75c}$$

$$= \frac{2(7200) + 2(1542)}{32}$$

$$= \frac{17{,}500}{32}$$

$$= 546 \text{ kips}$$

The ductility demand, based on the strength required of an elastic system, is developed as follows:

$$S_a = \omega_n S_v$$

$$= 2.9(44)$$

$$= 129.6 \text{ in/sec}^2$$

$$V_E = 0.8 S_a W \tag{2.4.76}$$

where 0.8 is an estimate of $\Gamma_1 \Sigma \phi_{1,i}$ (see reference 2.4, Eq. 4.7.16). Thus

$$V_E = 0.8 \frac{(129.6)}{386.4}(5880)$$

$$= 1577 \text{ kips}$$

$$\mu_v = \frac{V_E}{V_M}$$

$$= \frac{1577}{288}$$

$$= 5.48$$

The most critical plastic hinge region will be at the base of the compression pier. The axial load imposed on this pier would be

$$P_{CB} = P_D + C$$

$$= 1260 + 546$$

$$\cong 1800 \text{ kips}$$

An estimate of the yield moment would be developed as follows:

$$a = \frac{P_{CB}}{0.85 f_c' t_w}$$

$$= \frac{1800}{0.85(5)8}$$

$$= 53 \text{ in.}$$

$$
\begin{aligned}
c &= \frac{a}{\beta_1} \\
&= \frac{53}{0.8} \\
&= 66 \text{ in.}
\end{aligned}
$$

The strain imposed on this plastic hinge (the last to form) would be developed as follows. The free body diagram of the compression side pier is shown in Figure 2.4.37.

$$
\begin{aligned}
P_{CT} &= P_D + C \\
&= 120 + 546 \\
&= 666 \text{ kips} \\
P_{TT} &= -426 \text{ kips} \\
P_{CB} &= 1800 \text{ kips} \\
P_{TB} &= 714 \text{ kips} \\
a_{CT} &= \frac{P_{CT}}{0.85 f'_c t_w} \\
&= \frac{666}{0.85(5)(8)} \\
&= 19.6 \text{ in.}
\end{aligned}
$$

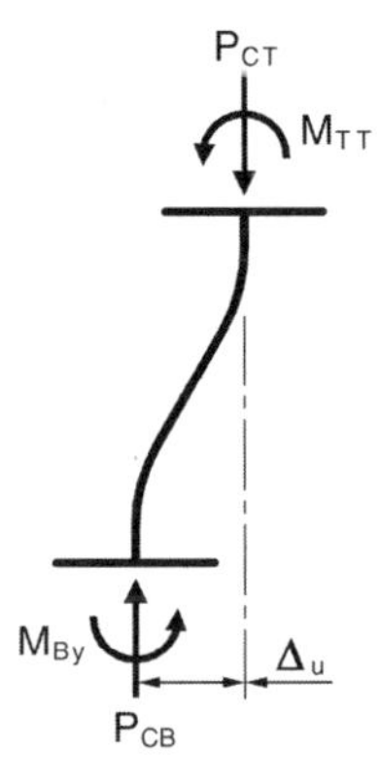

Figure 2.4.37 Partial free body diagram of the compression side pier (see Figure 2.4.35).

$$
\begin{aligned}
a_{CB} &= \frac{P_{CB}}{0.85 f_c' t_w} \\
&= \frac{1800}{0.85(5)(8)} \\
&= 53 \text{ in.} \\
a_{TB} &= 21 \text{ in.} \\
M_{y,CT} &= P_{CT}\left(\frac{h}{2} - \frac{a}{2}\right) + A_s f_y (d - d') \\
&= 666\left(\frac{26}{2} - 0.81\right) + 7200 \\
&= 8120 + 7200 \\
&= 15{,}320 \text{ ft-kips} \\
M_{y,CB} &= P_{CB}\left(\frac{h}{2} - \frac{a}{2}\right) + A_s f_y (d - d') \\
&= 1800\left(\frac{26}{2} - 2.4\right) + 7200 \\
&= 19{,}400 + 7200 \\
&= 26{,}600 \text{ ft-kips}
\end{aligned}
$$

Similarly, for the tension side pier,

$$
\begin{aligned}
M_{y,TB} &= P_{TB}\left(\frac{h}{2} - \frac{a}{2}\right) + M_{wy,f} \qquad (2.4.77) \\
&= 714(13 - 0.88) + 7200 \\
&= 8650 + 7200 \\
&= 15{,}850 \text{ ft-kips}
\end{aligned}
$$

$M_{y,TT}$ must consider the impact of a net tensile load of 426 kips.

$$
\begin{aligned}
P_{To} &= A_{st} f_y \\
&= 16.4(60) \\
&= 984 \text{ kips} \\
M_{y,TT} &= \frac{P_{To} - P_{TT}}{P_{To}} \left(A_s f_y (d - d')\right)
\end{aligned}
$$

$$= \frac{(984 - 426)}{984}(4.8)(60)(25)$$
$$= 4080 \text{ ft-kips}$$

Comment: To this point, discussion has centered about the global behavior of the coupled wall system (Figure 2.4.36*b*). Inertial forces are delivered to the floors so a reconciliation between pier force distribution and global distribution is sometimes required. The reconciliation may be accomplished with sufficient accuracy for design purposes as follows.

The shear at the base of the compression side pier is developed from the free body of Figure 2.4.38 which now includes the inertial floor loads (v_i):

$$M_{y,CB} - M_{y,CT} = \left(\sum v_i\right)(0.67)h_w = 0 \tag{2.4.78}$$

$$\frac{26{,}600 - 15{,}320}{0.67(210)} = \sum v_{i1}$$

$$\sum v_{i1} = 80.2 \text{ kips}$$

$$M_{y,TB} - M_{y,TT} = (\textstyle\sum v_i)(0.67)h_w$$

$$\frac{15{,}850 - 4080}{0.67(210)} = \sum v_{i2}$$

$$\sum v_{i2} = 83.7 \text{ kips}$$

The T/C couple described in Figure 2.4.36*b* will also translate into an inertial distribution on the coupled wall system.

$$T\ell = 546(32) \qquad \text{(see Figure 2.4.36}b\text{ and Eq. 2.4.75c)}$$
$$= 17{,}500 \text{ ft-kips}$$

$$\sum v_{i3} = \frac{T\ell}{0.67h_w}$$
$$= \frac{17{,}500}{0.67(210)}$$
$$= 124 \text{ kips}$$

and the total inertial load is

$$V = \sum v_{i1} + \sum v_{i2} + \sum v_{i3}$$
$$= 80.2 + 83.7 + 124$$
$$= 288 \text{ kips}$$

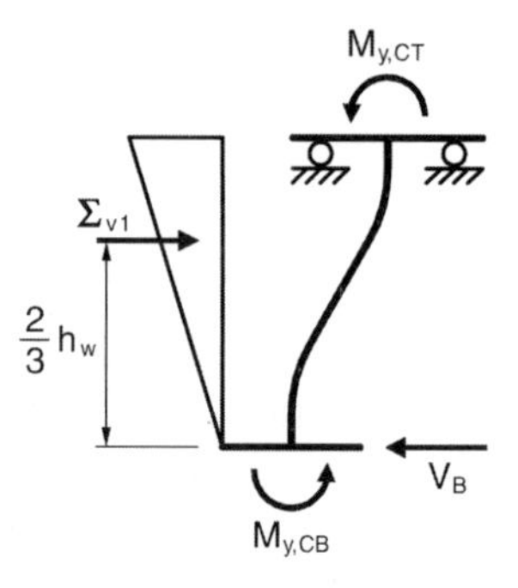

Figure 2.4.38 Free body diagram of the compression side pier.

This is the same conclusion as that developed from the mechanism shear that did not particularize the impact of the T/C couple on yield moments (Eq. 2.4.75b; $V_M = 288$ kips).

Step 6: Estimate the Probable Strain States. The strain imposed on the compression side pier at the base should now be determined. The deflection of this pier is comprised of an elastic and a postyield component as described in Figure 2.4.39.

In determining the elastic component of system drift, we may identify the probable point of inflection (h_T/h_B—see Figure 2.4.39) based on the yield moments or assume for convenience that the point of inflection is at midspan.

$$\begin{aligned}\phi_y &= \frac{0.0033}{\ell_w} \\ &= \frac{0.0033}{26(12)} \\ &= 0.0000106 \text{ rad/in.}\end{aligned}$$

$$\begin{aligned}\Delta_e &= 2\left(\frac{11}{40}\right)\phi_y\left(\frac{h_w}{2}\right)^2 \qquad \text{(see Eq. 2.4.17)} \\ &= 0.55(0.0000106)[105(12)]^2 \\ &= 9.2 \text{ in.}\end{aligned}$$

Comment: If $M_{y,CT}$ and $M_{y,CB}$ were used to locate the point of inflection, Δ_e would be 9.9 in. Remember that deflection compatibility (imposed at each floor) will not allow the point of inflection to move significantly.

$$\begin{aligned}\Delta_p &= \Delta_u - \Delta_e \\ &= 24.3 - 9.2 \\ &= 15.1 \text{ in.}\end{aligned}$$

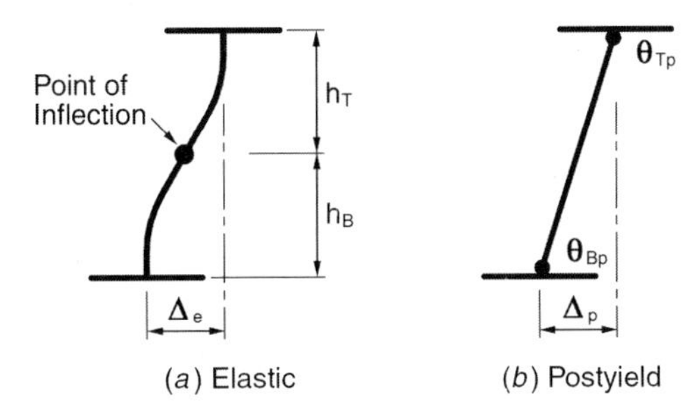

Figure 2.4.39 Components of pier deflection.

$$\theta_p = \frac{\Delta_p}{h_w - \ell_p}$$
$$= \frac{15.1}{(210 - 26)(12)}$$
$$= 0.0068 \text{ radian}$$
$$\phi_p = \frac{\theta_p}{\ell_p}$$
$$= \frac{\theta_p}{\ell_w/2}$$
$$= \frac{0.0068}{13(12)}$$
$$= 0.000044 \text{ rad/in.}$$

and this suggests a post (steel) yield concrete strain of

$$\varepsilon_{cp} = c\phi_p$$
$$= 66(0.000044) \qquad (a_{CB} = 53 \text{ in.}; \quad c_{CB} = 66 \text{ in.})$$
$$= 0.0029 \text{ in./in.}$$

The elastic component based on steel first yield is

$$\varepsilon_{cy} = \frac{\varepsilon_{sy}(kd)}{d - kd}$$
$$= \frac{0.002(0.33)}{0.67}$$
$$= 0.001 \text{ in./in.}$$

and the total strain at the extreme fiber in the plastic hinge region at the base of the compression pier at a system displacement of Δ_u is

$$\begin{aligned}\varepsilon_{cu} &= \varepsilon_{cy} + \varepsilon_{cp} \\ &= 0.001 + 0.0029 \\ &= 0.0039 \text{ in./in.}\end{aligned}$$

The axial load imposed on the base of the compression pier was 1800 kips. This corresponds to an axial load of $0.144 f_c' A_g$, which is well within acceptable limits.

The depth of the neutral axis is 66 in., and given our conservative limit state of $4t_w$, this suggests a wall thickness of 16.5 in., a wall width that we intuitively reject as being illogical. Recall that the T wall tested by Wallace (Figure 2.4.6) also violated this criterion ($c/t_w = 6.0$) with shear stresses $\left(6.5\sqrt{f_c'}\right)$ that were significantly higher than those imposed on this wall, and concrete strains that exceeded 0.01 in./in. It was noted in the analysis of the stem of the T section of Figure 2.4.6 that the confined core was capable of supporting the induced compression load on the concrete (C_c), and this might be a reasonable consideration here. The use of 6-ksi concrete also seems advisable. Observe that this will reduce the neutral depth to 55 in.

The introduction of an extended confined core (Figure 2.4.40) produces many positive side effects. The strain level at the outer fiber becomes smaller as the neutral axis depth is reduced and the yield strength of the wall is increased. The confined core should be capable of supporting the axial load.

$$\begin{aligned}f_\ell &= \frac{A_{sh} f_{yh}}{h_c s} \\ &= \frac{5(0.2)(60)}{(38)(3)} \\ &= 0.53 \text{ ksi} \qquad (0.088 f_c') \\ f_{cc}' &= f_c' + 4.1 f_\ell \\ &= 6.0 + 2.2 \\ &= 8.2 \text{ ksi} \\ C_{cc} &= f_{cc}' A_{cc} \\ &= 8.2(6)(38) \\ &= 1870 \text{ kips} > 1800 \text{ kips}\end{aligned}$$

Comment: Many simplifying assumptions afford significant insight into the behavior of this system. The designer should understand and accept, reject, or subliminally include the probable consequences of the adopted simplifications. Recall the moment

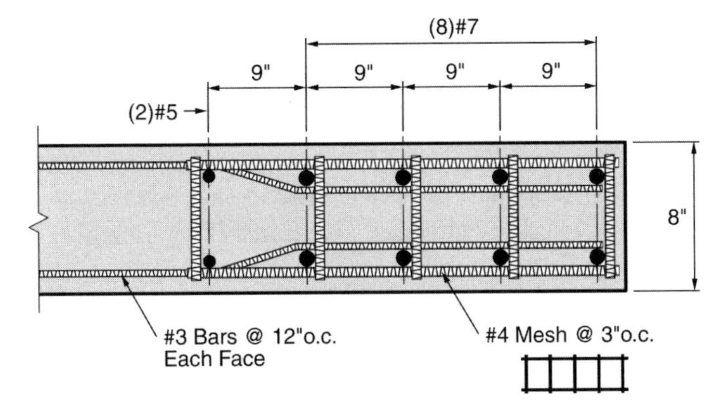

Figure 2.4.40 Boundary element—Levels 1 and 2.

curvature idealization used to develop the conceptual design. Idealized yield was assumed to be 9.2 in. ($\phi_y = 0.0000106$ rad/in.) and the mechanism base shear identified as being on the order of 288 kips. Clearly, these must be viewed as order of magnitude estimates.

The idealized behavior of the capped wall system is plotted on Figure 2.4.41 as is the base shear associated with the various hinge formations.

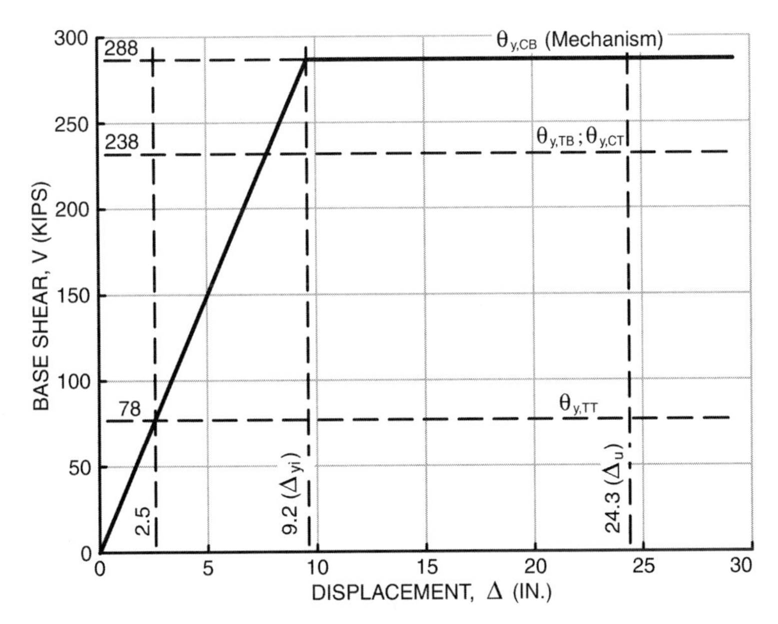

Figure 2.4.41 Moment displacement relationship for the capped wall described in Figure 2.4.35.

The estimate of concrete strain (0.0039 in./in.) was based on idealized behavior and should be fairly representative of the probable concrete strain imposed on the compression side pier at the base. The behavior of the plastic hinge at the top of the tension side pier would be almost exclusively in the postyield range. The consequences of this large a postyield displacement demand do not suggest damage that might prevent an acceptable behavior upon a reverse load cycle. The strain level in the concrete and steel at $M_{y,TT}$ might be estimated as follows

$$\theta_p = \frac{0.5\Delta_u - \Delta_y}{0.5(h_w - \ell_p)}$$

$$= \frac{0.5(24.3) - 2.5}{98.5(12)}$$

$$= 0.0082 \text{ radian}$$

$$\phi_p = \frac{\theta_p}{\ell_p}$$

$$= \frac{0.0082}{0.5(26)(12)}$$

$$= 0.000052 \text{ rad/in.}$$

The neutral axis depth will be on the centerline of the pier.

$$\varepsilon_{sp} = \frac{\phi_p \ell_w}{2}$$

$$= \frac{0.000052(26)(12)}{2}$$

$$= 0.0081 \text{ in./in.}$$

$$\varepsilon_s = \varepsilon_{sy} + \varepsilon_{sp}$$

$$= 0.0101 \text{ in./in.} \qquad (5\varepsilon_y)$$

In spite of the fact that yielding will occur very early at the top of the tension side pier, distress in this region should not be anticipated.

Step 7: Provide the Analysis Team with the Information Required to Confirm the Conceptual Design.

(1) System geometry—Figure 2.4.35

(2) Pier reinforcing:

Top: Figure 2.4.10*a* (Levels 19 and 20) $f'_c = 5000$ psi; $f_y = 60$ ksi

Bottom: Figure 2.4.40 (Levels 1 and 2) $f'_c = 6000$ psi; $f_y = 60$ ksi

Intermediate: Figure 2.4.10*b* (Levels 3–18) $f'_c = 5000$ psi; $f_y = 60$ ksi

(3) A stiffness criterion:
Pier: $t_w = 6$; $I_e = 0.5I_g$
Cap wall: $I_e = \infty$

(4) Load:
Tributary mass to system: 280 kips/floor
Tributary dead load to system: 120 kips/floor

2.4.2.6 Shear Walls with Randomly Placed Openings A random placement of openings in a shear wall is quite common, especially at the base of the wall where functional demands change as, for example, from residential to retail or parking. The location of the openings will want to be set by function, and usually only minor relocations are possible without impacting the functional objectives of the design. Accordingly, the behavior of a wall with randomly placed openings (Figure 2.4.12) must be understood by the conceptual designer and an acceptable load path must be developed during the conceptual design phase.

It is generally accepted that the load path will follow a strut and tie model, thereby creating a truss. This approach is discussed conceptually by Paulay and Priestley[2.5] for squat walls, but is also appropriately used in the design of tall walls in regions of discontinuity. By far the most troublesome opening location is one that occurs in the compression region near a boundary of a tall wall. Taylor, Cote, and Wallace[2.10] experimentally studied this problem by introducing an opening into the base of a wall whose overall configuration is that previously discussed and described in Figure 2.4.3*a* (RW2). This wall, which now contains an opening, and its behavior are shown in Figure 2.4.42. The attained level of drift and overall behavior (Figure 2.4.42) are essentially identical to that of RW2 (Figure 2.4.4*a*), leading the authors to conclude that "Slender structural walls with openings at the base can exhibit stable hysteretic behavior and significant ductility, even for the case where the opening is in the flexural compression zone." A stable load path and, in particular, an adequate compression flange and diagonal compression strut must be created.

The reinforcing program described in Figure 2.4.42*a* is complex, and to develop it into a strut and tie model requires many subjective decisions. Taylor, Cote, and Wallace[2.10] attempted such a development and monitored the forces in the tension ties; the monitored forces did not confirm the model developed. From a design perspective a simpler model can predict key forces to the order of magnitude required in the conceptual design process.

First, let us consider the case where the shear is directed toward the large pier (positive sense, Figure 2.4.42*d*). The complexities of the shear flow (Figure 2.4.43*a*) are simplified in this case to reflect a node at D that endeavors to collect the load that travels through the horizontal reinforcing in the pier from E, where it originates. to D. Quantitatively the load at D is developed from the mechanism strength of the wall ($\lambda_o M_n$). The design should create a compression node at E much as though it were the base of the wall and a tie between E and D to fully activate the strut DC. For conceptual design purposes, when the wall is tall, the design process might reasonably assume that the load at E was the same as that at C were the opening not to exist.

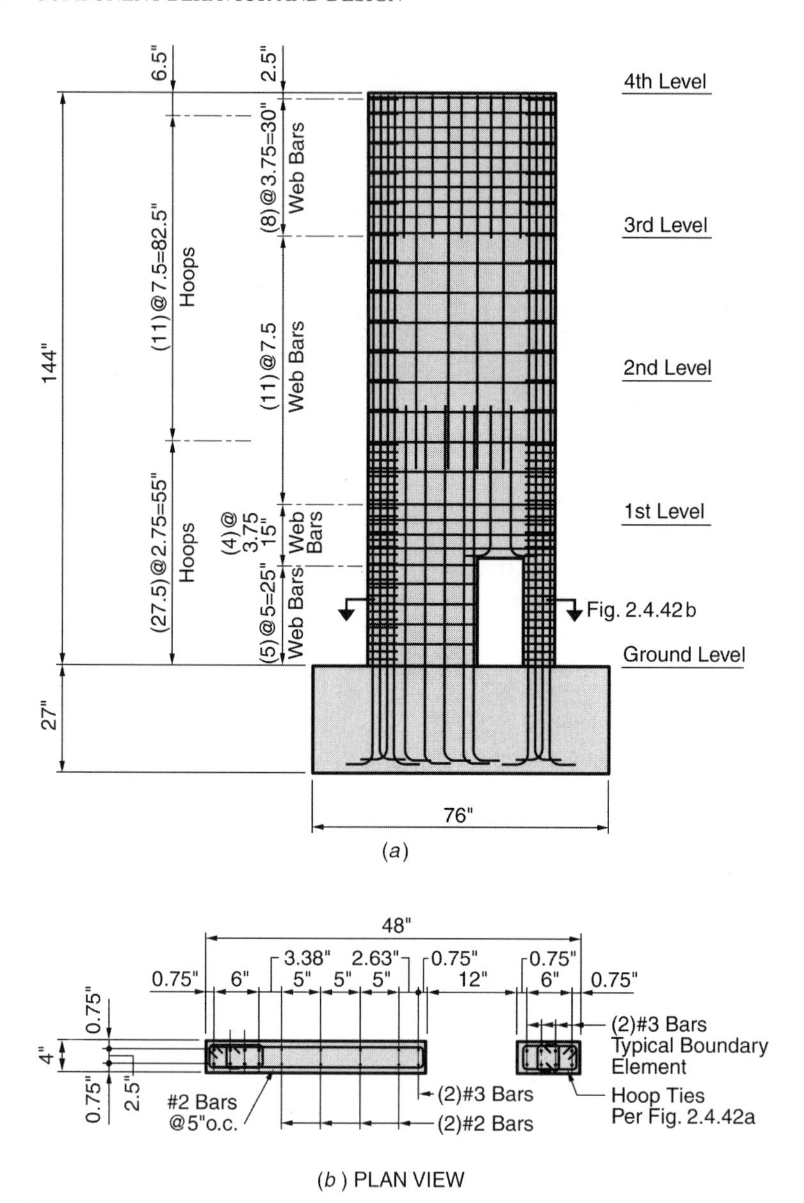

Figure 2.4.42 Tall thin wall model with an opening at the base. (*a*) Elevation view showing reinforcing. (*b*) Cross sections.

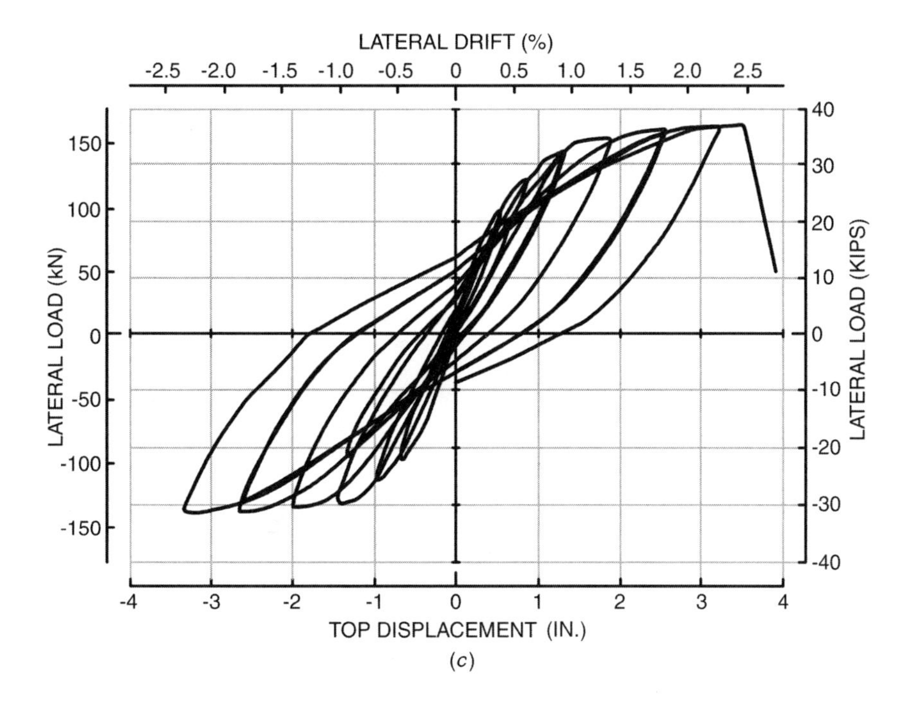

(*c*)

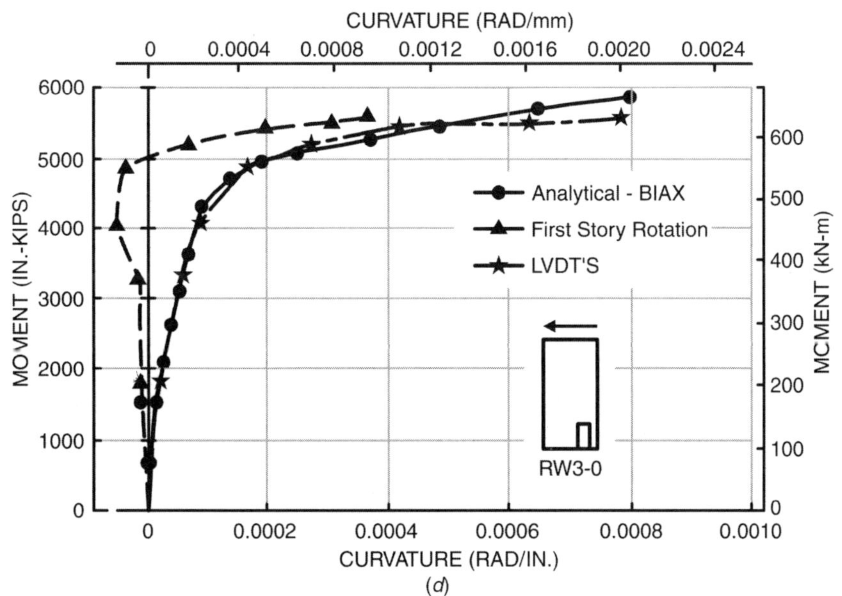

(*d*)

Figure 2.4.42 (*Continued*) (*c*) Lateral load versus top displacement. (*d*) Analytical versus experimental moment curvature response (positive displacement).

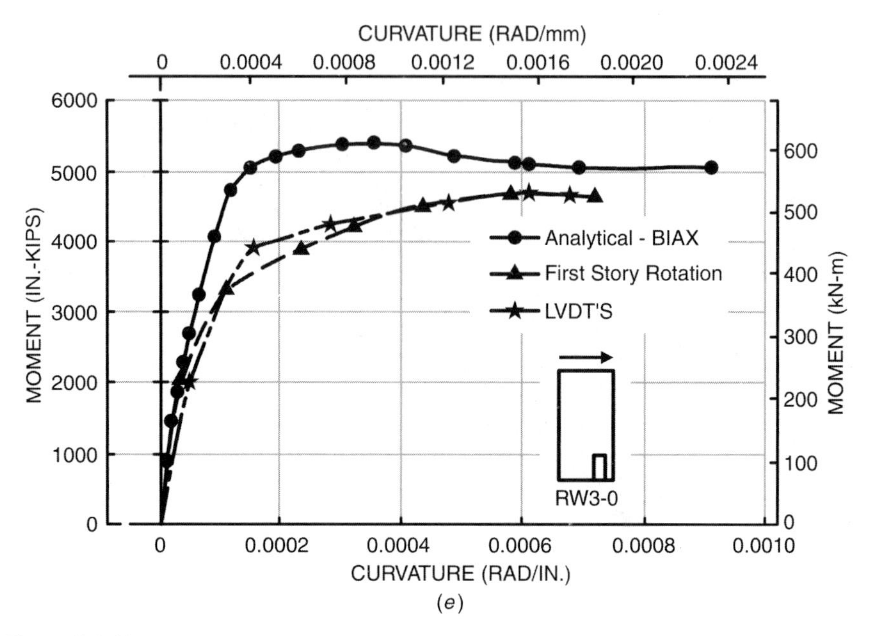

Figure 2.4.42 (*Continued*) (*e*) Analytical versus experimental moment-curvature response (negative displacement).

The design objective is to appropriately reinforce T_1 and T_2 and to ensure that the compression strut *CD* can carry the induced strut load. Two conditions are known, based on equilibrating the model described in Figure 2.4.43*a*. First,

$$M = Vh_w = P\ell_3 + T_1\ell_1 + T_2\ell_2 \tag{2.4.79}$$

Second, based on the load path developed in the pier,

$$V = \frac{\ell_2}{h_x}T_2 \tag{2.4.80}$$

Combining these relationships we can determine the relationship between T_1 and T_2 that will cause a balanced yielding in pier shear and wall flexure. This will allow us to decide how we wish the wall to behave in the postyield range.

$$\frac{\ell_2 h_w}{h_x}T_2 = P\ell_3 + T_1\ell_1 + T_2\ell_2 \tag{2.4.81a}$$

$$T_2 = \frac{P\ell_3 + T_1\ell_1}{(\ell_2 h_w)/h_x - \ell_2} \tag{2.4.81b}$$

Realizing that the geometry of the wall and the axial load will be known, and that the area of steel in the boundary element T_1 will probably be known because it has been developed based on attaining global design objectives, we can make a decision on how much strength should be provided at T_2.

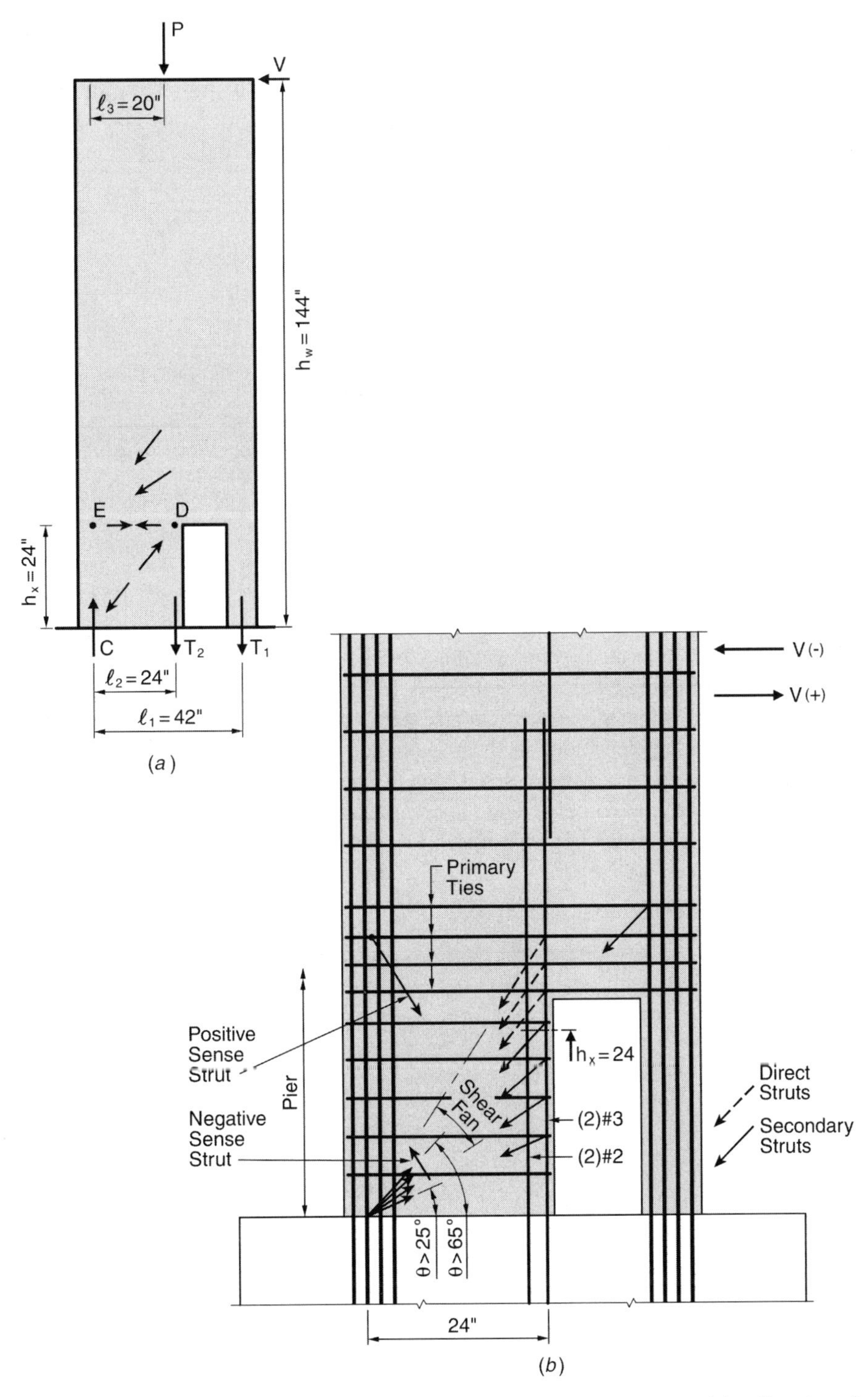

Figure 2.4.43 (*a*) Strut and tie model developed from Figure 2.4.42*a* (positive direction of load). (*b*) Detailed shear flow model developed from Figure 2.4.43*a*.

The dimensions and reinforcement of the Taylor, Cote, and Wallace[2.10] wall RW3-0 are shown on Figure 2.4.42*a* and *b*. The axial load appears to be 85 kips and the tensile capacity in the boundary appears to be

$$\begin{aligned} T_1 &= \lambda_o A_s f_y \\ &= 1.25(0.88)60 \\ &= 66 \text{ kips} \end{aligned} \tag{2.4.82}$$

If our objective were to develop a balancing shear load path in the pier, then

$$\begin{aligned} T_2 &> \frac{P\ell_3 + T_1\ell_1}{(\ell_2 h_w)/h_x - \ell_2} \qquad \text{(see Eq. 2.4.81b)} \\ &> \frac{85(20) + 66(42)}{(24)(144)/(24) - 24} \\ &> 37.2 \text{ kips} \end{aligned}$$

The measured load on the T_2 bars was 33.7 kips[2.10, Fig. 27] and this corresponds to a stress in the reinforcing bars of 105 ksi ($\lambda_o = 1.75$). Clearly the pier yielded (Figure 2.4.44) and some of the interior vertical pier steel shared in developing the load path.

The fact that the pier yielded in shear is not necessarily bad so long as the compression strut *CD* was not overstressed. Using the simplified model of Figure 2.4.43*a* we can estimate the force imposed on the compression diagonal.

The effective width (w) of the strut is

$$\begin{aligned} \ell_{CD} &= \sqrt{\ell_2^2 + h_x^2} \\ &= 34 \text{ in.} \end{aligned} \tag{2.4.83}$$

$$\begin{aligned} w &= 0.25\ell_{CD} \\ &= 8.5 \text{ in.} \end{aligned} \tag{2.4.84a}$$

$$\begin{aligned} F_{CD} &= \frac{\ell_{CD}}{h_x} T_2 \\ &= \frac{34}{24}(33.7) \qquad \text{(Measured conditions)} \\ &= 47.7 \text{ kips} \end{aligned} \tag{2.4.84b}$$

$$\begin{aligned} f_{CD} &= \frac{F_{CD}}{w t_w} \\ &= \frac{47.7}{8.5(4)} \\ &= 1.4 \text{ ksi} \qquad (0.35 f_c') \end{aligned}$$

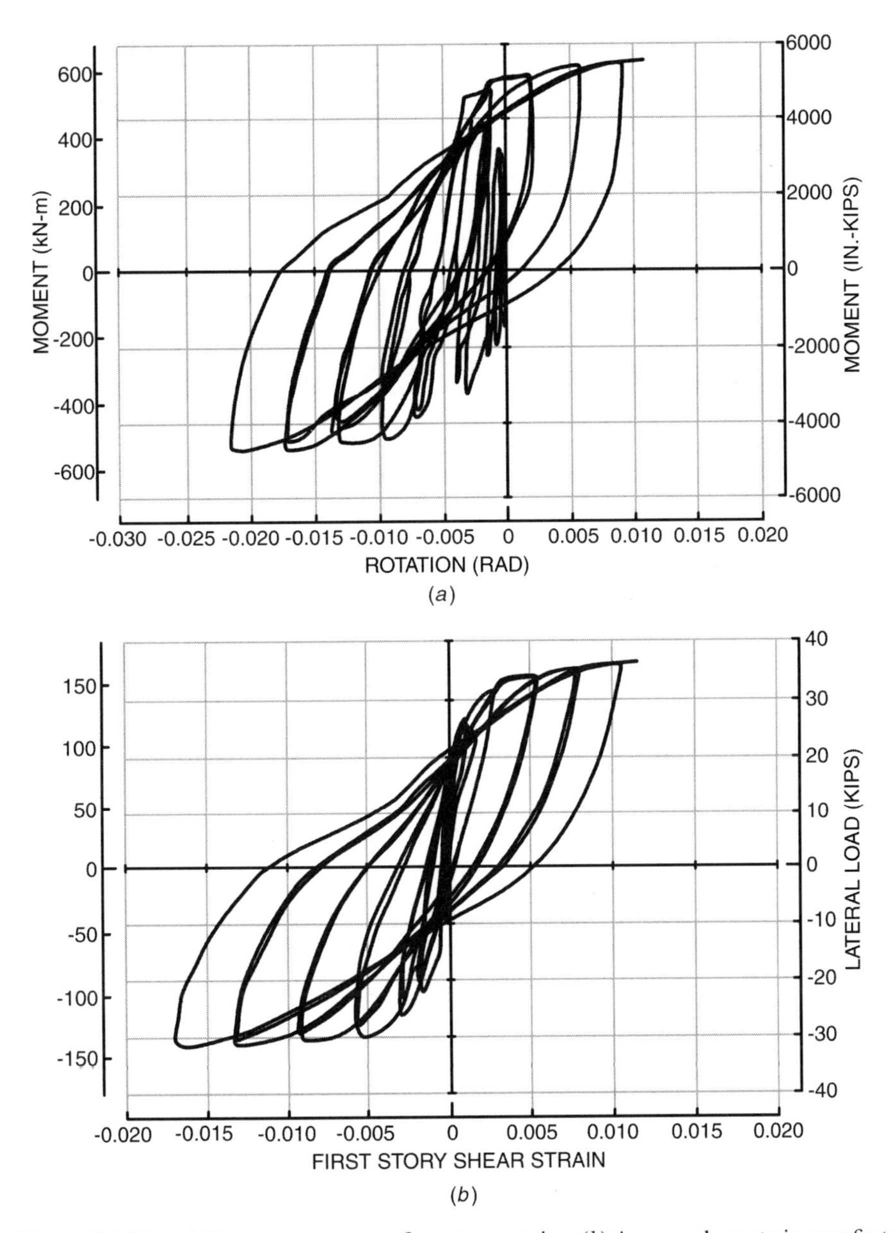

Figure 2.4.44 (*a*) Base moment versus first-story rotation. (*b*) Average shear strain over first story.

and this is reasonable (see Section 1.3.1). As in all engineering decisions the designer must realize that the strut model in this case tends to conservatively predict a concentration of stress in the pier, which probably occurs in the form of a shear fan (Figure 2.4.43*b*). Observe how the strut concentrations here (Figure 2.4.43*b*) because of the breadth of the shear fan will not be as severe as that imposed on the coupling beam diagonal (Figure 2.4.14*b*). Hence the strut strength limit state used for the coupling beam $\left(0.35 f_c'\right)$ would be conservative for this particular case.

The postyield deformation in the pier causes the boundary reinforcing to yield. Accordingly, significant ductility should be available. Figure 2.4.44*a* describes the hysteretic behavior of the shear pier.

It is interesting to note that Eq. 2.4.79 accurately predicts the attained level of moment in the wall.

The measured loads were

$$T_1 = 76.1 \text{ kips}$$
$$T_2 = 33.7 \text{ kips}$$
$$M = 5700 \text{ in.-kips}$$

We can also see that the developable pier shear (Eq. 2.4.80) appears to have established the limit state since the overstrength in the T_2 reinforcing was $1.75(\lambda_o)$ and only 1.44 in the T_1 reinforcing.

$$M = P\ell_3 + T_1\ell_1 + T_2\ell_2 \qquad \text{(see Eq. 2.4.79)}$$
$$= 85(20) + 66(42) + 37.2(24)$$
$$= 5400 \text{ in.-kips} \qquad \text{(see Figure 2.4.42}d\text{)}$$

Now consider shear flow in the opposite (negative or away from the pier) sense (Figure 2.4.45). The primary concern is with the compression force imposed on the column element (C_1). A conservative estimate of C_1 would be

$$C_{1,\max} = T + P \tag{2.4.85}$$
$$C_{1,\max} = 66 + 85$$
$$= 149 \text{ kips}$$

where T is $\lambda_o A_s f_y$.

This estimate is conservative because a part of T is expended in developing the shear transfer in the pier (Figure 2.4.45). Thus T may be viewed as $T_1 + T_2$ where T_1 provides flexural strength to the wall and T_2 activates the pier shear transfer mechanism. Proceeding as before in the development of Eq. 2.4.79, we find

$$M = Vh_w = P\ell_3 + T_1\ell_1 + T_2\ell_2 \tag{2.4.86}$$

$$V = \frac{\ell_2}{h_x} T_2 \tag{2.4.87}$$

$$T = T_1 + T_2 \tag{2.4.88}$$

We develop a relationship between T_1 and T_2:

$$T_1 = \frac{((h_w \ell_2)/h_x - \ell_2)\, T_2 - P\ell_3}{\ell_1} \tag{2.4.89}$$

Now the selection of h_x requires a review of the load path (Figure 2.4.45). The primary tie is above the opening, so h_x should be 34 in. Thus

$$T_1 = \frac{(144(24)/34 - 24)\, T_2 - 85(20)}{42}$$
$$= 1.83 T_2 - 40.5$$

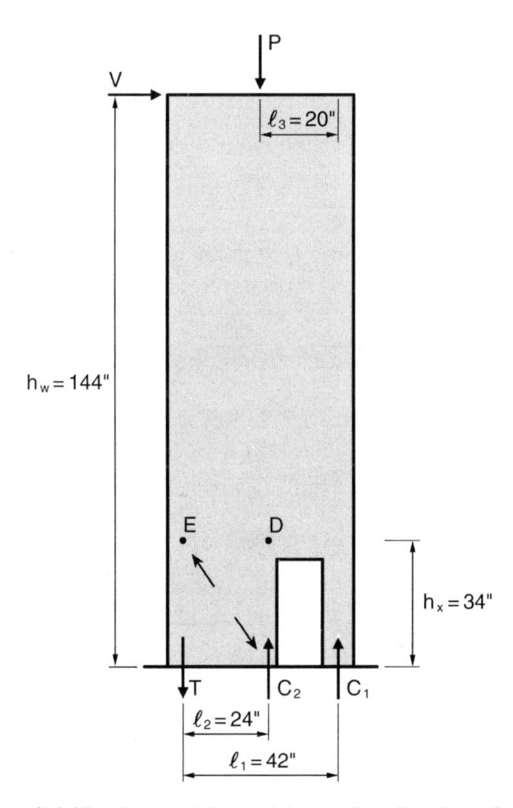

Figure 2.4.45 Strut and tie model (negative direction of load).

For $T = 66$ kips,

$$66 - T_2 = 1.83T_2 - 40.5$$
$$T_2 = 37.6 \text{ kips}$$
$$T_1 = 28.4 \text{ kips}$$
$$M = 3800 \text{ in.-kips}$$

and this slightly below the attained 4650 in.-kips (see Figure 2.4.42*e*).

The tensile load measured at T was 90 kips,[2.10, Fig. 28] and this corresponds to an overstrength factor (λ_o) of 1.7, which is consistent with the attained level of overstrength reported in the positive sense loading.

$$T_2 = 46 \text{ kips}$$
$$T_1 = 44 \text{ kips}$$
$$M = 4650 \text{ in.-kips}$$

The resultant load on the column element (C_1) is then

$$C_1 = P + T_1$$
$$= 85 + 44$$
$$= 129 \text{ kips}$$

or 20 kips less than the maximum value predicted by Eq. 2.4.85. The load on the compression strut C_2E (Figure 2.4.45) is developed as follows:

$$\ell_{C_2E} = \sqrt{34^2 + 24^2} \qquad \text{(see Eq. 2.4.83)}$$
$$= 41.6 \text{ in.}$$
$$F_{C_2E} = \frac{41.6}{34}(46) \qquad \text{(see Eq. 2.4.84b)}$$
$$= 56.3 \text{ kips}$$

The stress on the compression strut is

$$w = 0.25(41.6) \qquad \text{(see Eq. 2.4.84a)}$$
$$= 10.4 \text{ in.}$$
$$f_{C_2E} = \frac{56.3}{10.4(4)} \qquad \text{(see Eq. 2.4.36)}$$
$$= 1.35 \text{ ksi} \qquad (0.34 f_c')$$

Observe that for this direction of applied load, a strut will be developed much as it is in a beam-column joint—the truss mechanism (Figure 2.3.1*c*) being activated only after the primary truss mechanism breaks down (Figure 2.3.1*b*). The beam-column joint and the pier element (Figure 2.4.43*a*) represent discontinuity regions, a subject well presented by MacGregor.[2.3]

The capacity of the column element is

$$\begin{aligned} P_o &= 0.85 f_c' A_g + A_{st} f_y \qquad \text{(Eq. 2.2.4)} \\ &= 0.85(4)(4)(7.5) + 0.88(60) \\ &= 102 + 52.8 \\ &= 154.8 \text{ kips} \end{aligned}$$

and the postyield strength of the column, based on a confined core ($f_\ell \cong 500$ psi), is

$$\begin{aligned} f_{cc}' &= f_c' + 4.1 f_\ell \qquad \text{(Eq. 1.2.1)} \\ &= 4.0 + 4.1(0.5) \\ &= 6.05 \text{ ksi} \\ P_p &= f_{cc}' A_{\text{core}} + \lambda_o A_{st} f_y \\ &= 6.05(3)(6) + 1.25(0.88)(60) \\ &= 108.9 + 66 \\ &= 174.9 \text{ kips} \end{aligned}$$

Design Conclusions: When openings exist in shear walls, considerable care must be taken in selecting the reinforcing program. Too much reinforcing will cause the shear transfer element to break down and/or the compression area outboard of the opening to fail. The design developed by Taylor, Cote, and Wallace[2.10] recognized the load paths and limit states. As a consequence, they were able to accommodate the opening and attain the same level of story drift as those available in an identical solid panel. Compare Figure 2.4.4*a* (RW2) with Figure 2.4.42*c* (RW3-0).

Comment: Observe (Figures 2.4.42*d* and *e*) that analytical predictions of curvature based on strain state projections (BIAX) are consistent with recorded strains (LVDTs).

Load path development at the base of a tall wall that contains several openings (Figure 2.4.46) can be rather subjective in its development. The important issue is that a rational load path be developed in a consistent manner. The shear wall described in Figure 2.4.46 presents some modeling alternatives, each of which may be rationally developed. The overall wall design can be accomplished as though the openings at the base did not exist, and this will allow the development of the probable shear

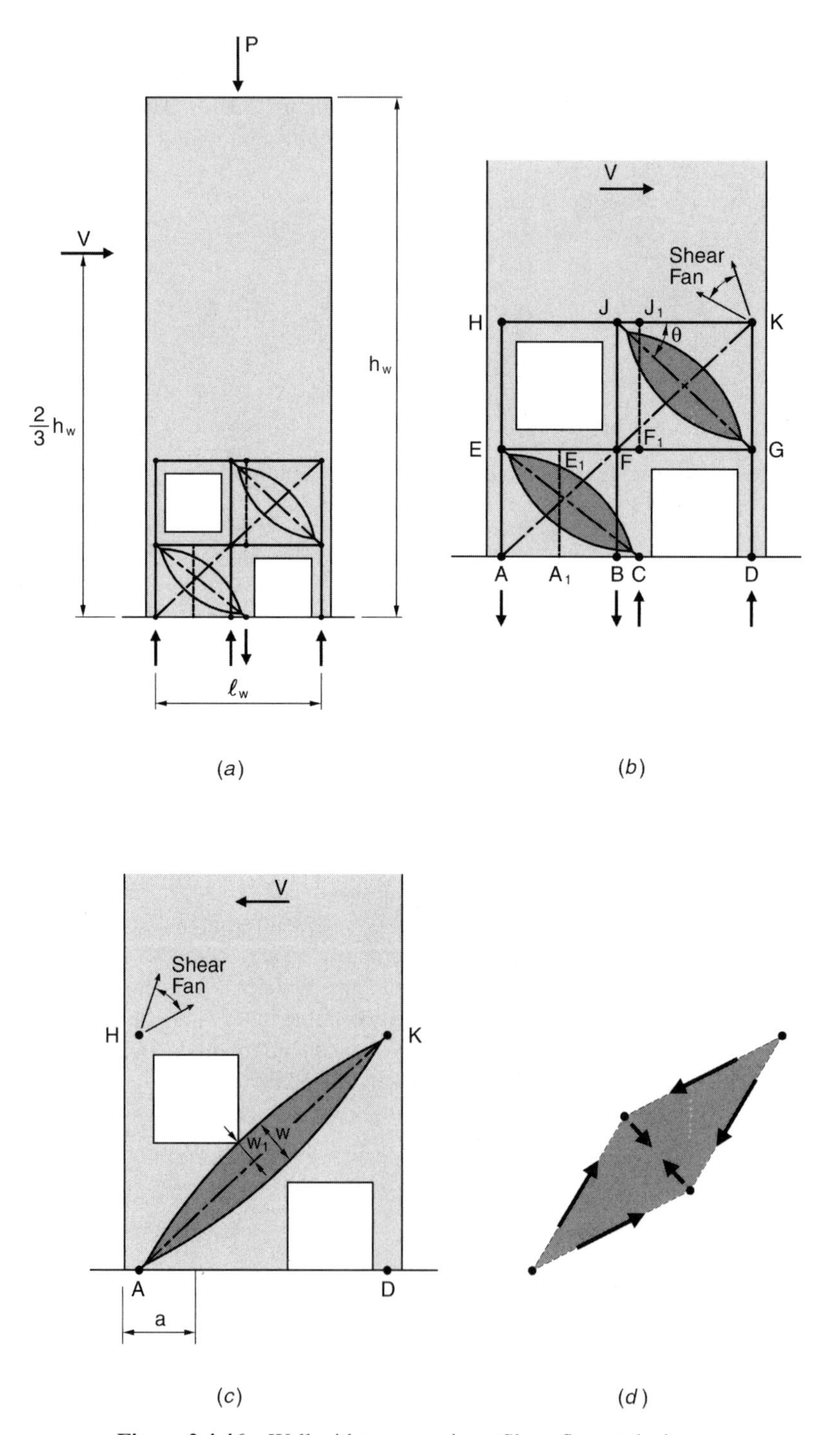

Figure 2.4.46 Wall with two openings. Shear flow at the base.

demand as well as the tension and compression demands at the base. This will allow a somewhat conservative sizing of the column element at D (Figure 2.4.46*b*) because at least some of the compression will go to the node at Point C (see Figure 2.4.46*b*). Once the design of the overall wall is complete, the shear load path must be developed at the base of the wall.

When the shear flow is from the left (Figure 2.4.46*b*), a shear fan will load the node at K which, of course, must be carefully developed. A tie capable of transferring the shear (V) back to node J must be created.

Now if the shape of pier $FGJK$ allows a load path development that fits within our trajectory limits (25° to 65°), a compression diagonal will be created between nodes J and G. A tie must now be provided to transfer the shear from G to E and a compression strut from E to C, unless this exceeds trajectory limits. When the trajectory limits are exceeded, an additional panel can be introduced (E_1A_1), or the load may be transferred by activating the vertical reinforcing in the pier as occurs naturally in a long short wall (see Figure 1.3.9). Observe that this option is not available in the upper pier ($FGJK$). Here, when the strut becomes too flat ($\theta < 25°$), an additional node must be created at F_1 and J_1. Designers must be careful not to introduce compression nodes over openings.

The conceptual design for the direction of loading described in Figure 2.4.46*b* might reasonably proceed as follows.

With loading from the left,

$$M_w = T_A(d - d') + P\left(\frac{\ell_w}{2} - \frac{a}{2}\right) \tag{2.4.90a}$$

where

$$T_A = \lambda_o A_s f_y$$

$$V = \frac{M_w}{0.67h_w}$$

Tie load KJ: $T_{HK} = V$

Tie load JB: $T_{EG} = V \tan\theta$

Strut load JG: $C_{JG} = V(\ell_{JG}/\ell_{FG})$

provided that

$$1.1 < \frac{\ell_{JG}}{\ell_{FG}} < 2.37 \qquad (\cos 25° = 0.9;\ \cos 65° = 0.42)$$

and

$$f_{c,JG} = \frac{C_{JG}}{0.25\ell_{JG}t_w} < 0.35f'_c \qquad \text{(see Eq. 2.4.36)}$$

Strut load *EC* is

$$C_{EC} = V\frac{\ell_{EC}}{\ell_{AC}}$$

provided that

$$1.1 < \frac{\ell_{EC}}{\ell_{AC}} < 2.37$$

Recall that the flexural strength for this direction of loading will be less than predicted by Eq. 2.4.90a because a portion of the tensile load at A (T_A) will be required to develop the shear transfer mechanism (strut *EC* or EA_1, as the case may be).

When the shear is from the right (Figure 2.4.46*c*), the shear load path is much more direct. The tie *HK* must be capable of developing the design shear and the strut *KA* may be capable of developing a singular load path in the lower part of the wall. The location of the compression node at A should be the center of the compressive stress block (a), a distance $a/2$ from the left end of the wall. The stress imposed on strut *AK* will determine whether or not the load path may be relied upon. The effective width of the strut will depend on w_1 (Figure 2.4.46*c*), and this is because the dilatational forces developed by expanding the load path (Figure 2.4.46d) must be equilibrated.

$$C_{AK} = V\frac{\ell_{AK}}{\ell_{AD}}$$

$$f_{c,AK} = \frac{C_{AK}}{2w_1 t_w} < 0.35f_c'$$

This limit state is conservatively set because of the critical nature of the load path developed. To the extent that alternative load paths (i.e., A_1F_1) are developed, this limit state ($0.35f_c'$) may be rationally increased to $0.5f_c'$.

Detailing of the base of a tall shear wall that contains openings is especially important. Nodes, particularly those that will be subjected to bidirectional tension loads (Node *E*—Figure 2.4.46*b*) must be carefully detailed, for the complete development of tie *EG* will be essential to the attainment of design objectives.

Clearly, many apparently subjective decisions are required in the development of a rational strut and tie model, and it is probably for this reason that shear stress limits are more typically used to design shear walls. Unfortunately, strut limit states are not prescribed, and this can lead to pier failures. Regardless of the design approach, the load path must be understood, as must the details that facilitate the load flow development. Considerable care must be taken so as not to overload compression struts, and this usually means resisting the temptation to overreinforce the wall.

2.4.3 Precast Concrete Shear Walls

Precast concrete shear walls can be effectively integrated into a building system, and this is especially true if the adopted vertical load carrying system is largely precast concrete. To accomplish this objective design procedures developed for cast-in-place shear wall construction must be abandoned because their prescriptive requirements endeavor to compensate for all possible shear load paths (Section 1.3), and this is not a constraint that must be imposed on a precast concrete system.

Precast concrete shear walls can be developed so as to produce virtually damage-free performance. The goal of this section is to demonstrate how this can be accomplished.

2.4.3.1 Experimental Efforts Rahman and Restrepo[2.47] tested two precast concrete shear wall systems, one connected to the foundation by post-tensioning only and the other connected by hybrid systems (post-tensioning supplemented by mild steel). This effort was followed by Holden,[2.48] who compared the postyield behavior of a precast concrete wall system that emulated cast-in-place construction in its connection to the foundation with that of a carbon fiber reinforced precast concrete hybrid panel designed to minimize damage at high levels of drift. Figures 2.4.47*a* and 2.4.47*c* describe the post-test condition of Holden's panels. Figure 2.4.47*b* describes Rahman and Restrepo's second hybrid wall at design levels of drift. The emulative system (Figure 2.4.47*a*) was capable of attaining a drift of only 2.5% (recall that this is higher than the cast-in-place systems discussed in Section 2.4.1), while Rahman and Restrepo's hybrid Unit 2 (Figure 2.4.47*b*) reached 4.5% without losing any strength. The carbon fiber system tested by Holden (Figure 2.4.47*c*) reached a drift of 6.2% before losing strength. The behaviors described in Figure 2.4.47 clearly demonstrate the behavior enhancement opportunities available through the appropriate use of yielding precast concrete systems.

Our focus will be on the behavior of the hybrid system (Figure 2.4.47*b*) because the carbon fiber system, though interesting, is one that would be expensive to build (Figure 2.4.48) and drifts in excess of 4% not expected. The reinforcing concept (Figure 2.4.48) should not be discarded, however, because the cost impact is only imposed on the lowermost panel, and the resultant reduction in damage is an important consideration given a performance-based objective (see Figure 2.4.47*c*).

The three units tested by Rahman and Restrepo[2.47] as well as those tested by Holden were constructed to the basic size described in Figure 2.4.49. Rahman and Restrepo Units 2 and 3 (the hybrid panels) contained milled mild steel bars located 90 mm (3.54 in.) on either side of the centerline of the panel. End confinement varied on Units 2 and 3 as described in Figure 2.4.50. An axial load of 45 kips was imposed on Unit 3.

The need for energy dissipation is clearly demonstrated by reviewing the behavior of a prestressed (only) wall assembly. The behavior of Rahman and Restrepo's Unit 1 (prestress only) was quite predictable. The two $\frac{1}{2}$-in. ϕ, 270-ksi prestressing strands clamp the precast wall to the foundation. Strand loads are described in Figure 2.4.51 for the eastward direction of drift (Figure 2.4.49). Drift controlled load displacement cycles varied in magnitude, so the envelope connecting load cycles 9, 12, 13, 17, and 21 describes behavior representative of monotonic loading. The applied shear load at cycle 21 is developed analytically as follows:

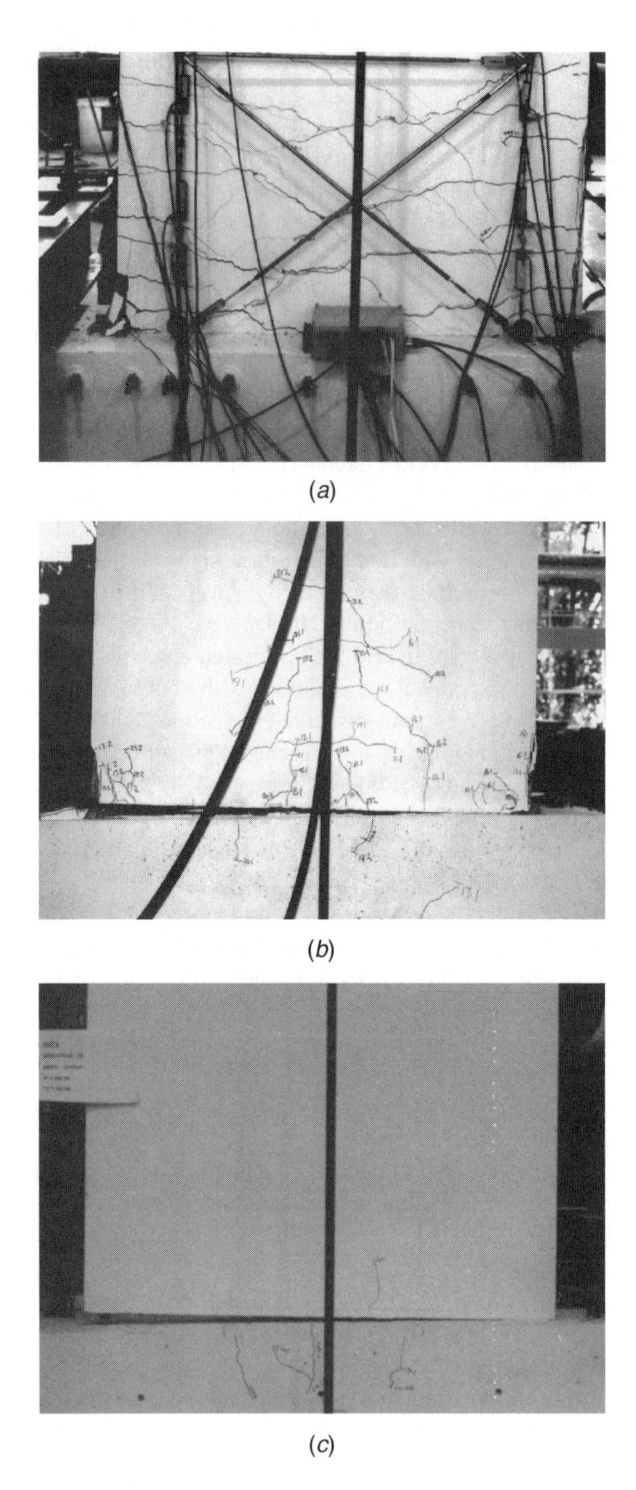

(a)

(b)

(c)

Figure 2.4.47 Cracking experienced in precast concrete wall panels at design level drifts.[2.48] (*a*) Emulative system at 2.5% drift (Holden[2.48]). (*b*) Hybrid system at 3% drift (Rahman and Restrepo[2.47, unit 2]). (*c*) Carbon fiber reinforced hybrid panel at 3% drift (Holden[2.48]).

Figure 2.4.48 Diagonal and longitudinal bars prior to being fillet welded to the base plate. Holden carbon fiber reinforced hybrid panel.[2.48]

$$F_{ps,W} = 35.3 \text{ kips} \qquad \text{(West strand)}$$

$$F_{ps,E} = 27.4 \text{ kips} \qquad \text{(East strand)}$$

$$\sum F_{ps} = 62.7 \text{ kips} \qquad \text{(240 psi)}$$

$$a = \frac{\sum F_{ps}}{0.85 f_c' t_w}$$

$$= \frac{62.7}{0.85(5.4)(4.92)} \qquad (f_c' = 5.4 \text{ ksi})$$

$$= 2.78 \text{ in.} \qquad \text{(70 mm)}$$

$$c = 3.5 \text{ in.} \qquad \text{(89 mm)}$$

Comment: The location of the neutral axis as projected from strand strains is shown on Figure 2.4.52. The analytically predicted depth of the compressive stress block (a) seems to provide an estimate of the location of the ultimate or gravitated neutral axis (75 mm at a DR of 2.7%) while c (89 mm) describes the location of the neutral axis at lower drift ratios (DR = 0.8%).

The liftoff (δ) described in Figure 2.4.52 at a wall displacement of 2.7% creates a rotation at the base of the wall of

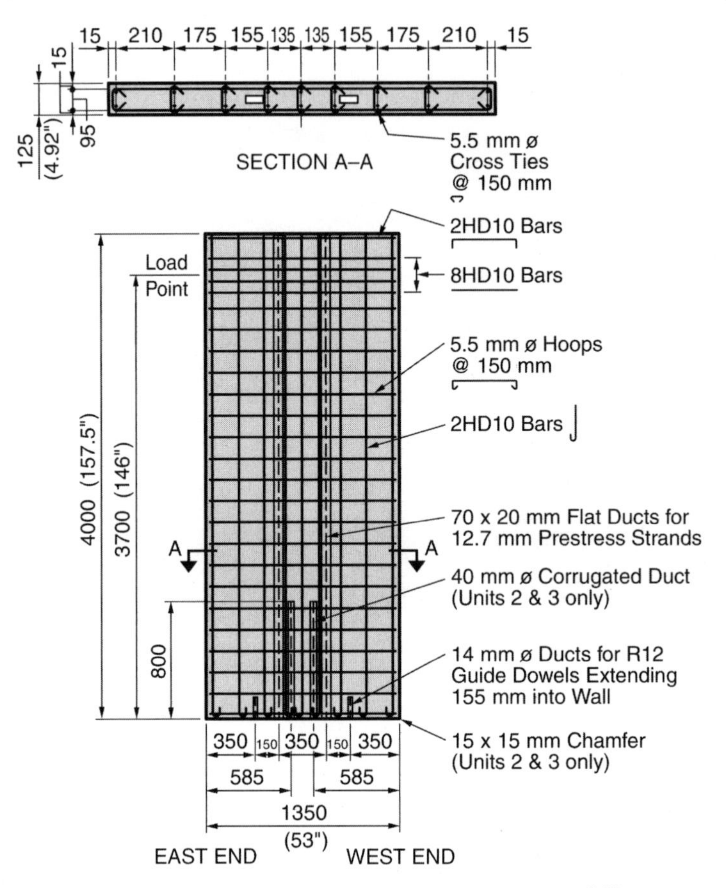

Figure 2.4.49 Reinforcement details of wall units.[2.47]

$$\theta_{\text{gap}} = \frac{\delta}{850 - 75}$$

$$= \frac{18}{775}$$

$$= 0.023 \text{ radian}$$

or about 85% of the drift angle Δ / h_w (0.027 radian). Hence, from a design perspective the panel may be viewed as a rigid body.

The strain imposed on the strand defines the induced level of strand stress. In post-tensioned systems strand elongation will occur over the unbonded length, and this usually results in a minimal increase in strand strain. From a design perspective this makes it difficult to estimate the attainable nominal moment capacity of the wall. The

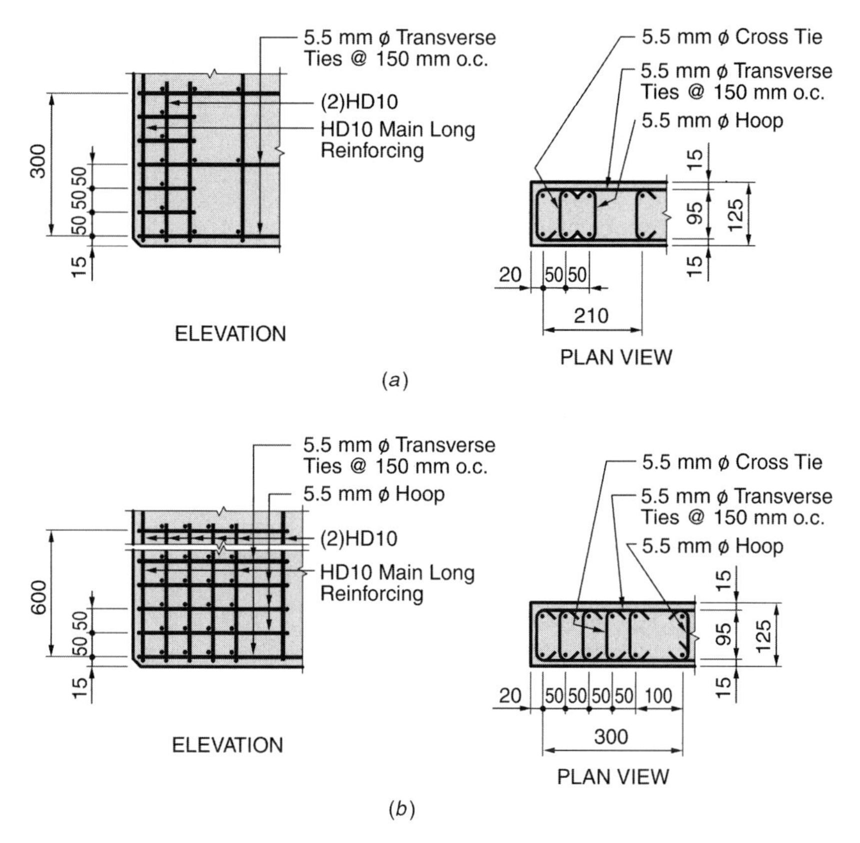

Figure 2.4.50 Details of confinement at the toes of the test units.[2.47] (*a*) Unit 2. (*b*) Unit 3.

usual design approach would be to assume the (nominal) stress in the strand (f_{ps}) to be around 175 ksi for an unbonded strand. The resultant nominal moment capacity is

$$M_n = A_{ps} f_{ps} \left(d - \frac{a}{2}\right) \tag{2.4.90b}$$

$$= 0.155(175)\,[(33.5 - 1.4) + (19.7 - 1.4)]$$

$$= 27.1\,(32.1 + 18.3)$$

$$= 1367 \text{ in.-kips}$$

$$V = 9.36 \text{ kips} \qquad (42 \text{ kN})$$

and this underestimates the test results (Figure 2.4.53*a*). The difference is attributed to the short strand length. Observe that the implied stress gain is quite large (Figure 2.4.51). The shear imposed on the wall, based on recorded strand forces, is

$$
\begin{aligned}
M &= F_{ps,W}(33.5 - 1.4) + F_{ps,E}(19.7 - 1.4) \\
&= 1634 \text{ in.-kips} \\
V &= 11.2 \text{ kips} \qquad (49.8 \text{ kN})
\end{aligned}
$$

The probable strain imposed on the strand is important because strand rupture must be avoided and overstraining of the strands will cause a loss in the effective level

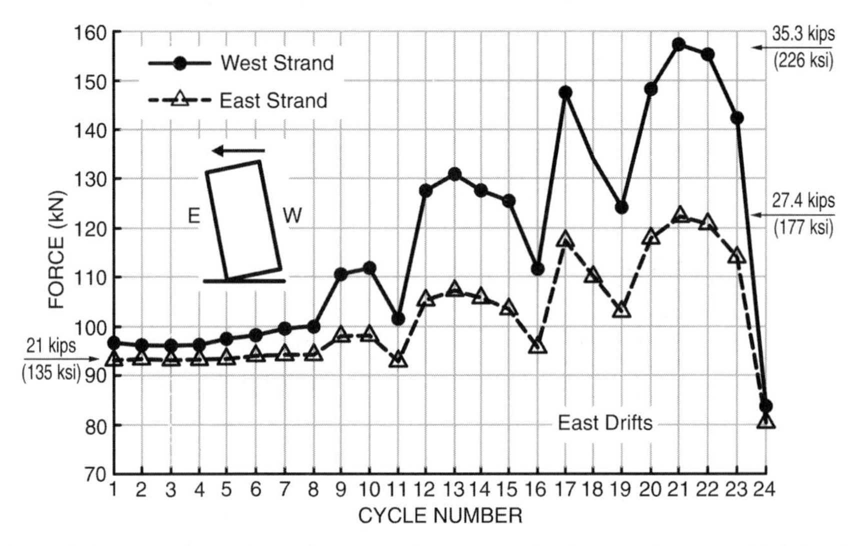

Figure 2.4.51 Maximum forces in prestressing strands of Rahman and Restrepo Unit 1 at the different loading cycles.[2.47]

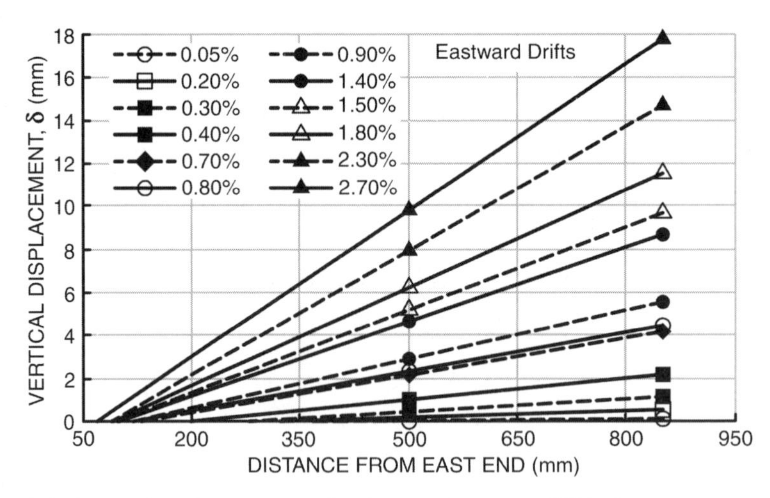

Figure 2.4.52 Vertical displacement of Rahman and Restrepo[2.47] Unit 1 at the construction joint corresponding to different lateral drift levels.[2.47]

of prestress. Observe (Figure 2.4.51) that a significant amount of the effective prestress has been lost at a drift ratio of 2.7%. Clearly this demonstrates the importance of maintaining a central pretensioning as opposed to pretensioning the wall at its boundaries. It is also important to create a long debond length for the strand.

The probable level of strain in the strand may be conservatively estimated by assuming that the gap opening (θ_{gap}) is the same as the drift ratio (Δ/h_w); hence the strand elongation is

$$\delta_{\text{pre}} = \theta_{\text{gap}}(d_{\text{pre}} - c) \tag{2.4.91}$$

This strain occurs after decompression, producing a total strain in the strand that is slightly higher than

$$\varepsilon_{ps} \cong \frac{f_{pse}}{E_s} + \frac{\theta_{\text{gap}}(d_{\text{pre}} - c)}{\ell_{ps}} \tag{2.4.92}$$

where ℓ_{ps} is the unbonded length of the prestressing strand. The conservative assumption regarding the gap angle (Eq. 2.4.91) will more than compensate for the decompression strain induced in the strand.

A force-based approach may also be used to control the strain in the strand. A force-based procedure was used in the design of the hybrid beam (Section 2.1.4.4). The ultimate force that an unbonded strand can sustain is a function of the anchoring device, and as was the case in the hybrid beam system, the objective should be to keep the force imposed on the strand to that associated with a stress of $0.85 f_{pu}$ or, for 270-ksi strand, 230 ksi. This corresponds to a strain of about 0.02 in./in. (see Figure 2.1.54). Observe that this strain limit state was not reached in the test as the measured stress in the west strand was 226 ksi. The flatness of the stress-strain relationship in this strain region suggests that a strain based design which adheres to strain limit states is the best guard against strand rupture.

$$\begin{aligned} \varepsilon_{psu} &= \frac{f_{pse}}{E_{ps}} + \frac{\theta_{\text{gap}}(d - c)}{\ell_{ps}} && \text{(Eq. 2.4.92)} \\ &\cong 0.005 + \frac{0.023}{157.5}(33.5 - 3.5) && (\theta_{\text{gap}} = 0.85(0.027)) \\ &\cong 0.0093 \text{ in./in.} \end{aligned}$$

Comment: Strand rupture must be guarded against in the detailing process. Holden[2.48] used a rectangular tube to contain the strand, and this was intended to prevent a kinking or fraying of the strand at the gap as the wall panel rocked. Alternatively, a splayed end to the duct (Figure 2.4.66*b*) would serve the same purpose. Detailing is also important at panel ends. A squared-off (no chamfer) corner was used by Rahman and Restrepo[2.47] on their nonhybrid unit, and this may have promoted a somewhat early spalling of the concrete in the toe region. A thin mortar bed was placed between the foundation and the wall to promote a more uniform bearing.

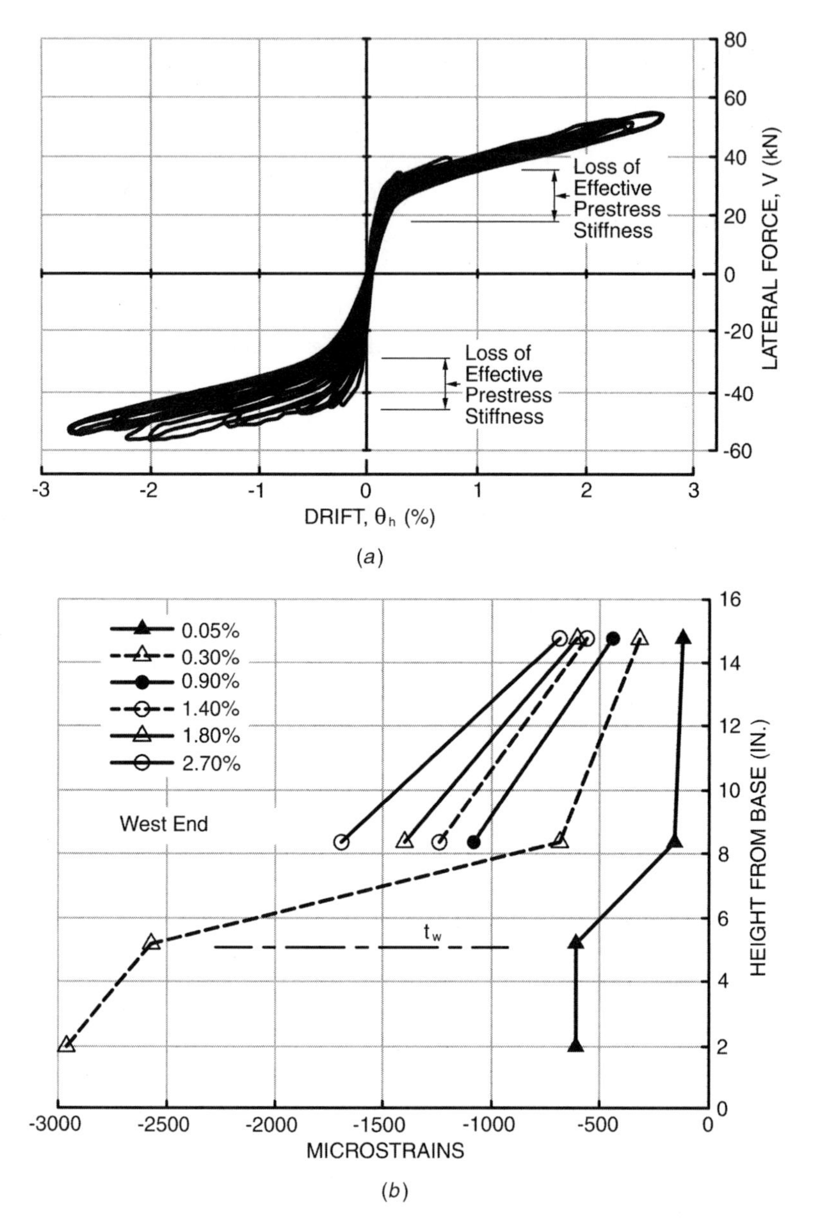

Figure 2.4.53 (*a*) Lateral force–drift response of Rahman and Restrepo Unit 1.[2.47] (*b*) Concrete strains at ends of Rahman and Restrepo Unit 1.[2.47]

Induced levels of concrete strain will define the deformation limit state of the wall. Strain states in the toe of precast wall panels are quite high, and this is because the strain gradient in the toe is steep. Figure 2.4.53*b* describes the strain gradient in the toe of Rahman and Restrepo Unit 1. The increase in measured strain between a point that is 5 in. above the base and one that is 8 in. above the base is dramatic. Strain measurements at drifts that exceeded 0.3% were not possible, but one might reasonably presume that peak strains are directly proportional to the rotated angle since this is the case for measured strains at lower drift angles. This means that the toe strain at a drift angle of 2.7% would be 9 times that reported for a drift angle of 0.3%. This implies a concrete strain in the toe region on the order of 0.027 in./in. This level of strain is consistent with the limit state suggested by other tests performed at the University of Canterbury. From a design perspective the indicated strain state may be developed from the total shortening required at the edge of the panel divided by a strain or hinge length that is equivalent to the thickness of the wall (t_w).

The introduction of an energy dissipating device at the base of a precast concrete wall panel may be accomplished in many ways. Clearly, it must accommodate the construction program. The Precast Seismic Structural Systems (PRESSS) Program,[2.49] for example, chose to construct a shear wall out of four precast wall panels (Figure 2.4.54), thereby creating both horizontal and vertical precast interfaces. A

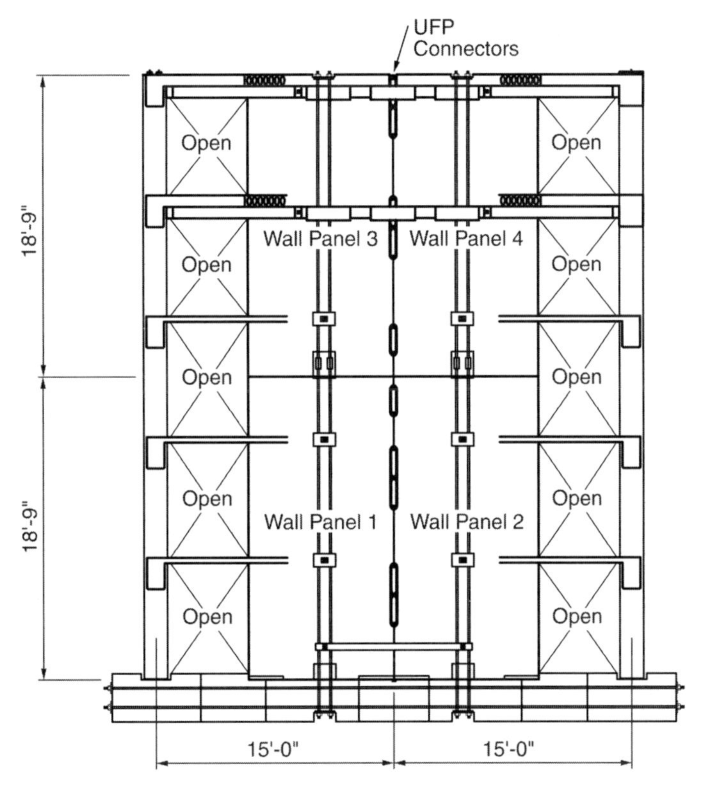

Figure 2.4.54 Jointed structural wall elevation.[2.47]

structural steel shape (acronym UFP), bent so as to allow significant postyield deformation as it transferred interface shears, provided the desired energy dissipation. Welding was required, as was the inclusion of embedded connections in wall panels, and this is not only expensive but requires a significant quality control effort. Most contractors prefer grouted assembly methods. Accordingly, the hybrid beam approach is more easily adapted to precast concrete wall construction.

Rahman and Restrepo's Unit 3 (second hybrid system) provides significant input into both behavior and system design. Construction of this panel is described in Figures 2.4.49 and 2.4.50*b*. The energy dissipater was a milled bar (Figure 2.4.55) whose milled portion was embedded in the foundation. Alternatives such as debonding (Section 2.1.4.4) or forged rods (Figure 2.1.57) may be used. A constant axial load of 45 kips was externally applied to Unit 3. The force-drift response of this unit is described in Figures 2.4.56*a* and 2.4.56*b*. The presence of the prestress force and applied axial load (45 kips) causes the wall to be initially somewhat stiffer than a wall that is not prestressed. The impact of prestressing on stiffness is easily recognized by comparing the response described in Figure 2.4.4*a* ($P/A = 0$) to that of Figure 2.4.56*a* ($P/A = 0.073 f_c'$) and Figure 2.4.2*a* ($P/A = 0.24 f_c'$). A drift of 4% is attained, and this produces strains of 6.5% in the energy dissipaters ($F_{\max} = 25$ kips; $f_s = 90$ ksi).

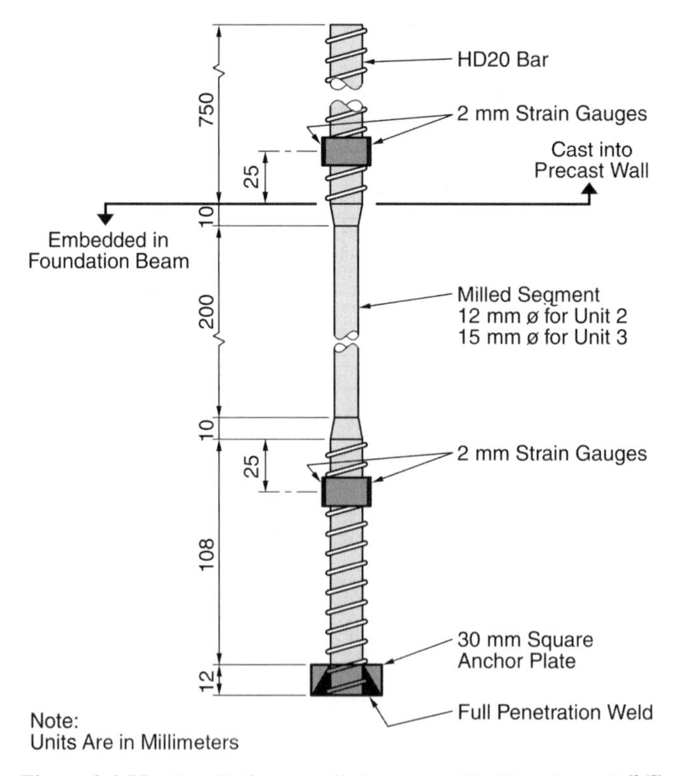

Figure 2.4.55 Detail of energy dissipater used in Units 2 and 3.[2.47]

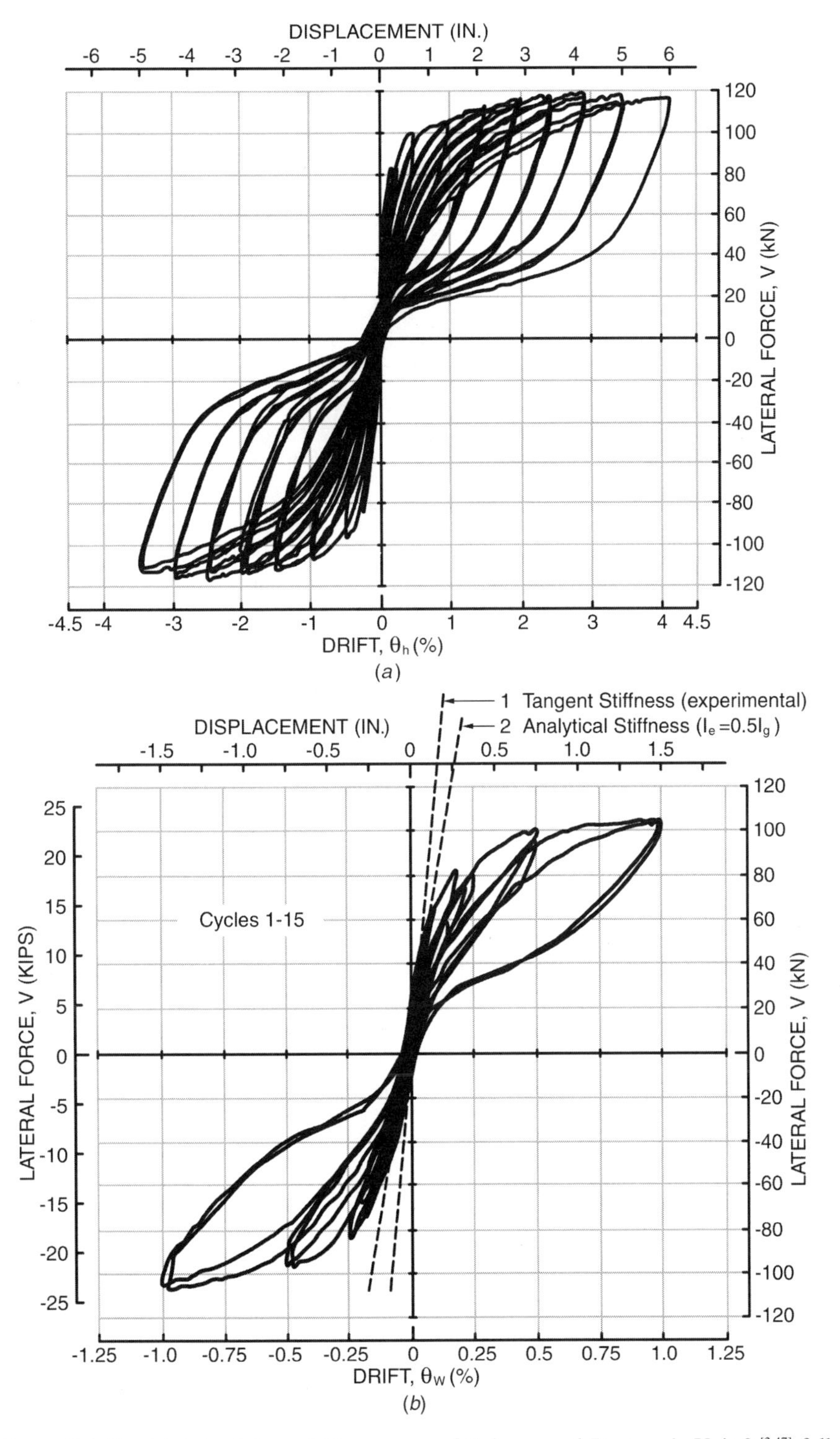

Figure 2.4.56 (*a*) Lateral force-drift response of Rahman and Restrepo's Unit 3,[2.47] full response. (*b*) Lateral force-drift response of Rahman and Restrepo's Unit 3,[2.47] cycles 1–15.

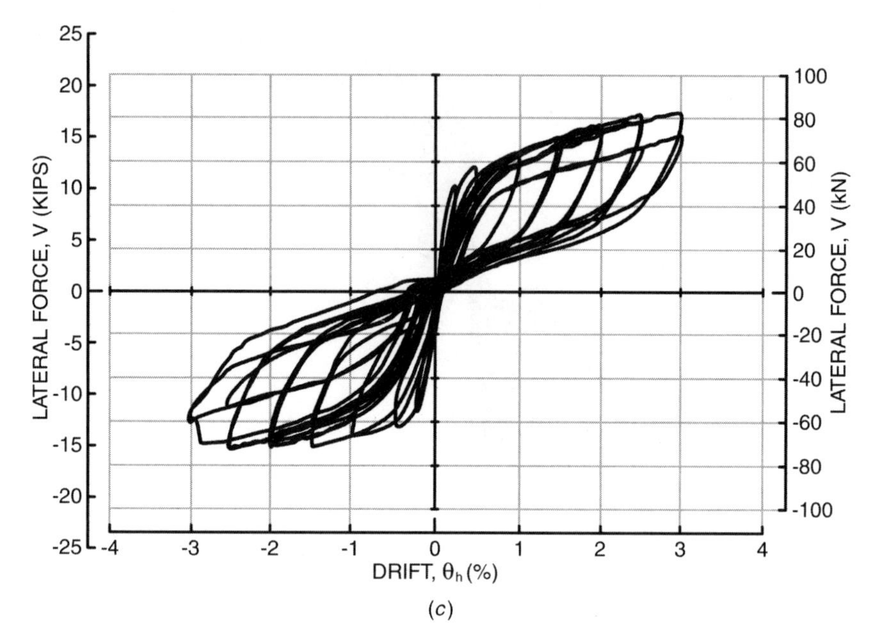

Figure 2.4.56 (*Continued*) (*c*) Lateral force-drift response of Rahman and Restrepo's Unit 2,[2.47] cycles 7, 11, 13, 16, and 23–31.

The initial force imposed on the prestressing strands was 20.6 kips. The resultant axial stress imposed on the precast wall panel, initial post-tension plus externally applied axial load, was 330 psi ($0.073 f'_c$).

Stiffness characterization in the elastic range is required for design purposes. A quantification of the elastic deflection component of wall drift is also required in order to evaluate strain states.

The idealization proposed for cast-in-place wall systems, defined as an effective moment of inertia of $0.5 I_g$, is reasonably used. The resultant elastic component of story drift for the experimental wall (Figure 2.4.49) would be

$$I_e = 30{,}500 \text{ in.}^4 \qquad (I_e = 0.5 I_g)$$

$$E = 3800 \text{ ksi} \qquad (f'_c = 4500 \text{ psi})$$

$$\theta_w = \frac{\Delta}{h_w}$$

$$= \frac{P h_w^2}{3 E I_e}$$

$$= \frac{22.5(146)^2}{3(3800)(30{,}500)}$$

$$= 0.00138 \text{ radian} \qquad (0.2 \text{ in.})$$

The analytically developed stiffness is compared with the experimental stiffness of the second hybrid wall system on Figure 2.4.56*b*. The behavior of Rahman and Restrepo Unit 2 is described in Figure 2.4.56*c*.

From the standpoint of strain calculations at significant drift angles ($\pm 2\%$), the elastic component of wall drift can be ignored, for it does not seem to be consistent with the accuracy of the analytical assumptions used to estimate ultimate concrete strain states.

Neutral axis depths for Unit 3 are developed in the standard manner. Presuming first that unconfined concrete in the toe region is capable of sustaining the predicted strains

$$
\begin{aligned}
F_{ps} &= 36 + 25 \qquad \text{(Test data)} \\
&= 61.8 \text{ kips} \\
F_s &= 2(22.5) \qquad \text{(Test data)} \\
&= 45 \text{ kips} \\
P &= 45 + 3.6 \qquad \text{(Includes panel weight)} \\
&= 48.6 \text{ kips} \\
C_c &= F_{ps} + F_s + P \\
&= 155.4 \text{ kips} \\
a &= \frac{155.4}{0.85(4.5)(4.92)} \qquad (f'_c = 4.5 \text{ ksi}) \\
&= 8.26 \text{ in.} \\
c &\cong 10 \text{ in.}
\end{aligned}
$$

and, alternatively, based on a confined core and spalled shell

$$
\begin{aligned}
a &= \frac{155.4}{0.85(7)(3.7)} \qquad \left(f'_{cc} = 7 \text{ ksi}\right) \\
&= 7.06 \text{ in.}
\end{aligned}
$$

to which must be added the cover,

$$
\begin{aligned}
c &= a + \text{cover} \\
&= 7.85 \text{ in.}
\end{aligned}
$$

The experimentally derived location of the neutral axis is about 7.9 in., and this is reasonable, for the concrete strain states at the panel foundation interface are undoubtedly above 0.007 in./in. Concrete strains are not reported for Unit 3, but they were reported for Rahman and Restrepo's Unit 2. Unit 2 differed from Unit 3 in the extent of provided confinement (Figure 2.4.50) and the absence of an axial load (the energy

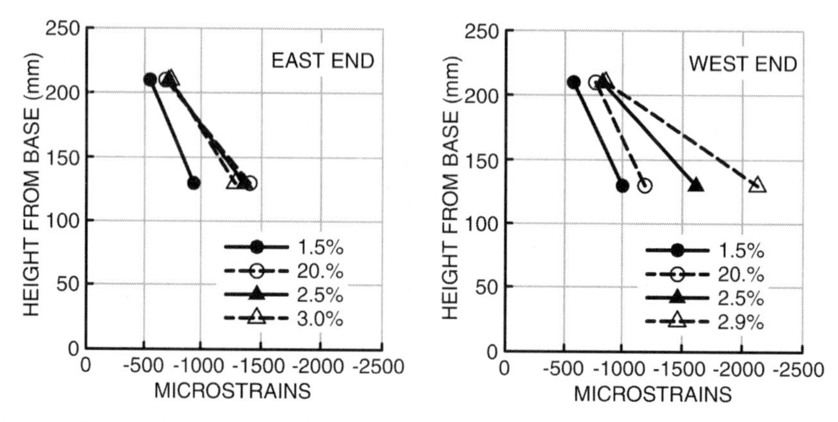

Figure 2.4.57 Microstrains developed in the toe region of Rahman and Restrepo's Unit 2.[2.47]

dissipater was also smaller). The measured strain gradient for Rahman and Restrepo's Unit 2 is shown in Figure 2.4.57. Compressive strains decrease rapidly with distance from the base of the wall. Observe that these strain gradients are consistent with those reported in Figure 2.4.53*b*. The strains in the ties reportedly reached 0.0009 in./in.

Design simplifications are clearly called for, and Rahman and Restrepo propose some that may be generalized to fit an even broader application. Figure 2.4.58 describes the pier in a rocked position. The concerns are the strain states in the concrete, post-tensioning, and energy dissipating bars. Rather than develop a closed form design solution to address the force on each post-tensioning assembly, the use of an average value is proposed by Rahman and Restrepo, based on a sensitivity study. This allows the quantification of the force provided by all of the post-tensioning:

$$F_{ps} = n \sum A_{ps} f_{ps} \tag{2.4.93}$$

where n is the averaging factor—assumed to be 0.9—and $A_{ps} f_{ps}$ is the force developed in the outermost strand. The force developed in the strands (F_{ps}) will, of course, depend on the strain induced in the strand, which for capacity design purposes may be conservatively assumed to be as high as $0.9 f_{psu} A_{ps}$. The force in the outermost strand of Unit 3 reached 36.8 kips or $0.88 f_{psu}$. This level of strain may not be reached in an assembled wall system, for the length over which the strand is strained will probably be significantly greater than the 12 ft of Unit 3, and this must at least be subliminally recognized in the conceptual design.

The nominal prestress force may be as high as

$$\begin{aligned} F_{nps} &= 0.9 \sum A_{ps} 0.9 f_{psu} \\ &= 0.81 \sum A_{ps} f_{psu} \end{aligned} \tag{2.4.94}$$

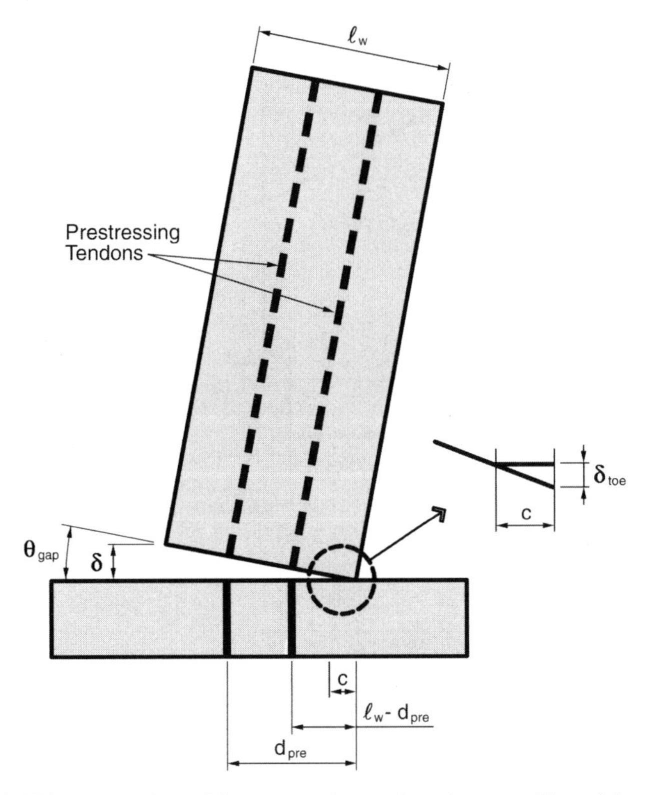

Figure 2.4.58 Elongation of the prestressing tendons due to rocking of the wall.[2.47]

The energy dissipater will be strained well into the strain hardening range and the force developed (F_s) will be significantly larger than the specified yield:

$$\begin{aligned} F_s &= \lambda_o f_y A_s \\ &= 1.5 f_y A_s \end{aligned} \tag{2.4.95}$$

At large levels of panel rotation, the unconfined shell will become ineffective, causing the compression and shear load to be transferred through the confined toe region (Figure 2.4.59). The mortar must also have an ultimate compressive strength greater than the confined strength of the concrete (f'_{cc}) in order to transfer confined concrete stresses. Accordingly, the developable concrete force will be

$$C_c = f'_{cc} b_e a \tag{2.4.96a}$$

The design process is simplified by presuming that the unconfined shell (c_c) will provide no strength and, as a consequence of this conservatism, c and a may be presumed to be essentially the same. Since the objective confining pressure will be

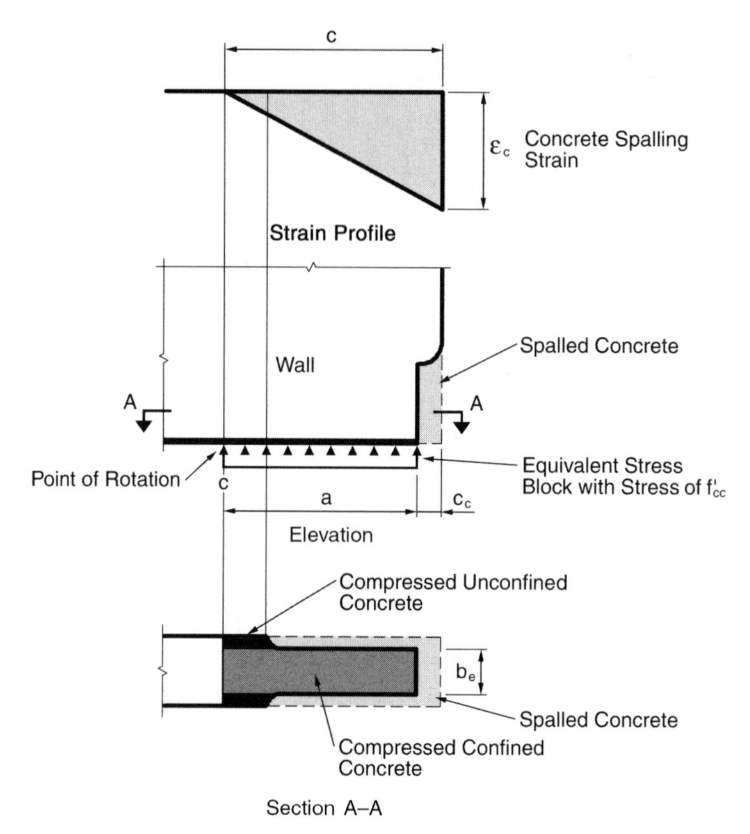

Figure 2.4.59 Distribution of compression stresses and strains at the toe of the wall during rocking.[2.47]

on the order of 600 psi, f'_{cc} can be assumed to be $f'_c + 4.1 f_\ell$ or $f'_c + 2.5$ ksi (Eq. 1.2.1), and the developed concrete force becomes

$$\begin{aligned} C_c &= 0.85 f'_{cc} b_e a \\ &= 0.85 \left(f'_c + 2.5\right) b_e a \end{aligned} \tag{2.4.96b}$$

Now the depth to the effective neutral axis or point of rotation becomes

$$a = \frac{P + F_{ps} + F_s}{0.85\left(f'_c + 2.5\right) b_e} \tag{2.4.97a}$$

and for Rahman and Restrepo's Unit 3, this becomes

$$\begin{aligned} a &= \frac{48.6 + 0.81(2)(0.155)(270) + 2(1.5)(0.31)(67)}{(0.85)(4.5 + 2.5)(3.75)} \\ &= 8 \text{ in.} \end{aligned} \tag{2.4.97b}$$

Figure 2.4.60 shows the deterioration experienced by Rahman and Restrepo's Unit 3 at an ultimate drift ratio of 4%. Clearly, the identified parameters used to develop Eq. 2.4.97a are somewhat speculative, but more design precision is not warranted.

The postyield strain induced in the outer strand at a drift ratio of 4% (see Figure 2.4.56*a*) is

$$\varepsilon_{psp} = \frac{\theta_u\left(d_{\text{pre}} - a\right)}{\ell_{ps}} \tag{2.4.98a}$$

where ℓ_{ps} is the unbonded length of the strand. Thus

$$\begin{aligned}\varepsilon_{psp} &= \frac{0.04(33.5 - 8)}{157.5}\\ &= 0.0064 \text{ in./in.}\end{aligned}$$

and

$$\begin{aligned}\varepsilon_{ps} &= \varepsilon_{pse} + \varepsilon_{psp} \\ &= 0.005 + 0.0064 \\ &= 0.0114 \text{ in./in.} \qquad < 0.02 \text{ in./in.}\end{aligned} \tag{2.4.98b}$$

The strain induced in the outermost energy dissipater should have been

$$\begin{aligned}\varepsilon_s &= \frac{\theta_u\,(d_s - a)}{\ell_p} \\ &= \frac{0.04(30 - 8)}{7.9} \qquad \text{(see Figure 2.4.55)} \\ &= 0.112 \text{ in./in.}\end{aligned} \tag{2.4.98c}$$

and this is less than the guaranteed ultimate elongation of 15% for the provided milled rod and the 20% available in a ductile rod (see Figure 2.1.41).

It should be recalled that the drift ratio (Δ / h_w) overestimates θ_{gap} and 85% of the drift ratio would be a better estimate of the gap created when the objective is to identify strain limit states in the concrete. The measured liftoff at the outermost strand was 0.63 in., while the theoretical liftoff based on $\theta_u d_{\text{pre}}$ of 4% would have been 0.77 in. Accordingly, the evaluation procedure might be modified or viewed as being at least somewhat conservative.

One of the advantages of the hybrid system is that the post-tensioning tends to restore the system to a displacement neutral position. Compare the displacement of the thin cast-in-place wall (Figure 2.4.4*a*) after the load is removed to that described in Figure 2.4.56. In the last displacement cycle for the cast-in-place wall, the residual drift is 1.5%, or 75% of the maximum drift. The residual displacement of the hybrid

(*a*)

(*b*)

Figure 2.4.60 (*a*) North face of Unit 3 at end of test.[2.47] (*b*) North face (west end) of Unit 3 at end of test.[2.47]

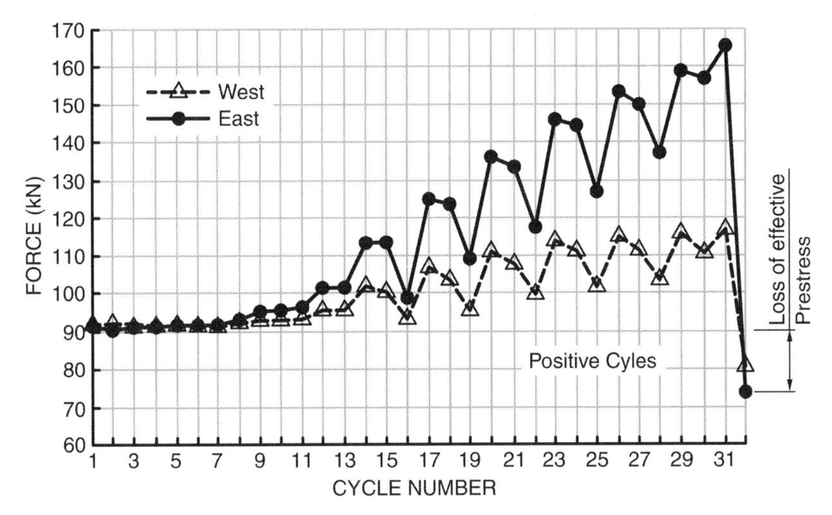

Figure 2.4.61 Maximum forces in prestressing strands of Unit 3 at the different loading cycles.[2.47]

wall system (Figure 2.4.56) after the load is released is essentially zero. Obviously this can only happen if the post-tensioning force, acting in conjunction with the axial load, can compress the energy dissipater (mild steel reinforcing). This goal will establish an objective relationship between the amount of post-tensioning and the overstrength developed in the energy dissipater. Accordingly,

$$F_{pse} + P_D > \lambda_o F_s \tag{2.4.99}$$

where, for Unit 3,

$$\begin{aligned}
F_s &= \sum A_s F_y \\
&= 0.62(60) \\
&= 37.2 \text{ kips} \qquad (167 \text{ kN}) \\
\lambda_o &= 1.5 \\
\lambda_o F_s &= 1.5(37.2) \\
&= 55.8 \text{ kips} \qquad (250 \text{ kN}) \\
F_{pse} + P_D &= 40 + 48.6 \\
&= 88.6 \text{ kips} > 55.8 \text{ kips}
\end{aligned}$$

The measured tensile force in one energy dissipater ($A_s = 0.31$ in.^{2}) at a reported peak strain of 6.7% (60% of the calculated ultimate of 11.2%—see Eq. 2.4.98c) was almost 120 kN, essentially confirming the use of an overstrength factor (λ_o) of 1.5.

The applied dead load was 48.6 kips and accordingly, the required restorative force would be only 7.2 kips. The effective post-tensioning load after the panel had been relieved of load (Figure 2.4.61) was approximately 153 kN (34 kips). Clearly, the measured force on the post-tensioning and measured strain in the energy dissipater demonstrate the level of conservatism inherent in the simplifications proposed as design procedures. More precise estimates of ultimate strains may, of course, be used, but the level of conservatism adopted in the proposed design procedures is warranted because of the consequences associated with the failure of either the post-tensioning or the energy dissipater.

Compressive strain states in the toe region define the behavior limit state for the wall. The behavior of Rahman and Restrepo's Unit 2 offers some insight into the strain gradient along the compression face. Approximately 5 in. above the base of the panel, the compressive concrete strain is reportedly 0.0021 in./in. (Figure 2.4.57) at the peak drift ratio of 3%. Nine inches above the base the reported strain drops off to 0.0008 in./in. These strain states should be viewed as being consistent with the conservative proposition that about 80% of the displacement is attributable to rocking at the base.

The post test condition of the toe of Unit 2 is essentially the same as that which occurred in Unit 3 (Figure 2.4.60). Spalling extends to a height of about 5 in. The concrete strain state in the toe region of Unit 2 can be quantified from the displacement of the panel at the maximum drift ratio of 3%.

Base rotation; $\theta_{\text{gap}} \cong 0.024$ radian (80%)

Depth to neutral axis $\cong$ 5 in. (extrapolated from measured strand strains)

Depth of plastic hinge region (ℓ_p) $\cong$ 5 in. (experimental observation and the measured strain state 5 in. above the base—$\varepsilon_e \cong 0.002$ in./in. at $\theta = 0.03$ radian—Figure 2.4.58)

Compression associated shortening in the toe region is

$$\begin{aligned} \delta_{\text{toe}} &= \theta_{\text{gap}} a \qquad \text{(see Figure 2.4.58)} \\ &= 0.024(6) \\ &= 0.144 \text{ in.} \end{aligned}$$

while strain in the toe region is

$$\begin{aligned} \varepsilon_c &= \frac{\delta_{\text{toe}}}{\ell_p} \\ &= \frac{0.144}{5} \\ &= 0.0288 \text{ in./in.} \end{aligned}$$

A similar analysis of the toe region of Rahman and Restrepo's Unit 3 supports conclusions extrapolated from Units 1 and 2. At a peak drift ratio of 4% (termination of cyclic loading program), the toe region had spalled (Figure 2.4.60*b*). Using the same analytical process, the probable strain state in the toe can be estimated.

Drift ratio = 4%
Base rotation, $\theta_{gap} \cong 0.032$ radian (80%)
$a \cong 8$ in. (extrapolated from strand strain data by analysis (Eq. 2.4.97*b*))

The shortening at the toe (δ_{toe}) in the plastic hinge region ($\ell_p = t_w = 5$ in.) is

$$\begin{aligned}\delta_{toe} &= \theta_{gap} a\\ &= 0.032(8)\\ &= 0.256 \text{ in.}\\ \varepsilon_c &= \frac{\delta_{toe}}{\ell_p}\\ &= \frac{0.256}{5}\\ &= 0.051 \text{ in./in.}\end{aligned}$$

In all likelihood the concrete in the shell had been ineffective for many of the preceding cycles. The strain state in the confined core would be

$$\begin{aligned}\delta_{cc} &= \theta_{gap}(a - c_c)\\ &= 0.032(7)\\ &= 0.224 \text{ in.}\\ \varepsilon_c &= \frac{\delta_{cc}}{\ell_p}\\ &= \frac{0.224}{5}\\ &= 0.045 \text{ in./in.}\end{aligned}$$

It is not realistic to use this strain as a design limit state, but it is clear that attainable levels of concrete strain are high. The fact that the strain gradient is steep (Figure 2.4.57) means that the presence of a confined core will permit the apparent strain in the core to reach strains on the order of 0.04 in./in.

Comment: It is entirely possible that a similar strain concentration occurs in the toe of cast-in-place walls, in which case average strain limits based on a plastic hinge region of $0.5\ell_w$ could be used to design precast wall systems. Following cast-in-place prediction methodologies we find

$$\begin{aligned}\phi &= \frac{\theta}{\ell_p} \\ &= \frac{0.04}{26.5} \qquad (\ell_p = 0.5\ell_w) \\ &= 0.0015 \text{ rad/in.} \\ \varepsilon_c &= \phi a \\ &= 0.0015(8) \\ &= 0.012 \text{ in./in.}\end{aligned}$$

which is consistent with observed limit states for the cast-in-place walls studied in Section 2.4.1.

From a design perspective it seems reasonable to assume that

The depth of the plastic hinge region in a jointed precast wall system is on the order of the thickness of the wall panel (t_w).

The depth to the neutral axis (a) may be presumed to be quantifiable analytically.

The attainable unconfined concrete strain state at the outer fiber should exceed 0.01 in./in.

The concrete in the confined core should be capable of exceeding a strain of 0.025 in./in.

Comment: The Holden test of the emulative precast wall system (Unit 1), because curvatures are reported, tends to confirm the selection of the depth of the plastic hinge region used in the design of cast-in-place shear wall systems ($\ell_p = 0.5\ell_w$). Figure 2.4.62*a* describes measured curvature above the base of Unit 1 at various drift ratios. Observe how the plastic hinge region lengthens as the drift ratio advances beyond 1.5%. At a drift ratio of 1.5% the reported curvature in the plastic hinge (Figure 2.4.62a) is 0.00039 rad/in. The procedures developed in Section 2.4.1 to estimate the curvature predict this behavior.

$$\begin{aligned}\phi_y &= \frac{0.0033}{\ell_w} \\ &= \frac{0.0033}{52.7} \\ &= 0.000063 \text{ rad/in.}\end{aligned}$$

A linear distribution of curvature over the height of the wall is used to estimate panel drift at yield because the applied load is concentrated at the top of the panel.

$$\Delta_y = \frac{\phi_y h_w^2}{3}$$

$$= \frac{0.000063(146)^2}{3}$$

$$= 0.45 \text{ in.} \qquad (11 \text{ mm})$$

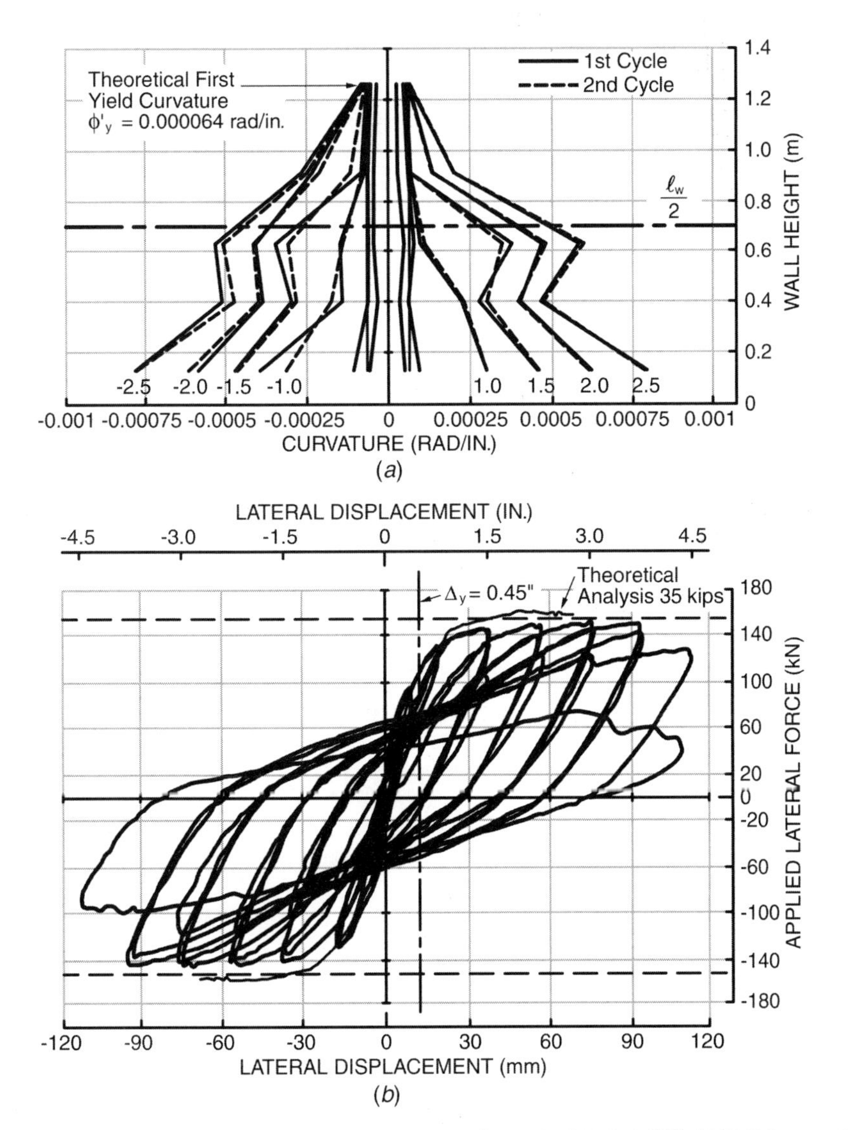

Figure 2.4.62 (*a*) Curvature distribution in the wall panel of Unit 1.[2.48] (*b*) Full hysteretic response of Holden's emulative wall.[2.48]

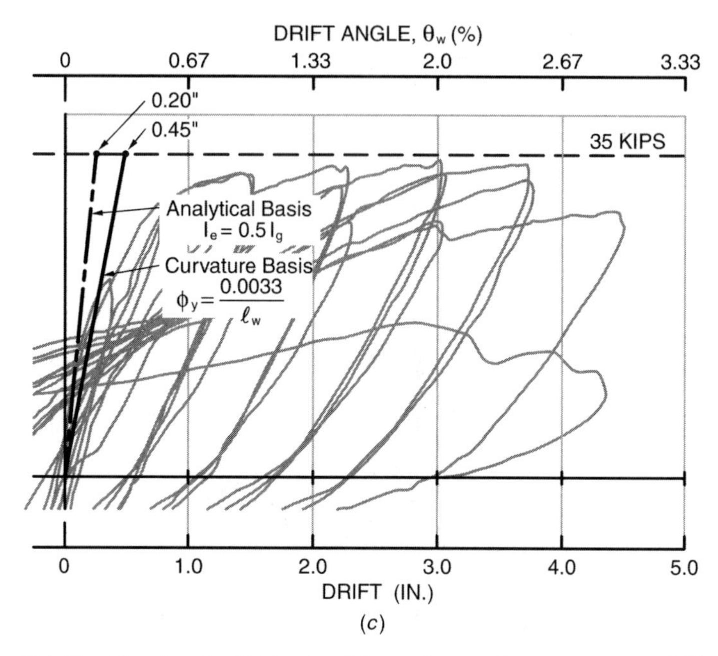

Figure 2.4.62 (*Continued*) (*c*) Elastic stiffness characterizations—Holden[2.48] Unit 1.

This is a reasonable idealization of the behavior reported by Holden (Figure 2.4.62*c*), given that the flexural component of drift was reportedly 75% of the measured drift. About 10% of the measured drift was reported as foundation rotation, 3% as shear deformation, and the rest was unidentified.

The postyield curvature at a drift of 1.5% is developed as follows:

$$\begin{aligned}\Delta_p &= \Delta_{1.5\%} - \Delta_y \\ &= 2.19 - 0.45 \\ &= 1.74 \text{ in.}\end{aligned}$$

$$\begin{aligned}\theta_p &= \frac{\Delta_p}{h_w - \ell_p/2} \qquad \text{(see Figure 2.1.10)} \\ &= \frac{1.74}{146 - 13} \\ &= 0.013 \text{ radian}\end{aligned}$$

$$\begin{aligned}\phi_p &= \frac{\theta_p}{\ell_p} \\ &= \frac{0.013}{26} \\ &= 0.0005 \text{ rad/in.}\end{aligned}$$

$$\begin{aligned}\phi_{1.5\%} &= \phi_y + \phi_p \\ &= 0.000063 + 0.0005 \\ &= 0.00056 \text{ rad/in.}\end{aligned}$$

Since 75% of this curvature estimate (0.00042 rad/in.) is attributable to flexure (by Holden), the agreement is quite good (see Figure 2.4.62*a*).

As the plastic hinge region deepens to the point at which the curvature at the outer edge of the plastic hinge region is that associated with first yield ($\phi_y = 0.000063$ rad/in.), either the procedures used to model behavior (Figure 2.1.10) or the linear curvature gradient described in Figure 2.4.62*a* predicts the flexural component of wall displacement with satisfactory accuracy.

$$\begin{aligned}\Delta_y &= \frac{\phi_y h_w^2}{3} \\ &= \frac{0.000063(146)^2}{3} \\ &= 0.45 \text{ in.} \\ \phi_{p,\text{avg}} &= 0.00064 \text{ rad/in.} \qquad \text{(Drift ratio} = 2.5\%\text{—Figure 2.4.62}a) \\ \Delta_p &= \phi_p \ell_p \left(h_w - \frac{\ell_p}{2}\right) \qquad \text{(Plastic hinge region)} \\ &= 0.00064(26)(146 - 13) \\ &= 2.2 \text{ in.} \\ \Delta_u &= 2.65 \text{ in.} \qquad \text{(Drift ratio} = 1.8\%)\end{aligned}$$

Again, this is consistent with the flexural component of drift (1.88%).

Were we to alternatively assume a uniform postyield strain gradient over a plastic hinge region of 48 in., (1.2 m, Figure 2.4.62*a*) the results would be

$$\begin{aligned}\Delta_p &= \frac{\phi_p}{2} \ell_p \left(h_w - \frac{\ell_p}{3}\right) \qquad \text{(Uniform curvature gradient)} \\ &= \frac{0.0008}{2}(48)(146 - 16) \\ &= 2.5 \text{ in.} \\ \Delta_u &= \Delta_y + \Delta_p \\ &= 2.95 \text{ in.}\end{aligned}$$

The projected strain state in the concrete, based on a probable depth to the neutral axis of 5 in. at a drift angle of 2.5%, would be developed as follows:

$$\phi_u = 0.00095 \text{ rad/in.} \qquad \text{(Extrapolated from Figure 2.4.62}a\text{)}$$

$$\begin{aligned} \varepsilon_c &= \phi_u c \\ &= 0.00095(5) \\ &= 0.0047 \text{ in./in.} \end{aligned}$$

which is low based on generally accepted limit states of 0.01 in./in. (Figure 2.4.4*c*).

Holden suggests that deterioration in the toe region is attributable to bar buckling, for he too projects strain states in the concrete that are much below spalling associated strains. Observe that the curvature ductility is at least 10.

$$\mu_\phi = \frac{\phi_y + \phi_p}{\phi_y} \tag{2.4.100}$$

$$= \frac{0.000063 + 0.00064}{0.000063}$$

$$= 11$$

Clearly, this test confirms previously developed design procedures for cast-in-place concrete walls (Section 2.4.1).

Shear transfer at the base of the tested precast walls produced no interface slip. The friction coefficient necessary to affect the shear transfer was only 17% (H/C_c), and this is well below the friction factor required to develop the hybrid beam system (Section 2.1.4.4) as well as the rather conservative shear friction factor suggested for a smooth interface (0.6). The shear stress across the effective bearing area (ab_e—Figure 2.4.59) in Rahman and Restrepo's Unit 3 approached 900 psi, and this is more than the limit state suggested in Figure 1.3.14 (700 psi). This also suggests that a limit state based on stress of 800 psi, as proposed in shear friction theory, is low.

Precast post-tensioned systems will tend to maintain their stiffness. Figure 2.4.56*b* describes early hysteretic behavior of Rahman and Restrepo's Unit 3. At a displacement of 0.20 in. (0.136%) the experimentally suggested secant stiffness is 112 kips/in., and this is consistent with the analytically suggested stiffness based on an effective moment of inertia of 50% of the gross moment of inertia (116 kips/in.). Observe that these projected early experimental cycles (Figure 2.4.56*b*) suggest a tangent stiffness on the order of 225 kips/in. and this suggests that the effective moment of inertia should be the gross moment of inertia given an "elastic" idealization similar to that used in cast-in-place concrete wall idealizations (Figure 2.4.4*a*). This perfectly elastic idealization ($I_e = I_g$) is certainly reasonable, if our objective is to compare alternatives (hybrid vs. cast-in-place) or to develop compatible design procedures.

The use of the same effective stiffness ($I_e = 0.5I_g$) to describe the cast-in-place wall whose behavior is described in Figure 2.4.4*a* and the precast wall whose behavior is described in Figure 2.4.56*b* is not logical. This conclusion is also supported by comparing the behavior of Holden's emulative panel (Figure 2.4.62*c*) with the hybrid panel (Figure 2.4.56b). This topic has been discussed in Section 1.1.5 and will be expanded on in Chapter 3 when we discuss the design of systems.

The secant stiffness at the probable drift ratio is important if damping-dependent, displacement-based design procedures are to be considered. If, for example, we were to presume that probable system drifts for a precast hybrid wall system were on the order of 1.5%, the corresponding decrease in analytical stiffness (see Figure 2.4.56*b*) would be

$$\delta_{1.5} = \frac{0.015(146)}{0.2}\delta_e$$

$$= 11\delta_e$$

and this suggests a secant stiffness of

$$k_{\text{sec}-1.5} = 0.091k_e$$

where k_e is the stiffness associated with the assumption that $I_e = 0.5I_g$.

Dynamic characteristics developed using the secant stiffness suggest that the period of a structure would increase 330% $\left(\sqrt{11}\right)$, and we see in Chapter 4 that this large a period shift does not materialize.

A relationship between viscous damping and the energy dissipated in postyield behavior cycles was developed in Section 1.1.3. Holden[2.48] develops comparative viscous damping ratios for the emulative precast wall and the hybrid system (Rahman and Restrepo—Unit 3). This comparison is reproduced in Figure 2.4.63. At a target drift ratio of 1.5%, the equivalent viscous damping in the emulative system (Figure 2.4.63*a*) is twice that in the hybrid system whose viscous damping is summarized in Figure 2.4.63*b* (17% versus 8.5%).

Structural damping, as a function of the hysteretic behavior of a member, has been analytically developed, but its extension to available system damping has yet to be demonstrated physically or analytically. Engineering intuition is the best design guide, at least at this writing. Engineering intuition is developed from a physical understanding of behavior.

Consider the hysteretic behavior described in Figure 2.4.53*a* for the post-tensioned (only) wall which did not contain energy dissipating devices. For a concentric prestress the forces acting on the wall are those described in Figure 2.4.64*a*. Figure 2.4.53*a* tells us that the post-tensioning yields when Fh_w is greater than $T_{py}(0.5\ell_w - a)$. This yield point is not well defined (Figure 2.1.53*a*), and as the post-tensioning is strained beyond theoretical yield, the sustainable load (F) increases until the displacement objective has been reached. Upon load release ($F = 0$) the system is restored by the post-tensioning force and this restoring force F^* (Figure 2.4.64*a*) is

$$F^* = \frac{T_{ps}(0.5\ell_w - a)}{h_w}$$

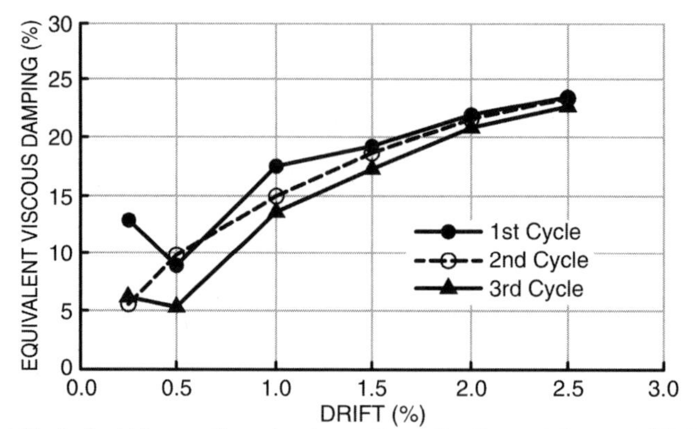

(*a*) Equivalent Viscous Damping for an Emulative Precast Concrete Wall System

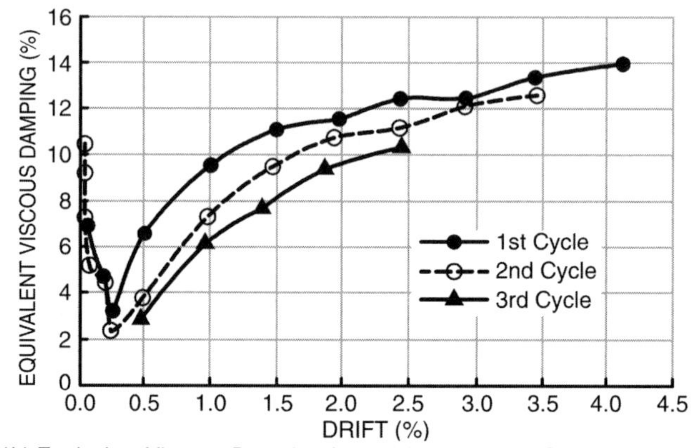

(*b*) Equivalent Viscous Damping for a Hybrid Precast Concrete Wall System

Figure 2.4.63 Comparison of equivalent viscous damping ratios.[2.47, 2.48]

The amount of postyield deformation experienced in the strand during each cycle will reduce the stiffness of the system as can be seen in Figure 2.4.53*a*.

When an energy dissipater is introduced into a system, the relationship between the force in the energy dissipater and that provided by the post-tensioning will define the shape of the hysteresis loop. Consider, for example, Figure 2.1.6, which describes the postyield behavior of a conventionally reinforced beam. Here the beam recovers the elastic portion of its deformation and, at no load ($F = 0$), a permanent displacement exists in the beam.

Now consider the hysteretic behavior described in Figure 2.4.56 for the hybrid system. Upon load release ($F = 0$), the elastic force in the energy dissipater is relieved and the restoring force provided by the post-tensioning will, if sufficient, overcome the postyield strain sustained by the energy dissipater. The hybrid system whose behavior is described in Figure 2.4.56 had as a design objective attaining a balance between the developable force in the energy dissipater and the post-tensioning. This objective was exceeded (see Eq. 2.4.99). The relationship between $\lambda_o F_y$ and

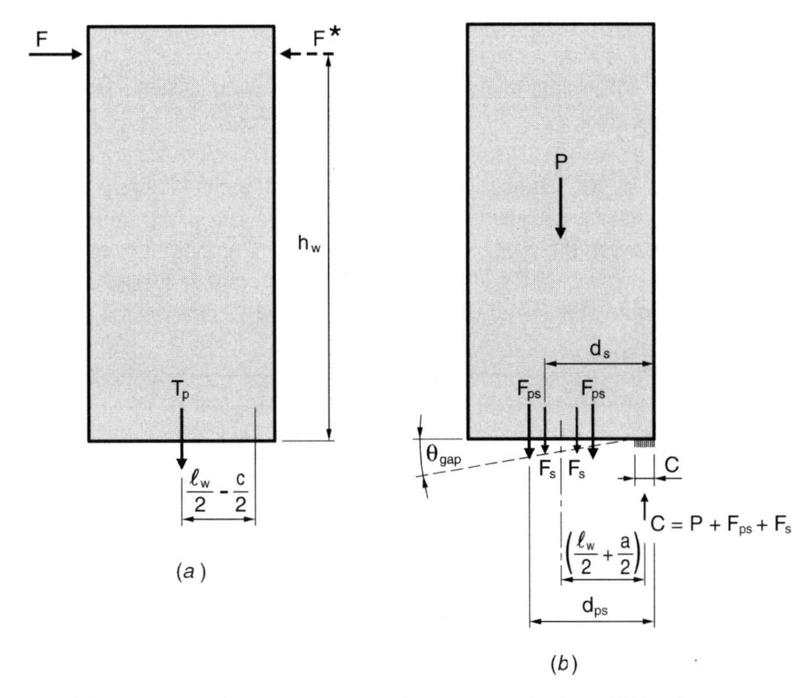

Figure 2.4.64 (*a*) Restoring force developed by post-tensioning. (*b*) Loads imposed on hybrid wall system.

$F_{pse} + F_D$ is reflected in the hysteretic behavior. A lesser strength in the energy dissipater will reduce the hysteretic area, and this is evident in Rahman and Restrepo's Unit 2, whose hysteretic behavior is described in Figure 2.4.56*c*. The same level of prestress was applied to each panel, but the area of the energy dissipater for Unit 2 was only 56% of that used in Unit 3. The change in hysteretic area is proportional to the overstrength force that can be sustained by the energy dissipater, and this is logical.

From a design perspective it appears reasonable to strive for the objective strength balance and utilize the damping values proposed in Figure 2.4.63*b*. If more structural damping is required, it may be attained by increasing the strength of the energy dissipater and interpreting linearly from the values suggested by Figures 2.4.63*a* and 2.4.63*b*. A significant imbalance ($\lambda_o F_y \gg F_{ps} + F_D$) will, however, result in an increase in probable permanent deformation of the system.

Accordingly, it is advisable to maintain a balance between the provided post-tensioning force and the strength of the energy dissipater.

$$F_{pse} \cong F_s \tag{2.4.101}$$

and, in order to restore the system, satisfy Eq. 2.4.99.

2.4.3.2 Experimentally Inferred Conclusions—Hybrid Precast Wall System

Postyield Behavior Postyield behavior is concentrated at the foundation/wall panel interface, thereby creating a gap (θ_{gap}).

System Stiffness A post-tensioned precast concrete wall panel system is stiffer than a comparable cast-in-place system because postyield rotation accumulates at the precast interface, leaving the panel virtually uncracked. The post-tensioned precast wall panel is about twice as stiff as the cast-in-place system (see Figure 2.4.56*b* and 2.4.4*a*), and this should at least subliminally be considered in the design process.

Concrete Strain Postyield strain in the toe region appears to be concentrated in a depth equivalent to the thickness of the wall (t_w). An ultimate strain of 0.01 in./in. seems reasonably conservative for design purposes for unconfined concrete. The toe region should be confined to the depth of the neutral axis and a height of at least $2t_w$. Attainable concrete strain in the confined core can be presumed to be at least 0.025 in./in.

Precast Interface Gap A gap will open at the precast interface (Figure 2.4.58). This gap (θ_{gap}) will be somewhat less than the panel rotation (θ) or drift ratio (Δ / h_w). A gap angle equivalent to 80% of the drift ratio (Δ / h_w) is appropriately assumed. The base of the panel will rotate as a plane about the neutral axis. The center of rotation at maximum sustainable rotations seems to coincide with the edge of the analytically quantified confined compressive stress block. Concrete and reinforcing strain states are reasonably estimated using this model.

Energy Dissipaters Energy dissipaters should be sized so as to be equivalent to the strength of the post-tensioning force. The restoring force that will include the effective dead load should exceed the overstrength of the energy dissipating device.

Detailing Post-tensioning strands and energy dissipating devices should be detailed so as to accommodate gap rotations.

The inclusion of panel reinforcement, similar to that described in Figure 2.4.48, has been shown[2.48] to increase the attainable drift ratio to 6%.

2.4.3.3 Design Procedures

Step 1: Determine the Design Approach. Alternative design approaches are strength and deflection based. See Sections 1.1.4 and 1.1.5 for a discussion of these approaches and Section 2.4.1.2 for their application to shear walls.

Step 2: Select the Number of Walls. In most cases this will evolve from building function and the proposed structural system.

Step 3: Select a Trial Thickness for the Wall. Precast walls should be as thin as possible, so the selected thickness should be between $h_x/25$ and $h_x/16$. The minimum thickness will be based typically on the development of the connection between the wall and the floor system, especially if a precast floor system is to be used.

The ability to provide confinement in the toe region should be maintained. The imposed axial demand $(P_D + P_L)$ imposed on the wall in the lowermost panel should not exceed $0.25 f'_c A_g$.

Step 4: Determine the Probable Fundamental Period of the Building.

Comment: If a strength-based approach is to be used, underestimate the period; if a displacement-based approach is to be used, overestimate the period. A good estimate for strength-based design purposes may be developed from the flexural stiffness of the wall, assuming that the effective stiffness is the uncracked stiffness $(I_e = I_g)$.

$$T_f = 0.09 \left(\frac{h_w}{\ell_w}\right)^2 \sqrt{\frac{w\ell_w}{Et_w}} \qquad \text{(see Eq. 2.4.10 and 2.4.11)}$$

When attaining deflection objectives is the design focus, the effective stiffness should be based on $0.5I_g$. Then,

$$T_f = 0.13 \left(\frac{h_w}{\ell_w}\right)^2 \sqrt{\frac{w\ell_w}{Et_w}}$$

where

E is expressed in kips per square in.
t_w, ℓ_w, and h_w are expressed in feet.
$w(W/h_w)$ is the tributary weight (mass), expressed in kips per foot, that must be supported laterally (restored) by the shear wall.

Step 5:

(a) Determine the Spectral Velocity. If spectral velocity (C_v) is defined by the building code, convert C_v to the spectral velocity for a single-degree-of-freedom system (Eq. 2.4.2b).

(b) Insure That the Structure Falls within the Velocity Constant Region.

$$T > \frac{C_v}{2.5C_a} \qquad \text{(Eq. 2.4.3)}$$

$$T_{\max} < 4 \text{ seconds} \qquad \text{(see Eq. 2.4.4)}$$

(c) Quantify the Angular Frequency (ω_n).

$$\omega_n = \frac{2\pi}{T} \qquad \text{(Eq. 2.4.5)}$$

(d) Determine the Spectral Displacement (S_d).

$$S_d = \frac{S_v}{\omega_n} \qquad \text{(Eq. 2.4.6)}$$

(e) Determine the Probable Ultimate Displacement of the Roof (Δ_u).

$$\Delta_u = \Gamma S_d \qquad \text{(Eq. 2.4.7)}$$

(f) Determine the Spectral Acceleration.

$$S_a = \omega_n S_v \qquad \text{(see Eq. 1.1.6b)}$$

Step 6: Determine the Yield Moment Required to Attain the Objective Displacement and Select a Displacement Ductility Factor (μ_Δ). Develop a base shear objective from the spectral acceleration using a system displacement ductility factor of at least 10 (see Figure 2.4.56*a*).

$$V_E = \frac{0.8 S_a W}{\mu_\Delta}$$

where 0.8 is an estimate of $\Gamma_1 \Sigma \phi_1$ [2.4, Eq. 4.7.16] and S_a is expressed as a percentage of g.

Establish a flexural strength objective:

$$M_{yi} = 0.67 h_w V_E$$

Step 7: Determine the Flexural Reinforcing Program Required to Attain System Design Objectives. Figure 2.4.64*a* describes the loads imposed on a hybrid wall system. From a design perspective, several generalizations will be helpful in terms of quickly quantifying an objective reinforcing program.

(a) The ultimate (average) strength provided by the post-tensioning is

$$F_{psu} = 0.81 \sum A_{ps} f_{ps} \qquad \text{(see Eq. 2.4.94)}$$

f_{ps} should reflect a best guess of the ultimate stress in the strand most distant from the toe. When the strand debond length is significant, an average prestress of 175 ksi is suggested for estimating the flexural strength of the wall.

$$F_{ps} = 175 \sum A_{ps} \qquad (2.4.102\text{a})$$

(b) The strength provided by the energy dissipater is

$$F_s = A_s f_y$$

(c) The flexural strength is satisfied provided

$$(P_D + F_{ps} + F_s)\left(\frac{\ell_w}{2} - \frac{a}{2}\right) > M_n \qquad (2.4.102b)$$

(d) The restoring objective is satisfied provided

$$P_D + F_{psu} > 1.5F_s \qquad \text{(see Eq. 2.4.99)}$$

(e) The forces (P_D, F_{ps}, F_s) act concentrically and, from a design perspective, $(\ell_w/2 - a/2)$ may be assumed to be on the order of $0.40\ell_w$ and, if need be, refined iteratively.

Introducing this approximation into Eq. 2.4.100 and Eq. 2.4.102b allows the quantification of $F_s + F_{ps}$:

$$P_D + F_{ps} + F_s > \frac{M_n}{0.4\ell_w} \qquad \text{(see Eq. 2.4.102b)}$$

$$F_{ps} + F_s > \frac{M_n}{0.4\ell_w} - P_D \qquad (2.4.103)$$

Step 8: Check the Stress Imposed on the Wall.

$$F_{pse} = \sum A_{ps} f_{pse}$$

$$f_c = \frac{F_{pse} + P_D}{A_g} < 0.25f'_c \qquad (2.4.104)$$

Step 9: Determine the Strain States in the Concrete and the Reinforcing Steel.

(a) Determine the probable gap, assuming the wall acts essentially as though it were a rigid body. Assume that the flexural and shear components of drift within the panel are 20% of the total drift.

$$\theta_{gap} = 0.8(\text{DR})$$

$$= 0.8\frac{\Delta_u}{h_w}$$

(b) Determine the effective depth of the neutral axis. This may be based on the presumption that the shell will spall (strength limit state) or on a performance basis whose strain objective would be the shell spalling limit state. The strength limit state would consider the additional strength provided by confined concrete strength (f'_{cc}), which may be assumed to be essentially the same over the entire confined compressed area:

$$f'_{cc} = f'_c + 2.5 \qquad \text{(see Eq. 2.4.96b)}$$

$$a = \frac{C_c}{0.85(f'_c + 2.5)b_e}$$

(c) Determine the shortening that must occur in the toe, either at the edge of the confined region (see Figure 2.4.58),

$$\delta_{cc} = \theta_{\text{gap}} a$$

or at the compression face of the panel,

$$\delta_{\text{toe}} = \theta_{\text{gap}}(a + c_c)$$

where c_c is the thickness of the cover.

(d) Determine the strain in the concrete, assuming that the plastic hinge length is t_w.

$$\varepsilon_c = \frac{\delta_{cc}}{t_w}$$

Given a shell performance objective, revise the design until the suggested or hypothetical strain is less than 0.01 in./in. The strain limit state would be,

$$\varepsilon_c = \frac{\delta_{\text{toe}}}{t_w}$$

(e) Determine the elongation and strain imposed on the post-tensioning strand at Δ_u,

$$\delta_{psp} = \theta_{\text{gap}}\left(d_{ps} - a\right) \qquad \text{(Figure 2.4.64}b\text{)} \qquad (2.4.105)$$

$$\varepsilon_{psp} = \frac{\delta_{psp}}{h_w} \qquad (2.4.106)$$

provided that the post-tensioning is unbonded over the entire height of the wall:

$$\varepsilon_{psu} \cong \varepsilon_{pse} + \varepsilon_{psp} \qquad (2.4.107)$$

Adjust the design until ε_{psu} is less than 0.02 in./in.

(f) Determine the strain state in the energy dissipaters at Δ_u.

$$\delta_{sp} = \theta_{\text{gap}}\left(d_s - a\right) \qquad \text{(Figure 2.4.64}b\text{)} \qquad (2.4.108)$$

$$\varepsilon_{sp} = \frac{\delta_{sp}}{\ell_{p,ed}} \tag{2.4.109}$$

where $\ell_{p,ed}$ is the length of the yielding portion of the energy dissipater. Adjust the design until the strain is within the capabilities of the energy dissipaters, usually on the order of 0.2 in./in.

Step 10: Detail the Panel System.

Step 11: Provide the Analysis Team with the Information Required to Effectively Evaluate the Conceptual Design.

2.4.3.4 Example Design—Ten-Story Shear Wall Design an assembled precast concrete shear wall to replace the thin wall system design developed in Section 2.4.1.4 (see Figure 2.4.9). Propose a hybrid system utilizing centrally located post-tensioning and energy dissipaters. Assume for the purposes of this design that the objective levels of deformability are the same as for the cast-in-place system. Hence

$$\Delta_u = 10.8 \text{ in.}$$

$$\text{Drift ratio} = \frac{10.8}{105(12)}$$

$$= 0.85\%$$

Step 1: Determine the Design Basis. A displacement design basis has been selected.

Step 2: Select the Number of Walls. This is an alternative design to the thin wall designed in Section 2.4.1.4. Accordingly, mass and design loads are as identified in that example.

Step 3: Select a Trial Thickness for the Wall. Assume that the floor will be constructed using a 6-in. cast-in-place slab. Try a 6-in. thick panel.

Steps 4 and 5: These tasks were completed as a part of the design of the thin wall system. See Section 2.4.1.4, Step 4. They produced the deflection criterion previously described.

Step 6: Determine the Yield Moment.

$$V_E = \frac{0.8 S_a W}{\mu_\Delta} \quad \text{(Eq. 2.4.103)}$$

$$= \frac{0.8(0.75)(1400)}{10}$$

$$= 84 \text{ kips}$$

$$M_{yi} = 0.67 h_w V_E$$

$$= 0.67(105)(84)$$

$$= 5900 \text{ ft-kips}$$

Comment: Observe that $\lambda_o \cong 1.0$ (see Figure 2.4.56*a*).

Step 7: Determine the Flexural Reinforcing Required to Attain System Design Objectives.

$$M_n = \frac{M_{yi}}{\phi}$$

$$= \frac{5900}{0.9}$$

$$= 6560 \text{ ft-kips}$$

$$F_{ps} + F_s > \frac{M_n}{0.40\ell_w} - P_D \qquad \text{(Eq. 2.4.103)}$$

$$> \frac{6560}{0.4(26)} - 600$$

$$> 630 - 600$$

$$> 30 \text{ kips}$$

Prestressing minimums will control the design. A minimum effective post-tensioning force of $0.125A_g$ is recommended. The effective post-tensioning force is

$$F_{pse,\min} = (0.125)A_g$$

$$= (0.125)6(26)(12)$$

$$= 234 \text{ kips}$$

Now the selection of reinforcing must consider the energy dissipater of interest. For example, if a forged ductile rod (Figure 2.1.57) were to be used, the alternatives would be to provide either two or three ductile rods (282 kips or 423 kips).

The function of the energy dissipater was to create structural damping in the system. The depth of the return or unloading cycle shown on Figure 2.4.56*a* was attributed to the strength of the energy dissipater. It was suggested that a balance be maintained between F_{ps} and F_s. Since the overstrength factors (λ_o) are about the same, a balance should be maintained between F_s and F_{pse}.

Conclusion: Select a three ductile rod configuration and provide twelve 0.6-in. ϕ strands.

$$F_s = 423 \text{ kips} \qquad \text{(3 DDCs)}$$

$$F_{pse} = N(A_{ps})f_{pse}$$

$$= 12(0.217)162 \qquad \text{(Twelve 0.6-in. } \phi \text{ strands)}$$

$$= 421 \text{ kips}$$

Before proceeding, it is advisable to check the depth of the comprehensive stress block assumed in the development of the flexural strength ($0.2\ell_w$).

$$\begin{aligned} C_c &= P_D + \lambda_o\left(F_{pse} + F_{sy}\right) \\ &= 600 + 1.5(421 + 423) \\ &= 1866 \text{ kips} \end{aligned}$$

Comment: It is unlikely that F_{pse} will be amplified by $\lambda_o = 1.5$ simply because the strands will be strained over a significant distance. A more probable overstrength force in the prestress would be

$$\begin{aligned} \lambda_o F_{ps} &= \lambda_o f_{pse} A_{ps} \\ &= (162 + 10)(2.6) \\ &\cong 450 \text{ kips} \end{aligned}$$

and then

$$\begin{aligned} C_c &\cong 1700 \text{ kips} \\ a &= \frac{C_c}{0.85 f_c' t_w} \\ &= \frac{1700}{0.85(5)(6)} \\ &= 67 \text{ in.} \qquad (0.2\ell_w = 62.4 \text{ in.}) \end{aligned}$$

Comment: The neutral axis will gravitate toward the toe as the wall displaces and the confined concrete stress (f_{cc}') is activated. A concrete strength of 6 ksi should be considered.

Step 8: Check the Stress Imposed on the Wall.

$$f_c = \frac{F_{pse} + P_D}{A_g} \tag{2.4.110}$$

$$\begin{aligned} &= \frac{421 + 600}{6(26)(12)} \\ &= 0.546 \text{ ksi} \qquad \left(0.109 f_c'\right) \qquad \text{(OK)} \end{aligned}$$

Step 9: Determine the Strain States in the Concrete and Reinforcing Steel.

(a) Determine the gap angle (θ_{gap}) created based on the presumption that the wall is uncracked and will rotate essentially as a rigid body. Compare the behavior of the precast hybrid system (Figure 2.4.56*a*) with that of the cast-in-place system (Figure 2.4.4*a*), and compare the developed crack pattern described in Figure 2.4.2*b* with that of the post-test precast wall panel (Figure 2.4.60*a*). Based on the proposition that the precast wall system is twice a stiff as the comparable cast-in-place alternative ($I_e = I_g$), the ultimate drift ratio might be as low as

$$\begin{aligned} \text{DR}_{\text{precast}} &= \frac{\text{DR}_{\text{cip}}}{\sqrt{2}} \\ &= \frac{0.85}{1.414} \\ &= 0.6\% \end{aligned}$$

The gap rotation becomes

$$\begin{aligned} \theta_{\text{gap}} &= 0.8(\text{DR}) \\ &= \frac{0.8(0.6)}{100} \\ &= 0.0048 \text{ radian} \end{aligned}$$

(b) Determine the effective depth of the neutral axis.

$$P_D = 600 \text{ kips} \qquad \text{(Given)}$$

$$F_{ps} = 175 \sum A_{ps} \qquad \text{(Eq. 2.4.101a)}$$

$$\begin{aligned} &= 175(2.6) \\ &= 455 \text{ kips} \end{aligned}$$

$$\begin{aligned} \lambda_o F_s &= \lambda_o \sum A_s f_y \\ &= 1.5(423) \\ &= 635 \text{ kips} \end{aligned}$$

$$\begin{aligned} C_c &= P_D + F_{psu} + \lambda_o F_{su} \\ &= 600 + 455 + 635 \\ &= 1690 \text{ kips} \end{aligned}$$

$$a = \frac{C_c}{0.85(f_c' + 2.5)(b_e)} \qquad \text{(see Eq. 2.4.96b)}$$

For $f_c' = 6$ ksi and $b_e = 4.5$ in.,

$$\begin{aligned} a &= \frac{1690}{0.85(8.5)(4.5)} \\ &= 52 \text{ in.} \end{aligned}$$

(c) Determine the shortening of the toe.

$$\begin{aligned}\delta_{\text{toe}} &= \theta_{\text{gap}}(a - c_c)\\ &= 0.0048(52 - 1)\\ &= 0.25 \text{ in.}\end{aligned}$$

(d) Determine the strain in the concrete.

$$\begin{aligned}\varepsilon_c &= \frac{\delta_{\text{toe}}}{t_w}\\ &= \frac{0.25}{6}.\\ &= 0.042 \text{ in./in.}\end{aligned}$$

Conclusion: A design revision is reasonably considered. Revise the concrete strength to 8 ksi $\left(f_c' = 8 \text{ ksi}\right)$ and the wall thickness to 8 in.

$$a = \frac{C_c}{0.85(f_c' + 2.5)(b_e)} \qquad \text{(see Eq. 2.4.96b)}$$

$$\begin{aligned}&= \frac{1690}{0.85(10.5)(6.5)}\\ &= 29 \text{ in.}\end{aligned}$$

$$\begin{aligned}\delta_{\text{toe}} &= \theta_{\text{gap}}(a - \text{cover})\\ &= 0.0048(29 - 1)\\ &= 0.134 \text{ in.}\end{aligned}$$

$$\begin{aligned}\varepsilon_c &= \frac{\delta_{\text{toe}}}{t_w}\\ &= \frac{0.134}{8}\\ &= 0.017 \text{ in./in.} \qquad \text{(OK)}\end{aligned}$$

(e) Determine the elongation and strain in the post-tensioning strand at Δ_u. Prepare sketch describing the probable location of the reinforcing at the base of the panel (Figure 2.4.65).

$$d_{ps} = 192 \text{ in.}$$

$$\delta_{psp} = \theta_{\text{gap}}\left(d_{ps} - a\right) \qquad \text{(Eq. 2.4.105)}$$

$$\begin{aligned}&= 0.0048(192 - 29)\\ &= 0.78 \text{ in.}\end{aligned}$$

$$\varepsilon_{psp} = \frac{\delta_{psp}}{h_w} \qquad \text{(see Eq. 2.4.106)}$$

$$= \frac{0.78}{105(12)}$$

$$= 0.00062 \text{ in./in.}$$

$$\varepsilon_{psu} \cong \varepsilon_{pse} + \varepsilon_{psp}$$

$$= 0.0058 + 0.00062$$

$$= 0.0064 \text{ in./in.} < 0.02 \qquad \text{(OK)}$$

(f) Determine the strain state in the energy dissipaters at Δ_u (see Figure 2.4.65).

$$\delta_{sp} = \theta_{\text{gap}}(d_s - a) \qquad \text{(see Eq. 2.4.108)}$$

$$= 0.0048(156 - 29)$$

$$= 0.61 \text{ in.}$$

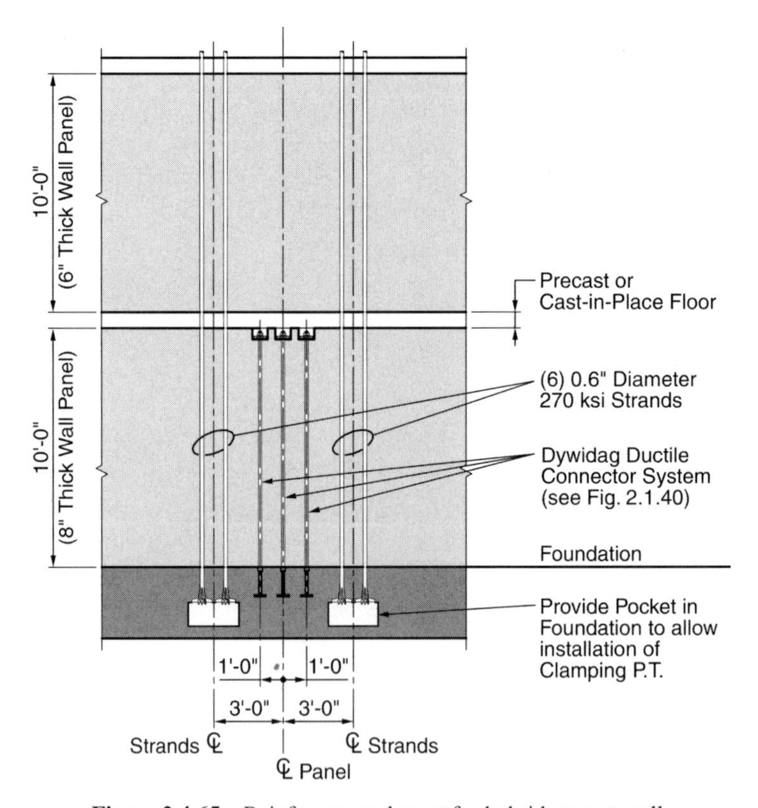

Figure 2.4.65 Reinforcement layout for hybrid precast wall.

$$\varepsilon_{sp} = \frac{\delta_{sp}}{\ell_{p,ed}} = \frac{0.61}{9} = 0.068 \text{ in./in.} < 0.2 \text{ in./in.} \qquad \text{(OK)}$$

Step 10: Detail the Precast Panel System. The DDCs should be embedded in the foundation or supporting beam. There must be enough reinforcing below the DDCs to allow them to be compressed on reverse cycle loading. The DDC anchorage in the foundation must develop the overstrength of the DDCs.

$$T_u = \lambda_o N(T_y) = 1.5(3)(141) = 635 \text{ kips}$$

$$A_s = \frac{T_u}{\phi f_y} = \frac{635}{0.9(60)} = 11.75 \text{ in.}^2$$

Conclusion: Propose fourteen #6 hoop ties. See Figure 2.4.66*a* to transfer the DDC anchorage load to the foundation or supporting beam. Observe that load transfer ties must be located well within the development cone ($\theta = 45°$) and that 135° hook enclosing an anchoring bar that should at least be 1 in. in diameter are required.

The post-tension anchorage must allow the tendons to be placed after the wall system has been assembled. Flat anchorages are probably most appropriate for precast wall construction (Figure 2.4.66*b*). Observe that the duct need only be 2 in. long and 1 in. thick. The foundation end will typically be a dead end. The stressing end is only 4 in. wide and will fit into a 6-in. thick wall panel. The development force should be that required to resist a yielding of the strand ($0.9 f_{psu} \Sigma A_{ps}$).

$$T_{ps} = 3(0.217)(0.9)(270) = 158 \text{ kips}$$

$$A_s = \frac{T_{ps}}{\phi f_y} = \frac{158}{0.9(60)} = 2.93 \text{ in.}^2$$

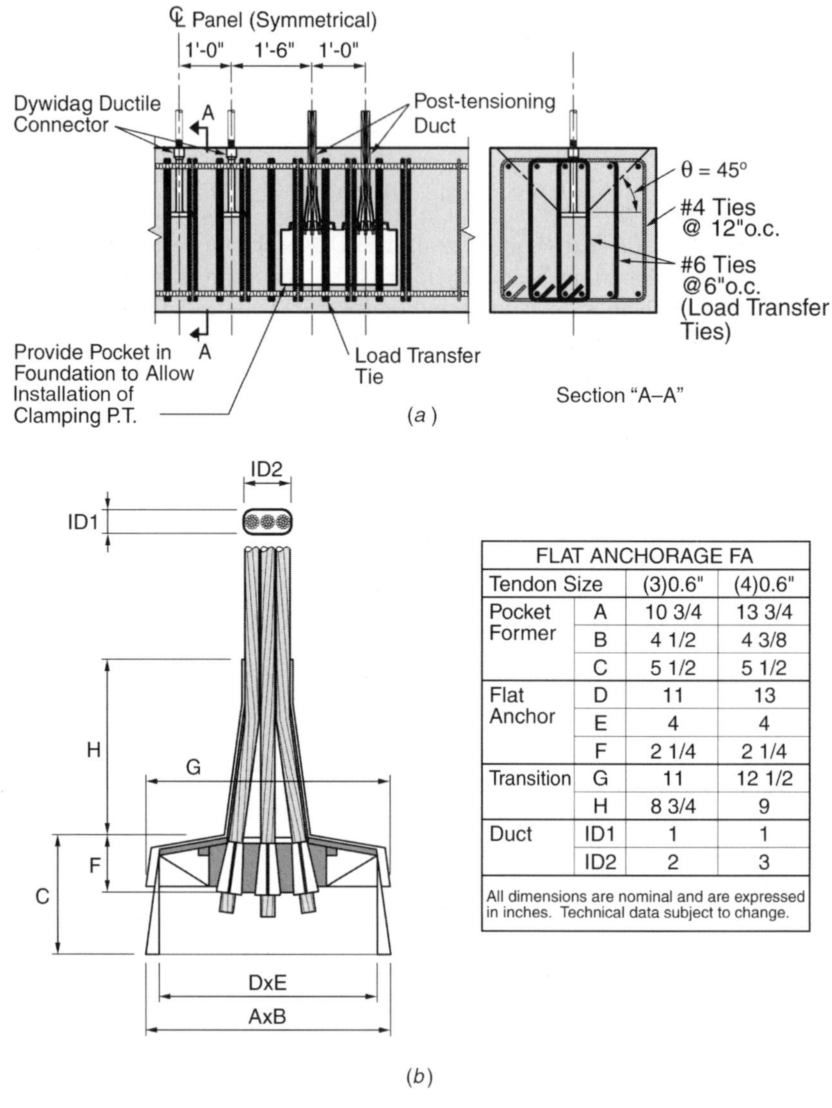

FLAT ANCHORAGE FA			
Tendon Size		(3)0.6"	(4)0.6"
Pocket Former	A	10 3/4	13 3/4
	B	4 1/2	4 3/8
	C	5 1/2	5 1/2
Flat Anchor	D	11	13
	E	4	4
	F	2 1/4	2 1/4
Transition	G	11	12 1/2
	H	8 3/4	9
Duct	ID1	1	1
	ID2	2	3
All dimensions are nominal and are expressed in inches. Technical data subject to change.			

Figure 2.4.66 (*a*) DDC anchorage. (*b*) Post-tensioning anchors.

An additional seven #6 hoop ties are required to anchor the two post-tensioning ducts (see Figure 2.4.66*a*).

The DDCs must be connected to thread bars or similar high-strength bars that can develop the overstrength of the DDCs (see Figure 2.1.40). A groutable sleeve must be placed in at least the first panel in order to develop the strength of the thread bar along with a load path that can resolve the delivered load. From a detailing perspective it is advisable to allow at least some rotational freedom for both the energy dissipaters and the prestressing strands as they enter the panel so as to minimize strain concentrations. Similar conditions have been extensively tested in the hybrid beam and not found to be a problem (see Section 2.1.4.4).

It is desirable to resolve the energy dissipater load transfer within the lowermost panel, provided the flexural strength of the precast wall system can be maintained. The strut mechanism described in Figure 2.4.67*a* is similar to that developed by Holden[2.48] in his Unit 2. This resolution of forces allows the panels above the base to be clamped entirely by post-tensioning because a boundary post-tensioning system has been added.

The boundary post-tensioning force (T_{psb}) is

$$
\begin{aligned}
T_{psb} &= \frac{\lambda_o F_{su}(d_s)}{d_{psb}} \\
&= \frac{\lambda_o F_{su}\,(\ell_w/2 - a/2)}{\ell_w - d' - a/2} \\
T_{psb} &= \frac{635(11.75)}{26 - 2.5} \\
&= 317.5 \text{ kips} \\
A_{psb} &= \frac{T_{psb}}{f_{psu}} \\
&= \frac{317.5}{\phi 0.9(270)} \\
&= 1.45 \text{ in.}^2
\end{aligned}
$$

Conclusion: Use seven 0.6-in. ϕ 270-ksi strands in the boundary. The force along the compression diagonal is

$$
F_{cd} = \frac{\lambda_o F_{su}}{2}\,\frac{\ell_{cd}}{h_x} \qquad \text{(see Figure 2.4.67}b\text{)}
$$

$$
\begin{aligned}
&= \frac{635}{2}\left(\frac{15.6}{10}\right) \\
&= 495 \text{ kips} \\
w &= 0.25\ell_{cd} \\
&= 0.25(15.6) \\
&= 3.9 \text{ ft} \\
f_{cd} &= \frac{F_{cd}}{w t_w}
\end{aligned}
$$

$$= \frac{495}{3.9(12)8}$$

$$= 1.32 \text{ ksi} < 0.5 f_c' = 4 \text{ ksi} \qquad \text{(OK)}$$

The thread bars anchoring the dissipating rods should be effectively anchored at the top so as to develop a node at the top of the lower panel. They need not be grouted. The area of the anchoring plate would be developed in a standard manner.

$$A_1 = \frac{\lambda_o A_b f_y}{0.85 f_c'} \qquad \text{(see Ref. 2.6, Sec. 10.17.1)}$$

$$= \frac{1.5(141)}{0.85(8)}$$

$$= 31 \text{ in.}^2$$

Conclusion: Use a 5 in. by 8 in. plate of the appropriate thickness.

Comment: Because of the high concentration of bearing stresses, confinement is reasonably provided.

Confinement in the toe region is required in the amount of $0.09 f_c'$ (see Section 1.2). If we select a transverse spacing of 6 in. (Figure 2.4.67*c*), the force that must be developed is

$$T_{sh} = 0.09(8000)6s$$

$$= 4320s$$

For a number 4, Grade 75 bar ($T_y = 0.2(75) = 15$ kips), the vertical spacing would be

$$s = \frac{15}{4.32}$$

$$= 3.5 \text{ in. on center}$$

Suggested reinforcing for the lowermost panel is shown on Figure 2.4.67*c*. Minimum horizontal and vertical wall reinforcing in the amount of $0.0025A_c$ should be provided. Observe that the wall reinforcing in this case is sufficient to resist the dilatational stresses required to develop the strut (Figure 2.4.67*b*). Since the remainder of the wall, above the lowermost panel, is not subjected to either large transfer stresses or unusual boundary stresses, a single curtain of reinforcing that satisfies shear strength and minimum steel requirements is sufficient.

Step 11: Provide Analysis Team with Data Necessary to Analyze the Wall. Most of this information is contained in Figures 2.4.65, 2.4.66, and 2.4.67.

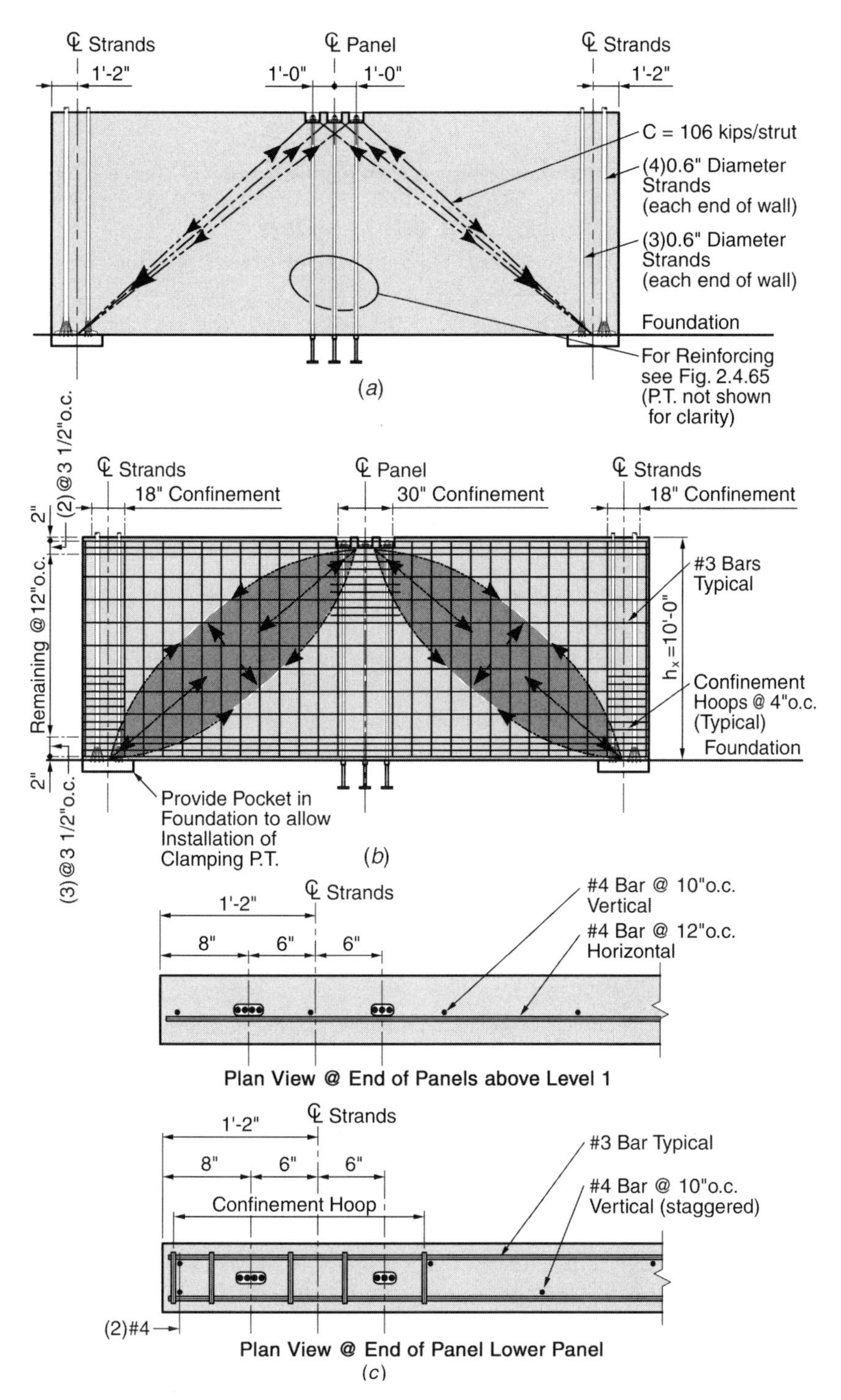

Figure 2.4.67 (*a*) Strut mechanism in lower hybrid precast wall panel. (*b*) Lower hybrid precast panel development—elevation. (*c*) Hybrid precast concrete wall reinforcing program.

SELECTED REFERENCES

[2.1] E. P. Popov, V. V. Bertero, and H. Krawinkler, "Cyclic Behavior of Three R.C. Flexural Members with High Shear," Report No. EERC 72-5, Earthquake Engineering Research Center, University of California, Berkeley, October 1972.

[2.2] R. Park and T. Paulay, *Reinforced Concrete Structures.* John Wiley & Sons, New York, 1975.

[2.3] J. G. MacGregor, *Reinforced Concrete: Mechanics and Design*, 2nd Edition. Prentice Hall, Englewood Cliffs, New Jersey, 1992.

[2.4] R. E. Englekirk, *Steel Structures: Controlling Behavior through Design.* John Wiley & Sons, New York, 1994.

[2.5] T. Paulay and M. J. N. Priestley, *Seismic Design of Reinforced Concrete and Masonry Buildings.* John Wiley & Sons, New York, 1992.

[2.6] American Concrete Institute, *Building Code Requirements for Structural Concrete (318-99) and Commentary (318R-99),* ACI 318-99 and ACI 318R-99, Farmington Hills, Michigan, June 1999.

[2.7] M. J. N. Priestley, "Overview of PRESSS Research Program," *PCI Journal,* Vol. 36, No. 1, July–August 1991, pp. 50–57.

[2.8] G. S. Cheok and H. S. Lew, "Model Precast Concrete Beam-to-Column Connections Subject to Cyclic Loading," *PCI Journal,* Vol. 38, No. 4, July–August 1993.

[2.9] International Conference of Building Officials, *Uniform Building Code*, 1997 Edition, Whittier, CA.

[2.10] C. P. Taylor, P. A. Cote, and J. W. Wallace, "Design of Slender Reinforced Concrete Walls with Openings," *ACI Structural Journal*, Vol. 95, No. 4, July–August 1998.

[2.11] M. J. N. Priestley, F. Seible, and M. Calvi, *Seismic Design and Retrofit of Bridges*, John Wiley & Sons, New York, 1995.

[2.12] G. H. Powell, "*DRAIN-2DX, Element Description and User Guide for Element Type 01, Type 02, Type 04, Type 06, Type 09, and Type 15*," Version 1-10, Report No. UCB/SEMM-93/18, Department of Civil Engineering, University of California at Berkeley, Berkeley, California, December 1993.

[2.13] Y. J. Park, A. M. Reinhorn, and S. K. Kunnath, "*DARC: Inelastic Damage Analysis of Reinforced Concrete Frames—Shearwall Structures*," Technical Report NCEER-87-0008, State University of New York at Buffalo, Buffalo, New York, 1987.

[2.14] R. Li and M. Pourzanjani, "Sensitivity of Building Response to Nonlinear Analysis Models," *The Structural Design of Tall Buildings.* John Wiley & Sons, 1999.

[2.15] G. Warcholik and M. J. N. Priestley, "Structural Systems Research Project: High Strength Concrete Joints Tests—Series 3," Report No. TR-98/12, University of California, San Diego, July 1998.

[2.16] J. B. Mander, M. J. N. Priestley, and R. Park, "Theoretical Stress-Strain Model for Confined Concrete," *Journal of Structural Engineering, ASCE,* Vol. 114, No. 8, August 1988, pp. 1804–1826.

[2.17] American Institute of Steel Construction (AISC), *Manual of Steel Construction—Load and Resistance Factor Design Specification for Structural Steel Building,* 3rd Edition. Chicago, IL, 2001.

[2.18] S. K. Ghosh and N.M. Hawkins, "Seismic Design Provisions for Precast Concrete Structures in ACI 318." *PCI Journal*, Vol. 46, No. 1, January–February 2001, pp. 28–32.

[2.19] A. H. Mattock and N. M. Hawkins, "Shear Transfer in Reinforced Concrete—Recent Research," *PCI Journal*, Vol. 21, No. 1, January–February 1976, pp. 20–39.

[2.20] H. R. Foerster, S. H. Rizkalla, and J. S. Heuvel, "Behavior and Design of Shear Connections for Load Bearing Wall Panels," *PCI Journal,* Vol. 34, No. 1, January–February 1989, pp. 102–119.

[2.21] E. G. Nawy, *Prestressed Concrete: A Fundamental Approach,* 3rd Edition. Prentice Hall, Upper Saddle River, New Jersey, 2000.

[2.22] C. M. Lin, J. I. Restrepo, and R. Park, "An Alternative Design Method for Interior Beam-Column Joints of Reinforced Concrete Moment Resisting Frames," NZNSEE Conference, Wairakes, Taupo, New Zealand, March 1997.

[2.23] R. E. Englekirk, "The Development and Testing of a Ductile Connector for Assembling Precast Beams and Columns," *PCI Journal*, Vol. 40, No. 2, March–April, 1995, pp. 36–51.

[2.24] Y. Xiao and H. W. Yun, "Full Scale Experimental Studies on High-Strength Concrete Columns," Report No. USC SERP 98/05, University of Southern California, Los Angeles, July 1998.

[2.25] R. E. Englekirk and M. Pourzanjani, "High Strength Concrete Applications in Regions of High Seismicity," *Proceedings of the Fifth Conference on Tall Buildings in Seismic Regions,* Los Angeles, California, May 5, 2000, pp. 201–217.

[2.26] V. Ciampi, R. Eligenhausen, V. Bertero, and E. Popov, "Analytical Model for Concrete Anchorages of Reinforced Bars Under Generalized Excitations," EERC Report No. UCA/eerc 82/83, University of California, Berkeley, November 1982.

[2.27] R. T. Leon, "Interior Joints with Variable Anchorage Lengths," *Journal of Structural Engineering, ASCE*, Vol. 115, No. 9, September 1989.

[2.28] G. Warcholik and M. J. N. Priestley, "Structural Systems Research Project: High Strength Concrete Joints Tests," Report No. TR-97/10, University of California, San Diego, July 1997.

[2.29] G. Warcholik and M. J. N. Priestley, "Structural Systems Research Project: High Strength Concrete Joints Tests," Report No. TR-98/01, University of California, San Diego, January 1998.

[2.30] D. F. Meinhelt and J. O. Jirsa, "Shear Strength of R/C Beam-Column Connections," *ASCI Journal of the Structural Division, Proceedings of the American Society of Civil Engineers,* Vol. 107, No. ST11, November 1981.

[2.31] J. Wallace, "New Methodology for Seismic Design of RC Shear Walls," *Journal of Structural Engineering, ASCE*, Vol. 120, No. 3, March 1994.

[2.32] J. Wallace, "Ductility and Detailing Requirements of Bearing Wall Buildings, *Journal of Structural Engineering, ASCE*, Vol. 118, No. 6, June 1992.

[2.33] J. Wallace, "Seismic Design of RC Structural Walls. Part I: New Code Format," *Journal of Structural Engineering, ASCE*, Vol. 121, No. 1, January 1995.

[2.34] J. W. Wallace and J. H. Thomsen, IV, "Seismic Design of RC Structural Walls. Part II: Applications," *Journal of Structural Engineering, ASCE*, Vol. 121, No. 1, January 1995.

[2.35] F. V. Yanez, R. Park, and T. Paulay, "Seismic Behavior of Reinforced Concrete Structural Walls with Regular and Irregular Openings," *Pacific Conference on Earthquake Engineering,* New Zealand, November 1991.

[2.36] Y. Zhang and Z. Wang, "Seismic Behavior of Reinforced Concrete Shear Walls Subjected to High Axial Loading," *ACI Structural Journal*, Vol. 97, No. 5, September–October 2000.

[2.37] T. Paulay, "The Design of Ductile Reinforced Concrete Structural Walls for Earthquake Resistance," *Earthquake Spectra,* Vol. 2, No. 4, 1986.

[2.38] J. M. Vallenas, V. V. Bertero, and E. P. Popov, Hysteretic Behavior of Reinforced Concrete Structural Walls, Report UCB/EERC-79/20, Earthquake Engineering Research Center, College of Engineering, University of California, Berkeley, August 1979, 234p.

[2.39] J. H. Thomsen, IV and J. W. Wallace, "Displacement-Based Design of RC Structural Walls: Experimental Studies of Walls with Rectangular and T-Shaped Cross Sections," Report No. CU/CEE-95-06, Department of Civil and Environmental Engineering, Clarkson University, Potsdam, New York, 1995.

[2.40] J. W. Wallace, "Reinforced Concrete Walls: Recent Research & ACI 318-2001," *Proceedings of the 6th U.S. National Conference on Earthquake Engineering.*

[2.41] Seismology Committee, Structural Engineers Association of California, *Recommended Lateral Force Requirements and Commentary*, 7th Edition. Sacramento, California, 1999.

[2.42] A. J. Chopra, K. Rakesh, and M. Goel, "Direct Displacement-Based Design: Use of Inelastic vs. Elastic Design Spectra," *Earthquake Spectra,* Vol. 17, No. 1, February 2001.

[2.43] L. Galano and A. Vignoli, "Seismic Behavior of Short Coupling Beams with Different Reinforcement Layouts," *ACI Structural Journal,* November-December 2000, pp. 876–885.

[2.44] K. A. Harries, D. Mitchell, W. D. Cook, and R. G. Redwood, "Seismic Response of Steel Beams Coupling Concrete Walls, *Journal of Structural Engineering, ASCE*, Vol. 119, No. 12, December 1993.

[2.45] American Institute of Steel Construction, Inc., *Seismic Provisions for Structural Steel Buildings*, Chicago, IL, April, 1997 and Supplement No. 2, September, 2000.

[2.46] K. A. Harries, G. Gong, and B. M. Shahrooz, "Behavior and Design of Reinforced Concrete, Steel, and Steel-Concrete Coupling Beams," *Earthquake Spectra*, Vol. 16, No. 4, November 2000.

[2.47] A. M. Rahman and J. I. Restrepo, *Earthquake Resistant Precast Concrete Buildings: Seismic Performance of Cantilever Walls Prestressed Using Unbonded Tendons*, Civil Engineering Research Report No. 2000-5, University of Canterbury, Christchurch, New Zealand, August 2000.

[2.48] T. J. Holden, *A Comparison of the Seismic Performance of Precast Wall Construction: Emulation and Hybrid Approaches,* Civil Engineering Research Report No. 2001-4, University of Canterbury, Christchurch, New Zealand, April 2001.

3

SYSTEM DESIGN

"Imagination is more important than knowledge."
—Albert Einstein

The objective of this chapter is to allow the designer to create an effective performance-based design procedure that is founded on an understanding of component behavior, building-specific characteristics, and relevant fundamentals of dynamics.

Performance-based design requires that building behavior be controlled so as not to create undesirable strain states in its components. Methodologies have been developed in Chapter 2 that allow component deformation states to be extended to induced strain states. Experimentally based conclusions were also extended to strain states so as to allow the designer to select component strain limit states consistent with performance objectives. An experimental basis does not exist for calibrating building design procedures, and this makes it difficult to assess the appropriateness of design presumptions. In Chapter 2 design approaches were studied and developed into examples. Some of these design approaches are extended into examples in this chapter and then tested analytically, primarily to allow the designer to evaluate both the design and evaluation process. The analytical testing of a design is not intended, however, to justify the design or design approach, for this must await the development of experimental procedures. With this approach in mind, I have adopted a standard forcing function presented in the form of an elastic design response spectrum (Figure 4.1.1) and a set of calibrated ground motions (Table 4.1.1). They are used in the examples of Chapters 3 and 4.

Building performance has been the basis for the design procedures I have used for over thirty years, and these procedures will be a focal topic. The thrust here is to demonstrate how the designer can understand the way a design is impacted by each decision and thereby maintain control of the design process and hopefully the behavior of the building as well. Computer-based analysis will be used to demonstrate how modern technology can be effectively introduced into the conceptual design process without losing design control or designer confidence.

Creativity is emphasized, both in the design procedures developed and in the selection of system components. Precast concrete alternatives demand a special emphasis on creativity because only a very few solutions have been extended to the constructed building domain. The systems developed herein and described in Chapter 2 lend themselves to creative adaptations, and the reader is encouraged to make such adaptations because performance objectives can be more easily attained in this way.

The first topic explored in this chapter is the design of shear wall braced buildings, and it is the first topic for two reasons. First, the response of a building to earthquake ground motions is most easily understood when its bracing system can be readily reduced to a single degree-of-freedom system. Second, system and component ductilities are the same for simple shear wall braced buildings. Alternative design procedures are reviewed and streamlined. From the simple shear wall braced buildings, we proceed to explore the relationship between component and system ductility by examining bracing systems of increasing complexity. The developed design procedures are used to design coupled shear walls, cast-in-place concrete frames, precast concrete frames, and diaphragms.

3.1 SHEAR WALL BRACED BUILDINGS

In Section 2.4.1.2 several design procedures for shear walls were presented as a part of exploring component design. The length of the wall was presumed to have been established by functional necessities. The various procedures proposed, be they strength- or displacement-based, had as an end product the identification of the appropriate strength of the shear wall. In this section we start by examining the shear wall design procedures proposed in Section 2.4.1.2, and then we investigate alternative design procedures that allow more design flexibility. The very simplicity of the shear wall system developed in Section 2.4.1.2 allows a comparison of the conclusions reached through the implementation of the alternative design procedures. In Chapter 4 we evaluate these various design conclusions using inelastic time history analyses. Our objective is to explore the consequences of design decisions and procedures as they relate to the effective sizing and detailing of the shear wall.

The methodologies proposed for the design of buildings braced by shear walls of equivalent stiffness will be extended to study more complex design problems.

Next the precast shear wall design of Section 2.4.3.3 will undergo the same scrutiny. The alternative design procedures explored for the cast-in-place shear wall will be tested on the precast walls, for they allow the inclusion of component postyield behavior characteristics inherent to precast concrete construction.

3.1.1 Shear Walls of Equivalent Stiffness

The seismic design of shear wall braced buildings, regardless of the approach used (displacement- or force-based), is often accomplished by reducing the building to a single-degree-of-freedom model that has an equivalent lumped (effective) mass operating at an effective height. This reduction is considerably simplified when the bracing elements are identical, for the appropriate amount of building mass can usually be allocated to each wall. The stiffness and postyield behavior of the shear wall can be modeled from experimental efforts by the appropriate selection of an effective stiffness and ductility. When the shear walls that brace a building are identical, the design process is further simplified by the fact that system and component (shear wall) ductilities are the same. This reduction from a multi-degree-of-freedom system to a single-degree-of-freedom system is further facilitated by the fact that the fundamental mode dominates system response insofar as the primary design interests are concerned—displacement and flexural strength (see Section 2.4.1.2). The resultant single-degree-of-freedom design model is described in Figure 3.1.1*a*. The selected effective mass and height coefficients are those developed for an elastic uniform cantilever tower.[3.1, Fig. 16.6.3] Observe that the model has an effective mass (M_e) operating at an effective height (h_e). The effective height and mass identified in Figure 3.1.1*a* can be developed so as to incorporate any mass or system stiffness distribution as well as the probable impact of the inelastic displacement shape.

The stiffness of the model described in Figure 3.1.1*a* can be characterized in a variety of ways. The initial stiffness (k_i), the rate of strength hardening (r), and the effective or secant stiffness (k_s) at the anticipated ultimate displacement (Δ_u)

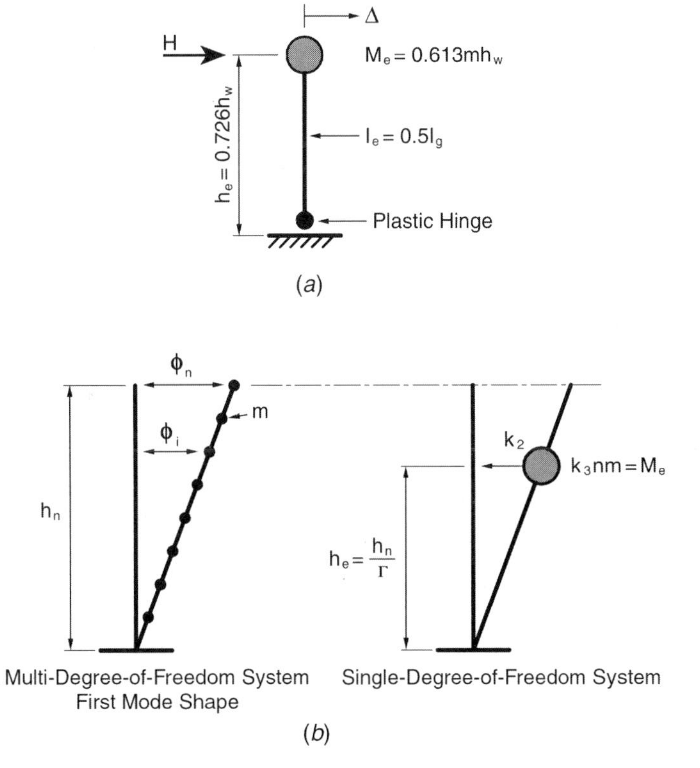

Figure 3.1.1 (*a*) Single-degree-of-freedom elastic model for shear wall braced building of uniform mass m.[3.1] (*b*) Multi-degree-of-freedom building system modeled as a single-degree-of-freedom system.

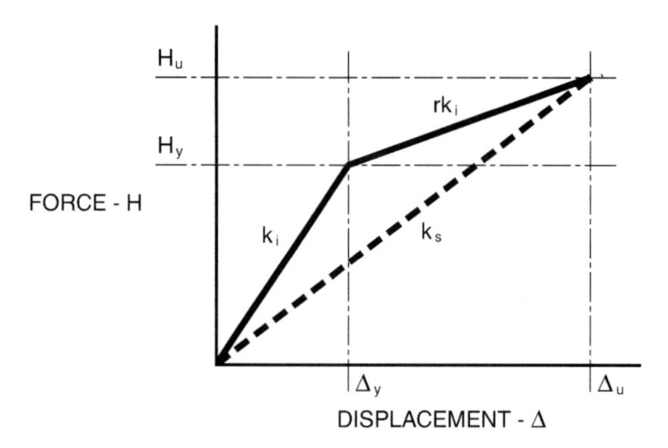

Figure 3.1.2 Alternative representations of system stiffness.

(Figure 3.1.2) can all be incorporated into the design process. These system stiffness describing characteristics can be taken directly from experimental work, as has been proposed in Chapter 2, and this allows the rational design of a variety of systems.

In Section 2.4.1.4 a cast-in-place shear wall design was developed. The wall configuration is described in Figure 2.4.9. The dimensions of the wall (h_w and ℓ_w) were pre-established because they had to satisfy functional requirements. The fundamental period (T) of the wall was based on the typical thickness of the wall (6 in.) and its idealized flexural stiffness ($I_e = 0.5I_g$) was developed from experimental efforts (Figure 2.4.4*a*). These procedures produce acceptable design conclusions, as we see in Chapter 4. They do not, however, allow any insight in terms of selecting the appropriate number and size of shear walls, nor do they allow the designer to account for the behavior characteristics inherent to alternative construction methods such as might be found, for example, in precast concrete construction.

3.1.1.1 Alternative Shear Wall Design Procedures Let us now explore force- and displacement-based design procedures that allow us to treat both the quantity (length and number) and construction type as design variables. Force-based design procedures will follow response spectra techniques that are commonly used today to design buildings.[3.1–3.3] Two generic classifications describe alternative displacement-based design methodologies. One is based on the presumption that inelastic displacements can be developed for any system regardless of the energy dissipated from displacements developed using an elastic model (equal displacement),[3.2, 3.4] and the other proposes that dissipated energy will significantly impact system displacement prediction (direct displacement).[3.1, 3.4, 3.5] The displacement-based design procedures that are explored in this section are currently referred to as equal displacement-based design (EBD) and direct displacement-based design (DBD).[3.4] The equal displacement-based design procedure is quite similar to the displacement-based procedures developed for the shear wall design of Section 2.4.1. The only major difference is that it allows for the sizing of the shear wall and the development of a compatible level of flexural strength. The direct displacement-based design (DBD) allows the same compatible development of stiffness (wall sizing) and flexural strength, plus the inclusion of system behavior characteristics; accordingly, the design presumably can account for postyield material behavior characteristics (cast-in-place or precast, steel, masonry, etc.), as well as the type of bracing system proposed (frame, shear wall, EBF, etc.).

The design objective will be to create a shear wall that will provide lateral support for the system described in Figure 2.4.9. The earthquake intensity used in the described design procedures is identified in the form of a global response spectrum adjusted to account for site soils. This site-specific response spectrum is described in Figure 4.1.1. Both the EBD and the DBD procedures propose target drifts that should produce the desired objective performance level, given the type of bracing system proposed. These procedures are new and, as a consequence, the target drifts will probably require some calibration, for the ultimate design objective is the control of postyield strain states and, accordingly, a comprehensive analytical review of the proposed conceptual design should be undertaken during the system development phase. A push-over analysis and spectral analysis of each design are used to evaluate the appropriateness of each design. Inelastic time histories will be performed in Chapter 4 to allow insight into the conclusions reached in the conceptual design phase.

The design response spectra (Figure 4.1.1) and the single-degree-of-freedom model described in Figure 3.1.1*a* can be used in a variety of ways to suggest the level of required strength of a structure. The classical *strength-based approach* follows.

$$\text{Tributary mass} = \frac{W}{g} \tag{3.1.1}$$

$$\text{Effective mass } (M_{\text{el}}) = \frac{0.613W}{g} \qquad \text{(see Figure 3.1.1}a\text{)} \tag{3.1.2}$$

The applied effective static elastic load ($H_{\text{el1}} - \mu F_{\text{max}}$ of Figure 1.1.1), as developed from the response spectrum of Figure 4.1.1 for a structure whose fundamental period is 0.96 second, (see Section 2.4.1.4) is

$$\begin{aligned} H_{\text{el1}} &= S_{a1} M_{e1} \\ &= 0.9g(0.613)\frac{W}{g} \\ &= 0.55W \end{aligned} \tag{3.1.3}$$

Comment: The spectral acceleration ($0.9g$) is taken directly from the design spectra of Figure 4.1.1. The subscript 1 (H_{el1}) indicates an elastic response in the first mode.

The yield moment associated with an elastic response (Figure 3.1.1*a*) is

$$M_{y1} = H_{\text{el1}} h_e \tag{3.1.4}$$

$$= 0.55W(0.726h_w) \qquad \text{(see Figure 3.1.1}a\text{)}$$
$$= 0.4Wh_w$$

M_y represents the contribution of the first mode only to the elastic flexural strength demand imposed on the multi-degree-of-freedom (MDOF), which Figure 3.1.1*a* describes. The contribution from higher modes would increase the base shear experienced by the multi-degree-of-freedom system significantly, but this increase in base shear would have little impact on the MDOF design yield moment for it can be developed with sufficient accuracy for design purposes directly from the single-degree-of-freedom model described in Figure 3.1.1*a*.

Comment: The effective modal mass and height for the second mode are $0.188mh_w$ and $0.209h_w$.[3.1, Fig. 16.6.3] The spectral acceleration (Figure 4.1.1) for the second mode, whose period is 0.15 second ($T_2 = T_1/6.27$), may be conservatively assumed to be $S_{a,\max}(1.45g)$. It follows that the associated effective inertial force (H_{el2}) is

$$H_{el2} = S_{a2}M_{e2} \tag{3.1.5}$$
$$= 1.45g(0.188)\frac{W}{g}$$
$$= 0.27W$$

and

$$M_{y2} = H_{el2}h_{e2} \tag{3.1.6}$$
$$= 0.27W(0.209)h_w$$
$$= 0.056Wh_w$$

Combining modal effects using a square root sum of the squares (SRSS) procedure produces the following multimode design parameters:

$$\hat{H} = \sqrt{H_{el1}^2 + H_{el2}^2} \tag{3.1.7}$$
$$= \sqrt{(0.55)^2 + (0.27)^2}(W)$$
$$= 0.61W \qquad (+11\%)$$

$$\hat{M} = \sqrt{M_1^2 + M_2^2} \tag{3.1.8}$$
$$= \sqrt{(0.4)^2 + (0.056)^2}(Wh_w)$$
$$= 0.404Wh_w \qquad (+1\%)$$

where $\hat{H}$ and $\hat{M}$ are MDOF design shears and moments that consider a combined response in both the first and second modes. The increase associated with the inclusion of the second mode from that developed using the first mode only is parenthetically identified.

The conclusions reached with regard to the dominance of the first mode in quantifying the flexural strength of a shear wall equally apply to predicting MDOF system displacement. This is because the period of higher modes decreases rapidly, and this means that spectral displacements will be quite small. Further, the roof displacement is obtained by taking the product of the participation factor (Γ_n) and the spectral displacement (S_{dn}). Thus the roof drift is a function of the first mode spectral displacement (S_{d1}). For a numerical example see Ref. 3.6, Example 4.7.3.

An *equal displacement-based procedure*, as developed in Section 2.4.1.2, would assume that the ultimate displacement of the inelastic system could be reasonably predicted from the elastic design spectrum.

$$\begin{aligned}\omega_n &= \frac{2\pi}{T}\\ &= \frac{2\pi}{0.96} \qquad (T = 0.96 \text{ second}) \text{ (see Section 2.4.1.4)}\\ &= 6.54 \text{ rad/sec}\end{aligned}$$

$$\begin{aligned}S_d &= \frac{S_v}{\omega_n} \qquad (3.1.9)\\ &= \frac{49}{6.54}\\ &= 7.5 \text{ in.}\end{aligned}$$

Comment: A spectral velocity of 49 in./sec is used to be consistent with the design spectrum of Figure 4.1.1 in the 1-second period range.

This corresponds to a roof displacement (Δ_n) of

$$\begin{aligned}\Delta_n &= \Gamma S_d \qquad (3.1.10)\\ &\cong 1.5(7.5)\\ &= 11.3 \text{ in.}\end{aligned}$$

Comment: The participation factor (Γ) is discussed and developed in Eq. 3.1.15 and, more extensively, in Ref. 3.6.

The associated elastic yield force may also be developed directly from the spectral velocity.

$$H_y = S_a W \qquad (3.1.11)$$

$$= \frac{\omega_n S_v}{g} W$$

$$= \frac{6.54(49)}{386.4} W$$

$$= 0.83W$$

which is, of course, essentially the same value as that read directly from the response spectra, the difference being attributable to the steepness of the acceleration response curve in this period range.

Comment: The spectral velocity is constant in this period range (see Figure 1.1.5), and it is for this reason that spectral velocity quantifications should be more commonly used in design procedures.

It follows that the elastic design moment, given the equivalent single-degree-of-freedom model described in Figure 3.1.1*a*, would be

$$M_{y1} = M_e S_{a1} g h_e \tag{3.1.12}$$

$$= 0.613(0.83)W(0.726)h_w$$

$$= 0.37Wh_w$$

To this point our focus has been on elastic response (see Figure 1.1.12). Our objective is to determine the appropriate design strength, in this case the objective moment demand (M_u). This design flexural strength objective would be developed directly from this spectral-based moment by including, albeit subjectively, a defined system ductility and overstrength factor (see Section 1.1.6).

$$M_u = \frac{M_y}{R_d R_o} \tag{3.1.13}$$

$$= \frac{0.37Wh_w}{\mu \Omega_o}$$

where R_d is the codified identification of the ductility factor of Figure 1.1.1 (μ), and Ω_o is an estimate of probable system overstrength, R_o in Figure 1.1.12.

Equal Displacement-Based Design The equal displacement-based design (EBD) procedure, as developed in the Blue Book,[3.4] endeavors to create the required system stiffness and design strength from displacement design objectives. Accordingly, the EBD design process will define not only the strength objective but also the required stiffness of the shear wall, thereby producing dynamic consistency.

The EBD procedure starts by selecting a target displacement (Δ_T) at the effective height of the equivalent single-degree of freedom system (Figure 3.1.1).

$$\Delta_T = \delta_1(h_e)(k_2) \tag{3.1.14a}$$

where

δ_1 is the objective interstory drift ratio.

k_2 is a "factor to relate the expected displaced shape function to a linear displaced shape function."

h_e is the (effective) height ($k_1 h_n$) of the (effective) mass.

The effective mass (M_e) is $k_3 M$. The resultant single-degree-of-freedom system (Figure 3.1.1*b*) varies from that described in Figure 3.1.1*a* only in terms of the modifying constants.

It follows that the target drift is

$$\Delta_T = \delta_1 k_1 k_2 h_w \tag{3.1.14b}$$

Objective interstory drift ratios (δ_1) are performance based. They strive to identify the level of performance desired, and this ultimately will be identified by the building owner or society. The performance level that is consistent with current strength-based code objectives is identified in the Blue Book[3.4] as performance category SP3: "Damage is moderate—extensive structural repairs are expected to be required."

Comment: Relating the drift ratio directly to structural damage requires considerable subjectivity for the extent of damage will clearly be a function of the postyield strain demand imposed on structural components. From a design perspective a target drift is essential to the rational development of a conceptual design, but the appropriateness of the adopted target drift must be confirmed early in the design process by evaluating the induced strain states imposed on structural components.

Prescriptive values for k_1, k_2, and k_3 are provided in the Blue Book,[3.4] but the designer is allowed to use a more detailed analysis to develop these factors.

Comment: The effective (modal) mass and height are used to develop a single-degree-of-freedom design model (Figure 3.1.1*b*) appropriate for use in the design of a multi-degree-of-freedom system. Several building-specific features allow for an easy reduction of a multistory building to a single-degree-of-freedom system.

The first of these building specific features is particular to shear wall braced buildings—the first or fundamental mode dominates the quantification of displacement and flexural strength and, as a consequence, allows a direct correlation between the multistory building and the adopted single-degree-of-freedom model because higher modes need not be considered. See, for example, the development of Eq. 3.1.8. Shear demand is a capacity based function of the provided level of flexural strength and therefore of no design interest when the task is to create a shear wall of the appropriate stiffness and strength.

The second simplification is not exclusively a characteristic of shear wall braced buildings. Mass and story heights are typically uniform, and this allows the adoption of a linear mode shape in spite of the fact that the elastic deflection is not linear. The linear behavior model is reasonable because elastic response does not dominate building response. Inelastic behavior in any concrete member starts long before idealized yield deflection (Δ_{yi}) is reached. Accordingly, when the induced moment reaches about one-half of the yield moment, rotation becomes concentrated at the base of the wall and the upper portion of the wall tends to rotate as a rigid body. The deformation of frame braced building follows a shear model, and this is linear.[3.6] Accordingly, a linear mode shape is not an unreasonable assumption. (Refer to Figure 2.1.7.)

Modal analysis is described in a variety of ways. The simplest form starts by relating spectral drift to the displacement of the top level in a multi-degree-of-freedom system.

$$\Delta_n = \Gamma S_d \qquad \text{(Eq. 3.1.10)}$$

The participation factor (Γ) and the spectral displacement (S_d) are both particular to possible mode shapes. Since our interest is exclusively with the first mode, subscripts designating the mode will be omitted.

The participation factor (Γ) is developed directly from the mode shape ϕ

$$\Gamma = \frac{\sum_{i=1}^{n} m_i \phi_i}{\sum_{i=1}^{n} m_i \phi_i^2} \qquad \text{(see Ref. 3.1, Eq. 13.2.3)} \quad (3.1.15)$$

where n is the number of stories.

The effective height (h_e) of the single-degree-of-freedom system is logically

$$h_e = \frac{h_w}{\Gamma} \qquad (3.1.16)$$

as suggested by Eq. 3.1.10.

The effective or modal mass (M_e) is developed so as to provide work equivalence between the single-degree-of-freedom model and the multi-degree-of-freedom system responding in its first mode (Figure 3.1.1*b*) or any mode for that matter.

$$h_e M_e = \sum m_i \phi_i h_w \qquad (3.1.17a)$$

This allows the development of an expression for the effective mass (M_e), presuming that all unit masses are the same (m).

$$\frac{h_w}{\Gamma} M_e = h_w m \sum_{i=1}^{n} \phi_i \qquad \text{(Eq. 3.1.17a)}$$

$$M_e = \frac{\Gamma M}{n} \sum_{i=1}^{n} \phi_i \qquad (3.1.17b)$$

Now if a linear mode shape and height distribution is reasonably adopted the series ($\Sigma\phi_i$ and $\Sigma\phi_i^2$) may be converted to normalized integer form and this allows both the participation factor and effective mass to be considerably simplified.

$$\sum \phi_i = \frac{n\,(n+1)}{2n} \tag{3.1.17c}$$

$$\sum \phi_i^2 = \frac{n\,(n+1)\,(2n+1)}{6n^2} \tag{3.1.17d}$$

Now

$$\Gamma = \frac{n\,(n+1)\,6}{2n\,(n+1)\,(2n+1)}\left(\frac{n^2}{n}\right) \qquad \text{(see Eq. 3.1.15)}$$

$$\boxed{\Gamma = \frac{3n}{2n+1}} \tag{3.1.18a}$$

and

$$M_e = \Gamma m \sum_{i=1}^{n} \phi_i \qquad \text{(Eq. 3.1.17b)}$$

$$= \left(\frac{3n}{2n+1}\right)\frac{M}{n}\left(\frac{n(n+1)}{2n}\right)$$

$$\boxed{M_e = \frac{1.5\,(n+1)}{2n+1}M} \tag{3.1.19a}$$

where n is the number of stories.

Compare conclusions developed from a linear mode shape with the comparable factors contained in Table APP1B-7 of the Blue Book.[3.4]

Ten-Story Linear System (Shape 2—Linear)

Blue Book Development[3.4]*: Effective Height*

$$k_1 k_2 = 0.7(1.0)$$
$$= 0.7$$

Modify the Blue Book[3.4] definition of $h_e(k_1 h_n)$ to make it consistent with Figure 3.1.1:

$$h_e = k_1 k_2 h_w \qquad \text{(see Eq. 3.1.14a and Eq. 3.1.14b)} \tag{3.1.18b}$$
$$= 0.7 h_w$$

$$\Gamma = \frac{1}{k_1 k_2} \qquad \text{(see Eq. 3.1.14b)} \quad (3.1.18c)$$

$$= 1.43$$

Linear Mode Shape Basis: Effective Height

$$\Gamma = \frac{3n}{2n + 1} \qquad \text{(Eq. 3.1.18a)}$$

$$= \frac{3(10)}{20 + 1}$$

$$= 1.428$$

$$h_e = \frac{h_w}{\Gamma} \qquad \text{(Eq. 3.1.16)}$$

$$= \frac{h_w}{1.428}$$

$$= 0.7h_w$$

Blue Book Development: Effective Mass

$$M_e = k_3 M \qquad (3.1.19b)$$

$$k_3 = 0.79$$

$$M_e = 0.79M$$

Linear Mode Shape Basis: Effective Mass

$$M_e = \frac{1.5(n + 1)}{2n + 1} M \qquad \text{(Eq. 3.1.19a)}$$

$$= \frac{1.5\,(n + 1)}{2n + 1} M$$

$$= \frac{1.5(11)}{21} M$$

$$M_e = 0.79M$$

The modeling factors developed by the Blue Book for a ten-story shear wall braced building whose aspect ratio (h_w/ℓ_w) and system ductility (μ) are each 2 will be slightly different from those developed for the linear model, which is not affected by aspect ratio or system ductility.

Blue Book Development[3.4] $(\mu = 2;\ h_w/\ell_w = 2)$

$$k_1 k_2 = 0.85(0.75)$$
$$= 0.64$$
$$k_3 = 0.85$$

Taken individually, the difference between Blue Book coefficients and those developed from the linear mode shape seem significant, but the product of the coefficients (k_1, k_2, k_3), which creates the design moment, is essentially the same.

Linear Mode Shape Basis

$$M_e h_e = 0.79M(0.7)h_w$$
$$= 0.55Mh_w$$

Blue Book Development[3.4]

$$k_1 k_2 k_3 M h_w = 0.64(0.85)M h_w$$
$$= 0.54M h_w$$

The major impact is on the relationship between the displacement at the top of the multistory building and the spectral displacement that is now associated with a participation factor of 1.56 (1/0.64). See also Figure 3.1.1*a*.

Conclusion: There seems to be little point in refining the single-degree-of-freedom model to this "apparent" degree of accuracy. The linear mode shape assumption is certainly reasonable for strength purposes and the participation factor fairly stable in the 1.5 range. Remember, the degree of overall accuracy must consider the probable level of ductility and the accuracy of the response spectrum.

The designer may also create a single-degree-of-freedom system from an assumed mode shape modeled after the proposed building by combining an elastic mode shape with any level of inelastic behavior.

Table 3.1.1 and Figure 3.1.3 describe the assumed relative deflected shapes for a shear wall braced building and the combined normalized mode shape. In this case the masses are equal, but this need not be the case. The effective mass and height of the equivalent single-degree-of-freedom system (Figure 3.1.1*a*) would be developed as follows:

$$\Gamma = \frac{\sum m_i \phi_i}{\sum m_i \phi_i^2} \qquad \text{(Eq. 3.1.15)}$$

$$= \frac{1.91}{1.475}$$

$$= 1.29$$

$$M_e = \frac{\Gamma}{h_w}\sum_{i=1}^{n} m_i\phi_i h_w \qquad \text{(see Eq. 3.1.17b)}$$

$$= \frac{\Gamma M}{n}(0.28 + 0.63 + 1.0)$$

$$= \left(\frac{1.29}{3}\right) M(1.91)$$

$$= 0.82M$$

and

$$h_e = \frac{h_w}{\Gamma} \qquad \text{(Eq. 3.1.16)}$$

$$= \frac{h_w}{1.29}$$

$$= 0.77h_w$$

TABLE 3.1.1 Elastic and Inelastic Mode Shapes—Three-Story Shear Wall Braced Building

Level	Normalized Elastic Deflection	Normalized Postyield Deflection ($\mu = 3$)	Combined Deflection	Normalized Inelastic Mode Shape
3	1.0	2.0	3.0	1.0
2	0.56	1.33	1.89	0.63
1	0.18	0.67	0.85	0.28

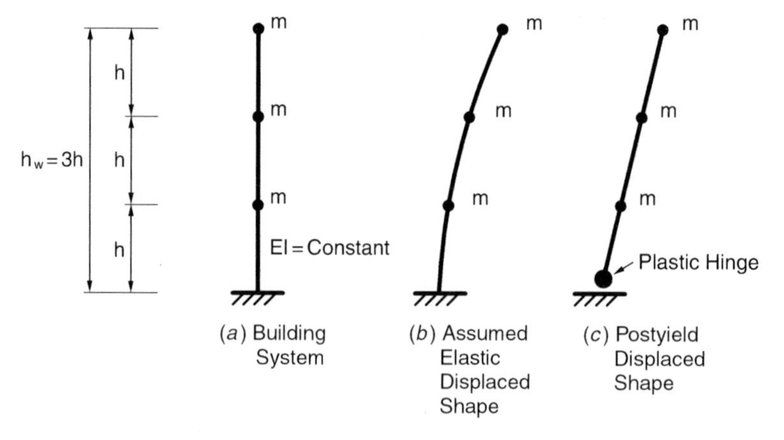

Figure 3.1.3 Assumed displacements describing the mode shape for a three-story shear wall braced building.

The resultant product is the objective flexural strength.

$$M_e h_e = 0.82M(0.77)h_w$$
$$= 0.64Mh_w$$

The prescriptive approach developed in the Blue Book[3.4] recognizes the impact of ductility on the displaced shape. The height (effective) of the mass (effective) is $k_1k_2h_w$; hence for a ductility factor of 3 and h_w/ℓ_w of 2,

$$h_e = k_1k_2h_w$$
$$= 0.82h_w \qquad \text{(see Ref. 3.4, Table App IB-7)}$$
$$M_e = k_3M$$
$$= 0.82M \qquad \text{(see Ref. 3.4, Table App IB-7)}$$
$$M_e h_e = 0.67h_w$$

Accordingly, the Blue Book[3.4] recommendation is slightly conservative, as it should be.

In the development of the example (Figure 2.4.9) using the equal displacement-based design (EBD) procedure, an objective or target drift ratio (δ_1) of 0.018 is assumed. The target drift (Δ_T) for $h_w/\ell_w = 4$ and $\mu = 4$ is

$$\Delta_T = \delta_1 k_1 k_2 h_n \tag{3.1.20}$$
$$= 0.018(0.77)(0.84)(1260) \qquad \text{(see Ref. 3.4, Table App IB-7-SP3)}$$
$$= 14.7 \text{ in.}$$

Now the initial objective period following the Blue Book[3.4] procedure is developed from an acceleration displacement response spectrum (ADRS) (Figure 3.1.4). For a site soils type S_D the spectral acceleration S_a, corresponding to a spectral displacement (S_d) of 14.7 in., is about $0.6g$. This relationship (see Eq. 1.1.6c) requires a fundamental building period of

$$T = 2\pi\sqrt{\frac{S_d}{S_a}} \tag{3.1.21}$$

$$= 2(3.14)\sqrt{\frac{14.7}{0.6(386.4)}}$$

$$= 1.58 \text{ seconds}$$

The preceding development can be considerably simplified if the response falls within the constant spectral velocity range (Figure 1.1.5). The spectral displacement is the target drift divided by the participating factor (Γ).

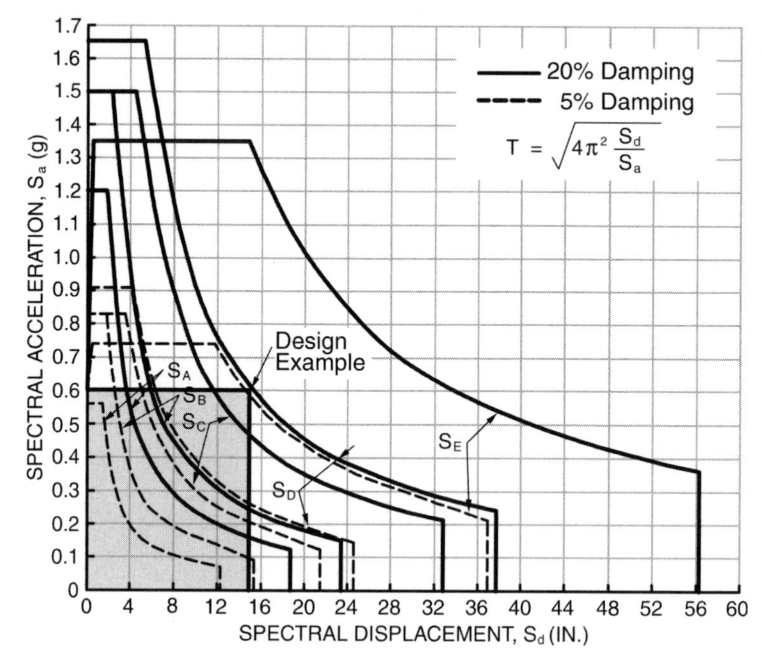

Figure 3.1.4 Acceleration-displacement response spectra.[3.4]

$$S_d = \frac{\delta_1 h_n}{\Gamma} \tag{3.1.22}$$

$$= \frac{0.018(1260)}{1.5}$$

$$= 15.1 \text{ in.}$$

Comment: The participation factor is assumed to be 1.5.

The (constant) spectral velocity can be developed from the acceleration response spectrum.

$$S_v = \frac{T}{2\pi} S_a \tag{3.1.23}$$

$$= \frac{2}{2\pi}(0.42)(386.4) \qquad \text{(see design spectrum of Figure 4.1.1, } T \cong 2 \text{ sec)}$$

$$= 52 \text{ in./sec}$$

$$\omega_n = \frac{S_v}{S_d} \qquad \left(\frac{\text{Constant spectral velocity}}{\text{Objective spectral displacement}}\right) \tag{3.1.24}$$

$$= \frac{52}{15.1}$$

$$= 3.44 \text{ rad/sec}$$

$$T = \frac{2\pi}{\omega_n} \tag{3.1.25}$$

$$= \frac{6.28}{3.44}$$

$$= 1.83 \text{ seconds}$$

and the associated spectral acceleration is

$$S_a = \omega_n S_v \tag{3.1.26}$$

$$= \frac{3.44(52)}{386.4} g$$

$$= 0.46g$$

Alternatively, the desired objective relationship between the period (T) and the spectral acceleration (S_a) can be developed from the site-specific response spectrum (Figure 4.1.1).

$$S_a = S_d \omega_n^2 \tag{3.1.27}$$

$$= S_d \left(\frac{4\pi^2}{T^2} \right)$$

$$T^2 S_a = \frac{S_d 4\pi^2}{386.4} \tag{3.1.28}$$

where S_a is expressed as a percentage of g as it appears in most response spectra.

Now following the spectra to a point where the product of T^2 and S_a is equal to the product generated by Eq. 3.1.28,

$$T^2 S_a = \frac{15.1(39.4)}{386.4}$$

$$= 1.54$$

This corresponds to a period of about 1.85 seconds and a spectral acceleration of about $0.45g$. The slope of the acceleration response spectrum (Figure 4.1.1) makes this quantification difficult, so it is easiest to work directly with the spectral velocity, and this may be extracted directly from an acceleration-based spectrum through the use of Eq. 3.1.23.

Since the procedure is iterative, select a trial initial period (T_i) of 1.85 seconds. Once a period has been selected, the initial stiffness (k_i) follows from the idealized model described in Figure 3.1.1a altered to reflect an effective mass of $0.75M$.[3.4]

$$k_i = \frac{4\pi^2 M_e}{T_i^2} \tag{3.1.29}$$

$$= \frac{39.4(0.75)1400}{(1.85)^2 386.4} \qquad (k_3 = 0.75)$$

$$= 31.2 \text{ kips/in.}$$

The corresponding (base) shear becomes

$$V = k_i \Delta_T \tag{3.1.30}$$

$$= 31.2(14.7) \qquad \text{(Eq. 3.1.20)}$$

$$= 459 \text{ kips} \qquad (0.33W)$$

Now the design shear or objective strength (F_o, Figure 1.1.1; V_s, Figure 1.1.12) is developed from this elastic prediction (μF_o, Figure 1.1.1) by dividing V by the system ductility factor (μ or R_d) and overstrength factor (Ω_o or R_o):

$$V_d = \frac{k_i \Delta_T}{\mu \Omega_o} \tag{3.1.31}$$

where the product $\mu\Omega_o$ is referred to as μ_Δ, and, according to the Blue Book (see Ref. 3.4, Table App. 1B-5), is a function of the objective performance level.

For our basic commercial design objective (SP-3), the suggested system ductility-overstrength factor is 4; hence

$$V_d = \frac{0.33W}{4}$$

$$= 0.0825W$$

This is clearly not consistent with the current strength-based criterion.[3.7]

Comment: I have intentionally chosen to refer to objective levels of strength as design strength objectives (V_d), as opposed to ultimate strength objectives (V_u), and this is because non–strength-based designs include provisions for member (λ_o) and system (Ω_o) overstrength. The selection of overstrength factors will depend on the choice of the design objective. Since Ω_o is not quantitatively established by the Blue Book, the choice of the relationship between M_d and M_n is deferred, and will be discussed as a part of the development of various design approaches.

The moment design criterion is

$$M_d = V_d h_e \tag{3.1.32}$$

$$= 0.0825W(0.65)h_w \qquad (k_1 k_2 \cong 0.65)^{[3.4]}$$

$$= 0.053Wh_w$$

Comment: The use of a system ductility factor of 4 is consistent with experimental efforts (Figure 2.4.4*a*) provided the axial load is not too high. Strain states must be used to determine the acceptability of the design. Observe that at this stage in the Blue Book development of the EBD procedure, system overstrength does not appear to have been considered, and this is inconsistent with the behavior idealizations described in Figure 1.1.1 or Figure 1.1.12.

The strength objective developed by the EBD approach is predicated on an assumption that the anticipated dynamic characteristics will in fact be attained. The period of the system described in Figure 2.4.9 is half of that assumed in developing the EBD strength objective ($M_d = 0.053Wh_w$). Accordingly, the system proposed in Figure 2.4.9 is incompatible with this EBD design.

The length of the shear wall that would be compatible with the objective period of 1.85 seconds may be developed from Eq. 2.4.22, rearranged to solve directly for ℓ_w.

$$\boxed{\ell_w = \left[\frac{0.13}{T}h_w^2\left(\frac{w}{Et_w}\right)^{0.5}\right]^{0.67}} \tag{3.1.33}$$

where units are in their most practical form:

ℓ_w, h_w, and t_w are in feet.
T is in seconds.
w is in kips per foot.
E is in kips per square inch.

$$\ell_w = \left[\frac{0.13}{1.85}(105)^2\left(\frac{13.3}{4000(0.5)}\right)^{0.5}\right]^{0.67}$$

$$= 16 \text{ ft}$$

Constant Spectral Velocity Method The constant spectral velocity method following the equal displacement (CVED) approach is the author's preference. It is demonstrated by designing the system described in Figure 2.4.9. The objective is to select a compatible wall length and strength based on the presumption that a system ductility-overstrength factor ($\mu\Omega_o$) of 5 will meet our strain-based performance objectives.

The adopted ductility-overstrength factor ($\mu\Omega_o = 5$) is somewhat less than one might conclude based on the experimental evidence reviewed in Section 2.4.1 but consistent with current suggested values[3.4] were they to include a system overstrength factor of 1.25.

Step 1: Create the Single-Degree-of-Freedom Model. For the purposes of development of a comparative design, use the constants previously (EBD) adopted to develop the design model of Figure 3.1.1*a*.

$$\begin{aligned} M_e &= k_3 M &&\text{(Eq 3.1.19b)}\\ &= 0.75M \qquad (k_3 = 0.75)\\ &= \frac{0.75W}{g}\\ &= \frac{0.75(1400)}{g}\\ &= 2.72 \text{ kip sec}^2/\text{in.} \end{aligned}$$

$$\begin{aligned} h_e &= k_1 k_2 h_w &&\text{(Eq. 3.1.16)}\\ h_e &= 0.65 h_w \qquad (\Gamma = 1.54;\ 1/0.65)\\ &= 0.65(105)\\ &= 68 \text{ ft} \end{aligned}$$

Step 2: Identify the Target Building Drift. This has been done in the previously developed displacement-based designs.

$$\begin{aligned} \Delta_u &= \delta_1 h_w\\ &= 0.018(105)(12)\\ &= 22.7 \text{ in.} \end{aligned}$$

Step 3: Determine the Objective Spectral Displacement (S_d). A system-specific participation factor (Γ) must be developed or assumed. Assume that $\Gamma = 1.5$ is reasonably accurate for a tall shear wall based building responding in the postyield range. Alternatively, use the value suggested by the Blue Book[3.4] ($\Gamma = 1.54$).

$$\begin{aligned} S_d &= \frac{\Delta_{un}}{\Gamma}\\ &= \frac{22.7}{1.5}\\ &= 15.1 \text{ in.} \end{aligned}$$

Step 4: Solve for the Objective Natural Frequency. A constant spectral velocity of 52 in./sec has been adopted for this example (see Figure 4.1.1) because it is consistent with the spectral velocity in the 2-second period range.

$$\omega_n = \frac{S_v}{S_d} \tag{3.1.34}$$

$$= \frac{52}{15.1}$$

$$= 3.44 \text{ rad/sec} \qquad (T = 1.83 \text{ seconds})$$

Step 5: Determine the Effective Stiffness.

$$k_e = \omega_n^2 M_e \tag{3.1.35}$$

$$= (3.44)^2(2.72)$$

$$= 32.2 \text{ kips/in.}$$

Step 6: Determine the Required Strength of the Shear Wall. An objective system ductility (μ_s) of 4 and overstrength factor (Ω_o) of 1.25 will be assumed.

$$M_{\text{max}} = \frac{k_e S_d}{\Omega_o \mu} h_e \qquad \text{(see Figure 1.1.11)} \tag{3.1.36}$$

$$= \frac{32.2(15.1)}{1.25(4)}(68)$$

$$= 6620 \text{ ft-kips} \qquad (0.045 W h_w)$$

Step 7: Size and Reinforce the Wall. The building period must be less than 1.83 seconds in order to meet our displacement objectives. The required length of the wall is developed from Eq. 3.1.33.

$$\ell_w = \left[\frac{0.13}{T} h_w^2 \left(\frac{w}{E t_w}\right)^{0.5}\right]^{0.67} \qquad \text{(Eq. 3.1.33)}$$

$$= \left[\frac{0.13}{1.83}(105)^2\left(\frac{13.3}{4000(0.5)}\right)^{0.5}\right]^{0.67}$$

$$= 16 \text{ ft}$$

Comment: One need not create a single-degree-of-freedom model to design a shear wall braced building to an equal displacement criterion. This alternative, in the author's opinion, is the most straightforward design approach.

Alternative CVED Follow Steps 2, 3, and 4 of the CVED method.

- Next use Step 7 to determine the required length of the wall. Adjust the period so as to match the proposed wall length should the wall length exceed the minimum value (use Eq. 2.4.22).
- Determine the displacement of the structure.

$$\Delta_u = \frac{\Gamma S_v T}{2\pi} \tag{3.1.37}$$

$$= \frac{1.5(52)(1.83)}{6.28}$$

$$= 22.7 \text{ in.}$$

- Determine the objective yield displacement.

$$\Delta_y = \frac{\Delta_u}{\Omega_o \mu} \tag{3.1.38}$$

$$= \frac{22.7}{1.25(4)}$$

$$= 4.54 \text{ in.}$$

- Assume a force distribution consistent with a first mode distribution to determine the design moment. Use a linear mode shape (triangular force distribution).

$$M_d = \frac{3.6 E I_e \Delta_y}{h_w^2} \tag{3.1.39}$$

$$= \frac{3.6(4000)(1{,}770{,}000)(4.54)}{(105)^2(144)(12)}$$

$$= 6200 \text{ ft-kips} \qquad (0.042\, W h_w)$$

The flexural reinforcement of the wall would be developed from Eq. 2.2.20b or more appropriately from Eq. 2.4.23 for $P_D = 600$ kip, $t_w = 6$ in., $\ell_w = 16$ ft.

$$a = \frac{P_D}{0.85 f_c' t_w}$$

$$= \frac{600}{0.85(5)(6)}$$

$$= 23.5 \text{ in.}$$

Consistent with current practice, it seems appropriate to associate M_d with M_u.

$$A_s = \frac{M_d - P_D\,(\ell_w/2 - a/2)}{\phi f_y (d - d')} \qquad \text{(see Eq. 2.2.20b)}$$

$$= \frac{6620 - 600(8 - 1)}{(0.9)60(14)}$$

$$= \frac{68{,}000}{11{,}200}$$

$$= 2.93 \text{ in.}^2 \qquad \text{(Five \#7 bars)}$$

Direct Displacement-Based Design The direct displacement-based design (DBD) as developed in the Blue Book[3.4] is more complex than it need be. I have adjusted it so as to include a constant spectral velocity (CVDD).

Step 1: Create a Substitute Structure That Emulates the Postyield Behavior of the System at the Level of Ductility That Is Consistent with the Design Objectives. Consider the behavior described in Figure 2.4.4*a*. The idealized behavior predicts a yield drift of 0.5 in. An ultimate drift of 2 in. seems certainly attainable. This corresponds to a system ductility of 4. The stiffness of the substitute structure ($k_{e,s}$) would be the idealized strength divided by the objective ultimate drift (Δ_u), in this case 2 in. It follows from Figure 3.1.2 that

$$k_s = \frac{k_i}{\mu} \tag{3.1.40}$$

Step 2: Estimate the Damping Available in the Substitute Structure.

Comment: Damping in the substitute structure is a combination of hysteretic structural damping and the traditional 5% of nonstructural viscous damping. The procedures discussed in Section 1.1.3 may be used for this purpose or an estimate of hysteretic energy absorption may be extracted from experimental efforts (i.e., Figure 2.4.63).

For partially full hysteretic behavior as might be found in cast-in-place concrete systems, the hysteretic structural component of damping is reasonably estimated by

$$\zeta_{\text{eq}} = \frac{\sqrt{\mu} - 1}{\pi\sqrt{\mu}} \qquad \text{(see Eq. 1.1.9d)} \tag{3.1.41}$$

and the total damping by

$$\hat{\zeta}_{\text{eq}} = \zeta_{\text{eq}} + 5 \qquad \text{(see Eq. 1.1.9a)} \tag{3.1.42}$$

all of which are expressed as a percentage of critical damping.

Step 3: Determine the Design Spectral Velocity for the Substitute Structure. The criterion 5% damped spectral velocity must now be reduced to account for the inclusion of structural damping. This may be accomplished through the use of the Newmark-Hall spectral development relationship (Eq. 1.1.8b).

$$S_v = \alpha_v \dot{d}_{max} \tag{3.1.43}$$

where $\dot{d}_{max}$ is the spectral velocity of the ground (see Figure 1.1.5), and α is the amplification factor (Eq. 1.1.8b)

$$\alpha_v = 3.38 - 0.67 \ln \zeta \tag{3.1.44a}$$

which for a 5% damped system is

$$\begin{aligned} \alpha_v &= 3.38 - 0.67 \ln 5 \\ &= 2.3 \end{aligned} \tag{3.1.44b}$$

If the design spectral velocity (response spectrum) is based on a 5% damped structure, the reduction factor ($\hat{R}$) for a system whose effective damping is $\hat{\zeta}$ may be determined from

$$\hat{R} = \frac{3.38 - 0.67 \ln \hat{\zeta}_{eq}}{2.3} \tag{3.1.45}$$

and the design spectral velocity for the substitute structure ($S_{v,s}$) is

$$S_{v,s} = \hat{R} S_v \tag{3.1.46}$$

Step 4: Establish the Objective Spectral Displacement. A building drift objective is first established. This is done by selecting a target drift ratio (δ_1) and ultimate building drift limit (Δ_u),

$$\Delta_u = \delta_1 h_w \qquad \text{(see Eq. 3.1.14b)}$$

and the spectral displacement that follows,

$$S_d = \frac{\Delta_u}{\Gamma} \qquad \text{(see Eq. 3.1.10)}$$

Since the building response is dominated by the postyield component of drift, which is essentially linear, a first mode participation factor of 1.43 based on a linearized first mode shape may be used.

$$\Gamma = \frac{3n}{2n+1} \qquad \text{(Eq. 3.1.18a)}$$

Comment: This will tend to be conservative for a shear wall braced structure.

$$\Gamma = \frac{3(10)}{2(10)+1} \qquad (\text{for } n = 10)$$

$$= 1.43 \qquad \left(\text{Blue Book: } \frac{1}{k_1 k_2} = 1.56\right)$$

Step 5: Determine the Objective Fundamental Frequency for the Substitute Structure.

$$\omega_s = \frac{S_{v,s}}{S_{d,s}} \qquad \text{(see Eq. 1.1.6a)}$$

Step 6: Determine the Effective Mass of the Substitute Structure ($M_{e,s}$). Assume a linear mode shape.

$$M = \frac{1.5(n+1)}{2n+1} M \qquad \text{(Eq. 3.1.19a)}$$

Step 7: Determine the Effective Stiffness of the Substitute Structure.

$$k_{e,s} = \omega_s^2 M \qquad (3.1.47)$$

Step 8: Determine the Mechanism Moment of the Structure (V_M—Figure 1.1.12).

Step 9: Determine the Design Moment for the Structure.

$$M_d = \frac{M_{\max}}{\Omega_o} \qquad \left(\frac{V_M}{R_o}, \text{Figure 1.1.12}\right) \qquad (3.1.48)$$

Step 10: Determine the Length of Wall Required to Provide the Desired Dynamic Characteristic.

$$T = \frac{2\pi}{\omega_s \sqrt{\mu}}$$

$$\ell_w = \left[\frac{0.13}{T} h_w^2 \left(\frac{w}{E t_w}\right)^{0.5}\right]^{0.67} \qquad \text{(Eq. 3.1.33)}$$

Now apply the CVDD method to the example building (Figure 2.4.9).

Step 1: Create a Substitute Structure. Select $\mu = 4$.

$$k_s = \frac{k_i}{\mu} \qquad \text{(Eq. 3.1.40)}$$

$$= 0.25 k_i$$

Step 2: Estimate the Damping Available in the Substitute Structure.

$$\zeta_{eq} = \frac{\sqrt{\mu} - 1}{\pi\sqrt{\mu}} \qquad \text{(Eq. 3.1.41)}$$

$$= \frac{\sqrt{4} - 1}{\pi\sqrt{4}}$$

$$= 16\%$$

$$\hat{\zeta}_{eq} = \zeta_{eq} + 5$$

$$= 21\% \qquad \text{(Eq. 3.1.42)}$$

Step 3: Determine the Design Spectral Velocity for the Substitute Structure.

$$\hat{R} = \frac{3.38 - 0.67 \ln \hat{\zeta}_{eq}}{2.3} \qquad \text{(Eq. 3.1.45)}$$

$$= \frac{3.38 - 0.67 \ln 21}{2.3}$$

$$= 0.59$$

$$S_{v,s} = \hat{R} S_v \qquad \text{(Eq. 3.1.46)}$$

$$= 0.59(52)$$

$$= 30.7 \text{ in./ sec}$$

Step 4: Establish the Objective Spectral Displacement.

$$\Delta_u = \delta_1 h_w \qquad \text{(see Eq. 3.1.14b)}$$

$$= 0.018(105)(12)$$

$$= 22.7 \text{ in.}$$

and the spectral displacement

$$S_d = \frac{\Delta_u}{\Gamma} \qquad \text{(see Eq. 3.1.10)}$$

$$= \frac{22.7}{1.43} \qquad \text{(Linear mode shape basis—see Eq. 3.1.18a)}$$

$$= 15.9 \text{ in.}$$

Step 5: Determine the Objective Fundamental Frequency for the Substitute Structure.

$$\omega_s = \frac{S_{v,s}}{S_d} \qquad \text{(see Eq. 1.1.6a)}$$

$$= \frac{30.7}{15.9}$$

$$= 1.93 \text{ rad/sec} \qquad (T = 3.25 \text{ seconds})$$

Step 6: Determine the Effective Mass of the Substitute Structure. Assume a Linear Mode Shape.

$$M_{e,s} = \frac{1.5(n+1)}{2n+1} M \qquad \text{(Eq. 3.1.19a)}$$

$$= \left(\frac{16.5}{21}\right) \frac{W}{g}$$

$$= \frac{0.78(1400)}{386.4}$$

$$= 2.84 \text{ kip sec}^2/\text{in.}$$

Step 7: Determine the Effective Stiffness of the Substitute Structure.

$$k_{e,s} = \omega_s^2 M_{e,s}$$

$$= (1.93)^2(2.84)$$

$$= 10.6 \text{ kips/in.}$$

Step 8: Determine the Flexural Strength Objective ($\Omega_o M_d$) for the Structure. The flexural strength objective corresponds to $F_{\max}$ (Figure 1.1.1)—or H_u (Figure 3.1.2).

$$H_u = k_{e,s} S_d$$

$$= 10.6(15.9)$$

$$= 168 \text{ kips}$$

$$\Omega_o M_d = H_u h_e$$

$$= \frac{H_u h_w}{\Gamma} \qquad \text{(Eq. 3.1.16)}$$

$$= \frac{168(105)}{1.43}$$

$$H_u h_e = 12{,}350 \text{ ft-kips}$$

Step 9: Determine the Design Moment.

$$M_d = \frac{H_u h_e}{\Omega_o}$$

$$= \frac{12{,}350}{1.25}$$

$$= 9880 \text{ ft-kips} \qquad (0.067Wh_w)$$

Step 10: Determine the Length of the Wall Required to Provide the Desired Idealized Dynamic Characteristic.

$$k_i = \mu k_s \qquad \text{(Eq. 3.1.40)}$$

$$\omega_i = \sqrt{\frac{k_i}{m}} \qquad \text{(Eq. 1.1.5)}$$

$$= \sqrt{\frac{\mu k_s}{m}}$$

$$= \omega_s \sqrt{\mu}$$

The period of the idealized structure is

$$T_i = \frac{2\pi}{\omega_i} = \frac{2\pi}{\omega_s \sqrt{\mu}}$$

$$= \frac{2\pi}{1.93\sqrt{4}}$$

$$= 1.6 \text{ seconds}$$

$$\ell_w = \left[\frac{0.13}{T} h_w^2 \left(\frac{w}{Et_w}\right)^{0.5}\right]^{0.67} \qquad \text{(Eq. 3.1.33)}$$

$$= \left[\frac{0.13}{1.6}(105)^2 \left(\frac{13.3}{4000(0.5)}\right)^{0.5}\right]^{0.67}$$

$$= 17.6 \text{ ft}$$

Conclusions:

$$\ell_w = 17.6 \text{ ft}$$

$$M_d = 0.067Wh_w$$

Table 3.1.2 summarizes the design parameters developed by the various approaches. The first equal displacement-based design (EBD 1) is that developed in section 2.4.1.4 for a specific wall length (26 ft). EDB2 is developed in this section following Blue Book procedures. The code-based criterion is not dependent on the adopted wall length, but the other design procedures are. The four displacement-based procedures

TABLE 3.1.2 Design Criterion Summary

Design Methodology	Wall Length (ℓ_w, ft)	M_d
Code[a]	(No requirements)	$0.125Wh_w$
Equal displacement-based design (EBD 1)[a]	26 (set)	$0.066Wh_w$
Equal displacement-based design (EBD 2)[b]	16	$0.053Wh_w$
Constant velocity equal displacement (CVED)	16	$0.042Wh_w$
Constant velocity direct displacement (CVDD)	17.6	$0.067Wh_w$

[a] Strength objectives have been adjusted to a spectral velocity of 49 in./sec and a system ductility factor of $5(R = 5)$ so as to be consistent with the design spectrum of Figure 4.1.1, which was used to develop the designs of this chapter.

[b] A system ductility factor of 4 was used in this design. Were it to have been 5, as was the case in the CVED design, the ultimate strength objective would have been $0.043Wh_w$.

predicate their strength criteria on the attainment of a particular stiffness objective quantified in Table 3.1.2 by defining the proper or compatible length of wall (ℓ_w).

3.1.1.2 Analyzing the Design Processes Two systems are proposed, a 26-ft long wall and one that is 16-ft long. The flexural strengths proposed for the two systems based on an equal displacement approach are consistent and clearly a function of the length of the wall. This is convenient from a design perspective because it allows a rapid conversion to any length of wall. We know, provided the spectral velocity is constant, that the associated design force level (spectral acceleration) is the product of the angular frequency (ω) and the spectral velocity. The angular frequency is a function of the square root of the stiffness, and the stiffness a function of the cube of the length of the wall and its thickness. Accordingly, we may convert a design strength without repeating a complete design. The strength criterion $\left(M_d^*\right)$ for the new wall $\left(\ell_w^*, t_w^*\right)$ is

$$M_d^* = \sqrt{\left(\frac{\ell_w^*}{\ell_w}\right)^2 \left(\frac{t_w^*}{t_w}\right)} M_d \tag{3.1.49a}$$

Equation 3.1.49a may be expanded to include spectral velocity changes.

$$M_d^* = \frac{S_v^*}{S_v} \left(\frac{\ell_w^*}{\ell_w}\right)^{1.5} \left(\frac{t_w^*}{t_w}\right)^{0.5} M_d \tag{3.1.49b}$$

The strength objective for the 26-ft wall (CVED) may be developed directly from that developed for the 16-ft wall if differences in ductility (4.27 versus 4.0) and concrete moduli are included in Eq. 3.1.49b.

$$M_{d-26}^* = \frac{S_v^*}{S_v} \left(\frac{\ell_w^*}{\ell_w}\right)^{1.5} \left(\frac{t_w^*}{t_w}\right)^{0.5} \left(\frac{\mu}{\mu^*}\right) \left(\frac{E^*}{E}\right)^{0.5} M_{d-16} \tag{3.1.49c}$$

$$= \frac{44}{52}\left(\frac{26}{16}\right)^{1.5}\left(\frac{6}{6}\right)^{0.5}\left(\frac{4.0}{4.27}\right)\left(\frac{3600}{4000}\right)^{0.5} 0.042Wh_w$$

$$= 0.85(2.07)(1.0)(0.94)(0.95)(0.042)Wh_w$$

$$= 0.066Wh_w$$

It is clear that a direct tie exists between stiffness and strength, but current[3.7] strength-based procedures do not make this connection because the design period is prescriptively established. The code strength objective (Table 3.1.2) would apply to walls of any quantity and length, whereas the strength objectives of the other design procedures are specifically tied to a particular wall configuration. The attainment of the appropriate stiffness/strength linkage is the objective and rationale behind all displacement-based design procedures; it should also be the objective of all strength-based designs.

Expending a significant effort in the development of an "accurate(?)" single-degree-of-freedom model is not warranted, especially in the conceptual design phase. This is because coefficients tend to be fairly stable and the extent of postyield participation not known. The probable mode shape (ϕ_i) will be a function of the "experienced" system ductility, a quantity that will not be known until the design is tested using sequential yield analysis procedures (Section 3.1.1.3) or inelastic time history procedures (Section 4.1.1).

Consider the modeling conclusions that would be developed from theory, a linear mode shape, and the Blue Book.[3.4] Theoretically Eq. 3.1.16a and 3.1.17b define the equivalent single-degree-of-freedom model for a lumped mass system whose primary or fundamental mode shape is known.

$$M_e = \frac{\Gamma}{h_w}\sum_{i=1}^{n} m_i\phi_i h_w \qquad \text{(see Eq. 3.1.17b)}$$

$$h_e = \frac{h_w}{\Gamma} \qquad \text{(Eq. 3.1.16)}$$

In the subsection that follows (Section 3.1.1.3), we explore several designs, one of which will approach our objective level of system ductility (4). The mode shape used in the designs is developed in Table 3.1.3. The level of experienced system ductility is 3.92. The plastic hinge in this analysis is assumed to have an effective point of rotation at Level 1; hence $\Delta_{p1} = 0$.

The effective mass is developed as follows:

$$\Gamma = \frac{\sum m_i\phi_i}{\sum m_i\phi_i^2} \qquad \text{(Eq. 3.1.15)}$$

$$= \frac{4.822}{3.354}$$

$$= 1.44$$

$$M_e = \frac{\Gamma M}{n} \sum_{i=1}^{n} \phi_i \qquad \text{(Eq. 3.1.17b)}$$

$$= \frac{1.44(4.82)}{10} M$$

$$= 0.69M$$

and the effective height is

$$h_e = \frac{h_w}{\Gamma} \qquad \text{(Eq. 3.1.16)}$$

$$= \frac{105}{1.44}$$

$$= 72.9 \text{ ft}$$

$$= 0.69h_w$$

TABLE 3.1.3 Modal Characteristics of the System Whose Behavior Is Described in Figure 3.1.5*a* ($\mu_\Delta = 3.92$)

Story	Elastic Spectral Displacement (in.)	Yield Displacement Δ_{yi} (in.)	Plastic Displacement Δ_p (in.)	Inelastic Spectral Displacement (in.)	Normalized Inelastic Mode Shape ϕ_i	ϕ_i^2	$m_i\phi_i$	$m_i\phi_i^2$
10th	22.84	5.83	17.02	22.84	1.000	1.000	1.000	1.000
9th	19.55	4.99	15.13	20.11	0.880	0.775	0.880	0.775
8th	16.29	4.16	13.23	17.39	0.761	0.580	0.761	0.580
7th	13.10	3.34	11.34	14.69	0.643	0.413	0.643	0.413
6th	10.07	2.57	9.45	12.02	0.526	0.277	0.526	0.277
5th	7.26	1.85	7.56	9.42	0.412	0.170	0.412	0.170
4th	4.79	1.22	5.67	6.89	0.302	0.091	0.302	0.091
3rd	2.75	0.70	3.78	4.48	0.196	0.038	0.196	0.038
2nd	1.25	0.32	1.89	2.21	0.097	0.009	0.097	0.009
1st	0.34	0.09	0.00	0.09	0.004	0.000	0.004	0.000
					4.822	3.354	4.822	3.354

The product $M_e h_e$ is

$$M_e h_e = 0.69M(0.69h_w)$$
$$= 0.48Mh_w$$

This is much lower than the Blue Book[2.4] recommendation for a shear wall (0.61 Mh_w) and the $0.55Mh_w$ developed from a linear elastic mode shape (EBD procedure, Section 3.1.1.1). Clearly the inelastic mode shape described in Table 3.1.3 is not linear but some of this divergence can be attributed to the selection of the effective centroid of postyield rotation as being coincident with the first floor ($\ell_p/2$).

We can conclude that the use of the modeling factors recommended in the Blue Book, though somewhat conservative, are as appropriate as are those suggested by the adoption of a linear mode shape given the possible considerations associated with defining a mode shape. The precision used in the development of an analytical model must recognize the degree of accuracy inherent to other assumptions such as the centroid of the plastic hinge and variations with regard to effective stiffness. The simplest approach (linearized behavior model) seems most appropriate for design.

3.1.1.3 Conceptual Design Review We are now in a position to decide which of the designed shear walls is most appropriate for the project described in Figure 2.4.9. Two alternatives appear to be reasonable:

- A 26-ft long wall whose flexural strength (M_u) is $0.066Wh_w$
- A 16-ft long wall whose flexural strength (M_u) is $0.042Wh_w$

Clearly many other alternatives are available, and they can be quickly developed through the use of Eq. 3.1.49b. Our objective here is to determine whether or not we have in fact attained our design objectives and, most importantly, if the selection of a system ductility factor (μ) of 4 and overstrength factor (Ω_o) of 1.25 resulted in strain demands that meet our performance objective.

Consider the 16-ft long wall. For the reasons discussed in Section 2.4.1, it is advisable to use an 8-in. wall in the plastic hinge region:

$$\frac{h_x}{t_w} = \frac{120}{6} = 20 > 16$$

The flexural reinforcement of the wall would be developed from Eq. 2.4.15 or directly from Eq. 2.2.23 for

$$M_d = 0.042Wh_w$$
$$= 0.042(1400)(105)$$
$$= 6174 \text{ ft-kips}$$
$$P_D = 600 \text{ kip}, t_w = 8 \text{ in.}$$

$$a = \frac{P_D}{0.85 f_c' t_w}$$

$$= \frac{600}{0.85(5)(8)}$$

$$= 17.6 \text{ in.}$$

Consider the design moment to be the ultimate moment ($M_d = M_u$).

$$A_s = \frac{M_u - P_D(\ell_w/2 - a/2)}{\phi f_y(d - d')} \qquad \text{(see Eq. 2.4.1)}$$

$$= \frac{6860 - 600(8 - 0.75)}{60(14.5)}$$

$$= 2.88 \text{ in.}^2$$

Two tasks are easily performed by the computer with available software—a response spectral analysis and a pushover or sequential yield analysis. Figure 3.1.5 describes the results of the pushover analysis performed on a 16-ft wall reinforced flexurally with six #6 bars at each end ($A_s = 2.64$ in.^{2}).

Steel first yield occurred at a drift ratio of 0.45% and a base shear of 89 kips (Figure 3.1.5*b*). The drift ratio predicted by our design idealization (Eq. 2.4.27 and 2.4.17) is 0.6%. The roof drift predicted by a spectral analysis is the same as our target drift (1.8%). This corresponds to a system ductility factor (μ) of 4 based on steel first yield or 3 based on idealized first yield.

Comment: Yield drift predictions and adopted system ductility factors must be compatible. In the development of this design, a system ductility and overstrength factor ($\Omega_o \mu_s$) of 5 was adopted (Eq. 3.1.38). The design strength was developed utilizing an idealized stiffness of 50% of I_g (Eq. 3.1.39). The bilinear behavior idealization shown in Figure 3.1.5*b* ($\lambda_o M_u$, Δ_{yi}) is clearly consistent with the pushover projection and the experimental data of Figure 2.1.1*b*.

The generated overstrength at spectrum predicted drift is 28% (109/85), but the steel yield strength used in the pushover analysis was 60 ksi and the vertical wall steel, which will exist between the chord reinforcing (six #6 bars), has not been included in this analysis. Accordingly, the use of an overstrength factor (Ω_o) of 1.25 is reasonable for developing the objective flexural strength but not appropriate for capacity considerations.

Strain states at our objective or target drift are reasonable and should not produce structural damage.

$$\varepsilon_{cu} = 0.0033 \text{ in./in.} \qquad (\text{DR} = 1.81\%; \text{ Figure 3.1.5}b)$$

$$\varepsilon_s = 0.03 \text{ in./in.} \qquad (\text{DR} = 2.5\%; \text{ Figure 3.1.5}a)$$

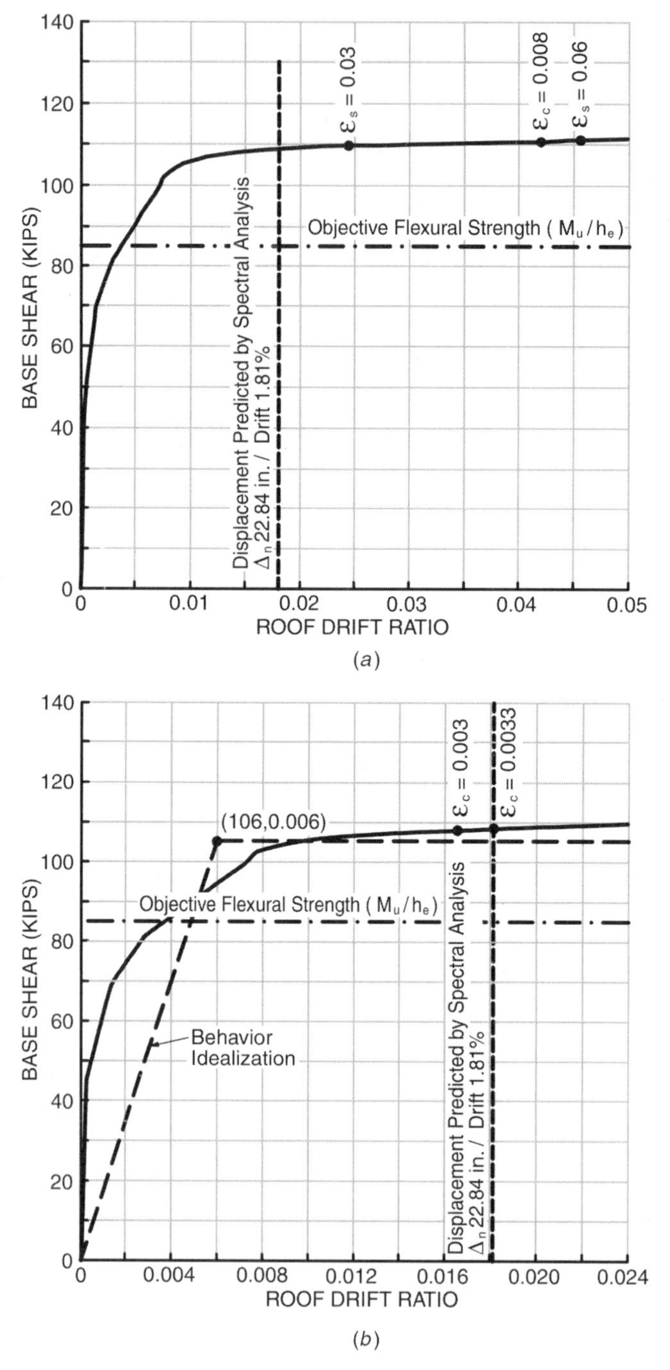

Figure 3.1.5 (*a*) Pushover results, capacity spectrum method—$\ell_w = 16$ ft; six #6 bars ($\rho = 0.25\%$). (*b*) Pushover results, capacity spectrum method—$\ell_w = 16$ ft; six #6 bars ($\rho = 0.25\%$).

At a drift ratio of about 4.5% ultimate strain limit states in both the concrete and steel are reached, but this corresponds to a displacement ductility of 7.5 based on the proposed behavior idealization (Figure 3.1.5*b*).

Conclusion: The design procedure produces the desired results provided the displacement of the system does not significantly exceed the target drift, and this is evaluated in Chapter 4 when we study the final design using time history procedures. At this point in the design process, we should have confidence in the proposed solution and, more importantly, in our ability to create the objective behavior.

Consider now the design proposed for the 26-ft long wall (Figure 2.4.10*a*). The results of a pushover analysis are presented on Figure 3.1.6. The spectral projection of roof drift is, of course, much lower than that of the 16-ft wall. Had a participation factor of 1.45 been used in the development of the (spectrum) predicted drift (see Eq. 3.1.10), the conclusion would have been consistent with the projected computer estimate. The system ductility (μ) demand is somewhat lower than the design objective of 4. The yield deflection predicted by the computer (Table 2.4.3 and Figure 3.1.6*b*) is on the order of 0.25% while that developed from Equations 2.4.17 and 2.4.27 are on the order of 0.37%. This variation is of no real consequence for this wall is clearly overreinforced in spite of the fact that it does not quite meet our suggested minimum reinforcing level ($0.0025bd$). Observe, however, that this has a significant impact on the provided level of system overstrength, Ω_o (1.71, Figure 3.1.6*a*), and it is for this reason that the use of a universal overstrength factor in order to attain capacity-based design objectives is dangerous.

The strain in the concrete is 0.0018 in./in at the roof drift ratio predicted by spectral analysis (0.84%). A concrete strain of 0.003 in./in. is not reached until the drift ratio exceeds 1.5%.

This wall is also capable of attaining a system ductility factor of 10 (DR = 2.5%) and an ultimate drift ratio of 4.5% at shell spalling. Structural damage projections for the 26-ft wall would be minimal, but not too dissimilar to those of the 16-ft wall. Obviously the displacement of a building braced by a 16-ft wall would be about twice that of a comparable building braced by a 26-ft long wall.

$$\Delta_{16} = \Delta_{26}\left(\frac{\ell_{26}}{\ell_{16}}\right)^{3/2} \tag{3.1.50}$$

System strength is commonly believed to be inversely related to structural damage caused by an earthquake. This inverse relationship is effectuated in strength-based designs through the introduction of an importance factor (I). The use of an importance factor of 1.25 is common in the design of essential facilities. The intent is to promote post-earthquake serviceability. To demonstrate that strength and performance are not necessarily related, the reinforcing proposed for the 26-ft wall was increased by 32%: eight #8s as opposed to eight #7s. The pushover analysis was repeated and its results presented in Figure 3.1.7. The flexural strength of the wall is increased by only 15% so we have not attained the code objective increase in strength ($1.25M_u$), but the

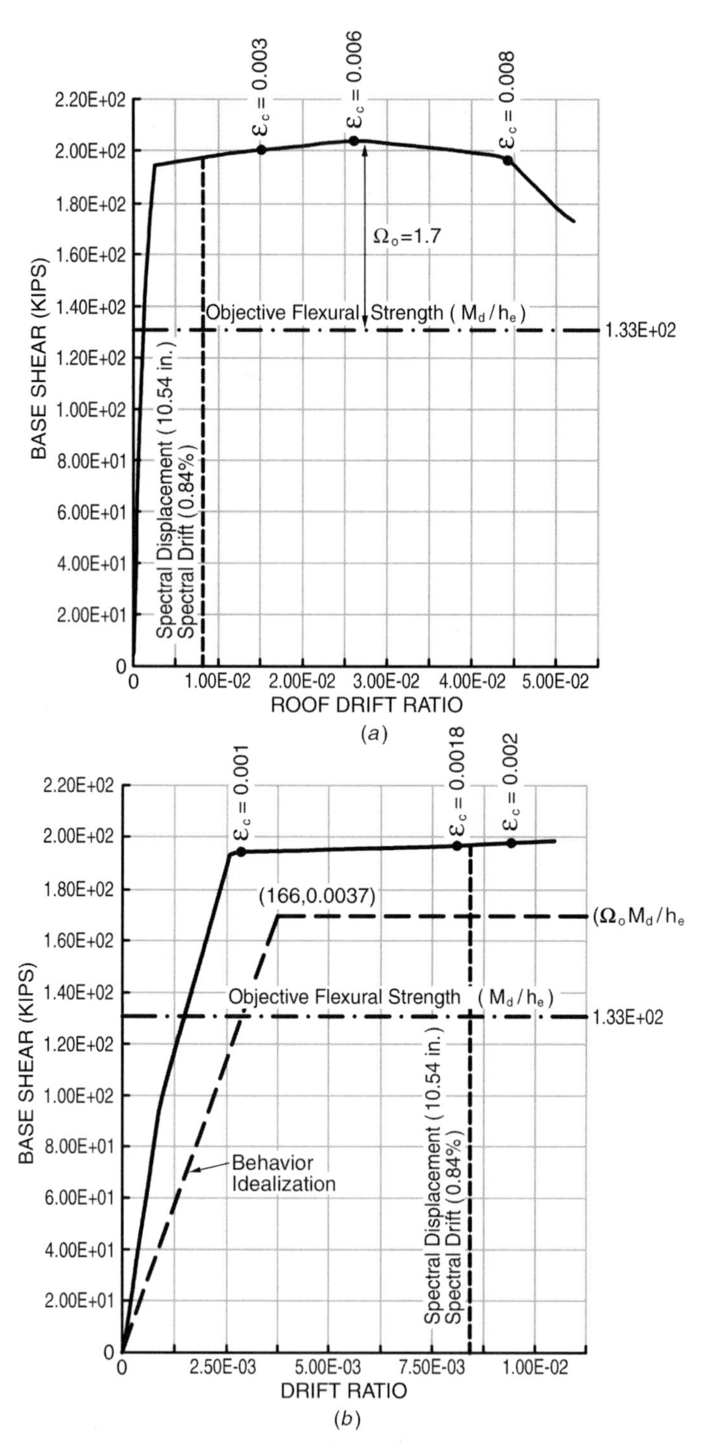

Figure 3.1.6 (*a*) Pushover results, capacity spectrum method, for wall of Figure 2.4.10*a*—ℓ_w = 26 ft, reinforcing = eight #7 bars ($0.0025A_c$). (*b*) Pushover results, capacity spectrum method, for wall of Figure 2.4.10*a*—ℓ_w = 26 ft, reinforcing = eight #7 bars ($0.0025A_c$).

design strength objective has been exceeded as has our attained level of overstrength. Spectral drift projections remain the same, of course, but induced strain states are little changed (Figure 3.1.7*b*).

This conclusion regarding the impact of strength on performance is also reached by parametrically studying the 16-ft wall. Figure 3.1.7*c* describes the concrete strain states induced in the 16-ft wall reinforced with varying levels of boundary reinforcement ($\rho = 0.0025$ to 0.0057). The strain states are essentially the same, and unless the actual displacement experienced by the stronger wall is significantly less than that experienced by the weaker wall, a topic deferred to Section 4.1.1, no apparent improvement in performance is to be expected. Logic suggests, however, that a reduced restoring force will in most cases produce lower drift demands (Section 1.1.1). One might reasonably conclude that strength objectives are not closely tied to the performance or damageability of shear wall braced buildings.

An obvious casualty of an increase in system strength is the expected level of acceleration. Following the spectral development discussed in Section 2.4.1.2, we may generally relate the level of peak roof acceleration $\left(A_{r,\max}\right)$, including higher mode effects, to the base shear by

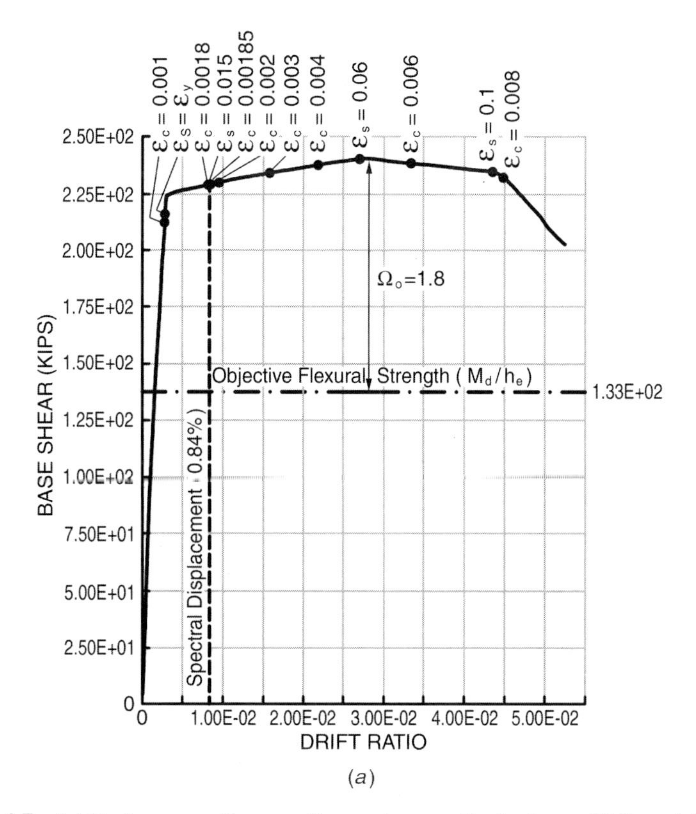

(*a*)

Figure 3.1.7 (*a*) Pushover results, capacity spectrum method—$\ell_w = 26$ ft—reinforcing = eight #8 bars.

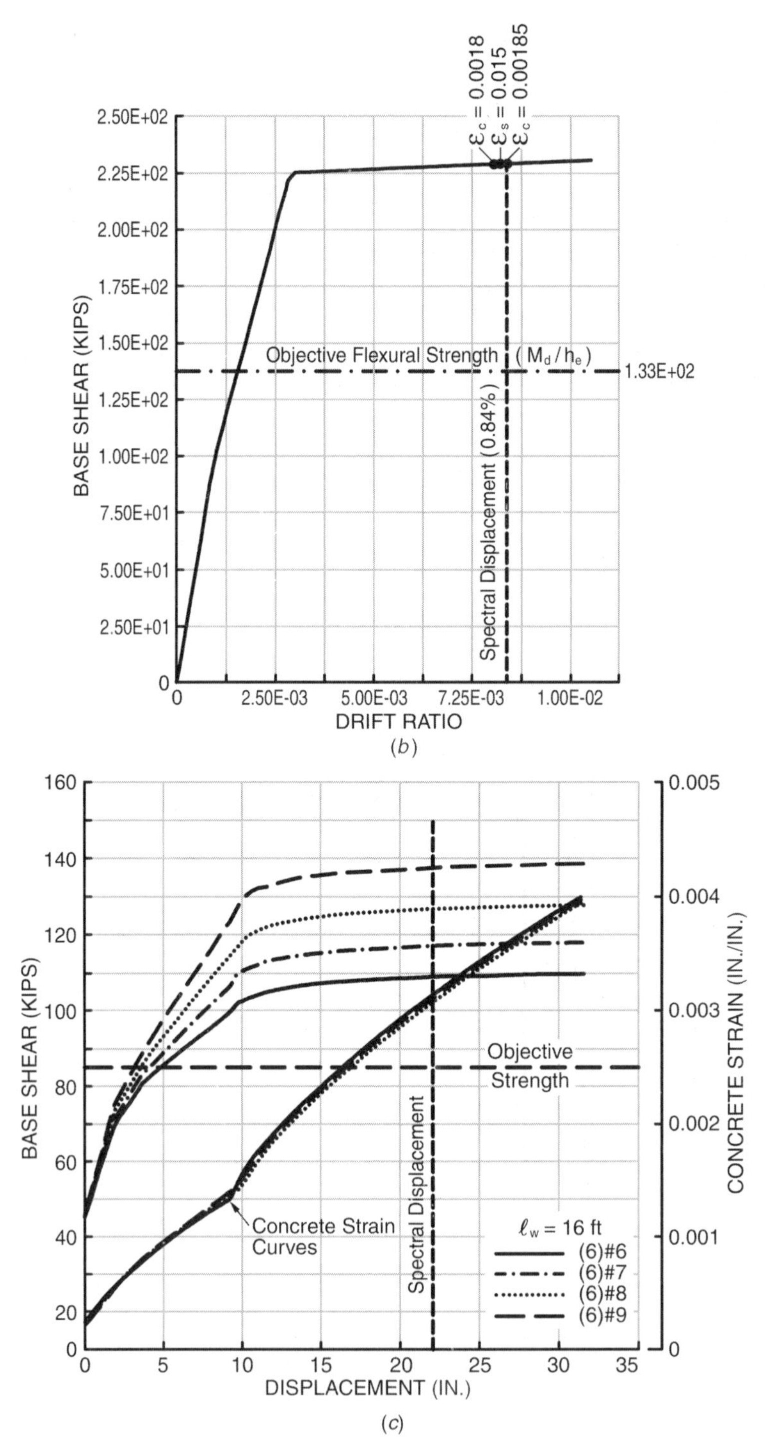

Figure 3.1.7 (*Continued*) (*b*) Pushover results, capacity spectrum method—$\ell_w = 26$ ft—reinforcing = eight #8 bars. (*c*) Pushover results, capacity spectrum method—$\ell_w = 16$ ft.

$$A_{r,\max} = \frac{\Gamma}{0.7}\frac{V}{W}g \qquad \text{(see Eq. 2.4.2a)}$$

$$\cong 2.0\frac{V}{W}g \qquad (3.1.51)$$

For the 16-ft shear wall reinforced with six #6 bars, this suggests an acceleration of

$$A_{r,\max} = 2.0\frac{(110)}{1400}g \qquad \text{(see Figure 3.1.5}a\text{)}$$

$$= 0.16g \qquad \text{(16-ft wall)}$$

For the 26-ft wall reinforced with eight #8 bars, the peak acceleration would be

$$A_{r,\max} = \frac{2.0(240)}{1400}g \qquad \text{(see Figure 3.1.7}a\text{)}$$

$$= 0.34g \qquad \text{(26-ft wall)}$$

Conclusion: Arbitrarily increasing system strength does not necessarily produce a building that will perform better.

Consider the overstrength factor. It would be convenient to adopt a universal overstrength factor for a category of bracing system, but this can be dangerous. Reflect on the overstrength factors of the two shear wall designs developed to provide lateral support for the system described in Figure 2.4.9. Strength-based designs[3.7] would propose an equivalent strength criterion along with an arbitrary overstrength factor, yet the 26-ft wall could easily be 2 to 3 times stronger than the 16-ft wall. If the shear wall load must proceed through a brittle member as, for example, a supporting column, either the universal overstrength factor must be extremely conservative or the produced design will be potentially dangerous.

Conclusion: A carefully considered sequential yield analysis (pushover) provides the best possible design insight. It also allows the designer to introduce the appropriate level of conservatism.

3.1.1.4 Summarizing the Design Process The general objective of the design process is to produce a shear wall that possesses a balance between wall stiffness and strength at the objective level of component (wall) ductility. The known factors are the mass of the system and the design spectral velocity, it being assumed that the structure falls within the response region where spectral velocity is constant (Figure 1.1.5).

The summaries that follow are categorized so as to deal with the identified design preconditions as efficiently as possible.

Procedure 1—Wall Characteristics Are a Precondition. The usual case is one where the wall dimensions are limited by functional or aesthetic considerations. In this case the design steps for a force-based procedure are as follows.

Step 1: Determine the Effective Stiffness and the Fundamental Period (T) for the Structure. Refer to Section 2.4.1.3, Step 3.

$$T_f = 0.13\left(\frac{h_w}{\ell_w}\right)^2\sqrt{\frac{w\ell_w}{Et_w}} \qquad \text{(Eq. 2.4.22)}$$

where E is expressed in ksi; t_w, ℓ_w, and h_w are in feet; and w is the tributary weight (mass) expressed in kips/ft that must be supported laterally by the shear wall (W/h_w).

Step 2: Determine the Spectral Acceleration.

$$S_a = \frac{S_v}{g}\left(\frac{2\pi}{T}\right) \qquad \text{(see Eq. 3.1.23)}$$

Step 3: Determine the Elastic Base Shear. (See Figure 3.1.1*b*).

$$V_{e1} = k_3 S_a W$$

where k_3 is the effective mass factor (Eq. 3.1.19b) and S_a is the spectral acceleration expressed in terms of g. k_3 may be estimated through the use of Eq. 3.1.19a or the following values developed from Eq. 3.1.19a.

$$n = 1; \qquad k_3 = 1$$
$$n = 3; \qquad k_3 = 0.86$$
$$n = 5; \qquad k_3 = 0.82$$
$$n = 10; \qquad k_3 = 0.79$$
$$n = 15; \qquad k_3 = 0.77$$
$$n = \infty; \qquad k_3 = 0.75$$

where n is the number of stories.

If mass distribution and/or story height vary use Eq. 3.1.15, 3.1.16, and 3.1.17a.

Step 4: Reduce the Elastic Base Shear to a Design or Ultimate Base Shear.

$$V_d = \frac{V_{\text{elastic}}}{\mu\Omega_o}$$

The product $\mu\Omega_o$ may reasonably be taken as 5 (see Figure 3.1.5*b*).

Step 5: Estimate the Design Moment.

$$M_d = V_d h_e$$

where h_e, the effective height, may be developed in a variety of ways (see Eq. 3.1.18a) but can reasonably be approximated by a triangular distribution of force ($h_e \cong 0.67h_w$).

Comment: This procedure ($h_e = 0.67h_w$) in effect creates an equivalent single-degree-of-freedom system (Figure 3.1.1*b*) whose effective weight (mass) and height are, for a ten-story building, $0.79W$ and $0.67h_w$. Observe that the product $M_e h_e$ is $0.53Mh_w$, as opposed to the $0.55Mh_w$ developed for the linear model described in Figure 3.1.1*b*.

Procedure 2—The Design Objective Is to Limit Building Drift (Δ_u)*—Ref. Figure 1.1.1.* An objective or target drift (Δ_T), as well as an objective level of system ductility (μ), are identified. The level of system damping is consistent with that used to develop the response spectra. An equal displacement-based approach is used to attain design objectives. No preconditions exist that define the length of the wall.

Step 1: Determine the Objective Period for the Structure.

$$T = \frac{2\pi S_d}{S_v} \qquad \text{(see Eq. 3.1.9)}$$

where S_d is the target drift divided by the participation factor Γ. The participation factor may be estimated through the use of Eq. 3.1.18a, which is developed from a linear mode shape. It may also be developed from Eq. 3.1.15 using an assumed mode shape that could reflect a combination of elastic and inelastic response. See Table 3.1.1.

Step 2: Calculate the Appropriate or Dynamically Compatible Length of Wall. The use of Eq. 3.1.33 facilitates this process.

$$\boxed{\ell_w = \left[\frac{0.13}{T}h_w^2\left(\frac{w}{Et_w}\right)^{0.5}\right]^{0.67}} \qquad \text{(Eq. 3.1.33)}$$

where units are in their most practical form:

ℓ_w, h_w, and t_w are in feet.

T is in seconds.

w is in kips per foot.

E is in kips per square inch.

Step 3: Determine the Displacement of the Structure If ℓ_w Is Greater Than That Developed in Step 2.

$$T = 0.13\left(\frac{h_w}{\ell_w}\right)^2\sqrt{\frac{w\ell_w}{Et_w}} \qquad \text{(Eq. 2.4.22)}$$

where units are those described in Step 2.

$$\Delta_u = \frac{\Gamma S_v T}{2\pi} \qquad \text{(Eq. 3.1.37)}$$

The author's preference is to assume a linear mode shape and mass distribution wherever possible.

$$\Gamma = \frac{3n}{2n+1} \qquad \text{(Eq. 3.1.18a)}$$

Step 4: Determine the Objective Yield Displacement.

$$\Delta_y = \frac{\Delta_u}{\Omega_o \mu} \qquad \text{(Eq. 3.1.38)}$$

Step 5: Determine the Design Flexural Strength (M_d). M_d corresponds to F_o in Figure 1.1.1.

$$M_d = \frac{3.6EI_e\Delta_y}{h_w^2} \qquad \text{(Eq. 3.1.39)}$$

Procedure 3—The Design Requires the Adoption of a Substitute Structure. The effect of structural damping has not been included in the forcing function (response spectrum). An objective target (Δ_T) drift and system ductility (μ_s) have been prescribed. A substitute structure is created (Figure 3.1.2). The adoption of a substitute structure could be especially helpful in dealing with more complex structures as well as those constructed of precast concrete.

Step 1: Adjust the Spectral Velocity to Account for the Inclusion of Structural Damping. The reduction coefficient $\hat{R}$ is developed as follows.

$$S_{v,s} = \hat{R}S_v \qquad \text{(Eq. 3.1.46)}$$

Quantify the structural damping:

$$\zeta_{eq} = \frac{\sqrt{\mu}-1}{\pi\sqrt{\mu}} \qquad \text{(Eq. 3.1.41)}$$

and the total damping:

$$\hat{\zeta}_{\text{eq}} = \zeta_{\text{eq}} + 5 \qquad \text{(Eq. 3.1.42)}$$

$$\hat{R} = \frac{3.38 - 0.67 \ln \hat{\zeta}_{\text{eq}}}{2.3} \qquad \text{(Eq. 3.1.45)}$$

Equation 3.1.45 assumes that the design spectrum is developed for a damping ratio of 5%.

Step 2: Determine the Objective Period (T_s) for the Substitute Structure.

$$T_s = \frac{2\pi}{\hat{R} S_v} \left(\frac{\Delta_T}{\Gamma} \right) \qquad (3.1.52a)$$

and for the real structure (T)

$$T = \frac{T_s}{\sqrt{\mu}} \qquad (3.1.52b)$$

Step 3: Determine the Dynamically Compatible Length of Wall. Use Eq. 3.1.33.

$$\boxed{\ell_w = \left[\frac{0.13}{T} h_w^2 \left(\frac{w}{E t_w} \right)^{0.5} \right]^{0.67}} \qquad \text{(Eq. 3.1.33)}$$

where units are in their most practical form.

ℓ_w, h_w, and t_w are in feet.
T is in seconds.
w is in kips per foot.
E is in kips per square inch.

Use Eq. 2.4.22 to determine the period of the structure if the adopted wall characteristics differ from those developed from Eq. 3.1.33.

Step 4: Calculate the Effective Stiffness of the Substitute Structure.

$$k_{e,s} = \omega_s^2 M_e \qquad (3.1.53)$$

$$= \frac{4\pi^2}{T_s^2} M_e$$

where M_e may be estimated using Blue Book[3.4] values, calculated using Eq. 3.1.17b in conjunction with an assumed mode shape or directly from Eq. 3.1.19a for systems that may reasonably be linearized.

Step 5: Determine the Flexural Strength Objective (F_o in Figure 1.1.1) for the Structure.

$$\Omega_o M_u = k_{e,s} S_{d,s} h_e \tag{3.1.54}$$

where h_e may be estimated using Blue Book[3.4] values, calculated using Eq. 3.1.15 and 3.1.16 and an assumed mode shape, or Eq. 3.1.16 and 3.1.18a for systems that may reasonably be linearized.

3.1.2 Shear Walls of Varying Lengths

The objective of this section is to explore design methodologies that might be used on a system that is a little more complex than the bracing system discussed in the preceding section—shear walls of equivalent stiffness. Even the casual reader of Section 3.1.1 will understand that no two designers are likely to arrive at the same design conclusion. Fortunately, a unique solution is not the objective—only one that satisfies performance objectives. In spite of this, the experimental data of Chapter 2 must be used to develop a solution for more complex problems and determine its appropriateness. After a review of design alternatives a design methodology will be developed, but each designer is encouraged to modify the approach to reflect his or her partiality.

3.1.2.1 Alternative Design Methodologies The design problem is described in Figure 3.1.8, wherein two walls of unequal length are tied together in a manner that assures displacement compatibility.

Current force-based design procedures do not offer satisfactory design solutions to this problem from a seismic performance perspective because they do not consider

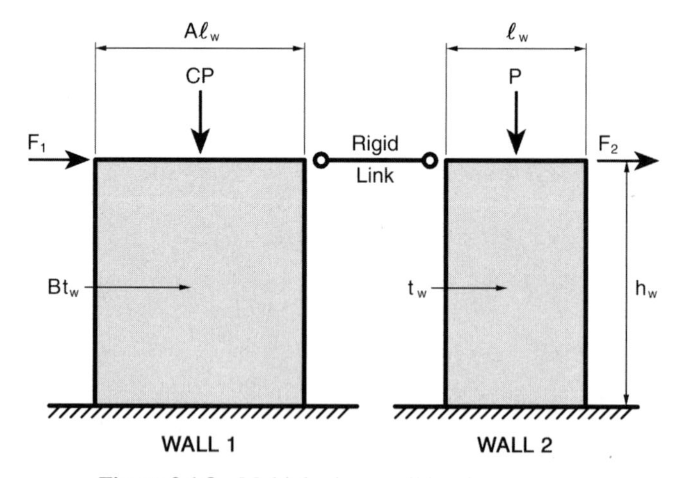

Figure 3.1.8 Multiple shear wall bracing program.

the consequences of postyield behavior. Consider the force-based procedure currently used in the design of shear wall braced buildings were it to be applied to a building braced by the walls shown in Figure 3.1.8. The two walls have different lengths and could have different thicknesses as well as imposed levels of axial load. If we assume that the shear deformation of the wall is small, we would follow elastic mandates and allocate the design forces in direct proportion to the flexural stiffness of each wall.

The distribution of the design load (ΣF) would be in direct proportion to the moment of inertia of each wall; hence

$$I_1 = \frac{Bt_w(A\ell_w)^3}{12} \tag{3.1.55a}$$

$$I_2 = \frac{t_w \ell_w^3}{12} \tag{3.1.55b}$$

and Wall 1 would need to be designed for BA^3 times the design lateral load imposed on Wall 2:

$$F_1 = BA^3 F_2 \qquad \text{(Elastic distribution)} \tag{3.1.55c}$$

The stiffness of the building would be developed from the force ($F_1 + F_2$) required to produce a subyield deformation in the stiffer wall. The strength provided by Wall 2 would, following elastic theory, be based on the subyield drift of Wall 1 ($\Delta_{yi,1}$ in Figure 3.1.9), and this would cause Wall 2 to be very ineffective. The ductility objective for the system would be that developed for Wall 1, and the ductility available in Wall 2 would be significantly more than required. In summary, for any significant variation in wall length, the economic value associated with adding a shorter wall becomes questionable when elastic behavior is a design constraint.

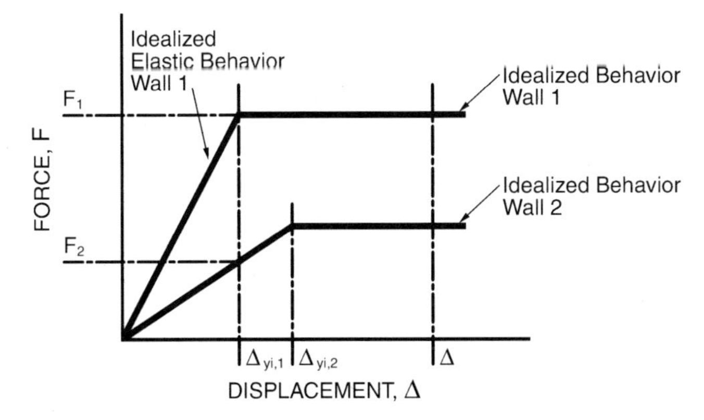

Figure 3.1.9 Force–displacement relationship for the bracing system described in Figure 3.1.8.

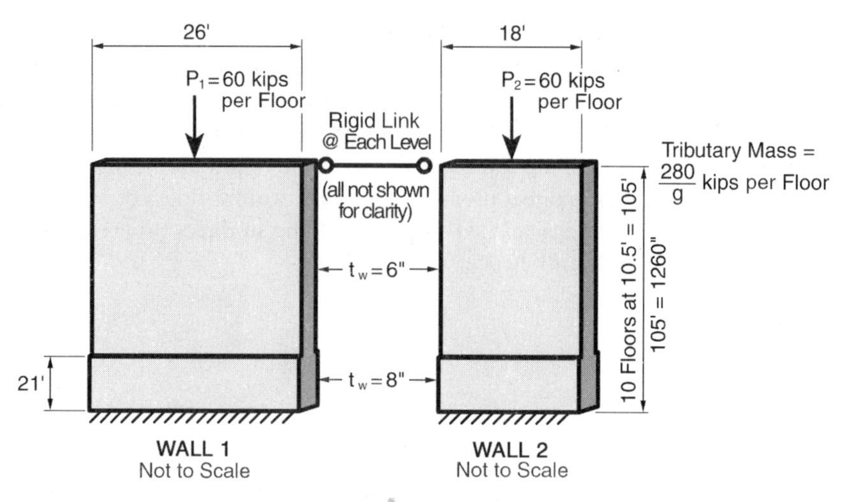

Figure 3.1.10 Multiple shear wall design example.

Consider for example the two walls described in Figure 3.1.10 where A is 1.44 (26/18). The elastic strength ratio is

$$F_1 = BA^3 F_2$$

$$\sum F = \left(1 + \frac{1}{BA^3}\right) F_1$$

$$= \left(1 + \frac{1}{1(1.44)^3}\right) F_1$$

$$= 1.33 F_1$$

$$F_1 = \frac{\sum F}{1.33}$$

$$= 0.75 \sum F \tag{3.1.55d}$$

So, Wall 1 would be designed to carry 75% of the design load and moment, or almost 4 times as much as Wall 2.

Advocates of a displacement-based approach would more effectively use Wall 2 by considering the strength it might provide at the ultimate or target drift. A logical design objective would be to create an equivalent ductility demand on each wall. Since available ductility is generally directly proportional to the percentage or reinforcing provided, the design should have as an objective the development of two walls whose reinforcement ratios are the same. This will be impacted by the level of axial load, but it does represent a reasonable starting point.

The first step in the design process is to evaluate a system whose walls were of equivalent stiffness following the procedures developed in Section 3.1.1. Given the similarities of the proposed design (Figure 3.1.10) and the example design of Section 3.1.1 (Figure 2.4.9), we might reasonably conclude that minimum wall reinforcing was an appropriate starting point in the design process for both walls (1 and 2 of Figure 3.1.10). Based on the strength developed for a 26-ft wall (EBD), the strength objective for the system should be on the order of $0.066Wh_w$ (see Table 3.1.2).

The strength of Wall 1, given the minimum flexural reinforcement objective (0.0025 A_c), is

$$\begin{aligned} a &= \frac{P}{0.85 f_c' t_w} \\ &= \frac{600}{0.85(5)8} \\ &= 17.6 \text{ in.} \\ A_s &= 0.0025(6)(26)(12) \\ &= 4.68 \text{ in.}^2 \end{aligned}$$

Comment: An 8-in. thick wall is provided at the base to facilitate the placement of confinement reinforcing and to satisfy wall thickness objectives in the plastic hinge region. It is otherwise disregarded in the design process.

$$M_n = P\left(\frac{\ell_w}{2} - \frac{a}{2}\right) + A_s f_y (d - d') \qquad \text{(see Eq. 2.2.20a)}$$

$$\begin{aligned} &= 600\left(156 - \frac{17.6}{2}\right) + 4.68(60)[24(12)] \\ &= 88{,}300 + 80{,}900 \\ &= 169{,}200 \text{ in.-kips} \qquad (14{,}000 \text{ ft-kips}) \end{aligned}$$

The strain in the concrete is

$$\phi_y = \frac{0.0033}{\ell_w} \qquad \text{(Eq. 2.4.27)}$$

$$\begin{aligned} &= \frac{0.0033}{26(12)} \\ &= 0.0000106 \text{ rad/in.} \\ \varepsilon_{cy} &= \phi_y kd \\ &= 0.0000106(0.33)(312) \\ &= 0.00109 \text{ in./in.} \end{aligned}$$

$$\Delta_y = \frac{11}{40}\phi_y h_w^2 \tag{3.1.56}$$

$$= 0.275(0.0000106)(1260)^2$$

$$= 4.63 \text{ in.} \qquad (\text{DR} = 0.37\%)$$

If a target drift ratio of 1.8% is adopted, then

$$\Delta_u = \Delta_T$$

$$= 0.018h_w$$

$$= 22.7 \text{ in.}$$

The postyield component of drift becomes (see Figure 2.1.11)

$$\Delta_p = \Delta_u - \Delta_y$$

$$= 22.7 - 4.6$$

$$= 18.1 \text{ in.}$$

$$\theta_p = \frac{\Delta_p}{h_w - \ell_p/2} \qquad \text{(see Figure 2.1.11 and Eq. 2.1.8)}$$

$$= \frac{18.1}{1182}$$

$$= 0.015 \text{ radian}$$

$$\phi_p = \frac{\theta_p}{\ell_p}$$

$$= \frac{0.015}{156}$$

$$= 0.0001 \text{ rad/in.}$$

The curvature ductility demand is

$$\mu_\phi = \frac{\phi_p + \phi_y}{\phi_y}$$

$$= \frac{0.0001 + 0.0000106}{0.0000106}$$

$$= 10.4$$

$$c = \frac{a}{\beta_1}$$

$$= \frac{17.6}{0.8}$$
$$= 22 \text{ in.}$$

The strain imposed on the concrete is

$$\varepsilon_{cp} = \phi_p c$$
$$= 0.0001(22)$$
$$= 0.0022 \text{ in./in.}$$
$$\varepsilon_c = \varepsilon_{cy} + \varepsilon_{cp}$$
$$= 0.00109 + 0.0022$$
$$= 0.0033 \text{ in./in.}$$

The strain imposed on the steel in Wall 1 would be

$$\varepsilon_{sp} = \phi_p(d - c)$$
$$= 0.0001(300 - 22)$$
$$= 0.0278 \text{ in./in.}$$
$$\varepsilon_{su} = \varepsilon_{sp} + \varepsilon_{sy}$$
$$\cong 0.03 \text{ in./in.}$$

A similar analysis performed on Wall 2 (Figure 3.1.10) would suggest the following:

$$\phi_y = \frac{0.0033}{\ell_w} \qquad \text{(Eq. 2.4.27)}$$
$$= \frac{0.0033}{216}$$
$$= 0.0000153 \text{ rad/in.}$$
$$\varepsilon_{cy} = \phi_y kd$$
$$= 0.0000153(0.33)216$$
$$= 0.00109 \text{ in./in.}$$
$$\Delta_y = \frac{11}{40}\phi_y h_w^2 \qquad \text{(Eq. 3.1.56)}$$
$$= 0.275(0.0000153)(1260)^2$$
$$= 6.6 \text{ in.} \qquad (\text{DR} = 0.52\%)$$

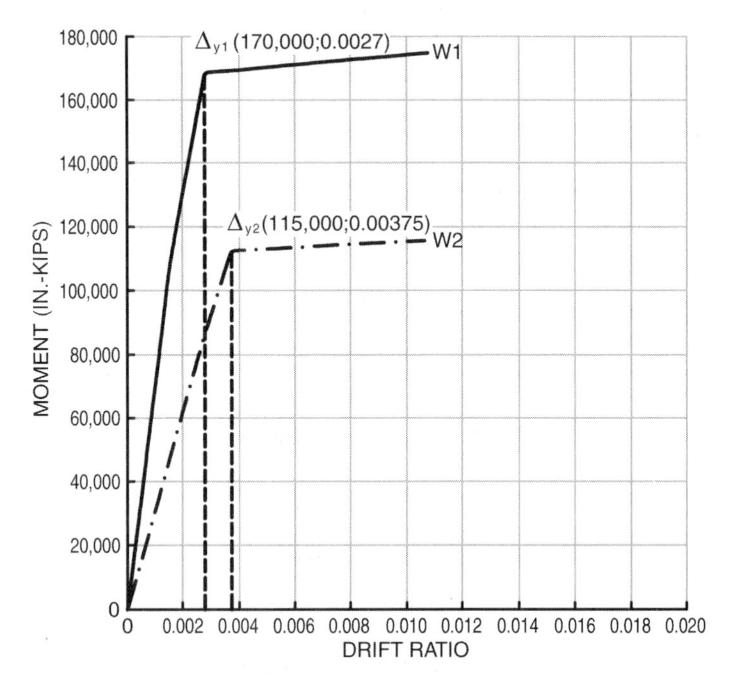

Figure 3.1.11 Pushover analysis of each wall described in Figure 3.1.10—flexural strength corresponding to a reinforcement ratio of 0.0025; flexural stiffness based on $I_w = 0.43I_g$.

Comment: Observe that in this case the yield drift (0.52%) ratio predicted by Equation 3.1.56 is significantly higher than that predicted by the computer (0.375%, Figure 3.1.11). This divergence is a characteristic of lightly reinforced members ($\rho = 0.0025$). The neutral axis depth (kd) is only $0.17d$ as opposed to the $0.33d$ assumed in the development of Equation 2.4.27. The yield curvature (ϕ_y) predicted by each of these alternatives is

$$kd = 0.33d \qquad \phi_y = 0.0000153 \text{ rad/in.}$$

$$kd = 0.17d \qquad \phi_y = 0.0000117 \text{ rad/in.}$$

As discussed in Section 3.1.1 the design must realize that the establishment of yield displacements is fairly subjective (Figure 2.4.4*b*), and this relegates the use of system ductilities to the design process and design acceptance must become a function of induced strain states.

$$\begin{aligned}\Delta_p &= \Delta_u - \Delta_y \\ &= 22.7 - 6.6 \\ &= 16.1 \text{ in.}\end{aligned}$$

$$\theta_p = \frac{\Delta_p}{h_w - \ell_p/2} \qquad \text{(see Figure 2.1.11 and Eq. 2.1.8)}$$

$$= \frac{16.1}{1260 - 54}$$

$$= 0.0133 \text{ radian}$$

$$\phi_p = \frac{\theta_p}{\ell_p}$$

$$= \frac{0.0133}{108}$$

$$= 0.000124 \text{ rad/in.}$$

The curvature ductility demand is

$$\mu_\phi = \frac{\phi_p + \phi_y}{\phi_y}$$

$$= \frac{0.00014}{0.0000153}$$

$$= 9.1$$

Observe that the curvature ductility demand imposed on Wall 2 is slightly less than the 10.4 imposed on Wall 1. The strain imposed on the concrete of Wall 2 is

$$a = \frac{P}{0.85 f_c' t_w}$$

$$= \frac{600}{0.85(5)(8)}$$

$$= 17.6 \text{ in.}$$

$$c = \frac{a}{\beta_1}$$

$$= \frac{17.6}{0.8}$$

$$= 22 \text{ in.}$$

$$\varepsilon_{cp} = \phi_p c$$

$$= 0.000124(22)$$

$$= 0.0027 \text{ in./in.}$$

$$\varepsilon_{cu} = \varepsilon_{cy} + \varepsilon_{cp}$$

$$= 0.00109 + 0.0027$$
$$= 0.0038 \text{ in./in.}$$

Had the yield drift ratio of 0.375% ($\phi_y = 0.0000117$ rad/in.) been adopted the predicted concrete strain state would be 0.0033 in./in.

Observe that the peak concrete strain imposed on Wall 2 is essentially the same as that suggested for Wall 1 (0.0033 in./in.).

The strain imposed on the steel in Wall 2 is

$$\varepsilon_{sp} = \phi_p(d - c)$$
$$= 0.000124(204 - 22)$$
$$= 0.023 \text{ in./in.}$$
$$\varepsilon_{su} = \varepsilon_{sp} + \varepsilon_{sy}$$
$$= 0.025 \text{ in./in.}$$

and this is slightly less than the strain imposed on the steel in Wall 1 (0.03 in./in.).

From a performance perspective, the objective balance seems to have been attained.

The proposed level of flexural strength for Wall 2 is

$$M_n = P\left(\frac{\ell_w}{2} - \frac{a}{2}\right) + A_s f_y (d - d') \qquad \text{(see Eq. 2.2.20a)}$$
$$= 600(108 - 8.8) + 0.0025(6)(18)(12)(60)(16)(12)$$
$$= 59{,}500 + 37{,}300$$
$$= 96{,}800 \text{ in.-kips} \qquad \text{(8100 ft-kips)}$$

The flexural strength or restoring force provided by the two walls described in Figure 3.1.10 is

$$M_1 + M_2 = 169{,}200 + 96{,}800$$
$$= 266{,}000 \text{ in.-kips}$$
$$M_u = \phi\,(M_1 + M_2)$$
$$= 0.9(266{,}000)$$
$$= 239{,}000 \text{ in.-kips}$$

Restated in terms of Wh_w, the objective design strength is

$$\frac{\phi(M_1 + M_2)}{Wh_w} = \frac{239{,}000}{2800(105)(12)}$$
$$= 0.068$$

and this exceeds our objective capacity as developed for a stiffer single wall system [two 26-ft walls in Section 3.1.1 (see Table 3.1.2)].

Two things should be clear:

- Wall 2 can safely carry considerably more shear than an elastic distribution would propose, in this case 36% as opposed to 21%.
- Ductility demands and strain states imposed on both walls are comparable, and induced concrete strain states are within accepted limit states.

An analysis of the proposed design confirms these design conclusions. Figure 3.1.11 summarizes the results of a pushover analysis performed on each wall, while Figure 3.1.12 identifies projected strain states based on a pushover analysis of the wall system described in Figure 3.1.10. A computer-generated spectral based drift projection of 13.5 in. (1.07%) is identified, but this spectral projection is developed from an elastic model of the system.

The system behavior described in Figure 3.1.12*a* combines the two walls and accordingly describes their behavior as being essentially elastic up to the point at

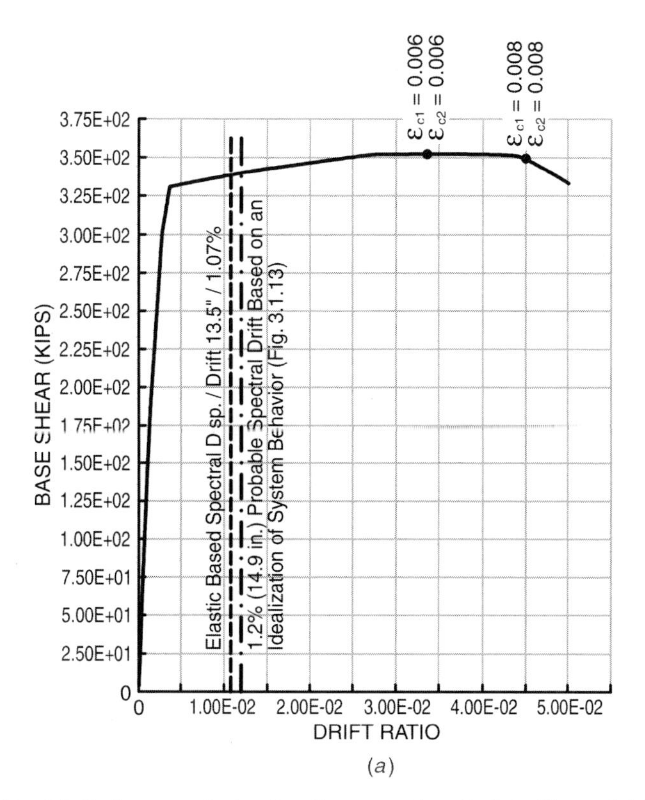

Figure 3.1.12 (*a*) Pushover analysis of wall system described in Figure 3.1.10—flexural strength corresponding to a reinforcement ratio of 0.0025.

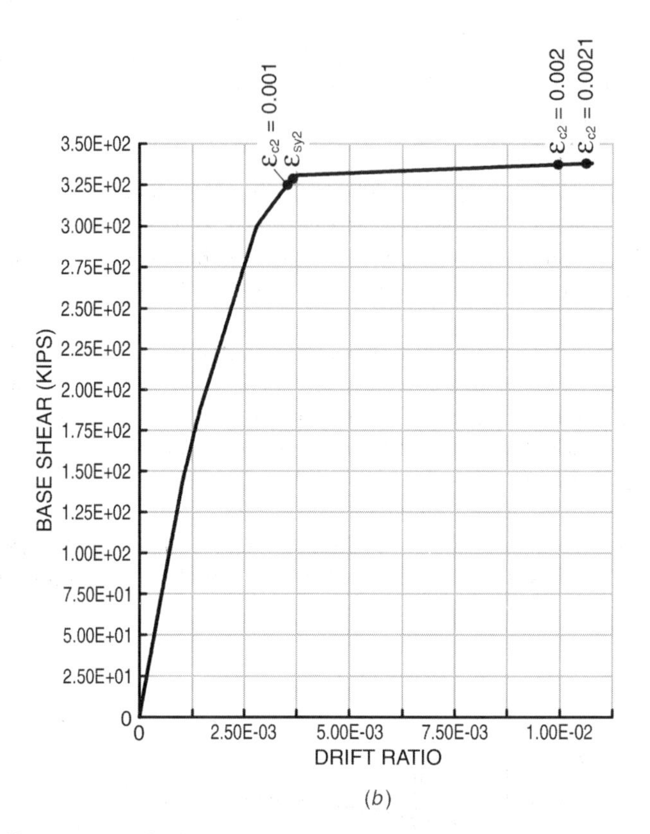

(*b*)

Figure 3.1.12 (*Continued*) (*b*) Pushover results, capacity spectrum method for the wall system of Figure 3.1.10, flexural reinforcement ratio 0.0025.

which Wall 2 yields (DR = 0.00375). This combined system will be softer than the elastic system because Wall 1 will have yielded before the system reaches a drift ratio of 0.375%. From a design perspective it is convenient to develop a comparable idealized elastic system and then use the design procedures developed in Section 3.1.1 to estimate the probable spectral-based drift of the system. Alternatively, one could use the elastic-based computer analysis procedures discussed in this subsection and realize that the spectral response projections must be adjusted to reflect the effective or idealized behavior of the system as described in Figure 3.1.13. Observe that this is no different than the idealization of experimental efforts developed to predict the behavior of a thin wall (see Figure 2.4.2*a* and Figure 2.4.4*b*).

The Design Process The relationship between the elastic and idealized or secant stiffnesses described in Figure 3.1.13 is most easily understood by extrapolating data from the pushover analyses presented in Figures 3.1.11 and 3.1.12, though it may also be extracted from the preceding development of the conceptual design used to predict strain states.

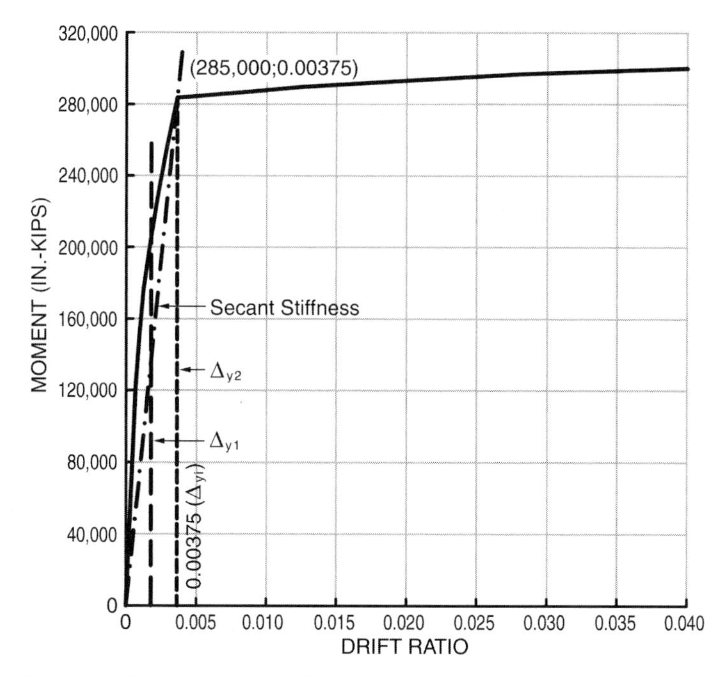

Figure 3.1.13 System model for the shear walls described in Figure 3.1.10.

Consider Figure 3.1.11. The elastic stiffness of the system described in Figure 3.1.10 would be developed from the strength of both walls, provided the selected level of displacement did not exceed Δ_y for Wall 1 (DR $\cong$ 0.0027). At a drift of $0.0027h_w$ the moments induced in the walls would be

$$M_1 \cong 170{,}000 \text{ in.-kips}$$

$$M_2 \cong 85{,}000 \text{ in.-kips}$$

$$\sum M = 255{,}000 \text{ in.-kips}$$

Elastic Stiffness, k_e

$$\Delta_y = 0.0027(1260) \qquad \text{(see Figure 3.1.11, } \Delta_{y1}\text{)}$$

$$= 3.4 \text{ in.}$$

$$k_e = \frac{\sum M}{h_e S_d}$$

$$= \frac{\sum M_{e\ell}\Gamma}{h_e \Delta_y}$$

$$= \frac{255{,}000(1.44)}{875(3.4)}$$

$$= 123.4 \text{ kips/in.}$$

Consider the combined behavior described in Figure 3.1.12*b*. The base shear at an idealized yield drift ratio of 0.00375 is 330 kips. If we assume that the force or base shear operates at an effective height of h_w/Γ or 875 in. ($\Gamma = 1.44$), the stiffness of the elastic system and the inelastic or idealized systems can be compared.

Idealized or Secant Stiffness, k_s

$$\Delta_y = 0.00375(1260) \qquad \text{(see Figure 3.1.12}b\text{)}$$

$$= 4.7 \text{ in.}$$

$$k_s = \frac{V}{S_d}$$

$$= \frac{V\Gamma}{\Delta_y}$$

$$= \frac{330(1.44)}{4.7}$$

$$= 101 \text{ kips/in.}$$

A similar conclusion could be extrapolated from the behavior of each wall as described in Figure 3.1.11

$$M_{y1} = 170{,}000 \text{ in.-kips}$$

$$M_{y2} = 115{,}000 \text{ in.-kips}$$

$$\sum M_y = 285{,}000 \text{ in.-kips}$$

$$k_y = \frac{\sum M_y \Gamma}{h_e \Delta_y}$$

$$= \frac{285{,}000(144)}{875(4.7)}$$

$$= 100 \text{ kips/in.}$$

The ratio of stiffnesses is

$$\frac{k_s}{k_e} = \frac{101}{123.4}$$

$$= 0.82$$

and this would cause the period of the inelastic structure to be

$$T_i = \sqrt{\frac{k_e}{k_s}} T_e \tag{3.1.57}$$

$$= \sqrt{\frac{1}{0.82}} T_e$$

$$= 1.1 T_e$$

The anticipated spectral drift of the system described in Figure 3.1.10 would be 10% higher than the elastic projection, or about 1.2% (Figure 3.1.12*a*)

Comment: Since the period of a system is inversely proportional to the square root of the change in stiffness ($\sqrt{k_2/k_1}$), as shown in Eq. 3.1.57, the designer may easily consider the consequences of alternative assumptions.

Suppose, for example, that you wish to estimate the probable inelastic or idealized response of the system described in Figure 3.1.10 and you have been provided the computer output that identifies the elastic period of the system (T_e). Previously developed data are summarized as

$$M_{y1} = 169{,}000 \text{ in.-kips}$$

$$\Delta_{y1} = 4.76 \text{ in.}$$

$$M_{y2} = 96{,}800 \text{ in.-kips}$$

$$\Delta_{y2} = 6.6 \text{ in.} \qquad (\text{DR} = 0.52\%)$$

The stiffness of the system in the elastic behavior range may for comparison purposes be developed as

$$M_2 = M_{y2}\left(\frac{\Delta_{y1}}{\Delta_{y2}}\right)$$

$$= 96{,}800\left(\frac{4.76}{6.6}\right)$$

$$= 69{,}800 \text{ in.-kips}$$

$$k_e^* = \frac{\sum M_{\text{el}}}{\Delta_{y1}}$$

where the superscript * identifies the stiffness as being relative to a similar development. Thus

$$k_e^* = \frac{169{,}000 + 69{,}800}{4.76}$$

$$= 50{,}170 \text{ in.-kips/in.}$$

The stiffness of the idealized (secant) system (Figure 3.1.12*b*) is

$$k_s^* = \frac{\sum M_y}{\Delta_{y2}}$$

$$= \frac{169{,}000 + 96{,}800}{6.6}$$

$$= 40{,}300 \text{ in.-kips/in.}$$

The stiffness ratio is

$$\sqrt{\frac{k_e}{k_s}} = \sqrt{\frac{50{,}170}{40{,}300}}$$

$$= 1.1$$

Therefore,

$$T_i = 1.1T_e$$

Comment: Observe that this system development and its component walls are more flexible than that developed by the computer model. This, as previously discussed, is a direct consequence of the selection of Δ_y (Eq. 2.4.27 and 3.1.56) and the modeling process.

The period of this idealized system, based upon its secant stiffness and a system yield drift of 6.6 in. can be developed as follows. A linear mode shape is adopted.

$$\Gamma = \frac{3n}{2n + 1} \quad \text{(Eq. 3.1.18a)}$$

$$= 1.43$$

The effective mass (M_e) is

$$M_e = \frac{1.5(n + 1)}{2n + 1}M \quad \text{(Eq. 3.1.19a)}$$

$$= 0.79\left(\frac{2800}{386.4}\right)$$

$$= 5.72 \text{ kip-sec}^2/\text{in.}$$

$$k_i = \frac{\sum M_y}{h_e S_d}$$

$$= \frac{\sum M_y}{h_w \Delta_y} \Gamma^2 \qquad \text{(see Eq. 3.1.16a and Eq. 3.1.10)}$$

$$= \frac{(169{,}000 + 96{,}800)(1.43)^2}{1260(6.6)}$$

$$= 65.4 \text{ kips/in.}$$

$$\omega_n = \sqrt{\frac{k_i}{M_e}}$$

$$= \sqrt{\frac{65.4}{5.72}}$$

$$= 3.38 \text{ rad/sec}$$

$$T = \frac{2\pi}{\omega}$$

$$= 1.86 \text{ seconds}$$

Reconciliation with Computer Model The computer model is elastic and, as a consequence, would be reduced to a single-degree-of-freedom system following the model of Figure 3.1.1. The computer period was reported as 1.09 seconds. This can be developed using the stiffness extracted from the pushover analysis.

$$M_{yi} = 255{,}000 \text{ in.-kips} \qquad \text{(see Figure 3.1.11)}$$

$$\Delta_{yi} = 3.4 \text{ in.} \quad (\text{DR} = 0.0027) \qquad \text{(see Figure 3.1.11)}$$

$$M = \frac{W}{g}$$

$$= \frac{2800}{386.4}$$

$$= 7.25 \text{ kip-sec}^2/\text{in.}$$

$$M_e = 0.613M \qquad \text{(see Figure 3.1.1)}$$

$$= 4.44 \text{ kip-sec}^2/\text{in.}$$

$$h_e = 0.726h_w \qquad \text{(see Figure 3.1.1)}$$

$$= 915 \text{ in.}$$

$$\Delta = k_1 k_2 \Delta_y \qquad \text{(see Eq. 3.1.14b)}$$

$$= 0.726(0.81)(3.4) \qquad \text{(see Figure 3.1.1}a\text{ and Blue Book}^{[3.4]}\text{)}$$

$$= 2 \text{ in.}$$

$$k_e = \frac{M_{yi}}{h_e \Delta}$$

$$= \frac{255{,}000}{915(2)}$$

$$= 139.4 \text{ kips/in.}$$

$$\omega = \sqrt{\frac{k_e}{M_e}}$$

$$= \sqrt{\frac{139.4}{4.44}}$$

$$= 5.6 \text{ rad/sec}$$

$$T = 1.12 \text{ seconds} \cong 1.09 \text{ seconds}$$

$$\Delta_u = \frac{T}{2\pi} S_v \Gamma$$

$$= \frac{1.12}{6.28}(49)(1.6) \qquad (S_v = 49 \text{ in./sec at } T \cong 1 \text{ second})$$

$$= 14.0 \text{ in.} \cong 13.5 \text{ in.}$$

Impact on Performance (If $T = 1.86$ seconds) This period might reasonably describe an inelastic system whose stiffness is developed from a secant model similar to that portrayed in Figure 3.1.13.

$$\Delta_u = \frac{T}{2\pi} S_v \Gamma$$

$$= \frac{1.86}{6.28}(49)1.43$$

$$= 20.8 \text{ in.} \qquad (\text{DR} = 0.0165)$$

and this is still less than the target drift of 22.7 in.

Conclusion: Estimates of maximum drift will be sensitive to the adopted model. Selected stiffness estimates should be based on experimental conclusions (see Figure

2.4.4*a*). Modeling considerations should be parametrically studied to develop a range from which to make a design judgement.

Concluding Comments: From a design perspective it is reasonable to presume that a designer would not opt to use walls of different lengths. Obviously the design engineer would not select the wall system described in Figure 3.1.10, especially if he or she knew that either two 26-ft long walls or two 18-ft walls would adequately brace the building. Shear wall or wall locations are based on functional requirements, aesthetic considerations, or construction practicalities. The wall system described in Figure 3.1.10 is not that uncommon, however, and it may result from the fact that units on either side of the corridor are of a different size or configuration—or that mechanical ventilation or plumbing considerations require that one wall be shorter than the other. Suffice it to say that walls of differing lengths will be a condition often imposed on a design and their lengths will be known. System characteristics will then become dependent on objectives of the design.

3.1.2.2 Suggested Design Approach For walls of different lengths or configurations, the most convenient design approach will be to idealize system behavior (Figure 3.1.2 and Figure 3.1.13) and proceed to use any of the design procedures developed in Section 3.1.1. This may be accomplished as follows.

Step 1: Establish a Strength Relationship between the Two Walls. The designer, understanding the lengths of the two walls, must arbitrarily select the desired relationship between the two walls in terms of strength. Alternatives include selecting a common reinforcement ratio. In the example development of the system described in Figure 3.1.10, a flexural reinforcement ratio of 0.25% was used for each wall. This resulted in a strength ratio of about 66% ($M_{y2} = 0.66M_{y1}$—see Figure 3.1.11).

Step 2: Determine the System Deflection at Yield (Δ_y). This will be the yield deflection of the more flexible wall—Wall 2 in this case. Δ_{yi} is developed through the use of Eq. 2.4.17 and Eq. 2.4.27.

$$\Delta_{y2} = \frac{11}{40}\phi_y h_w^2 \qquad \text{(Eq. 3.1.56)}$$

$$= 0.275\left(\frac{0.0033}{\ell_w}\right)h_w^2 \qquad \text{(see Eq. 2.4.27)}$$

$$= (0.00091)\frac{h_w^2}{\ell_w}$$

$$= 6.7 \text{ in.} \qquad (\text{DR} = 0.0053)$$

Comment: Figure 3.1.11 identifies the yield displacement of Wall 2 as being 0.37% as opposed to 0.53%. A design decision is required. Consider the behavior of the test specimen of Figure 2.4.3 as described in Figure 2.4.4*a*. Both procedures have a logical basis as previously discussed.

Conclusion: Select the larger estimate of deflection ($DR_y = 0.0053$) if the analysis is to be deflection based and the lesser ($DR_y = 0.0037$) if the design is to be strength based.

Step 3: Develop an Equivalent Single-Degree-of-Freedom Model (Figure 3.1.1). This requires the development of an effective mass (M_e) and an effective height (h_e). A variety of procedures for developing M_e and h_e were presented in Section 3.1.1 (see Eq. 3.1.16 and 3.1.19a). For this example, we adopt the following values

$$\begin{aligned} M_e &= 0.78M \\ &= \frac{0.78(2800)}{386.4} \\ &= 5.65 \text{ kip-sec}^2/\text{in.} \\ h_e &= 0.69h_w \\ &= 0.69(1260) \\ &= 869 \text{ in.} \end{aligned}$$

Step 4: Develop the Effective Stiffness (k_e) of the System (Figure 3.1.2).

$$\begin{aligned} \Delta_{yi} &= \frac{h_e}{h_w}\Delta_{y2} \qquad \text{(Single degree of freedom system)} \\ &= 0.69(6.7) \\ &= 4.6 \text{ in.} \end{aligned}$$

$$H_{yi} = \frac{M_{yi}}{h_e} \qquad \text{(Figure 3.1.1}a\text{)}$$

where M_{yi} is the idealized flexural strength of the system (Figure 3.1.13). Thus

$$\begin{aligned} H_{yi} &= \frac{285{,}000}{869} \\ &= 328 \text{ kips} \\ k_s &= \frac{H_{yi}}{\Delta_{yi}} \\ &= \frac{328}{4.6} \\ &= 71 \text{ kips/in.} \end{aligned}$$

Step 5: Determine the Period of the System.

$$\omega = \sqrt{\frac{k_i}{M_e}}$$

$$= \sqrt{\frac{71}{5.65}}$$

$$= 3.5 \text{ rad/sec}$$

$$T = \frac{2\pi}{\omega}$$

$$= 1.78 \text{ seconds}$$

Step 6: Using the Equal Displacement-Based Assumption, Find the Ultimate Spectral Drift of the System.

$$S_d = \frac{S_v}{\omega}$$

$$= \frac{52}{3.5}$$

$$= 14.9 \text{ in.}$$

where S_v is the design objective spectral velocity (52 in./sec) in the 1.8-second range (Figure 4.1.1).

This corresponds to a system ductility of 3.2 (14.9/4.6), which is below our objective system ductility of 4. The probable drift at the top of the wall is

$$\Delta_n = \Gamma S_d$$

$$= 1.44(14.9)$$

$$= 21.4 \text{ in.} \qquad (\text{DR} = 0.017)$$

and this, as should be expected, is higher than the elastic- or secant-based conclusion identified in Figure 3.1.12*a*. The target drift is 22.7 in. Accordingly, the system should perform to our expectations.

Step 7: Examine the Strain State at a Drift of 21.4 in. (Anticipated Range) in the Longer Wall.

$$\Delta_u = 21.4 \text{ in.}$$

$$\phi_y = \frac{0.0033}{312} \qquad \text{(Eq. 2.4.27)}$$

$$= 0.00001 \text{ rad/in.}$$

$$\Delta_y = \frac{11}{40}\phi_y h_w^2 \qquad \text{(Eq. 3.1.56)}$$

$$= 4.6 \text{ in.}$$

$$\Delta_p = \Delta_u - \Delta_y$$

$$= 21.4 - 4.6$$

$$= 16.8 \text{ in.}$$

$$\theta_p = \frac{\Delta_p}{h_w - \ell_w/4} \qquad \text{(see Figure 2.1.11 and Eq. 2.1.8)}$$

$$= \frac{16.8}{1260 - 78}$$

$$= 0.0142 \text{ radian}$$

$$\phi_p = \frac{\theta_p}{\ell_p} \qquad \text{(see Eq. 2.1.9)}$$

$$= \frac{0.0142}{156}$$

$$= 0.000091 \text{ rad/in.}$$

At this level of drift the strain in the steel will be on the order of 0.02 in./in. and it is reasonable to include an allowance for overstrength in the tension reinforcing (Figure 2.1.54) but not in the compression reinforcing.

$$C_c = P + (\lambda_o - 1)A_s f_y$$

$$= 600 + 0.25(4.68)(60)$$

$$= 670 \text{ kips}$$

$$a = \frac{C_c}{0.85 f_c' t_w}$$

$$= \frac{670}{0.85 f_c' t_w}$$

$$= \frac{670}{0.85(5)(8)}$$

$$= 19.8 \text{ in.}$$

$$c = \frac{a}{\beta}$$

$$
\begin{aligned}
&= \frac{19.8}{0.8} \\
&= 25 \text{ in.} \\
\varepsilon_{sp} &= \phi_p(d - c) \\
&= 0.000091(275) \\
&= 0.025 \text{ in./in.} \\
\varepsilon_s &= 0.027 \text{ in./in.} \\
\varepsilon_{cp} &= \phi_p c \\
&= 0.000091(25) \\
&= 0.0023 \text{ in./in.} \\
\varepsilon_c &= \varepsilon_{cy} + \varepsilon_{cp} \\
&= 0.0033 \text{ in./in.}
\end{aligned}
$$

Conclusion: In spite of the lack of apparent exactness in any of the design procedures, it seems clear that the proposed wall configuration (Figure 3.1.10) is reasonable and consistent with performance-based design objectives. The sole purpose of the design process is to make this decision.

3.1.3 Coupled Shear Walls—Design Confirmation

Section 2.4.2.4 developed the conceptual design of a coupled shear wall (Figure 2.4.31). Many assumptions were required in order to produce what appeared to be a reasonable bracing program. It should be clear that the behavior of the coupled wall system will be difficult to predict using elastic-based procedures, and this is because many, if not all, of the components will yield before the strength of the system is reached. Accordingly, the design procedures summarized in Section 3.1.1.4 should offer more insight into system behavior and allow the designer to decide if the system is appropriate. The focus of this section will be to confirm the conceptual design proposed for the coupled shear walls described in Figure 2.4.31.

Either a force- or a displacement-based approach might be used, but the behavior of the system must consider the fact that postyield behavior will be imposed on many of the components of the bracing system before it reaches a displacement that reasonably describes idealized yield. Accordingly, a displacement-based approach is best suited for the task at hand.

Step 1: Determine the Moment Required to Yield the System. Our objective is to create an idealized behavior representation of the system similar to that described in Figure 3.1.13. The idealized moment capacity of the system will be reached when all of the coupling beams yield. The tension compression couple associated with coupling beam yield (V_p) can be developed from Figure 2.4.32. For design purposes assume that all coupling beams will be steel beams (W18 × 130), and that the wall reinforcement will be as shown on Figure 2.4.33.

In Section 2.4.2.4 it was demonstrated that a W18 × 130 spanning 5 ft would yield in shear. Hence

$$V_p = 0.6F_y t_w \left(d - 2t_f\right) \qquad \text{(see Eq. 2.4.47b)}$$

$$= 339 \text{ kips}$$

$$\sum V_p = nV_p$$

$$= 15(339)$$

$$= 5085 \text{ kips}$$

The axial load imposed on the base of each pier is the dead load P_D plus the weight of the pier itself.

$$P_D = P_D^1 + \text{Pier weight}$$

$$= 15(40) + 1260 \qquad \text{(see Figure 2.4.31)}$$

$$= 1860 \text{ kips}$$

To this must be added the coupling beam shear.

The tensile strength of each pier (T_o) is

$$T_o = A_{st} f_y \qquad \text{(see Figure 2.4.33 and 2.2.2)}$$

$$= 2(18)(1.56)(60)$$

$$= 3370 \text{ kips}$$

At system yield the flexural strength provided by the tensile side pier will be minimal and in this case it may be neglected. To estimate the flexural strength provided by the tension side pier, consider the interaction diagram of Figure 2.2.2. If the applied axial tension load (P_T) is T_o, the associated flexural strength is zero. Observe that the variation in the level of provided flexural strength between $P = 0$ and T_o is linear. In the example T_o is 3370 kips, while the applied load, P_T, is

$$P_T = \sum V_p - P_D = 5085 - 1860$$

$$= 3225 \text{ kips} \qquad \text{(Tension)}$$

The nominal moment capacity of the pier, absent an applied axial load (M_o), is

$$M_o = A_s f_y (d - d')$$

$$= 18(1.56)(60)(27)$$

$$= 45{,}500 \text{ ft-kips}$$

The flexural strength of the tensile side pier is

$$M_{yT} = \left(1 - \frac{P_T}{P_o}\right) M_o$$

$$= \left(1 - \frac{3225}{3370}\right) 45{,}500$$

$$= 2000 \text{ ft-kips}$$

and accordingly negligible.

The compression side pier will yield once it reaches the flexural strength suggested by Eq. 2.2.3, which is

$$a = \frac{P_D + \sum V_p}{0.85 f_c' t_w}$$

$$= \frac{1860 + 5085}{0.85(5)(16)}$$

$$= 102 \text{ in.}$$

$$M_{yC} = A_s f_y (d - d') + \left(P_D + \sum V_p\right)\left(\frac{h}{2} - \frac{a}{2}\right) \quad \text{(see Eq. 2.2.3)}$$

$$= 18(1.56)(60)(27) + 6945(15 - 4.25)$$

$$= 45{,}500 + 74{,}500$$

$$= 120{,}000 \text{ ft-kips}$$

When the last coupling beam yields, the total moment resisted by the system will be

$$M_{yi} = M_{yT} + M_{yC} + \sum V_p(\ell_w + \ell_b) \tag{3.1.58}$$

$$= 2{,}000 + 120{,}000 + 5085(35)$$

$$= 300{,}000 \text{ ft-kips}$$

The deflection of the system at idealized system yield will usually coincide with yielding of the compression side pier.

$$\phi_y = \frac{0.0033}{\ell_w} \quad \text{(Eq. 2.4.27)}$$

$$= \frac{0.0033}{30(12)}$$

$$= 0.0000092 \text{ rad/in.}$$

$$\Delta_{yi} = \frac{11}{40}\phi_y h_w^2 \qquad \text{(Eq. 3.1.56)}$$

$$= 0.275(0.0000092)(210)^2(144)$$

$$= 16 \text{ in.} \qquad (0.64\%)$$

The idealized yield point ($\Delta_{yi} = 0.0064h_w$) and idealized base shear are shown on Figure 3.1.14*a*, which also describes the results of a sequential yield or pushover analysis of the coupled wall system.

$$V_{yi} = \frac{M_{yi}}{h_e}$$

$$= \frac{M_{yi}\Gamma}{h_w}$$

$$= \frac{300{,}000(1.45)}{210}$$

$$= 2075 \text{ kips}$$

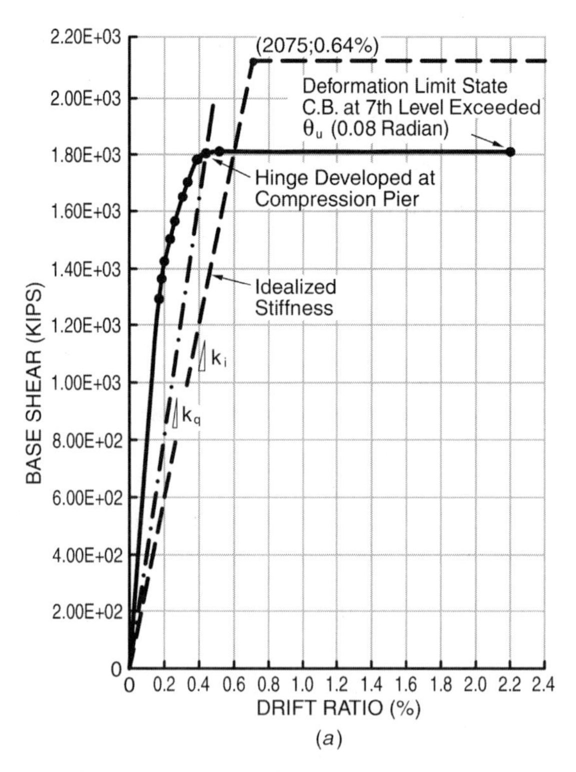

Figure 3.1.14 (*a*) Pushover results, capacity spectrum method—coupled wall system of Figures 2.4.31 and 2.4.33—W18 × 130 coupling beams.

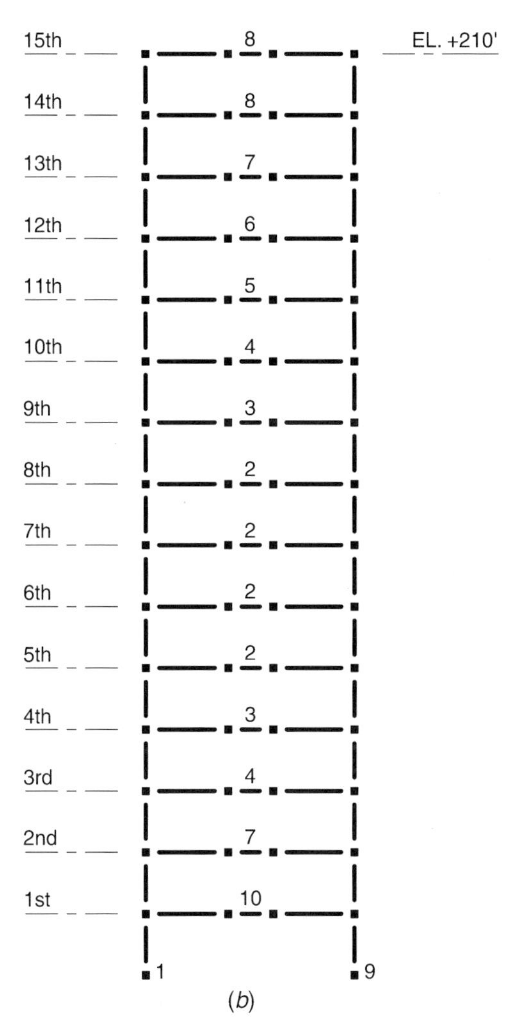

Figure 3.1.14 (*Continued*) (*b*) Sequence of hinge formation—coupled wall system whose behavior is described in Figure 3.1.14*a*.

The projected idealized yield deflection (Δ_{yi}) and strength (V_{yi}) are somewhat higher than that suggested by the sequential yield analysis. Nevertheless for a complex system such as this one, the estimate is accurate enough for conceptual design purposes. The sequence of hinge formation is described in Figure 3.1.14*b*. Observe that hinge formation starts in the middle levels of the system and then moves outward. This is caused by the tensile straining of the tension side pier, for this tends to relieve the deformation demand on the upper level coupling beams (see Figure 2.4.30).

In this case the coupling beams are almost all activated before the compression side pier yields, and this should provide significant energy dissipation. The coupling beam at Level 7 yields first but it does not exceed its deformation limit state (0.09 radian) until the story drift reaches 2.2% (Figure 3.1.14*a*).

In Section 2.4.2.4 a methodology for extending component characteristics (a single pier) into that of the coupled system (Figure 2.4.32) was proposed. It was suggested that the period of the coupled system could be found by considering the increase in system flexural strength provided by the coupling beams (see Eq. 2.4.67 through 2.4.70), and, accordingly, that the period of the coupled wall system would be

$$T_{cw} = \frac{T_w}{\sqrt{A}} \qquad \text{(Eq. 2.4.71)}$$

where

$$A = 1 + \frac{M'}{M_{yi}} \qquad \text{(Eq. 2.4.70)}$$

M' is the flexural strength provided to the system by the coupling beams (Eq. 2.4.67).
M_{yi} is the idealized strength of the piers.
T_w is the period of the uncoupled system.
T_{cw} is the period of the coupled system.

The flexural strength provided by the coupling beam (M') is

$$\begin{aligned} M' &= \sum V_p(\ell_w + \ell_b) \qquad \text{(Eq. 2.4.67)} \\ &= 5085(30 + 5) \\ &= 180{,}000 \text{ ft-kips} \end{aligned}$$

The strength provided by the piers is 122,000 ft-kips $\left(M_{yT} + M_{yC}\right)$.

$$\begin{aligned} A &= 1 + \frac{M'}{M_{yi}} \qquad \text{(Eq. 2.4.70)} \\ &= 1 + \frac{180{,}000}{122{,}000} \\ &= 2.48 \end{aligned}$$

The period (T_w) of a 30-ft pier that provided lateral support for half of the mass tributary to the coupled wall system described in Figure 2.4.31 was 2.81 seconds (see Section 2.4.2.4). This period ($T = 2.81$ seconds) must be adjusted to reflect the actual condition, and this is most easily accomplished in stages. First, realize that the tension pier provides little or no strength ($M_{yT} \cong 0$). Absent the contribution of the coupling beams the mass that must be supported by the acting compression side pier would be doubled. As a consequence, the period of the acting ($k_{Tw} = 0$) uncoupled system (T_{Cw}) would be

$$T_{Cw} = \sqrt{2}T_w$$
$$= 3.91 \text{ seconds}$$

Next the coupling effect must be considered.

The period of the coupled wall system (T_{cw}) is

$$T_{cw} = \frac{T_{Cw}}{\sqrt{A}} \qquad \text{(Eq. 2.4.71)}$$
$$= \frac{3.91}{\sqrt{2.48}}$$
$$= 2.48 \text{ seconds}$$

Step 2: Create a Single-Degree-of-Freedom Model. The development of a single-degree-of-freedom model is somewhat speculative, especially in the conceptual design phase of a complex system. The simplest approach is to adopt a linear displacement shape.

$$\Gamma = \frac{3n}{2n+1} \qquad \text{(Eq. 3.1.18a)}$$
$$= \frac{3(15)}{2(15)+1}$$
$$= 1.45$$

$$h_e = \frac{h_w}{\Gamma} \qquad (0.69h_w) \qquad \text{(Eq. 3.1.16)}$$
$$= \frac{210}{1.45}$$
$$= 144.7 \text{ ft}$$

$$M_e = \frac{1.5(n+1)}{2n+1}M \qquad \text{(Eq. 3.1.19a)}$$
$$= \frac{1.5(15+1)}{2(15)+1}M$$
$$= 0.77M$$
$$= \frac{0.77W}{g}$$

$$= \frac{0.77(15)(1000)}{g}$$

$$= 29.9 \text{ kips-sec}^2/\text{in.}$$

Comment: Table 3.1.4*a* describes the normalized shape of the structure as developed from the displaced shape extracted from the sequential yield analysis (Figure 3.1.14*a*) at compression pier yield ($\mu = 1$). Tables 3.1.4*b* and *c* describe normalized mode shapes given displacement ductilities of 2 and 4. Observe that there is little change in the mode shape as the system advances into the postyield range. The shape is not linear; for example, at the fifth floor a linear normalization of the modal displacement would be 0.33, not 0.283. As the system moves from a ductility of 1 to 2 to 4, the normalized displacement of the fifth node goes from 0.283 to 0.284 to 0.292. Regardless, both of our key parameters (Γ and M_e) remain fairly stable, and we might reasonably adopt a participation factor (Γ) of 1.45 and an effective mass factor (k_3) of 0.75.

TABLE 3.1.4*a*　Elastic Behavior (Extracted from Sequential Yield Analysis—Figure 3.1.14*a*, DR = 0.45%)

Story	Normalized Elastic Mode Shape $\phi_i = \phi_e$	ϕ_i^2	m_i	$m_i\phi_i$	$m_i\phi_i^2$	h_i	$h_i m_i \phi_i$
15	1.000	1.00	1	1.000	1.000	210	210.00
14	0.923	0.852	1	0.923	0.852	196	180.95
13	0.862	0.743	1	0.862	0.743	182	156.85
12	0.787	0.619	1	0.787	0.619	168	132.22
11	0.720	0.518	1	0.720	0.518	154	110.83
10	0.646	0.417	1	0.646	0.417	140	90.42
9	0.578	0.334	1	0.578	0.334	126	72.22
8	0.500	0.250	1	0.500	0.250	112	56.03
7	0.425	0.181	1	0.425	0.181	98	41.65
6	0.353	0.125	1	0.353	0.125	84	29.66
5	0.283	0.080	1	0.283	0.080	70	19.80
4	0.214	0.046	1	0.214	0.046	56	11.98
3	0.148	0.022	1	0.148	0.022	42	6.20
2	0.089	0.008	1	0.089	0.008	28	2.48
1	0.039	0.001	1	0.039	0.001	14	0.54
Σ	7.565	5.196	15.0	7.565	5.196		1122.38
Participation factor		Γ	1.46				
Effective mass factor		k_3	0.76				
Effective mass (kip-sec²/in.)		M_e	29.5				

TABLE 3.1.4*b* Inelastic Behavior $\mu = 2$ (DR = 0.9%—Figure 3.1.14*a*)

Story	Yield Displacement Δ_{yi} (in.)	Plastic Displacement Δ_{pi} (in.)	Inelastic Spectral Displacement (in.)	Inelastic Mode Shape $\phi_{i=\text{in.}}$	ϕ_i^2	$m_i\phi_i$	$m_i\phi_i^2$	$h_i m_i \phi_i$
15	10.10	10.10	20.19	1.000	1.000	1.000	1.000	210.00
14	9.32	9.37	18.69	0.926	0.857	0.926	0.857	181.48
13	8.70	8.65	17.35	0.859	0.739	0.859	0.739	156.42
12	7.95	7.93	15.88	0.786	0.618	0.786	0.618	132.11
11	7.27	7.21	14.48	0.717	0.514	0.717	0.514	110.41
10	6.52	6.49	13.01	0.644	0.415	0.644	0.415	90.21
9	5.83	5.77	11.60	0.574	0.330	0.574	0.330	72.38
8	5.05	5.05	10.10	0.500	0.250	0.500	0.250	56.01
7	4.29	4.33	8.62	0.427	0.182	0.427	0.182	41.82
6	3.57	3.61	7.17	0.355	0.126	0.355	0.126	29.83
5	2.86	2.88	5.74	0.284	0.081	0.284	0.081	19.90
4	2.16	2.16	4.32	0.214	0.046	0.214	0.046	11.99
3	1.49	1.44	2.93	0.145	0.021	0.145	0.021	6.10
2	0.90	0.72	1.62	0.080	0.006	0.080	0.006	2.24
1	0.39	0.00	0.39	0.019	0.000	0.019	0.000	0.27
Σ				7.533	5.187	7.533	5.187	1121.9
Participation factor			Γ	1.45				
Effective mass factor			k_3	0.75				
Effective mass (kip-sec²/in.)			M_e	29.0				

TABLE 3.1.4*c* Inelastic Behavior $\mu = 4$ (DR = 1.8%—Figure 3.1.14*a*)

Story	Yield Displacement Δ_{yi} (in.)	Plastic Displacement Δ_{pi} (in.)	Inelastic Spectral Displacement (in.)	Inelastic Mode Shape ϕ_i	ϕ_i^2	$m_i\phi_i$	$m_i\phi_i^2$	$h_i m_i \phi_i$
15	10.07	30.20	40.26	1.000	1.000	1.000	1.000	210.00
14	9.42	28.04	37.46	0.930	0.866	0.930	0.866	182.36
13	8.70	25.88	34.58	0.859	0.738	0.859	0.738	156.31
12	8.01	23.72	31.73	0.788	0.621	0.788	0.621	132.40
11	7.33	21.57	28.90	0.718	0.515	0.718	0.515	110.53
10	6.63	19.41	26.04	0.647	0.418	0.647	0.418	90.56
9	5.93	17.25	23.18	0.576	0.332	0.576	0.332	72.56
8	5.21	15.10	20.31	0.504	0.254	0.504	0.254	56.49
7	4.50	12.94	17.44	0.433	0.188	0.433	0.188	42.46
6	3.80	10.78	14.58	0.362	0.131	0.362	0.131	30.42
5	3.11	8.63	11.74	0.292	0.085	0.292	0.085	20.41
4	2.42	6.47	8.89	0.221	0.049	0.221	0.049	12.36
3	1.76	4.31	6.07	0.151	0.023	0.151	0.023	6.33

continued

TABLE 3.1.4*c* Inelastic Behavior μ = 4 (DR = 1.8%—Figure 3.1.14*a*) *(Continued)*

Story	Yield Displacement Δ_{yi} (in.)	Plastic Displacement Δ_{pi} (in.)	Inelastic Spectral Displacement (in.)	Inelastic Mode Shape ϕ_i	ϕ_i^2	$m_i\phi_i$	$m_i\phi_i^2$	$h_i m_i \phi_i$
2	1.12	2.16	3.28	0.081	0.007	0.081	0.007	2.28
1	0.53	0.00	0.53	0.013	0.000	0.013	0.000	0.18
Σ				7.575	5.226	7.575	5.226	1125.65
Participation factor			Γ	1.45				
Effective mass factor			k_3	0.75				
Effective mass (kip-sec²/in.)			M_e	29.0				

Our projected effective mass moment is

$$h_e M_e = \frac{h_w}{\Gamma} M_e$$

$$= 0.69(0.75) h_w M$$

$$= 0.52 h_w M$$

which is essentially the same conclusion as that reached using the linearized mode shape.

$$h_e M_e = 0.69 h_w (0.77) M \qquad \text{(Linear-mode shape)}$$

$$= 0.53 h_w M$$

Step 3: Develop the Effective Stiffness (k_s) of the Single-Degree-of-Freedom Model (Figure 3.1.2).

$$H = \frac{M_{yi}}{h_e} \qquad \text{(see Figure 3.1.1)}$$

$$= \frac{300{,}000}{144.7} \qquad \text{(see Eq. 3.1.58)}$$

$$= 2073 \text{ kips}$$

$$\Delta = \frac{\Delta_y}{\Gamma}$$

where Δ is the deflection of the single-degree-of-freedom model (Figure 3.1.1), and Δ_y is the yield drift at level n of the compression side pier (see Step 1). Thus

$$\Delta = \frac{16}{1.45}$$

$$= 11 \text{ in.}$$

$$k_i = \frac{H}{\Delta}$$

$$= \frac{2073}{11}$$

$$= 188 \text{ kips/in.}$$

Comment: The stiffness of the system may also be developed directly from the pushover analysis (Figure 3.1.14*a*). The strength of the system reaches its effective peak when the compression side pier yields (Figure 3.1.14*a*). The secant stiffness (k_q) associated with first yield of the compression side pier is

$$k_q = \frac{V}{\Delta}$$

where Δ is the deflection at the center of mass (h_e) and V is the base shear H.

$$k_q = \frac{V\Gamma}{\text{DR}h_w}$$

$$= \frac{1800(1.45)}{0.0043(210)(12)} \qquad \text{(see Figure 3.1.14}a\text{)}$$

$$= 241 \text{ kips/in.}$$

Step 4: Determine the Period of the System. Since the design confirmation process is displacement based consider the softer system.

$$\omega = \sqrt{\frac{k_i}{M_e}}$$

$$= \sqrt{\frac{188}{29.9}}$$

$$= 2.5 \text{ rad/sec}$$

$$T_{cw} = \frac{2\pi}{\omega}$$

$$= 2.5 \text{ seconds}$$

Observe that this is the same conclusion reached using Eq. 2.4.71 (see Step 1).

Comment: The period of the structure based upon secant stiffness (k_q) is

$$T_{\text{sec}} = 2\pi\sqrt{\frac{M_e}{k_q}}$$

$$= 2\pi\sqrt{\frac{29.9}{241}}$$

$$= 2.2 \text{ seconds} \qquad (\omega = 2.85 \text{ rad/sec})$$

Step 5: Using the Equal-Displacement-Based Assumption, Find the Ultimate Drift of the System. The spectral displacement, given the criterion spectral velocity of 48 in./sec, is

$$S_d = \frac{S_v}{\omega} \qquad \text{(Eq. 2.4.6)}$$

$$= \frac{48}{2.5}$$

$$= 19.2 \text{ in.}$$

and the anticipated drift of the system becomes

$$\Delta_u = \Gamma S_d \qquad \text{(Eq. 2.4.7)}$$

$$= 1.45(19.2)$$

$$= 27.8 \text{ in.} \qquad (1.1\%)$$

Both piers have clearly yielded (Figure 3.1.14*a*)

Comment: The displacement of the system if modeled by the k_q stiffness would be

$$\Delta_u = \frac{\Gamma S_v}{\omega}$$

$$= \frac{1.45(48)}{2.85}$$

$$= 24.4 \text{ in.} \qquad (0.97\%)$$

Step 6: Examine the Strain States at Ultimate Displacement (Δ_u). Computational accuracy may be possible but predicting displacements and strain states is clearly not done with any degree of accuracy. The designer should focus on answering one question: "Is the proposed design reasonable?"

The best way to feel comfortable with a design is to evaluate a series of hypothetical situations. The sequential yield analysis described in Figure 3.1.14*a* suggests that the coupling beam at Level 7 will start to degrade at a drift ratio of 2.2%. Our spectral analysis suggests that peak displacements should be on the order of 1.1%. Accordingly, a level of comfort should exist with regard to coupling beam behavior. Strain states in the compression pier should be estimated. The deflection states of interest might include 1.1% (spectral estimate) and 2.2% (coupling beam degradation). Accordingly,

$$\begin{aligned}
\Delta_u &= 0.011 h_w \\
&= 0.011(210)(12) \\
&= 27.8 \text{ in.} \\
\Delta_y &= 0.0043 h_w \qquad \text{(see Figure 3.1.14}a\text{)} \\
&= 10.8 \text{ in.} \\
\Delta_p &= \Delta_u - \Delta_y \\
&= 17.0 \text{ in.} \\
\theta_p &= \frac{\Delta_p}{h_w - \ell_p/2} \qquad \text{(see Eq. 2.1.8)} \\
&= \frac{\Delta_p}{h_w - 0.25\ell_w} \qquad (\ell_p = 0.5\ell_w) \\
&= \frac{17.0}{210(12) - 90} \\
&= 0.007 \text{ radian} \\
\phi_p &= \frac{\theta_p}{0.5\ell_w} \qquad \text{(see Eq. 2.1.9)} \\
&= \frac{0.007}{180} \\
&= 0.000039 \text{ rad/in.}
\end{aligned}$$

The depth of the compressive stress block might conservatively be developed as follows:

$$\begin{aligned}
P_D &= 1860 \text{ kips} \\
\sum V_p &= 5085 \text{ kips} \\
(\lambda_o - 1)\, A_s f_y &= 0.25(18)(1.56)(60) \qquad (\varepsilon_s \text{ assumed to be} > 0.02 \text{ in./in.}) \\
&= 420 \text{ kips} \\
C_c &= P_D + \sum V_p + (\lambda_o - 1) A_s f_y
\end{aligned}$$

$$
\begin{aligned}
&= 1860 + 5085 + 420 \\
&= 7365 \text{ kips} \\
a &= \frac{C_c}{0.85 f'_c t_w} \\
&= \frac{7365}{0.85(5)(16)} \\
&= 108 \text{ in.} \\
c &= \frac{a}{\beta_1} \\
&= \frac{108}{0.8} \\
&= 135 \text{ in.} \\
\varepsilon_{cp} &= \phi_p c \qquad \text{(Eq. 2.1.10a)} \\
&= 0.000039(135) \\
&= 0.0052 \text{ in./in.} \\
\varepsilon_{sp} &= \phi_p(d - c) \\
&= 0.000039(348 - 135) \\
&= 0.0083 \text{ in./in.}
\end{aligned}
$$

to which must be added ε_{cy}, the concrete strain at yield (± 0.001 in./in.).

$$
\begin{aligned}
\varepsilon_{cu} &\cong \varepsilon_{cy} + \varepsilon_{cp} \qquad \text{(Eq. 2.1.10b)} \\
&\cong 0.006 \text{ in./in.}
\end{aligned}
$$

Conclusion: Shell spalling in the toe region is not anticipated at a drift ratio of 1.1%. At a story drift of 2.2% the shell would clearly have spalled ($\varepsilon_{cp} \cong 0.016$ in./in.), and confined concrete in the toe must be relied upon to maintain system strength.

$$
\begin{aligned}
f'_{cc} &= f'_c + 2.5 \qquad \text{(Eq. 1.2.1—}f_\ell \cong 600 \text{ psi)} \\
&= 7.5 \text{ ksi} \\
a &= \frac{C_c}{0.85 f'_{cc} b_e} \\
&= \frac{7365}{0.85(7.5)(14)} \\
&= 82.5 \text{ in.}
\end{aligned}
$$

$$
\begin{aligned}
c &\cong \frac{a}{\beta_1} \\
&= \frac{82.5}{0.80} \\
&= 103 \text{ in.} \\
\Delta_p &= (0.022 - 0.0043)h_w \\
&= 44.6 \text{ in.} \\
\theta_p &= \frac{\Delta_p}{h_w - \ell_p/2} && \text{(see Eq. 2.1.8)} \\
&= \frac{44.6}{2430} \\
&= 0.0184 \text{ radian} \\
\phi_p &= \frac{\theta_p}{\ell_p} && \text{(Eq. 2.1.9)} \\
&= \frac{0.0184}{180} \\
&= 0.0001 \text{ rad/in.} \\
\varepsilon_{cp} &= \phi_p c && \text{(Eq. 2.1.10a)} \\
&= 0.0001(103) \\
&= 0.01 \text{ in./in.} \\
\varepsilon_{cu} &\cong 0.011 \text{ in./in.} && \text{(see Eq. 2.1.10b)}
\end{aligned}
$$

Conclusion: The compression pier will not strength degrade at a drift of 2.2%. The boundaries of each pier should be confined within the plastic hinge region extending to a height of $0.5\ell_w$.

The confined region should extend to the point where the concrete strain is on the order of 0.003 in./in. In this case, based on a displacement objective of 1.1%, this would be to a distance of about 68 in. from the outside face of the wall. Observe that the inside face of the wall in this case and in most cases will seldom experience high compressive strains.

Concluding Comments:

- *Analytical Procedures.* Care must be taken to appropriately model a coupled wall system. Nodes must be introduced at the wall/coupling beam interface so as to properly introduce changes in the behavior state of the coupling beam. A compatibility check should be performed during the development of a design that is predicated on a sequential yield analysis to insure that the analytical model describes probable component behavior.

- *Conceptual Design Procedures.* Methods were developed in Section 2.4.2.4 to estimate the probable level of rotation a coupling beam might experience (Eq. 2.4.66). These estimates will tend to be conservative because the rotation imposed on a coupling beam will be relieved by the axial deformations experienced by the piers. Once the strength limit state of the system has been reached, the rotation experienced in the coupling beam will become directly proportional to the rotation of the wall (Eq. 2.4.63).

Consider the impact on coupling beam behavior of axial strain experienced in the piers in the example design. The strain gradient imposed on the tensile pier at compression pier first yield in Figure 3.1.14*a* would vary from what we shall refer to as ε_o at the base to ε_m at the mid-height of the wall.

$$\begin{aligned}\varepsilon_o &= \frac{\sum V_p - P_D}{A_{st} E_s} \\ &= \frac{5085 - 1860}{2(18)(1.56)(29{,}000)} \qquad \text{(see Figure 2.4.32)} \\ &= 0.00198 \text{ in./in.}\end{aligned}$$

$$\begin{aligned}\varepsilon_m &= \frac{1}{2}\varepsilon_o \\ &= 0.001 \text{ in./in.}\end{aligned}$$

The elongation of the tensile pier at mid-height (δ_{tm}) would be

$$\begin{aligned}\delta_{tm} &= \frac{\varepsilon_o + \varepsilon_m}{2}\left(\frac{h_w}{2}\right) \\ &= 0.0015(105)(12) \\ &= 1.9 \text{ in.}\end{aligned}$$

while the shortening of the compression side pier at mid-height (δ_{cm}) would be

$$\begin{aligned}f_{co} &= \frac{\sum V_p + P_D}{A_c} \\ &= \frac{5085 + 1860}{30(12)16} \qquad \text{(see Figure 2.4.32)} \\ &= 1.2 \text{ ksi}\end{aligned}$$

$$\varepsilon_{co} = \frac{f_{co}}{E_c}$$

$$= \frac{1.2}{4000}$$

$$= 0.0003 \text{ in./in.}$$

$$\varepsilon_{cm} = 0.00015 \text{ in./in.}$$

$$\delta_{cm} = 0.000225(105)(12)$$

$$= 0.28 \text{ in.}$$

The relaxation of coupling beam rotation (θ'_{bx}) would be

$$\theta'_{bm} = \frac{\delta_{tm} + \delta_{cm}}{\ell_b}$$

$$= \frac{1.9 + 0.28}{60}$$

$$= 0.035 \text{ radian}$$

The W18 × 130 coupling beam has a yield rotation of

$$\theta_{by} = \gamma_y \qquad \text{(see Figure 2.4.25)}$$

$$= \frac{v_{sy}}{G}$$

$$= \frac{30}{0.4(29{,}000)}$$

$$= 0.0026 \text{ radian}$$

and an objective maximum rotation of

$$\theta_u - 0.08 \text{ radian}$$

The rotation imposed on the coupling beam at Level 7 and compression pier first yield (Figure 3.1.14*a*) would be on the order of 0.053 radian based on the development proposed by Eq. 2.4.65. Pier elongations would relieve this rotation to 0.018 radian or approximately $7\theta_{by}$.

Accordingly, allowing the tension pier to elongate will relieve the rotation demand imposed on the coupling beam. The designer who is aware of the impact of design alternatives should be in a better position to optimize system behavior.

Observe that the rotation limit state (0.08 radian) is exceeded in the coupling beam at Level 7 when the building drift ratio reaches 2.2% (Figure 3.1.14*a*).

- *Dynamic Response Prediction.* Two levels of stiffness have been proposed, the idealized stiffness (k_i) and that developed from the pushover analysis at strength limit state k_q (Figure 3.1.14*a*). An elastic analysis based on the effective moment of inertia of system components would suggest still another value ($k_{el} \cong 460$ kips/in.). The corresponding fundamental periods are

$$T_i = 2.5 \text{ seconds} \qquad \text{(see Step 4)}$$

$$T_{sec} = 2.2 \text{ seconds} \qquad \text{(see Step 4)}$$

$$T_{el} = 1.6 \text{ seconds}$$

The building code[3.7] suggests a period of

$$T = C_t h_n^{3/4}$$

$$= 0.02(210)^{0.75}$$

$$= 1.1 \text{ seconds}$$

or 1.43 seconds when amplified by the currently allowed 30%.

Associated with each of these approaches to establishing the dynamic characteristics of the system described in Figure 2.4.31 are the strength and deformation objectives identified in Table 3.1.5.

The proposed design appears to satisfy all but the code-suggested[3.7] design objectives. The base shear (V_u) for flexural design purposes should be

$$V_u = \phi V_M$$

$$= 0.9(1800)$$

$$= 1620 \text{ kips}$$

However, this ultimate strength is developed from the mechanism strength of the system, not the first element to yield (Figure 3.1.14*a*). First yield occurs in the tension pier at a base shear of 1200 kips (see Figure 2.1.14*b*). Clearly, tension pier first yield does not represent a reasonable definition of the strength limit state for this system.

TABLE 3.1.5 Design Objectives—Coupled Wall System Described in Figure 2.4.31 (S_v = 48 in./sec)

Fundamental Period (seconds)	Objective (Design) Base Shear (V_u)[a]	Probable Drift (in.) (DR—%)
1.43	2340[b]	15.8 (0.63)
1.6	1462	17.7 (0.7)
2.2	1064	24.4 (0.97)
2.5	936	27.7 (1.1)

[a] Response spectra—$\mu\Omega_o = 1.25(4) = 5$; code strength—$R = 5.5$.

[b] Base shear spectra = $S_a/0.7$.

3.1.4 Precast Concrete Shear Walls

Precast concrete wall systems are especially cost effective when they satisfy aesthetic objectives, for virtually any type of finish may be created in the plant. Figure 3.1.15 shows how precast bearing/shear walls may be effectively incorporated into a residential project. Figure 3.1.16 describes a variety of finishes available to the design team.

Section 2.4.3 has discussed recent experimental work in the field of assembled precast wall systems. In addition to these experimental efforts, a significant amount of comparative analysis has been performed on assembled precast wall systems.[3.8–3.10]

Design methodologies have been proposed for a variety of precast concrete wall systems, most of which include energy dissipation devices along panel interfaces. Basically these diverse assembly processes demonstrate the options available to the creative or imaginative designer. Regardless of the assembly process a ductile load path that can dissipate energy is created, and this common thread joins the various design processes and stresses the importance of the basic concepts discussed in this section. The treatment that follows endeavors to extend the example of Section 2.4.3.4 and describe how damping-sensitive design approaches may be used to create systems that will outperform comparable cast-in-place shear wall systems.

Precast shear wall location will typically follow function. As was the case with the cast-in-place shear wall designs, a designer will usually understand the length constraints imposed on a shear wall. Precast concrete shear walls have a further constraint; specifically, they must accommodate the construction constraints, in terms of both how they blend into the vertical load carrying system and how they are to be installed. For example, if the floor system is to be constructed using precast plank or a precast slab floor, a multistory wall panel is likely to be more difficult to incorporate into the building than a single-story wall. If, on the other hand, the floors are to be cast in place and conventionally reinforced, significant advantages may be gained by the use of wall systems two or more stories high.

From the standpoint of panel manufacture and erection, panel size and weight limitations need to be carefully considered. When the wall panels are cast off site, they must be transportable, and this usually limits one dimension to 12 ft.

Creativity is the key element in the design process, and starting with a predisposition toward a particular connector approach will seldom result in a successful design. Accordingly, many factors must be considered in the conceptual design phase, and the result of this process should be an economical system that can be quickly assembled and that will perform well when subjected to earthquake ground motions. Connection costs are always an important consideration, so the designer must be capable of quickly developing a concept that can be evaluated by a constructor. The delivery system best suited to the development of innovative systems is some version of the design/build approach. Often, the precaster is selected based on the acceptance of a budget developed from conceptual plans. This allows the precaster to participate in the development of construction details.

To demonstrate how a precast system might evolve, let us consider the precast alternatives that might be considered in the development of the fifteen-story residential

Figure 3.1.15 Precast concrete residential building. (Courtesy of Clark-Pacific.)

Figure 3.1.16 Concrete finishes. Precast concrete tames. (Courtesy of Pankow Builders.)

Figure 3.1.16 (*Continued*) Concrete finishes. Details incorporated into precast concrete frame beams. (Courtesy of Pankow Builders.)

Figure 3.1.16 (*Continued*) Concrete finishes. Limestone on precast concrete panels. (Courtesy of Clark-Pacific.)

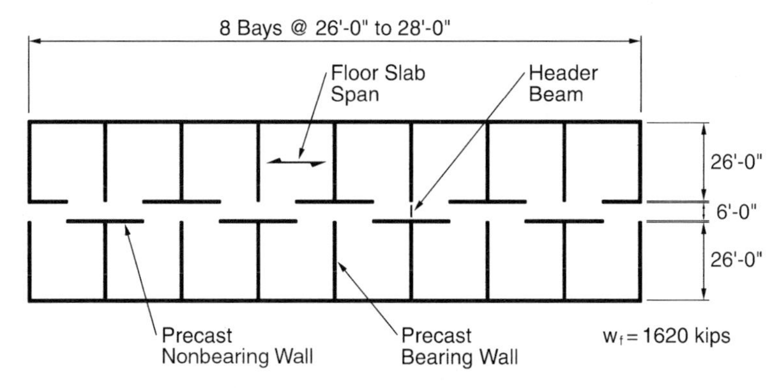

Figure 3.1.17 Residential building floor plan.

building whose basic plan is described in Figure 3.1.17. Slab spans will depend on both function and the capabilities of the floor framing system. Bedrooms in residential applications will be on the order of 12 ft wide, whereas common rooms can be in the 16-ft range. Accordingly, the first set of design options must deal with floor framing alternatives. Two basic options, each of which contain a variety of alternatives, are:

Option 1: Floor-Slab Span in the 26- to 28-ft Range. This will allow every other wall to be nonbearing, and this is functionally attractive because the nonbearing (nonconcrete) wall may be used to accommodate plumbing and a majority of the electrical delivery systems. Precast/prestressed floors are the logical structural system, and while hollow core plank is the most frequently selected alternative, cast-in-place post-tensioned floor slabs might also be considered. Conventionally reinforced floor slabs are too heavy, and the required reshoring process is time consuming. As a consequence, conventionally reinforced slabs are seldom used in long span precast bearing wall systems. From an erection perspective, it is desirable to reduce the number of floor units. Fewer floor units are required in a long span floor option than in comparable shorter span systems. The 28-ft span is within the reach of an 8-in. pretopped system. Shear wall systems for this long span bearing wall option would be most cost effective if they were single-story panels extending from the exterior enclosure to the corridor and 8 to 12 ft high, depending on the desired ceiling (floor-to-floor) height. Figure 3.1.18 demonstrates the use of this construction alternative.

Option 2: Floor-Slab Span 16 ft. This option places a shear wall at every demising wall. Walls of varying lengths could be a program requirement in order to accommodate plumbing or mechanical penetrations. From an engineering perspective mechanical openings in floors can be accommodated easily and economically by using walls of different lengths, as we saw in Section 3.1.2 when

Figure 3.1.18 MGM Grand. (Photo courtesy of PCL Construction Services, Inc.)

we dealt with cast-in-place walls of varying lengths. Normal to the orientation of the bearing walls, shear wall location and dimension are always constrained by function (Figure 3.1.17). Methodologies for providing support for the floor must be carefully considered. Floor construction options might include large precast panels that are conventionally reinforced or cast-in-place alternatives. Precast panels may be two (or more) stories in height (see Photo 2.9) with floor slab reinforcing placed in slots or voids incorporated in the panels. The principal advantage of two-story construction lies in the ability to cast two levels of floor at a time or in a rapid sequence, an option little explored to date, but especially attractive to builders who are willing to invest in the formwork, because construction speed is a key component of building cost. In the multistory option, the wall dimension could be limited to a transportable dimension and this, of necessity, introduces a vertical joint in the completed wall.

If the function of the building is not residential, the story heights are likely to exceed 12 ft. This too will require a panel seam along the vertical axis. Now the construction options must consider structural design objectives and connection cost. Panels are most economically joined through the use of a late cast wet joint, and this from a design perspective, emulates a solid wall panel. Alternatively, friction dampers may be installed along the vertical seam, as they were in the PRESSS shear wall shown in Figure 2.4.54. Late cast wet joints and friction dissipating connectors must be concealed if the finished product is to be exposed (Figure 3.1.15).

From a seismic design perspective it is reasonable to assume that available shear wall locations will be dictated by function and the requirements of the vertical load carrying system. Thus, the only real designer options in terms of the bearing wall itself will be thickness of the wall and the type of connection system. It is then reasonable to explore from a design perspective several logical wall construction alternatives.

3.1.4.1 Hybrid Wall System—Equal Displacement-Based Design (EBD, Section 3.1.1)

Design Considerations

- *Prestressing Force.* The design process must anticipate tendon overstrength conditions. The experimental efforts summarized in Section 2.4.3 assumed that the ultimate strength provided by the post-tensioning would reach an average stress on the order of $0.81 f_{psu}$ (see Eq. 2.4.94). This was reasonable because the length of the strand was 13 ft. If the unbonded length of the post-tensioning strand is on the order of 100 ft, however, this level of overstrength or ultimate strength will not materialize. The designer must select an ultimate strand stress that is consistent with the envisioned construction program. In the examples in this section, a tendon stress ($\lambda_o f_{pse}$) of 172 ksi will be assumed in the design process (see Section 2.4.3.4, Step 7).

- *Energy Dissipater.* Ductile rods (DDCs) will be used in the examples of this section because the strength and overstrength characteristics have been established by test. Alternative energy dissipaters can be used, but their behavior characteristics must be established and quality assured. An overstrength factor of 1.5 is used for the ductile rod.
- *System Overstrength* (Ω_o). Given the nature of the assumptions used to develop design strengths of the reinforcing components, a system overstrength factor of up to 1.5 is reasonably used in the design of hybrid precast concrete wall systems.
- *Post-Tensioning Minimums.* Most designers feel that minimum stressing levels are required if the behavior characteristics attributed to prestressed concrete are to be attained. 125 psi is used in the designs of this section as a minimum level of post-tensioning.

A residential building whose proposed floor plan is described in Figure 3.1.17 will be built using precast concrete bearing walls and precast prestressed floors. It will be assumed that the tributary dead load/floor to a bearing wall is 90 kips per floor, both from a mass and applied axial load perspective. The building is fifteen stories high and each story height will be 10 ft. Precast hollow core floor planks (pretopped) will span between bearing walls. The design criterion spectral velocity is 50 in./sec. Hence

$$h_w = 150 \text{ ft} \qquad \text{(see Figure 3.1.19}a\text{)}$$

$$\ell_w = 26 \text{ ft}$$

$$t_w = 8 \text{ in.} \qquad \text{(see detail in Figure 3.1.19}d\text{)}$$

$$W_f = 90 \text{ kips} \qquad \text{(Tributary mass and vertical load/floor)}$$

The structural system to be considered is a vertically post-tensioned assembly of precast walls with energy dissipaters at the base. Conclusions are developed based on the constant velocity equal displacement (CVED) method (see Section 3.1.1.1).

Follow the design procedures developed in Section 2.4.3.3 and the example developed in Section 2.4.3.4.

CVED

Step 1: Select a Displacement-Based Design.

Step 2: Select the Number of Walls. The number of walls is a precondition (see Figure 3.1.17).

Step 3: Check the Panel Thickness in the Plastic Hinge Region. Minimum wall thickness in the plastic hinge region should be the lesser of

$$t_{w,\min} = \sqrt{\frac{P}{3.4 f'_c \beta_1}} \qquad \text{(Eq. 2.4.21b)}$$

$$= \sqrt{\frac{1350}{3.4(5)0.8}} \qquad (f'_c = 5000 \text{ psi})$$

$$= 10 \text{ in.}$$

$$t_{w,\min} = \frac{h_n}{16} \qquad \text{(see Section 2.4.1.1)}$$

$$= \frac{120}{16}$$

$$= 7.5 \text{ in.}$$

Conclusion: Minimum wall thickness should be 7.5 in. A confinued core is suggested (see Section 2.4.1.3, Step 2). Consider increasing the strength of the concrete in the plastic hinge region.

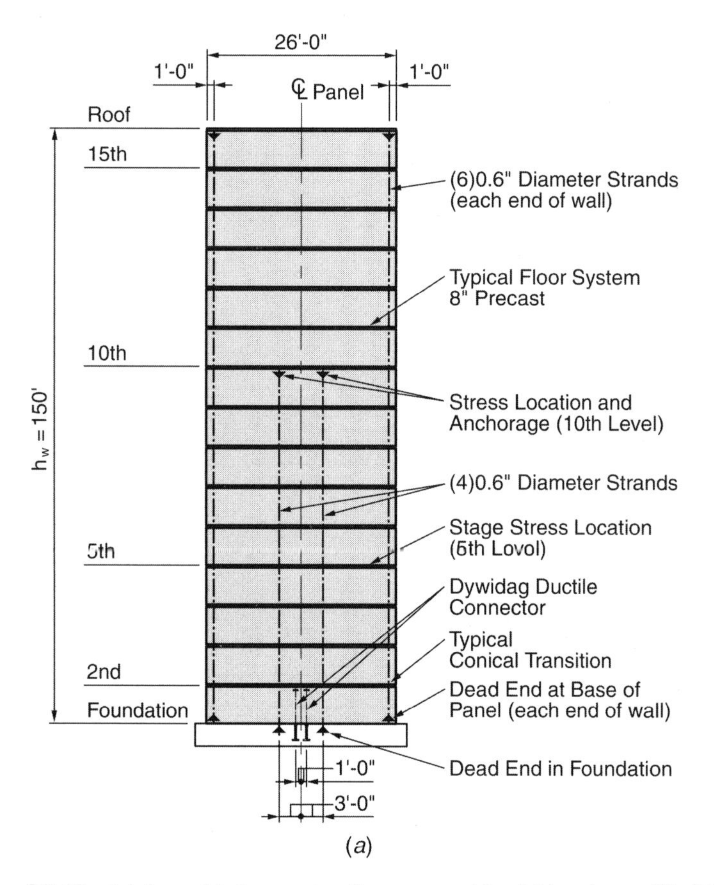

Figure 3.1.19 (*a*) Assembled precast wall system, residential bearing wall building.

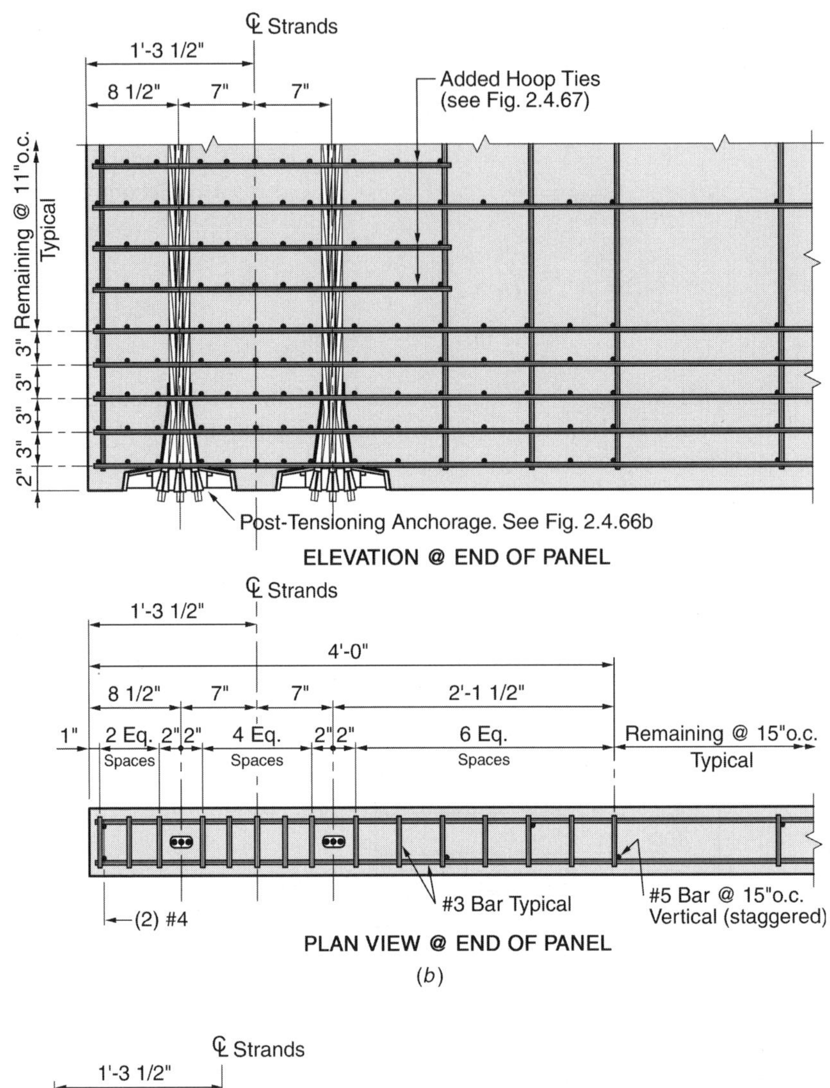

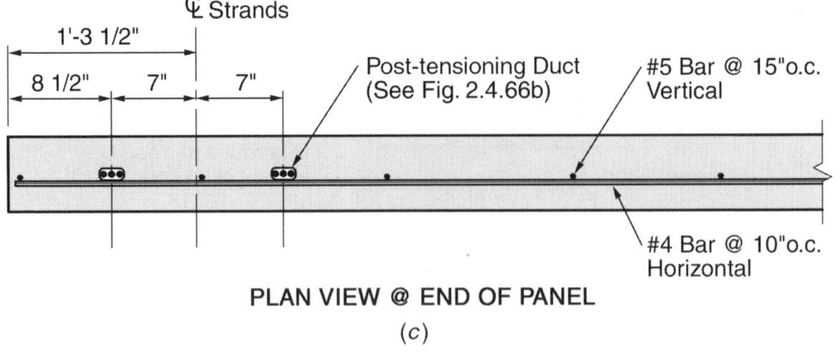

Figure 3.1.19 (*Continued*) (*b*) Detailing the toe region—lowermost panel. (*c*) Detailing intermediate panels.

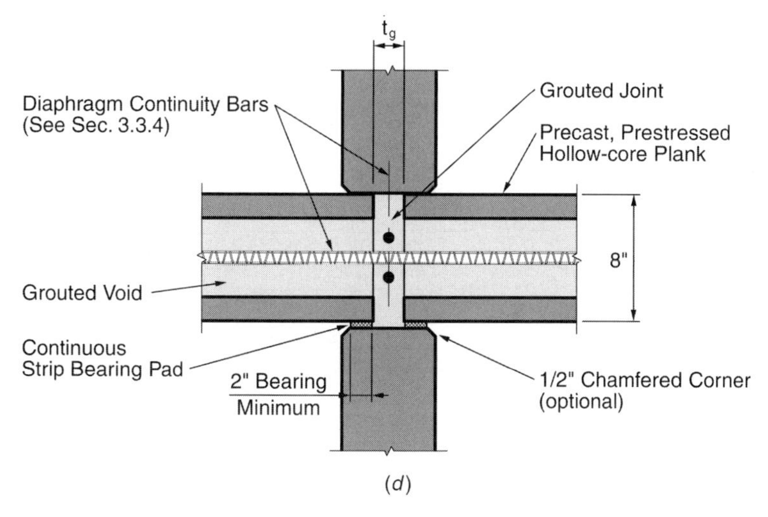

Figure 3.1.19 (*Continued*) (*d*) Joint at floor slab.

Step 4: Determine the Probable Period of the Building.

$$w = \frac{W_f}{h_x} \qquad \text{(Uniformly distributed weight (mass))}$$

$$= \frac{90}{10}$$

$$= 9 \text{ kips/ft}$$

$$T_f = 0.13\left(\frac{h_w}{\ell_w}\right)^2 \sqrt{\frac{w\ell_w}{Et_w}} \qquad \text{(Eq. 2.4.22)}$$

$$= 0.13\left(\frac{150}{26}\right)^2 \sqrt{\frac{9(26)}{4000(0.67)}} \qquad (t_w = 8 \text{ in.})$$

$$= 1.28 \text{ seconds}$$

Step 5:

(a) and (b) See Section 2.4.3.3, Step 5 a and b.

(c) Quantify the Angular Frequency.

$$\omega_n = \frac{2\pi}{T}$$

$$= \frac{6.28}{1.28}$$

$$= 4.9 \text{ rad/sec}$$

(d) Determine the Spectral Displacement (S_d).

$$S_d = \frac{S_v}{\omega_n} \qquad \text{(Eq. 2.4.6)}$$

$$= \frac{50}{4.9}$$

$$= 10.2 \text{ in.}$$

(e) Determine the Ultimate Displacement Objective of the Wall System.

$$\Delta_u = \Gamma S_d \qquad \text{(Eq. 2.4.7)}$$

$$= 1.5(10.2) \qquad \text{(Conservatively—see Step 6)}$$

$$= 15.3 \text{ in.}$$

Step 6: Determine the Yield Moment Required to Attain the Displacement Objective. Use the single-degree-of-freedom model (Figure 3.1.1) and assume that the inelastic mode shape is linear. This is especially appropriate for a hybrid precast wall panel system because, as we saw in Section 2.4.3, the postyield rotation is accumulated at the base of the wall where a gap was created.

$$\Gamma = \frac{3n}{2n+1} \qquad \text{(Eq. 3.1.18a)}$$

$$= \frac{3(15)}{2(15)+1}$$

$$= 1.45$$

$$h_e = \frac{h_w}{\Gamma} \qquad \text{(Eq. 3.1.16)}$$

$$h_e = \frac{150}{1.45}$$

$$= 103 \text{ ft}$$

$$M_e = \frac{1.5(n+1)}{2n+1} M \qquad \text{(Eq. 3.1.19a)}$$

$$= \frac{1.5(16)}{31}\left(\frac{1350}{386.4}\right)$$

$$= 2.7 \text{ kips-sec}^2/\text{in.}$$

$$k_e = \omega^2 M_e$$

$$= (4.9)^2(2.7)$$

$$= 64.8 \text{ kips/in.}$$

$$\Delta_{u,\text{SDOF}} = \frac{\Delta_u}{\Gamma} \qquad (\text{or, } \Delta_{u,\text{SDOF}} = S_d)$$

$$= \frac{15.3}{1.45} \qquad (\text{conservative})$$

$$= 10.6 \text{ in.}$$

$$\Delta_{y,\text{SDOF}} = \frac{\Delta_{u,\text{SDOF}}}{\mu}$$

$$= \frac{10.6}{4} \qquad (\mu = 4, \text{ see Figure 2.4.56}b)$$

$$= 2.65 \text{ in.}$$

$$H_u = k_e \Delta_{y,\text{SDOF}} \qquad (\text{see Figure 3.1.1})$$

$$= (64.8)2.65$$

$$= 171 \text{ kips}$$

$$\Omega_o M_{d,\text{SDOF}} = H_u h_e$$

$$= 171(103)$$

$$= 17{,}600 \text{ ft-kips}$$

Comment: M_d refers to the design moment. The designer may wish to use the ultimate moment (M_u) or the nominal moment (M_n). Either case is reasonable provided the choice of the adopted ductility factor (μ) and overstrength factor (Ω_o) is consistent with the selected objective. Both are viewed as being conservative (see Figure 2.4.56*b*).

Adopt an overstrength factor of 1.25

$$M_d = \frac{17{,}600}{1.25}$$

$$= 14{,}080 \text{ ft-kips}$$

Step 7: Determine the Flexural Reinforcing Program Required to Attain System Design Objectives.

$$F_{psu} + F_s \geq \frac{M_n}{0.4\ell_w} - P_D \qquad \text{(Eq. 2.4.102a)}$$

$$\geq \frac{169{,}000}{0.4(26)(12)} - 1350 \qquad (M_n = M_d)$$

$$\cong 0$$

Objective reinforcement minimums are

$$F_{pse} = 0.125A_g \qquad \text{(Design consideration, see Section 3.1.4.1)}$$

$$= 0.125(26)(12)(8)$$

$$= 312 \text{ kips}$$

$$F_s \cong F_{pse} \qquad \text{(see Eq. 2.4.101)}$$

$$\cong 312 \text{ kips}$$

$$A_{ps} = \frac{F_{pse}}{f_{pse}}$$

$$= \frac{312}{162}$$

$$= 1.93 \text{ in.}^2$$

Use two 4-strand (0.6-in. ϕ) sets (see Figure 2.4.66*b*).

$$A_{ps} = 8(0.213)$$

$$= 1.7 \text{ in.}^2$$

$$F_{pse} = 1.7(162)$$

$$= 276 \text{ kips}$$

$$A_s = \frac{276}{60}$$

$$= 4.6 \text{ in.}^2$$

Use two ductile rods (Figure 2.1.57), $F_y = 141$ kips each.

Step 8: Check the Stress Imposed on the Wall.

$$f_c = \frac{F_{pse} + P_D}{A_g} \qquad \text{(Eq. 2.4.104)}$$

$$= \frac{276 + 1350}{2496}$$

$$= 0.65 \text{ ksi} \qquad \left(0.13 f_c' < 0.25 f_c'\right) \qquad \text{(OK)}$$

Step 9: Determine the Strain States in the Concrete and Reinforcing Steel.

(a) Determine the gap angle (θ_{gap}).

$$\theta_{\text{gap}} = 0.8 \left(\frac{\Delta_u}{h_w}\right) \qquad \text{(see Section 2.4.3.2)}$$

$$= (0.8)\frac{15.3}{150(12)}$$

$$= 0.0068 \text{ radian}$$

Comment: Eq. 2.4.22, which in essence establishes θ_{gap}, assumes that the effective moment of inertia of the wall is $0.5I_g$. The effective stiffness of the wall will increase as the axial load and prestress level are increased. Compare, for example, the stiffness of the wall described in Figure 2.4.2*a* with that in Figure 2.4.4*a*. This obvious increase in stiffness is also apparent in the pretensioned walls (Figure 2.4.56*b*). The effective stiffness of the wall whose behavior is described in Figure 2.4.2*a* might reasonably be described as twice that of a comparable lightly loaded cast-in-place wall (Figure 2.4.4*a*). The resultant reduction in the created gap could be as much as 30%.

$$\theta_{\text{gap,pr}} = \frac{\theta_{\text{gap}}}{\sqrt{2}} \qquad (I_e = I_g)$$

$$= 0.0048 \text{ radian}$$

(b) Determine the effective depth of the neutral axis based on the presumption that the core will be confined.

Comment: Since our objective is to identify worst case conditions, the average ultimate or attainable (F_{psu}) strength in the prestressing is used.

$$F_{psu} = 0.81 \sum A_{\text{pre}} f_{psu} \qquad \text{(Eq. 2.4.94)}$$

$$= 0.81(8)(0.213)270$$

$$= 373 \text{ kips}$$

$$\lambda_o F_{sy} = 1.5(2)(141)$$

$$= 423 \text{ kips}$$

$$P = 1350 \text{ kips}$$

$$C_c = F_{psu} + \lambda_o F_{sy} + P$$
$$= 373 + 423 + 1350$$
$$= 2146 \text{ kips}$$

$$a = \frac{C_c}{0.85\left(f_c' + 2.5\right) b_e} \qquad \text{(see Eq. 2.4.97a)}$$

$$= \frac{2146}{0.85(7.5)(6.5)}$$

$$c \cong 52 \text{ in.} \qquad (c \cong a\text{—Figure 2.4.52})$$

$$\lambda_o M_n = C_c \left(\frac{h}{2} - \frac{c}{2}\right)$$

$$= 2146\left(\frac{26}{2} - \frac{52}{24}\right)$$

$$= 25{,}400 \text{ ft-kips}$$

(c) Determine the shortening at the edge of the confined toe (δ_{core}).

$$\delta_{\text{core}} = \theta_{\text{gap}}(c - \text{cover})$$
$$= 0.0068(52 - 1)$$
$$= 0.347 \text{ in.}$$

(d) Determine the strain in the concrete.

$$\varepsilon_{cu} = \frac{\delta_{\text{core}}}{t_w} \qquad \left(\ell_p \cong t_w\right)$$

$$= \frac{0.347}{8}$$

$$= 0.043 \text{ in./in.}$$

The strain state is reasonable based on the experimental evidence reviewed in Section 2.4.3.1. Remedial considerations might include increasing the strength of the concrete, thickening the panel, or recognizing that the predicted drift is based on the softer cast-in-place wall model. Were the designer to increase strength of concrete to 8 ksi, a significant reduction of concrete strain would be realized.

$$c = \frac{2146}{0.85(10.5)(6.5)} \qquad \text{(see Step 9b)}$$

$$= 37 \text{ in.}$$

$$\delta_{\text{core}} = 0.0068(37 - 1) \qquad \text{(see Step 9c)}$$

$$= 0.245 \text{ in.}$$

$$\varepsilon_{cu} = \frac{\delta_{\text{core}}}{\ell_p} \qquad (\ell_p \cong t_w)$$

$$= \frac{0.245}{8}$$

$$= 0.03 \text{ in./in.}$$

The primary consequence of increasing the strength of the concrete is the fact that confinement reinforcing will be increased by 33% because required confining pressure is directly proportional to concrete strength $(f_\ell = 0.09 f_c')$.

The designer might opt to use 6000-psi concrete and place some reliance on the impact of increased system stiffness on displacement. For example, if

$$\theta_{\text{gap,pr}} = 0.0048 \text{ radian} \qquad \text{(see Step 9a)}$$

$$f_c' = 6 \text{ ksi}$$

then

$$c = \frac{2146}{0.85(8.5)(6.5)} \qquad \text{(see Step 9b)}$$

$$\cong 46 \text{ in.}$$

$$\delta_{\text{core}} = \theta_{\text{gap,pr}}(c - \text{cover}) \qquad \text{(see Step 9c)}$$

$$= 0.0048(46 - 1)$$

$$= 0.215 \text{ in.}$$

$$\varepsilon_{cu} = \frac{0.216}{8}$$

$$= 0.027 \text{ in./in.}$$

Comment: The design decision should not be made in a vacuum. The designer should understand the degree of conservatism used to develop the forcing function (S_v) and in the reduction of the building to the design model. It is for this reason that it is so difficult to codify the design process, for the codification process must typically select a series of conservative postures, the sum of which is likely to produce an unnecesarily conservative design. A series of conservative postures as adopted in force-based codifications will tend to produce an overly strong building, which is not likely to attain performance objectives. Consider, for example, the impact on performance of doubling the reinforcement in this wall.

Conclusions: For the purposes of this design example, it will be assumed that we will use 6000-psi concrete in the lowermost panel ($c = 46$ in., $\varepsilon_{cu} = 0.027$ in./in.).

(e) Determine the strain in the post-tensioning strand. The principal variable here will be the length of the tendon. The designer must first decide how the system will be built. In Section 2.4.3.4, boundary strands were introduced to attain sufficient strength in the wall construction program above the plastic hinge region (Figure 2.4.67*a*). This strength may also be used to replace the centrally loaded strands at a convenient intermediate point where building stability during construction or intermediate stressing requires that the centrally located post-tensioning be activated. The adopted program is described in Figure 3.1.19*a*.

$$d_{\text{pre}} = 16 \text{ ft} \qquad \text{(see Figure 3.1.19}a\text{)}$$

$$\delta_{\text{pre},p} = \theta_{\text{gap}}(d_{\text{pre}} - a) \qquad \text{(Eq. 2.4.105)}$$

$$= 0.0068(192 - 46)$$

$$= 0.99 \text{ in.}$$

Single ended pulls that exceed 100 ft are not advisable. Accordingly, intermediate stressing is required. One reasonable option would be to terminate the central strand group at level ten. Hence

$$\varepsilon_{psp} = \frac{\delta_{\text{pre},p}}{10h_x} \qquad \text{(Eq. 2.4.106 and Figure 3.1.19}a\text{)}$$

$$= \frac{0.99}{10(10)(12)}$$

$$= 0.00083 \text{ in./in.}$$

$$\varepsilon_{ps} = \varepsilon_{pse} + \varepsilon_{psp}$$

$$= \frac{162}{E_{ps}} + 0.00083$$

$$= \frac{162}{28{,}000} + 0.00083$$

$$= 0.0066 \text{ in./in.} \qquad \text{(OK, see Figure 2.1.54)}$$

$$f_{ps} = 185 \text{ ksi}$$

(f) Determine the strain in the energy dissipaters. Since only two DDCs are proposed,

$$d_s = \frac{\ell_w}{2} + 6 \text{ in.} \qquad \text{(see Figure 3.1.19}a\text{)}$$

$$= 162 \text{ in.}$$

$$\delta_{sp} = \theta_{\text{gap}}(d_s - a) \qquad \text{(Eq. 2.4.108)}$$

$$= 0.0068(162 - 46)$$

$$= 0.788 \text{ in.}$$

$$\varepsilon_{sp} = \frac{\delta_{sp}}{\ell_{p,ed}}$$

$$= \frac{0.788}{9} \qquad \text{(see Figure 2.1.57)}$$

$$= 0.088 \text{ in./in.} \qquad \text{(OK, see Figure 2.1.41)}$$

Step 10: Detail the Precast Panel System. Details of the foundation connection are developed in Section 2.4.3.4, as were the load transfers required in the lowermost panel (Figure 2.4.67). Our focus here is on developing a constructible wall system.

Consider the strength required at level one (Figure 3.1.19*a*). It should be at least the overstrength provided at the base of the wall. Only the exterior and central post-tensioning cross this plane, and it is unlikely that these strands will develop a stress that significantly exceeds f_{pse} (see Step 9e—$f_{ps} = 185$ ksi). From a conceptual design perspective it is reasonable to replace the flexural overstrength provided by the energy dissipaters with that provided by the exterior post-tensioning.

$$\lambda_o M_{n,ed} = \lambda_o T_s \left(d_s - \frac{a}{2}\right)$$

$$= 1.5(2)(141)(162 - 23)$$

$$= 58{,}800 \text{ in.-kips}$$

$$A_{\text{pre},2} = \frac{\lambda_o M_{n,ed}}{\left(d_{\text{pre},2} - \frac{c}{2}\right) f_{pse}}$$

where the subscript pre, 2 refers to the strand farthest from the compression face of the wall. Therefore,

$$A_{\text{pre},2} = \frac{58{,}800}{(300 - 23)(162)}$$

$$= 1.31 \text{ in.}^2 \qquad \text{(Six 0.6-in. } \phi \text{ strands)}$$

Consider the confinement required at the end of the lowermost panel.

$$c = 46 \text{ in.}$$

$$f_\ell = 0.09 f_c' \qquad \text{(see Eq. 1.2.7)}$$

$$= 0.09(6)$$

$$= 0.540 \text{ ksi}$$

$$A_{sh} = \frac{0.54sh_c}{f_{yh}}$$

$$sh_c = \frac{A_{sh} f_{yh}}{0.54}$$

For #3 bars,

$$sh_c = \frac{0.11(60)}{0.54}$$

$$= 12.2 \text{ in.}$$

Conclusion: A 4 in. by 3 in. grid of #3 bars is required.

Consider shear reinforcing requirements.

$$C_c = F_{ps} + \lambda_o F_{sy} + P \qquad \text{(see Step 9b)}$$

$$= 2146 \text{ kips}$$

$$c = 46 \text{ in.} \qquad (a \cong c, \text{ see Eq. 2.4.97a})$$

$$\lambda_o M_n = C_c \left(\frac{\ell_w}{2} - \frac{c}{2} \right)$$

$$= 2146(13 - 1.92)$$

$$\cong 23{,}800 \text{ ft-kips}$$

$$V_u = \frac{\lambda_o M_n}{h_e}$$

$$\cong \frac{23{,}800}{103} \qquad \text{(see Step 6)}$$

$$\cong 231 \text{ kips}$$

$$v \cong \frac{V_u}{t_w d}$$

$$= \frac{231}{8(25)(12)}$$

$$\cong 0.1 \text{ ksi}$$

Caution: If the extension of the energy dissipaters is continued over the full height of the wall and the edge post-tensioning is not provided, the shear flow will occur over only half of the panel. In any case the shear stress will typically be low. Since the panel will not be cracked, the usual case will be to provide minimum wall reinforcement quantities.

Current wall reinforcing minimums require that v_s be 150 psi ($\rho_v = \rho_h = 0.0025$) for Grade 60 reinforcing ($f_{yh} = 60$ ksi). Prescriptive detailing requirements include the need for two curtains of reinforcing when the in-plane factored shear force assigned to the wall exceeds $2A_{cv}\sqrt{f'_c}$. This requirement is based on "the observation" that maintaining a single layer of reinforcement under ordinary construction situations will not be possible, as well as on a desire to "inhibit fragmentation of the concrete in the event of severe cracking." Neither of these concerns apply to precast post-tensioned hybrid wall panel systems (Figure 2.4.47*b*), so construction practicalities should be the primary concern.

For the example 8-in. thick panel at the lowermost level, mesh type reinforcing could provide the confinement required at the toe ($\ell_p \cong t_w$). This would allow for post-tensioning and for energy dissipaters being effectively placed. See Figure 3.1.19*b*. Bar selection is accomplished as follows:

$$A_h = A_v = 0.0025 t_w s$$

$$s = \frac{A_v}{0.0025 t_w}$$

$$= \frac{0.22}{0.02} \qquad \text{Two \#3 bars}$$

$$= 11 \text{ in. on center}$$

It is advisable to use larger vertical bars. Maximum spacing considerations are appropriate—$3t_w$ and 18 in.

$$A_v = 0.0025 t_w (18)$$

$$= 0.36 \text{ in.}^2$$

Provide a #5 bar at 15 in. on center.

Above the lowermost panel a similar reinforcing program may be used. However, an eccentrically placed mesh might also be used (Figure 3.1.19*c*), especially if the energy dissipaters are terminated in the lowermost panels.

Consider the joint between panels. Stresses must be transferred across the intersection between the floor slabs and the wall panels. Figure 2.4.57 suggested that high stresses and strains dissipated rapidly with distance from the toe region. Figure 2.4.47 described upper panel rotation as being essentially that of a rigid body. If we assume (conservatively—see Step 9e) that the clamping force provided by the post-tensioning is developed from a stress of 172 ksi, the force imposed on the panel joint described in Figure 3.1.19*d* may be quantified as

$$F_{ps} = A_{ps} f_{ps}$$

$$= A_{ps}(172)$$

For the proposed 20 strands that clamp the panels together (Figure 3.1.19*a*),

$$F_{ps} = 20(0.213)172$$
$$= 734 \text{ kips}$$
$$P \cong 1350 \text{ kips}$$
$$F_{ps} + P \cong 2084 \text{ kips}$$

The overstrength moment developed at the base was about 23,800 ft-kips. This places the axial load at an eccentricity of

$$e = \frac{\lambda_o M_n}{F_{ps} + P}$$
$$= \frac{23{,}800}{2084}$$
$$= 11.4 \text{ ft}$$

The bearing stress at the edge of the panel ($f_{c,\max}$) acting on an effective panel width of t_{we} is

$$f_{c,\max} = \frac{2(F_{ps} + P)}{3(\ell_w/2 - e)t_{we}} \qquad \text{(see Figure 3.1.20)}$$
$$= \frac{2(2084)}{3(13 - 11.4)(12)t_{we}}$$
$$= \frac{4168}{(57.6)(t_{we})}$$
$$= \frac{72.4}{t_{we}}$$
$$= \frac{72.4}{8}$$
$$= 9.05 \text{ ksi}$$

Observe that the depth of the compressive stress block is approaching the depth of the Whitney stress block (46 in.) and, even if the joint is fully effective in transferring the load ($t_{we} = t_w$), the stress distribution will become nonlinear accordingly, the stress in end of the panel will be somewhere between 4.5 and 9 ksi. However, this does not mean that the strain will necessarily reach 0.003 in./in. (see Figures 1.2.3 and 1.2.4). Regardless of the level of stress and strain reached

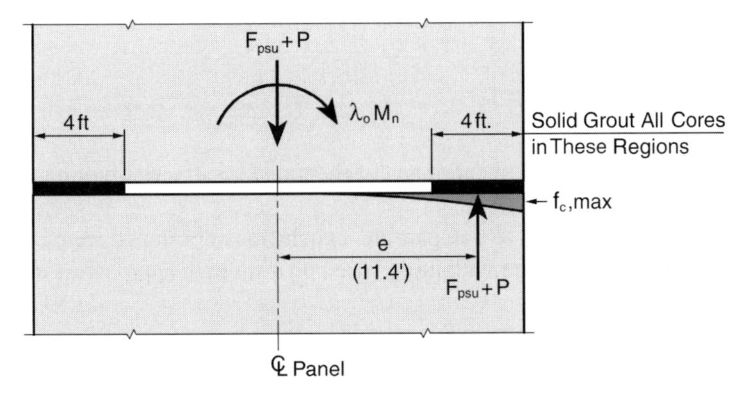

Figure 3.1.20 Resolution of forces at Level 1 (see Figure 3.1.19*a*).

in the ends of the panel, it seems prudent in this case to provide an effective load transfer mechanism in the outermost 4 ft of the panel. This can be accomplished by providing continuity bars in the hollow-core voids and insuring the placement of effective grout dams in this region (see Figure 3.1.19*d*).

Vertical load transfer between the end regions can be accomplished through the grout core (t_g), which at a minimum is, for this development, 3 in. wide. The stress on the grout (f_{cg}) is

$$f_{cg} = \frac{F_{psu} + P}{\ell_w t_g}$$

$$= \frac{2084}{26(12)3}$$

$$= 2.23 \text{ ksi}$$

The allowable bearing stress[3.11] is

$$f_{cg} = \phi 0.85 f'_c$$

$$= 0.7(0.85) f'_c$$

$$= 0.6 f'_c$$

Grout strengths and ductilities are typically high, and since 6-ksi grout is advisably placed at the ends of the wall, the typical vertical load transfer mechanism need not rely on the floor planks.

Observe that the shear transfer mechanism need only develop a friction factor of

$$f = \frac{V}{F_{ps} + P}$$

$$= \frac{231}{2084}$$

$$= 0.11$$

Accordingly, special treatment of any of the transfer surfaces is not required.

Comment: It is interesting to compare the conclusions developed from a displacement-based design procedure with those based on a strength approach as dictated by the building code.[3.7]

First, the basic prescriptive code period is

$$T = C_t(h_n)^{3/4} \qquad \text{(see Ref. 3.7, Eq. 30-9)}$$

$$= 0.02(150)^{3/4}$$

$$= 0.85 \text{ second}$$

This, when amplified by the currently allowed 30%, is 1.1 seconds. The period used in the displacement-based design was 1.28 seconds.

The design force level would be developed as follows:

$$C_v = \frac{S_v}{61.5} \qquad \text{(see Eq. 2.4.2b)}$$

$$= \frac{50}{61.5}$$

$$= 0.81$$

$$R = 4.5 \qquad \text{(Bearing wall[3.7])}$$

$$V = \frac{C_v I}{RT} W \qquad \text{(see Ref. 3.7, Eq. 30-4)}$$

$$= \frac{0.81(1)}{4.5(1.1)} W$$

$$= 0.164W$$

$$M_u - Vh_e$$

$$= 0.164(1350)(0.67)(150)$$

$$= 22{,}140 \text{ ft-kips}$$

$$M_n = \frac{M_u}{\phi}$$

$$= \frac{22{,}140(12)}{0.9}$$

$$= 295{,}200 \text{ in.-kips} \qquad (24{,}600 \text{ ft-kips})$$

The objective nominal moment capacity for the displacement-based design was 156,500 in.-kips, the variance is attributable to the period restriction and the conservative transformation of the single-degree-of-freedom response spectrum to a base shear spectrum (see Section 2.4.1.2, Eq. 2.4.2a).

The nominal capacity of the provided system would be

$$\begin{aligned} T_{ps} &= f_{ps} A_{ps} \\ &= (162 + 10)(8)(0.213) \qquad \text{(see Ref. 3.4, Eq. 18-4)} \\ &= 293 \text{ kips} \\ T_s &= 282 \text{ kips} \qquad \text{(Two DDCs)} \\ P &= 1350 \text{ kips} \\ a &= \frac{T_{ps} + T_s + P}{0.85 f_c' t_w} \\ &= \frac{293 + 282 + 1350}{0.85(6)(8)} \\ &= 47 \text{ in.} \\ M_n &= \left(T_{ps} + T_s + P\right)\left(\frac{\ell_w}{2} - \frac{a}{2}\right) \\ &= 1925(13 - 1.96) \\ &= 21{,}252 \text{ ft-kips} \end{aligned}$$

which is 86% of the code required nominal capacity (24,600 ft-kips).

3.1.4.2 Hybrid Wall System—Direct Displacement Design Procedure

The objective of this section is to demonstrate how important imagination is to the effective development of a design. The design must follow a loose form that allows alternatives to be quickly considered and previous design decisions rapidly adjusted.

The residential building described in Figures 3.1.17 and 3.1.19 is braced in the longitudinal direction by a series of nonbearing walls placed along the interior corridor. The design objective will usually be to provide the appropriate number of walls. The length of each wall must be compatible with building function. Function will usually dictate a wall length that is typical as, for example, down the corridor of the example building (Figure 3.1.17). Society, however, is increasingly unwilling to accept the rectangular blocks produced by functionally efficient units of the type shown in Figure 3.1.17. Also, from a leasing or sales perspective, a mixture of unit sizes is important, and this means that the use of a standard wall dimension will not be possible or will be structurally inefficient. In this case, unequal wall lengths, the concepts explored in Section 3.1.2 need to be incorporated into the design process.

In this section we explore the use of direct displacement design procedures (CVDD) and force-based design procedures in the search for a conceptual design solution for

Photo 3.1 Hybrid precast wall panel test program, PRESSS Program. (Courtesy of Englekirk Partners, Inc.)

the longitudinal bracing program. The goal is to select a precast system that will satisfy the objective strain limit states developed in Section 2.4.3.

Precast Hybrid Walls of Uniform Length First, follow Procedure 3 of Section 3.1.1.4, for it allows us to differentiate between the behavior of precast and cast-in-place systems.

Step 1: Adjust the Spectral Velocity to Account for the Inclusion of Structural Damping. Since we propose to use a hybrid system and our design objective drift is on the order of 1.5%, adopt a structural damping (ζ_{eg}) of 9% (see Figure 2.4.63*b*). Observe that the damping in a comparable cast-in-place system would be almost twice this level (Figure 2.4.63*a*). Total system damping for the precast system is

$$\begin{aligned} \hat{\zeta}_{\text{eq}} &= \zeta_{\text{eq}} + 5 \qquad \text{(Eq. 3.1.42)} \\ &= 9 + 5 \\ &= 14 \end{aligned}$$

Since the design spectral velocity assumes 5% nonstructural damping, use Eq. 3.1.45 to determine the spectral reduction factor $\hat{R}$,

$$\hat{R} = \frac{3.38 - 0.67 \ln \hat{\zeta}_{eq}}{2.3} \qquad \text{(Eq. 3.1.45)}$$

$$= \frac{3.38 - 0.67 \ln 14}{2.3}$$

$$= 0.7$$

and the spectral velocity of the substitute structure ($S_{v,s}$) is

$$S_{v,s} = \hat{R} S_v \qquad \text{(Eq. 3.1.46)}$$

$$= 0.7(50)$$

$$= 35 \text{ in./sec}$$

Step 2: Determine the Objective Period for the Substitute Structure. Adopt an objective target drift ratio of 1.8%.

$$\Delta_n = 0.018 h_n \qquad \text{(for the structure)}$$

$$= 0.018(150)(12)$$

$$= 32.4 \text{ in.}$$

For the substitute single-degree-of-freedom system, the spectral displacement may be determined in a variety of ways (see Eq. 3.1.15 through Eq. 3.1.19a). Use Equation 3.1.10 and a conservative participation factor of 1.5.

$$S_d = \frac{\Delta_u}{\Gamma} \qquad \text{(Eq. 3.1.10)}$$

$$= \frac{32.4}{1.5}$$

$$= 21.6 \text{ in.}$$

and the period for the substitute structure should be

$$T_s = \frac{2\pi S_d}{\hat{R} S_v} \qquad \text{(see Eq. 3.1.9 or 3.1.52a)}$$

$$= \frac{2\pi(21.6)}{35}$$

$$= 3.88 \text{ seconds}$$

For the real structure, assuming a system ductility factor (μ_s) of 4, the period is

$$T = \frac{T_s}{\sqrt{\mu}} \qquad \text{(see Eq. 3.1.52b)}$$

$$= \frac{3.88}{\sqrt{4}}$$

$$= 1.94 \text{ seconds}$$

Step 3: Determine the Dynamically Compatible Length of the Wall. The process becomes iterative, and designer understanding of functional constraints is important to a rapid conversion. Assume that the maximum wall length is 30 ft (functional constraint) and that you would like to use a 6-in. thick panel at least above the base (constructibility consideration—panel weight). Start the iterative process by assuming five panels that are 6 in. thick.

Use Eq. 3.1.33 to expedite the iterative process.

$$\ell_w = \left\{ \frac{0.13}{T} h_w^2 \left(\frac{w}{E t_w} \right)^{0.5} \right\}^{0.67} \qquad \text{(Eq. 3.1.33)}$$

$$= \left\{ \frac{0.13}{1.94} (150)^2 \left(\frac{w}{4000(0.5)} \right)^{0.5} \right\}^{0.67}$$

$$= \left\{ 1508 \sqrt{\frac{w}{2000}} \right\}^{0.67}$$

Now the only variable is w, which is the tributary weight (mass) per foot of height expressed in kips per foot. Since the total weight of each floor (W_f) is 1620 kips (Figure 3.1.17), the following values of $w(W_f/h_x)$ might reasonably be considered.

Five Walls

$$w = \frac{1620}{5(10)}$$

$$= 32.4 \text{ kips/ft}$$

$$\ell_{w5} = \left(1508 \sqrt{\frac{32.4}{2000}} \right)^{0.67}$$

$$= 33.8 \text{ ft}$$

This exceeds our space limitation of 30 ft.

Six Walls

$$w = \frac{1620}{6(10)}$$

$$= 27 \text{ kips/ft}$$

$$\ell_{w6} = \left(1508\sqrt{\frac{27}{2000}}\right)^{0.67}$$

$$= 31.8 \text{ ft}$$

Seven Walls

$$w = \frac{1620}{7(10)}$$

$$= 23.1 \text{ kips/ft}$$

$$\ell_{w7} = \left(1508\sqrt{\frac{23.1}{2000}}\right)^{0.67}$$

$$= 30 \text{ ft}$$

Conclusion: Propose seven walls that are 30 ft long.

Comment: It is always a good idea to periodically check your conclusions to catch math errors. Equation 2.4.11 provided an easy means of estimating the period of a structure. Use it to check the conclusion reached.

$$I_e = \frac{0.5 t_w \ell_w^3}{12} \qquad (I_e = 0.5 I_g)$$

$$= \frac{0.5(0.5)(30)^3}{12}$$

$$= 563 \text{ ft}^4$$

$$w = 23.1 \text{ kips/ft}$$

$$\Delta_f = \frac{w h_w^4}{8 E I_e}$$

$$= \frac{23.1(150)^4}{8(4000)(144)(563)}$$

$$= 4.5 \text{ ft}$$

$$T_f = 0.89(\Delta_f)^{0.5} \qquad \text{(Eq. 2.4.11)}$$

$$= 1.89 \text{ seconds} \cong 1.94$$

Step 4: Calculate the Effective Stiffness of the Substitute Structure. For the purposes of this example, assume a linear mode shape and use Eq. 3.1.19b. Since we propose to use seven walls,

$$M = \frac{W}{7g}$$

$$= \frac{15(1620)}{7(386.4)}$$

$$= 9 \text{ kip-sec}^2/\text{in.}$$

$$M_e = \frac{1.5(n+1)}{2n+1} M \qquad \text{(Eq. 3.1.19a)}$$

$$= 0.77(9)$$

$$= 7 \text{ kip-sec}^2/\text{in.}$$

The effective angular frequency of the substitute structure (ω_s) is

$$\omega_s = \frac{2\pi}{T_s}$$

$$= \frac{6.28}{3.88}$$

$$= 1.62 \text{ rad/sec}$$

$$k_{e,s} = \omega_s^2 M_e \qquad \text{(Eq. 3.1.47)}$$

$$= (1.62)^2(7)$$

$$= 18.3 \text{ kips/in.}$$

Step 5: Determine the Flexural Strength Required of Each Wall.

$$\Omega_o M_d = k_{e,s} S_{d,s} h_e \qquad \text{(Eq. 3.1.54)}$$

$$= 18.3(21.6)\left(\frac{h_w}{\Gamma}\right) \qquad \text{(see Step 2)}$$

$$= 18.3(21.6)\left(\frac{150}{1.5}\right)$$

$$= 39{,}500 \text{ ft-kips}$$

Step 6: Determine the Design Moment.

$$M_d = \frac{\Omega_o M_u}{\Omega_o}$$

$$= \frac{39{,}500}{1.5}$$

$$= 26{,}350 \text{ ft-kips}$$

The reinforcement program for the wall would be developed from Eq. 2.4.113, which was developed from

$$M_n = \left(P + F_{ps} + F_s\right)\left(\frac{\ell_w}{2} - \frac{a}{2}\right) \qquad \text{(Eq. 2.4.102b)}$$

where P is essentially the weight of the wall (340 kips), and

$$F_{ps} + F_{s,ed} = \frac{M_n - P(0.4\ell_w)}{0.4\ell_w} \qquad \text{(see Eq. 2.4.103a)}$$

$$= \frac{26{,}350 - 340(12)}{12}$$

$$= 1833 \text{ kips}$$

Following our objective relationship between F_{ps} and F_s, where $F_{ps} \cong F_s$ suggests

$$F_{ps} = F_s \cong 915 \text{ kips}$$

this does not seem like a reasonable solution, since it requires a prestress force on the order of 420 psi $(915/A_g)$.

An equal-displacement-based approach, given the seven-wall solution, would proceed as follows (see Procedure 2, Section 3.1.4.4):

$$T \cong 1.9 \text{ seconds}$$

$$\omega = 3.3 \text{ rad/sec}$$

$$S_a = \frac{S_v \omega}{g} g \qquad \text{(see Eq. 3.1.26)}$$

$$= \frac{50(3.3)}{386.4} g$$

$$= 0.42g$$

$$M_{\max} = S_a M_e h_e g \qquad \text{(see Figure 1.1.1)}$$

$$= 0.42(7)\left(\frac{150}{1.5}\right)(386.4)$$

$$= 155{,}600 \text{ ft-kips}$$

$$M_u = \frac{M_{\max}}{\mu_s \Omega_o}$$

$$= \frac{155{,}600}{4(1.5)}$$

$$= 25{,}900 \text{ ft-kips}$$

Conclusion: The suggested strength of each of the seven walls is on the order of 26,000 ft-kips, and this requires eight DDCs and a post-tensioning stress in excess of 400 psi. Alternative solutions must be investigated.

Alternative designs include:

- Increasing the number of walls—nine or ten walls might be used, and this would change the dynamic response.
- The transverse wall might reasonably be connected to form T sections. This would activate the post-tensioning in these walls as well as the dead load imposed on them.
- A vertical joint could be introduced into the panel and friction dampers or energy dissipater used to connect the walls.
- The strain states in the seven-wall solution could be studied to determine if higher levels of displacement ductility could be tolerated.

Increasing the number of walls would be accomplished using design procedures explored in Step 3. The T wall alternative explores new design concepts, and is developed next. Behavior enhancement opportunities available in the vertical joining of wall segments are discussed and demonstrated in Section 3.1.4.3.

Hybrid T *Wall Solutions* Longitudinal walls may be easily connected to transverse walls with a wet cast joint. The resultant shape is described in Figure 3.1.21. Both the transverse and the longitudinal designs would be impacted by this joining of the walls. The behavior of cast-in-place T walls was discussed in Section 2.4.1. Figure 2.4.7 describes a common behavior characteristic, specifically, how these walls tend to be much stronger in one direction than the other. From the standpoint of the typical residential building that contains a central corridor, as described in Figure 3.1.17, this is definitely an asset, for the strength of the total system will be increased. Since the reinforcing required was developed from minimums, we can assume that the design proposed in Figure 3.1.19*a* is appropriate for the joined walls of Figure 3.1.21*a*. In any case, the design of the longitudinal wall must precede any review of the transverse bracing program.

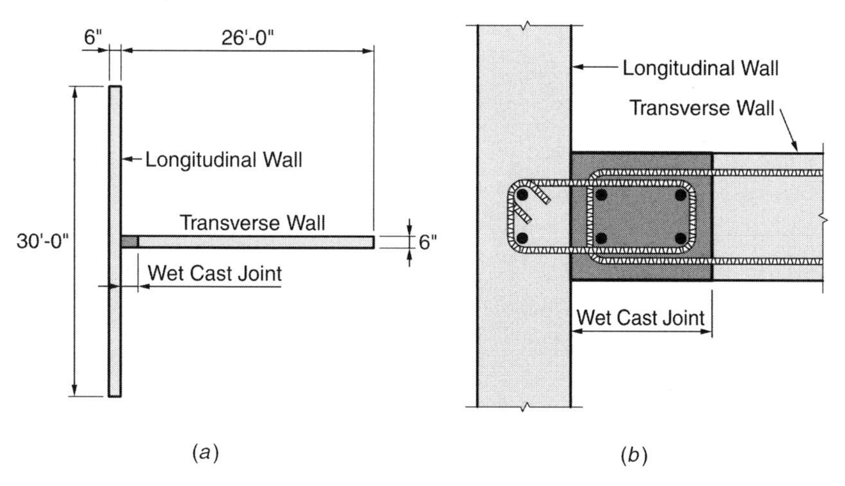

Figure 3.1.21 (*a*) Precast T wall. (*b*) Detail of wet cast joint.

Consider first the impact the transverse wall will have on the longitudinal wall from a design perspective. We know that the longitudinal wall will, at its strength limit state, open a gap about a neutral axis located at some distance from its toe. This means that the joined end of the transverse wall must also be raised. A force equal to half of the dead load, post-tensioning load, and energy dissipater reaction will be required to raise the transverse wall. Hence, the activated load at the middle of the longitudinal wall would include

$$P' = \frac{1}{2}\left(P_D + F_{ps} + F_{s,ed}\right) \tag{3.1.59}$$

$$= 0.5(1350 + 8(0.213)(162) + 2(141))$$

$$= 954 \text{ kips}$$

This significantly reduces the strength demand imposed on the post-tensioning and energy dissipaters to be installed in the longitudinal wall.

$$F_{ps} + F_{s,ed} = \frac{M_n - \left(P_D + P'\right)(0.4\ell_w)}{0.4\ell_w} \qquad \text{(see Eq. 2.4.103)}$$

$$= \frac{26{,}350 - (340 + 954)(12)}{12}$$

$$= 900 \text{ kips}$$

where P_D is the dead load supported by the longitudinal wall, its own weight in this case.

Consider reinforcement minimums.

$$F_{pse} = 0.125\ell_w t_w \qquad \text{(see Design Considerations, Section 3.1.4.1)}$$
$$= 0.125(30)(12)(6)$$
$$= 270 \text{ kips}$$

It appears as though two DDCs ($T_y = 141$ kips) are reasonably considered,

$$F_{s,ed} = 2(141)$$
$$= 282 \text{ kips}$$

as are eight 0.6-in. ϕ strands,

$$F_{psu} = 8(0.213)172$$
$$= 293 \text{ kips}$$

The nominal moment capacity of this system is developed as follows. Assume that the lowermost panel will be 8 in. thick, constructed using 6-ksi concrete, and confined at the end.

$$C_c = F_{ps} + F_{s,ed} + P_T + P' \qquad \text{(see Eq. 3.1.59)}$$
$$= 293 + 282 + 340 + 954$$
$$= 1869 \text{ kips}$$

$$c = \frac{C_c}{0.85 f'_{cc} b_e} \qquad (a \cong c)$$
$$= \frac{1869}{0.85(8.5)(6.5)}$$
$$= 40 \text{ in.}$$

$$M_n = C_c \left(\frac{\ell_w}{2} - 2 - c_c \right) \qquad \text{(see Figures 2.4.64}a\text{ and 2.4.59)}$$
$$= 1857(180 - 20 - 1)$$
$$= 297{,}000 \text{ in.-kips} \qquad (24{,}800 \text{ ft-kips})$$

Conclusion: The objective moment (M_u) was about 26,000 ft-kips (Section 3.1.4.2, Step 6). The reinforcement in both the transverse and longitudinal walls should be increased. The next design iteration should be of a balanced (equally prestressed) T wall section ($f_{ps} \cong 140$ psi).

3.1.4.3 Vertically Jointed Wall Panels Often it is impractical or impossible to construct a shear wall by stacking precast walls in the manner described in Figure

3.1.19*a*. The hybrid system may still be used with joined walls as was proposed Section 3.1.4.2, where a T section was developed by joining two panels with a wet joint (Figure 3.1.21*b*). The longer the created wall length becomes, the greater the strain imposed on the energy dissipater, and this increases the chance for rupture.

An energy dissipating device can be installed on the vertical panel joints using readily available materials, including plates or perforated plates similar to those shown in Figure 3.1.22. The objective of the plate design is to create a plate that will yield in shear, thereby creating a link beam (Section 2.4.2.1). Shear yielding should dominate when

$$0.8V_p\frac{\ell_p}{2} < M_p \qquad \text{(see Figure 3.1.22)}$$

$$0.8(0.6)F_y t_w h\frac{\ell_p}{2} < \frac{t_w h^2}{4}F_y \qquad \text{(see Eq. 2.4.47)}$$

$$0.96\ell_p < h \qquad (3.1.60)$$

Accordingly, shear yielding should occur provided the length of the plate (ℓ_p) is less than 96% of its height (h).

Connected panel system behavior is described in Figure 3.1.23. The deformation imposed on the energy dissipater will correspond to a vertical displacement of

$$\delta_{ed} = \theta_w \ell_w \qquad (3.1.61)$$

$$\ell_p = \frac{\theta_w}{\theta_{u,ed}}\ell_w$$

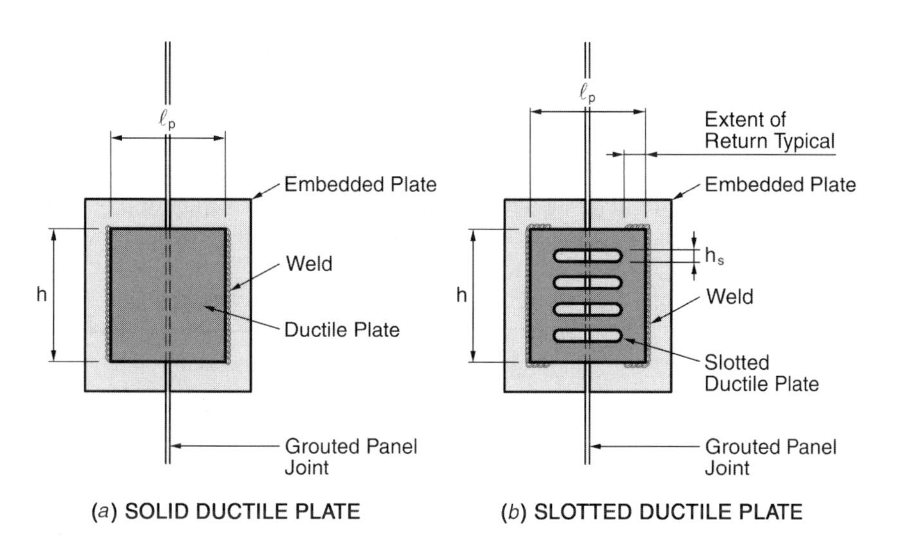

Figure 3.1.22 Energy dissipater—vertical panel joint.

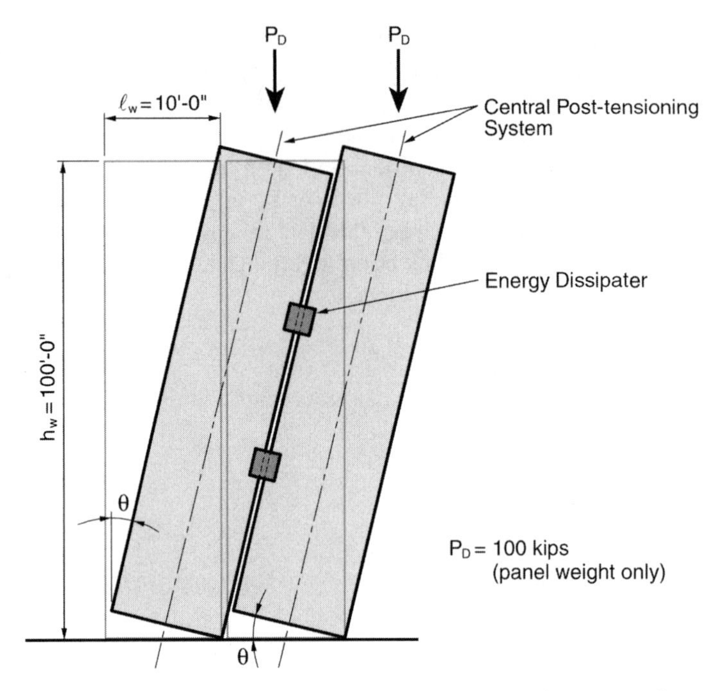

Figure 3.1.23 Vertically joined precast concrete wall panels, two-panel system.

And it follows that

$$\theta_{ed} = \frac{\delta_{ed}}{\ell_p} \qquad \left(\lambda_p, \text{Figure 2.4.25}\right)$$

$$= \frac{\theta_w \ell_w}{\ell_p} \tag{3.1.62}$$

The deformation angle (θ_{ed}) should be capable of attaining rotation of 0.08 radian. The design approach generally follows that developed for the coupled shear wall system (Sections 2.4.2.4 and 3.1.3).

Equation 3.1.62 tells a lot about probable component sizes. Assume that, because the created system will tend to be quite stiff, the drift ratio (θ, Figure 3.1.23) will probably be small, possibly less than 0.01 radian. If we assume that

$$\theta_{u,ed} = 0.08 \text{ radian}$$

and

$$\theta_{wu} = 0.01 \text{ radian} \qquad \text{(design objective)}$$

then

$$\ell_p > \frac{\ell_w}{9} \qquad \text{(see Eq. 3.1.61)}$$

Now if the panel length is 10 ft (recall that 12 ft is a practical maximum from a perspective of transportability), the system and component geometry are established.

$$\ell_p = \frac{120}{9}$$

$$= 13.33 \text{ in.}$$

Select a plate dimension of 14 in. by 16 in. ($\ell_p \times h$).

Consider next the behavior limit states of the coupled walls described in Figure 3.1.23. If these wall panels are 8 in. thick and post-tensioned to minimum levels, the behavior states may be developed as follows.

$$F_{pse} = 0.125\ell_w t_w \qquad \text{(Design Consideration, Section 3.1.4.1)}$$

$$= 0.125(120)(8)$$

$$= 120 \text{ kips}$$

$$A_{ps} = \frac{F_{pse}}{f_{pse}}$$

$$= \frac{120}{162}$$

$$= 0.74 \text{ in.}^2$$

Consider a four-strand prestressing program ($A_{ps} = 0.85$ in.^{2}).

$$F_{pse} = 4(0.213)(162)$$

$$= 138 \text{ kips}$$

The design objective is to create a restoring force as well as energy dissipation; hence the strength of the energy dissipaters must be less than that of the post-tensioning plus the activated dead load (P_D).

$$F_{ed} < F_{pse} + P_D$$

Consider using two energy dissipaters.

$$V_p < \frac{138}{2} + \frac{P_D}{2}$$

$$= 69 + \frac{P_D}{2}$$

Since the geometry of the dissipater has been established, solve for the thickness of the plate.

$$V_p = 0.6F_y t_w h \qquad \text{(see Eq. 2.4.48)}$$

$$t_p = \frac{V_p}{0.6F_y h}$$

$$= \frac{69 + P_D/2}{0.6(50)(16)}$$

Comment: P_D is limited to the dead load tributary to the energy dissipater; accordingly it may be neglected. Therefore,

$$t_p = 0.081 \text{ in.} \qquad \left(\tfrac{1}{12} \text{ of an inch}\right)$$

Practically, this is too thin. Revise the number of connectors and reverse the design process.

Structural steel codes[3.11] identify the minimum thickness required to maintain the shear capacity of a panel zone under cyclic loading as

$$t_p > \frac{d_z + w_z}{90} \tag{3.1.63}$$

but this assumes that the plate will be stabilized on all edges.

Stiffener spacing in a shear link is determined through the use of

$$s = 30t_w - \frac{d}{5} \tag{3.1.64}$$

A $\frac{3}{8}$-in. thick plate would satisfy Eq. 3.1.63, while a $\frac{5}{8}$-in. thick plate would be required by Eq. 3.1.64. Shear buckling is not a catastrophic condition, so one might adopt a $\frac{1}{2}$-in. thick plate and confirm behavior by testing.

$$V_p = 0.6F_y t_w h \qquad \text{(see Eq. 2.4.47b)}$$

$$= 0.6(50)(0.5)(16)$$

$$= 240 \text{ kips}$$

Now size the post-tensioning to produce this shear on the energy dissipater. Twelve strands produce an effective prestress of

$$F_{pse} = 12(0.213)(162)$$

$$= 414 \text{ kips} \qquad (575 \text{ psi})$$

The behavior of the system is described in Figure 3.1.23.

$$F_{pse} + P_D = 12(0.213)(162) + 100$$
$$= 514 \text{ kips}$$

Conclusion: Provide two energy dissipaters ($2V_p = 480$ kips)

System strength will be provided by two mechanisms, the system working as a unit and each panel acting individually (Figure 3.2.24*a*). The idealized yield strength would be that associated with dissipater yield or strand decompression.

$$M_n = \sum V_p \left(\ell_w + \frac{\ell_w}{2} - 0.1\ell_w \right) + \left(F_{ps} + P_D \right) \left(\frac{\ell_w}{2} - 0.1\ell_w \right)$$
$$+ \left(F_{ps} + P_D - \sum V_P \right) \left(\frac{\ell_w}{2} - 0.1\ell_w \right)$$
$$= \sum Vp \left(\ell_w \right) + \left(2F_{ps} + 2P_D \right) (0.4\ell_w)$$
$$= 480(10) + 1080(4)$$
$$= 9120 \text{ ft-kips} \qquad \text{(at energy dissipater yield)}$$

A balancing of the post-tensioning and dissipater yield shear will cause decompression to occur before first yielding of the energy dissipater.

Incipient lift-off $f_c = 0$ (see Figure 3.1.24*b*)

$$M = \left(2F_{pse} + 2P_D \right) \left(\frac{2\ell_w}{6} \right)$$
$$= 1028(3.33)$$
$$= 3423 \text{ ft-kips}$$

Decompression $f_c = 0$ at tension strand

$$C_c = 2F_{pse} + 2P_D$$
$$= 2(514)$$
$$= 1028 \text{ kips}$$

The centroid of the compression force at decompression must be located $0.5\ell_w$ from the compression face of the wall system (Figure 3.1.24*b*).

$$M_n = 0.5\ell_w C_c$$
$$= 5(1028)$$
$$= 5400 \text{ ft-kips} \qquad \text{(at strand decompression)}$$

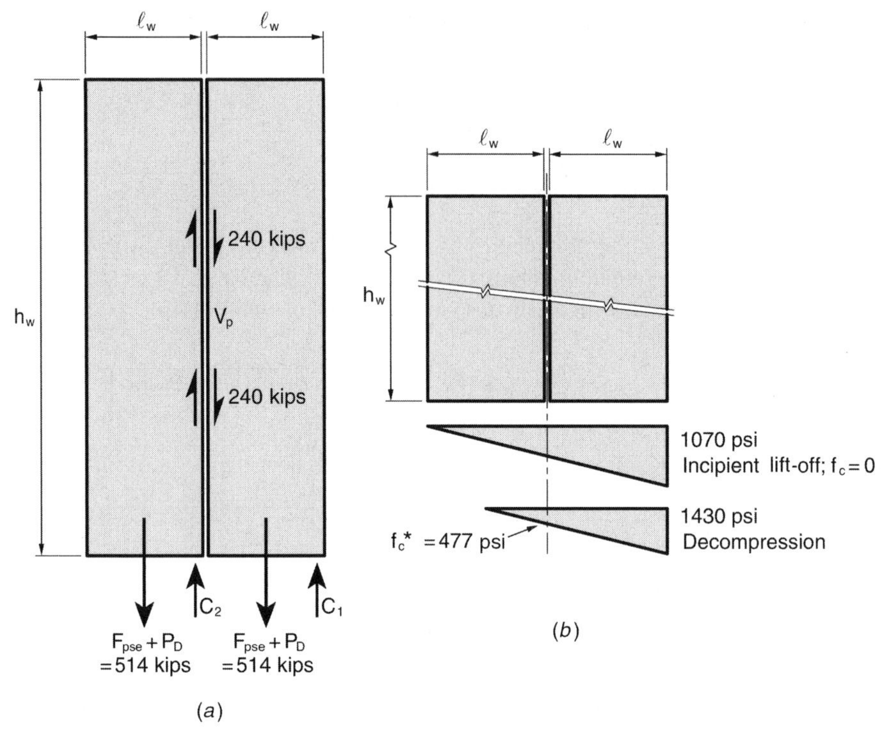

Figure 3.1.24 (*a*) Vertical force distribution on two-panel system at system strength. (*b*) Panel stresses at various load stages.

$$f_c = \frac{2C_c}{1.5\ell_w t_w} \qquad \text{(see Figure 3.1.24}b\text{)}$$

$$= \frac{2(1028)}{1.5(10)(12)(8)}$$

$$= 1.43 \text{ ksi}$$

$$\varepsilon_c = \frac{f_c}{E_c}$$

$$= \frac{1.43}{4000}$$

$$= 0.000357 \text{ in./in.}$$

$$\phi_y = \frac{\varepsilon_c}{1.5\ell_w}$$

$$= \frac{0.000357}{1.5(10)(12)}$$

$$= 0.000002 \text{ rad/in.}$$

$$\Delta_y = \frac{\phi_y h_w^2}{3} \qquad \text{(Conjugate Beam—linear strain gradient)}$$

$$= 0.96 \text{ in.}$$

The load on the energy dissipaters at decompression is

$$V_{ed} = F_{pse} + P_D - f_c^* \frac{0.5\ell_w t_w}{2} \qquad \text{(see Figure 3.1.24}b\text{)}$$

$$= 514 - \frac{0.477(5)(12)(8)}{2}$$

$$= 400 \text{ kips}$$

Conclusion: The energy dissipaters will remain elastic until the strain in the post-tensioning begins to increase. The idealized behavior of the wall system is reasonably associated with $M_n = 9120$ ft-kips and $\Delta_y \cong 1.0$ in. (incipient dissipater yield).

The postyield deformation of the system will assume the form described in Figure 3.1.23 and the compression loads imposed on the toe of each panel (Figure 3.1.24*a*) will be the larger of

$$C_1 = F_{ps} + P_D + \lambda_o \sum V_p$$

$$= 540 + 1.25(480)$$

$$= 1140 \text{ kips}$$

$$C_1^* = 2\left(F_{ps} + 2P_D\right)$$

$$= 1080 \text{ kips}$$

$$C_2 = F_{ps} + P_D - \lambda_o \sum V_p$$

$$\cong 0$$

The probable deflection of the system can be estimated in a variety of ways. The system may be reduced to a single-degree-of-freedom system of effective mass (M_e) and idealized stiffness developed either from M_n (M_{yi}) and Δ_{yi} or directly from behavior idealizations.

If, for example, the tributary mass at each floor (W_f) is 150 kips, then

$$M = \frac{nW_f}{g}$$

$$= \frac{10(150)}{386.4}$$

$$= 3.9 \text{ kips-sec}^2/\text{in.}$$

$$M_e = \frac{1.5(n+1)}{2n+1} M \qquad \text{(Eq. 3.1.19a)}$$

$$= \frac{1.5(11)}{21} M$$

$$= 0.79M$$

$$= 3.06 \text{ kips-sec}^2/\text{in.}$$

The effective stiffness of the substitute single-degree-of-freedom structure is developed as follows:

$$M_n = M_{yi}$$

$$= 9120 \text{ ft-kips}$$

$$\Gamma = \frac{3n}{2n+1} \qquad \text{(Eq. 3.1.18a)}$$

$$= \frac{30}{21}$$

$$= 1.43$$

$$h_e = \frac{h_w}{\Gamma} \qquad \text{(Eq. 3.1.16)}$$

$$= \frac{100}{1.43}$$

$$= 70 \text{ ft}$$

$$H_y = \frac{M_n}{h_e} \qquad \text{(see Figure 3.1.1)}$$

$$= \frac{9120}{70}$$

$$= 130.3 \text{ kips}$$

$$k_e = \frac{H_y}{\Delta_{yi}}$$

$$= \frac{130.3}{0.96}$$

$$= 135.7 \text{ kips/in.}$$

$$\omega = \sqrt{\frac{k_e}{M_e}}$$

$$= \sqrt{\frac{135.7}{3.06}}$$

$$= 6.66 \text{ rad/sec}$$

$$T = \frac{2\pi}{\omega}$$

$$= 0.94 \text{ second}$$

Following the equal displacement method (CVED) and assuming a spectral velocity of 50 in./sec, we get

$$S_d = \frac{S_v}{\omega} \qquad \text{(Eq. 2.4.6)}$$

$$= \frac{50}{6.66}$$

$$= 7.5 \text{ in.}$$

$$\Delta_u = \Gamma S_d \qquad \text{(Eq. 2.4.7)}$$

$$= 1.43(7.5)$$

$$= 10.74 \text{ in.}$$

$$\Delta_p = \Delta_u - \Delta_y$$

$$= 10.74 - 0.96$$

$$= 9.8 \text{ in.}$$

The two panels will have become uncoupled (Figure 3.1.23) and the rotation of each is

$$\theta = \frac{\Delta_p}{h_w}$$

$$= \frac{9.8}{100(12)}$$

$$= 0.008 \text{ radian}$$

and the rotation imposed on the link or energy dissipater becomes

$$\theta_{ed} = \frac{\theta_w \ell_w}{\ell_p} \qquad \text{(Eq. 3.1.62)}$$

$$= \frac{0.008(120)}{14}$$

$$= 0.07 \text{ radian}$$

and this is less than the proposed allowable of 0.08 radian.

From a detailing perspective it is important to insure that the overstrength of the plate can be developed into the panel. This will be impossible given the solid plate described in Figure 3.1.22*a* unless its end section is reinforced in some manner, as, for example, by the slotting described in Figure 3.1.22*b*.

A better design approach would be to use four energy dissipating plates that are slotted or drilled like the one shown in Figure 3.1.25. Drilled sections have been successfully tested in the development of ductility in steel assemblies and represent a good solution for energy dissipation in precast concrete.

Multiple panel assemblies can be created following the same general approach (see Figure 3.1.26). The most practical and behavior effective approach is to create an energy dissipater that yields in shear at a load that is less than the effective prestress, as was done in the two-panel system described in the preceding development. This will cause the units to rotate individually, minimizing the strain imposed on the strand and dissipating a maximum amount of energy. The strength of the system is developed logically from Figure 3.1.26 for as many panels (n) as might be provided.

$$M_n = \sum V_p(n-1)\ell_w + n\left(T_{ps} + P_D\right)0.4\ell_w \quad (3.1.65a)$$

$$C_1 = \sum V_p + T_{ps} + P_D \quad \text{(End panel, compression side)} \quad (3.1.65b)$$

$$C_i = T_{ps} + P_D \quad \text{(Interior panel)} \quad (3.1.65c)$$

$$C_2 = T_{ps} + P_D - \sum V_p \quad \text{(End panel, tension side)} \quad (3.1.65d)$$

The preceding development, most importantly, describes how equilibrium combined with an objective postyield behavior state can be effectively used to create energy absorbing precast concrete systems.

Example Design The design example developed in Section 3.1.4.2 for the longitudinal walls of the residential building described in Figures 3.1.17 and 3.1.19 is now extended into a series of vertically and horizontally connected precast panels similar to the one tested in the PRESSS program (Figure 2.4.54). The purpose of the design is to allow a contractor to evaluate the cost effectiveness of this design. Accordingly, only key elements need to be developed quantitatively.

From the designs developed in Section 3.1.4.2 the following parameters provide a reasonable starting point.

Number of walls: 7

Flexural strength objective (M_n): 26,500 ft-kips

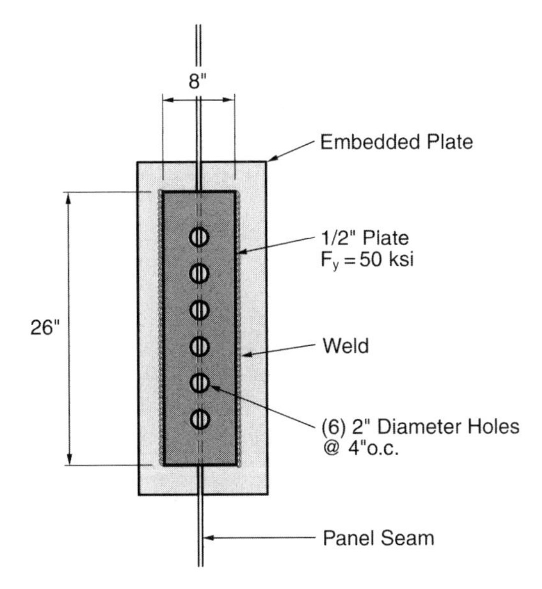

Figure 3.1.25 Energy dissipating plate.

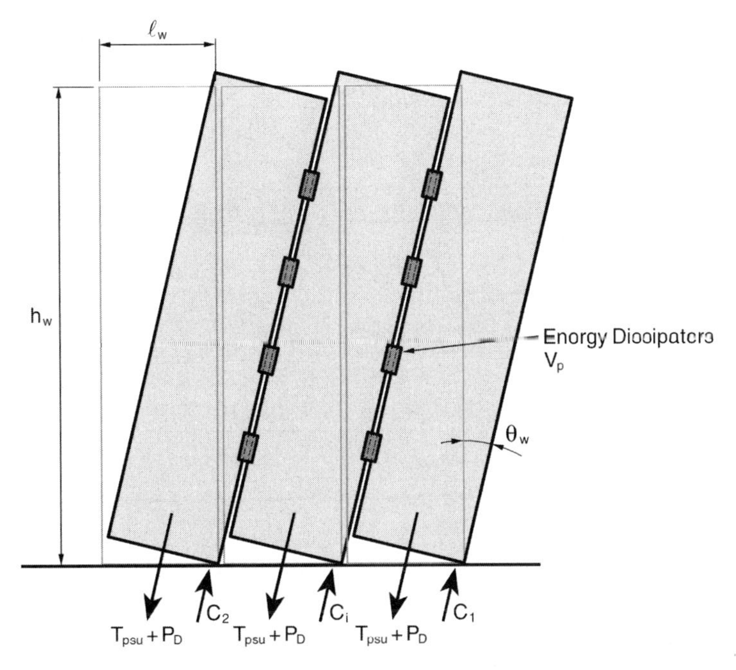

Figure 3.1.26 Multiple-panel wall system.

Drift objective: $0.018h_w$
Postyield panel rotation (Figure 3.1.26): $0.8(0.018h_w)$
Panel dimensions: 10 ft by 50 ft (Figure 3.1.19*a*)

Two things should be obvious based on the conceptual designs developed in Section 3.1.4.2.

- Most of the strength will need to be provided by the energy dissipaters so the quantification of the post-tensioning will be fairly arbitrary.
- Concrete strain levels in the toe will be large and confinement will be required—the panels should be at least 8 in. thick.

Step 1: Select the Post-Tensioning Reinforcement. Propose an effective prestress of about 150 psi.

$$\begin{aligned} F_{pse} &= 0.15\ell_w t_w \\ &= 0.15(120)(8) \\ &= 144 \text{ kips} \end{aligned}$$

Conclusion: Provide six strands (Figure 2.4.66*b*)

$$\begin{aligned} F_{psu} &= A_{ps} f_{ps} \\ &= 6(0.213)(175) \\ &= 224 \text{ kips} \end{aligned}$$

Step 2: Determine the Number of Energy Dissipaters Required.

$$M_n = \sum V_p(n-1)\ell_w + n\left(T_{ps} + P_D\right)0.4\ell_w \quad \text{(Eq. 3.1.65a)}$$

$$\begin{aligned} \sum V_p &= \frac{M_n - n\left(T_{ps} + P_D\right)0.4\ell_w}{(n-1)\ell_w} \\ &= \frac{26{,}500 - 3(244 + 158)0.4(10)}{2(10)} \\ &= 1083.8 \text{ kips} \end{aligned}$$

Conclusion: Propose 5 energy dissipaters similar to the one developed in the preceding example ($V_p = 240$ kips).

Step 3: Check to Insure That Concrete Strains Are within Performance Objectives.

$$C_1 = \sum V_p + T_{ps} + P_D \quad \text{(Eq. 3.1.65b; Figure 3.1.26)}$$

$$= 5(240) + 224 + 158$$

$$= 1582 \text{ kips}$$

$$c = \frac{C_c}{0.85 f'_{cc} b_e} \qquad \text{(Figure 2.4.59)}$$

$$= \frac{1582}{0.85(8.5)(6.5)} \qquad (f'_c = 6 \text{ ksi})$$

$$= 33.7 \text{ in.}$$

$$\theta_{wp} = 0.8(0.018)$$

$$= 0.0144 \text{ rad}$$

$$\phi_p = \frac{\theta_{wp}}{\ell_p} \qquad (\ell_p = t_w)$$

$$= \frac{0.0144}{8}$$

$$= 0.0018 \text{ rad/in.}$$

$$\varepsilon_{cp} = \phi_p c$$

$$= 0.0018(33.7)$$

$$= 0.06 \text{ in./in.}$$

Since this considerably exceeds our objective maximum of 0.025 in./in. consider higher-strength concrete $\left(f'_c = 12 \text{ ksi}\right)$ and a thicker lowermost wall ($t_w = 10$ in.).

$$c = \frac{1582}{0.85(14.5)(8.5)}$$

$$= 15.1 \text{ in.}$$

$$\phi_p = \frac{0.0144}{10}$$

$$= 0.00144$$

$$\varepsilon_{cp} = 0.00144(15.1)$$

$$= 0.0217 \text{ in./in.}$$

Design Conclusion

- Use 10-in. thick wall panels at the lowermost lift, $f'_c = 12$ ksi.
- Use 8-in. thick wall panels from level 5 thru 15, $f'_c = 5$ ksi.
- Use six 0.6-in. ϕ strands at the center of each panel.
- Provide five energy dissipaters at each wall interface (Figure 3.1.26), ten total.

3.2 FRAME BRACED BUILDINGS

The behavior and design of the components of a ductile frame have been studied in Sections 2.1, 2.2, and 2.3. Designs were based on an assumed strength objective, and this is the usual case. In Section 3.1 we have explained various means of deciding how strong shear walls should be and selecting the appropriate number and length. In this section, we endeavor to decide how many frames are required to brace a building. As was the case in the design of shear walls, functional and aesthetic objectives will usually limit our choices, so we are not entirely free to select optimal spans, beam depths, and frame locations. Further, attaining or optimizing frame ductility will impose another set of constraints on the design of a frame—beam span to depth ratios must be maintained, reinforcement ratios cannot be too high, induced levels of shear on beams, columns, and beam-column joints must be controlled and limited, and so on. Accordingly, the usual case will be that frame geometry is fairly well established and the design problem is to determine how many frames or frame bays are required.

Another feature or characteristic of concrete construction, be it cast-in-place or precast, is that member sizes, once established, will not change over the height of the building. Only the level of provided strength will change. In other words, beams will not usually be deeper in the lower levels than they are higher up in the building, and the same applies to columns. Changing the formwork to accommodate a change in beam size from one level to another is expensive, and the resulting structure takes longer to build. Time and forming costs are more expensive than a few cubic feet of concrete. In columns a transition in column sizes not only requires a new set of forms, but it creates an expensive transition in the column reinforcing program, one that must be carefully programmed in the design details and shop drawing development. Obviously, in buildings in the thirty-plus-story range, a change in member size may reasonably be considered; otherwise adjustments in column strength should be limited to the strength of the concrete and the amount and strength of the reinforcing.

These constraints tend to make the design task easier, and the design procedures developed in Section 3.2.1 are usually based on an adopted set of constraints.

Design procedures that are force based will be described in Section 3.2.2. Displacement-based procedures will be developed in Section 3.2.3. Damping-sensitive approaches are developed in Section 3.2.4, which explores the design of precast concrete frames.

3.2.1 Design Objectives and Methodologies

Strength- and displacement-based design procedures are both appropriately used to design frames. Regardless of the approach, system dynamics and strength must be compatible. This means that system stiffness and strength limit states must be understood during the conceptual design phase. The required reconciliation of system stiffness, strength, and dynamic response can be considerably simplified through insight into system behavior.

The deformed shape of a frame-braced structure is dominated by a sliding of planes or shear type deformation. Frame flexure can influence the displacement shape of the building, but its contribution will be small if the aspect ratio of the frame (frame height/frame length) is three or less. Methodologies for estimating building periods for frames that have large aspect ratios are contained in Reference 3.6, Sec. 4.7.4(b). Here we presume the dominance of shear deformation.

The fundamental period of a frame structure dominated by a shear type deformation[3.6 and 3.12] is

$$T = 0.288(\Delta_s)^{0.5} \tag{3.2.1}$$

where Δ_s is the deflection of the frame in inches when loaded by its tributary mass (weight).

The shear deformation (Δ_s) will be a function of the stiffness of the subassemblies that make up the frame. Procedures for developing estimates of subassembly drift were also developed in Reference 3.6, Sec. 4.2. Specifically, the deflection of an interior subassembly is

$$\Delta_i = \frac{V_i h_x^2}{12E}\left[\frac{\ell_c}{I_b}\left(\frac{\ell_c}{\ell}\right)^2 + \frac{h_c}{I_c}\left(\frac{h_c}{h_x}\right)^2\right] \tag{3.2.2a}$$

where

h_x is the story height in inches.
E is the modulus of elasticity in kips per square inch.
ℓ is the beam length—node to node.
ℓ_c is the clear span of the beam in inches.
h_c is the clear span of the column in inches.
I_b is the effective moment of inertia of the beam in in.^{4}.
I_c is the effective moment of inertia of the column in in.^{4}.

Similarly, the deflection of an exterior subassembly may be ascertained.

$$\Delta_e = \frac{V_e h_x^2}{6E}\left[\frac{\ell_c}{I_b}\left(\frac{\ell_c}{\ell}\right)^2 + \frac{h_c}{2I_c}\left(\frac{h_c}{h_x}\right)^2\right] \tag{3.2.2b}$$

NOTE: In the development of Eq. 3.2.2b the length of the exterior beam is the full length of the exterior beam (node to node). See Reference 3.6, Fig. 4.2.8.

Comment: In Section 2.1.1 the stiffness of a beam was developed from experimental efforts (Figure 2.1.2). It was concluded that the effective moment of inertia for beams should be between 0.25 and $0.35I_g$. From a design perspective the ACI recommendation of $0.35I_g$ seems reasonable. The stiffness of a column was evaluated in Section 2.2.2 and the adopted effective moment of inertia was $0.7I_g$.

In Section 2.3.2 the results of a subassembly (Figure 2.3.7) test were presented and conclusions relative to the capacity of the beam-column joint developed. The behavior of the test subassembly described in Figure 2.3.8 allows us to calibrate our subassembly stiffness relationship (Eq. 3.2.2a).

$$I_{bg} = \frac{bh^3}{12}$$

$$= \frac{20(30)^3}{12} \qquad \text{(see Figure 2.3.7}b\text{)}$$

$$= 45{,}000 \text{ in.}^4$$

$$I_{be} = 0.35 I_{bg}$$

$$= 15{,}750 \text{ in.}^4$$

$$I_{cg} = \frac{bh^3}{12}$$

$$= \frac{24(30)^3}{12} \qquad \text{(see Figure 2.3.7}b\text{)}$$

$$= 54{,}000 \text{ in.}^4$$

$$I_{ce} = 0.7 I_{cg}$$

$$= 37{,}800 \text{ in.}^4$$

$$h_x = 197 \text{ in.} \qquad \text{(see Figure 2.3.7}a\text{)}$$

$$h_c = 167 \text{ in.}$$

$$\ell = 240 \text{ in.}$$

$$\ell_c = 210 \text{ in.}$$

$$E = 4400 \text{ ksi}$$

$$\Delta_i = \frac{V_i h_x^2}{12E}\left[\frac{\ell_c}{I_{be}}\left(\frac{\ell_c}{\ell}\right)^2 + \frac{h_c}{I_c}\left(\frac{h_c}{h_x}\right)^2\right] \qquad \text{(Eq. 3.2.2a)}$$

$$= \frac{V_i(197)^2}{12(4400)}\left[\frac{210}{15{,}750}\left(\frac{210}{240}\right)^2 + \frac{167}{37{,}800}\left(\frac{167}{197}\right)^2\right]$$

$$= 0.74 V_i (0.01 + 0.003)$$

$$= 0.0096 V_i$$

A beam load (Figure 2.3.8) of 165 kips would be equilibrated by a story shear of

$$V_i = V_b\left(\frac{\ell}{h_x}\right)$$
$$= 165\left(\frac{240}{197}\right)$$
$$= 200 \text{ kips}$$

This suggests a story drift of

$$\Delta_i = 0.0096V_i \qquad \text{(see Eq. 3.2.2a)}$$
$$= 0.0096(200)$$
$$= 1.92 \text{ in.}$$

and a subassembly rotation of

$$\theta_y = \frac{\Delta_i}{h_x}$$
$$= \frac{1.92}{197}$$
$$= 0.0097 \text{ radian}$$

This corresponds to a beam deflection of

$$\delta_b = \theta\frac{\ell}{2}$$
$$= 0.0097(120)$$
$$= 1.17 \text{ in.}$$

And this seems to be consistent with the reported deflection pattern described in Figure 2.3.8.

Strength limits for the components of the frame were developed in Section 2.3. Those that provide design guidance appropriate to the frame sizing process are:

For the interior subassembly,

$$V_{cu} = 0.22A_j \qquad \text{(Office buildings)} \qquad \text{(Eq. 2.3.22a)}$$
$$V_{cu} = 0.18A_j \qquad \text{(Residential buildings)} \qquad \text{(Eq. 2.3.22b)}$$

where A_j is the gross area of the joint expressed in square inches.

For an exterior subassembly joint, shear limit states will typically not control the design. The column sizing process must satisfy

$$A_j = 96A_s \qquad \text{(Eq. 2.1.17)}$$

where A_s is the flexural reinforcing proposed for the frame beam. The shear strength provided by a subassembly should be based on an objective maximum beam reinforcement program of no more than 1.5% (see Section 2.1.2.5).

$$V_{cu} = \frac{0.015bd(d - d')\phi f_y}{h_x} \qquad \text{(Exterior subassembly)} \qquad (3.2.3a)$$

$$\cong \frac{0.015b(0.9)h(0.8)h(0.9) f_y}{h_x}$$

$$= \frac{0.01bh^2 f_y}{h_x} \qquad \text{(Exterior subassembly)} \qquad (3.2.3b)$$

$$= \frac{0.02bh^2 f_y}{h_x} \qquad \text{(Interior subassembly)} \qquad (3.2.3c)$$

where b and h are the out-to-out dimensions of the frame beam.

The strength provided by a ductile frame designed to a capacity-based criterion is most easily developed using mechanism-based concepts. We have used this approach in Chapter 2 to develop beam design strength objectives (see Figure 2.1.25 and 2.2.9).

The shear strength provided by the frame at any level is related to the flexural strength of frame beams.

$$V_M = \frac{\sum M_{bn}(\ell/\ell_c)}{h_x} \qquad (3.2.4a)$$

where V_M is the story mechanism shear and the amplification factor (ℓ/ℓ_c) transfers the beam strength to the centerline of the column from the face of the column; it can be treated as a constant in the design process.

The mechanism approach is often questioned because it seems to rely on the occurrence of a point of inflection in the column, which is presumed to be constant, hence separated by h_x. This assumption is especially a concern when it is used to develop the strength of a two-story frame. The concern has no basis, however. Consider the two-story frame described in Figure 3.2.1 subjected to a primary mode shape distribution of shears. A frame mechanism will occur in a frame designed to a capacity-based criterion when all of the plastic hinges shown form. This allows the following relationships to be developed:

$$AM_p + DM_p = 2V_e h_{x2} \qquad (3.2.4b)$$

$$BM_p + EM_p = 3V_e h_{x1} \qquad (3.2.4c)$$

$$D + E = 1$$

$$c = \frac{h_{x2}}{h_{x1}}$$

Rearranging, we get:

Exterior Column

$$A + D = \frac{2V_e c h_{x1}}{M_p} \qquad \text{(see Eq. 3.2.4b)}$$

$$B + E = \frac{3V_e h_{x1}}{M_p} \qquad (3.2.4d)$$

$$A + 1 - E = \frac{2V_e c h_{x1}}{M_p} \qquad (3.2.4e)$$

Combining Eq. 3.2.4d and 3.2.4e results in

$$A + B + 1 = (2c + 3)\frac{h_{x1} V_e}{M_p}$$

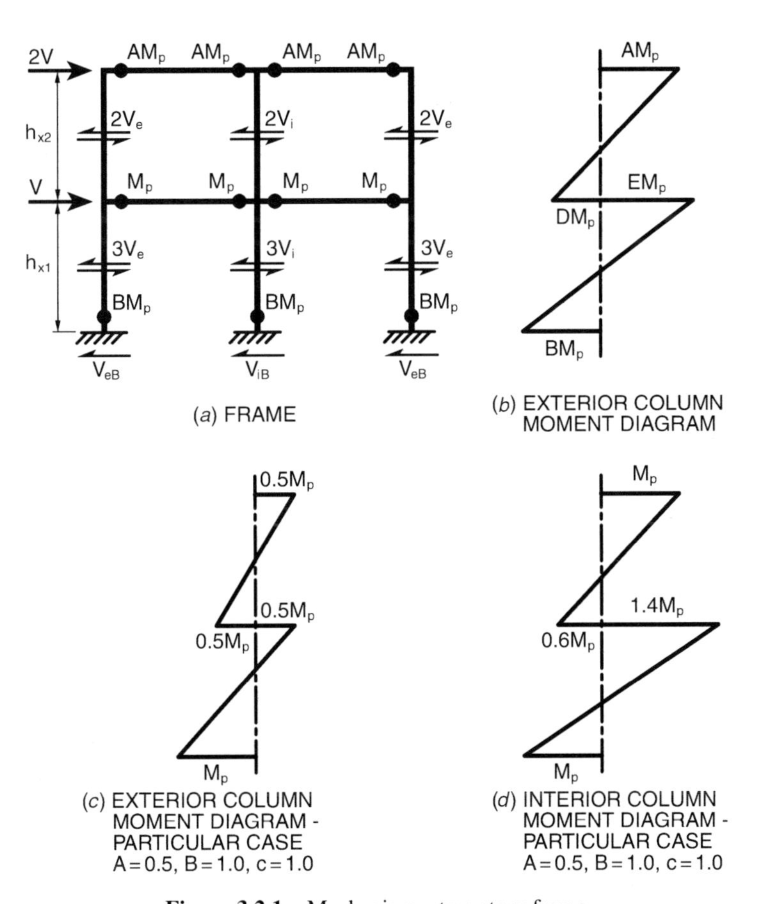

Figure 3.2.1 Mechanism—two story frame.

Now

$$V_e = \left(\frac{A + B + 1}{2c + 3}\right) \frac{M_p}{h_{x1}} \tag{3.2.4f}$$

and, for the case where $A = \frac{1}{2}$, $B = 1$, and $c = 1$,

$$V_e = \frac{2.5}{5} \frac{M_p}{h_{x1}}$$

and

$$D = \frac{2V_e h_{x1}}{M_p} - A \qquad \text{(see Eq. 3.2.4b)}$$

$$= 2\left(\frac{1}{2}\right) - \frac{1}{2}$$

$$= 0.5$$

and the shear at the base of an exterior column is

$$V_{eB} = 3V_e$$

$$= 3\left(\frac{2.5}{5}\right) \frac{M_p}{h_{x1}}$$

$$= 1.5 \frac{M_p}{h_{x1}}$$

Observe that this is 50% more than the shear that would be developed from Eq. 3.2.4a.

Similarly, for an interior column, it follows that

$$V_i = \left(\frac{2A + B + 2}{2c + 3}\right) \frac{M_p}{h_{x1}} \tag{3.2.4g}$$

And, for the particular case where $A = \frac{1}{2}$, $B = 1$, and $c = 1$,

$$V_i = 2.4 \frac{M_p}{h_{x1}} \tag{3.2.4h}$$

In addition, the mechanism shear for the frame of Figure 3.2.1, given the particular conditions ($A = 0.5$, $B = 1.0$, $c = 1.0$), becomes

$$V_M = 2V_e + V_i \tag{3.2.4i}$$

$$= (3 + 2.4) \frac{M_p}{h_{x1}}$$

$$= 5.4\frac{M_p}{h_{x1}}$$

Had we used Eq. 3.2.4a to estimate the mechanism base shear, we would have concluded that

$$V_M = \frac{4M_p}{h_{x1}} \qquad \text{(see Eq. 3.2.4a)}$$

and the mechanism strength prediction would have been conservative by 35% (5.4/4). If, however, the objective were to predict capacity-based demands, Eq. 3.2.4a would be nonconservative and one would need to use equations similar to those developed in the preceding example or a sequential yield analysis to predict shears.

3.2.1.1 How to Avoid Lower Level Mechanisms When the base of a column is rigidly supported by the substructure, it will yield as a part of the development of a mechanism in the lower part of a frame braced building. This cannot be avoided. The best a designer can hope to accomplish is to delay the onset of postyield rotation in the column as long as possible. This delaying action can be accomplished by allowing the column some rotational freedom at the base or by providing a yield strength that will force the inception of the mechanism into the stories immediately above the lowermost story. Consider the impact of $B = 2.0$ on Eq. 3.2.4g.

The problem is usually further complicated by the fact that the lowermost floor is often taller than the typical floor. In concrete construction frame beam sizes are constrained by the forming system (Section 3.2.1), so it is not a simple matter of providing a deeper, stronger frame beam. In precast concrete construction the connector size will usually establish the beam size and, as for example in the DDC system (Section 2.1.4.5), increasing the strength of the frame will not be practical. In order to promote a more uniform mechanism, the point of inflection in the lowermost column (Figure 3.2.1) must be driven upwards so as to at least reach a height of $(h_{x1} - h_{x2}/2)$. Observe that this was accomplished in the "particular case" (Figure 3.2.1*c*) by making the yield moment strength of the exterior column equal to the strength of the frame beam. The flexural strength of the interior column would need to be significantly greater than M_p in order to avoid early hinging, and this will usually be the case because the axial load imposed on it will be high.

The need to precisely identify strength objectives at the base is not pressing, for sequential yield analyses performed early on in the development of the design will allow these strengths to be appropriately adjusted. Realize for now that the mechanism approach developed Eq. 3.2.4a will tend to be conservative, and that the variables identified in Figure 3.2.1 should be at least subliminally factored into the development of column strengths.

3.2.2 Force- or Strength-Based Design Procedures

The design criterion used in this section is the response spectrum of Figure 4.1.1. The spectral velocity is, for the example building, presumed to be constant (50 in./sec).

Photo 3.2 Cast-in-place concrete ductile frame construction, jump form framing. (Courtesy of Englekirk Partners, Inc.)

System ductility is as suggested by the Blue Book for seismic performance category SP-3 ($\mu = 6.2$). An objective overstrength factor of 1.29 is adopted for convenience, for it produces an overall strength reduction factor ($\Omega_o\mu$) of 8 (6.2 × 1.29).

The example building is that described by the floor plan of Figure 3.1.17. The objective is to design a perimeter frame in the longitudinal direction. A design that proceeds without client input is a waste of time and effort. The following functional and aesthetic constraints are proposed. The maximum height of the beam should not exceed 36 in. The number of frame bays is a design option; however, the width of the frame beam should not be greater than 20 in. The architect has expressed a desire to have the column protrude no more than 8 in. from the face of the beam. After a first pass at the design, the impact of functional and aesthetic constraints should be reviewed and accepted or appropriately modified.

Given the aesthetic and functional objectives, proceed to develop the design of the frame.

Step 1: Adopt a Conceptual Design Configuration for the Frame.

$$h_x = 120 \text{ in.}; \qquad h_c = 84 \text{ in.}; \qquad \left(\frac{h_c}{h_x}\right)^2 = 0.49$$

$$\ell = 312 \text{ in.}; \qquad \ell_c = 276 \text{ in.}; \qquad \left(\frac{\ell_c}{\ell}\right)^2 = 0.78$$

$$b_b = 20 \text{ in.}$$

$$h_b = 36 \text{ in.}$$

Column size is not constrained. Select a 36 in. by 36 in. column.

$$f_y = 60 \text{ ksi}$$

$$f'_c = 5 \text{ ksi}$$

Tributary weight (mass) = 1620 kips/floor (see Figure 3.1.17)

Step 2: Characterize the Stiffness of the Frame.

$$I_{be} = 27{,}200 \text{ in.}^4 \qquad (0.35 I_g)$$

$$I_{ce} = 98{,}000 \text{ in.}^4 \qquad (0.70 I_g)$$

$$\Delta_i = \frac{V_i h_x^2}{12E}\left[\frac{\ell_c}{I_{be}}\left(\frac{\ell_c}{\ell}\right)^2 + \frac{h_c}{I_{ce}}\left(\frac{h_c}{h_x}\right)^2\right] \qquad \text{(Eq. 3.2.2a)}$$

$$= \frac{V_i(120)^2}{12(4000)}\left[\frac{276}{27{,}200}(0.78) + \frac{84}{98{,}000}(0.49)\right]$$

$$= 0.3V_i(0.0079 + 0.0004)$$

$$\Delta_i = 0.0025 V_i \qquad (3.2.5a)$$

$$k_i = 400 \text{ kips/in.}$$

$$\Delta_e = \frac{V_e h_x^2}{6E}\left[\frac{\ell_c}{I_{be}}\left(\frac{\ell_c}{\ell}\right)^2 + \frac{h_c}{I_{ce}}\left(\frac{h_c}{h_x}\right)^2\right] \qquad \text{(Eq. 3.2.2b)}$$

$$= 0.6V_e(0.0079 + 0.0002)$$

$$\Delta_e = 0.0049 V_e \qquad (3.2.5b)$$

Comment: A reduction in the effective panel zone dimension to 60% of the provided size will result in a significant reduction in subassembly stiffness.

$$h_x = 120 \text{ in.}; \qquad h_c = 98 \text{ in.}; \qquad \left(\frac{h_c}{h_x}\right)^2 = 0.67$$

$$\ell = 312 \text{ in.}; \qquad \ell_c = 290 \text{ in.}; \qquad \left(\frac{h_c}{\ell}\right)^2 = 0.86$$

$$\Delta_i = \frac{V_i(120)^2}{12(4000)}\left[\frac{290}{27{,}200}(0.86) + \frac{98}{98{,}000}(0.67)\right]$$

$$= 0.00295V_i \tag{3.2.5c}$$

$$k_i' = 338 \text{ kips/in.} \qquad (85\% \text{ of } 400 \text{ kips/in.})$$

Consider the implications associated with the preceding stiffness calculations:

(a) Columns are very stiff and minor changes in the size of columns will not affect design decisions.

(b) Beams control the drift and one might reasonably conclude that interior columns will experience twice the load resisted by the exterior columns.

Step 3: Select a Trial Frame. Start by considering a three-bay frame. The shear resisted by an interior column will be

$$V_i = \frac{V}{(j-1) + 2\,(V_e/V_i)} \tag{3.2.6a}$$

where V is the shear resisted by the frame and j is the number of frame bays. For the example three-bay frame described in Figure 3.2.2,

$$V_i = \frac{V}{3} \tag{3.2.6b}$$

Step 4: Calculate the Deflection of the Frame Subjected to Its Own Weight and the Associated Fundamental Period of the System (Δ_s—Eq. 3.2.1). The deflection of each subassembly will be a function of the shear imposed on it. The tributary mass to each frame per floor is 810 kips, and the shear imposed on each floor will be the sum of the masses above the floor. Calculate the displacement of a typical interior subassembly (V_i) when loaded by a unit weight. The unit shear imposed on an interior subassembly is

$$V_i = \frac{810}{3} \qquad \text{(see Eq. 3.2.6a)}$$

$$= 270 \text{ kips}$$

$$\Delta_i = 0.0025V_i \qquad \text{(Eq. 3.2.5a)}$$

$$= 0.0025(270)$$

$$= 0.68 \text{ in.}$$

Then use a series reduction to predict the total drift. The sum of the n integers times the unit story drift (δ_i) provides an estimate of the shear deformation of the frame loaded by its tributary mass (Figure 3.2.2*b*).

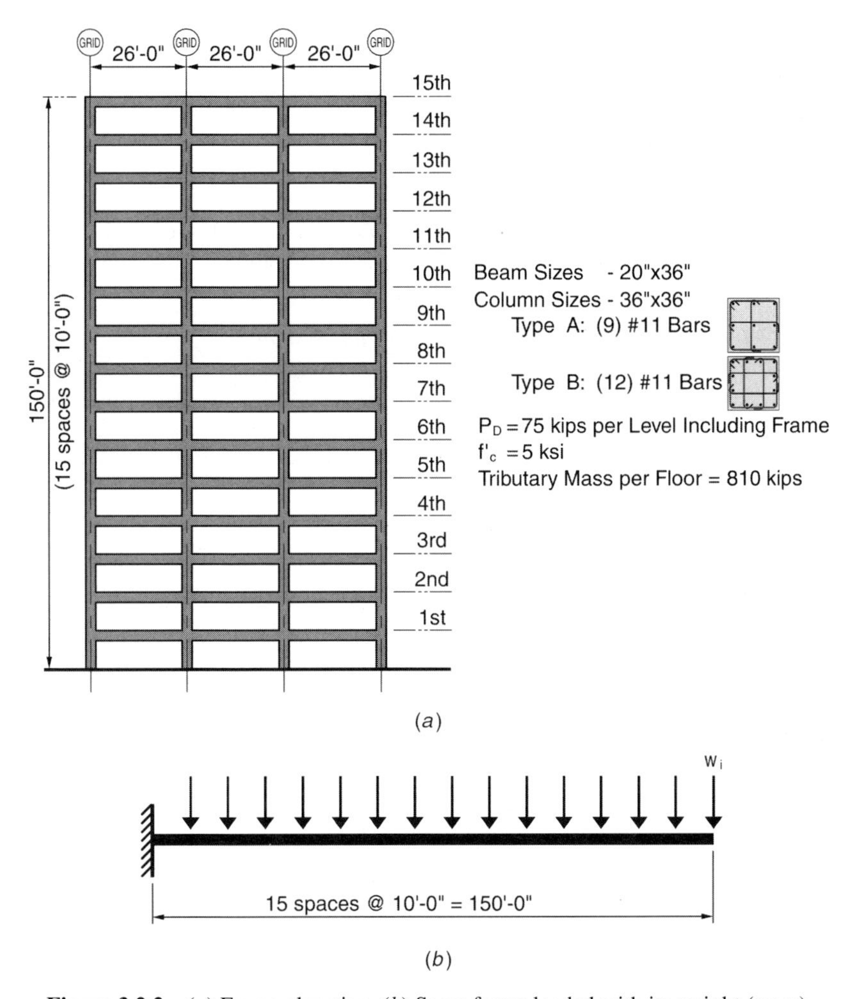

Figure 3.2.2 (*a*) Frame elevation. (*b*) Same frame loaded with its weight (mass).

$$\Delta_s = \left(\frac{n+1}{2}\right) n\Delta_i \tag{3.2.7a}$$

where n in this case is the number of stories. Thus

$$\Delta_s = \left(\frac{16}{2}\right) 15(0.68)$$

$$= 81.6 \text{ in.}$$

The fundamental period of a beam vibrating in shear is

$$T_s = 0.288(\Delta_s)^{0.5} \qquad \text{(Eq. 3.2.1)}$$

where Δ_s is the shear deflection of the roof in inches. Hence

$$T_s = 0.288(81.6)^{0.5} \quad (3.2.7b)$$
$$= 2.6 \text{ seconds}$$

Comment: Computer-determined period is 2.63 seconds. If, however, the stiffness were based on the 30% effective joint area, the period would be 2.82 seconds.

Step 5: Determine the Spectral Acceleration of the System.

$$S_a = \omega S_v \quad \text{(see Eq. 1.1.6b)}$$

$$S_a = \frac{2\pi S_v}{T}$$
$$= \frac{6.28(50)}{2.6}$$
$$= 120.8 \text{ in./sec}^2 \qquad (0.31g)$$

Step 6: Determine the Base Shear. The effective mass, as developed from a fundamental mode response given a linear mode shape, is

$$M_e = \frac{1.5(n+1)}{2n+1}M \quad \text{(Eq. 3.1.19a)}$$
$$= \frac{1.5(16)}{31}M$$
$$= 0.77M$$

This will underestimate the base shear because higher modes will contribute to the base shear, as was pointed out in comments contained in Section 3.1.1. Recall also that it was for this reason that the code adopted the single-degree-of-freedom spectrum as the base shear spectrum. For design purposes use at least 80% of the mass.[3.6]

The base shear is then

$$V_u = S_a M_e$$
$$= \frac{0.31g0.8W}{\Omega_o \mu g}$$
$$= \frac{0.31(0.8)(15)810}{8}$$
$$= 376.7 \text{ kips}$$

Step 7: Determine Whether or Not the Shear Imposed on an Interior Subassembly Exceeds Its Capacity.

$$V_i = \frac{V}{3} \qquad \text{(Eq. 3.2.6a)}$$

$$= \frac{376.7}{3}$$

$$= 125.6 \text{ kips}$$

The subassembly shear associated with beam reinforcing of 1.5% is defined by Eq. 3.2.3c.

$$V_c = \frac{0.02(20)(36)^2(60)}{120} \qquad \text{(Interior column)} \qquad \text{(Eq. 3.2.3c)}$$

$$= 259.2 \text{ kips}$$

A 36 in. × 36 in. column is capable of sustaining this level of shear given our design objective shear stress of $5\sqrt{f'_c}$.

$$V_{cu,\text{allow}} = 5\sqrt{f'_c}bd \qquad \text{(see Section 2.1.2.4)}$$

$$= 5\sqrt{5000}(36)(33)$$

$$= 420 \text{ kips} \qquad \text{(OK)}$$

Conclusion: A three-bay frame will satisfy our design objectives

The minimum size of the interior column must also consider beam-column joint limit states. These can be developed directly from Eq. 2.3.22b.

$$A_j = \frac{V_{cu}}{0.18} \qquad \text{(Eq. 2.3.22b)}$$

$$= \frac{125.6}{0.18}$$

$$= 698 \text{ in.}^2 \qquad \text{(27 in. by 27 in.)}$$

Step 8: Determine the Required Interior Flexural Reinforcement Program.

$$M_{bu} = \frac{V_i h_x}{2}\left(\frac{\ell_c}{\ell}\right)$$

$$= \frac{125.6(10)}{2}\left(\frac{23}{26}\right)$$

$$= 556 \text{ ft-kips}$$

$$A_s = \frac{M_{bu}}{(d - d')\phi f_y}$$
$$= \frac{556(12)}{30(0.9)60}$$
$$= 4.11 \text{ in.}^2 \qquad (0.6\%)$$

The required level of reinforcing is minimal. The probable strength of this beam based on the experimental data described in Figure 2.1.2 would be

$$\frac{\lambda_o M_u}{\phi} = \frac{1.5(556)}{0.9}$$
$$= 927 \text{ ft-kips}$$

Comment: Building codes[3.7] prescriptively describe the period of a building. For this structure the prescriptive period would be

$$T = C_t h_n^{3/4}$$
$$= 0.03(150)^{3/4}$$
$$= 1.29 \text{ seconds}$$

This may be increased by 30%[3.7] so long as the increase may be justified analytically.

$$T_{\text{code}} = 1.7 \text{ seconds}$$

And this is considerably less than the period of the proposed three-bay frame ($T_3 = 2.6$ seconds). If the prescriptive period were used to design the three-bay frame, the increase in base shear would be 24% $\left(\sqrt{T_3/T_{\text{code}}}\right)$.

Conclusion: The frame described in Figure 3.2.2 represents the solution proposed to brace the building whose plan is described in Figure 3.1.17 in the longitudinal direction. Beam strength objectives are described in Table 3.2.1 for the strength-based approach, and the design procedures will be developed in the sections that follow. A design evaluation of this and the other solutions will be undertaken in Section 3.2.5 using a sequential yield analysis approach. An inelastic time history analysis comparison will be made in Chapter 4.

A four-bay alternative can be easily evaluated, provided bookkeeping has been done in an organized manner.

$$V_i = \frac{V}{(j - 1) + 2(V_e/V_i)} \qquad \text{(Eq. 3.2.6a)}$$

$$= \frac{V}{(4-1)+2(0.0025/0.0049)} \qquad \text{(see Eqs. 3.2.5a, b)}$$

$$= \frac{V}{4}$$

Repeat Step 4.

$$V_i = \frac{810}{4}$$

$$= 202 \text{ kips}$$

and this is 75% of the unit load (mass) imposed on the three-bay frame. Accordingly,

$$\Delta_s = 0.75(81.6)$$

$$= 61.2 \text{ in.}$$

$$T_s = 0.288(61.2)^{0.5} \qquad \text{(see Eq. 3.2.1)}$$

$$= 2.25 \text{ seconds}$$

Comment: Computer determined period is 2.27 seconds.

Step 9: Determine the Spectral Acceleration.

$$S_a = \frac{2\pi S_v}{T}$$

$$= \frac{6.28(50)}{2.25}$$

$$= 139.6 \text{ in./sec}^2 \qquad (0.36g)$$

TABLE 3.2.1 Alternative Frame Designs

Design Approach	Beam Flexural Strength ($\lambda_o M_n$) ft-kips			Column Type		
	Levels 1–5	Levels 6–10	Levels 11–15	1/5	6/10	11/15
1. Strength—Spectra	927	816	510	B	A	A
2. Strength—Code[a]	1150	1012	633	B	A	A
3. Equal displacement	1000	880	550	B	A	A
4. Equal displacement (SDOF)	800	700	440	B	A	A
5. Hybrid precast direct displacement	2000	1760	1100	B	A	A
6. DDC direct displacement	910	910	910	B	A	A

[a]Accounts for prescriptive period only.

Step 10: Determine Base Shear.

$$\begin{aligned} V &= S_a M_e \\ &= \frac{0.36g0.8W}{\Omega_o \mu_s g} \\ &= \frac{0.36(0.8)(15)(810)}{8} \\ &= 438 \text{ kips} \end{aligned}$$

Step 11: Calculate the Shear Imposed on an Interior Subassembly.

$$\begin{aligned} V_i &= \frac{V}{4} \\ &= 110 \text{ kips} \end{aligned}$$

Observe that the process may be shortened by realizing that the response of the building will be increased by the square root of the change in stiffness and the ratio of base shear allocation. Hence, the interior subassembly demand resulting from increasing the three-bay frame to a four-bay frame would be

$$V_{i4} = AV_{i3} \tag{3.2.8}$$

where V_{i3} and V_{i4} are the ultimate interior subassembly shear demands for three-bay and four-bay frames, respectively.

$$A = \frac{\sqrt{r_4/r_3}}{r_4/r_3} \tag{3.2.9}$$

where

$$r_j = (j-1) + 2\left(\frac{V_e}{V_i}\right) \quad \text{(see Eq. 3.2.6a)} \tag{3.2.10}$$

where j is the number of bays in the example.

$$\begin{aligned} r_3 &= 3 \\ r_4 &= 4 \\ \frac{r_4}{r_3} &= 1.33 \end{aligned}$$

$$\begin{aligned} A &= \frac{\sqrt{1.33}}{1.33} \quad \text{(see Eq. 3.2.9)} \\ &= 0.87 \end{aligned}$$

$$\frac{V_{i,4}}{V_{i,3}} = \frac{110}{125.6}$$
$$= 0.87$$

The reinforcement ratio can be determined by proportion also.

$$\rho_s = \frac{V_{i,4}}{V}(0.006)$$
$$= \frac{110}{125.6}(0.006)$$
$$= 0.0053$$

Regardless of the design approach used, it will be advisable to vary beam strengths throughout the height of the building so as to promote the distribution of postyield behavior. Since displacement demands at peak building drift follow a primary mode shape, a linear force distribution is appropriately selected as a design objective. Changing beam strengths at every level is an overreaction; it will, in addition to costing more money to detail and place, become a source of error. A five- or six-floor pattern for beam adjustments is typically used.

The shear at any level may be developed as follows[3.7]:

$$V_{i/(i+1)} = V\left(\frac{2na - a^2 + a}{n^2 + n}\right) \tag{3.2.11a}$$

where

i is the floor of interest.

$V_{i/(i+1)}$ is the story shear.

V is the base shear.

n is the number of stories.

a is $n - i$.

This relationship may also be used to develop beam strengths from that required at the lowermost level. For example, the flexural strengths required in level 6 of the frame of Figure 3.2.2 would be:

Level 5: $a = 10, n = 15$

$$M_5 = M_1\left(\frac{2an - a^2 + a}{n^2 + n}\right) \tag{3.2.11b}$$
$$= M_1\left(\frac{2(10)(15) - (10)^2 + 10}{(15)^2 + 15}\right)$$
$$= 0.875M_1$$

3.2.3 Displacement-Based Design

Most displacement-based design approaches[3.4] reduce multi-degree-of-freedom systems to single-degree-of-freedom systems. This reduction unnecessarily clouds design issues.

Two basic approaches are developed in this section. One works directly with the multi-degree-of-freedom system (building model), while the other utilizes the single-degree-of-freedom (SDOF) model before reverting to the multi-degree-of-freedom system.

Building Model	SDOF Model
(1) Establish a criterion.	(1) Establish a criterion.
(2) Identify objective dynamic characteristics.	(2) Create SDOF system.
(3) Size frame to meet objective dynamic characteristics.	(3) Determine SDOF objective stiffness.
(4) Develop the dynamic characteristics of the provided frame.	(4) Determine yield strength of SDOF system.
(5) Establish objective yield for the provided frame.	(5) Convert SDOF strength objective to a building objective strength and stiffness.
(6) Select member strengths based on objective yield criterion.	

The chief advantage of the building model lies in the fact that a simple building model is created early in the design process and iterations are minimized, as is the potential for error created by shifting from one model to another. Both approaches are explored and applied to the design problem (Figure 3.1.17 and Figure 3.2.2*a*).

3.2.3.1 Building Model

(1) Criterion Development The first issue that must be dealt with is the adoption of an appropriate system ductility factor. This was discussed briefly in Section 1.1.6. In the design of a frame for a tall building, the reduction from component ductility to system ductility can take advantage of an understanding of system behavior characteristics. In the preceding section we found that the column would typically be much stiffer than the beam, especially if story heights are small compared to beam spans. Further, a capacity-based design approach will force inelastic behavior into the beam. This being the case, if all of the beams experience the same amount of postyield rotation, the ductility of the system and the beam will be about the same ($\mu \cong \mu_b$).

Unfortunately, not all frame beams will undergo the same level of postyield rotation. Rather, most of the displacement ductility demand at peak drifts will tend to accumulate in the lower levels of tall buildings. A reasonable projection for the extent of the effective region of plastification is the depth of the frame.[3.6, Sec. 5.5.2] The ductile region in the three-bay frames developed in the preceding section would extend about

Photo 3.3 Placing a two-story column cage, The Remington, Westwood, CA. (Courtesy of Morley Construction Co.)

seven or eight floors, or one-half of the building height. This understood, the system ductility might reasonably be assumed to be one-half of that which could be developed in the frame beam. Figure 2.1.2 suggests that drift ratios of 4.7% ($\delta = 3.7$ in.) are attainable in frame beams. This level of deformation will produce a significant level of structural damage (Figure 2.1.9) but not collapse; hence it is consistent with seismic performance level SP-3.

The displacement ductility of the beam evaluated in Section 2.1.1 was developed by a variety of means. The design procedures we plan to use, as well as our reinforcement objectives, must be considered in the selection of a member ductility factor. If we adopt an idealized strength associated with that developed by Eq. 2.1.1, the yield load will be about 100 kips. The yield deflection using an effective moment of inertia of $0.35I_g$ will be on the order of 0.37 in. (Figure 2.1.3). The deflection associated with Point 77 (Figure 2.1.2) is 3.7 in., and this corresponds to a displacement ductility of 10. This beam was reinforced to 1.5% (ρ_b), and this was our suggested reinforcing limit state. Were our design to result in a reinforcement ratio of 1%, then the level of displacement ductility would be at least 15 $((0.015/0.01) \cdot (\Delta_u/\Delta_{y,1.5\%}))$ because the yield drift would be lower and the attainable level of drift probably higher. Accordingly, for frame design purposes adopt a system reduction factor ($\Omega_o \mu$—because of the selection of δ_y) of about 8. This is consistent with current recommendations[3.7] ($R = 8.5$; $R_d R_o$—Figure 1.1.12).

Comment: Observe that if the reinforcement ratio (ρ) of the proposed frame beam is significantly greater than the 1% used as a basis for establishing the system ductility factor the design process may need to be revisited.

A forcing function must be adopted. For consistency use a spectral velocity of 50 in./sec.

(2) Identify Objective Dynamic Characteristics

Step 1: A Drift Objective or Target Drift for the Structure (Δ_T) Is First Established.

$$\begin{aligned}\Delta_T &= \delta_1 h_n \\ &= 0.018(150)(12) \\ &= 32.4 \text{ in.}\end{aligned} \tag{3.2.12}$$

where δ_1 is the target criterion drift ratio and 0.018 the drift ratio suggested by the Blue Book[3.4] for performance level SP-3.

Step 2: Determine the Objective Spectral Displacement. Availing ourselves of an understanding of the linear, shear related, deformation characteristics of the frame-braced structure, we can transform this drift objective into a frame size.

$$\begin{aligned}\Gamma &= \frac{3n}{2n+1} \qquad \text{(Eq. 3.1.18a)} \\ &= \frac{45}{31} \\ &= 1.45\end{aligned}$$

$$\begin{aligned}S_d &= \frac{\Delta_T}{\Gamma} \qquad \text{(see Eq. 3.1.10)} \\ &= \frac{32.4}{1.45} \\ &= 22.3 \text{ in.}\end{aligned}$$

Step 3: Determine the Maximum Fundamental Period for the Structure. Since our objective spectral velocity is 50 in./sec, we may define an objective natural frequency.

$$\omega_n = \frac{S_v}{S_d} \qquad \text{(see Eq. 3.1.9)}$$

$$= \frac{50}{22.3}$$

$$= 2.24 \text{ rad/sec}$$

and a maximum fundamental period

$$T = \frac{2\pi}{\omega_n}$$

$$= 2.8 \text{ seconds}$$

(3) Size the Frame to Meet Objective Dynamic Characteristics

Step 4: Convert the Objective Period to an Interior Subassembly Stiffness Criterion. Equation 3.2.1 allows us to quantify the shear deformation (Δ_s) of the structure when loaded by its own weight (mass).

$$T = 0.288(\Delta_s)^{0.5} \qquad \text{(Eq. 3.2.1)}$$

$$\Delta_s = \frac{T^2}{0.083}$$

$$= \frac{(2.8)^2}{0.083}$$

$$= 94.4 \text{ in.}$$

In Section 3.2.2, Step 1, a frame configuration was adopted. It was consistent with program functional and aesthetic objectives. In Section 3.2.2, Step 2, frame stiffness characteristics were developed. This included the development of a relationship between the stiffness of an interior and an exterior subassembly. Given this established relationship, we are now in a position to work exclusively with the established characteristics of an interior subassembly. The stiffness of an interior subassembly was assessed at

$$k_i = 400 \text{ kips/in.} \qquad \text{(see Eq. 3.2.5a)}$$

This allows us to determine how much weight (mass) can be supported by an interior subassembly.

$$w_i = k_i \Delta_i \qquad \text{(see Eq. 3.2.5a)}$$

where w_i is the unit of weight applied to an interior subassembly (Figure 3.2.2*b*) and Δ_i is the resultant story drift.

The shear deflection of each level (Figure 3.2.2*a*) would be the (shear) load imposed on that level times the unit deflection Δ_i. The total shear deflection is

$$\Delta_s = \sum n \Delta_i \qquad (3.2.13a)$$

$$= \left(\frac{n+1}{2}\right) n \left(\frac{w_i}{k_i}\right) \qquad \text{(see Eq. 3.2.7)}$$

$$= 8(15)(0.0025)w_i$$

$$= 0.3w_i$$

$$w_i = \frac{\Delta_s}{0.3}$$

$$= \frac{94.4}{0.3}$$

$$w_i = 314.7 \text{ kips}$$

Step 5: Determine the Minimum Number of Bays Required to Accomplish Stiffness Objectives. The tributary unit floor weight (W_f) is 810 kips. The number of required frame bays (j) can be developed from Eq. 3.2.6a.

$$(j-1) + 2\left(\frac{V_e}{V_i}\right) = \frac{W_f}{w_i} \qquad \text{(see Eq. 3.2.6a)}$$

The ratio of V_e to V_i for the example frame is 0.5 (see Eq. 3.2.5a and 3.2.5b). Accordingly,

$$j \geq \frac{W_f}{w_i} \tag{3.2.13b}$$

$$\geq \frac{810}{314.7}$$

$$\geq 2.57$$

Conclusion: A three-bay frame meets our performance-based objective. It remains to establish the strength of the frame.

Comment: The process can be reordered so as to seek a required interior subassembly stiffness.

Step C1: For a Selected Number of Bays, Solve for w_i. If $j = 3$, then

$$w_i = \frac{V_f}{3}$$

$$w_i = 270 \text{ kips}$$

Step C2: Use Eq. 3.2.13a and Eq. 3.2.5 ($w_i = k_i \Delta_i$) to Solve for k_i.

$$k_i = \frac{\sum n w_i}{\Delta_s} \tag{3.2.13c}$$

$$= (8)(15)\frac{w_i}{\Delta_s}$$

$$= 120\left(\frac{270}{94.4}\right)$$

$$= 343 \text{ kips/in.}$$

Step C3: Use Eq. 3.2.2a to Solve for the Required Effective Moment of Inertia of the Beam, Realizing That the Column Is Essentially Rigid.

$$I_{be} = \frac{k_i h_x^2}{12E}\ell_c\left(\frac{\ell_c}{\ell}\right)^2$$

$$= \frac{325(120)^2}{12(4000)}(276)(0.78)$$

$$= 21{,}000 \text{ in.}^4$$

Conclusion: Provide a 20 in. by 33 in. beam—$I_e = 21{,}000$ in.^{4}

(4) Develop the Dynamic Characteristics of the Provided Frame

Step 6: Determine the Period of the Proposed Bracing System. The frame that is proposed will be stiffer than our objective frame; hence the maximum drift will be less than the target drift of 32.4 in.

Use Eq. 3.2.13a to determine the period of the proposed frame (Figure 3.2.2).

$$\Delta_s = \frac{\sum n w_i}{k_i} \qquad (\Delta_s = \Delta_n) \qquad \text{(Eq. 3.2.13c)}$$

$$= \left(\frac{n+1}{2}\right) n \left(\frac{w_i}{k_i}\right) \qquad \text{(see Eq. 3.2.7a)}$$

$$= 8(15)\frac{270}{400} \qquad \left(w_i = \frac{W_f}{3}\right)$$

$$= 81 \text{ in.}$$

$$T = 0.288(\Delta_s)^{0.5} \qquad \text{(Eq. 3.2.1)}$$

$$= 0.288(81)^{0.5}$$

$$= 2.6 \text{ seconds}$$

(5) Establish Objective Yield for the Provided Frame

Step 7: Determine the Spectral Displacement of the System.

$$\omega = \frac{2\pi}{T}$$

$$= \frac{6.28}{2.6}$$

$$= 2.42 \text{ rad/sec}$$

$$S_d = \frac{S_v}{\omega} \qquad \text{(Eq. 3.1.9)}$$

$$= \frac{(50)}{2.42}$$

$$= 20.7 \text{ in.}$$

Step 8: Determine the Objective Drift at Idealized Yield.

$$\Delta_n = \Gamma S_d \qquad \text{(Eq. 3.1.10)}$$

$$= 1.45(20.7)$$

$$= 30 \text{ in.}$$

$$\Delta_y = \frac{\Delta_n}{\mu \Omega_o}$$

$$= \frac{30}{8}$$

$$= 3.75 \text{ in.}$$

Comment: Consider the consequences on the design of the adoption of a 60% effective beam-column joint (see Figure 2.3.8, Section 2.3.2).

$$\left(\frac{k_i}{k_i'}\right) = \left(\frac{400}{338}\right) \qquad \text{(see Eqs. 3.2.5a, c)}$$

$$= 1.18$$

$$\left(\frac{k_i}{k_i'}\right)^2 = 1.09$$

First, our drift objectives would be met (see Step 3).

$$T' = T\left(\frac{k_i}{k_i'}\right)^2$$

$$= 2.6(1.09)$$

$$= 2.8 \text{ seconds} \qquad \text{(Maximum period: 2.8 seconds, Step 3)}$$

A three-bay frame would experience a drift of

$$\begin{aligned}\omega &= \frac{2\pi}{T}\\ &= \frac{2\pi}{2.8}\\ &= 2.24 \text{ rad/sec}\\ S_d &= \frac{S_v}{\omega}\\ &= \frac{50}{2.24}\\ &= 22.3 \text{ in.}\\ \Delta_R &= \Gamma S_s\\ &= 1.45(22.3)\\ &= 32.3 \text{ in.}\end{aligned}$$

This is our objective drift ratio (1.8%) and the drift at yield would be

$$\begin{aligned}\Delta_y &= \frac{\Delta_R}{\mu\Omega_o}\\ &= \frac{32.3}{8}\\ &= 4.04 \text{ in.}\end{aligned}$$

(6) Select Member Strengths Based on Objective Yield Criterion

Step 9: Determine the Base Shear (V) Required to Produce the Required Yield Roof Drift. The force distribution at peak drift will be consistent with (and exclusively in) the primary mode shape; accordingly, it will be linear as described by

$$F_m = \frac{mV_i}{(n(n+1))/2} \tag{3.2.14}$$

where

F_{mi} is the force applied to an interior subassembly at level m.
V_i is the portion of the base shear resisted by an interior subassembly.
n is the number of stories.

This force will cause each subassembly to deflect in accordance with the relationship developed in Eq. 3.2.2a, provided, of course, that the beams are designed so as to remain in the idealized elastic domain. If we sum the story drifts (recalling that they are shear induced and represent a sliding of planes), we can express the deflection at the roof in terms of a series summation.

$$\Delta_m = \sum F_m m k_i \tag{3.2.15}$$

where k_i is the story stiffness as developed in Eq. 3.2.2a.

$$k_i = \frac{12E}{h_x^2((\ell_c/I_{be})(\ell_c/\ell)^2 + (h_c/I_{ce})(h_c/h_x)^2)} \tag{3.2.16}$$

or in our example, $k_i = 1/0.0025$ or 400 kips/inch (see Eq. 3.2.5a).

Combining Eq. 3.2.14 and 3.2.15 produces a direct relationship between the design base shear (at yield) allocated to an interior subassembly and the displacement at the roof.

$$\Delta_m = \frac{V_i k_i^{-1}}{(n(n+1))/2} \left\{ \sum_{m=1}^{n} m^2 \right\} \tag{3.2.17a}$$

$$= \frac{V_i}{k_i} \left(\frac{m(m+1)(2m+1)}{3n(n+1)} \right) \tag{3.2.17b}$$

Our interest is in the displacement of the roof or nth level; hence

$$\Delta_n = \frac{V_i}{k_i} \left(\frac{n(n+1)(2n+1)}{3(n)(n+1)} \right)$$

which, recalling that $3n/(2n+1)$ is the normalized participation factor, reduces to

$$\Delta_n = \frac{nV_i}{\Gamma_1 k_i} \tag{3.2.18}$$

where Γ_1 is the normalized participation factor for the fundamental mode and V_i is the total force or base shear resisted by each interior subassembly.

At yield the base shear of an interior subassembly becomes

$$V_i = \frac{\Delta_n \Gamma_1 k_i}{n} \qquad \text{(see Eq. 3.2.18)}$$

$$= \frac{\Delta_y \Gamma_1 k_i}{n} \tag{3.2.19}$$

$$= \frac{3.75(1.45)(400)}{15}$$

$$= 145 \text{ kips}$$

Comment: The alternative model (60% effective panel zone) would require an objective interior subassembly base shear of

$$V_i = \frac{\Delta_n \Gamma_1 k_k'}{n} \qquad \text{(see Eq. 3.2.18)}$$

$$= \frac{4.04(1.45)(338)}{15}$$

$$= 132 \text{ kips}$$

which is less than the proposed objective of 145 kips.

The associated base shear (V) for the frame is

$$V = \left[(j-1) + 2\left(\frac{V_e}{V_i}\right)\right] V_i \qquad \text{(Eq. 3.2.6a)}$$

$$= (2+1)V_i \qquad \text{(see Eq. 3.2.5a, b)}$$

$$= 3(145)$$

$$= 435 \text{ kips} \qquad (0.036W)$$

The design moments for the displacement-based approach might reasonably be established as follows:

$$M_{bu} = \frac{V_{iu} h_x}{\ell}\left(\frac{\ell_c}{2}\right) \qquad (3.2.20a)$$

$$= \frac{V_{iu}}{12}\left(\frac{120}{312}\right)\left(\frac{276}{2}\right)$$

$$= 4.4V_{iu} \qquad (3.2.20b)$$

$$= 4.4(145)$$

$$= 638 \text{ ft-kips}$$

Again, reverting to our component design model (Figure 2.1.2), an overstrength factor of 1.5 seems appropriate based on a tip design load of 100 kips.

$$\lambda_o M_d = 1.5(638)$$

$$\cong 1000 \text{ ft-kips}$$

3.2.3.2 Single-Degree-of-Freedom (SDOF) Model

(1) Criterion Development Refer to Section 3.2.3.1, Subsection on Criterion Development.

(2) Objective Dynamic Characteristics Steps 1 through 3 are the same as those developed in this design phase in Section 3.2.3.1, Subsection on Objective Dynamic Characteristics.

(3) Determine SDOF Objective Stiffness.

Step 4: Determine Effective Mass.

$$M_e = \frac{1.5(n+1)}{2n+1}M \qquad \text{(Eq. 3.1.19a)}$$

$$= \frac{1.5(16)}{31}\left(\frac{W}{g}\right)$$

$$= 0.77\left(\frac{12{,}150}{386.4}\right)$$

$$= 24.2 \text{ kips-sec}^2/\text{in.}$$

The objective maximum period is 2.8 seconds (Sec. 3.2.3.1—Step 3). The effective stiffness must be

$$k_e = \omega^2 M_e$$

$$= (2.24)^2(24.2) \qquad (\omega = 2\pi/2.8)$$

$$= 121.4 \text{ kips/in.}$$

(4) Determine Yield Strength of the SDOF System.

Step 5: Solve for H_y. (See Figures 3.1.1*a* and *b*.)

$$H_y = \frac{k_e S_d}{\mu \Omega_o} \qquad \text{(see Figure 3.1.2)}$$

$$= \frac{121.4(22.3)}{8}$$

$$= 335 \text{ kips}$$

(5) Convert SDOF Strength to a Building Objective Strength and Stiffness.

Step 6: Adjust the Base Shear So As to Be Compatible with the Provided Building. ($T = 2.6$ seconds; see Section 3.2.3.1, Step 6).

The stiffness developed in Step 4 was based on the attainment of the maximum period T_{max}. The period of the proposed building is 2.6 seconds (T).

$$V_y = \frac{T_{max}}{T} H_y$$

$$= \frac{2.8}{2.6}(335)$$

$$= 360 \text{ kips}$$

Step 7: Determine Objective Beam Moments. Use mechanism methods.

$$M_{bu} = \frac{V_M\,(\ell_c/\ell)\,h_x}{2n} \qquad \text{(see Eq. 3.2.4a and Figure 2.2.9)}$$

where n is the number of bays. Thus for a three-bay frame,

$$M_{bu} = \frac{360(376/312)(10)}{6}$$

$$= 531 \text{ ft-kips}$$

Step 8: Establish the Overstrength of the Beam Consistent with the Design Model. (See Figure 2.1.2.) Select $\lambda_o = 1.5$.

$$\lambda_o M_d = 1.5(531)$$

$$\cong 800 \text{ ft-kips}$$

Comment: The SDOF procedure is shear or force dependent and, accordingly, some consideration of higher mode effects is appropriate. Given the introduction of a multimode factor the system strengths as developed from either procedure would be the same. The building model approach is based on building drift and as a consequence is not sensitive to higher-mode contributions—see Section 3.1.1.1. The author's preference is to use the system-based design approach developed in Section 3.2.3.1.

3.2.4 Precast Concrete Frame—Direct Displacement-Based Design

Precast concrete frames may be designed by either the strength-based procedure developed in Section 3.2.2 or the displacement-based procedure of Section 3.2.3. The hysteretic behavior described for the hybrid system (Figure 2.1.47*a*) and the DDC® System (Figure 2.3.17) differ from that of the cast-in-place beam (Figure 2.1.2), in terms of both energy dissipated and available ductility. This suggests that energy dissipation might reasonably be included in the design process.

The design of frame-braced structures by either strength- or displacement-based methods requires the adoption of a system ductility factor (μ), and this was discussed in Section 1.1.6. The direct displacement-based approach also requires that estimates of component damping be extended to reflect an estimate of effective structural damping. In Section 3.2.3 an effective system ductility factor of 5.33 (8/1.5) was used in the design. In this section we test the sensitivity of the direct displacement-based

Photo 3.4 Nine-story precast concrete column, Getty Center Parking Structure, Brentwood, CA. (Courtesy of Englekirk Partners, Inc.)

design process to system ductility assumptions by using system ductility factors of 2, 3, 4, and 8.

Step 1: Develop the Design Response Spectrum. This step starts by evaluating component damping using the procedures developed in Section 1.1.3. Figure 3.2.3 describes the energy dissipated in one response cycle at a displacement ductility of 6 for

A cast-in-place beam: Figure 3.2.3*a*

A DDC subassembly: Figure 3.2.3*b*

A hybrid subassembly: Figure 3.2.3*c*

These, when compared to a complete or full hysteresis loop (Figure 1.1.6) suggest that structural damping provided by the cast-in-place frame and DDC systems is about 33% effective while that of the hybrid system is about 16% effective. The former ratio (33%) is also confirmed by Eq. 1.1.9b and c. For a component ductility of 6 the effective damping is as follows:

Full Hysteresis Loop

$$\zeta_{eq} = \frac{2(\mu - 1)}{\pi \mu} \qquad \text{(Eq. 1.1.9b)}$$

$$= \frac{2(5)}{\pi 6}$$

$$= 0.53$$

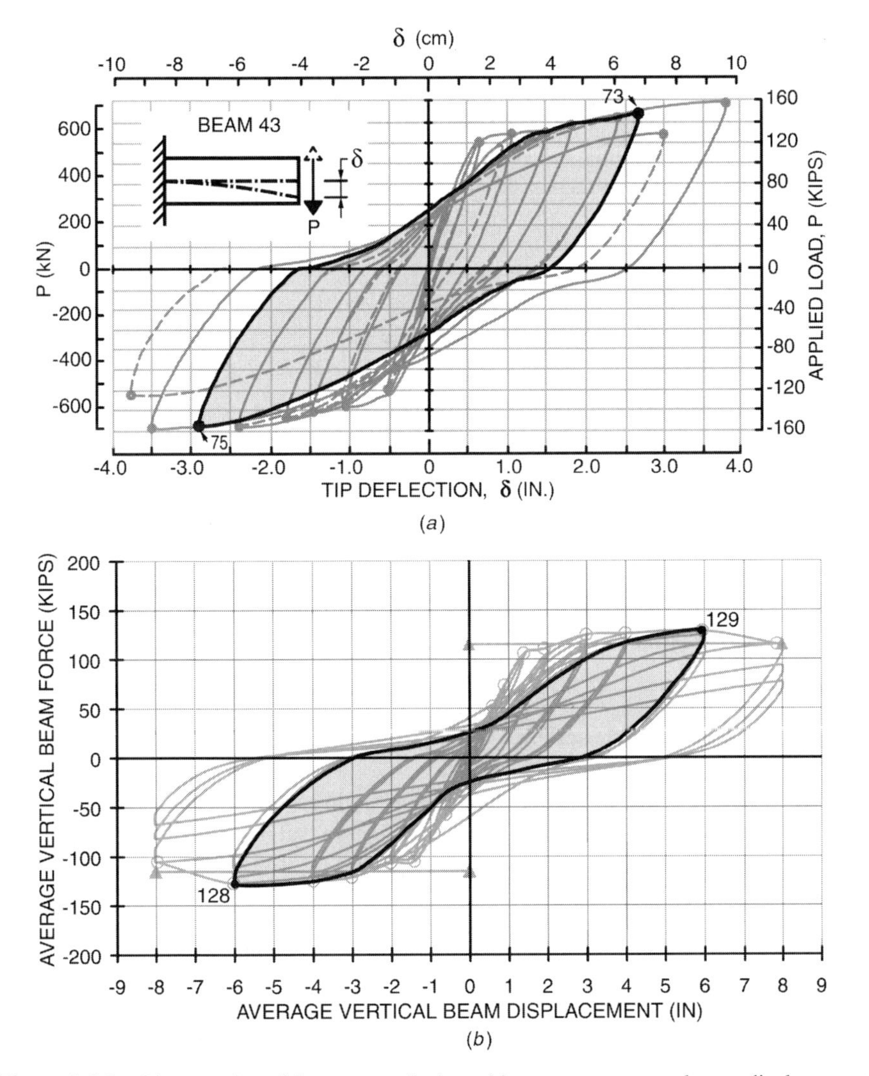

Figure 3.2.3 Diagram describing energy dissipated in one response cycle at a displacement ductility of 6. (*a*) Cast-in-place beam (Figure 2.1.2). (*b*) DDC® beam (Figure 2.3.17).

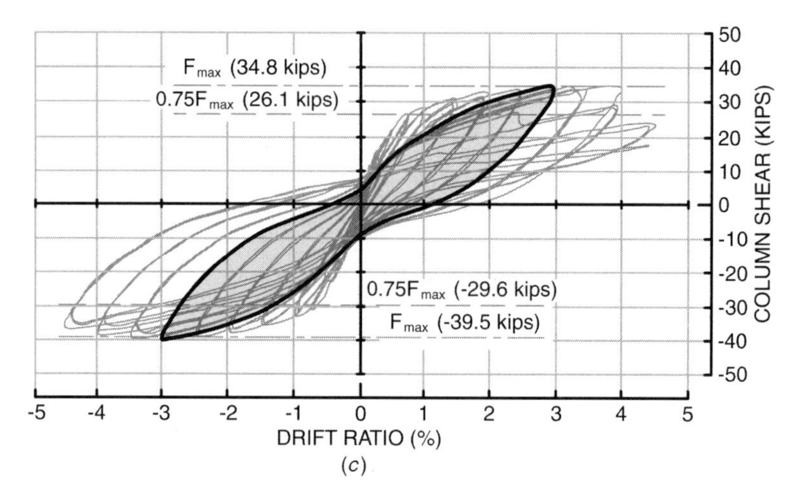

Figure 3.2.3 (*Continued*) (*c*) Hybrid beam (Figure 2.1.47*a*).

Pinched Hysteresis Loop Described in Figure 3.2.3b.

$$\zeta_{eq} = \frac{\sqrt{\mu} - 1}{\pi\sqrt{\mu}} \qquad \text{(Eq. 1.1.9d)}$$

$$= \frac{\sqrt{6} - 1}{\pi\sqrt{6}}$$

$$= 0.175$$

Conclusion: The structural damping component provided by a concrete frame beam is 33% effective (17.5/53).

It seems appropriate to use Eq. 1.1.9d to develop designs for the cast-in-place frame as well as the DDC system.

3.2.4.1 DDC Frame Proceed to develop a direct displacement-based design (CVDD, Section 3.1.1.1) for a DDC system. Observe that it would also be appropriate for a cast-in-place system.

Step 1: Develop the Spectrum Velocities. Spectrum reduction factors ($\hat{R}$) must be calculated for the various levels of assumed system ductility (2, 3, 4 and 8).

$$\hat{R} = \frac{3.38 - 0.67 \ln \hat{\zeta}_{eq}}{2.3} \qquad \text{(Eq. 3.1.45)}$$

where

$$\hat{\zeta} = \frac{\sqrt{\mu} - 1}{\pi\sqrt{\mu}} + 5 \qquad \text{(see Eqs. 1.1.9a, d)}$$

$$\begin{array}{lll} \mu = 2 & \hat{\zeta} = 14 & \hat{R} = 0.7 \\ \mu = 3 & \hat{\zeta} = 19 & \hat{R} = 0.62 \\ \mu = 4 & \hat{\zeta} = 21 & \hat{R} = 0.59 \\ \mu = 8 & \hat{\zeta} = 25 & \hat{R} = 0.53 \end{array}$$

The reduced spectral velocities, based on a criterion 5% damped spectral velocity of 50 in./sec^2, are

$$\begin{array}{ll} S_v = 35 \text{ in./sec} & \mu = 2 \\ S_v = 31 \text{ in./sec} & \mu = 3 \\ S_v = 29.5 \text{ in./sec} & \mu = 4 \\ S_v = 26.5 \text{ in./sec} & \mu = 8 \end{array}$$

Comment: At some point the spectral velocity will no longer be constant, for the structure will enter the displacement-sensitive region. Should this be the case, conclusions based on a constant spectral velocity will be conservative. Alternately, the design process may be adjusted so as to incorporate displacement maxima.

Step 2: Determine Spectral Displacement Objective for the Substitute Structure (Figure 3.1.2). The previously established target drift for the structure was 32.4 in. (see Eq. 3.2.12). The associated spectral drift, given a participation factor of 1.45 (Section 3.2.3.1, Step 2) is 22.3 in.

Step 3: Determine the Maximum Period Required of the Various Substitute Structures.

$$T = \frac{2\pi S_d}{S_{v,s}} \qquad \text{(see Eq. 3.1.24)}$$

$$= \frac{2\pi(22.3)}{35}$$

$$\begin{array}{ll} = 4 \text{ seconds} & \mu = 2 \\ = 4.5 \text{ seconds} & \mu = 3 \\ = 4.75 \text{ seconds} & \mu = 4 \\ = 5.28 \text{ seconds} & \mu = 8 \end{array}$$

Step 4: Determine the Maximum Period Required of the Real Structure.

$$T = \frac{T_s}{\sqrt{\mu}}$$

$$= \frac{4}{\sqrt{2}}$$

$$= 2.8 \text{ seconds} \qquad \mu = 2$$

$$= 2.6 \text{ seconds} \qquad \mu = 3$$

$$= 2.38 \text{ seconds} \qquad \mu = 4$$

$$= 1.87 \text{ seconds} \qquad \mu = 8$$

Step 5: Determine the Number of Frame Bays Required to Provide a Building of the Required Stiffness. The design steps are the same as Steps 4 and 5 of the displacement-based design developed in Section 3.2.3.1. This example adopts the design constraints that led to the development of the frame described in Figure 3.2.1. Where this is not the case, the stiffness of the components may be developed as they were in Step 5 of Section 3.2.3.1.

The period of a frame structure was developed based on self weight (see Eq. 3.2.1). Accordingly,

$$T = 0.288\sqrt{\Delta} \qquad \text{(Eq. 3.2.1)}$$

$$\Delta_s = \frac{T^2}{0.083}$$

$$= \frac{(2.8)^2}{0.083}$$

$$= 94.5 \text{ in.} \qquad \mu = 2$$

$$= 81.4 \text{ in.} \qquad \mu = 3$$

$$= 68.2 \text{ in.} \qquad \mu = 4$$

$$= 42.1 \text{ in.} \qquad \mu = 8$$

Then the stiffness of an interior subassembly was used in conjunction with the deflection to determine the tributary weight (mass) that could be supported by one interior subassembly (w_i).

$$\Delta_s = \sum n\Delta_i \qquad \text{(Eq. 3.2.13a)}$$

$$= \left(\frac{n+1}{2}\right) n \frac{w_i}{k_i} \qquad \text{(see Eq. 3.2.7a)}$$

$$= \frac{8(15)w_i}{400}$$

$$w_i = \frac{\Delta_s}{0.3}$$

$$
\begin{aligned}
w_i &= \frac{94.5}{0.3} \\
&= 315 \text{ kips} \qquad \mu = 2 \\
&= 271 \text{ kips} \qquad \mu = 3 \\
&= 227 \text{ kips} \qquad \mu = 4 \\
&= 140 \text{ kips} \qquad \mu = 8
\end{aligned}
$$

Conclusion: A three-bay frame may be used to brace a structure whose effective system ductility is at least 3, 4, or 8.

$$
\begin{aligned}
j &> \frac{W_f}{w_i} \qquad \text{(see Eq. 3.2.13b)} \\
&= \frac{810}{271} \\
&= 3 \qquad \mu = 3
\end{aligned}
$$

The beam strength required to limit the ductility demand to 2 exceeds our objective maximum ($\rho_b > 1.5\%$).

$$
\begin{aligned}
M_{bu} &= 4.4V_{id} \qquad \text{(see Eq. 3.2.20b)} \\
&= 4.4(315) \\
&= 1386 \text{ ft-kips} \\
A_s &= \frac{M_{bu}}{\phi f_y (d - d')} \\
&= \frac{1386}{0.9(60)(2.5)} \\
&= 10.26 \text{ in.}^2 \qquad (\rho = 1.56\%)
\end{aligned}
$$

Comment: The selection of a four-bay frame for a structure of assumed system ductility of 2 presumes that the structure will experience a system drift that is 2 times Δ_y. This will probably not be the case, as we see in Chapter 4. Accordingly, the procedures used to define the number of frame bays only serve to help the designer make this key decision.

Step 6: Determine the Strength Required of the Frame. Now consider the characteristics of the provided frame and the drift levels expected, given the adopted reduced spectrum velocity ($\hat{R}S_v$). The period of the three-bay structure of Figure 3.2.1 is 2.6 seconds. The associated secant periods T_s would be

$$
\begin{aligned}
T_s &= \sqrt{\mu} T \\
&= \sqrt{2} T \\
&= \sqrt{2}(2.6) \\
&= 3.67 \text{ seconds} \qquad \mu = 2 \\
&= 4.5 \text{ seconds} \qquad \mu = 3 \\
&= 5.2 \text{ seconds} \qquad \mu = 4 \\
&= 7.36 \text{ seconds} \qquad \mu = 8
\end{aligned}
$$

Comment: Drift limits will be reached, and they must be considered in the evaluation of strength objectives. Consider Figure 1.1.5 and our criterion spectral velocity of 50 in./sec. The displacement constant region starts at a period of 4 seconds. Accordingly,

$$
\begin{aligned}
S_d &= \frac{T}{2\pi} S_v \\
&= \frac{4}{2\pi}(50) \\
&\cong 32 \text{ in.}
\end{aligned}
$$

This is converted to a peak ground displacement by Eq. 1.1.8c.

$$
\begin{aligned}
d_{\max} &= \frac{S_d}{2.73 - 0.45 \ln \zeta} \qquad (3.2.21) \\
&= \frac{32}{2.73 - 0.45 \ln 5} \\
&= 16 \text{ in.}
\end{aligned}
$$

Now the spectral displacement maxima are

$$
\begin{aligned}
S_{d,\max} &= (2.73 - 0.45 \ln \hat{\zeta}) d_{\max} \qquad \text{(Eq. 1.1.8c)} \quad (3.2.22) \\
&= (2.73 - 0.45 \ln 14)(16) \qquad (\text{see Step 1 for } \hat{\zeta}) \\
&= 24.6 \text{ in.} \qquad \mu = 2 \\
&= 22.0 \text{ in.} \qquad \mu = 3 \\
&= 21.8 \text{ in.} \qquad \mu = 4 \\
&= 21.8 \text{ in.} \qquad \mu = 8
\end{aligned}
$$

The spectral displacement will be the lesser of S_d and $S_{d,\max}$, where

$$
S_{d,s} = \frac{T_s S_v}{2\pi} \qquad (3.2.23)
$$

$$= \frac{3.67(35)}{2\pi}$$

$$= 20.4 \text{ in.} \qquad \mu = 2$$

$$= 22.2 \text{ in.} \qquad \mu = 3 \qquad (S_{d,\max})$$

$$= 24.4 \text{ in.} \qquad \mu = 4$$

$$= 31.0 \text{ in.} \qquad \mu = 8$$

Thus the yield drift of the real structure must be

$$\Delta_y = \frac{\Delta_u}{\mu}$$

$$= \frac{\Gamma S_d}{\mu}$$

$$= \frac{1.45(20.4)}{2}$$

$$= 14.8 \text{ in.} \qquad \mu = 2$$

$$= 10.6 \text{ in.} \qquad \mu = 3$$

$$= 7.9 \text{ in.} \qquad \mu = 4$$

$$= 3.95 \text{ in.} \qquad \mu = 8$$

The strength or base shear (V_i) required of an interior subassembly was developed by Eq. 3.2.19.

$$V_i = \frac{\Gamma k_i}{n} \Delta_y \qquad \text{(Eq. 3.2.19)}$$

$$= \frac{1.45(400)}{15} \Delta_y$$

$$= 38.7 \Delta_y$$

and the design base shears required of an interior subassembly to attain the respective levels of system ductility are

$$V_i = k_s \Delta_y$$

$$= 38.7(14.8)$$

$$= 573 \text{ kips} \qquad \mu = 2$$

$$= 410 \text{ kips} \qquad \mu = 3$$

$$= 306 \text{ kips} \qquad \mu = 4$$

$$= 154 \text{ kips} \qquad \mu = 8$$

Comment: The use of an overstrength factor of 1.29 is intended to attain comparability between the equal displacement procedure developed in Section 3.2.3.1 and the direct displacement design developed in this section. The Blue Book[3.4] recommended value for μ is 6.2, hence ($\Omega_o = 8/6.2 = 1.29$).

For the entire frame the design base shear, assuming an overstrength factor of 1.29, would be

$$V = \left((j-1) + 2\left(\frac{V_e}{V_i}\right)\right) V_i \qquad \text{(Eq. 3.2.6)}$$

$$= 3V_i$$

$$V_d = \frac{V}{\Omega_o}$$

$$= \frac{3V_i}{1.29}$$

$$= 2.33\ V_i$$

$$= 1355 \text{ kips} \quad (0.11W) \quad \mu = 2$$

$$= 955 \text{ kips} \quad (0.078W) \quad \mu = 3$$

$$= 713 \text{ kips} \quad (0.059W) \quad \mu = 4$$

$$= 359 \text{ kips} \quad (0.03W) \quad \mu = 8$$

Figure 2.1.65 describes the behavior of a precast DDC assembly. A ductility-overstrength factor of 8 is certainly reasonable.

$$M_{bu} = 4.4V_{iu} \qquad \text{(Eq. 3.2.20a)}$$

$$= 4.4(154) \quad (\mu = 8)$$

$$= 678 \text{ ft-kips}$$

$$\lambda_o M_{bu} = 1.29(678)$$

$$= 875 \text{ ft-kips}$$

3.2.4.2 Hybrid Frame The structural damping in the hybrid frame might be presumed to be half of that used in the DDC design. Our objective system ductility factor should be at least 8. Refer to Section 3.2.3.1 for a description of the process.

Step 1:

$$\hat{\zeta}_{eq} = \frac{\sqrt{\mu} - 1}{2\pi\sqrt{\mu}} + 5 \qquad \text{(see Eqs. 1.1.9a, d)}$$

For $\mu = 8$, the result is

$$\hat{\zeta}_{\text{eq}} = \frac{\sqrt{8}-1}{2\pi\sqrt{8}} + 5$$

$$= 15$$

$$\hat{R} = \frac{3.38 - 0.67 \ln \hat{\zeta}_{\text{eq}}}{2.3} \qquad \text{(Eq. 3.1.45)}$$

$$= 0.68$$

$$S_{v,s} = \hat{R} S_v$$

$$= 0.68(50)$$

$$= 34 \text{ in./sec}$$

Step 2:

$$S_d = 24.2 \text{ in.} \qquad \text{(Eq. 3.2.21 and 3.2.22)}$$

Step 3:

$$T_s = \frac{2\pi S_{d,s}}{S_{v,s}} \qquad \text{(see Eq. 3.1.22)}$$

$$= \frac{2\pi(24.2)}{34}$$

$$= 4.5 \text{ seconds}$$

Step 4:

$$T = \frac{T_s}{\sqrt{\mu}}$$

$$= \frac{4.5}{\sqrt{8}}$$

$$= 1.59 \text{ seconds}$$

Step 5:

$$\Delta_s = \frac{(T)^2}{0.083} \qquad \text{(see Eq. 3.2.1)}$$

$$= \frac{(1.59)^2}{0.083}$$

$$= 30.5 \text{ in.}$$

$$w_i = \frac{\Delta_s}{0.3} = \frac{30.5}{0.3} = 101.7 \text{ kips}$$

Conclusion: Three bays are suggested. The adopted three-bay structure has a fundamental period of 2.6 seconds.

Step 6:

$$T_s = \sqrt{\mu} T = \sqrt{8}(2.6) = 7.36 \text{ seconds}$$

$$S_{d,\max} = 24.2 \text{ in.} \qquad (d_{\max} = 16 \text{ in., see Eq. 1.1.8c})$$

Step 7:

$$\Delta_y = \frac{\Delta_u}{\mu} = \frac{\Gamma S_d}{\mu} = \frac{1.45(24.2)}{8} = 4.4 \text{ in.}$$

and this is consistent with the previous example ($V_i = 170$ kips). Hence,

$$\lambda_o M_{bu} = 965 \text{ ft-kips}$$

3.2.4.3 Precast Frame Beam Designs Given the breadth of design options developed in Table 3.2.1 (800 ft-kips to 1150 ft-kips), it seems appropriate to adopt a pragmatic member design solution. Size constraints were identified and used to develop estimates of dynamic actions. Hardware constraints for the DDC and design objectives for the hybrid system were discussed in Chapter 2. It remains to develop a practical design for each system and review the impact of the design on system and component ductility demand.

DDC Beam Design The 20-in. by 36-in. beam size described in Figure 3.2.2 will allow the placement of one set of DDCs at the top and bottom of the beam. Following the concepts developed in Section 2.1.4.5, the nominal strength of a DDC reinforced beam is

$$\begin{aligned} M_n &= NT_n(d - d') \qquad \text{(see Eq. 2.1.73)} \\ &= (1)(282)(2.5) \\ &= 705 \text{ ft-kips} \end{aligned}$$

Based on Figure 3.2.3*b* and Figure 2.1.65 it seems reasonable to revert to an overstrength of 1.29.

$$\begin{aligned} \lambda_o M_n &= 1.29(705) \\ &= 910 \text{ ft-kips} \end{aligned}$$

This is consistent with a system ductility overstrength factor of 8.

The mechanism shear for the system would be

$$V_M = \frac{2jM_n\,(\ell/\ell_c)}{h_x} \tag{3.2.24}$$

where

j is the number of bays.
ℓ is the beam length.
ℓ_c is the clear span of the beam.
h_x is the story height.

$$\begin{aligned} V_M &= \frac{2(3)(705)(26/23)}{10} \\ &= 478 \text{ kips} \qquad (0.04W) \end{aligned}$$

Hybrid Beam Design The procedures developed in Section 2.1.4.4 may be used, but a looser design is also reasonably adopted. The beam size is 20 in. × 36 in. We know (Section 2.1.4.4) that an efficient design is one in which the level of post-tensioning is about 800 psi and that the mild steel should provide a moment capacity of about 80% of that provided by the post-tensioning.

$$\begin{aligned} T_{pse} &= f_{pse}bh \qquad (3.2.25) \\ &= 0.8(20)(36) \\ &= 576 \text{ kips} \end{aligned}$$

$$T_{ps} = \frac{f_{ps}}{f_{pse}} T_{pse} \tag{3.2.26}$$

$$\cong \frac{172}{162}(576)$$

$$= 612 \text{ kips}$$

$$M_{psn} = T_{ps}\left(\frac{h}{2} - \frac{a}{2}\right) \tag{3.2.27}$$

$$\cong (612)(1.33)$$

$$\cong 814 \text{ ft-kips}$$

$$M_{sn} \cong 0.8 M_{ps} \tag{3.2.28}$$

$$= 650 \text{ ft-kips}$$

$$M_n = M_{ps} + M_s$$

$$= 814 + 650$$

$$= 1464 \text{ ft-kips}$$

Based on Figure 3.2.2*c* it seems reasonable to revert to an overstrength factor of 1.29.

$$\lambda_o M_n = 1.29(1464)$$

$$= 1900 \text{ ft-kips}$$

$$V_M = \frac{2jM_n(\ell/\ell_c)}{h_x} \qquad \text{(Eq. 3.2.24)}$$

$$= \frac{2(3)(1900)(26/23)}{10}$$

$$= 993 \text{ kips} \qquad (0.082W)$$

Conclusion: A three-bay system seems to be the best choice. The strength of the hybrid system should be greater than that of the cast-in-place system and the DDC *if* credence is given to energy dissipation differentials. Since a series of parametric studies are undertaken in Section 3.2.6 and Chapter 4, adopt a system strength that is efficient and a reasonable increment higher than those proposed for the cast-in-place alternative and the DDC system (see Table 3.2.1). Select a three-bay system and adopt a beam overstrength ($\lambda_o M_n$) on the order of 2000 ft-kips.

3.2.5 Irregular Frames

To this point our discussion of frames has been restricted to regular frames in the sense that bays and story heights have been the same throughout. Fortunately, the concepts developed in the preceding section can easily be adapted to irregular structures.

Photo 3.5 Precast DDC frame tower, Hollywood Highland Project, Hollywood CA. (Courtesy of Englekirk Partners, Inc.)

Consider first the irregular frame described in Figure 2.1.28 where it was assumed that residential function required alternating bays of 16 and 24 ft. Apply this constraint to the fifteen-story building design of Sections 3.2.2 through 3.2.4 (Figure 3.2.2*a*). The choices are 16/24/16 ft bays and 24/16/24 ft bays, the presumption being that a three-bay frame is adequate and this we will confirm or refute. Function and aesthetics will participate at least to the extent of proposing the most desirable of the alternatives. The engineer who can anticipate functional and aesthetic objectives will reduce the design time required to produce an acceptable solution.

The pattern 16/24/16 would be developed from two 12-ft wide bedrooms and two 16-ft wide common rooms. The spandrel beam is usually less offensive in a bedroom than in a common room. The 24/16/24 frame also creates a deeper frame, 64 ft as opposed to 56 ft. The only negative is the 8 additional feet of frame beam. From a design perspective, absent any specific aesthetic or functional input, try the 24/16/24 frame first.

The first step is to evaluate system dynamic characteristics starting with Eq. 3.2.2a and 3.2.2b. The presumption that the point of inflection will occur at midspan is valid if the flexural strength of the beam is the same on both ends so as to create antisymmetrical double curvature. A procedure for dealing with subassemblies that contain unequal spans is discussed in Reference 3.6, Sec. 4.2(d), where it is demonstrated that an average span length or the distance between points of inflection could be used without adversely impacting stiffness conclusions.

The effective length (ℓ_e) of an irregular subassembly is

$$\ell_e = 0.5\,(\ell_L + \ell_R) \tag{3.2.29}$$

and, for this example,

$$\begin{aligned}\ell_e &= 0.5(24 + 16)\\ &= 20 \text{ ft}\end{aligned}$$

The dimensions and properties used to develop stiffness characteristics are

$$\begin{aligned}b_b &= 20 \text{ in.}\\ h_b &= 36 \text{ in.}\\ b_c &= 36 \text{ in.}\\ h_c &= 36 \text{ in.}\end{aligned}$$

For the exterior subassembly ($\ell = 24$ ft), we have

$$\ell = 288 \text{ in.} \qquad \ell_c = 252 \text{ in.} \qquad \left(\frac{\ell_c}{\ell}\right)^2 = 0.77$$

while for the interior subassembly ($\ell = 20$ ft), we have

$$\begin{aligned}&\ell = 240 \text{ in.} && \ell_c = 204 \text{ in.} && \left(\frac{\ell_c}{\ell}\right)^2 = 0.72\\ &h_x = 120 \text{ in.} && h_c = 84 \text{ in.} && \left(\frac{h_c}{h_x}\right)^2 = 0.49\\ &I_{be} = 27{,}100 \text{ in.}^4 && (0.35 I_g)\\ &I_{ce} = 98{,}000 \text{ in.}^4 && (0.7 I_g)\\ &E_c = 4000 \text{ ksi}\end{aligned}$$

With a tributary mass of 810 kips/floor,

$$\Delta_i = \frac{V_i h_x^2}{12E}\left[\frac{\ell_c}{I_{be}}\left(\frac{\ell_c}{\ell}\right)^2 + \frac{h_c}{I_{ce}}\left(\frac{h_c}{h_x}\right)^2\right] \qquad \text{(Eq. 3.2.2a)}$$

$$\begin{aligned}&= \frac{V_i(120)^2}{12(4000)}\left[\frac{204}{27{,}100}(0.72) + \frac{84}{98{,}000}(0.459)\right]\\ &= 0.3V_i(0.0054 + 0.0004)\\ &= 0.00174V_i \qquad (k_i = 574 \text{ kips/in.})\end{aligned} \tag{3.2.30}$$

$$\Delta_e = \frac{V_e h_x^2}{6E}\left[\frac{\ell_c}{I_{be}}\left(\frac{\ell_c}{\ell}\right)^2 + \frac{h_c}{2I_{ce}}\left(\frac{h_c}{h_x}\right)^2\right] \qquad \text{(see Eq. 3.2.2a)}$$

$$= \frac{V_e(120)^2}{6(4000)}\left[\frac{252}{27{,}200}(0.77) + \frac{84}{2(98{,}000)}(0.49)\right]$$

$$= 0.6V_e(0.0071 + 0.0002)$$

$$= 0.00438V_e \qquad (k_e = 228 \text{ kips/in.})$$

$$(j-1) + 2\left(\frac{V_e}{V_i}\right) = (3-1) + 2\left(\frac{228}{574}\right) \qquad \text{(see Eq. 3.2.6a)}$$

$$= 2 + 0.8$$

$$V_i = \frac{V}{2.8} \qquad \text{(see Eq. 3.2.6a)}$$

Now proceed to determine the deflection of an interior subassembly loaded by its own weight (mass) as described in Figure 3.2.2*b*.

$$\sum w_i = \frac{W}{2.8}$$

$$= \frac{810}{2.8}$$

$$= 289 \text{ kips}$$

$$\delta_n = 0.00174\sum w_i \qquad \text{(see Eq. 3.2.30)}$$

$$= 0.00174(289)$$

$$= 0.5 \text{ in.}$$

The deflection of the frame (in shear) is

$$\Delta_s = \left(\frac{n+1}{2}\right)n\delta_n \qquad \text{(see Eq. 3.2.7a)}$$

$$= 8(15)(0.5)$$

$$= 60 \text{ in.}$$

and the period of the structure is

$$T_s = 0.288(\Delta_s)^{0.5} \qquad \text{(see Eq. 3.2.1)}$$

$$= 0.288(60)^{0.5}$$

$$= 2.23 \text{ seconds} \qquad \text{(Computer-generated period—2.26 seconds)}$$

In the displacement-based design (Section 3.2.3) it was determined that the period of the structure must be less than 2.8 seconds to attain our target drift objective. This, restated in compliance form, is

$$S_v = 50 \text{ in./sec} \qquad \text{(Design criterion)}$$

$$\begin{aligned} S_d &= \frac{TS_v}{2\pi} \\ &= \frac{2.23(50)}{6.28} \\ &= 17.8 \text{ in.} \\ \Delta_u &= \Gamma S_d \\ &= 1.45(17.8) \\ &= 25.8 \text{ in.} < \Delta_T = 32.4 \text{ in.} \qquad \text{(see Eq. 3.2.12)} \end{aligned}$$

If we adopt an objective system ductility factor of 6 so as to recognize the potential for early hinging in the interior bay, we may determine the strength required of an interior subassembly.

$$\begin{aligned} \Delta_y &= \frac{\Delta_u}{\mu\Omega_o} \\ &= \frac{25.8}{6} \\ &= 4.27 \text{ in.} \end{aligned}$$

and the shear strength required at yield would be developed as before.

$$\begin{aligned} V_i &= \frac{\Delta_y \Gamma k_i}{n} \qquad \text{(Eq. 3.2.19)} \\ &= \frac{4.27(1.45)(574)}{15} \\ &= 237 \text{ kips} \end{aligned}$$

The shear imposed on the exterior subassembly at interior subassembly yield will be in direct proportion to the stiffnesses. The first yield strength objective for the frame is

$$\begin{aligned} V_y &= \left\{ (j-1) + 2\left(\frac{V_e}{V_i}\right) \right\} V_{yi} \\ &= 2.8(237) \\ &= 663 \text{ kips} \qquad (0.055W) \end{aligned}$$

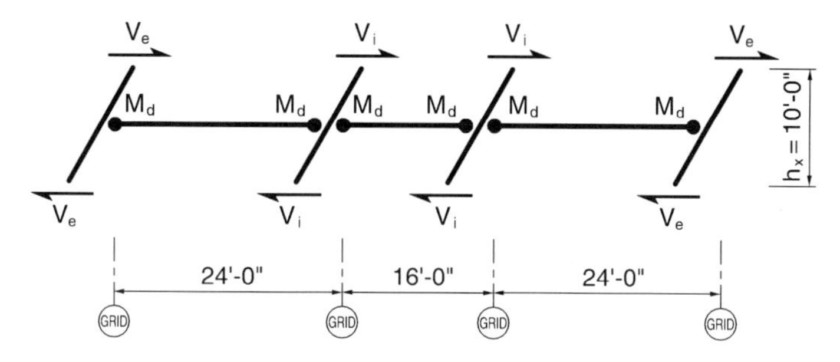

Figure 3.2.4 Mechanism shear, irregular frame.

The irrationality of an elastic distribution of shears to any combination of dissimilar bracing elements has been discussed in Section 3.1.2. The elastic analysis used to this point was only used to establish a design objective. The uniform nature of the proposed frame (24/16/24), coupled with the disproportionate stiffness of the columns (Eq. 3.2.30) suggests that we adopt a uniform design moment for both the exterior and interior beams. This should be the objective for cast-in-place construction, and an especially important consideration in precast concrete construction, where it will be difficult to alter the flexural strengths of adjoining beams.

A story mechanism approach should be used to develop the reinforcing program. The shear distribution, given this mechanism objective (Figure 3.2.4), will vary from that developed from an elastic distribution.

The mechanism shear, given an equivalent exterior subassembly strength, will be $3V_{yi}$ ($V_M = 3 \times 237 = 711$ kips) if a uniform reinforcing program is adopted (see Eq. 3.2.24).

Clearly a pragmatic approach to establishing the design strength of an irregular frame is called for. The mechanism moment demand, were the objective a mechanism story shear of 663 kips, is developed from Figure 3.2.4 as

$$6M_d = h_x V_M$$

$$= 10(663)$$

$$M_d = \frac{6630}{6}$$

$$= 1105 \text{ ft-kips}$$

M_d in this case is the moment at the centerline of the column. A reduction to the face of the column is appropriate. The lesser reduction would occur on the long beam where

$$V_b = \frac{2M_d}{\ell}$$

$$= \frac{2(1105)}{24}$$

$$= 92 \text{ kips}$$

$$\Delta M_d = V_b \frac{h}{2}$$

$$= 92\left(\frac{36}{2}\right)\left(\frac{1}{12}\right)$$

$$= 138 \text{ ft-kips}$$

The design moment should be

$$M_{bu} = M_d - \Delta M_d$$

$$= 1105 - 138$$

$$= 967 \text{ ft-kips}$$

The steel required to attain this design strength would be

$$A_s = \frac{M_u}{\phi f_y (d - d')}$$

$$= \frac{967}{0.9(60)(2.5)}$$

$$= 7.16 \text{ in.}^2 \qquad (\rho = 1.1\%)$$

and this is within our objective limit of 1.5%.

Were the irregular frame beam system (Figure 3.2.4) designed using a strength-based approach (Section 3.2.2), the suggested mechanism shear would be developed as follows.

Comment: The period used will be that developed in this section for the elastic irregular frame ($T = 2.23$ sec). The development of the dynamic characteristics of a system was the objective of Steps 1 thru 4. Accordingly we start with Step 5.

Step 5: (Section 3.2.2)

$$S_a = \frac{2\pi}{T} S_v$$

$$= \frac{6.28}{2.23}(50)$$

$$= 140.8 \text{ in./sec}^2 \qquad (0.36g)$$

Comment: Observe that the stiffness of an irregular-coupled system (secant stiffness, Figure 3.1.13) will be softer than that of the elastic based period ($T = 2.23$ seconds) and this will produce a lesser strength objective (S_a).

Step 6: (Section 3.2.2)

$$M_e = 0.8M$$

$$\begin{aligned} V_d &= S_a M_e \\ &= \frac{0.36g0.8W}{\Omega_o \mu g} \\ &= \frac{0.36(0.8)(15)810}{6} \\ &= 582 \text{ kips} \end{aligned}$$

Conclusion: A three-bay frame provides an acceptable solution. The design base shear objective is 582 kips based on a strength-based design methodology and the adopted level of system ductility.

$$\begin{aligned} V_i &= \frac{V}{3} \qquad \text{(mechanism approach)} \\ &= \frac{582}{3} \\ &= 194 \text{ kips} \end{aligned}$$

$$\begin{aligned} M_{bu} &= \frac{V_i h_x}{2}\left(\frac{\ell_c}{\ell}\right) \\ &= \frac{194(10)}{2}\left(\frac{21}{24}\right) \\ &= 849 \text{ ft-kips} \end{aligned}$$

Observe that this is consistent with the displacement/mechanism design conclusion of 967 ft-kips.

Design Suggestion: Use a mechanism approach and generally neglect the unequal beam lengths in the development of the design. The consequences of this approach will be explored in Section 3.2.6.5.

3.2.6 Frame Design Evaluation by Sequential Yield Analysis[3.6, Sec. 5.5.1]

The frame design procedures presented in the preceding sections varied with regard to how they considered the impact of ductility on the design process. The fact that, regardless of the approach, a three-bay frame seemed to be the logical selection for the size of the frame given the specific constraints imposed on the system (column spacing, story height, beam size, and so on) is encouraging, at least to the designer who assumes the responsibility for the design.

The resultant objective frame strength is clearly a function of system ductility assumptions, and these assumptions are far more difficult to quantify for a frame than for a shear wall. The design strength objectives summarized in Table 3.2.1 vary by as much as 100%, and this presents a dilemma to the responsible engineer. The easiest choice is to follow code mandates and assume that various code committees knew more about the building being designed than you, the engineer responsible for the design. Alternatively, you could arbitrarily select a strength that is significantly higher than suggested by any of the design procedures. It's quite possible, however, that neither of these procedures will be in the best interest of the building and its occupancy, even with construction cost aside. This is because stronger buildings will create larger accelerations and place a larger demand on the more brittle components of the structure. Accordingly, a decision regarding the appropriate strength of the frame should consider more than which of the preceding design alternatives is believed to be most appropriate. Fortunately, system response is not too sensitive to system strength, and this will be explored in Chapter 4. The question becomes one of understanding the consequences of strength-related decisions on structural and nonstructural damage.

A sequential yield analysis is well suited to making performance-related design decisions. In this section we explore its use on the various designs summarized in Table 3.2.1. In Chapter 4 these designs are evaluated using inelastic time history procedures. Again, the attempt will be to describe how these design confirmation procedures might be used to gain confidence in a proposed design, not to support one design procedure over another. The basic building model is described in Figure 3.2.5. Before we discuss the behavior of the frame described in Figure 3.2.5 it is appropriate to understand our behavior objectives and the analytical modeling process.

3.2.6.1 What Constitutes Good Behavior? The most difficult issue faced in the development of the design of a frame is the selection of the appropriate level of system ductility (μ). System ductility is clearly a function of the ductility available in the frame's yielding components as well as the distribution of this yielding throughout the structure. Accordingly, a good design will be one that has a closely grouped, well-distributed hinging pattern. A good hinge formation pattern is described in Figure 3.2.8 where most of the beams hinge between a building drift of 0.3% and 0.5%.

The tendency is to focus on the first hinge to form and to delay its occurrence as long as possible especially if it occurs in a column.

First yield will almost always occur at the base of the tension side column of the frame (Figure 3.2.8*b*). The design penchant will be to strengthen this column and repeat the analysis, but this is not usually a good idea. Concern over the behavior of the tension side column should not cause it to be overreinforced, for this is likely to produce large shear loads on this column when it is subjected to compression loading (see Section 2.2.4). Before you change your design, consider the fixity that will actually be provided. Is it real or merely perceived? Can the fixity be relieved? Remember,

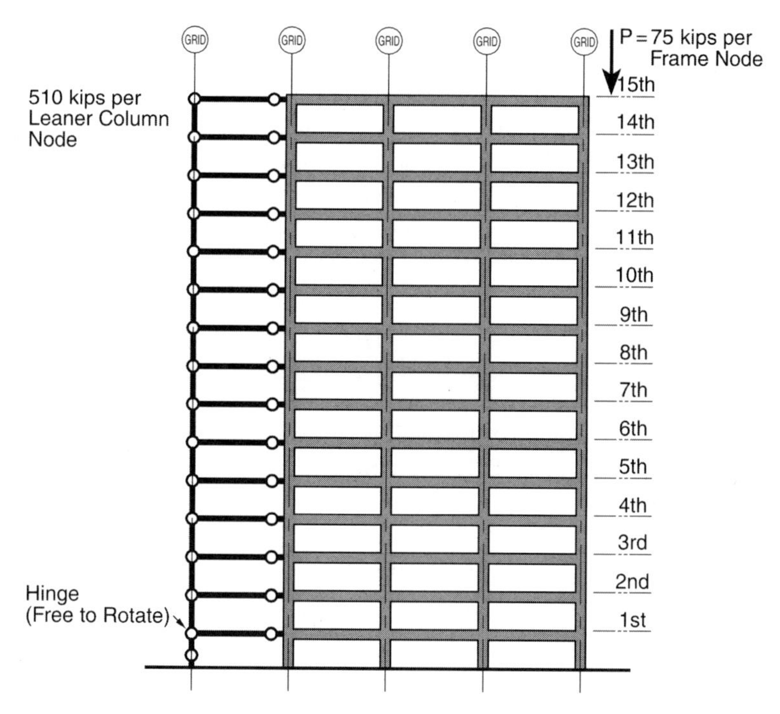

Figure 3.2.5 Frame model for $P\Delta$ effects.

this member when subjected to tension is quite ductile. Direct your concerns to the critical components—the compression side columns and the frame beams.

The compression side column will be critical and it must be designed in accordance with the concepts developed in Chapter 2. Strain states in all members at projected levels of building drift should be reviewed to insure that performance objectives have been attained. The postyield rotation demand imposed on beams must be reviewed to insure that strength degradation is not to be anticipated, for strength degradation will adversely impact behavior and your analysis.

Accomplishing these objectives should produce a frame that will perform well and when subjected to an inelastic time history analysis allow its confident adoption.

3.2.6.2 P*Δ *Concerns and Modeling Assumptions Modeling options include a consideration of strength hardening (Figure 3.1.2). The amount of strength hardening in beams (r, Figure 3.1.2) depends on the selection of the idealized flexural strength. The cyclic behavior of the frame beam described in Figure 2.1.2 shows an increase in load carrying capacity of 30 kips between Points 33 and 73, and this amounts to about 4% of the initial stiffness or rate of strength gain. The nominal strength of this beam was 645 ft-kips (based on the use of Eq. 2.1.1), and this corresponds to a tip load of 99 kips. Accordingly, the strength hardening rate of 4% must be used with a consistent definition of idealized yield strength. We have chosen (Table 3.2.1) to adopt an overstrength characterization of the idealized flexural strength, $M_{yi}(\lambda_o M_n)$. Accordingly, a strength hardening rate of 4% seems appropriate.

$P\Delta$ loads will have a strength degrading effect on the restoring force (Section 1.1.2). Most computer programs have a $P\Delta$ "switch" that presumably accounts for this degrading action. Unfortunately, most (if not all) computer programs do not properly account for $P\Delta$ effects. The usual case is one in which the axial loads imposed on the frame columns are considered but not the load imposed on the leaner or nonframe columns (see Figure 1.1.4). The $P\Delta$ induced shear on the frame whose behavior is described in Figure 3.2.5 is developed as follows:

$$V_{P\Delta} = \frac{P\Delta_x}{h_x} \qquad \text{(see Eq. 1.1.7)}$$

where P in this case should be the total weight (mass) tributary to the frame and Δ_x/h_x is the story drift ratio (DR). $P\Delta$ effects will be greater on the lower floors. For example, at Level 1,

$$P = 810(15)$$
$$= 12{,}150 \text{ kips}$$

Thus the $P\Delta$ shear is

$$V_{P\Delta} = P(\text{DR}) \qquad (3.2.29)$$

At a story drift ratio of 0.4%, the story shear will be increased by

$$V_{P\Delta} = P(0.004)$$

and this increase in story shear will increase the story drift proportionately ($V_{P\Delta}/V$).

In the example frame we know the relationships between story shear and story displacement:

$$\Delta_i = 0.0025V_i \qquad \text{(Eq. 3.2.5a)}$$
$$V_i = 400\Delta_i$$

And since we know the relationship between the story shear (V) and the shear imposed on an interior subassembly V_i,

$$V = 3V_i \qquad \text{(Eq. 3.2.6b)}$$

we may develop a story stiffness relationship

$$V = 1200\Delta_i$$

Now convert this to a relationship between story shear and drift ratio:

$$V = 1200\left(\frac{\Delta_i}{h_x}\right)h_x$$
$$= 1200(120)\ \text{DR}$$
$$= 144{,}000\ \text{DR}$$

and then the relationship between the axial load P, laterally braced by the frame and the drift ratio becomes

$$0.004P = 144{,}000\ \text{DR} \qquad \text{(see Eq. 3.2.29)}$$
$$\text{DR} = 0.000000027P$$

At the lowermost level this corresponds to an increase in drift ratio (DR) of

$$\Delta\text{DR}_1 = 0.000000027(15)(810)$$
$$= 0.000338$$

and this assumes that the story drift ratio caused by the lateral load will be on the order of 0.4%.

If we assume that the drift ratio is fairly constant at each level and that it will be increased at every level in proportion to the axial load at that level, a series summation can be developed that will allow us to estimate the increase in roof drift caused by the $P\Delta$ effect.

$$\sum_1^{15} \Delta\text{DR} = 0.000000027(P_f)\sum n$$
$$= 0.000000027(810)\left(\frac{n(n+1)}{2}\right)$$
$$= 0.000022\left(\frac{15(16)}{2}\right)$$
$$= 0.0026$$

which means that the roof drift would be more than 50% greater than that based on lateral loads acting in the absence of $P\Delta$ effects.

Consider what the pushover analysis tells us, realizing that the axial load considered by the $P\Delta$ analysis was only the vertical load imposed on the columns; hence the considered load (P) was 300 kips, not 810 kips. Accordingly,

$$\sum_{1}^{15} \Delta\text{DR} = 0.000000027(300)(120)$$
$$= 0.00097$$

In Section 3.2.3.1 a relationship between base shear and roof displacement ($V_y = 475$ kips; $\Delta = 3.75$ in.) was developed (Eq. 3.2.18). For the frame developed in the example, the projected stiffness is described in Figure 3.2.6.

Adjusting the drift rate increase $\Sigma_1^{15} \Delta\text{DR}$ to reflect the drift rate at $V = 725$ kips, we find the increase to be

$$\sum_{1}^{15} \Delta\text{DR} = \frac{0.0033}{0.0040}(0.00097) \qquad \text{(see Figure 3.2.6)}$$
$$= 0.0008$$

which when added to the basic drift ratio (no $P\Delta$) suggests the drift ratio developed in the sequential yield analysis:

$$\text{DR}_{P\Delta+V} = 0.0033 + 0.0008$$
$$= 0.0041 \cong 0.0042$$

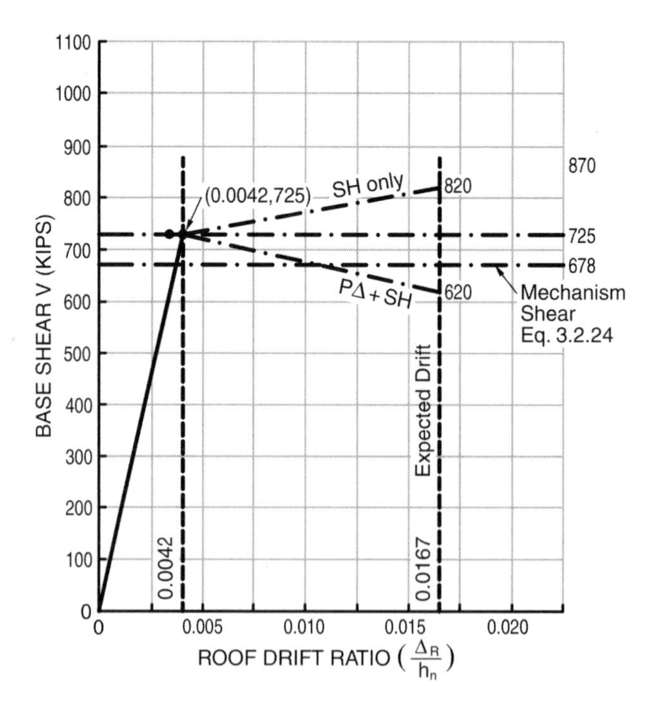

Figure 3.2.6 Idealized sequential yield analysis for three-bay frame—$\lambda_o M_n = 1000$ ft-kips.

Comment: The preceding evaluation assumes that the story drift ratio is constant in the elastic range, and this will not be the case. Nevertheless, it does demonstrate the impact of $P\Delta$ on system characteristics.

In the postyield behavior range the increase in story shear (V_{sh}) is associated with strength hardening.

$$V_{sh} = \frac{\sum M_p}{h_x} \frac{r\theta_p}{\theta_y} \tag{3.2.30}$$

$$= \frac{V_M r \theta_p}{\theta_y}$$

where V_{sh} is the increase in base shear provided by system strength hardening.

The net effect or effective mechanism strength (V_{eff}) will be a function of the shear that can be developed at the lower level.

$$V_{p\Delta} = P(\text{DR})$$

$$V_{sh} = \frac{V_M r \left(\text{DR} - \text{DR}_y\right)}{\text{DR}_y}$$

$$V_{\text{eff}} = \left(\frac{V_M}{P} + \frac{V_M}{P}(r)(\mu - 1) - \text{DR}\right)P$$

At a building drift ratio of 0.016 ($\mu = 4$) for the system described in Figure 3.2.5, the net impact would be developed as follows:

$$V_{\text{eff}} = \left(\frac{V_M}{P} + \frac{V_M}{P}(r)(\mu - 1) - \text{DR}\right)P$$

$$= \left(\frac{725}{12{,}150} + \frac{725}{12{,}150}(0.04)3 - 0.016\right)P$$

$$= (0.0597 + 0.007 - 0.016)P$$

$$= 0.051P \qquad \text{(85\% effective)}$$

Comment: This assumes a uniform distribution of postyield drift, and this is not likely to be the case, for postyield deformations will tend to concentrate in the lower floors.

At a displacement ductility of 4, the restoring force has been reduced by at least 15%. It is important to realize that this reduction in restoring force will not tend to increase the response of a structure (Section 1.1.2), as we shall see in Chapter 4.

Conclusion: $P\Delta$ effects can significantly reduce the restoring force provided by a bracing system. The impact of $P\Delta$ effects on response is not reliably estimated using approximate techniques. Accordingly, $P\Delta$ effects must be appropriately considered in the analysis phase of the design process.

An analytical model that correctly accounts for $P\Delta$ effects is described for this example frame in Figure 3.2.5. In essence a dummy column is tied to the frame and loaded with the excess tributary weight (mass).

Behavior models are many. Figure 3.2.6 may be used to describe several.

- An elastic/perfectly plastic model (725 kips). This is of little use unless it is adjusted to account for strain hardening in which case it describes the strength of the system for capacity-based design purposes (820 kips; 0.0167). This is because $P\Delta$ loads may reduce the available restoring force, but they do not reduce the moment demand imposed on the base of the frame (see Figure 3.2.7).
- A strength degrading model (620 kips). This model may be used to study the response of a single-degree-of-freedom model. Alternatively, this degrading stiffness can be replaced by a bilinear model (678 kips).

Design suggestions.

- Use the appropriate analytical model (Figure 3.2.5)
- Insure that a column mechanism is avoided (Section 2.2.3.3)
- Consider increasing the mechanism strength of the frame to compensate for $P\Delta$ effects if strength hardening does not appear to adequately compensate for this loss of strength.

3.2.6.3 Behavior Review—Frame 3 (Table 3.2.1) This section reviews the design proposed for a three-bay frame (see Figure 3.2.2*a*). Focal issues include those of

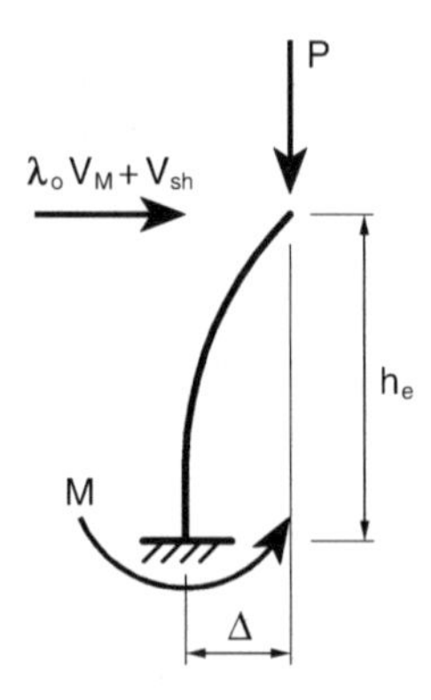

Figure 3.2.7 Analytical model including $P\Delta$ effects.

concern from a performance evaluation perspective as well as those of an academic interest.

From a performance evaluation perspective, major items of concern include

- Hinge propagation or distribution.
- Induced strain states in frame beams.
- Induced strain states in columns where plastic hinging is to be expected.

From an academic perspective, the issues are:

- The accuracy of the proposed mechanism prediction methodologies.
- System overstrength.
- Single-degree-of-freedom modeling procedures.
- Column load prediction methodologies.

The first item to review in any computer solution is the consistency between the computer solution and the design parameters that were used to generate the computer solution.

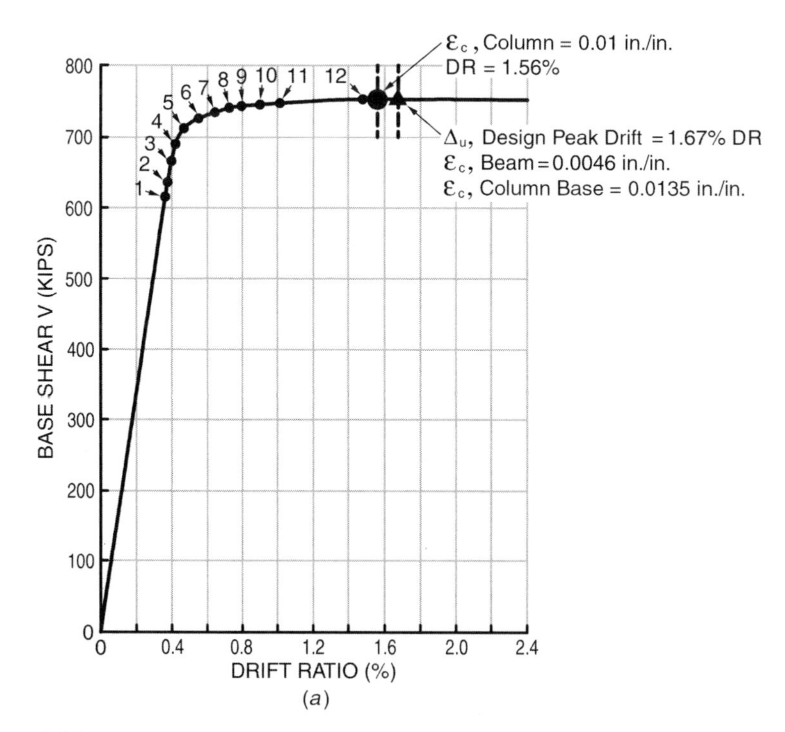

Figure 3.2.8 (*a*) Sequential yield analysis—Frame 3 (Table 3.2.1). (This frame will be referred to as 3–2 in a later discussion.)

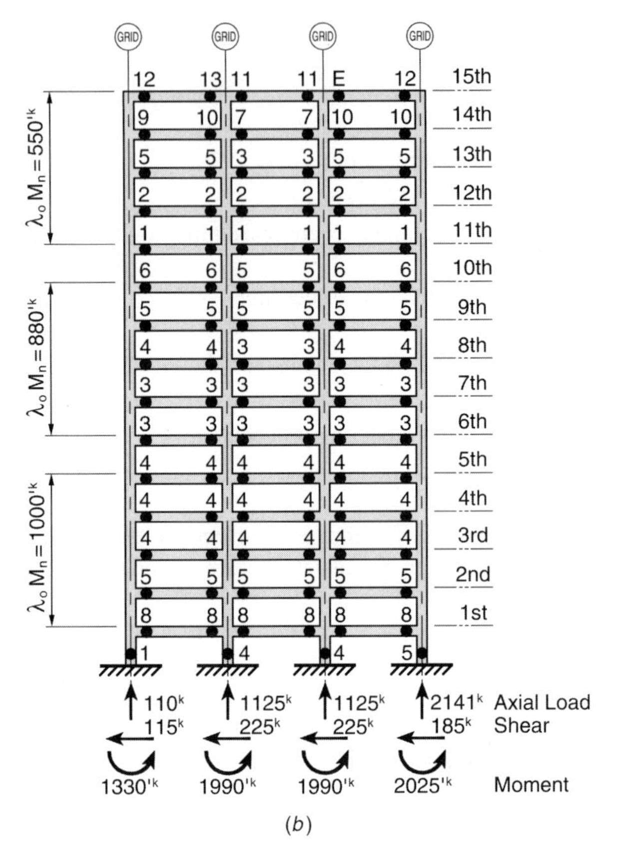

Figure 3.2.8 (*Continued*) (*b*) Hinge formation sequence.

Check mechanism strength. The behavior of Frame 3 as developed from a sequential yield (pushover) analysis is described in Figure 3.2.8*a*. $P\Delta$ effects and strain or strength hardening were not considered in the development of Figure 3.2.8*a*. System strength is projected at 750 kips. Figure 3.2.9 describes an idealization of this behavior extended so as to include $P\Delta$ and strength hardening. Observe that the restoring force at a drift of 1.67% is on the order of 550 kips.

Equation 3.2.24 developed the story mechanism shear. For Frame 3 the predicted mechanism shear is

$$\lambda_o V_M = \frac{2j\lambda_o M_n(\ell/\ell_c)}{h_x} \qquad \text{(Eq. 3.2.24)}$$

$$= \frac{2(3)(1000)(26/23)}{10}$$

$$= 678 \text{ kips}$$

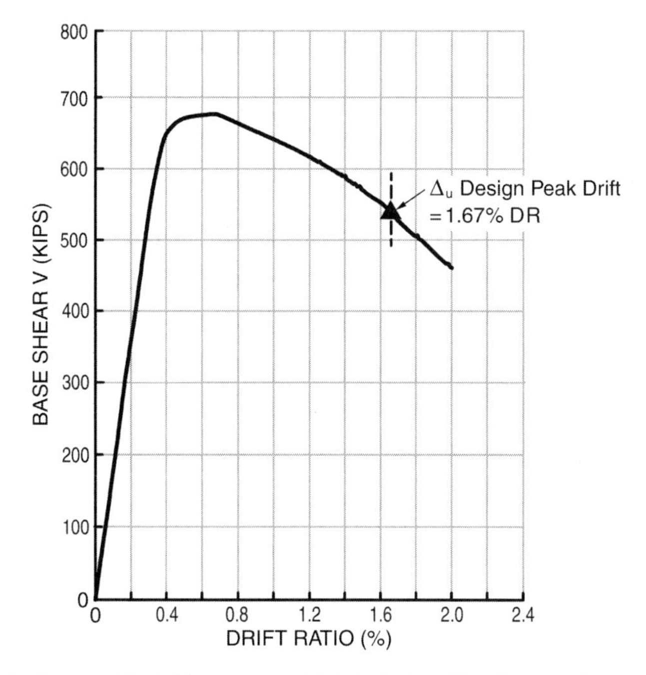

Figure 3.2.9 Sequential yield analysis which includes $P\Delta$ effects and member strength hardening.

This suggests a system overstrength factor (Ω_o) of 1.1 (750/678), but Eq. 3.2.24 did not include component strength hardening.

At the predicted level of displacement demand or drift ratio (DR = 1.67%), the restoring force, based on a 4% strength hardening rate (r), will be on the order of

$$V_{sh} = \frac{V_M r\,(\mathrm{DR})}{\mathrm{DR}_y} \qquad \text{(see Eq. 3.2.20)}$$

$$= \frac{725(0.04)(0.0167 - 0.004)}{0.004} \qquad \text{(see Figure 3.2.6)}$$

$$= 95 \text{ kips}$$

The restoring force lost as a consequence of $P\Delta$ effects will be on the order of

$$V_{P\Delta} = P(\mathrm{DR})$$

$$= 12{,}150(0.0167)$$

$$= 200 \text{ kips}$$

The restoring force (R) acting on the mass at a drift ratio of 1.67% should be on the order of

$$\begin{aligned} R &= \Omega_o \lambda_o V_M + V_{sh} - V_{p\Delta} \\ &= 725 + 95 - 200 \\ &= 620 \text{ kips} \end{aligned}$$

while that predicted by the pushover analysis (Figure 3.2.9) is only 550 kips (R/W = 0.045).

Comment: The 70-kip variant is probably attributable to the fact that component strength hardening does not translate directly to system strength hardening.

The actual force or shear imposed on the components of the frame will include $P\Delta$ effects (see Figure 3.2.7); hence from a component design perspective the shear load which the lowermost subassembly ($\Sigma V_i + \Sigma_I V_e$, Figure 3.2.4) must sustain is

$$\begin{aligned} \sum V_i + \sum V_e &= \Omega_o \lambda_o V_M + V_{sh} \qquad \text{(see Figure 3.2.4)} \\ &= 725 + 95 \\ &= 820 \text{ kips} \end{aligned}$$

As it relates to the nominal moment capacity of the beam (M_{bn}) proposed for frame 3 of Table 3.2.1 the overstrength factor to be used in the capacity-based design process would be 1.7.

$$\begin{aligned} M_n &= \frac{M_u}{\phi} \\ &= \frac{638}{0.9} \\ &= 709 \text{ ft-kips} \\ V &= \frac{3}{4.4}(709) \qquad \text{(see Eq. 3.2.20b)} \\ &= 483 \text{ kips} \\ \frac{\sum V_i + \sum V_e}{V} &= \frac{820}{483} \\ &= 1.7 \end{aligned}$$

The distinction between restoring force and the demand imposed on the components of the frame described in Figure 3.2.8b and as idealized in Figure 3.2.7 is

$$M = (\Omega_o \lambda_o V_M + V_{sh})\,h_e + P\Delta \qquad \text{(Moment demand)}$$

$$V = \Omega_o \lambda_o V_M + V_{sh} \qquad \text{(Shear demand)}$$

$$R = \Omega_o \lambda_o V_M + V_{sh} - \frac{P\Delta}{h_e} \qquad \text{(Restoring force)}$$

Comment: To this point we have not actually designed the frame beam. Once the frame beam is designed the choice of overstrength factor must be made so as to be consistent with the adopted level of both member and system overstrength. In this instance the use of a member overstrength factor of 1.5 seems appropriate based on Figure 2.1.2 and the beam design methodology, but it does not consider system overstrength.

One might then rationally use several values to account for overstrength, one for design purposes and the others for capacity considerations.

Insofar as the design procedures developed in the preceding sections are concerned, the overstrength factor was combined with an estimate of system ductility to produce an overall reduction factor and this is reasonable. For capacity design purposes the adopted overstrength factor should be member specific and appropriately include both member (λ_o) and system (Ω_o) overstrength considerations. Based on what we know about this frame and its components the following might be used to design system components.

System overstrength, $\Omega_o = 1.25$
Member overstrength, $\lambda_o = 1.5$
Column design, $\lambda_o \Omega_o = 1.88$
Beam and beam-column joint, $\lambda_o = 1.50$

And this is based on the design moment being the nominal moment capacity developed using Eq. 2.1.1.

Comment: The component designs in Chapter 2 used the customary member overstrength factor of 1.25, and this was consistent with the experienced behavior, prediction methodology, and adopted limit states. It was appropriate for that purpose but not necessarily for assessing system behavior.

The system overstrength factor should be confirmed by performing a pushover analysis which does not include $P\Delta$ effects. It need not include strength hardening if this is appropriately considered in the development of λ_o.

In Chapter 4 we will discuss stability and its impact on design decisions, and this will add yet another dimension to the decision making process as it relates to quantifying overstrength.

Next, it is advisable to check equilibrium to confirm the validity of the computer solution and understand the source of the strength exceedance. The base moments, shears, and axial loads are shown on Figure 3.2.8*b*. Deducting the dead load axial force of 1125 kips per column, we find the total overturning moment at the base is

$$\begin{aligned}\text{OTM} &= (2141 - 1125)(3)(26) + \sum M_c \\ &= 79{,}248 + 7335 \\ &= 86{,}583 \text{ ft-kips}\end{aligned}$$

and this corresponds to an effective height (Figure 3.1.1*a*) of

$$\begin{aligned}h_e &= \frac{\text{OTM}}{V_M} \qquad \left(\frac{V_M h_e}{V_M}\right) \\ &= \frac{86{,}583}{750} \\ &= 115 \text{ ft} \qquad (0.77h_n)\end{aligned}$$

Observe that this is higher than would be predicted by equivalent linear single-degree-of-freedom design methodologies (0.69—Eq. 3.1.16 and 3.1.18a). It is also considerably more than the 0.62 suggested by the Blue Book.[3.4]

Next, review hinge formation patterns and adjust strength as may be suggested.

The first hinge to form is in the tension side column at the base. Given the low level of axial load (110 kips—Figure 3.2.8*b*), the ductility available in this member should be essentially the same as that available in a beam. Initially hinges form in frame beams at Levels 11 and 12. Consideration might be given to increasing their strengths slightly, but frame beams in the lower levels are subjected to the largest postyield excursions. In general, the produced pattern seems acceptable.

Consider the load imposed on the compression side column, for its ability to withstand imposed levels of axial load and rotation will be critical to the survival of the frame. The stress induced on the column is

$$\begin{aligned}f_c &= \frac{P}{A_g} \\ &= \frac{2141}{(36)^2} \\ &= 1.65 \text{ ksi}\end{aligned}$$

For 5000-psi concrete, this represents a stress ratio of

$$\begin{aligned}\frac{f_c}{f'_c} &= \frac{1.65}{5.0} \\ &= 0.33\end{aligned}$$

and this approaches the balanced load. It seems advisable to reduce this stress ratio to something around $0.25f'_c$.

$$f'_c = \frac{f_c}{0.25} = \frac{1.65}{0.25} = 6.6 \text{ ksi}$$

Conclusion: Increase concrete strength in lower level columns to 7 or 8 ksi.

Check shear stress ratios in columns. The shear distribution in the postyield range will differ slightly from the elastic distribution, and if the mechanism strength is high, the redistribution of shears may cause undesirably high shear stresses.

$$V_i = 225 \text{ kips} \qquad \text{(see Figure 3.2.8}b\text{)}$$

$$v_i = \frac{V_i}{bd} = \frac{225}{36(33)} = 0.19 \text{ ksi} \qquad 2.7\sqrt{f'_c}$$

and this is well within our objective shear stress limit states.

Check strain states imposed on the compression side column.

$$P = 2141 \text{ kips} \qquad \text{(see Figure 3.2.8}b\text{)}$$

$$M = 2025 \text{ ft-kip}$$

$$A_s = A'_s = 9.36 \text{ in.}^2 \qquad \text{(see Figure 3.2.1)}$$

$$f'_c = 8 \text{ ksi} \qquad \text{(See previous consideration)}$$

Assume

$$C_c \cong \lambda_o A_s f_y - A'_s f_y = 0.25(9.36)(60) = 140 \text{ kips}$$

$$a = \frac{P + C_c}{0.85 f'_c b}$$

$$= \frac{2281}{0.85(8)(36)}$$

$$= 9.32 \text{ in.}$$

$$c = \frac{9.32}{0.65}$$

$$= 14.3 \text{ in.}$$

The postyield rotation reported at the projected building displacement (1.67%) was 0.0135 radian. Several conditions need to be considered. First, the hinge length in a heavily loaded column seemed to be at least $0.75h$ (see Figure 2.2.8). Second, the full fixity assumed in the computer analysis at the base of the columns will probably not be realized. Accordingly, the design that follows might be viewed as being conservative.

$$\ell_p = 0.75h$$

$$= 0.75(36)$$

$$= 27 \text{ in.}$$

$$\phi_p = \frac{\theta_p}{\ell_p}$$

$$= \frac{0.0135}{27}$$

$$= 0.0005 \text{ rad/in.}$$

$$\varepsilon_{cp} = \phi_p c$$

$$= 0.0005(14.3)$$

$$= 0.007 \text{ in./in.}$$

$$\varepsilon_{cu} = \varepsilon_{cp} + \varepsilon_{cy}$$

$$= 0.007 + 0.001$$

$$= 0.008 \text{ in./in.}$$

Conclusion: The compression side column will probably be at incipient shell spalling.

Review curvature demands imposed on frame beams. The maximum postyield rotation (0.018 radian) occurred simultaneously at Levels 5 and 6.

Comment: Observe that the postyield rotation of the frame beam used as our design model was 0.035 radian (3.5%, Eq. 2.1.8). If 1.8% is the maximum level of postyield rotation, the 4% strength hardening rate is too high (see Point 49, Figure 2.1.2).

Further, the strength hardening predicted for the system will not materialize because it is based on an average postyield rotation of 1.63%, and this was confirmed by the pushover analysis (Figure 3.2.8*a*).

$$\theta_p = 1.8\%$$

$$d = 33 \text{ in.}$$

$$\phi_y = \frac{\varepsilon_{sy}}{2kd}$$

$$\cong \frac{0.002}{22}$$

$$= 0.0001 \text{ rad/in.}$$

$$\ell_p = 0.5h$$

$$= 18 \text{ in.}$$

$$\phi_p = \frac{\theta_p}{\ell_p}$$

$$= \frac{0.018}{18}$$

$$= 0.001 \text{ rad/in.}$$

$$\mu_\phi = \frac{\phi_p + \phi_y}{\phi_y}$$

$$= \frac{0.0011}{0.0001}$$

$$= 11$$

Observe that this curvature ductility approaches experimental limits. As we have discussed in Chapter 2, an assessment of concrete strain states will require that an estimate of the effectiveness of the compression steel be made.

$$T = \frac{M}{d - d'}$$

$$= \frac{1000}{2.5} \qquad \text{(Level 5)}$$

$$= 400 \text{ kips}$$

$$a = \frac{400}{0.85(5)20}$$

$$= 4.7 \text{ in.}$$

$$c = \frac{a}{\beta_1}$$

$$= 5.9 \text{ in.}$$

$$\varepsilon_{cp} = \phi_p c$$

$$= 0.00109(5.9)$$

$$= 0.0064 \text{ in./in.}$$

Even discounting the compression steel entirely, the strain in the concrete would be on the order of 0.0075 in./in. and a neutral axis depth of 6 in. would cause the compression steel to yield. A second iteration should suggest the probable peak strain state.

Assume

$$c = 4 \text{ in.}$$

$$\varepsilon_{sc} = (c - d')\phi_p + kd\phi_y$$

$$= (1)0.001 + 8(0.0001)$$

$$= 0.0018 \text{ in./in.}$$

$$f_{sc} = \varepsilon_{sc} E_s$$

$$= 52 \text{ ksi}$$

Since the tension force (400 kips) is probably developed from a stress of 75 ksi, the compression force would, assuming a neutral axis depth of 4 in., be

$$C_s = T\left(\frac{52}{75}\right)$$

$$= 400(0.69)$$

$$= 277 \text{ kips}$$

This results in neutral axis depth of

$$a = \frac{400 - 277}{0.85(5)(20)}$$

$$= 1.45 \text{ in.}$$

$$c = 1.8 \text{ in.}$$

From these two iterations we can reasonably conclude that the neutral axis depth is about 3.5 in.

This suggests a concrete strain state of

$$\begin{aligned}\varepsilon_{cp} &= \phi_p c \\ &= 0.0010(3.5) \\ &= 0.0035 \text{ in./in.} \\ \varepsilon_{cu} &= \varepsilon_{cp} + \varepsilon_{cy} \\ &= 0.0035 + 0.001 \\ &= 0.0045 \text{ in./in.}\end{aligned}$$

Conclusion: Shell spalling should not be anticipated in the frame beams. Observe that both of these estimates of the concrete strain state are consistent with those predicted by the computer analysis (Figure 3.2.8*a*).

3.2.6.4 Frame 3—Consequences of Alternative Strengths This section considers the consequences associated with using alternative system strengths on the behavior of the frame described in Figure 3.2.2. The selected alternative system strengths are two-thirds to twice the strength of Frame 3 whose behavior was reviewed in the

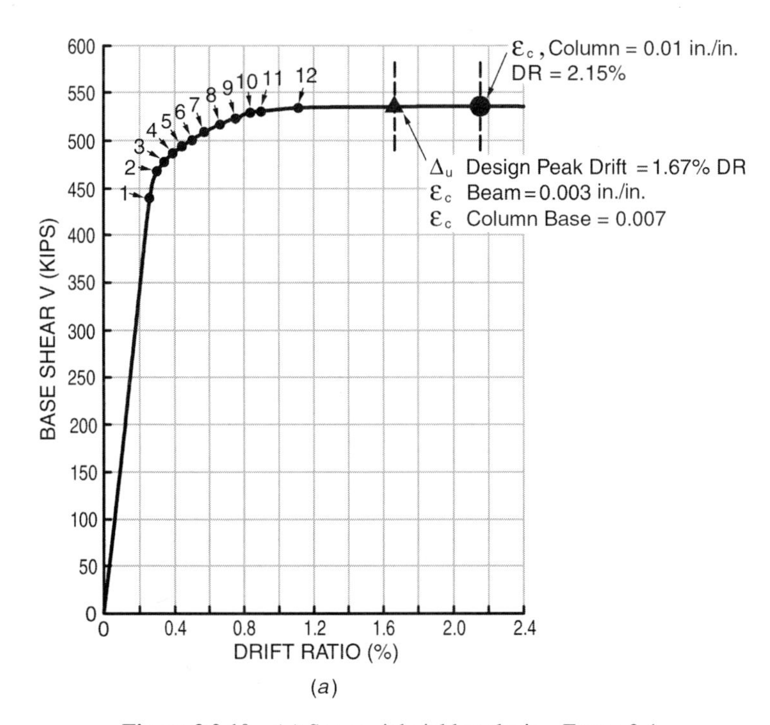

Figure 3.2.10 (*a*) Sequential yield analysis—Frame 3-1.

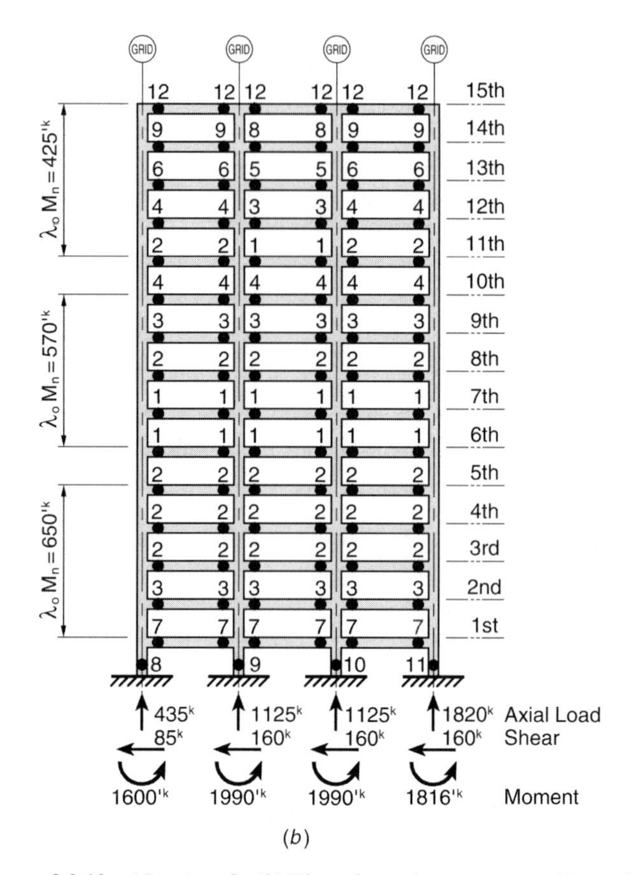

Figure 3.2.10 (*Continued*) (*b*) Hinge formation sequence—Frame 3-1.

preceding (Section 3.2.6.3). These frames will be identified as 3-1 and 3-3, while the original frame will be 3-2. The evaluated strengths of Frames 3-1, 3-2, and 3-3 based on overstrength of 1.5 would correspond to system ductilities of 8.2, 6.33, and 2.67, respectively.

Comment: The baseline frame (Frame 3-2—Table 3.2.1) was designed to a system ductility/overstrength factor of 8 ($\lambda_o \mu$). Subsequently a member overstrength factor of 1.5 was adopted (see Section 3.2.3.1). The member overstrength of 1000 ft-kips presumed that 18.7% of this system overstrength (1.5/8) was attributable to the overstrength of the member. Hence the design system ductility factor of 6.33. In Section 3.2.6.2 the relationship between probable system strength, as determined by sequential yield analysis (Figure 3.2.8*a*) and an analytical evaluation (Eq. 3.2.24), based on a member overstrength ($\lambda_o M_n$) of 1000 ft-kips, was found to be approximately 11% (750/678). Further, the sequential yield analysis (Figure 3.2.8*a*) suggests an idealized system ductility of 4 (0.0167/0.0042). These factors suggest that Frame 3-2 is best described as having an effective system ductility of 4. Frames 3-1 and 3-3 are now analyzed. Beam overstrengths of 650 ft-kips and 2000 ft-kips are used respectively. Accordingly, the system ductilities of Frames 3-1 and 3-3 might better be classified as 6 and 2.

First consider the behavior predicted by the sequential yield analysis performed on the system whose system ductility factor is 6 (Frame 3-1, Figure 3.2.10*a*). Observe that hinging first occurs in the beams (Figure 3.2.10*b*) as opposed to the tension side column as it did in Frame 3-2 (Figure 3.2.8*b*). Beam hinging is also closely grouped, occurring between drift ratios of 0.25% and 0.6%. Column hinging started with the tension side column at a drift ratio of 0.65% and this was quickly followed by hinging in the other columns. At the anticipated drift (1.67%) concrete strains in the frame beams were on the order of 0.003 in./in. and in the compression side column 0.007 in./in., based on $f_c' = 5$ ksi. Concrete strains in the compression side column reached 0.01 in./in. at a drift ratio of about 2.15%.

Conclusion: This frame should perform better than Frame 3-2 provided ultimate drifts are the same for both frames and this will be explored in Chapter 4.

The behavior of Frame 3-3 ($\mu_3 = 2$) is described in Figure 3.2.11*a*. Mechanism shear occurs when most of the frame beams, as well as the compression side column,

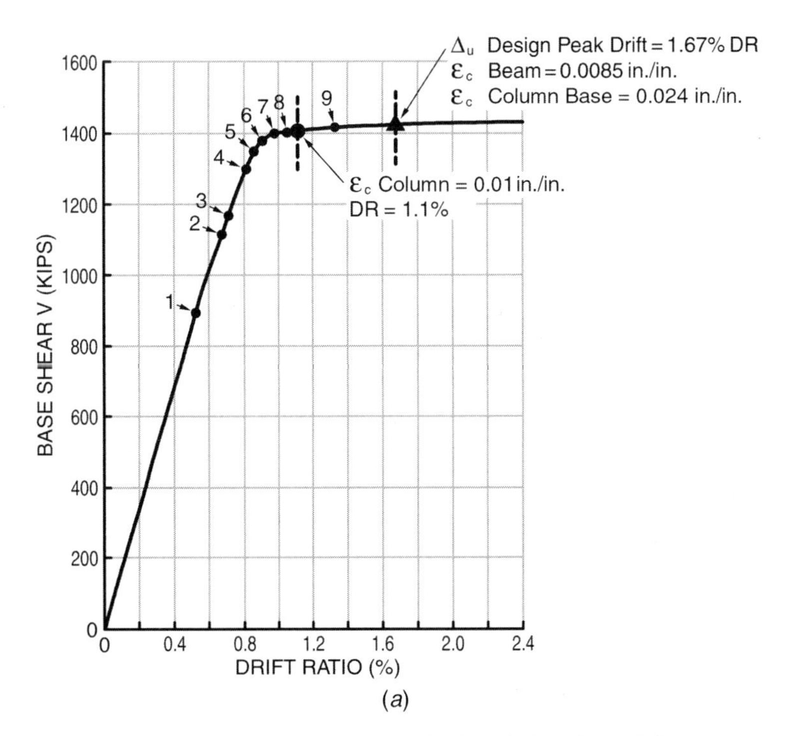

Figure 3.2.11 (*a*) Sequential yield analysis—Frame 3-3.

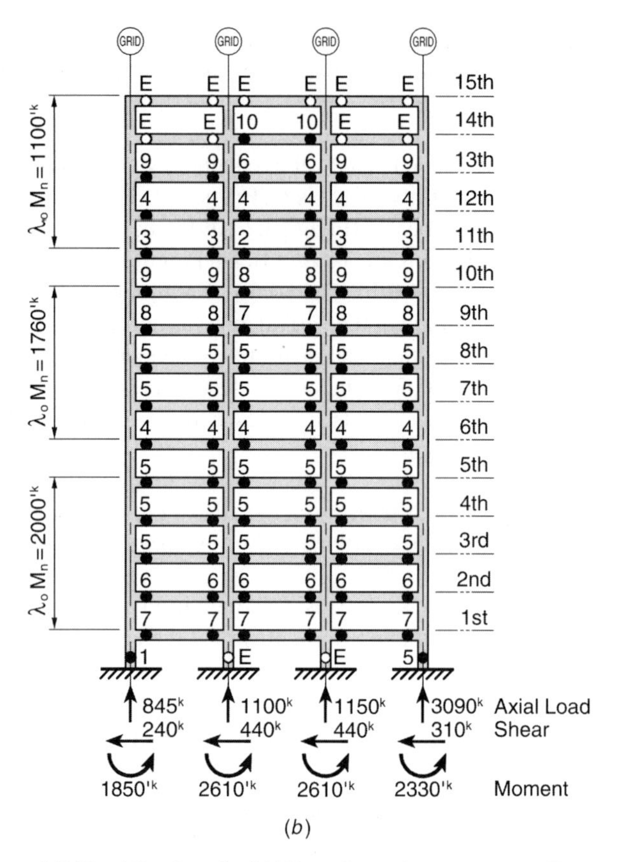

Figure 3.2.11 (*Continued*) (*b*) Hinge formation sequence—Frame 3-3.

yield (Figure 3.2.11*b*). Mechanism shear is associated with a drift ratio of 0.85%. At a drift ratio of 1.1% the concrete strain in the compression side column is expected to reach 0.01 in./in. At the anticipated drift (1.67%), the strain in the compression side column is expected to reach 0.024 in./in. and that in the frame beam to reach 0.008 in./in. Clearly less damage would be expected of Frame 3-1 or 3-2.

Conclusion: The three-bay frame described in Figure 3.2.2*a* should not be used if mechanism shears on the order of 1356 kips are required to satisfy an adopted design criterion.

Comment: Would the designer have been alerted to the problem? A beam reinforcement ratio (ρ_s) of less than 1.5% would have produced the overstrength moment of 2000 ft-kips. Accordingly, this alert would be insufficient.

The column alert is easily recognized:

$$\begin{aligned} P_E &= \frac{V_M h_e}{3\ell} \\ &= \frac{1356(100)}{3(26)} \\ &= 1738 \text{ kips} \end{aligned}$$

The sequential analysis suggested a seismic demand of 1975 kips ($\Omega_o = 1.14$).

$$\begin{aligned} P &= P_E + P_D \\ &= 1975 + 1125 \\ &= 3100 \text{ kips} \end{aligned}$$

The balanced axial load, based on a concrete strength of 5 ksi, is

$$\begin{aligned} P_b &\cong 0.35 f_c' A_g \qquad \text{(Maximum probable value)} \\ &\cong 0.35(5)(36)^2 \\ &\cong 2268 \text{ kips} \end{aligned}$$

Concrete strength must be significantly increased so as to provide ductility, but the behavior of a 12-ksi column would not be good, and this can be quickly checked because the postyield rotation of the column is 2% based on the sequential yield analysis given the frame drift objective of 1.67%.

$$\begin{aligned} P &= 3100 \text{ kips} \\ A_{st} &= 37.44 \text{ in.}^2 \qquad (2.9\%) \\ A_s &= 12.48 \text{ in.}^2 \\ C_s &\cong 0.75 f_y A_s \qquad \text{(Conservatively adopted, minor impact)} \\ C_c &= P + \lambda_o f_y A_s - 0.75 f_y A_s \\ &= 3100 + 0.5(60)(12.48) \\ &= 3475 \text{ kips} \\ a &\cong \frac{3475}{0.85(12)(36)} \\ &\cong 9.5 \text{ in.} \\ c &\cong \frac{9.5}{0.65} \\ &\cong 14.5 \text{ in.} \end{aligned}$$

$$\phi_p = \frac{\theta_p}{\ell_p}$$

$$= \frac{0.02}{0.75h}$$

$$= \frac{0.02}{27}$$

$$= 0.00074 \text{ rad/in.}$$

$$\varepsilon_{cp} = \phi_p c$$

$$= 0.00074(14.5)$$

$$\cong 0.011 \text{ in./in.}$$

To this must be added ε_{cy} and at best we can conclude that the shell of the column would spall.

3.2.6.5 *Behavior Review—Irregular Frame* Figure 3.2.12*a* describes the behavior of the irregular frame designed in Section 3.2.5. The projected (idealized) yield

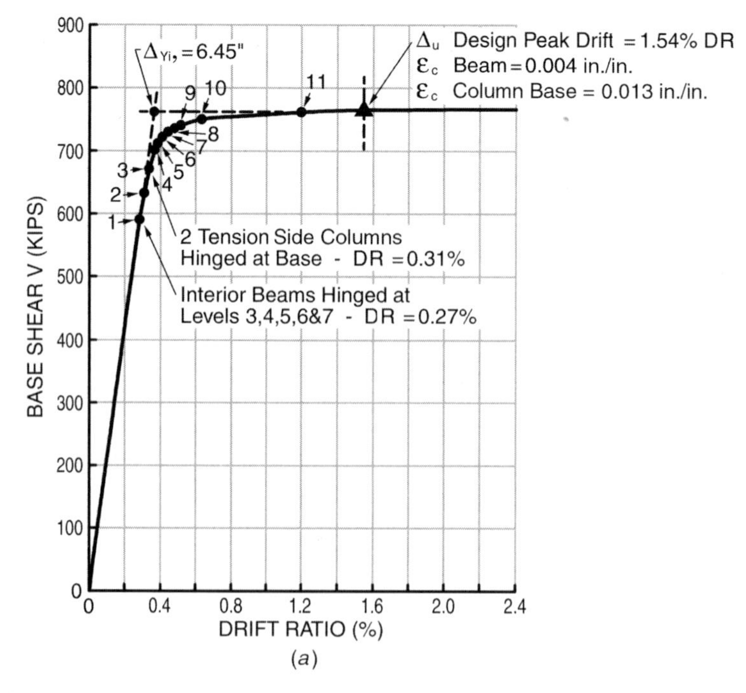

Figure 3.2.12 (*a*) Sequential yield analysis—Irregular frame (see Section 3.2.5).

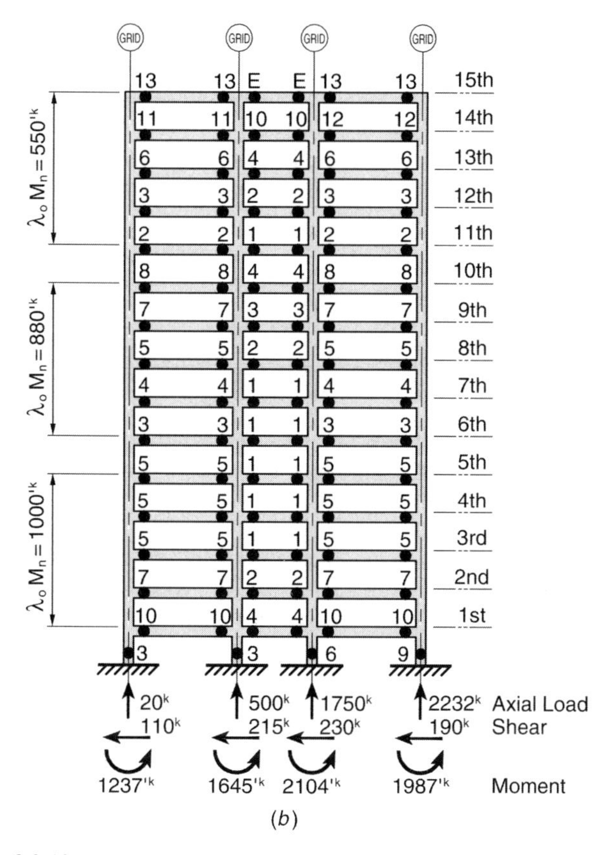

Figure 3.2.12 (*Continued*) (*b*) Hinge formation sequence—Irregular frame.

displacement is 6.45 in. (0.36%). The predicted mechanism strength, based on a member overstrength ($\lambda_o M_n$) of 1000 ft-kips, is

$$V_M = \frac{6\lambda_o M_n}{h_x}\left(\frac{\ell_{\text{eff}}}{\ell_c}\right)$$

$$= \frac{6(1000)}{10}\left(\frac{20}{17}\right)$$

$$= 706 \text{ kips}$$

The sequential yield prediction was 760 kips.

Observe (Figure 3.2.12*b*) that hinges formed early in the interior beams (0.27%) and that corresponding exterior hinges formed a little later (0.33%). Most of the beams hinged before the compression side column hinged.

At the predicted spectrum-based drift of 1.54%, the concrete strain in the compression side column had reached 0.013 in./in. and spalling should be expected. The maximum beam strain at a drift level of 1.54% was 0.004 in./in.

One might reasonably conclude that the provided strength is higher than it should be because the axial load imposed on the compression side column will cause it to spall before the design level drift is reached. The axial load imposed on the column was 2232 kips (Figure 3.2.12*b*), and this corresponds to

$$\frac{P}{f_c' A_g} = \frac{2232}{5(32)^2}$$

$$= 0.344$$

An increase in concrete strength would significantly improve the behavior of this column. Given this mitigation, the behavior seems reasonable. Observe how the interior columns shared the overturning load.

3.2.6.6 *Behavior Review—Precast Frame Systems* The design proposed for the hybrid system envisioned a beam strength ($\lambda_o M_n$) of 1900 ft-kips (see Section 3.2.4.3). The anticipated behavior would be essentially the same as that reported for Frame 3-3 (see Figure 3.2.11*a*).

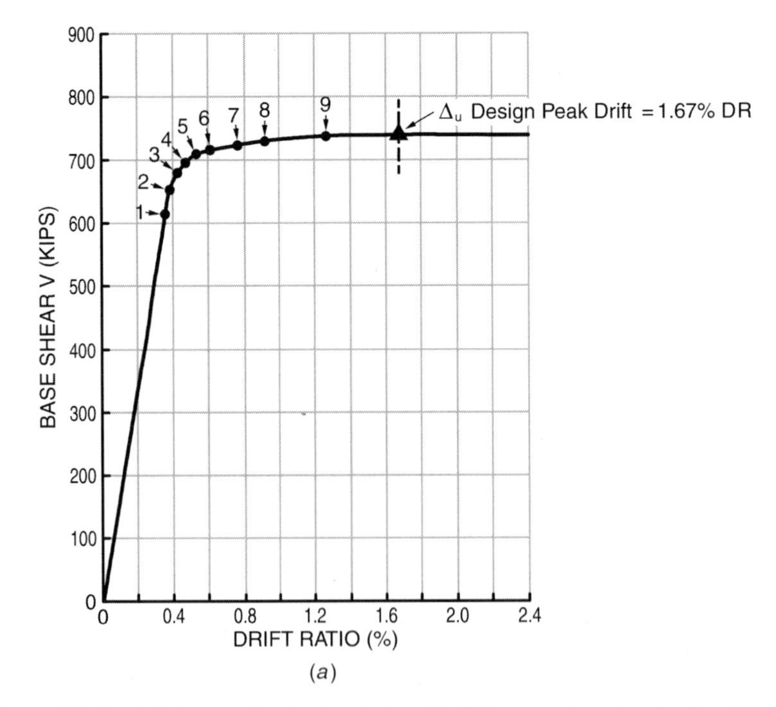

Figure 3.2.13 (*a*) Sequential yield analysis—DDC frame system (see Section 3.2.4.3).

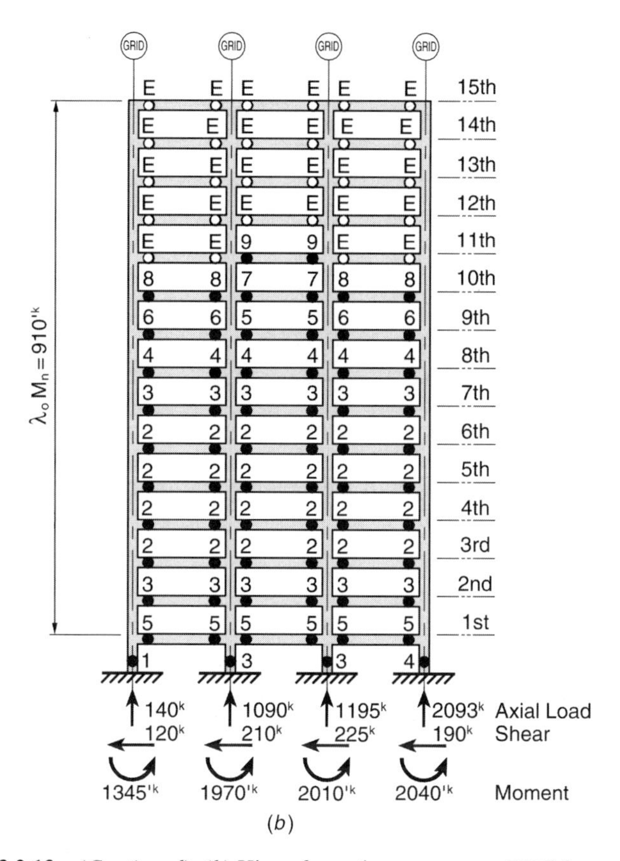

Figure 3.2.13 (*Continued*) (*b*) Hinge formation sequence—DDC frame system.

The behavior of the DDC frame demonstrates the consequences of maintaining the same beam strength over the entire height of the frame. Figures 3.2.13*a* and 3.2.13*b* show how the hinging is restricted to the lower ten floors. This appears to impose significantly more postyield rotation demand on the compression side column ($\theta_p = 2.25\%$ at 1.67%). This is more than twice the rotation imposed on Frame 3-1, whose behavior is described in Figure 3.2.10, and 25% more than that generated in the compression side column of Frame 3-2 (Figure 3.2.8*a*).This suggests that plastic hinge distribution throughout the frame will be important, a subject explored in Chapter 4. Mitigating features that might be considered since beam strengths cannot be reduced are:

- Reduce the number of provided frame bays above Level 10.
- Provide DDC connectors at the column bases

The strain imposed on a ductile rod at a plastic rotation of 2.25% is developed as follows:

$$\delta_{\text{rod}} = \frac{\theta_p(d - d')}{2}$$

$$= \frac{0.0225(30)}{2}$$

$$= 0.337 \text{ in.}$$

$$\varepsilon_p = \frac{\delta_{\text{rod}}}{\ell_p}$$

$$= \frac{0.337}{9}$$

$$= 0.037 \text{ in./in.}$$

where ℓ_p in this case is the length of the yielding portion of the ductile rod (Figure 2.1.57) and this is well within the capabilities of the forged rod (see Figure 2.1.63).

3.3 DIAPHRAGMS

Complex methods of analysis have no place in the design of diaphragms because there are too many uncertainties involved. Uncertainties are, of course, true to a lesser extent in all structural systems. Still, when one considers the fact that for a diaphragm the loading function (ground motion) is filtered through the primary bracing system of the building before it reaches the diaphragm, one can only accept the futility of any design approach that is not reduced to the simplest of forms. The development of this simplest form is the objective of this section.

3.3.1 Design Approach

Let us first consider the peculiarities of diaphragms that allow for design simplification.

- *System Decoupling:* Dynamically activated systems can feed off of each other or respond in an independent manner. Fortunately, the dynamic characteristics of the diaphragm and of the primary bracing system of the building are usually quite different. This will typically allow them to be treated separately. This separation occurs when the fundamental period of the two systems varies by a factor of 2 or more.[3.12]

- *Diaphragm Forcing Function:* Ground motion will be filtered as it proceeds through the building's bracing system, and an acceleration response will be created at the diaphragm/bracing system interface. This diaphragm acceleration input will be created from a spring/mass system (Figure 3.1.1*a*) or the model of a building subjected to the design earthquake ground motion. For example, if the bracing system for the building were rigid, the acceleration input to the diaphragm would be that of the ground. Otherwise, amplification or softening would occur. In any case, the nodal acceleration record for the building can be used to develop a nodal or support input acceleration to the diaphragm $\ddot{d}$, $\dot{d}$, d (see Figure 1.1.5). The diaphragm will, of course, respond to this motion in accord with its dynamic characteristic and damping (Eq. 1.1.8). This so-created elastic or inelastic response spectrum (Figure 1.1.7) can be used to design the diaphragm.

- *Diaphragm Behavior Characteristics:* Shear was the focus of Section 1.3. Behavior models for members with short shear spans were developed (Figures 1.3.6 and 1.3.7). The behavior of these short shear span models is extremely complex and does not lend itself to a rigorous analysis. Fortunately, accepting the characteristic behavior of the arch and the probable dynamic behavior range of diaphragms allows us to simplify the design processes considerably.

Given this intuitive understanding we may proceed to simplify the design process.

First, the response region for a diaphragm will be the acceleration constant region (Figure 1.1.5). As a consequence, the fundamental period of the diaphragm will have no impact on the inertial loading. Should the period of the diaphragm force its response into the velocity control region ($T > 0.5$ second), assumptions based on an acceleration constant response will be conservative.

Second, diaphragm displacements will tend to be small and conservative (high) estimates of the fundamental period of the diaphragm can be reasonably used to predict displacement ductility demands (see Figure 1.1.7).

Third, the fundamental frequency of a diaphragm can be estimated from

$$\omega = \frac{\pi^2}{\ell^2}\sqrt{\frac{EI_e}{m}} \tag{3.3.1}$$

Fourth, the effective moment of inertia (I_e) can be conservatively developed from a displacement prediction perspective (see Figure 2.4.17) from the experimental work on coupling beams (Section 2.4.2.1).

$$I_e = \frac{0.2I_g}{1 + 3(h/\ell)^2} \qquad \text{(see Eq. 2.4.38a)}$$

Fifth, from a force prediction perspective the stiffness assumed for cracked walls might be adopted ($I_e = 0.35I_g$). Recall, however, that when the probable period of the diaphragm is less than 0.5 second, conclusions relative to inertial forces will be the same regardless of the assumed stiffness of the diaphragm, and this clearly supports the use of a force-based criterion.

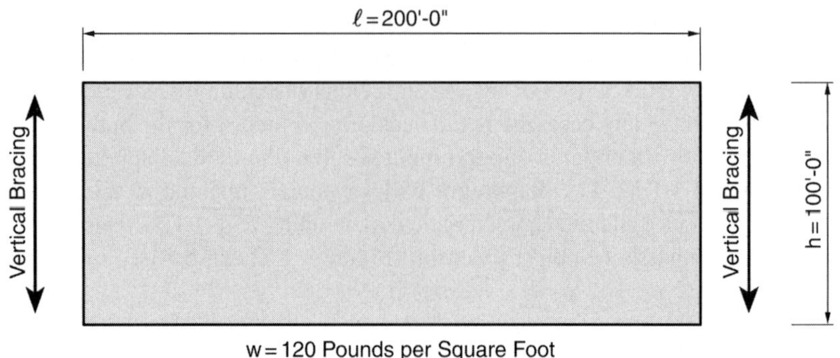

Figure 3.3.1 Diaphragm characteristics.

3.3.2 Estimating Diaphragm Response

The basic steps are

(1) Determine the dynamic characteristics of the diaphragm.
(2) Compare these with the dynamic characteristics of the building.
(3) Using the nodal response for the level of interest, develop the effective input acceleration for the diaphragm.
(4) Create a response spectrum for the design of the diaphragm.
(5) Estimate the probable maximum inertial force and displacement of the diaphragm.
(6) Design the diaphragm.

The execution of these steps is best described by example. The diaphragm shown in Figure 3.3.1 is reasonably representative of what might exist in an office building braced by a perimeter frame. Alternative diaphragm types will be discussed in the next sections. For now assume that the effective slab thickness is 6 in. and that the weight of the system is 120 lb/ft^2.

Step 1: Determine the Dynamic Characteristics of the Diaphragm.

$$I_e = 0.35 I_g \qquad \text{(Cracked shear wall model—Figure 2.4.4}a\text{)}$$

$$= \frac{0.35 t h^3}{12}$$

$$= \frac{0.35(6)(1200)^3}{12}$$

$$= 303{,}000{,}000 \text{ in.}^4$$

$$m = \frac{wh}{g} \qquad \text{(see Figure 3.3.1)}$$

$$= \frac{0.120(100)}{12(386.4)}$$

$$= 0.0026 \text{ kips-sec}^2/\text{in.}$$

$$f'_c = 4000 \text{ psi}$$

$$E_c = 3600 \text{ ksi}$$

$$\omega = \frac{\pi^2}{\ell^2}\sqrt{\frac{EI_e}{m}} \qquad \text{(Eq. 3.3.1)}$$

$$= \frac{9.86}{(200)^2(144)}\sqrt{\frac{3600(303{,}000{,}000)}{0.0026}}$$

$$= 35 \text{ rad/sec}$$

$$T = 0.18 \text{ second}$$

Step 2: Compare the Dynamic Characteristics of the Diaphragm with Those of the Building. The frame braced buildings had periods of 2 seconds or more, while the shear wall braced buildings had periods in the range of 0.9 to 1.5 seconds.

Conclusion: The building design and that of the diaphragm may be dealt with separately.

Step 3: Develop the Effective Acceleration Input to the Diaphragm. The location of the diaphragm of interest and some knowledge of the characteristics of the building's bracing system are required. Assume that the diaphragm that is being designed is at the roof of a fifteen-story building and that the building has the following characteristics:

$$W_f = 1200 \text{ kips}$$

$$T = 2.25 \text{ seconds}$$

$$V_M = 1500 \text{ kips} \qquad \text{(Mechanism shear at system and component overstrength)}$$

From these, we can determine the spectral acceleration and the acceleration at the roof ($n = 15$).

$$M_e = \frac{1.5(n+1)}{2n+1}M \qquad \text{(Eq. 3.1.19}a\text{)}$$

$$= \frac{0.77W}{g}$$

$$S_a = \frac{H}{M_e} \qquad \text{(see Figure 3.1.1}a\text{)}$$

Since the bracing system is ductile, the maximum spectral acceleration will be that associated with mechanism shear (V_M). See Eq. 1.1.3a.

$$\begin{aligned} S_a &= \frac{V_M}{M_e} \\ &= \frac{1500(g)}{0.77W} \\ &= \frac{1500(386.4)}{0.77(15)(1200)} \\ &= 41.8 \text{ in./sec}^2 \end{aligned}$$

The acceleration at level 15 associated with a response in the first mode is

$$\begin{aligned} a_{15} &= \Gamma S_a \qquad \text{(See Figure 3.1.1}b\text{)} \qquad (3.3.2) \\ &= 1.45(41.8) \\ &\cong 60 \text{ in./sec}^2 \end{aligned}$$

The contributions of higher modes must be considered since our concern is with acceleration (see Eq. 3.1.7). This multimode amplification can reasonably be accomplished by increasing the nodal acceleration by about 25%. An estimate of the acceleration at level 15 is

$$\begin{aligned} a_{15} &= 1.25\Gamma_1 S_a \qquad (3.3.3) \\ &= 1.25(60) \\ &= 75 \text{ in./sec}^2 \qquad (\cong 0.2g) \end{aligned}$$

Step 4: Create a Design Response Spectrum for the Diaphragm. Use the concepts briefly described in Section 1.1.3 to create a response spectrum for the diaphragm. This requires the adoption of an effective damping ratio and member ductility factor (μ_D) for the diaphragm. The ductility factor can become a design and detailing objective (see Figure 1.3.12), but the damping will be subjectively established and based to a certain extent on the vertical load-carrying system selected. For example, damping should be lower for a post-tensioned floor slab than for a floor slab that is likely to have experienced significant cracking. For a cast-in-place floor slab which has a significant number of shrinkage cracks one might estimate the impact of structural damping (Eq. 1.1.9d) or subjectively adopt the following:

$$\zeta = 10\% \qquad \text{(see Eq. 1.1.9b)}$$

$$
\begin{aligned}
\mu_D &= 4 \\
S_a &= \alpha_a \ddot{d}_{\max} && \text{(Eq. 1.1.8}a\text{)} \\
&= (4.38 - 1.04 \ln \zeta)\ddot{d}_{\max} \\
&= (4.38 - 1.04 \ln 10)75 \\
&= 156 \text{ in./sec}^2 \qquad (\cong 0.4g)
\end{aligned}
$$

Step 5: Estimate the Probable Maximum Inertial Force and Displacement of the Diaphragm. Figure 3.3.2 describes the various spectral values used in the design of the diaphragm. The maximum diaphragm displacement, assuming that $I_e = 0.35I_g$, should be on the order of

$$
\begin{aligned}
S_d &= \frac{S_a}{\omega^2} \frac{\mu_D}{\sqrt{2\mu_D - 1}} && \text{(see Figure 1.1.7)} \\
&= \frac{156}{(35)^2} \frac{4}{\sqrt{2(4) - 1}} \\
&= 0.19 \text{ in.} \\
\Delta_D &= \Gamma S_d && \text{(3.3.4a)} \\
&= 1.5(0.19) \\
&= 0.29 \text{ in.}
\end{aligned}
$$

The use of a participation factor of 1.5 is based on the similarity between the deflected shape of the cantilever (shear wall) and the half sine curve.

Given the uncertainties used to develop the preceding conclusion, it is advisable to make a worst-case prediction of diaphragm response in order to establish an upper bound displacement and ductility demand. Consider the stiffness assumption used to develop a force-based criterion ($S_a \cong 0.4g$).

$$I_e = 0.35I_g$$

Members with shorter shear spans are likely to be more flexible—as, for example, was the case for coupling beams (Section 2.4.2.3).

$$
\begin{aligned}
I_{e2} &= \frac{0.2I_g}{1 + 3(h/\ell)^2} && \text{(see Eq. 2.4.38a)} \\
&= \frac{0.2}{1.75} I_g \\
&= 0.11I_g
\end{aligned}
$$

and this is about one-third of the effective stiffnesses assumed ($0.35I_g$) for the force-based design.

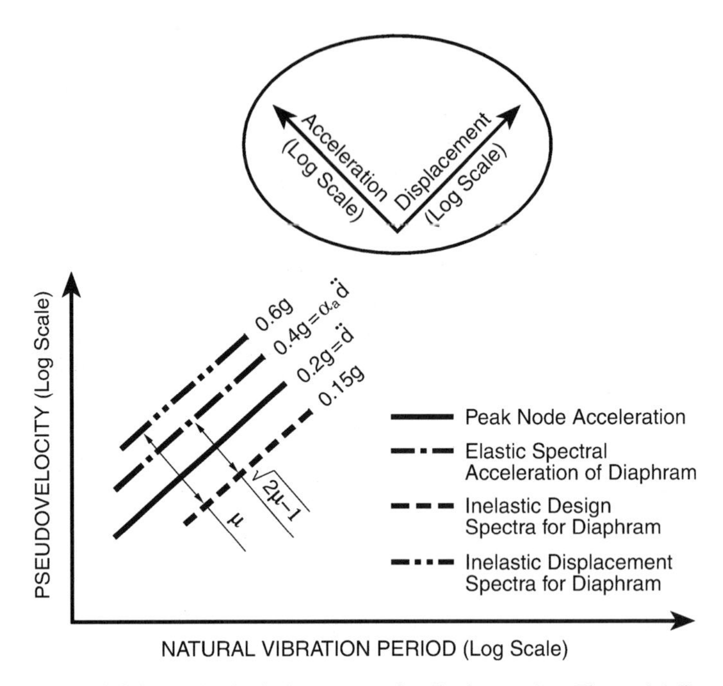

Figure 3.3.2 Inelastic design spectra for diaphragm (see Figure 1.1.5).

The calculations need not be repeated, for the conclusions previously reached are easily modified. The revised frequency (ω_r) is

$$\omega_r = \frac{\omega}{\sqrt{3}} \qquad (T = \sqrt{3}(0.18) = 0.31 \text{ second})$$

The calculated displacement, based on $I_e = 0.35 I_g$, should be adjusted upward by the square of the period shift. This produces the following estimate of diaphragm drift:

$$\begin{aligned} \Delta_{D2} &= \frac{I_{e1}}{I_{e2}} \Delta_{D1} \qquad (3.3.4b) \\ &= 3(0.29) \\ &= 0.87 \text{ in.} \end{aligned}$$

Adjustments may also be made to reflect more conservative estimates of damping and ductility.

Step 6: Design the Diaphragm. Having identified a range of displacements, we are now in a position to select an objective level of diaphragm strength. The configuration of the load caused by inertial forces is that shown in Figure 3.3.3.

Inertial loads follow the deflected shape, and the deflected shape is described by a sine function.

We seek a relationship between the projected elastic displacement and the flexural strength of the diaphragm. Given the confidence level one might reasonably have in the many assumptions used in the development of the design approach, it seems reasonable to adopt a triangular loading pattern, and this results in the following moment deflection relationship.

If W_i is the total inertial load (Figure 3.3.3), then

$$M_{\max} = \frac{W_i \ell}{6} \tag{3.3.5}$$

and

$$\Delta_{\max} = \frac{W_i \ell^3}{60 E I_e} \tag{3.3.6a}$$

$$= \frac{M_{\max} \ell^2}{10 E I_e} \tag{3.3.6b}$$

In the example then, for a maximum displacement of 0.29 in.,

$$M_{\max} = \frac{10 E I_e \Delta_{\max}}{\ell^2} \tag{3.3.7}$$

$$= \frac{10(3600)(303{,}000{,}000)(0.29)}{(200)^2 1728}$$

$$= 45{,}800 \text{ ft-kips}$$

This corresponds to a chord force of about 600 kips or an area of steel on the order of 12 in.^{2} if $d_{\text{eff}} \cong 0.8h$. Observe that the moment deflection relationship will not be affected by changes in stiffness assumptions (see Eq. 3.3.4b). Intuitively this is reasonable because the response falls within the acceleration constant response region (Figures 1.1.5 and 3.3.2).

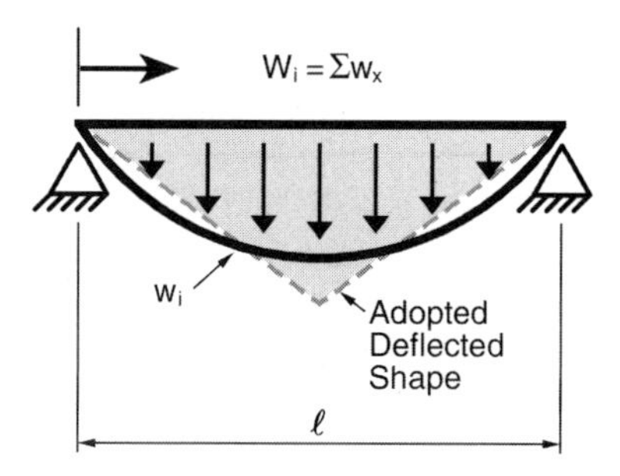

Figure 3.3.3 Inertial load imposed on diaphragm.

It remains then to introduce appropriate overstrength and ductility factors into the expression developed for the flexural strength objective M_{max}. The selection of the appropriate $\Omega_o \mu_\Delta$ factor will depend on how the designer plans to develop the strength of the diaphragm, along with the confidence he or she might have in the selection of the forcing function. A significant overstrength factor will probably not materialize, as was evident in the behavior of coupling beams (Figure 2.4.15*a*). For the purposes of the design example, adopt an overstrength factor of unity ($\Omega_o = 1$) and a diaphragm ductility factor of 4 ($\mu_D = 4$). This suggests that chord reinforcing should be provided in the amount of ± 3 in.^{2} and that the maximum displacement of the diaphragm will or could be on the order of 1 in. ($\ell/2400$), based on the use of coupling beam data.

Comment: Observe that a similar strength objective can be developed from a rigid diaphragm assumption (constant acceleration). The input base acceleration ($\ddot{d}_o$—Figure 3.3.2) is $0.2g$. This produces a design moment of

$$M_{max} = \frac{w\ell^2}{8}$$

$$= \frac{0.2(0.120)(100)(200)^2}{8}$$

$$= 12{,}000 \text{ ft-kips}$$

The deflection is, however, unrealistically low, but this is consistent with the rigid diaphragm modeling assumption.

$$\Delta_D = \frac{5w\ell^4}{384EI_e}$$

$$= \frac{5(2.4)(200)^4(1728)}{384(3600)(303{,}000{,}000)} \qquad (I_e = 0.35I_g)$$

$$= 0.08 \text{ in.}$$

3.3.3 Establishing the Strength Limit State of a Diaphragm

One thing should be clear at this point—predicting loads and deflections of diaphragms is not accomplished with a great degree of precision. Experimental efforts other than those performed on shear walls and coupling beams that can be used to support a design do not exist. Since the behavior of shear walls is closer to beam behavior, the use of shear wall data is probably appropriate only when the shear span ratio (a/h) is greater than 1.5. For short shear spans ($a/h \cong 1$), the use of coupling beam data seems more appropriate. This uncertainty makes it especially important to develop a reliable ductile load path in the diaphragm, one that recognizes conditions that will exist.

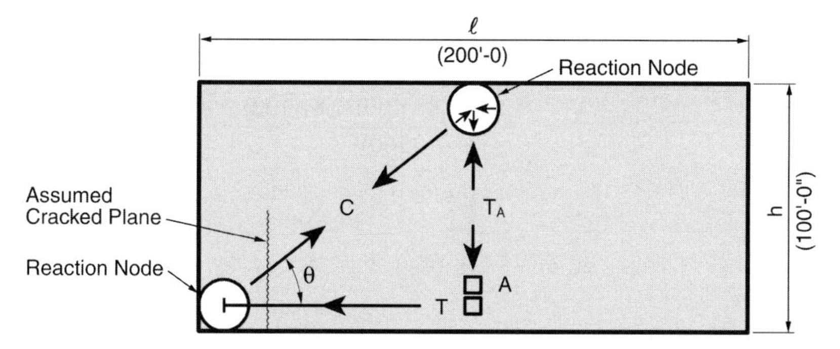

Figure 3.3.4 Development of arch action in a diaphragm.

Figure 1.3.12 described how the strength limit state of a short span diaphragm is a function of the tensile reinforcing provided at midspan. The distribution of load has little impact on the load transfer mechanisms (see Figure 1.3.6), except for the fact that inertial loads must be delivered to the struts. A strut and tie mechanism was used in Figure 1.3.7 to describe the arch action that will want to develop. This fundamental load path is repeated in Figure 3.3.4 so as to describe how it can be activated. Consider the load path of unit mass A as it travels to the reaction node. The inertial force imposed on unit mass A must travel up to the compressive strut where it is equilibrated. The maximum tension load acting on the tension tie T_A is a function of the flexural strength of the diaphragm.

$$M = T d_{\text{eff}} \tag{3.3.8}$$

In our example diaphragm (Figure 3.3.1), the tie force T was developed from the adopted load model described in Figure 3.3.3 as follows:

$$M_{\max} = \frac{W_i \ell}{6} \qquad \text{(Eq. 3.3.5)}$$

$$T d_{\text{eff}} = \frac{W_i \ell}{6}$$

Accordingly, the load path to the compression strut will not be too difficult to develop.

The load imposed on the compression strut (C in Figure 3.3.4), provided the reaction node is fully developed, will be

$$W_i = \frac{6\lambda_o A_s f_y d_{\text{eff}}}{\ell} \tag{3.3.9a}$$

$$w_{\max} = \frac{2W_i}{\ell} \qquad \text{(see Figure 3.3.3)}$$

The maximum tensile strength required to deliver the inertial loads by the compression diagonal to the truss mechanism is

$$T_A = \frac{12\lambda_o A_s f_y d_{\text{eff}}}{\ell^2} \tag{3.3.9b}$$

$$= \frac{12(1.25)(3)(60)(0.8)(100)}{(200)^2}$$

$$= 5.4 \text{ kips/ft}$$

$$A_s = \frac{T_A}{f_y}$$

$$= \frac{5.4}{60}$$

$$= 0.09 \text{ in.}^2/\text{ft}$$

The force activated along the compression diagonal is

$$C = \frac{T}{\cos\theta}$$

and in the example case this corresponds to

$$C = \frac{\lambda_o A_s f_y}{\cos 45}$$

$$= \frac{1.25(3)(60)}{0.707}$$

$$= 318 \text{ kips}$$

The spread of this load may conservatively assumed to be 15° (Figure 1.3.6) starting from the reaction node. Accordingly, any assistance in transferring load across a cracked plane that might exist must consider the detail development of the node, and this is done when pretopped precast diaphragms are considered in Section 3.3.4.

If we were to entirely discount the friction load path activated by the compression strut, then a shear friction or mechanical shear transfer load path would need to be developed. The shear force acting along a vertical crack would be

$$V = \frac{T}{\tan\theta} \tag{3.3.10a}$$

and in the example case,

$$V = \frac{\lambda_o A_s f_y}{\tan 45} \tag{3.3.10b}$$

$$= \frac{1.25(3)(60)}{1.0}$$

$$= 225 \text{ kips}$$

To develop this nominal force in shear friction would require

$$A_{sh} = \frac{V}{f_y} \qquad \text{(see Eq. 1.3.19)}$$

$$= \frac{225}{60}$$

$$= 3.75 \text{ in.}^2$$

or 0.0375 in.²/ft if the resistance is distributed evenly along the depth of the diaphragm.

The preceding approach does not appear to be too dissimilar from that developed by a beam analogy except for the fact that reinforcement is required normal to the direction suggested by beam theory (truss analogy). Observe that any minimum slab reinforcing program would satisfy the needs of this shear transfer mechanism and the tensile demand T_A in a 6 in. thick cast-in-place slab.

Comment: It is not reasonable to discount the resistance provided by friction. Observe that if the static friction is 700 psi (Figure 1.3.14), very little area of concrete is required to activate a shear resistance of 225 kips.

$$A_{c,sh} = \frac{V}{0.7} \qquad (3.3.11)$$

$$= \frac{225}{0.7}$$

$$= 321 \text{ in.}^2$$

For a 6-in. thick slab, only 4.5 ft of slab is required to resist the shear force. Accordingly, slippage along precracked planes should not be expected.

The fact that tension ties or chord bars are overstrained during earthquakes was well documented during the Northridge Earthquake. Figure 3.3.5 describes the post-earthquake condition of a chord bar, and conditions like this were not uncommon. The reverse cycle deformation imposed on the chord bars shown in Figure 3.3.5 has caused the overstrained bars to buckle and destroy the concrete that originally encased it.

In Section 1.3.1 an expression that describes the yield displacement was developed for a deep beam (Eq. 1.3.16b). This development presumed that either the arch action described in Figure 1.3.7 (or the beam model) would be achieved. The tension tie, in either case, would be debonded, and the deflection would be reasonably modeled by the rotation of two rigid bodies (Figure 1.3.12). In the example diaphragm in Figure 3.3.1, the yield displacement would be

$$\Delta_y = \frac{\varepsilon_{sy} a \ell}{2} \qquad \text{(Eq. 1.3.17)}$$

$$= \frac{0.002(1)(200)(12)}{2}$$

$$= 2.4 \text{ in.}$$

Clearly, the debonding of the chord reinforcing is not advisable because the stiffness of the so-created system would be much less than assumed in the development of the dynamic response. Compare, for example, the displacement of the debonded diaphragm with that assumed in the response calculations using the coupling beam model ($I_e = 0.11I_g$). Figure 3.3.7 describes the two models.

Pure Arch Action

$$\Delta_y = 2.4 \text{ in.} \qquad \text{(Eq. 1.3.17)}$$

$$W_i = \frac{6M}{\ell} \qquad \text{(see Figure 3.3.3 and Eq. 3.3.5)} \qquad (3.3.12)$$

$$= \frac{6A_s(d_{\text{eff}})f_y}{\ell}$$

Figure 3.3.5 Buckling of chord bars in a diaphragm.

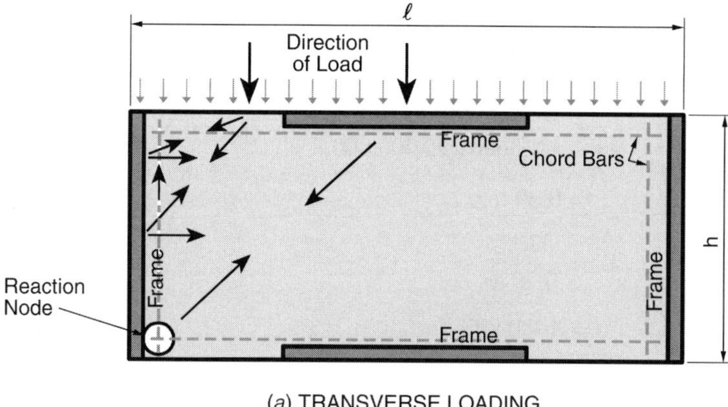

(*a*) TRANSVERSE LOADING

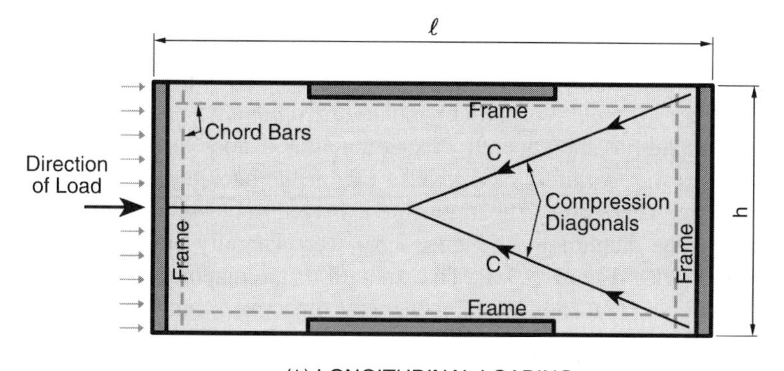

(*b*) LONGITUDINAL LOADING

Figure 3.3.6 Development of diaphragm load paths.

For the example design (Figure 3.3.1)

$$W_i = \frac{6(4.0)(80)}{200}60$$
$$= 576 \text{ kips}$$

and the stiffness using the arch analogy is

$$k_a = \frac{W}{\Delta} \tag{3.3.13}$$
$$= \frac{576}{2.4}$$
$$= 240 \text{ kips/in.}$$

The comparable stiffness used in the design was developed from pseudobeam theory.

$$\Delta = \frac{W_i \ell^3}{60 E I_e} \tag{3.3.14}$$

$$= \frac{576(200)^3(1728)}{60(3600)(95{,}000{,}000)} \qquad (I_e = 0.35 I_g)$$

$$= 0.39 \text{ in.}$$

$$k_b = \frac{576}{0.29}$$

$$= 1480 \text{ kips/in.}$$

This is six times stiffer than that of the debonded arch model. Clearly the debonded arch model suggests a much softer system than the coupling beam model used in the development of the response.

The question of concentrated strain in debonded reinforcing bars was dealt with in the development of the hybrid system in Section 2.1.4.4. Intentionally debonding is an option here also, but it is probably not required nor advisable because the hinge region will expand as the concrete cracks much as it does in a coupling beam or wall. It is, however, certainly advisable to inhibit the buckling described in Figure 3.3.5 in regions where postyield strains are expected to be significant. Suppose, for example, that the diaphragm of Figure 3.3.1 were laterally braced by 100-ft long frames on all sides (Figure 3.3.6). The strength of the diaphragm in the transverse direction, assuming that chord yielding were the objective, would be a function of the strength of the tension tie between the frame and the reaction node (Figure 3.3.6*a*). The only difference from a performance perspective would be that the yielding of the tension tie would probably take place just outside the longitudinal frame. Observe that the development of the reaction node could be easily accomplished in this case by developing the chord reinforcing in the transverse frame.

The longitudinal response of the diaphragm (Figure 3.3.6*b*) would undoubtedly be elastic and, because of the provided stiffness, the diaphragm would behave as a rigid body subjected to the unamplified input accelerations of the vertical bracing system $\ddot{d}$ of Figure 3.3.2.

To this point our focus has been on the arch action described in Figure 3.3.4. The development of this load path relies on the creation of an effective reaction node that in effect must transfer, according to the adopted load path, the maximum chord force T. Design practice today follows the beam model, and, as a consequence, the chord reinforcing is reduced as the beam moment decreases. Does this mean that diaphragms that do not develop reaction nodes will fail? Certainly not! The supporting rationale is largely analogous to the beam-column joint behavior mechanisms described in Figure 2.3.1 where a strut and truss mechanism combine to create a complex, redundant set of load paths. Observe that the inclination of the compression strut at the reaction node in Figure 2.3.1*c* is quite steep and that a significant amount of load has been relieved from the tension tie (T) before it reaches the reaction node. The existence of these truss nodes, in essence, validates the beam behavior model, and this can easily be

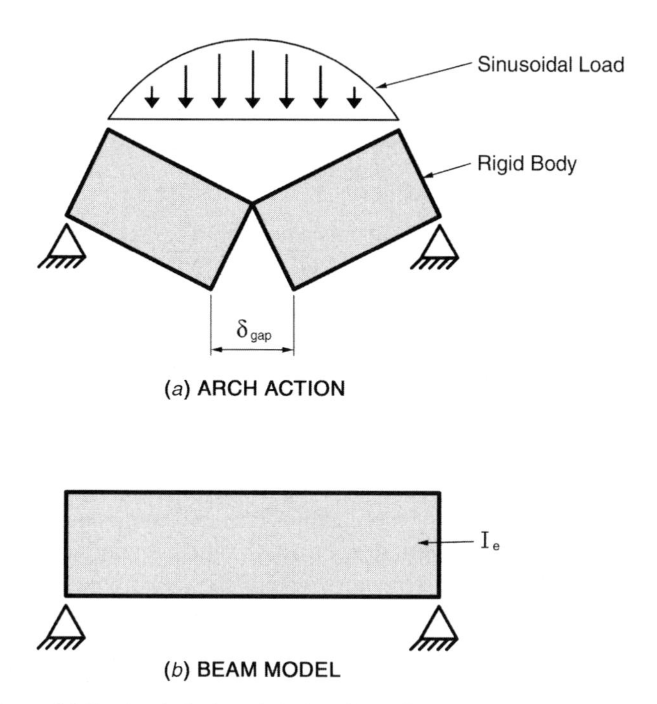

Figure 3.3.7 Analytical models describing flexural behavior alternatives.

supported analytically. Defining and understanding the consequences of alternative load paths will be essential in the development of precast concrete diaphragms.

3.3.4 Precast Concrete Diaphragms

Concrete code prescribed practice effectively requires that precast concrete diaphragms to be topped with a cast-in-place topping slab. Exceptions are typically allowed by design codes to prescriptive requirements but, these exceptions place the burden on the designer to " . . . demonstrate by experimental evidence and analysis that the proposed system will have strength and toughness equal to or exceeding those provided by a comparable monolithic reinforced concrete structure. . . ."[3.13, Sec. 21.2.1.5]

In this section we touch briefly on the design of precast concrete diaphragms that are topped with a cast-in-place, composite slab, but the focus is on untopped or, as they are frequently referred to, pretopped diaphragms.

3.3.4.1 Composite Diaphragms Composite action is naturally created between the precast member and the cast-in-place topping slab. A nominal shear strength of 80 psi is accepted for intentionally roughened surfaces[3.13, Sec. 17.5.2.1] but this can be increased significantly if the amplitude of the roughening is increased to $\frac{1}{4}$ in. and ties are provided across the interface.

Comment: The definition of roughness has been debated over the years. An unfinished surface as, for example, the top of an extruded plank or a double T qualifies as intentionally roughened. Research done in support of these requirements is identified as that performed by the Portland Cement Association (PCA) in the late 1950s to support the design of composite concrete bridge beams.

Shear flow in a diaphragm is generally considered as being only active in the topping slab but this is very conservative especially when the joint between the precast members is grouted. Modeling consideration have been discussed in the preceding sections and load paths developed. The concern usually raised in the development of composite precast concrete diaphragms is with the natural coincidence of topping slab cracks and joints in precast members. This concern has little, if any, real basis, for the width of the resultant pattern of cracks is essentially the same as one would find in a conventional cast-in-place slab and generally less than would be found in a metal deck/cast-in-place concrete slab diaphragm system.

The transfer of shear across precracked joints is usually accomplished analytically by developing shear friction. This, as previously discussed, is a conservative approach because arch action passively precompresses these surfaces (Figure 3.3.4).

The development of a composite precast concrete diaphragm follows the theory discussed in the preceding section (3.3.3). The essential difference lies in the development of effective details at node points and where buckling of chord bars are anticipated. Figure 3.3.8 describes how the reaction node might be effectively developed while Figure 3.3.9 describes how chord bars may be supported in plastic hinge regions. The details in Figure 3.3.9 may also be used to improve behavior at reaction nodes.

3.3.4.2 Pretopped Precast Concrete Diaphragms The concepts developed thus far in Section 3.3.4 as well as those developed in Section 3.3.3 are applicable to the

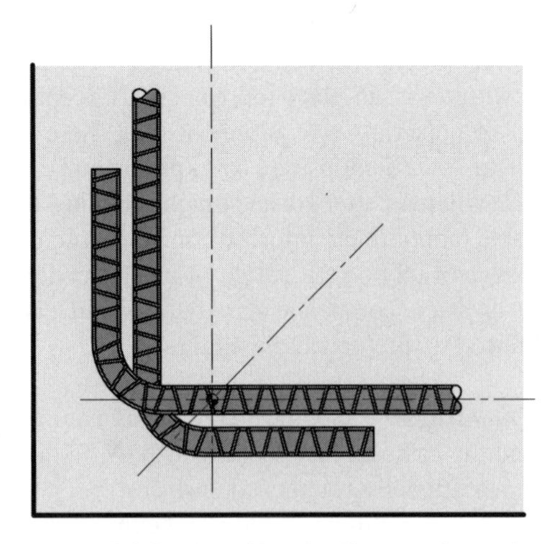

Figure 3.3.8 Chord bar detail at reaction node.

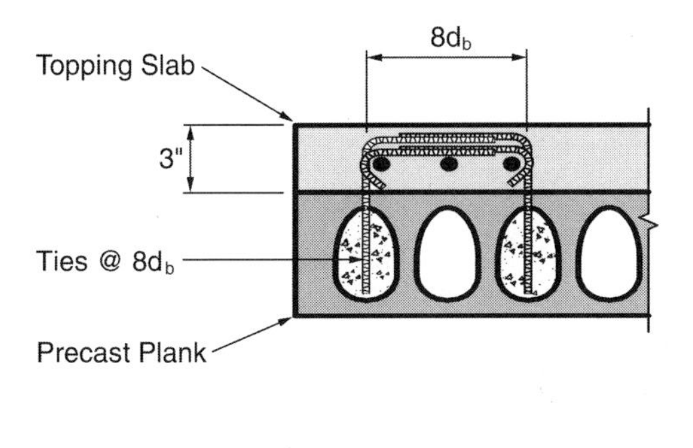

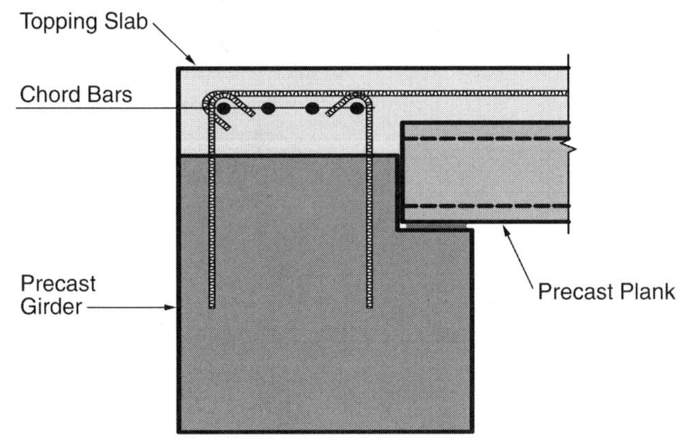

Figure 3.3.9 Stabilizing detail—chord bars in plastic hinge region.

design of precast concrete diaphragms that do not include a topping slab. The key to the design process is the development of a rational load path. Diaphragm load paths will depend, to a large extent, on the type of vertical load carrying system adopted and the flexibility available in the construction of the precast components. Typically, a precast concrete floor will be developed from primary members (girders) and secondary members (planks or tees). The resulting arrangement is generically described in Figure 3.3.10*a*.

Any number of diaphragm load paths may be developed in a precast diaphragm system as described in Figure 3.3.10*a* and *b*. Clearly, the primary program described in Figure 3.3.4 is possible so long as the nodes may be developed. Secondary load paths will also be developed as long as the requisite struts (C_1) and ties (T_1 and T_s) are created. One might reasonably conclude that the load path described in Figure 3.3.10*b* could be significantly altered by the considered selection of tension elements following the plastic truss analogy (Figure 1.3.2). The alternative load paths described in Figure 1.3.6 may also be developed as well as the truss mechanism of Figure 2.3.1*c*.

From an analytical perspective the easiest load path to describe involves the introduction of a series of bounded panel elements similar to those described in Figure 3.3.10*c*. Now, regardless of the direction of load a truss is developed that imposes a

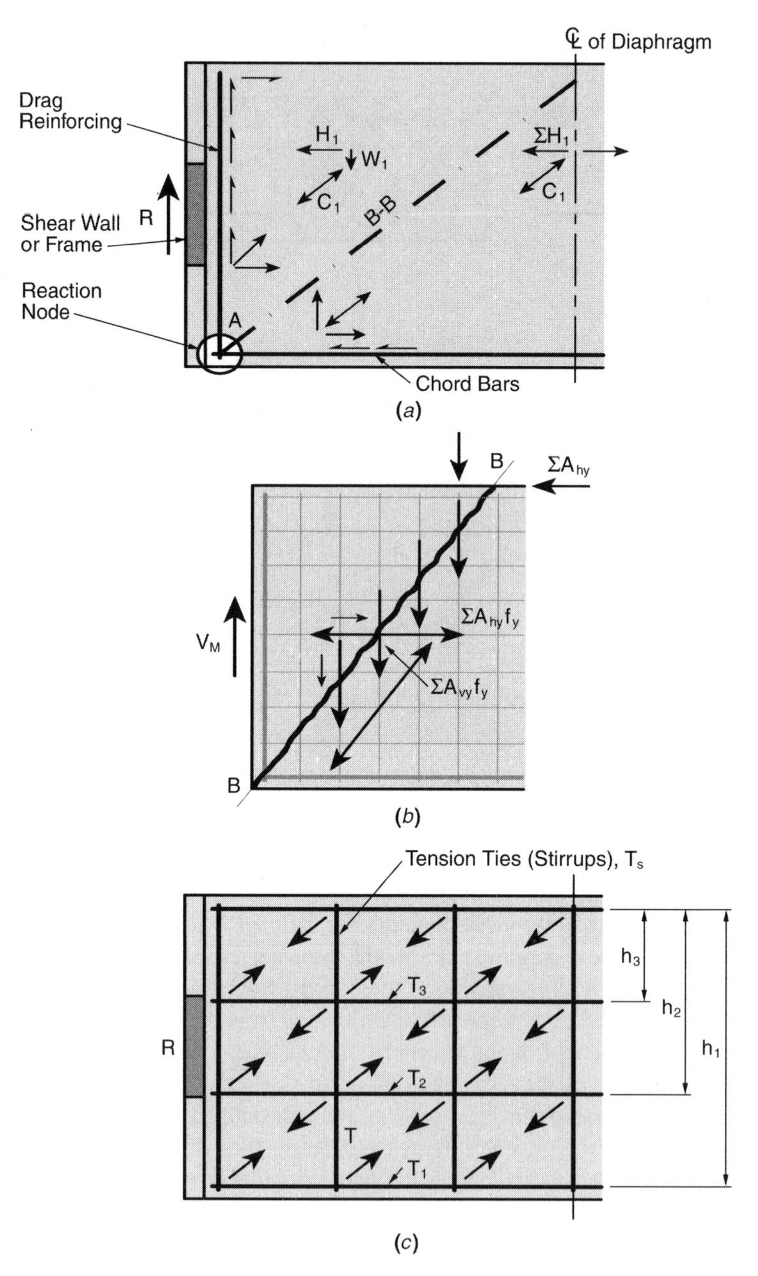

Figure 3.3.10 Alternative load paths in a reinforced diaphragm.

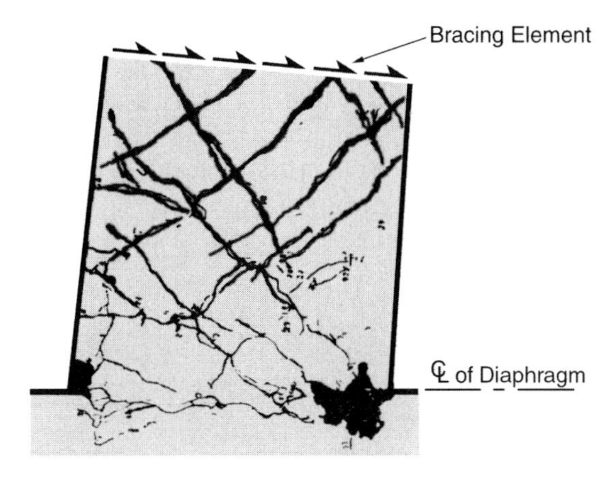

Figure 3.3.11 Postyield distress in a diaphragm.

compression only load on a panel. The tension loads are carried along the bounding elements. The magnitude of the load, R, will be a function of the capacity of the ties (stirrups), T_s, or the flange elements, T_1, T_2, and T_3. The load that the stirrup tie, T_s, or boundary elements, T_i, must resist can be developed using capacity-based concepts.

$$M_{\max} = \lambda_o \left(T_1 h_1 + T_2 h_2 + T_3 h_3\right) \qquad \text{(see Figure 3.3.10}c\text{)}$$

$$V_{\max} = 3\lambda_o T_s \qquad \text{(see Figure 3.3.10}c\text{)}$$

For example, the boundary tie T_s must develop the system overstrength if the intent is that is not be ductile. Alternatively, T_s can become the load limiter.

Observe that the load on the reaction node has been considerably reduced from that associated with a direct load path similar to that described in Figure 3.3.4. Regardless of the developed load path, care must be taken to reasonably create and bound panel elements. Effective detailing is especially important in the development of pretopped precast concrete diaphragms.

The postyield behavior described in Figure 3.3.11 is representative of what one might expect in a diaphragm which has been subjected to cyclic load reversals. Clearly, the details developed in Figures 3.3.8 and 3.3.9 are important.

Conclusion: The alternative approaches to the design of precast concrete diaphragms are many. The key to a successful design will be the development of a rational load path that includes a ductile component capable of limiting the load imposed on less ductile load transfer points, i.e., nodes and cracked or weakened planes.

3.4 DESIGN PROCESS OVERVIEW

While it is not my intention to attempt to regulate the design process, it seems important to identify certain key issues and discuss how a designer might approach key design decisions. It has always been clear to me that strength is not a good gauge of performance, for once stability is assured, increments of strength may lead to inferior performance and this is especially true of concrete structures where the true measure of performance is a function of induced strain states.

Strain states are a function of imposed stress levels and the deformation state; they can only be estimated after a structure has been designed. The issues discussed in this section are intended to help the designer develop a structure that will attain strain objectives.

3.4.1 System Ductility

It would be nice to identify a universal system ductility factor. Codes[3.4, 3.7] attempt this and in concrete structures two or three values are proposed. Displacement-based procedures are recommended. Optimal ductility factors (R_d, Figure 1.1.12) are

$$\mu = 6 \qquad \text{(Ductile frame braced structures)}$$

$$\mu = 4 \qquad \text{(Shear wall braced structures)}$$

These optimal values presume the following:

Frame Braced Structures

- Member ductilities are on the order of 8 to 10 based on the adopted idealization of yield (see Figure 2.1.3). Optimal ductilities should be associated with a beam reinforcement ratio in the 1% to 1.5% range.
- Inelastic behavior will be well distributed over the height of the frame (see Figure 3.2.8*b*).
- System yield will be developed from idealized behavior (see Figure 3.2.12*a*).
- The probable axial load imposed on the compression side column will be controlled so as to limit the imposed capacity-based load to $0.3 f_c' A_g$.

Shear Wall Braced Structures

- The axial load is relatively low, $\pm 0.15 f_c' A_g$.
- The flexural reinforcing is limited to $0.005 A_g$.
- The capacity-based shear stress is limited to $5\sqrt{f_c'}$.

Where these design objectives cannot be met, the level of assumed system ductility should be reduced to levels suggested by experimental efforts.

3.4.2 Capacity Considerations

Both system and member overstrengths must be considered in the design. Universal values are entirely inappropriate. Member test data must be coordinated with the adopted member design methodology and appropriate material stress. The same is true when it comes to evaluating system overstrength. A sequential yield analysis is the most effective tool for assessing system overstrength, but member overstrengths must be incorporated into the analytical model for it to be effective.

3.4.3 Recommended Design Approach

The following approach assumes that the building design will impose certain constraints on the development of the bracing program.

Step 1: Establish the Probable Design Drift for the Structure. The fundamental period of the structure must be consistent with reasonable drift objectives. If, as is most often the case, the intensity of the earthquake can be stated in terms of a constant spectral velocity (S_v), a relationship exists between the drift ratio (DR) and the fundamental period of the building.

$$\frac{TS_v\Gamma}{2\pi h_n} = \text{DR} \tag{3.4.1a}$$

The participation factor can be reasonably quantified for most buildings as 1.5 for buildings over ten stories and in the range of 1.3 for 3 to ten-story buildings (see Eq. 3.1.18a).

For example: The objective building drift ratio (DR) is 0.02, $h_n = 150$ ft, $S_v = 50$ in./sec. Then

$$T \leq \frac{2\pi \text{ DR } h_n}{S_v\Gamma} \tag{3.4.1b}$$

$$< \frac{6.28(0.02)150(12)}{50(1.5)}$$

$$< 3.0 \text{ seconds}$$

Conclusion: The period of the structure must be less than 3 seconds if the objective drift limit is to be met.

Step 2: Size the Bracing Component to Meet the Established Period Limit State. Use relationships developed in this chapter.

Shear walls: Eq. 3.1.33
Frames: Eq. 3.2.1 and 3.2.2a
System model:

$$I_{be} = 0.35 I_{bg}$$

$$I_{ce} = 0.7 I_{cg}$$

When the wall is prestressed or subjected to a significant axial load (Figure 2.4.2*a*).

$$I_{we} = 0.5 I_{wg}$$

Otherwise (Figure 2.4.4*a*).

$$I_{we} = 0.35 I_{wg}$$

This establishes a direct link between the system and its dynamic characteristics.

Step 3: Establish the Objective System Drift at Yield. This is a two step process.

- First, the dynamic characteristics of the proposed system must be established. In essence the calculations performed in Step 2 and Step 1 must be reversed now that the bracing system has been proposed. The result of this process will be a quantification of Δ_u.
- Second, determine the objective yield displacement Δ_y.

$$\Delta_y = \frac{\Delta_u}{\Omega_o \mu} \qquad \text{(Eq. 3.1.38)} \quad (3.4.2)$$

Step 4: Determine the Yield Strength of the Structure. The characteristics of the structure have been established. The load distribution must be consistent with the fundamental or first mode shape, and this is reasonably accomplished by a uniformly varying load, which is a maximum at the top of the structure (inverse triangle).

For a shear wall braced building,

$$M_d = \frac{3.6 E I_e}{h_w^2} \Delta_y \qquad \text{(Eq. 3.1.39)} \quad (3.4.3)$$

For a frame braced building, this is done by the sequential application of Eq. 3.2.6*a* and 3.2.16.

$$V_i = \frac{\Delta_y \Gamma k_i}{n} \qquad \text{(see Eq. 3.2.18)} \quad (3.4.4)$$

where

V_i is the yield base shear of an interior subassembly.
k_i is the stiffness of an interior subassembly as developed from Eq. 3.2.16.
n is the number of levels.
Γ is the participation factor associated with the first mode (Eq. 3.1.18a).

$$V_d = \left\{ (j-1) + 2\left(\frac{k_e}{k_i}\right) \right\} V_i \qquad \text{(see Eq. 3.2.6a)} \quad (3.4.5)$$

where

V_d is the design level base shear.
k_e is the stiffness of the exterior subassembly (Eq. 3.2.2b).
j is the number of bays.

For example: $k_i = 400$ kips/in., $k_e = 200$ kips/in., $\Delta_y = 7.5$ in., $n = 15$, $j = 3$, and $\Gamma = 1.5$.

$$V_i = \frac{\Delta_y \Gamma k_i}{n} \qquad \text{(Eq. 3.4.4)}$$

$$= \frac{7.5(1.5)(500)}{15}$$

$$= 300 \text{ kips}$$

$$V_d = \left\{ (j-1) + 2\left(\frac{k_e}{k_i}\right) \right\} V_i \qquad \text{(Eq. 3.4.5)}$$

$$= \left\{ (3-1) + 2\left(\frac{200}{400}\right) \right\} 300$$

$$= 900 \text{ kips}$$

Step 5: Develop Ductile Component Designs. For a shear wall this amounts to sizing the flexural reinforcing in the wall. Following established procedures, treat M_d as M_u. Since the behavior is flexural, use a ϕ factor of 0.9 so as not to create an excessive overstrength.

$$A_s = \frac{M_d/\phi - P_D\,(\ell_w/2 - a/2)}{f_y(d - d')} \qquad \text{(see Eq. 2.4.23)} \quad (3.4.6)$$

For frame braced structures, design the frame beam to the mechanism shear:

$$M_d = \frac{V_d h_x}{2\,\{(j-1) + 2\,(k_e/k_i)\}} \qquad \text{(Eqs. 3.2.6a and 3.2.20a)} \quad (3.4.7)$$

and then convert this to a beam strength objective:

$$M_u = M_d \left(\frac{\ell_c}{\ell}\right) \qquad (3.4.8)$$

Step 6: Design Remaining Members to a Capacity-Based Criterion. Several steps are involved.

(a) Estimate ductile component strengths based on provided reinforcement and probable material strength.

(b) Estimate system overstrength. Use either prior analytically based data (Figure 3.2.8*a* and 3.1.6*a*) or a system specific sequential yield analysis.

The net result should be a conservative estimate of the force that components are likely to experience. In the case of the frame the critical element will be the compression side column. In the case of the shear wall it will be the shear stress imposed on the wall and the strain imposed on the toe region.

SELECTED REFERENCES

[3.1] A. K. Chopra, *Dynamics of Structures: Theory and Applications to Earthquake Engineering,* 2nd Edition, Prentice Hall, Upper Saddle River, New Jersey, 2001.

[3.2] J. A. Blume, N. M. Newmark, and L. H. Corning, "Design of Multistory Reinforced Concrete Buildings for Earthquake Motions," Portland Cement Association, Skokie, Illinois, 1961.

[3.3] R. E. Englekirk, G. C. Hart, and the Concrete Masonry Association of California and Nevada, *Earthquake Design of Concrete Masonry Buildings*, Volume 1. Prentice Hall, Englewood Cliffs, New Jersey, 1982.

[3.4] Seismology Committee, Structural Engineers Association of California, *Recommended Lateral Force Requirements and Commentary*, 7th Edition. Sacramento, California, 1999.

[3.5] M. J. N. Priestley and M. J. Kowalsky, "Direct Displacement-Based Seismic Design of Concrete Buildings," *Bulletin of the New Zealand Society for Earthquake Engineering,* Vol. 33, No. 4, December 2000.

[3.6] R. E. Englekirk, *Steel Structures: Controlling Behavior Through Design*. John Wiley & Sons, New York, 1994.

[3.7] International Conference of Building Officials, *Uniform Building Code*, 1997 Edition, Whittier, California.

[3.8] Y. Kurama, R. Sause, S. Pessiki, and L. W. Lu, "Lateral Load Behavior and Seismic Design of Unbonded Post-Tensioned Precast Concrete Walls," *ACI Structural Journal*, Vol. 96, No. 4, July–August, 1999, pp. 622–632.

[3.9] Y. Kurama, R. Sause, S. Pessiki, and L. W. Lu, "Seismic Behavior and Design of Unbonded Post-Tensioned Precast Concrete Walls," *PCI Journal*, Vol. 44, No. 3, May–June, 1999, pp. 72–89.

[3.10] Y. Kurama, "Seismic Design of Unbonded Post-Tensioned Precast Walls with Supplemental Viscous Damping," *ACI Structural Journal*, Vol. 97, No. 4, July–August, 2000, pp. 648–658.

[3.11] American Institute of Steel Construction, Inc., *Seismic Provisions for Structural Steel Buildings*, Chicago, IL, April 1997 and Supplement No. 2, September 2000.

[3.12] J. M. Biggs, *Introduction to Structural Dynamics,* McGraw-Hill Book Company, New York, 1964.

[3.13] American Concrete Institute, *Building Code Requirements for Structural Concrete (318–99) and Commentary (318R0–99)*, ACI 318–99 and ACI 318R–99, Farmington Hills, Michigan, June 1999.

4

DESIGN CONFIRMATION

"It's what you learn after you know it all that counts."
—Earl Weaver

Design procedures must introduce many simplifying assumptions into the process. The validity of these simplifications must be tested analytically if they are to be used with any degree of confidence. This testing is done through the use of elastic and inelastic time history analyses. This chapter endeavors to analyze some of the component and system designs developed in the preceding chapters. A confirmation of any of the design procedures is not the objective. The focus is to demonstrate the application of the procedures that might be used to confirm a design. The designer who wishes to improve his or her design skills will study completed designs to determine how the next design might reasonably be improved.

Several issues were raised in Chapter 3 that have a direct bearing on the design process. These issues can be grouped into major categories as follows:

- Equal displacement.
- System ductility.
- Design strength.
- Modeling considerations.

The time history evaluations undertaken in this chapter are limited to frame and shear wall designs developed in Chapter 3. The focus of each analysis is to probe these identified focal issues. Similar types of analyses performed over the last 35 years have led the author to accept the recommended design approaches developed in Chapter 3.

Implicit in the confirmation of a design by time history analysis is the comparability of the design spectrum and the ground motions used to evaluate the design. Two means of developing comparable representations of an earthquake are being used today. One alters the ground motion so as to match the spectrum at every period (matched spectrum) while the other amplifies (only) an actual ground motion record to fit the design spectrum (scaled ground motion) in a particular period range.

Both methods are used today but I believe that the latter is more appropriate provided the selected ground motions are appropriate and a sufficient number of earthquake records are used to analyze the building. This is because earthquakes have a dominant frequency content that can be a benefit insofar as mitigating the impact of the earthquake on the response of ductile structures, and this will be an observed characteristic. The design spectra and those that were produced from the scaled ground motion used to perform the analyses of this chapter are described in Figure 4.1.1. A set of matched spectra are shown in Figure 4.1.2 to describe the inherent difference in the two approaches.

The selected ground motions are identified in Table 4.1.1. An attempt has been made in the scaling process to match the ground motion spectrum to the design spectrum in the response range of the studied structures. For example, Earthquakes 1 and 5 match the design spectrum in the period range of the shear wall structure ($T \cong 0.95$ second) while Earthquakes 2 and 3 match the design spectrum in the period range of the frame braced structure. Earthquake 4 was included as a curiosity because it contained a large single pulse (Figure 4.1.3*a*); it had no effect on any of the studied structures.

The scaled ground motion for Earthquake 1 is shown in Figure 4.1.3*b*. Compare this to the scaled and altered ground motion of this event used to develop the matched spectrum of Figure 4.1.2*b*, which is shown in Figure 4.1.3*c*. Observe that the accelerations used to develop the matched spectrum (Figure 4.1.3*c*) are significantly increased in the post 10-second range and that the peak acceleration for the matched ground motion exceeds that of the scaled ground motion. This in spite of the fact that the spectral velocity defined by the matched spectrum is only 42 in./sec, as opposed to the 51 in./sec defined by the scaled spectrum.

Comment: One caveat should be reaffirmed as it relates to the use of time history analyses. Building designs are not in themselves academic efforts. Designs are performed and buildings produced to satisfy a perceived societal need in a manner consistent with the extant standard of care. The designer who performs a time history analysis of a proposed design does so only to answer one question—Is the produced design likely to meet the design objective? Academic interests may be explored in the process but this is only to improve future designs; not to "perfect" a produced design. Introducing design changes at this stage in the design process will create chaos, and this usually results in a less effective design. Design changes should be resisted unless the need is unquestionable.

4.1 RESPONSE OF SHEAR WALL BRACED BUILDINGS TO GROUND MOTION

The objective of this section is to review and analyze inelastic time history analyses of the shear wall braced buildings designed in Section 3.1. The equal displacement hypothesis and the impact of system strength on response and behavior are the focal issues.

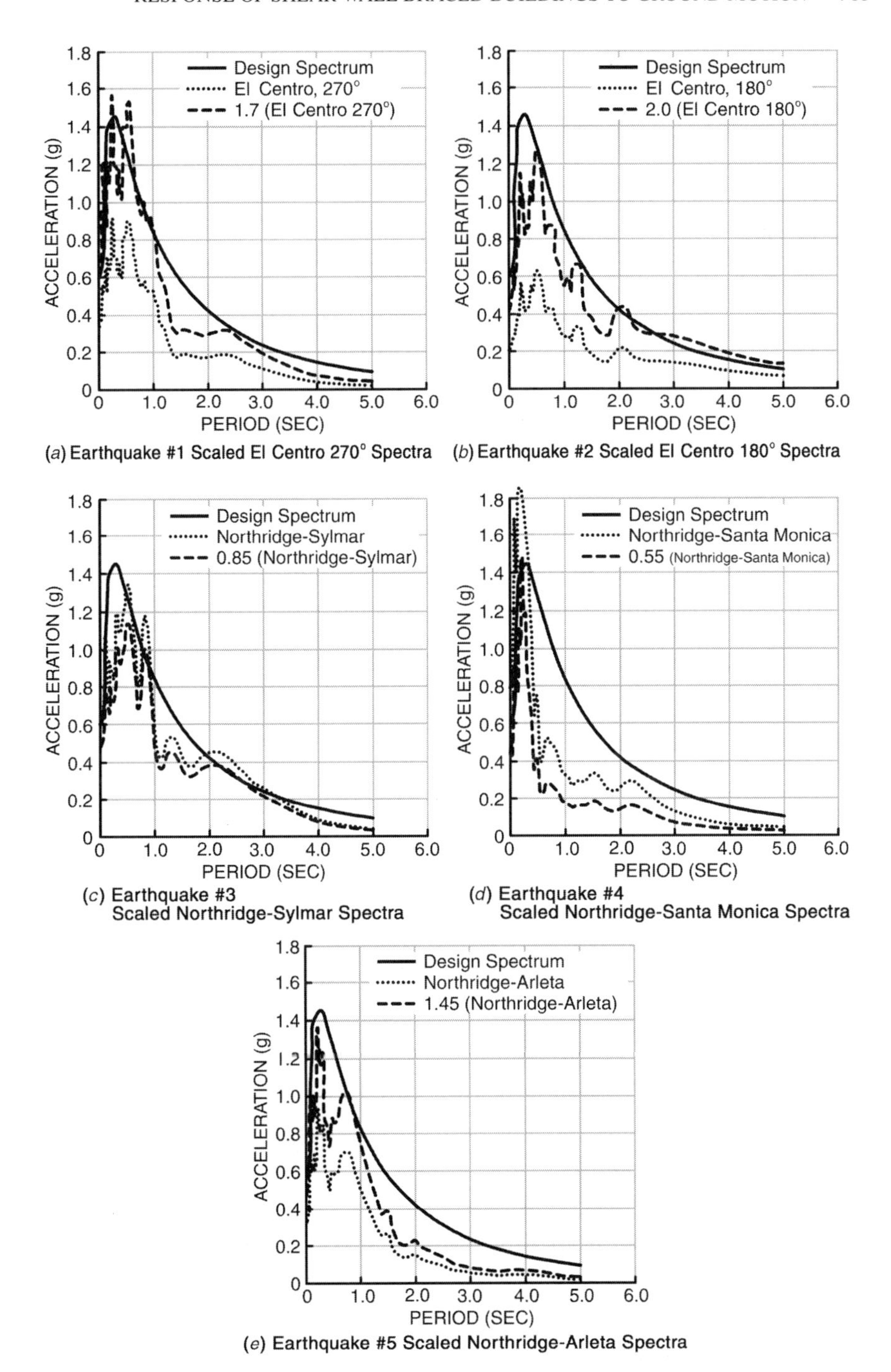

Figure 4.1.1 Design response spectrum and response spectra developed from scaled ground motions.

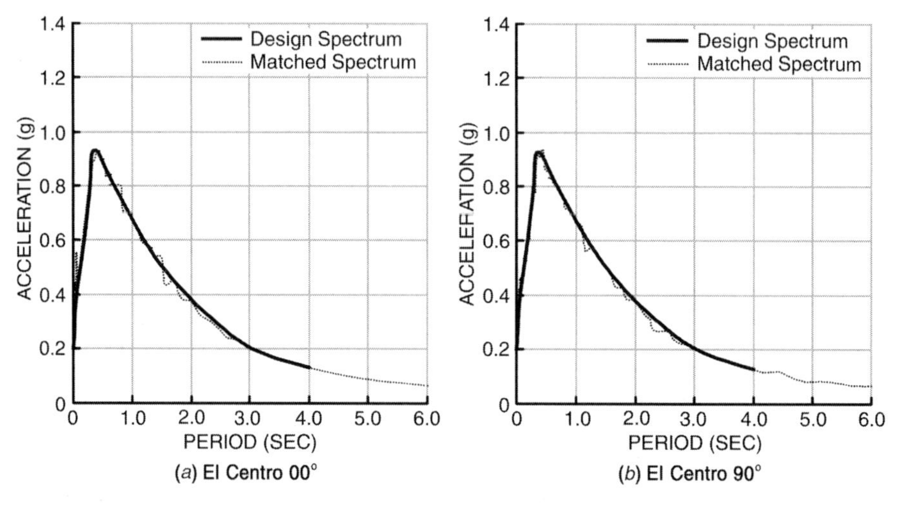

Figure 4.1.2 Matched spectra ($S_v = 42$ in./sec).

TABLE 4.1.1 Earthquake Ground Motions

	Earthquake Information		Scaling to Site Spectra	
Earthquake Number	Earthquake Record	File Name	Case	Scale Factor
1	Imperial Valley Earthquake—El Centro May 18, 1940, 20:37 PST Corrected Accelerogram, 270°, Caltech IIA001 ISEE, UC Berkeley, California	IMPVAL1.ACC	ELC270	1.70
2	Imperial Valley Earthquake—El Centro May 19, 1940, 20:37 PST Corrected Accelerogram, 180° Caltech IIA001 ISEE, UC Berkeley, California	IMPVAL1.ACC	ELC180	2.00
3	Northridge Earthquake—Sylmar County Hospital January 17, 1994, 04:31 PST Corrected Accelerogram, Channel (90°) CDMG QN94A514 ISEE, UC Berkeley, California	NRIDGE1.ACC	NR1	0.85

(Continued)

TABLE 4.1.1 Earthquake Ground Motions *(Continued)*

Earthquake Information			Scaling to Site Spectra	
Earthquake Number	Earthquake Record	File Name	Case	Scale Factor
4	Northridge Earthquake—Santa Monica City Hall Grounds January 17, 1994, 04:31 PST Corrected Accelerogram, Channel 1, 90° CDMG QN94A538 ISEE, UC Berkeley, California	NRIDGE2.ACC	NR2	0.55
5	Northridge Earthquake—Arleta and Nordhoff Fire Station January 17, 1994, 04:31 PST Corrected Accelerogram, Channel 1, 90° CDMG QN94A087 ISEE, UC Berkeley, California	NRIDGE3.ACC	NR3	1.45

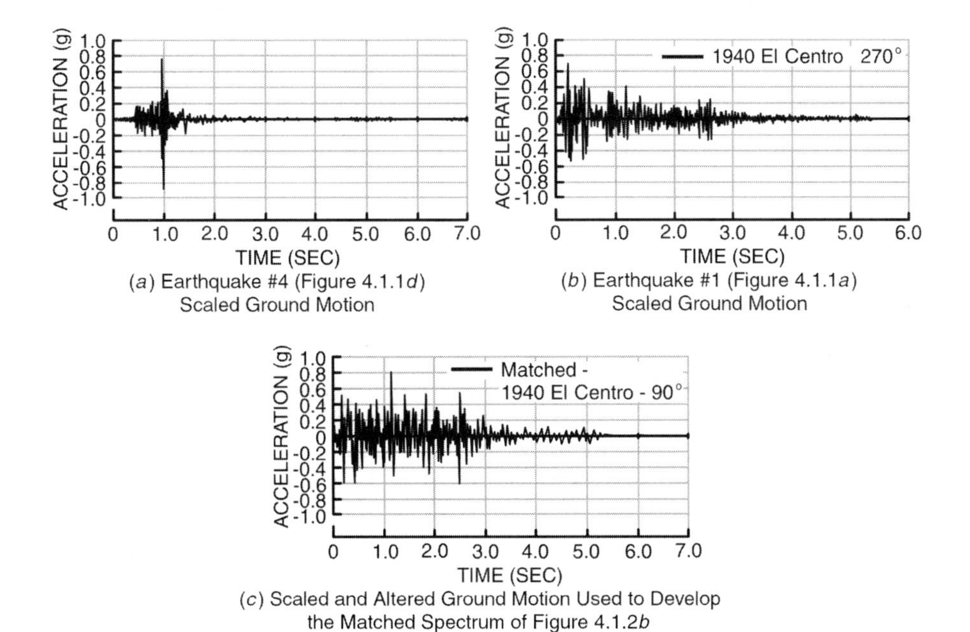

(*a*) Earthquake #4 (Figure 4.1.1*d*) Scaled Ground Motion

(*b*) Earthquake #1 (Figure 4.1.1*a*) Scaled Ground Motion

(*c*) Scaled and Altered Ground Motion Used to Develop the Matched Spectrum of Figure 4.1.2*b*

Figure 4.1.3 Selected scaled ground motions used in the various time history analyses.

4.1.1 Testing the Equal Displacement Hypothesis

A 26-ft long shear wall was designed using several methodologies in Section 2.4.1.4 and Section 3.1.1.1. The wall system and its reinforcing are described in Figures 2.4.9 and 2.4.10. This shear wall braced system was subjected to the ground motions used to develop the response spectra of Figure 4.1.1. Roof displacement history for Earthquake 1 is shown in Figure 4.1.4. The magnitude of the response of the wall was essentially the same for Earthquakes 3 and 5, and this is to be expected, for each of the scaled spectra closely match the design spectrum (Figures 4.1.1*a*, *c*, *e*) in the period range of the system ($T \cong 0.95$ second).

The response spectral projection of building drift should match the elastic time history analysis. The analysis that follows is repeated here for the reader's convenience.

The design spectral velocity in the 1-second period range is

$$S_v = \frac{TS_a}{2\pi} \tag{4.1.1}$$

$$= \frac{1.0(0.83)(386.4)}{6.28} \quad \text{(see Figure 4.1.1)}$$

$$= 51 \text{ in./sec}$$

The spectral displacement, based on a spectral velocity of 51 in./sec, would be

$$S_d = \frac{TS_v}{2\pi} \tag{4.1.2}$$

$$= \frac{0.95(51)}{6.28} \quad (T \cong 0.95 \text{ second})$$

$$= 7.71 \text{ in.}$$

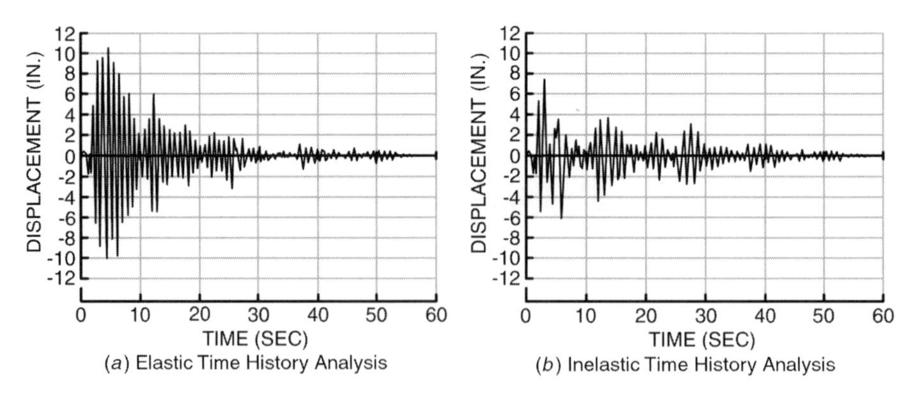

Figure 4.1.4 Response of 26-ft shear wall to Earthquake 1, Table 4.1.1 (building described in Figures 2.4.9 and 2.4.10).

and the displacement at the roof (Δ_n) would be

$$\Delta_n = \Gamma S_d \tag{4.1.3}$$
$$= 1.45(7.71) \qquad (\Gamma = 1.45)$$
$$= 11.2 \text{ in.}$$

This is quite comparable to the peak value predicted by the elastic time history analysis ($\Delta_{n,\max} = 10.66$ in. in Figure 4.1.4*a*).

An inelastic model of the building was also subjected to the ground motions described in Table 4.1.1. The peak displacement was slightly less than 8 in. and this warrants some study.

The elastic and inelastic responses of the wall system to Earthquake 1 are described in greater detail in Figure 4.1.5. Observe how the periodicity of the inelastic response is lengthening. The behavior pattern follows the characteristic yielding response described in Section 1.1.1. The structure yields 2+ seconds into the earthquake ($\Delta_y = 3.2$ in. in Figure 3.1.6*b*) and continues to displace to a drift of 5 in. in the elastic model and slightly more in the inelastic model. This corresponds to a displacement ductility demand of about 1.56.

The associated period shift is 1.25 ($\sqrt{1.56}$) times. The period of the yielding structure should be 1.2 seconds (1.25 (0.95)), and this is what is predicted by the time history analysis (Figure 4.1.5*b*). Observe that the response of the inelastic model after 3 seconds (Figure 4.1.5*a*) never exceeds 6 in. ($t = 6$ seconds) and thereafter is limited to 4 in.—essentially elastic. Understanding this behavior requires a review of the characteristics of the driving earthquake. Observe (Figure 4.1.1*a*) that the response drops significantly in the now period of interest ($T = 1.25$ seconds) range. The spectral acceleration response in the 1.3- to 1.5-second period range is only one-third of that in the 1-second range. In essence we have a new or different structure ($T = 1.2$ seconds) responding to a different earthquake ground motion ($S_v \cong 30$ in./sec). Spectral projections based on the new structure and the response spectrum

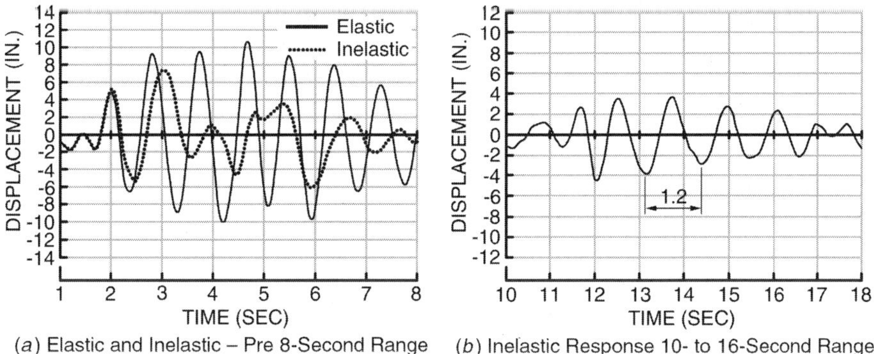

Figure 4.1.5 Enlarged time history responses of the 26-ft shear wall to Earthquake 1.

for Earthquake 1 ($\Delta_u \cong 8.6$ in.) are not likely to confirm the time history projections ($\Delta_{\max} = 7.6$ in.) because a portion of the strong ground motions ($t = 2$ seconds) have passed (Figure 4.1.3*b*)—and it was these larger ground motions ($0.75g$) that drove the initial structure ($T = 0.95$ second).

The drift experienced in the next cycle ($t = 3.5$ seconds) is 9.5 in. for the elastic model but only 7.6 in. for the inelastic model. Now the ductility demand is on the order of 2.4 (7.6/3.2) and another period shift might reasonably be anticipated. The period shift should amount to $\sqrt{\mu_s}$, or 1.55 times the elastic period. This magnitude of period shift is not confirmed in the response because the subsequent ground motions appear to excite higher modes in the structure. Figure 4.1.6 describes behavior in the plastic hinge region in the pre 8-second range (Figure 4.1.6*a*) and in the 12- to 14-second response range (Figure 4.1.6*b*). Observe how the stiffness in the 12- to 14-second range is consistent with the observed period change.

$$\sqrt{\frac{k_{12}}{k_8}} \cong 1.3 \qquad \text{(see Figure 4.1.6)}$$

$$T_{12} = 1.3T_8$$

$$T_{12} \cong 1.3(0.95)$$

$$\cong 1.2 \text{ seconds}$$

Comment: The response of this system to ground motion identifies an important attribute of a ductile structure. Specifically, the inherent shift in system period makes

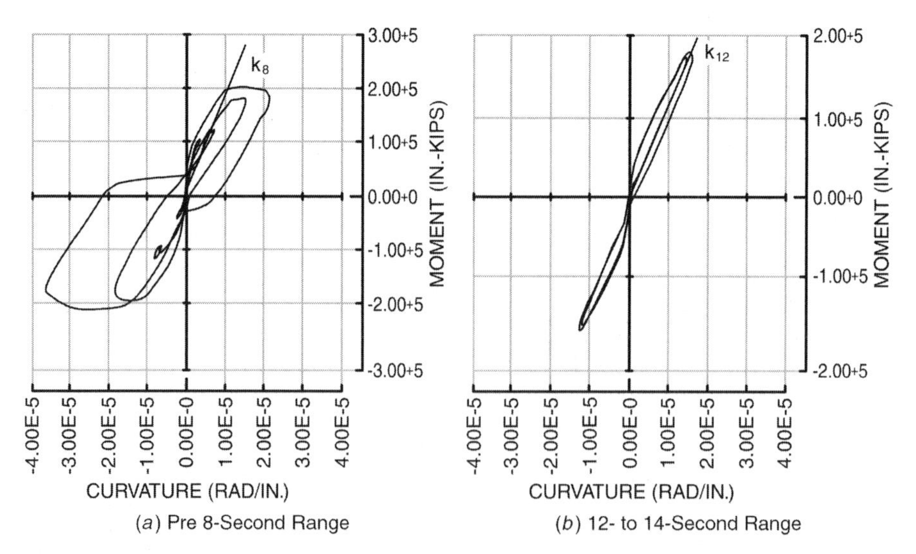

Figure 4.1.6 Hysteretic response in the plastic hinge region of the 26-ft shear wall (inelastic time history of Figure 4.1.3*a*).

it virtually impossible for the inelastic system to resonate with the ground motion. This will logically make it unlikely that the inelastic system will ever reach the response projections of the elastic system, which are, in their peak ranges ($T = 1$ second for Earthquake 1), largely attributable to resonant effects.

Observe that in Earthquakes 1, 3, and 5 (Figure 4.1.1) a significant reduction of response is apparent in the period range of 1.2 to 1.5 seconds. Obviously care must be taken to select earthquakes that reasonably describe potential ground motion characteristics. This argues for the adoption of an artificial ground motion similar to the ones described in Figure 4.1.2.

The matched design spectra described in Figure 4.1.2 are of a lesser intensity than the design spectrum described in Figure 4.1.1. The spectral velocity in the 1-second range is

$$S_v = \frac{TS_a}{2\pi} \qquad \text{(Eq. 4.1.1)}$$

$$= \frac{(1.0)(0.68)(386.4)}{6.28}$$

$$= 42 \text{ in./sec}$$

The ultimate displacement (Δ_u) of the 26-ft shear wall braced system ($T = 0.95$ second) would be

$$S_d = \frac{TS_v}{2\pi} \qquad \text{(Eq. 4.1.2)}$$

$$= \frac{0.95(42)}{6.28}$$

$$= 6.35 \text{ in.}$$

$$\Delta_u = \Gamma S_d \qquad \text{(Eq. 4.1.3)}$$

$$= 1.45(6.35)$$

$$= 9.2 \text{ in.}$$

The elastic time history analyses, using the ground motions that produced the spectra of Figure 4.1.2, were consistent in their prediction of a peak displacement response of slightly more than 8 in. Of particular interest is the response of the system to the altered El Centro 90° ground motion described in Figure 4.1.3*c*. The elastic response is slightly less than the spectral prediction of 9.2 in., and the early peaks ($t = 5$ seconds) developed for the scaled El Centro 90° ground motion (Figure 4.1.4*a*) have been suppressed. The displacement peaks now occur at $t = 12$ seconds (Figure 4.1.7*a*). The inelastic response (Figure 4.1.7*b*) describes a similar shift in the peak response (to $t = 12$ seconds), but in this case the displacement response of the inelastic response is essentially the same as the elastic response. This tends to confirm

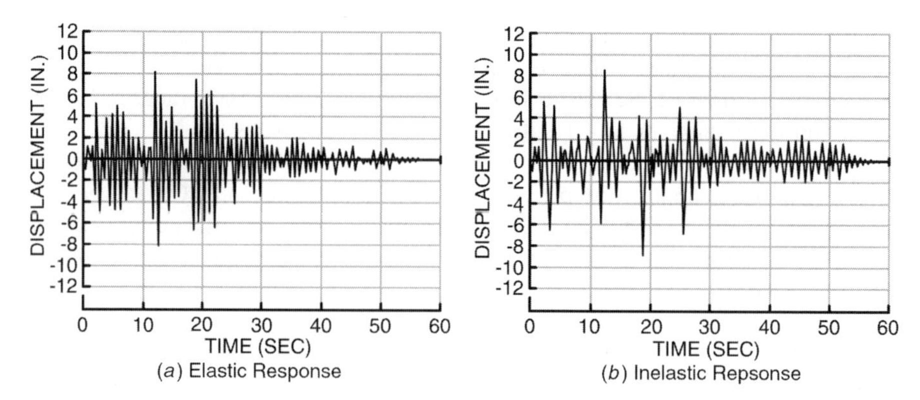

Figure 4.1.7 Response of the 26-ft shear wall to the (matched) ground motion described in Figure 4.1.3*c*.

the fact that the large reductions observed in the response to the scaled ground motion (Figure 4.1.4) are entirely attributable to the characteristics of the ground motion. Clearly, this analysis supports the equal displacement proposition.

Comment: Clearly, the selection of ground motions used in the analysis of this system raises several issues. Is an artificial earthquake that creates a constant spectral intensity reasonably assumed? If not, the designer must study the proposed ground motions prior to performing a time history analysis to insure that anomalies do not exist in the period range of interest. This latter approach seems more reasonable. In either case the structural engineer and the geotechnical consultant must work together to insure that the proposed system is reasonably tested.

The elastic and inelastic response of this structure (Figure 2.4.9) to the ground motion of Earthquake 2 (Figure 4.1.1*b*) is interesting. The elastic response (Figure 4.1.8*a*)

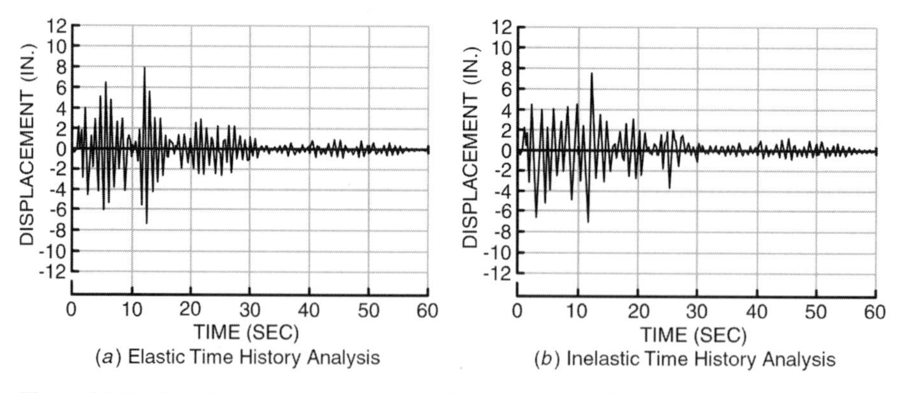

Figure 4.1.8 Roof displacement response of the 26-ft shear wall to Earthquake 2, Table 4.1.1.

suggests two ground pulses, one in the 3- to 7-second range and another in the 12-second range. Figure 4.1.8*b* shows the inelastic response and identifies a period shift quite similar to that caused by Earthquake 1 (see Figure 4.1.5*b*). Observe that the maximum inelastic and elastic displacement responses are the same. The behavior patterns, however, are quite different. The first inelastic excursion in the positive sense provokes a slightly larger inelastic response, while the negative sense inelastic response is significantly greater (50%) than the elastic response and the periodicity of the response has increased.

The response spectra for Earthquake 2 (in Figure 4.1.1*b*) is quite different from the spectra produced by ground motions 1, 3, and 5. Instead of suggesting a reduction in response, the response actually increases. When the second pulse arrives ($t = 12$ seconds), the spectral acceleration imposed on the inelastic structure ($T = 1.2$ seconds) is about $0.67g$, while that imposed on the elastic model ($T = 0.93$ second) is on the order of $0.6g$.

The displacement response for these models suggested by a spectral analysis is developed as follows. First combine Eq. 4.1.1, 4.1.2, and 4.1.3 to create

$$\Delta_n = \Gamma S_d \qquad \text{(Eq. 4.1.3)}$$

$$= \frac{\Gamma T^2 S_a}{4\pi^2} \qquad (4.1.4a)$$

$$= \frac{1.45T^2(386.4)S_{ag}}{39.4}$$

which for $\Gamma = 1.45$ becomes

$$\Delta_n = 14.2T^2 S_{ag} \qquad (4.1.4b)$$

where S_{ag} is the spectral acceleration express in terms of g.

For the elastic structure the predicted displacement at the roof is

$$\Delta_n = 14.2(0.93)^2(0.6) \qquad \text{(see Eq. 4.1.4b)}$$

$$= 7.4 \text{ in.}$$

and, given the steepness of the response spectrum in this period range, this confirms the elastic time history prediction of 8 in.

The response of the inelastic structure, using the same procedures, should be

$$\Delta_n = 14.2T^2 S_{ag} \qquad \text{(Eq. 4.1.4b)}$$

$$= 14.2(1.2)^2(0.67)$$

$$= 13.7 \text{ in.}$$

and this is significantly greater than the 8 in. predicted by the inelastic time history analysis (Figure 4.1.8*b*).

The reduced displacement suggests that energy dissipation may indeed have reduced the response. Consider the curvature ductility suggested by Figure 4.1.6*a*, which reflects plastic hinge behavior at inelastic displacements in the 8-in. range.

$$\mu_\phi = \frac{0.000035}{0.00001}$$
$$= 3.5$$

The structural component of system damping is

$$\zeta_{\text{eq}} = \frac{\sqrt{\mu} - 1}{\sqrt{\mu}\pi} \qquad \text{(Eq. 1.1.9d)}$$
$$= \frac{\sqrt{3.5} - 1}{\sqrt{3.5}\pi}(100)$$
$$= 15$$
$$\hat{\zeta}_{\text{eq}} = \zeta + \zeta_{\text{eq}} \qquad \text{(Eq. 1.1.9a)}$$
$$= 20$$
$$\hat{R} = \frac{3.38 - 0.67}{2.3} \ln \hat{\zeta}_{\text{eq}} \qquad \text{(Eq. 3.1.45)}$$
$$= 0.6$$
$$\Delta_{n,\text{eq}} = \Delta_n \hat{R}$$
$$= 13.8(0.6)$$
$$= 8.3 \text{ in.}$$

and this is quite consistent with the reported peak inelastic displacement of slightly less than 8 in. It seems that structural damping is impacting the inelastic system response, at least for this single-degree-of-freedom system whose system ductility and dissipated energy are reasonably quantified (Figure 4.1.6*a*).

The curvature response described in Figure 4.1.6*a* is acceptable from the standpoint of suggested strain states. The maximum curvature is slightly less than 0.00004 rad/in. and Table 2.4.3 suggests that induced material strains will be

$$\varepsilon_c = 0.0016 \text{ in./in.}$$
$$\varepsilon_s = 0.01094 \text{ in./in.}$$

From a design acceptance perspective, this meets our objectives. From a design procedures perspective, it merits some analysis.

Computer programs adopt one of two possible element models. One models the end of an element as a rotational spring; the other creates a fiber model. The spring model is consistent with the model used in the design; specifically one that assumes a constant curvature in the prescribed plastic hinge region. The fiber model creates a linearly changing curvature from the face of the support to the inception of elastic behavior, provided this point coincides with a node. In other words, it develops the hinge length for the user. The distinction is important because our design basis was developed from testing. Test results were reduced to strain states using the plastic hinge model (Figure 2.1.8). A plastic hinge length of 13 ft ($0.5\ell_w$) was used in the analysis of the example 26-ft long wall.

The time history analysis was performed on a fiber model. Nodes were established at each floor level (Figure 2.4.9), and the computer adopted a plastic hinge length of 10.5 ft or one floor. The conclusions reported in Figure 4.1.4 are for an average curvature in the plastic hinge region and, as a consequence, are consistent with the curvature model described in Figure 2.1.10. The curvatures developed in the design process follow the curvature model described in Figure 2.1.11. The two are easily compared.

$$\Delta_u = 7.6 \text{ in.} \qquad \text{Figure 4.1.5}a\text{—}(t = 3 \text{ seconds})$$

$$\Delta_y = 3.2 \text{ in.} \qquad (\text{Figure 3.1.6}b)$$

$$\Delta_p = 4.4 \text{ in.}$$

$$\theta_p = \frac{\Delta_p}{h_w - \ell_p/2} \qquad \text{(see Figure 2.1.11)}$$

$$= \frac{4.4}{1182}$$

$$= 0.0037 \text{ radian}$$

$$\phi_p = \frac{\theta_p}{\ell_p} \qquad \text{(Eq. 2.1.9)}$$

$$= \frac{0.0037}{156}$$

$$= 0.000024 \text{ rad/in.}$$

$$\phi_y = \frac{0.0033}{\ell_w} \qquad \text{(Eq. 2.4.27)}$$

$$= 0.00001 \text{ rad/in.}$$

$$\phi_u = \phi_y + \phi_p$$

$$= 0.00001 + 0.000024$$

$$= 0.000034 \text{ rad/in.}$$

The reported curvature is clearly consistent with that developed for conceptual design purposes. The computer model adopted a plastic hinge length of 126 in. This is more conservative than the hinge length adopted in the conceptual design. The impact on the comparative analysis would not change our confirmation of the behavior predicted by the computer analysis for

$$\theta_p = \frac{4.4}{1197}$$
$$= 0.00367 \text{ radian}$$
$$\phi_p = \frac{0.00367}{126}$$
$$= 0.000029 \text{ rad/in.}$$
$$\phi_u = \phi_y + \phi_p$$
$$= 0.000039 \text{ rad/in.}$$

and this is consistent with the curvatures reported in Figure 4.1.6*a*.

A fiber model will be used to analyze the 16-ft shear walls in the next section. The computer analysis will adopt the same plastic hinge length ($h_x = 126$ in.), while the design opted for a plastic hinge length of 96 in. ($0.5\ell_w$). The conceptual design prediction of curvature, in this case, will be greater than that developed from the computer analysis.

From a design perspective, it is probably more rational to select a plastic hinge length that coincides with a story height as opposed to one that is a function of the length of the wall, but the impact should not alter conclusions drawn from the analysis.

Several conclusions may be drawn:

- Time history based conclusions must be developed with care and rationally explained.
- Ground motions used to describe inelastic behavior should be compatible with anticipated levels of spectral intensity in the pre- and postyield behavior ranges.
- The "equal displacement" hypothesis is reasonably used to design structures of this type.
- It seems unlikely that the ultimate displacement of this system will exceed 11 in. (0.9%). Given this and the strains predicted by the pushover analysis (Figure 3.1.6*b*), there seems no reason to believe that performance objectives will not be met.

4.1.2 Impact of Design Strength on Response

The selection of a 16-ft long wall was based on the attainment of displacement objectives and the balancing of dynamic response and strength, given the presumed system displacement ductility (μ_Δ) of 4. The selected reinforcing program proposed a flexural reinforcement ratio of 0.25% (six #6 bars). The elastic period of this wall is slightly less than 2 seconds. Predicted drifts were

$$\Delta_y = 7.5 \text{ in.} \qquad \left(\frac{11}{40}\phi_y h_w^2\right)$$

$$\Delta_u = 22.86 \text{ in.} \qquad \text{(Response spectra)}$$

$$\mu_\Delta = 4 \qquad \text{(Objective)}$$

$$\mu_\Delta = 3 \qquad \text{(Attained)}$$

The issue now addressed is the advantage that might result from an increase in flexural strength. Accordingly, the response of the proposed design, six #6 bars ($\rho_s = 0.25\%$), is compared with an otherwise identical wall reinforced with six #9 bars ($\rho_s = 0.57\%$).

These walls were subjected to the ground motion of Earthquake 3 (Table 4.1.1) whose response spectrum is described in Figure 4.1.1*c*. Observe that the ground motion based spectrum is less than the design spectrum at the probable elastic building period of 1.92 seconds, but exceeds the design spectrum in the 2.1- to 2.7-second period range. Accordingly, any period shift is likely to produce an increase in displacement.

The results of the time history analyses are presented in Figure 4.1.9. The peak elastic displacement occurs at 9 seconds and corresponds to almost 18 in. This is consistent with the response predicted using the design spectrum (22.86 in.). The peak inelastic displacement is 15 in. (Figure 4.1.9*b*) for the proposed design (six #6 bars) but almost 18 in. (Figure 4.1.9*d*) for the stronger wall (six #9 bars). This is explained following the concepts discussed in Section 1.1.1. Observe that the stronger wall (six #9 bars) did not yield at the 6-second excursion while the design wall (six #6 bars) did. This resulted in a reduced displacement demand on the design wall in subsequent cycles (Figure 4.1.9*b*) while the stronger wall had a deferred first plastic excursion that coincided with a larger pulse. Accordingly, there may often be an advantage associated with early yielding. This is not a conclusion that can be applied to the design of a wall, but it is not an uncommon occurrence.

Several observations are made:

- The proposed 16-ft wall design ($\rho = 0.25\%$) meets our performance objectives.
- The experienced deflection should be in the 15-in. range (1.2%), and this is twice that expected of the 26-ft wall.
- Concrete strain states will also be higher in the 16-ft long wall ($\varepsilon_c = 0.0023$ in./in. see Figure 3.1.7*c*) than those in the 26-ft long wall ($\varepsilon_c = 0.0016$ in./in.—see Figure 3.1.7*a*). The deflection predicted for the stronger wall was 18 in. This suggests a concrete strain of 0.0028 in./in. (Figure 3.1.7*c*). Structural damage should not be expected in either case.

Several generalizations are drawn from these observations:

- A stronger wall does not necessarily reduce the displacement demand.
- The inelastic displacement demand may reasonably be predicted using elastic response spectral techniques.

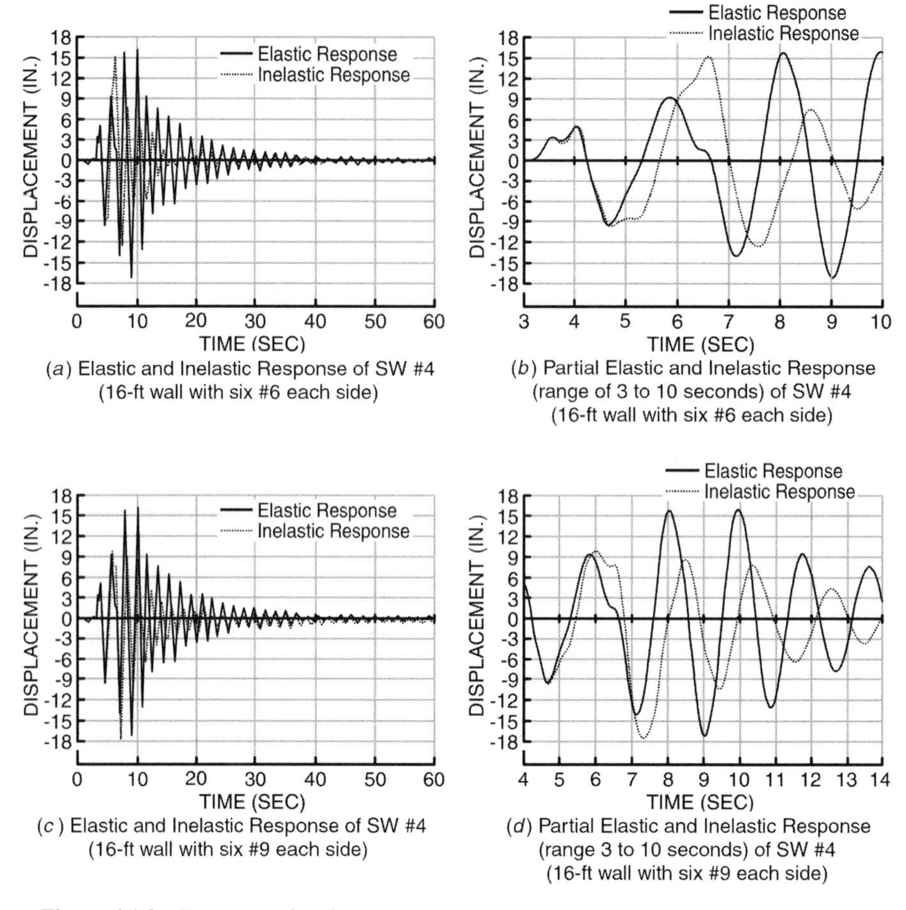

(*a*) Elastic and Inelastic Response of SW #4 (16-ft wall with six #6 each side)

(*b*) Partial Elastic and Inelastic Response (range of 3 to 10 seconds) of SW #4 (16-ft wall with six #6 each side)

(*c*) Elastic and Inelastic Response of SW #4 (16-ft wall with six #9 each side)

(*d*) Partial Elastic and Inelastic Response (range 3 to 10 seconds) of SW #4 (16-ft wall with six #9 each side)

Figure 4.1.9 Response of various strength shear wall braced structures to Earthquake 3.

- Concrete strain states are not directly related to wall strength (Figure 3.1.7*c*) but rather to displacement demand and, as a consequence, in shear wall braced buildings performance will not be improved by increasing strength.
- In near fault locations wall designs should be studied by subjecting them to pulse type excitations.
- Design procedures developed in Chapters 2 and 3 reasonably predict behavior and seem to produce systems that will survive the criterion earthquake.

The predicted level of spectral acceleration at the roof will increase significantly as the building is stiffened and strengthened.

$$\Gamma S_a = \frac{\Gamma S_d 4\pi^2}{T^2 \mu_\Delta} \qquad \text{(see Eq. 4.1.1 and 4.1.2)} \qquad (4.1.5)$$

The proposed designs are:

(A) 26-ft shear wall; $\Delta_u = 11.2$ in., $T \cong 1$ second.
(B) 16-ft shear wall, six #9; $\Delta_u = 18$ in., $T \cong 2$ seconds.
(C) 16-ft shear wall, six #6; $\Delta_u = 15$ in., $T \cong 2$ seconds.

The anticipated acceleration at the uppermost levels is

Wall A

$$\mu_\Delta = \frac{\Delta_u}{\Delta_y} = \frac{11.2}{3.2} = 3.5$$

Wall B

$$\mu_\Delta = \frac{18}{7.5} = 2.4$$

Wall C

$$\mu_\Delta = \frac{15}{7.5} = 2.0$$

The acceleration at the roof or nth level (A_n) would be

$$A_n = \frac{\Delta_u 4\pi^2}{T^2 \mu_\Delta} \qquad \text{(see Eq. 4.1.5)} \qquad (4.1.6)$$

Wall A

$$A_n = \frac{11.2(4)(3.14)^2}{(1)^2(3.5)} = 126 \text{ in./sec}^2 \qquad (0.33g)$$

Wall B

$$A_n = \frac{18(4)\pi^2}{(2)^2(2.4)} = 74 \text{ in./sec}^2 \qquad (0.19g)$$

Wall C

$$A_n = \frac{15(4)\pi^2}{(2)^2(2.0)}$$

$$= 74 \text{ in./sec}^2 \qquad (0.19g)$$

Conclusion: There appears to be little reason to select Walls A or B in spite of the fact that they are stiffer (A) or stronger (B).

4.2 FRAME BRACED BUILDINGS

The objective of this section is to study the behavior of the frames designed in Section 3.2. Analytical models start with the simplest form. Models are then modified so as to describe the peculiarities associated with precast systems. The impact model changes could have on predicted response is then discussed.

4.2.1 Impact of Design Strength on Performance

Strength has traditionally been viewed as having a significant impact on building performance. This proposition can only be supported when the increase in strength results in a lesser building drift and/or lower strain states in ductile components. Absent a fundamental change in the bracing program, this did not turn out to be the case

Photo 4.1 Precast clad concrete frame, Mariners Island, San Mateo, CA. (Courtesy of Englekirk Partners, Inc.)

in the shear wall examples reviewed. A similar conclusion is reached for frame braced structures. The intuitive argument suggests that an increase in strength will cause a more violent response, leading to larger displacements and induced levels of strain.

The frames designed in Section 3.2 were subjected to inelastic time history analyses. Component behavior was modeled using a bilinear elastic perfectly plastic representation of behavior—alternative behavior models will be considered in Section 4.2.2. Two frame types, each with three different levels of strength, are considered: the three-bay frame of Figure 3.2.2*a* and the four-bay alternative. Frame characteristics are summarized in Table 4.2.1 and identified as to number of bays and level of strength (for example, 3-1 refers to three bays—strength level 1).

Earthquakes 2 and 3 were selected because their spectral response was essentially the same as that of the design spectrum in the 2- to 3-second period range (see Figures 4.1.1*b* and 4.1.1*c*). The responses of the various models to the selected ground motions are presented in Table 4.2.2.

It seems clear that a generalization of behavior is not possible; however, intuitive logic tends to be supported by the responses described. Consider the following propositions:

- An increase in the strength of the frame will not guarantee a reduction in the level of experienced drift.
- Response is a function of the characteristics of the ground motion.

The impact of ductility on response was discussed in Section 1.1.1, and it should be obvious that the nature of the earthquake can affect the peak displacement more than the strength of the system.

The response spectrum for Earthquakes 2 and 3 identify essentially identical spectral velocities in the period range of interest, yet the range of predicted peak displacement response is significantly higher on average for Earthquake 2 than for Earthquake 3.

Figures 4.2.1*a*, through 4.2.1*c* describe the responses of the various three-bay frames to the ground motion of Earthquake 2. For Frame 3-1 the first inelastic excursion is the largest drift and the impact of a ground displacement pulse at $t \cong 4$ seconds

TABLE 4.2.1 Frame Descriptions (See Figure 3.2.2)

Designation Bay-Strength Category	Beam Overstrength, Levels 1–5 ($\lambda_o M_n$, ft-kips)	Mechanism Base Shear (kips)	V_M/W
3-1	650	535	0.044
3-2	1000	750	0.062
3-3	2000	1420	0.117
4-1	650	700	0.058
4-2	1300	1300	0.107
4-3	2600	2500	0.206

TABLE 4.2.2 Displacement Response Comparisons—Elastic/Perfectly Plastic Model (All Displacements Are in Inches)

Designation Bay-Strength Category	V_M/W	Δ_y	Predicted Drift	Earthquake 2 Max. Drift	(μ)	Earthquake 2 Residual Drift	Earthquake 3 Max. Drift	(μ)	Earthquake 3 Residual Drift
3-1	0.044	5.5	30	22.5	(4.1)	11.5	15.5	(2.8)	7
3-2	0.062	7.8	30	28.5	(3.7)	22.0	17.5	(2.2)	6
3-3	0.117	18	30	31.1	(1.7)	16.5	19.5	(1.1)	5.5
4-1	0.058	5	27	24.5	(4.9)	17	13.5	(2.7)	6
4-2	0.107	10	27	29.6	(3)	15.5	13.3	(1.3)	1.8
4-3	0.206	20	27	23.1	(1.3)	4.5	24.7	(1.2)	6.3

is significantly reduced. Compare this with the response of Frame 3-3, where the first positive pulse produces a response in the elastic range. Now the pulse at $t \cong 4$ seconds produces the first inelastic excursion, the consequences of which are a series of elastic excursions about a new baseline created by a permanent or residual drift of 16.5 in.

These anomalies aside, it seems reasonable to imagine that peak displacements of a ductile frame would become larger as the strength of the frame is increased. In Section

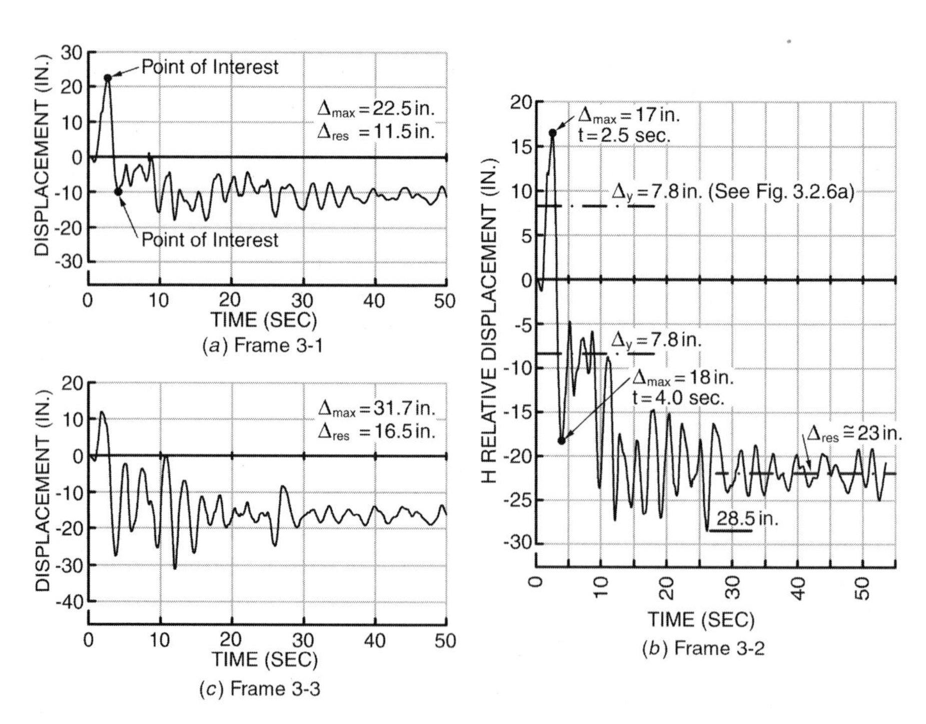

Figure 4.2.1 Response of the various three-bay frames to the ground motion of Earthquake 2.

1.1.1 it was pointed out that it was the strength of the spring that connects the mass to the ground that drives the mass and, as a consequence, a weakening of the spring tends to reduce the displacement response on all but the first inelastic excursion. This is the essence of base isolation so it is reasonable to assume that the displacement response would increase as the linkage between the ground and the mass is increased.

The addition of one bay does not seem to materially impact the level of experienced roof drift. Figure 4.2.2 describes the response of the four-bay frame (4-2) to Earthquake 2 and, as can be seen, it is quite similar in character to the response of Frames 3-2 and 3-3, which bracket it in terms of strength. The encouraging aspect from a design perspective is that peak drifts seem to be consistent with spectral projections.

Residual drifts should logically be reduced by an increase in frame strength, but this is not a reliable consequence. Observe (Table 4.2.2) that the residual drift of the various three-bay frames to Earthquake 3 and the various four-bay frames to Earthquake 2 exhibited the logical reduction in residual drift, while the other residual drift patterns were very irregular. Design objectives were discussed in Chapter 3 regarding residual displacements, but it seems clear that an increase in strength will not guarantee reduced residual drifts.

From a performance perspective, damage and failure potential can only be related to induced concrete strains. If it is assumed that the drift of Frames 3-2 and 3-3 will be essentially the same, then Frame 3-3 will experience much more damage than Frame 3-2 because, absent a change in member size, an increase in strength will reduce the available ductility in its components. Table 4.2.3 describes critical strain states

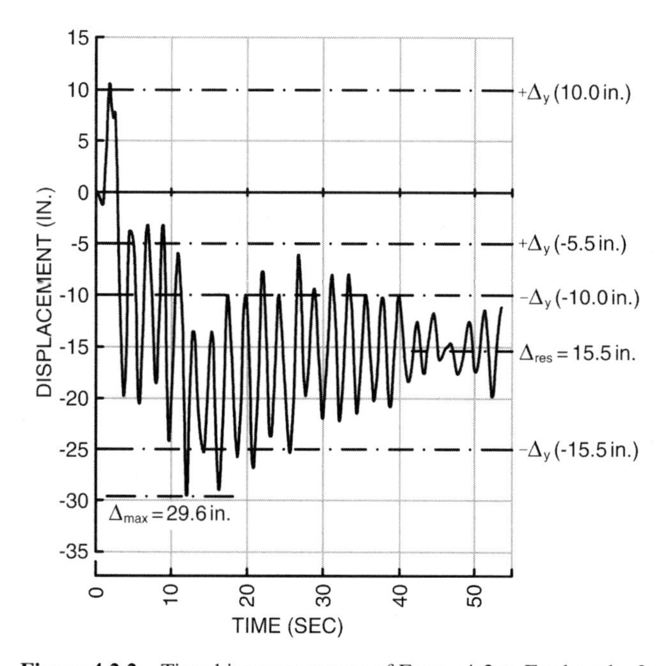

Figure 4.2.2 Time history response of Frame 4-2 to Earthquake 2.

TABLE 4.2.3 Strength versus Performance

Frame Identification			Maximum Induced Concrete Strains (in./in.)	
Designation	Strength (V_M/W)	Predicted Drift (in.)	Compression Side Column	Beam
3-2	0.062	30	0.0135	0.0046
3-3	0.117	30	0.024	0.008
4-2	0.107	27	0.014	0.007

in several frames. There seems to be no advantage from a performance perspective associated with adopting mechanism strength levels on the order of $0.11W$.

Several conclusions can be drawn from this summary:

- The "equal displacement" hypothesis is reasonably supported.
- System drifts can be predicted by response spectral procedures.
- System strength has a direct tie to displacement. As strength is increased it appears as though displacements will increase until they reach the elastic displacement response, which appears to be the upperbound.
- A quantifiable tie between frame strength and residual drift does not seem to exist; however, it is clear and logical that as frame strength is increased residual drift will asymptotically be reduced to zero (elastic behavior).
- Strength will have a direct bearing on strains imposed on components. The strength limit state of a frame should be based on component strain limit states.

4.2.2 Impact of Modeling Assumptions

The behavior idealization of the beams in the analyzed frames whose behavior is described in Figures 4.2.1 and 4.2.2 was elastic/perfectly plastic. This is the simplest and most commonly used element model. Member behavior in the plastic hinge region is described by a series of parallelograms whose stiffness is constant. The area within the parallelogram quantifies the energy dissipated during the particular cycle (see Figure 1.1.6).

This elastic/perfectly plastic behavior model is described in Figure 4.2.3*a*. In addition to energy dissipation, component stiffness will degrade with each cycle (see Figure 2.1.2), and energy dissipated will be less than that proposed by the elastic/perfectly plastic behavior model, especially if pinching becomes a significant factor. Further, $P\Delta$ effects cause the system to strength degrade. The impact of these considerations on behavior is discussed in this section.

Consider the three-bay alternatives reviewed in Section 4.2.1. The design ductilities, based on a predicted building drift of 30 in., were 5.5, 3.8, and 1.7 (see Table 4.2.2). The associated period shifts would be 2.3, 1.95, and 1.3 ($\sqrt{\mu}$—Figure 3.1.2). Period shifts do not materialize in the time history analyses described in Figures 4.2.1

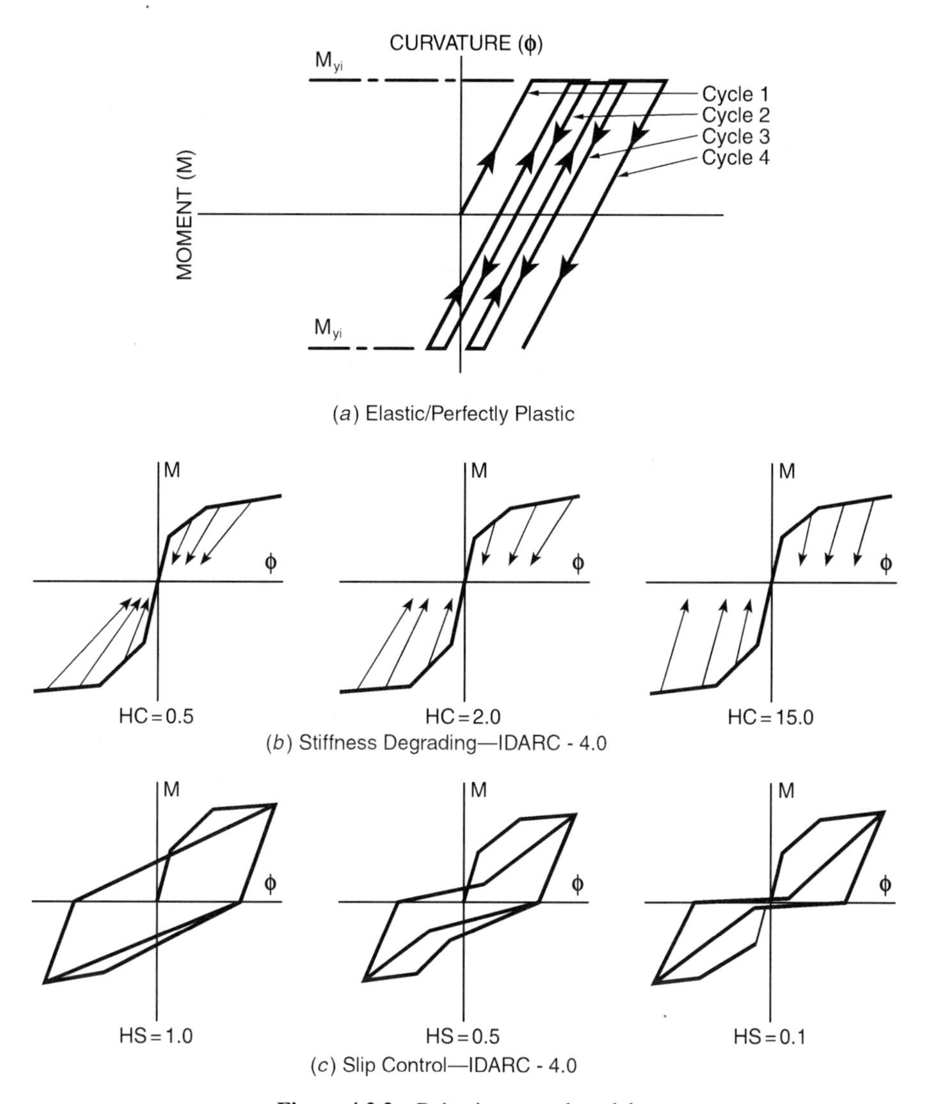

(*a*) Elastic/Perfectly Plastic

(*b*) Stiffness Degrading—IDARC - 4.0

(*c*) Slip Control—IDARC - 4.0

Figure 4.2.3 Behavior control models.

and 4.2.2 because the stiffness of the member is presumed to be constant (Figure 4.2.3*a*).

Stiffness degradation should generally serve to reduce the displacement response of a structure. In order to appreciate this fact, consider the response spectrum for Earthquake 3 (Figure 4.1.1*c*). A structure whose fundamental period is in the 0.5- to 1-second range will be tuned to this earthquake, and resonance would be a logical concern. When, however, the structure is forced into the inelastic range, the periodicity of the response will also change and the consequences of resonance will be avoided. Logically, this suggests that a reasonable level of ductility should be included in the design of any structure. The exception to this behavior enhancement characteristic might be experienced by a structure responding to Earthquake 2, for a period shift from 1.8 seconds to 2 seconds would increase the displacement response as was observed in Section 4.1. Fortunately, these resonant peaks tend to be transient in terms of their impact, for the structure will continue to soften and the displacement response to change. So long as the design spectrum reasonably captures spectral peaks, the drift of the structure should be predictable.

A stiffness degrading model was used to predict the behavior of structure types 3-2, 3-3, and 4-2 (Table 4.2.1). The response predicted for each system was less in all cases when the ground motion for Earthquake 2 was used. This pattern was reversed when Earthquake 3 was used, but this is probably a consequence of the low ductility demands and the characteristics of the ground motion. Observe, however, that the time history prediction of maximum displacement is in all cases less than the spectral projection. These results are summarized in Table 4.2.4.

$P\Delta$ effects were discussed in Chapter 1 where it was hypothesized that $P\Delta$ effects would tend to reduce the displacement response of a system. The three frame types (3-2, 3-3, and 4-2) were subjected to an inelastic time history analysis. A stiffness degrading component model was used and $P\Delta$ effects included as described in Figure 3.2.5. Peak displacement responses were typically reduced (see Table 4.2.4). The results obtained from the frame analyses used as examples are reasonably representative of similar comparative analyses and have caused most analysts to revert to the simpler bilinear component model, which includes some strength hardening.

Comparative time history analyses that consider variations in the hysteretic behavior of components are time consuming and require a significant amount of monitoring to insure that the behavior of the model is in fact consistent with the proposed objectives. A three-level, two-bay frame was adopted as a model to compare the response of a cast-in-place frame to a hybrid frame. Beam and column sizes were those used for the frames of Table 4.2.1 and the basic overstrength ($\lambda_o M_n$) of the frame beams was 1000 ft-kips (Frame 3-2). An elevation of the frame is described in Figure 4.2.4. IDARC 4.0 software was used to perform the comparative analysis.

Stiffness degradation and slip characteristics were used to develop the component models. Stiffness degradation and slip control parameters are described in Figures 4.2.3*b* and *c*. The two systems were modeled using the following parameters:

	CAST-IN-PLACE	HYBRID
Stiffness degradation (HC)	1.0	0.5
Slip control (HS)	1.0	0.2

Before proceeding with the time history analyses, both the cast-in-place and hybrid building models were subjected to a static displacement controlled response and member behavior was checked to insure that plastic hinge behavior objectives had been attained. Building displacement responses are shown in Figure 4.2.5. Observe that building responses reflect the characteristics of the components. Component behaviors are described in Figure 4.2.6. Compare the experimental response described in Figure 2.1.2 for the cast-in-place beam with the analytical model described in

TABLE 4.2.4 Summary of Time History Analysis (All Displacements in Inches)

Modeling Considerations					Earthquake 2			Earthquake 3		
Frame Type	Elastic Period (seconds)	Spectral Drift	$P\Delta + 3\%$ Strength Hardening	Stiffness Degradation	Δ_{max}	Residual Drift	Inelastic Period (seconds)	Δ_{max}	Residual Drift	Inelastic Period (seconds)
3-2	2.63	30	No	No	28.5	22	—	17.5	6	—
3-2	2.63	30	No	Yes	22.2	7	4.2	20.2	8	2.63
3-2	2.63	30	Yes	Yes	18.0	2.5	4.2	19	8.5	3.57
3-3	2.63	30	No	No	31.1	16.5	—	19.5	5.5	—
3-3	2.63	30	No	Yes	29.5	11.0	3.18	20.5	4	2.76
3-3	2.63	30	Yes	Yes	29.0	8.5	3.14	19.0	2	2.75
4-2	2.27	27	No	No	29.6	15.5	—	13.3	1.8	—
4-2	2.27	27	No	Yes	22.5	10.0	3.1	16.5	0.3	2.79
4-2	2.27	27	Yes	Yes	21.5	8.0	3.05	16.7	1.5	2.71

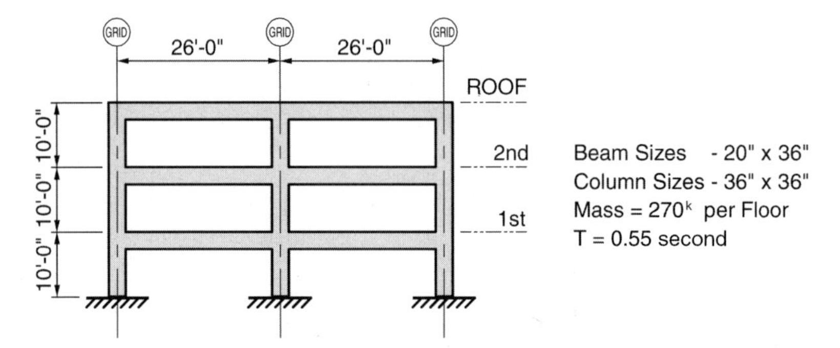

Figure 4.2.4 Frame elevation—example analysis—Section 4.2.2.

Figure 4.2.6*a*. Figures 4.2.6*b* and *c* describe how several experimental cycles compare with analytical cycles. Similarly, compare the experimentally determined hysteretic behavior of the hybrid system (Figures 2.1.47*a* and 2.1.47*b*) with the developed analytical model (Figure 4.2.6*d*). Observe that the "slip" feature is reasonably captured. The hybrid beam dissipates about 60% of the energy dissipated by the cast-in-place beam.

Both models were analyzed using the ground motion for Earthquake 3 (Figure 4.1.1*c*), which predicts a fairly constant response in the period range of interest (0.55 second). Building responses are shown in Figure 4.2.7. The response of the hybrid system (Figure 4.2.7*b*) is greater than that predicted for the cast-in-place system (Figure 4.2.7*a*), as might be expected.

It is interesting to note that the magnitude of the increase in response is consistent with that predicted by Eq. 3.1.45. Consider the following:

Cast-in-Place System

$$\mu_\Delta = \frac{\Delta_{\max}}{\Delta_y}$$

$$= \frac{7.0}{1.5} \qquad \text{(see Figure 4.2.7}a\text{)}$$

$$= 4.7$$

$$\zeta_{eq} = \frac{\sqrt{\mu} - 1}{\pi\sqrt{\mu}} \qquad \text{(Eq. 1.1.9d)}$$

$$= 0.17$$

$$\hat{\zeta}_{eq} = 17 + 5 \qquad \text{(Eq. 1.1.9a)}$$

$$= 22$$

$$\hat{R} = \frac{3.38 - 0.67 \ln \hat{\zeta}_{eq}}{2.3} \qquad \text{(Eq. 3.1.45)}$$

$$= 0.57$$

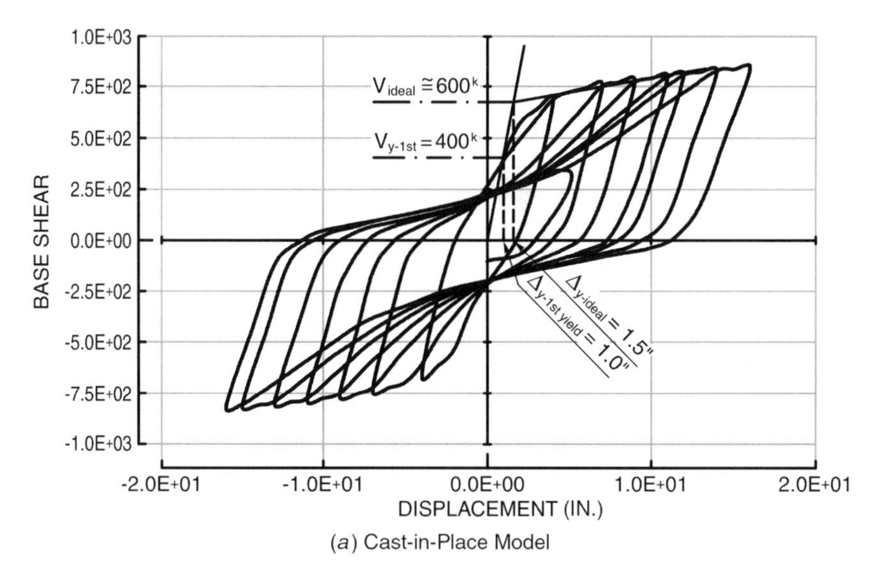

(*a*) Cast-in-Place Model

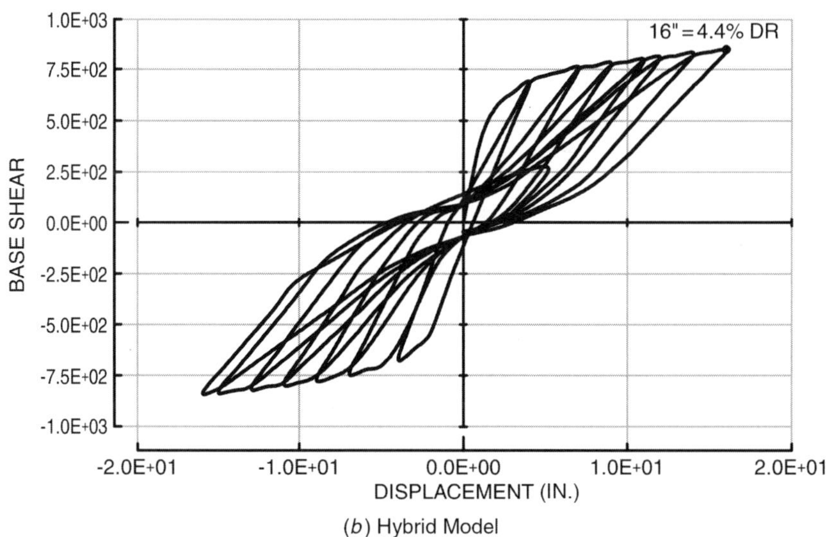

(*b*) Hybrid Model

Figure 4.2.5 Static displacement response—frame of Figure 4.2.4.

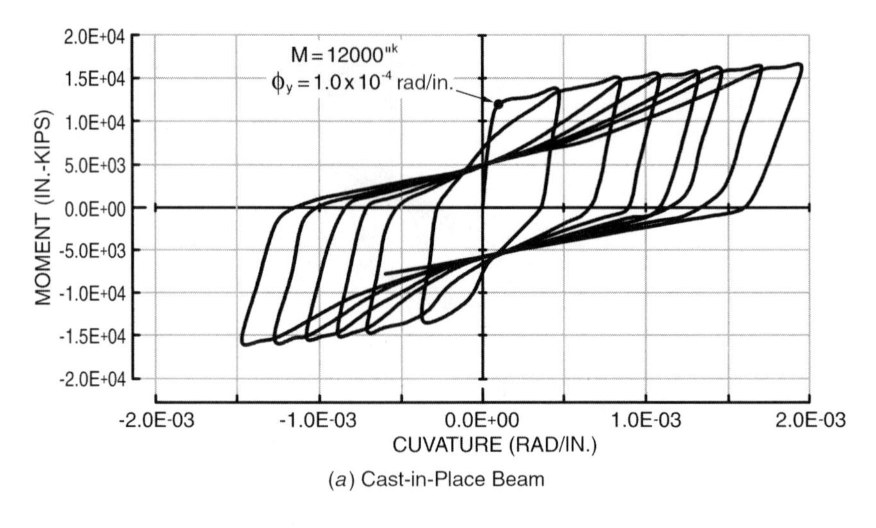

(a) Cast-in-Place Beam

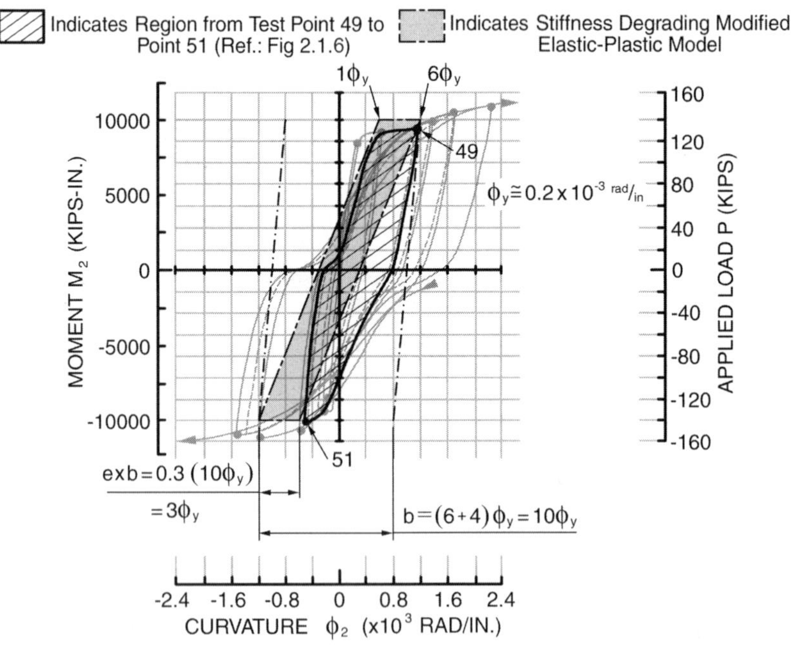

(b) Modified Elastic Plastic Beam Model @ μ_t = 6.0 (e = 0.3)

Figure 4.2.6 Beam behavior models static displacement controlled behavior.

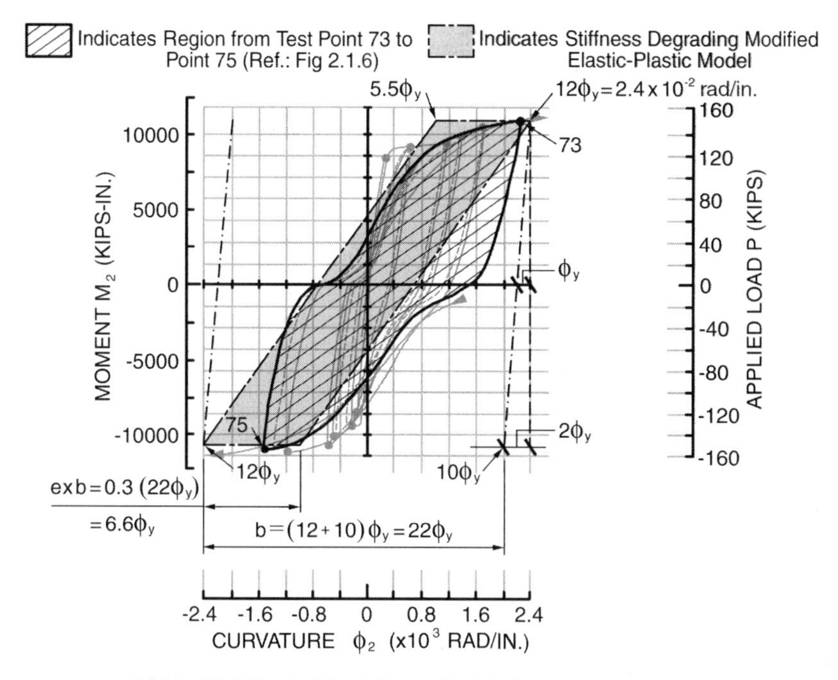

(*c*) Modified Elastic Plastic Beam Model @ μ_f = 12.0 (e = 0.3)

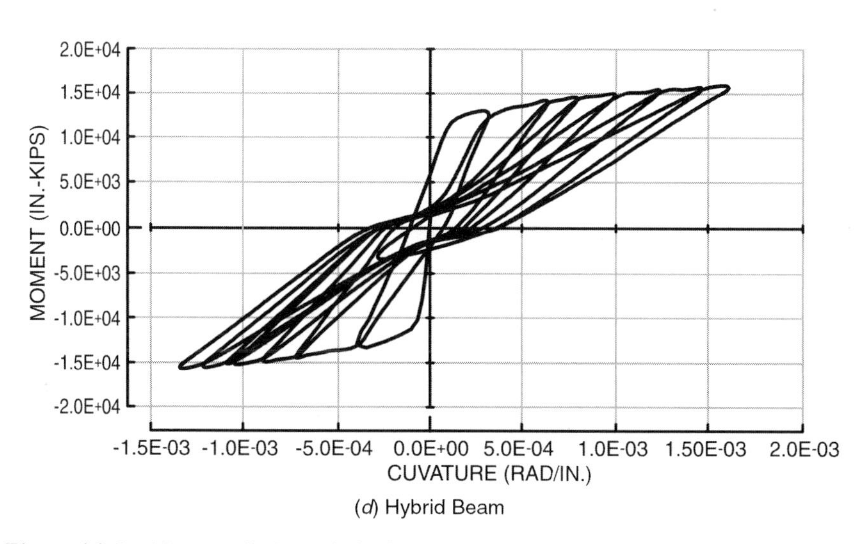

(*d*) Hybrid Beam

Figure 4.2.6 (*Continued*) Beam behavior models static displacement controlled behavior.

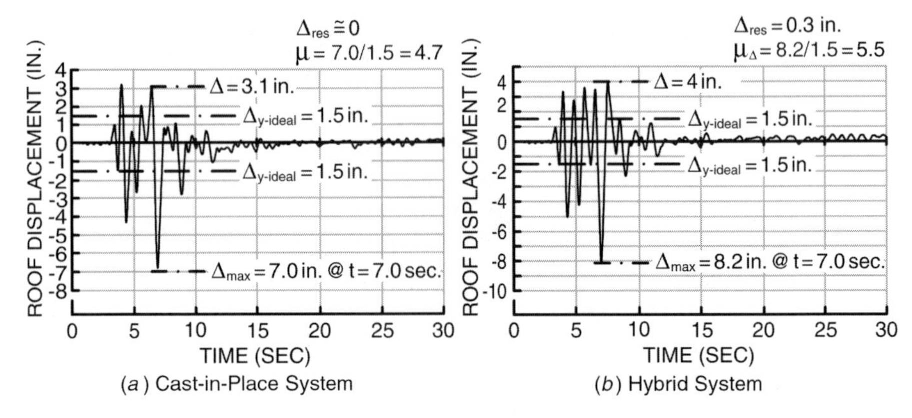

Figure 4.2.7 Response of structure described in Figure 4.2.4 to Earthquake 3.

Hybrid System

$$\mu_{\Delta} = \frac{\Delta_{\max}}{\Delta_y}$$

$$= \frac{8.2}{1.5} \qquad \text{(see Figure 4.2.7}b\text{)}$$

$$= 5.5$$

$$\zeta_{\text{eq}} = \frac{\sqrt{\mu} - 1}{\pi \sqrt{\mu}} \qquad \text{(Eq. 1.1.9d)}$$

$$= 0.18$$

Now, adjust the equivalent damping to account for the reduction in energy absorbed (Figure 1.1.8). Presuming that the hybrid system is 60% effective,

$$\zeta_{\text{eq}}' = 0.6\zeta_{\text{eq}}$$

$$= 0.11$$

$$\hat{\zeta}_{\text{eq}} = 11 + 5 \qquad \text{(Eq. 1.1.9a)}$$

$$= 16$$

$$\hat{R} = \frac{3.38 - 0.67 \ln \hat{\zeta}_{\text{eq}}}{2.3} \qquad \text{(Eq. 3.1.45)}$$

$$= 0.66$$

The resultant increase in displacement should be on the order of

$$\frac{\hat{R}_{\text{hybrid}}}{\hat{R}_{\text{cast-in-place}}} = \frac{0.66}{0.57}$$

$$= 1.16$$

The experienced increase in drift was

$$\frac{\Delta_{\text{max-hybrid}}}{\Delta_{\text{max-cast-in-place}}} = \frac{8.2}{7.0}$$

$$= 1.17$$

The recorded moment curvatures for the cast-in-place beam and the hybrid beam are presented in Figure 4.2.8. Observe that the curvature demand imposed on the hybrid beam is almost 40% greater than that imposed on the cast-in-place beam.

Conclusions

- Stiffness degradation should typically be beneficial, and it will usually result in lower displacement demands.

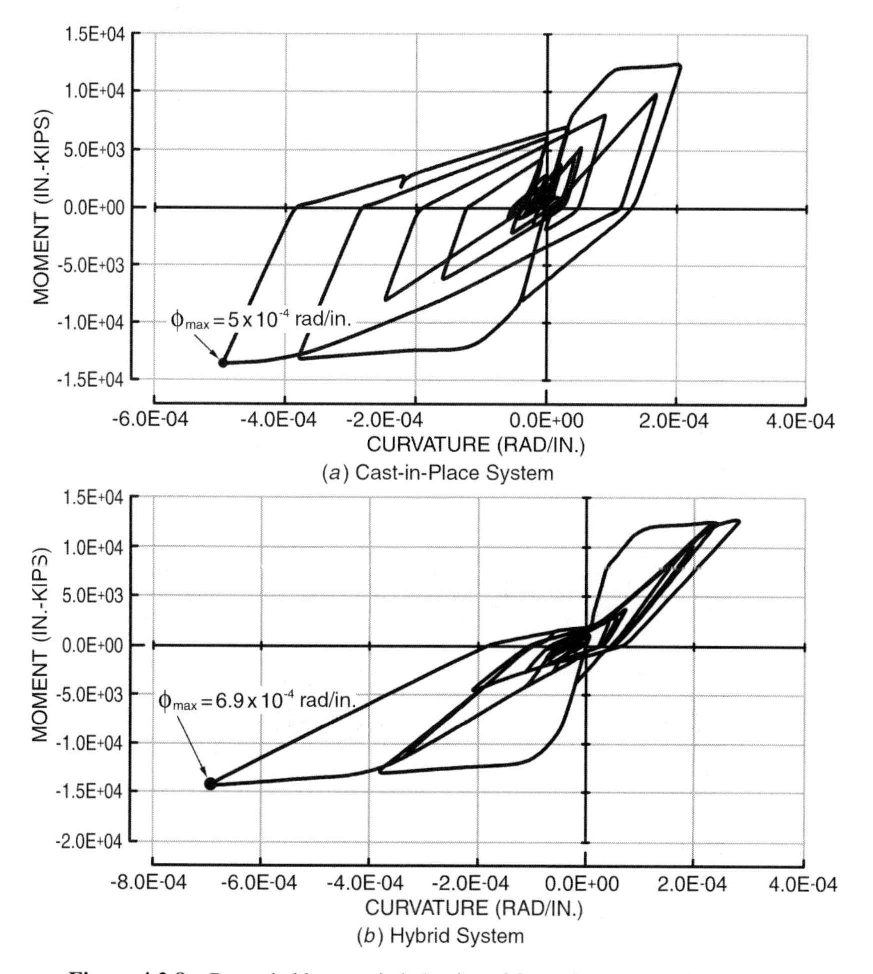

Figure 4.2.8 Recorded hysteretic behavior of frame beam (see Figure 4.2.7).

- The inclusion of $P\Delta$ effects will tend to reduce system displacement.
- Systems that exhibit significant pinching or slip in their behavior will probably displace more than those systems that do not exhibit this characteristic.

4.2.3 Distribution of Postyield Deformations

Our interest in this section is to review and analyze how postyield behavior propagates throughout a frame so that we may be in a better position to determine how to establish a link between component and system ductility.

Figure 4.2.9 provides a picture of the postyield rotation demand experienced in the frame beams of frame 3-2 (Table 4.2.1) when responding to Earthquake 2 (see Figure 4.2.1*b*). This postyield portrayal is taken from the first postyield excursion at

Photo 4.2 Precast concrete clad poured-in-place concrete ductile frame braced condominium, The Remington, Los Angeles, CA, 2002. (Courtesy of Magee Architects, Inc.)

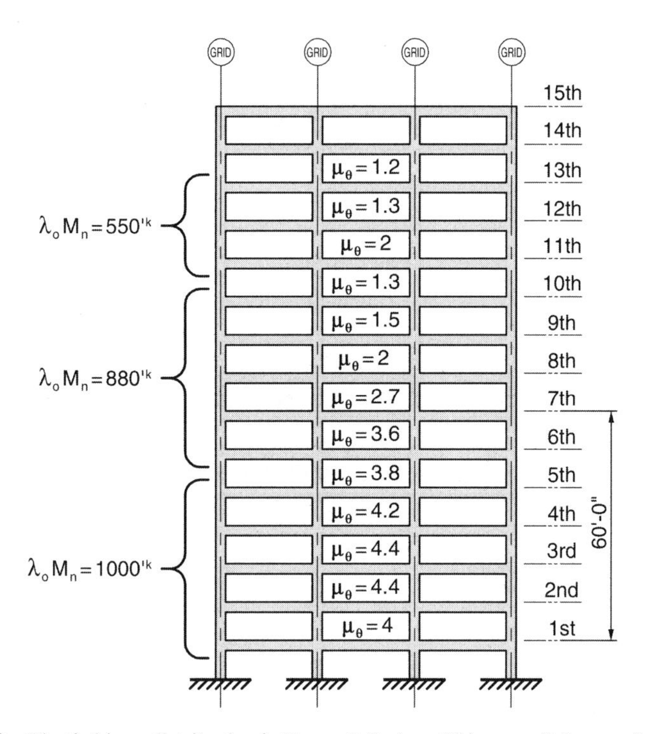

Figure 4.2.9 Plastic hinge distribution in Frame 3-2, $\Delta = 17$ in., $t = 2.5$ seconds (see Figure 4.2.1*b*).

$t = 2.5$ seconds. As a consequence, Figure 4.2.9 describes how a response in the first mode excursion might distribute inelastic actions.

First, note that the rotational ductility demand is concentrated in the lowermost part of the frame, about six to seven floors extending over a height that corresponds to the depth of the frame. Accordingly, these lower floors might be viewed as an effective plastic hinge region for the frame. The contrast is even more pronounced when described in terms of the postyield rotation demand. The postyield rotation demand, θ_p, in the lower floors (2 and 3) is 6.5 times that imposed on level 11.

The real concern is with the probable level of strain induced in the concrete. This can be estimated fairly easily.

The range of postyield rotation (θ_p) is between 0.014 radian ($\Delta = 17$ in.) and 0.031 radian at a building drift of 30 in. This corresponds to a range of postyield curvature demands of between 0.00078 and 0.0017 rad/in. One can either resort to computer developments similar to those whose conclusions are presented in Table 2.2.1 or use shorthand methodologies to identify the range of strain states.

$$\lambda_o M_n = 1000 \text{ ft-kips}$$

$$h = 36 \text{ in.}$$

$$d - \frac{a}{2} \cong 30 \text{ in.}$$

$$\begin{aligned} C_c &= \frac{\lambda_o M_n}{d - a/2} \\ &= \frac{1000}{2.5} \\ &= 400 \text{ kips} \end{aligned}$$

$$\begin{aligned} a &= \frac{C_c}{0.85 f_c' b} \\ &= \frac{400}{0.85(5)(20)} \\ &= 4.7 \text{ in.} \end{aligned}$$

$$\begin{aligned} c &= \frac{a}{\beta_1} \\ &= \frac{4.7}{0.8} \\ &= 5.9 \text{ in.} \end{aligned}$$

This would be a conservative estimate because it does not account for the compression steel. The depth to the neutral axis would be reasonably assumed to be on the order of 5 in. The induced level of postyield concrete strains would be estimated as follows:

$$\begin{aligned} \phi_y &= \frac{\lambda_o M_n}{E I_e} \qquad \text{(Computer model)} \\ &= \frac{1000(12)}{4000(27{,}200)} \\ &= 0.00011 \text{ rad/in.} \end{aligned}$$

$$\begin{aligned} \theta_y &= \frac{\phi_y \ell_c}{6} \qquad \text{(Conjugate beam)} \\ &= \frac{0.00011(276)}{6} \\ &= 0.005 \text{ radian} \end{aligned}$$

$$\frac{\theta_p}{\theta_y} + 1 = 4.4 \qquad \text{(see Figure 4.2.9)}$$

$$\theta_p = 3.4\theta_y$$
$$= 3.4(0.005)$$
$$= 0.017 \text{ radian}$$

$$\phi_p = \frac{\theta_p}{\ell_p} \quad \text{(Eq. 2.1.9)}$$
$$= \frac{0.017}{18}$$
$$= 0.00096 \text{ rad/in.}$$

$$\varepsilon_{cp} = \phi_p c$$
$$= 0.00096(5)$$
$$= 0.0048 \text{ in./in.}$$

$$\varepsilon_{cu} = \varepsilon_{cp} + \varepsilon_{cy}$$
$$= 0.0048 + 0.003$$
$$= 0.0078 \text{ in./in.}$$

Based on this, one might conclude that spalling was not likely to occur in the lower level frame beams.

In summary, it seems reasonable to conclude that

- System ductility is not equivalent to component ductility.
- The region of postyield behavior in a frame will extend over a height equal to the depth of the frame.
- The height of the building and the depth of the frame must be considered when one assesses the relationship between member (μ_b) and system (μ) ductility. The ratio (μ/μ_b) will be significantly different in taller buildings, where the height to depth ratio is typically larger than midrise buildings similar to the example building ($H/D \cong 2$—see Figure 2.2.10).

4.2.4 Design/Behavior Reconciliation

In Section 3.2.3 a displacement-based design procedure was used to develop Frame 3-2 (Table 4.2.1). A key element in the design process was the selection of a system ductility/overstrength factor of 8. The development of the system ductility component (μ) has intentionally been delayed so as to allow a reconciliation with the inelastic time history analysis.

The time history analysis demonstrated that inelastic behavior would be concentrated over the lower levels (Figure 4.2.9). In Section 3.2.3.1 (Step 8) yield building drifts of 3.75 to 4.04 in. were predicted for Frame 3-2. The predicted yield base shear was 435 kips. In Section 3.2.6 a sequential yield analysis was performed on Frame 3-2 (Figure 3.2.8*a*). Postyield behavior did not start until the building drift reached almost 7 in., (DR = 0.39%) and the base shear associated with first yield ($\lambda_o M_n = 1000$ ft-kips) was over 600 kips. This apparent contradiction is a direct consequence of modeling assumptions used to describe the behavior of the beam (see Figure 2.1.3). The analysis process used to generate Figure 3.2.8*a* adopted a frame behavior model that is quite different from that used to predict postyield beam strain states in Section 3.2.3.1. The difficulty encountered using the sequential yield model (Figure 3.2.8*a*) lies in the fact that concrete strain states are hard to predict, for behavior at idealized yield has been in the postyield domain for some time (see Section 2.1.1). This difficulty is overcome when we adopt steel first yield ($\varepsilon_s = \varepsilon_y$) as a point of departure.

The following yield baseline was developed in Section 3.2.3.1.

$$\phi_y = \frac{\varepsilon_{sy}}{0.67d} \qquad \text{(Eq. 2.1.7a)}$$

$$= \frac{0.002}{0.67(33)}$$

$$= 0.00009 \text{ rad/in.}$$

$$\theta_y = \frac{\phi_y \ell_c}{6} \qquad \text{(Conjugate beam)}$$

$$= \frac{0.00009(276)}{6}$$

$$= 0.00414 \text{ radian}$$

$$\Delta_y \cong 4 \text{ in.} \qquad \text{(Section 3.2.3.1, Step 8)}$$

Our performance-based objective is to identify the system ductility factor associated with a concrete strain state of 0.01 in./in.—incipient spalling of the shell. The assumption is that postyield behavior will be uniformly distributed over eight floors (80 ft $\cong$ 78 ft, which is the depth of the frame).

Each floor will slide—deform in shear—until the concrete strain in the frame beam reaches 0.01 in./in. The rotation experienced in the plastic hinge region will be modeled after Figure 2.1.11 wherein ϕ_y is associated with a steel strain of ε_y and a concrete strain of 0.001 in./in. (ε_{cy}).

$$\phi_p = \frac{\varepsilon_{cp}}{c}$$

$$c \cong 5 \text{ in.} \qquad \text{(see Section 4.2.3)}$$

$$\phi_p = \frac{0.01 - 0.001}{5}$$

$$= 0.0018 \text{ rad/in.}$$

$$\theta_{pb} = \phi_p \ell_p$$
$$= 0.0018(18)$$
$$= 0.0324 \text{ radian}$$

The rotation at the center of the column (story drift) is

$$\theta_j = \theta_{pb}\left(\frac{\ell_c - \ell_p}{\ell}\right)$$
$$= 0.0324\left(\frac{21.5}{26}\right)$$
$$= 0.027 \text{ radian}$$

The postyield story drift experienced in a story is

$$\delta_p = \theta_j h_x$$
$$= 0.027(120)$$
$$= 3.24 \text{ in.}$$

For the eight stories of assumed postyield response, this corresponds to a postyield component of building drift of

$$\Delta_p = 8(3.24)$$
$$= 26 \text{ in.}$$
$$\Delta_u = \Delta_y + \Delta_p$$
$$= 4 + 26$$
$$= 30 \text{ in.}$$
$$\mu = \frac{\Delta_u}{\Delta_y}$$
$$= \frac{30}{4}$$
$$= 7.5$$

Observe that this is reasonably consistent with the strain states conclusions developed from the time history analysis (Section 4.2.3), given the fact that no postyield rotation was presumed above level 8.

Conclusion: We should be convinced that the selected system will meet our performance objectives. Further, that the use of a system ductility/overstrength factor of 8 was appropriate.

4.2.5 Postyield Beam Rotations

Postyield beam rotations peaked at about 0.02 radian regardless of the frame type and strength. For example, the beams in Frame 3-2 ($\rho_s \cong 0.8\%$) experienced a postyield rotation of 1.7%, while those in Frame 3-3 ($\rho_s \cong 1.6\%$) experienced a postyield rotation of 2%, undoubtedly caused by the larger drift experienced by this frame coupled with a more concentrated zone of postyield plastification—six floors. The reported postyield rotation of the beams in Frame 4-2 ($\rho_s \cong 1.0\%$) was also 1.7%. Accordingly, there seems to be little incentive from a beam ductility demand perspective to adopt either the stronger three-bay frame (3-3), which will obviously be much more likely to experience damage than Frame 3-2 (see Section 4.2.3), or the stiffer and stronger four-bay frame (4-2).

The design confirmation phase is the only point in the design process where beam ductility demands can be realistically evaluated. This is because inelastic behavior and, in particular, component ductility demands are sensitive to strength distribution and the characteristics of an earthquake. The designer should be concerned with identifying large inelastic rotation demands and any accumulation of such demands as, for example, might tend to create a soft story or soft region in the building. Clearly, if a region is likely to reach displacements that suggest strength degradation, the situation should be mitigated and where this is not possible, the design approach should be changed.

Conclusions

- Inelastic time history analyses are most effectively used to confirm the existence of a well-distributed hinging pattern, one that avoids soft regions and strength degradation.
- The temptation to fine-tune beam strengths at this point in the design process should be resisted unless the need is clearly identified.

4.2.6 Evaluating Column Behavior

In all concrete members subjected to flexural load, the most critical element is the compression flange or face which, in the case of a frame, is the compression side column. The maximum axial load and shear imposed on this member are well identified in the sequential yield analysis, and the column size and axial capacity should have been adjusted in the conceptual design phase so as to assure a significant amount of ductility (Section 2.2). That having been the case, there should be no surprises in the design confirmation phase, though clearly the behavior of the compression side column is a, if not the, major concern.

A review of the strain states predicted by the inelastic time history analysis suggests how sensitive the frame design is to mechanism shear or provided strength. Four frames will be reviewed (see Table 4.2.2).

Frame 3-1

$$f'_c = 5 \text{ ksi}$$

$$P = 1820 \text{ kips}$$
$$M = 2097 \text{ ft-kips}$$
$$\theta_p = 0.01 \text{ radian}$$
$$\varepsilon_c = 0.007 \text{ in./in.}$$
$$\frac{P}{A_g f_c'} = 0.28$$

Frame 3-2

$$f_c' = 5 \text{ ksi}$$
$$P = 2141 \text{ kips}$$
$$M = 2025 \text{ ft-kips}$$
$$\theta_p = 0.0135 \text{ radian}$$
$$\varepsilon_c = 0.0135 \text{ in./in.}$$
$$\frac{P}{A_g f_c'} = 0.33$$

Frame 3-3

$$f_c' = 5 \text{ ksi}$$
$$P = 3100 \text{ kips}$$
$$M = 2330 \text{ ft-kips}$$
$$\theta_p = 0.02 \text{ radian}$$
$$\varepsilon_c = 0.024 \text{ in./in.}$$
$$\frac{P}{A_g f_c'} = 0.48$$

Frame 4-1

$$f_c' = 5 \text{ ksi}$$
$$P = 1820 \text{ kips}$$
$$M = 1903 \text{ ft-kips}$$
$$\theta_p = 0.012 \text{ radian}$$
$$\varepsilon_c = 0.0095 \text{ in./in.}$$
$$\frac{P}{A_g f_c'} = 0.28$$

Frame 4-2

$$f'_c = 5 \text{ ksi}$$
$$P = 2430 \text{ kips}$$
$$M = 2300 \text{ ft-kips}$$
$$\theta_p = 0.0165 \text{ radian}$$
$$\varepsilon_c = 0.0155 \text{ in./in.}$$
$$\frac{P}{A_g f'_c} = 0.375$$

Clearly, the concrete strengths used in Frames 3-2, 3-3, and 4-2 need to be increased so as to reduce the level of concrete strain and improve the level of available ductility, but this should have been discovered much earlier in the design process.

Conclusion: Frame strength must not exceed the reasonable postyield deformation capability of the compression side column.

4.2.7 Response of Irregular Frame

The irregular frame (3-2), whose design was developed in Section 3.2.5, was reinforced so as to produce a beam strength identical to Frame 3-2. The response of this frame to Earthquake 2 is shown on Figure 4.2.10*a*. Observe that the behavior is essentially the same as that of regular Frame 3-2 (see Figure 4.2.1*b*). The response of the irregular frame to Earthquake 3 was somewhat less, at 14.3 in., than the response of Frame 3-2 (17.5 in.).

Plastic hinge propagation for the irregular frame whose design was developed in Section 3.2.5 is described in Figure 4.2.10*b*. Inelastic behavior is concentrated in the lower six floors, as is expected. The rotation ductility imbalance between interior and exterior bays is as to be expected from the postyield deformation of the subassembly suggested by the sequential yield analysis described in Figure 3.2.10*b*.

Rotation ductility demand does not describe performance in this case. Postyield rotations will be essentially the same, as is clearly described in Figure 4.2.11. Rotation ductility is based on θ_y, and θ_y is a function of length and strength of the beam. The development of the displacement ductility factor follows:

$$\phi_y = \frac{M_y}{EI_e} \qquad \text{(Reasonably elastic behavior)}$$

$$\theta_y = \frac{M_y}{EI_e}\frac{\ell}{6}$$

For M_y, E, and I_e constant, this becomes

$$\theta_y = C_1 \ell$$

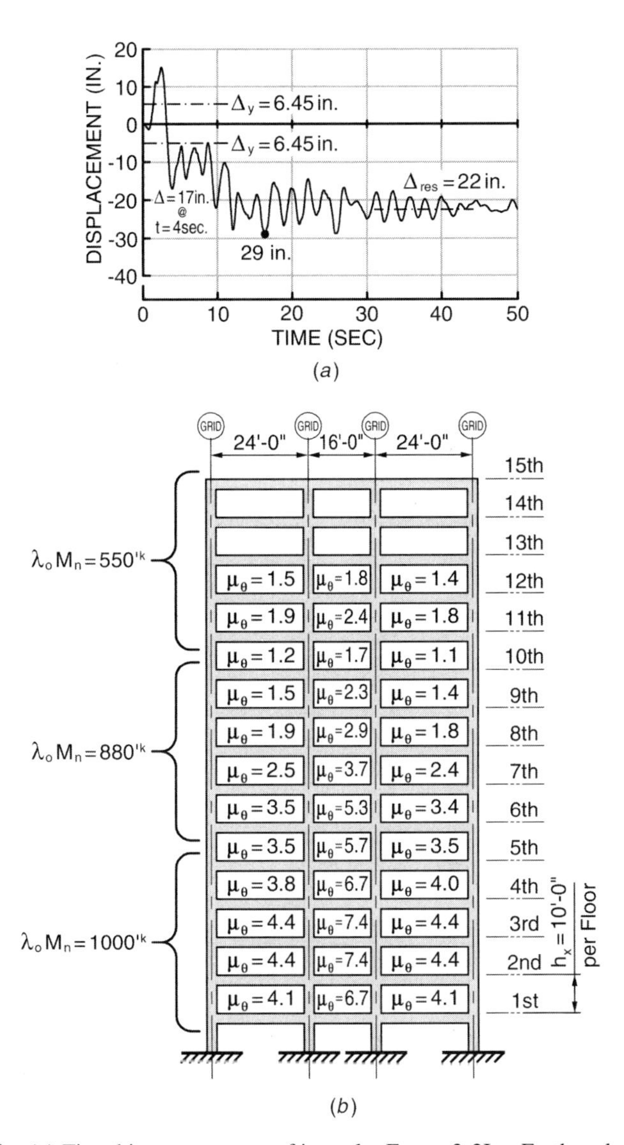

Figure 4.2.10 (*a*) Time history response of irregular Frame 3-2I to Earthquake 2. (*b*) Hinge propagation—irregular frame of part *a*.

Hence the rotation at yield for members of equal strength but unequal spans becomes

$$\theta_{y1} = \theta_{y2}\frac{\ell_1}{\ell_2}$$

when the dimension of the column is not a factor or is not considered.

The postyield rotation of the column (θ_{cp}) and beam are equal provided the column dimension is ignored. Hence

$$\mu_1 = \frac{\theta_{cp}}{\theta_{by1}}$$

$$\mu_2 = \frac{\theta_{cp}}{\theta_{by2}}$$

The ratio of rotation ductilities is

$$\frac{\mu_{\theta 1}}{\mu_{\theta 2}} = \frac{\theta_{cp}}{\theta_{by2}} \frac{\theta_{by1}}{\theta_{cp}}$$

$$= \frac{\ell_1}{\ell_2}$$

which, for the example case with the ratio of rotation ductility factors producing similar behavior ($\phi_y + \phi_p$), is 1.5 (24/16). The recorded values are slightly higher because the depth of the column is considered and the rotation is assumed to take place at the face of the column. Hence

$$\frac{\mu_{\theta 2}}{\mu_{\theta 1}} = \frac{\ell_{c1}}{\ell_{c2}}$$

$$= \frac{21}{13}$$

$$= 1.61$$

The relationship is further exacerbated when the centroid of beam postyield rotation moves away from the face of the column.

$$\frac{\mu_{\theta 1}}{\mu_{\theta 2}} = \frac{\ell_{c1} - \ell_p}{\ell_{c2} - \ell_p}$$

$$= \frac{23 - 1.5}{13 - 1.5}$$

$$= 1.87$$

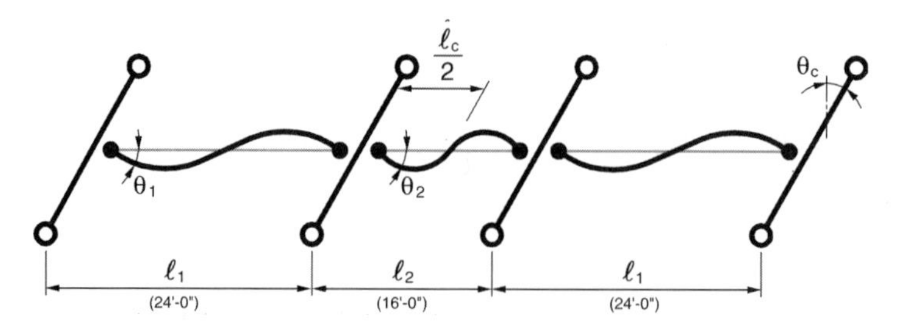

Figure 4.2.11 Story mechanism—irregular frame.

An obvious mitigation would be to increase the strength of the interior beam, and this can be done if it does not increase the reinforcement ratio to undesirable levels. Recall that an identical reinforcement program in adjoining beams was viewed as being desirable from a constructibility perspective, so a clear need should be perceived before adopting a dissimilar reinforcing program.

Consider the strain states imposed on the plastic hinge region on either side of the interior column if the postyield component of story drift is 1.8%.

$$\theta_{jp} = 0.018 \text{ radian}$$

$$\theta_{b1p} = \theta_j \left(\frac{\ell_1}{\ell_{c1}}\right)$$

$$= 0.018 \left(\frac{24}{21}\right)$$

$$= 0.021 \text{ radian}$$

$$\phi_{b1p} = \frac{\theta_{pb1}}{\ell_p}$$

$$= \frac{0.021}{18}$$

$$= 0.00117 \text{ rad/in.}$$

$$\varepsilon_{cp} = \phi_{pb1} c$$

$$\cong 0.00117(5)$$

$$= 0.0058 \text{ in./in.}$$

$$\varepsilon_{cu} = \varepsilon_{cy} + \varepsilon_{cp}$$

$$= 0.003 + 0.0058$$

$$= 0.0088 \text{ in./in.}$$

Comment: ε_{cy} is assumed to be 0.003 in./in. so as to be consistent with the computer model of θ_{by} $(\lambda_o M_n / E I_e)$.

$$\theta_{b2p} = \phi_{jp} \left(\frac{\ell_2}{\ell_{c2}}\right)$$

$$= 0.018 \left(\frac{16}{13}\right)$$

$$= 0.022 \text{ radian}$$

$$\phi_{b2p} = \frac{\theta_{b2p}}{\ell_p}$$

$$= \frac{0.022}{18}$$

$$= 0.0012 \text{ rad/in.}$$

$$\varepsilon_{cp} = \phi_{b2p} c$$

$$= 0.0012(5)$$

$$= 0.006 \text{ in./in.}$$

$$\varepsilon_{cu} = \varepsilon_{cy} + \varepsilon_{cp}$$

$$= 0.003 + 0.006$$

$$= 0.009 \text{ in./in.}$$

Observe that postyield concrete strain states are essentially the same on both sides of the interior column. Hence, the rotation ductility demands described in Figure 4.2.10*b* tend to deceive, and an increase in the strength of the interior beam will only tend to promote a more uniform yielding of the system, but not significantly alter induced postyield strain states.

Conclusions

- The "equal displacement" hypothesis is supported—$\Delta_u = 29$ in.—and this is consistent with the projected 27.7 in. given that the spectral coordinate exceeds the design spectrum (Figure 4.1.1*b*) at the building period of 2.25 seconds.
- The region of significant postyield demand is confined to a height equal to the depth of the frame.
- Rotation ductility demands are not a good measure of performance.

4.2.8 Response of Precast Concrete Frames—DDC®

The basic difference between the behavior of the hybrid and cast-in-place frames centered on the impact that reduced levels of structural damping might have on the response of the system to ground motion. This was explored in Section 4.2.2.

The basic difference between the DDC frame and Frame 3-2 was the distribution of the flexural strength over the height of the structure. The displacement responses of the cast-in-place and DDC frames to Earthquakes 2 and 3 are shown in Figure 4.2.12. In each case the maximum drift and residual drift of the DDC frame are less than those of the comparable cast-in-place frame (Figure 4.2.1*b*). This is probably attributable in part to the slightly lesser strength of the DDC frame (91%) and to the fact that the beam strength at the sixth floor did not allow it to yield prematurely. The upper part of the DDC frame remained elastic throughout the response.

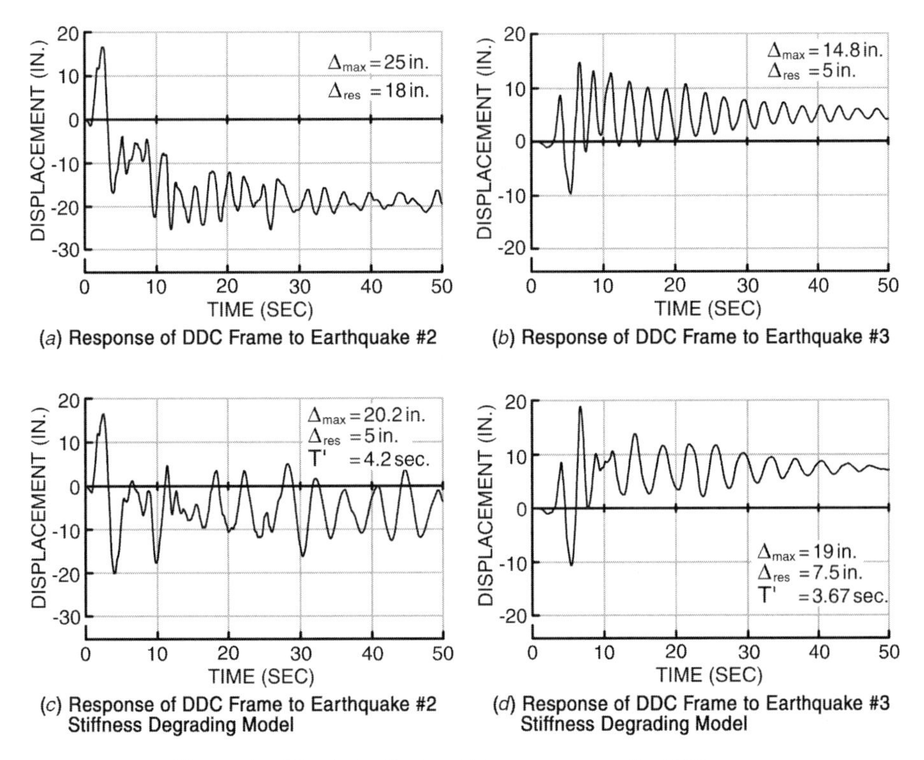

Figure 4.2.12 Inelastic time history response of DDC frame.

4.3 BEHAVIOR IMPONDERABLES

Stability and torsion are a major concern of the structural engineer. Both topics are reasonably well understood for systems whose behavior is elastic but the extrapolation to the inelastic behavior range is conjectural. A pragmatic overview is the goal of this section.

4.3.1 System Stability Considerations

Stability concerns have been addressed by some but conclusions that might be used in the design process are quite speculative at this point in time and are likely to remain so. The focus of this section is on those issues that should be of concern to the designer and on practical mitigation measures.

Instability occurs when the restoring force or mechanism shear, minus $P\Delta$ effects ($V_{P\Delta}$) and other strength degrading characteristics (V_D), reaches zero. See Figure 4.3.1. This definition is probably not satisfactory when the design consideration is earthquake safety. In this case, it seems appropriate to select a minimum resistance or restoring force that is capable of sustaining aftershocks.

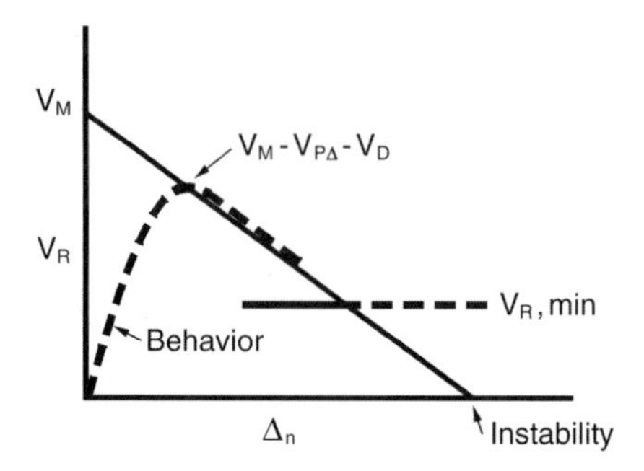

Figure 4.3.1 Available restoring force as a function of displacement.

The most rational discussion and development of a stability criterion is contained in Paulay and Priestley.[4.1] They follow the "stability index" approach, which essentially adds strength to a bracing program to counter the $P\Delta$ effect.

From a design perspective this is certainly a rational approach, so long as it is accomplished in a considered manner.

The frames studied in Section 4.2 all experienced a residual drift, but none of them approached negative levels of restoring force. Hence, the presumption is that they would still be stable were they to have been subjected to the design earthquakes. The establishment of a stability criterion must consider post-earthquake occupancy requirements. To consider a reoccurrence of the design seismic event is not rational. The continued use of the facility must be based on a lesser event and the occupancy time frame. For example, it might take 20 years or more to fund, design, and build a new hospital but significantly less to replace a police station or school.

Consider the story forces described in Figure 4.3.2. This story will, when deformed, have a spring force equal to $\Sigma M_p h_x$. The $P\Delta$ forces will counter this spring force as discussed previously. The $P\Delta$ force will be a function of the mass or weight tributary (P_{tr}) to the frame above the level under consideration, while the story drift will, of course, depend on the displacement or story displacement at peak building drift. This means that, in the case of the frame oscillating about a new baseline (see Figure 4.2.1*b*), at least half of the time $P\Delta$ effects will be adding resistance or

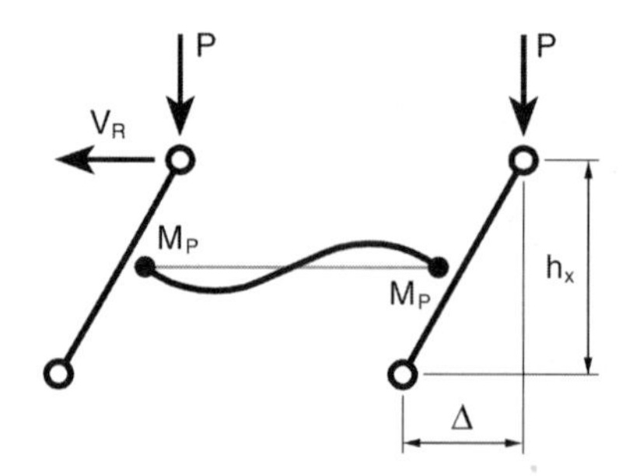

Figure 4.3.2 Forces acting on a story.

restoring force, and this complicates any reduction of the problem to a simple design rule or procedure.

The form adopted for the stability index is the ratio of the $P\Delta$ shear to the mechanism shear:

$$Q = \frac{P_{\text{tr}}\Delta}{V_M h_x} \tag{4.3.1}$$

where P_{tr} is the sum of the tributary mass above the level being considered and V_M is the mechanism shear which, for our purposes, is

$$V_M = \frac{\sum \lambda_o M_n}{h_x} \tag{4.3.2}$$

Through the years the stability index (Q) has been related to the elastic displacement associated with a static lateral load criterion (Δ_y) and Q^* to the inelastic displacement Δ_u or $\mu\Delta_y$. Paulay and Priestley[4.1] adopt the inelastic approach and define a consideration limit state for Q^* of 8.5%.

$$Q^* > 0.085 \tag{4.3.3}$$

In other words, a design should mitigate $P\Delta$ effect when the basic design strength V_M (Eq. 4.3.2) is not enough to cause Q^* to be less that 8.5%. In this case V_M should be increased by $(1 + Q^*)$. Hence

$$V_M^* = (1 + Q^*)V_M \tag{4.3.4}$$

where V_M^* is the objective design mechanism strength based on the mechanism strength developed absent $P\Delta$ considerations.

The identified limit state (8.5%) and mediation were developed from analytical work done after the San Fernando Earthquake (1971), which suggested that $P\Delta$ effects substantially increased building drifts and residual drifts. This is not consistent with the analytical efforts reported in this chapter or the author's experience. Further, inelastic drifts were, in the stability index approach, typically assumed to be a function of the system ductility factor amplified to account for the distribution of inelastic demand in the lower stories, a subject discussed in Section 4.2.

The compensating strength ($V_M^* - V_M$) has typically been based on an energy balancing approach where the actual increase in strength is half of that required to compensate for $P\Delta$ effects at the anticipated displacement or ductility demand. In the example building, this would amount to doubling the design ductility factor to account for a concentration of inelastic displacement in the lower part of the structure and then halving it to balance the energy dissipated.

Now apply these concepts to the frame designs of Chapter 3.

Working loosely so as to identify the potential level of additional or $P\Delta$ compensating shear on the three-bay frame (3-2), assume that we have concluded that the following appropriately describes the action anticipated in the lower levels:

$$V_M = 600 \text{ kips} \qquad (6\lambda_o M_n / h_x)$$

$$\Delta_y = 0.35 \text{ in.}$$

$$\mu_\Delta = 3 \qquad (15/5)$$

$$P_{\text{tr}} = 12{,}150 \text{ kips}$$

$$Q^* = \frac{\mu_\Delta \Delta_y P_{\text{tr}}}{V_M h_x} \qquad \text{(see Eq. 4.3.1)}$$

$$= \frac{3(0.35)(12{,}150)}{600(120)}$$

$$= 0.18$$

According to the criterion discussed, an increase of 18% in the design strength is suggested, and this would apply to the lower levels of the frame.

Whereas this increase seems to have a logical basis, it must be considered in the context of the building design. First, strength did not seem to have a major impact on behavior, so a fine-tuning of the strength will probably not necessarily result in a "better" building. Second, an increase in the strength of the lower part of the building will cause the upper floors to yield first, and this was observed in the sequential yield analyses (Figure 3.2.8*b*) given only a 12% reduction in strength between Levels 5 and 6. Clearly, pushing the inelastic behavior region higher into the building will produce generally better behavior, but a prescriptive approach based on the limited data currently available does not seem advisable.

Clearly, the $P\Delta$ issue is one that could benefit from a focused research effort. I have not seen any discussion of dynamic stability even as it relates to the simplest of structures and this, after all, is the primary concern.

4.3.2 Torsion

Few analytical procedures are more futile than an elastic torsional analysis to demonstrate compliance with prescriptive drift objectives. To confirm this consider the discussion regarding the design of unequal shear walls or irregular frames. A time history treatment of torsion must consider inelastic behavior if it is to have meaning, and this is not possible now nor is it likely to be in the near future. The intent of this section is to describe how the impact of torsion in the inelastic range can rationally be mitigated.

Paulay[4.2] presented an excellent discussion of inelastic torsion, and the essence of his solution was to provide an orthogonal bracing program that would remain elastic and thereby mitigate the consequence of unbalanced torsion in the postyield behavior range. This is certainly one solution, but it is not usually available to the design team. I see no need to repeat Paulay's arguments, and instead approach the problem from a pragmatic position and offer mitigating measures that can be easily developed in the design process.

Why does inelastic behavior impact torsion adversely? To understand the concern consider the spring supported beam described in Figure 4.3.3*a* and the spring

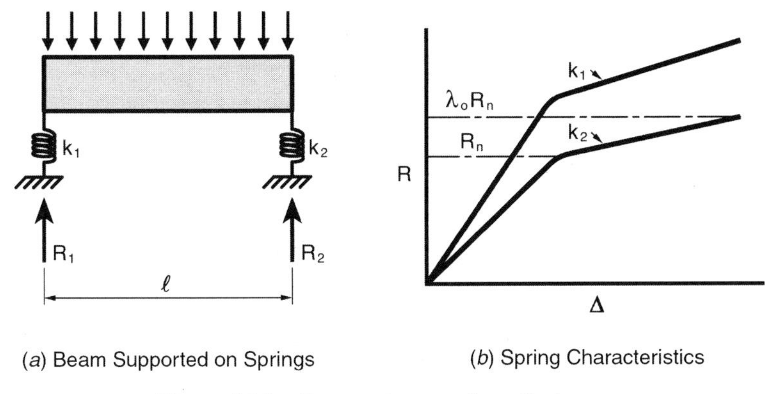

Figure 4.3.3 Force acting on a floor diaphragm.

characteristics described in Figure 4.3.3*b*. Equilibrium requires that the reactions R_1 and R_2 be identical so long as the forcing function, which in this case is the inertia load, remains uniformly distributed. This means that the strength of R_2 will dictate the resisting force provided by the bracing program. Once $R_{2\,\max}$ is reached, the center of resistance will move to the left hand support (R_1, Figure 4.3.3). Observe that even if both springs have identical stiffnesses but different strengths, the center of rigidity or resistance will move to the stronger support. The existence of an orthogonal bracing system would only impact this behavior if it were to remain in the elastic behavior range—Paulay's recommendation. This design objective is not usually possible given that most buildings are rectangular in plan and that the angle of earthquake attack is likely to cause an inelastic response on both axes.

The consequences of this action are many and none of them are good. Once the center of rigidity migrates to one end of the building, the displacement on the other end will increase as its strength hardens to $\lambda_o R_n$, and this displacement (Δ_2) may be many times the displacement of R_1 (see Figure 4.3.3*b*). This torsional response will further impact the dynamic response of the system and cause an even larger displacement at R_2. Observe that the resistance provided by this bracing program will be significantly different from the cumulative resistance provided by the unequal wall condition discussed in Section 3.1.2. It goes without saying that the likelihood of spring 2 reaching its displacement limit state is increased and that the performance of the structure will be adversely impacted.

Consider the structure whose plan is described in Figure 4.3.4. If bracing element 2 is weaker than 1 (i.e., Figure 4.3.3*b*), the center of rigidity will move toward wall 1 to the extent allowed by walls 3 and 4. If walls 3 and 4 are very stiff and strong, the center of rigidity will remain at the center of the building and the restoring force provided by walls 1 and 2 would be developed as it was in Section 3.1.2 for walls of varying length. Recall that this is not the conclusion that would be reached by an extension of an elastic analysis.

Consider the consequences of an earthquake attack along a 45° angle. Since designs assume that the yield strength of the bracing system will be significantly exceeded, the orthogonal bracing system (E-W) will also yield. If the strength imbalance

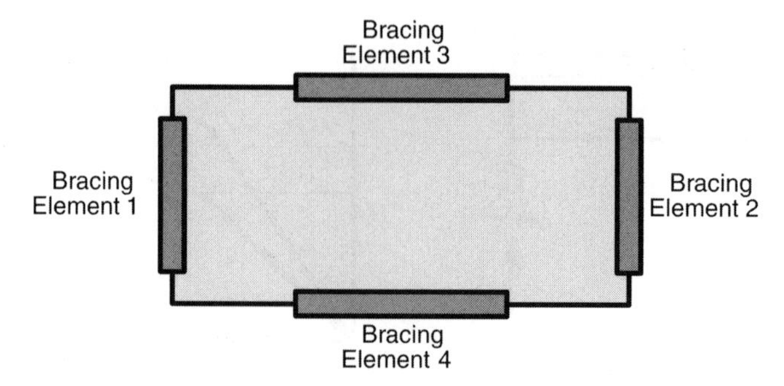

Figure 4.3.4 Planar arrangement of bracing elements.

between Systems 3 and 4 is significant, the center of rigidity will move to a corner of the building and the behavior will revert to the planar condition developed for the beam described in Figure 4.3.3*a*.

Eventually someone will develop an inelastic three-dimensional computer program that will consider the dynamic impact of inelastic behavior, but in the mean time and even given a computational assist, a practical design solution is required. My design approach has been to balance the strength of the bracing systems so as to match the inelastic behavior of each system to the extent possible. This is most simply done by providing identical bracing elements. Unfortunately, this is more often than not an unacceptable solution from an aesthetic or functional perspective (see Figure 4.3.5). Given a condition where bracing systems must be different, they must be strength balanced to the extent possible, even if it means that strength or prescriptive mandates be violated.

A sequential yield analysis is most effectively used to balance bracing systems. The behavior of two significantly different wall systems is described in Figure 4.3.3. The strength of shear wall 1 must be significantly reduced in order to match that of wall 2. The plan aspect ratio (length to width) of the building described in Figure 4.3.4 is nearly 3, and this would make the impact of any orthogonal bracing system on torsional response unlikely. Observe that this balancing of bracing system strengths should minimize torsional response even if the stiffness of the two systems is incorrectly assumed. This is fortunate because, as should be clear from the material

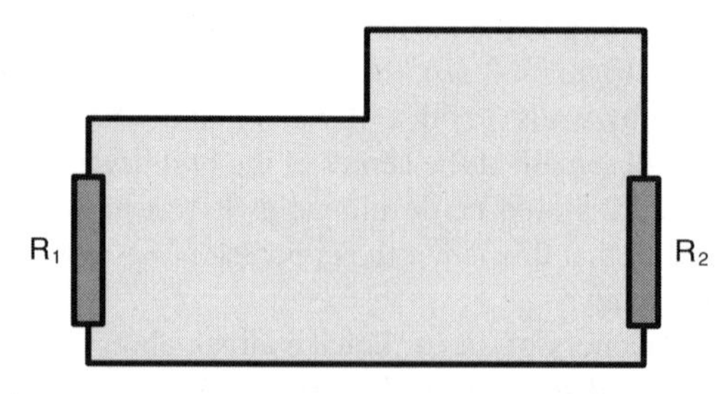

Figure 4.3.5 Irregular floor plan.

developed in Chapter 2, stiffness characteristics are more difficult to describe than are strength limit states, especially since stiffnesses continuously change with the imposition of postyield displacement cycles (see Figures 2.1.2, 2.1.6, and 2.3.8).

This concept, equal or compatible strength, should be extended to irregular bracing programs as well. Consider the plan described in Figure 4.3.5. The logical design approach is to view the bracing program as though it were a simple beam (Figure 4.3.3*a*). The strength provided by each reaction should be based on the mass (weight) that is tributary to it. The introduction of intermediate bracing elements complicates the problem but does not alter the objectives. Consider the bracing program described in Figure 4.3.6. The objective should be to minimize induced torsional moments when the system responds in the inelastic displacement range. The development assumes that all bracing components have been pushed into their inelastic range (Δ_u—Figure 4.3.6*b*). The relationship of strengths should strive to balance induced torsional moments such that

$$(R_1 - R_2)\frac{\ell}{2} - R_3 a\ell \cong 0$$

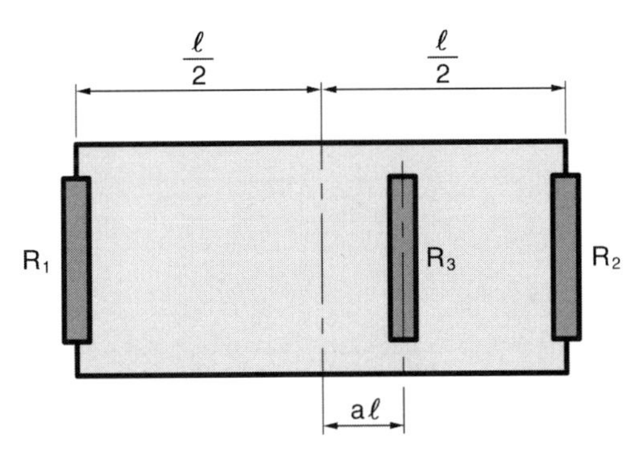

(*a*) Unbalanced Bracing Program

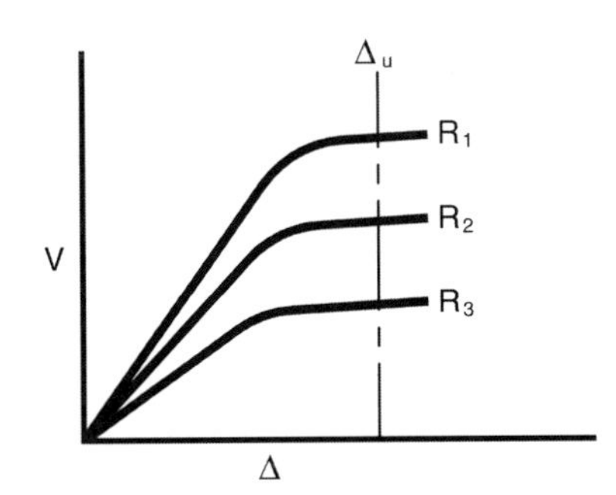

(*b*) Force Displacement Relationship—Bracing Elements Described in Figure 4.3.6*a*

Figure 4.3.6 Floor plan containing multiple bracing elements.

Clearly, one cannot expect to entirely eliminate the impact of torsion on the response of a building, but a strength-based approach will produce much better results than any elastic stiffness-based alternative.

SELECTED REFERENCES

[4.1] T. Paulay and M. J. N. Priestley, *Seismic Design of Reinforced Concrete and Masonry Buildings,* John Wiley & Sons, New York, 1992.

[4.2] T. Paulay, "Are Existing Seismic Torsion Provisions Achieving the Design Aims?" *Earthquake Spectra*, Vol. 13, No. 2, p. 249.

INDEX